Klystrons, Traveling Wave Tubes, Magnetrons, Crossed-Field Amplifiers, and Gyrotrons

For a listing of recent titles in the *Artech House Microwave Library*,
turn to the back of this book.

Klystrons, Traveling Wave Tubes, Magnetrons, Crossed-Field Amplifiers, and Gyrotrons

A. S. Gilmour, Jr.

ARTECH HOUSE
BOSTON | LONDON
artechhouse.com

Library of Congress Cataloging-in-Publication Data
A catalog record for this book is available from the U.S. Library of Congress.

British Library Cataloguing in Publication Data
A catalogue record for this book is available from the British Library.

ISBN-13 978-1-60807-184-5

Cover design by Vicki Kane

685 Canton Street
Norwood, MA 02062

10 9 8 7 6 5 4 3 2 1

Contents

Preface

This book is based on material assembled for courses, short courses, and seminars on microwave tubes that I have given over 100 times at various U.S. Navy, Air Force, Army, and NASA facilities, Varian Associates (now CPI), Hughes (now L3) Electron Dynamics Division, Teledyne Electronic Technologies, Litton (now L3) Industries, L3 Defense Systems Division, Texas Instruments, the French Ministry of Defense, the University of California at Los Angeles, The University at Buffalo, and other locations. There have been over 2,000 students with backgrounds ranging from third-class Navy electronic technicians through microwave tube experts to individuals at the chief scientist level.

This book contains a balance of practical and theoretical material so that those new to microwave tubes as well as experienced microwave tube technicians, engineers, and managers can benefit from its use. To a large extent, each chapter builds on material from earlier chapters and so, for the newcomers, it is advisable to work through the book starting at the beginning. For those who have microwave tube experience, most of the chapters can be used without referring to others. For those who are interested in further study and research, the references to the works of over 500 authors dating from the invention of the microwave tube to the present time should be of value.

Contributions to this book have come from many sources. First of all, Thomas Bekker of the Port Hueneme Division of the Naval Surface Warfare Center provided the encouragement that led me to give my first 5-day course on microwave tubes in 1982 and to publish the book *Microwave Tubes* in 1986. Then, with the encouragement and support of Eugene (Joe) Dutkowski of the Crane Division of the Naval Surface Warfare Center, I developed an 8-day course on traveling wave tubes that led to the publication in 1994 of *Principles of Traveling Wave Tubes*. Subsequently, I developed 8- and 10-day courses, which contained in-depth material on klystrons and crossed-field devices as well as traveling wave tubes. Recently, I was asked to prepare lecture material on fast wave devices. In this book, I have integrated the new material on klystrons, crossed-field devices, and fast wave devices with revised material from my previous two books.

Especially valuable technical contributions have come from the following:

Jon Christensen, James Hansen, and Dr. Ivo Tammaru of Hughes (now L3); David Zavidil, Phil Lally, and Robert Dipple of Teledyne; Dr. Richard True, Dr. Robert Symons, and Dr. Rodney Vaughan of Litton (now L3); Dr. George Farney, Dr. Armand Staprans, and George Miram of Varian (now CPI); Dr. Stanley Kaisel; and Dr. Marvin Chodorow. I am also indebted to the many course participants who made substantial recommendations and contributions to the course material and to my books.

Finally, I again thank my wife for her patience and understanding.

Chapter 1

Introduction

Microwave tubes are vacuum electron devices used for the generation and amplification of radio frequencies in the microwave frequency range. Microwave frequencies are generally considered to be from somewhat less than 1 GHz to over 100 GHz. The millimeter-wave frequency region is part of the microwave range and is generally considered to extend from about 30 GHz to 300 GHz.

1.1 THE MICROWAVE SPECTRUM

The microwave spectrum is divided into several frequency bands. The two most commonly used band designations are shown in Figure 1-1. The old band designation was established during World War II. The new designation, established to provide a more logical system, has not been widely accepted. The old band system is most commonly used in the microwave tube industry and is the one used in this book.

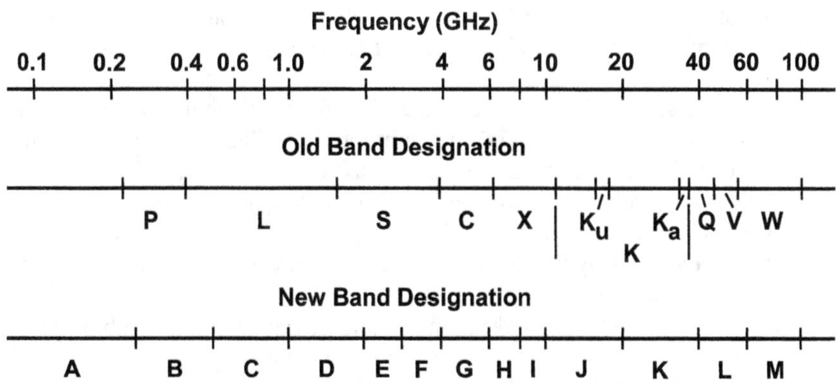

Figure 1-1 Microwave band designations.

1.2 THE DOMAIN OF MICROWAVE TUBES

Before the advent of solid-state devices, microwave tubes were the only active devices available for use at all power levels throughout the entire microwave range. Now, as indicated in Figure 1-2, vacuum devices (microwave tubes) predominate in high-power, high-frequency applications and solid-state devices are used mostly at relatively low powers and frequencies.

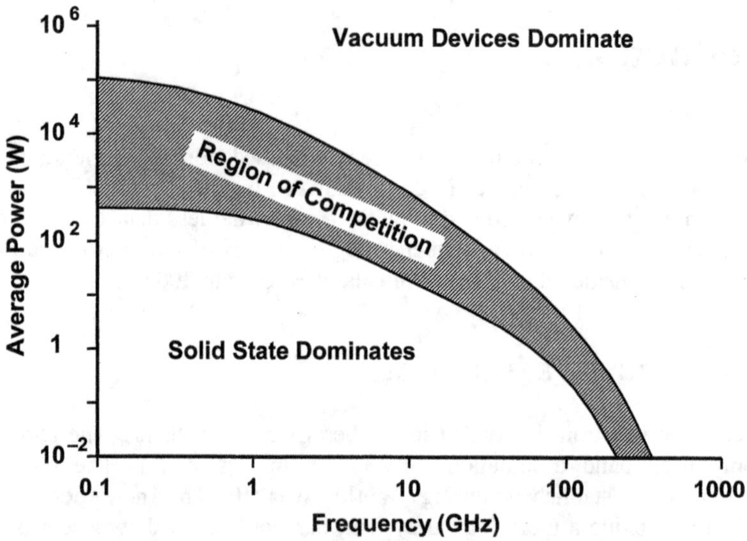

Figure 1-2 Ranges of application of microwave tubes and solid-state devices.

There is no clear dividing line between the domains of tubes and solid state because several factors (see [1] for a relevant discussion) beside average power and frequency must be considered in most applications. Some of these factors are:

1. In general, microwave tubes are far more efficient than solid-state devices. With the use of the appropriate collector technology, tube efficiencies can approach or, in some cases, exceed 70%.
2. Microwave tubes can operate at much higher temperatures than solid-state devices. This, coupled with higher efficiency, means that tubes can be smaller and lighter than solid-state devices and still eliminate waste heat.
3. Because of their extremely high reliability, traveling wave tubes are the amplifiers of choice for many satellite applications.
4. A factor of great importance in many applications is the large bandwidth for some microwave tubes. This can be over 2 octaves for some helix-type traveling wave tubes (TWTs).

So we see that even though there is an overlap in capabilities on a power versus frequency chart, tubes may be selected to obtain the required system capability.

1.3 CLASSICAL MICROWAVE TUBE TYPES

Most microwave tubes were invented prior to World War II [2–4]. Two types, the magnetron and the reflex klystron, were the subjects of intense development efforts during the war for use in radar. During and after the war, numerous other microwave tube types and configurations were invented. Most did not find application for any of several reasons (low efficiency, low gain, complexity). Today, five types of microwave tubes predominate. Four are conventional devices and the fifth is the gyrotron. These are the subjects of this book:

1. Klystrons;
2. Traveling wave tubes;
3. Magnetrons;
4. Crossed-field amplifiers;
5. Gyrotrons (gyro-monotrons and gyro-amplifiers).

These can be divided into the three categories, linear-beam, crossed-field, and fast-wave tubes, shown in Figure 1-3. The reasons for this division become clear when the configurations of the tubes and the electric and magnetic fields in them are considered.

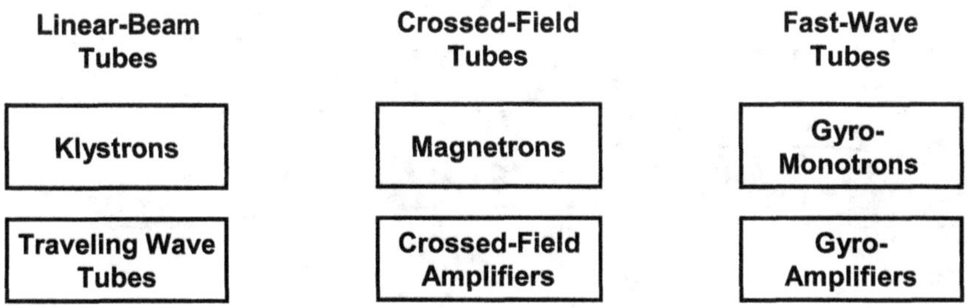

Figure 1-3 Major categories of microwave tubes.

Figures 1-4 and 1-5 show the basic configurations of klystrons and traveling wave tubes. In each case, after being formed in an electron gun, the electron beam travels linearly through the RF circuit to a collector. In the klystron, the RF circuit consists of resonant cavities. There is no electromagnetic coupling between cavities. The RF input signal accelerates and decelerates electrons in the beam. As the beam drifts, faster electrons catch up with slower electrons and this results in the formation of bunches of electrons and an RF current in the beam that grows as the beam moves along. This current is coupled first to intermediate cavities (only one is shown) and then to the output cavity. At each intermediate cavity, the RF beam current induces a signal, which, in turn, produces a field that enhances the bunching process in the beam. Finally, the coupling of the intense RF beam current to the output cavity produces the RF output power of the device. The gain of a klystron can be very large, 60 dB or more, but the bandwidth is usually limited to a few percent to about 10%. Output power levels can be tens of megawatts or more.

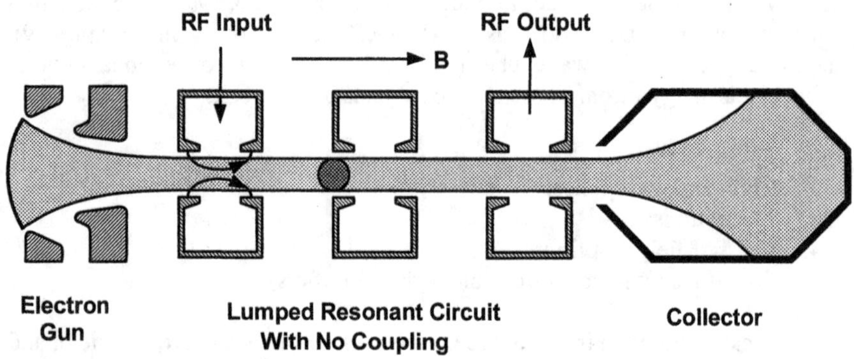

Figure 1-4 Basic configuration of a klystron.

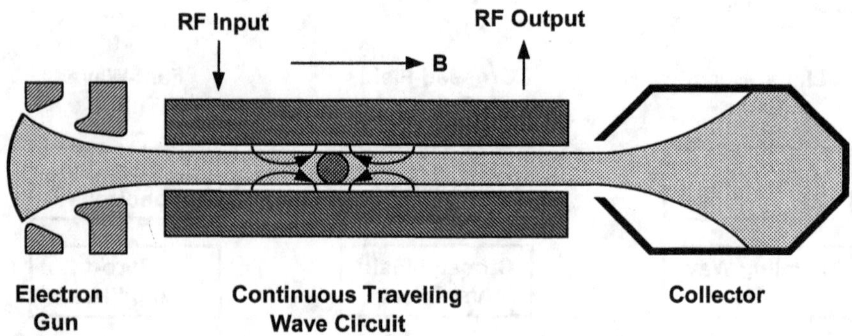

Figure 1-5 Basic configuration of a traveling wave tube.

In the traveling wave tube (TWT) the RF circuit is continuous so that a signal can travel along the circuit much like it does on a transmission line. The circuit is designed so that the velocity of the signal is nearly the same as the velocity of the electron beam passing through it. Also, the circuit is designed so that the electric fields from a signal on the circuit penetrate into the beam as indicated in Figure 1-5. These fields accelerate some electrons and decelerate others so that bunches of electrons are formed. As these bunches pass by the nearby circuit, they induce an RF current flow in the circuit and this causes the amplitude of the RF field on the circuit to increase. The increased field, in turn, increases the bunching action in the beam. As a result, as the bunches in the beam and the field on the circuit travel along (at the same velocity), the bunching action in the beam increases. As the bunching becomes more intense, the RF current in the beam and the field on the circuit become larger. Eventually, the amplified signal is coupled off from the circuit. The resulting gain of a TWT is usually in the 30–50-dB range and the bandwidth can be from 20% to over 2 octaves. Output power levels range from tens of watts for extremely broadband devices to hundreds of kilowatts to megawatts for narrowband devices.

The basic configurations of magnetrons and crossed-field amplifiers are shown in Figures 1-6 and 1-7. In these devices, the cathode is a cylindrical emitter in the center. In general, electron flow is radially outward towards the RF circuit, which also serves as the anode. Because a magnetic field is used, which is perpendicular to the cathode-to-anode electric field and to the direction of electron flow, electrons are forced to move in paths that circle about the cathode. When an RF field is present, the electrons are bunched into spoke-like configurations.

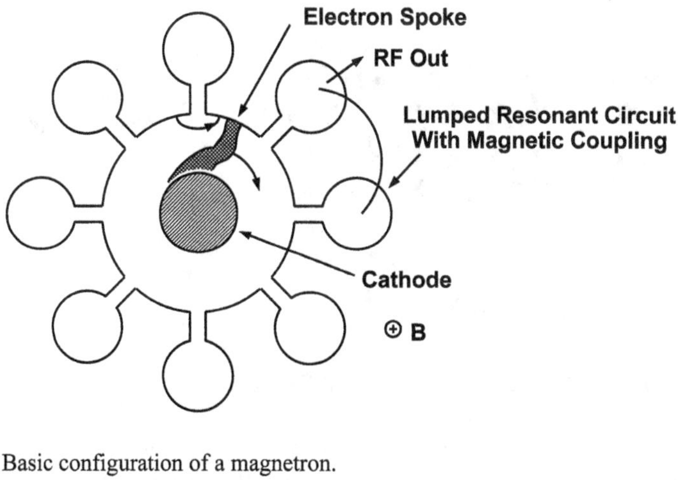

Figure 1-6 Basic configuration of a magnetron.

The magnetron is an oscillator in which the RF circuit consists of resonant cavities. The cavities are arranged so that the RF magnetic field generated in each cavity is coupled to adjacent cavities. As a result, ideally the entire cavity structure resonates at a single frequency with the RF electric field in adjacent cavities being 180° out of phase. As the RF field oscillates, the field pattern across cavity gaps rotates about the cathode. If an electron spoke circles about the cathode in synchronism with the rotating field pattern on the anode, then the spoke can arrive at each gap at a time when it will induce a current in the cavity that reinforces the oscillation. The resulting power out of the magnetron can be in the multimegawatt range and efficiencies reported to be as high as 88% are possible.

The operation of a crossed-field amplifier (CFA) (Figure 1-7) is similar to that of a traveling wave tube. The RF circuit is designed so that a signal can travel along it in synchronism with electrons that are moving next to it. In the TWT, electron bunches are formed that move along with the traveling wave. In a CFA, electron spokes are formed and these circle about the cathode in synchronism with the wave traveling on the circuit. The electric fields from the circuit wave enhance the bunching action in the spokes and the currents induced in the circuit by the spokes enhance the electric fields. As a result, the wave on the circuit grows as it travels from input to output. In some CFAs, the circuit is designed so that a backward wave interaction can occur and produce amplification. The gain of CFAs is usually less than 20 dB, but power outputs of tens of megawatts are possible.

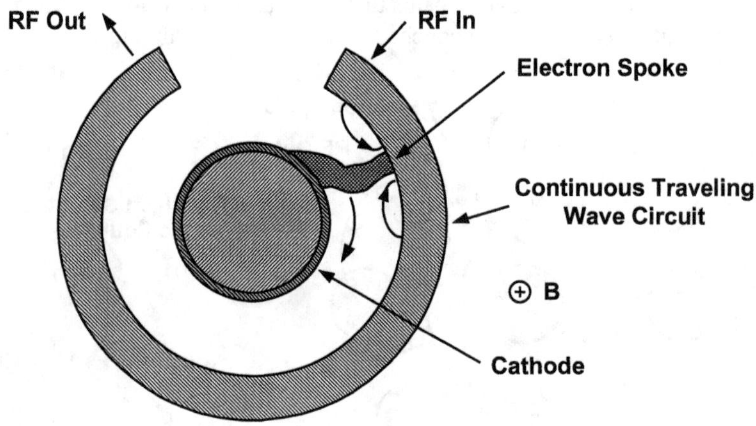

Figure 1-7 Basic configuration of a crossed-field amplifier.

In fast-wave devices, the beam-wave interaction mechanism is completely different than in linear beam and crossed-field devices. As Figure 1-8 indicates, in linear beam and crossed-field devices, the frequency of operation is determined by the RF circuit and so circuit dimensions scale with wavelength. As frequency is increased, power capability is decreased. In gyro-devices, the frequency of operation is determined by the cyclotron frequency and by the magnetic field. Frequency is independent of circuit dimensions, so very high powers can be generated at very high frequencies.

As Figure 1-9 indicates, the electron beam in gyro devices is hollow. Electrons in the beam spiral about magnetic flux lines in near synchronism with a transverse RF electric field. The transverse velocities of the electrons are normally 1.5 to 2 times the axial velocities. It is the transverse energy of the electrons that is used to amplify the electric field.

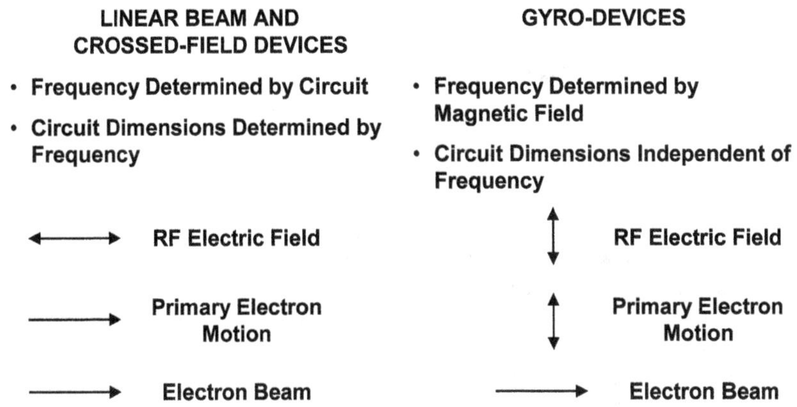

Figure 1-8 Comparison of RF electric field configurations and electron motions in linear beam crossed-field and gyro-devices.

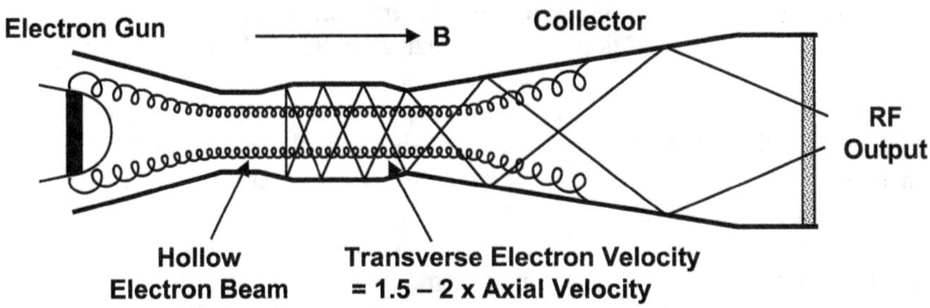

Figure 1-9 Basic configuration of a gyrotron oscillator.

From the preceding discussions, it is apparent that the basic operating principles of all microwave tubes are the same. An electron beam is passed through the tube with a circuit containing an electromagnetic field. The electron beam and the field interact to produce amplification or oscillation.

Hidden in this simplistic description are some clues to the extremely complex nature of a microwave tube. For example, the fact that an electron beam is used makes it necessary to evacuate the interior of a microwave tube to a pressure that may be over 11 orders of magnitude below atmospheric pressure. To achieve and maintain this pressure over the life of a tube, it is necessary to use special materials and fabrication techniques.

As another example, because of the high electron density required in the electron beam, the source of the electrons and the beam in most microwave tubes are very sophisticated structures. As we look further into the operation of a microwave tube, it is found that nearly every aspect of construction and operation is extremely complex.

1.4 OVERVIEW OF THIS BOOK

In selecting the subject matter for this book, a serious attempt has been made to include discussions of all the technologies that are essential to an understanding of microwave tube operation. The order of presentation of the material has been selected so that the reader, in working through the book from the beginning, can develop, in a logical sequence, a working knowledge of microwave tubes.

After a review of electron dynamics, cathodes and electrodes for forming electrons into linear beams are discussed. Next, the behavior of the beam after being subjected to a localized RF electric field is analyzed. This leads to discussions of the operation of klystrons followed by the analysis of the interaction of the beam with a traveling RF electric field. Perhaps the most significant result of this analysis is the importance of the relative velocities of the traveling RF field and the electron beam to the amplification of the RF field. This leads directly to a general discussion and analysis of the variation of velocity with RF frequency (dispersion) of TWT circuits. With an understanding of the significance of dispersion characteristics, it is possible to analyze in some detail helix TWTs and then coupled-cavity TWTs.

Many modern klystrons and TWTs use sophisticated electron beam collectors. These enhance the efficiency but do not directly affect the interaction of the beam with the circuit. Because of the importance of this subject, all of Chapter 16 is devoted to collector technology.

The operation of crossed-field devices is still not as well understood as that of linear-beam tubes, even though the magnetron was developed before linear-beam tubes were. Still, an attempt is made in Chapters 18, 19, and 20 to describe the most important aspects of the operation of these devices. The crossed-field device cathodes and the secondary emission process used in the cathodes of crossed-field tubes are discussed in some detail and then magnetrons and crossed-field amplifiers are described.

The gyrotron, which was invented in 1964 in Russia, is relatively new to the microwave tube field (klystrons and magnetrons were invented in the 1930s and the TWT was invented in 1940). There are many configurations for gyrotrons (gyro-oscillators, gyro-klystrons, gyro-twystrons, gyro-TWTs), and these and their applications are described.

The remainder of this book is devoted to subjects not related directly to the electromagnetic interactions that occur in microwave tubes but that, nevertheless, are important to the understanding of the total operation of a tube. For example, all microwave tubes use output windows that permit the RF power generated within the tube to be used outside the tube. Noise, nonlinearities, and distortion are important in most applications of tubes. When tubes fail, it sometimes results from degradation of the vacuum environment in the tube, and so a discussion of vacuum requirements is included.

A degradation in vacuum environment along with many other factors can lead to electrical breakdown in a tube, and so breakdown phenomena along with requirements for protecting tubes against the effects of breakdown are discussed.

REFERENCES

[1] R. S. Symons, "Modern microwave sources," *IEEE AESS Systems Magazine*, January 2002, pp.19–26.

[2] H. A. H. Boot, and J. T. Randall, "Historical notes on the cavity magnetron," *IEEE Trans. Electron Devices*, Vol. ED 23, No. 7, July 1976, pp. 724–729.

[3] G. R. Kilgore, "Recollections of pre World War magnetrons and their applications," *IEEE Trans. Electron Devices*, Vol. ED-31, No. 11, November 1984, pp. 1593–1595.

[4] R. H. Varian, and S. F. Varian, "A high frequency oscillator and amplifier," *Jour. App. Phys.*, Vol. 10, No. 5, May 1939. pp. 321–327.

Chapter 2

Static Fields Produced by Electrons

2.1 ELECTRIC FIELD

At many points in the analysis of electron motion in microwave tubes, it is necessary to include the effects of the static electric fields produced by the electrons. Those effects can be very large as the examples in this chapter demonstrate. Before examining two of those effects in electron guns, it is enlightening to look at the example in Figure 2-1. The effect of 10^6 electrons on the potential distribution in a small sphere is shown. With the shell of the sphere grounded (at zero potential), each electron that is placed within the sphere causes the potential at the center of the sphere to decrease a little over 1 μV. With 10^6 electrons, the potential at the center decreases to –1.44V. The density of electrons in the electron beam in a microwave tube is two to three orders of magnitude above that indicated here, so potential depressions produced by electrons are normally hundreds to thousands of volts.

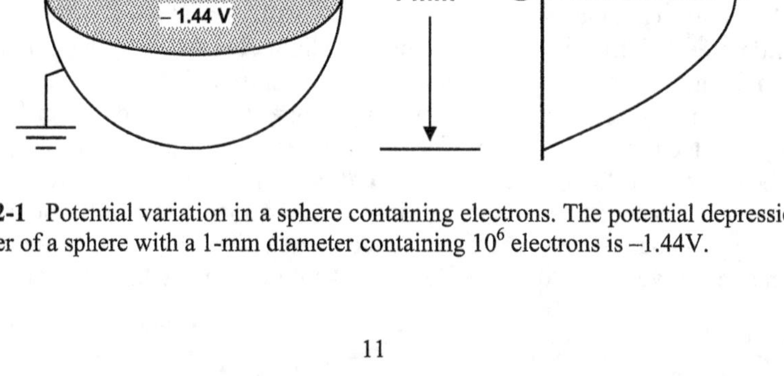

Figure 2-1 Potential variation in a sphere containing electrons. The potential depression in the center of a sphere with a 1-mm diameter containing 10^6 electrons is –1.44V.

It was pointed out by one of the author's professors many years ago that, in nearly all of the operations that must be carried out in designing, fabricating, testing, and manufacturing a microwave tube, "Mother Nature is there to trip you up." In the case of electron emission from the cathode of an electron gun in a TWT or klystron (Region 1 in Figure 2-2), it turns out that Mother Nature (and the very dense cloud of electrons near the cathode surface) is there to help us. In Chapter 5, this phenomenon is discussed in detail. Here we indicate why the electron cloud near the cathode is so important.

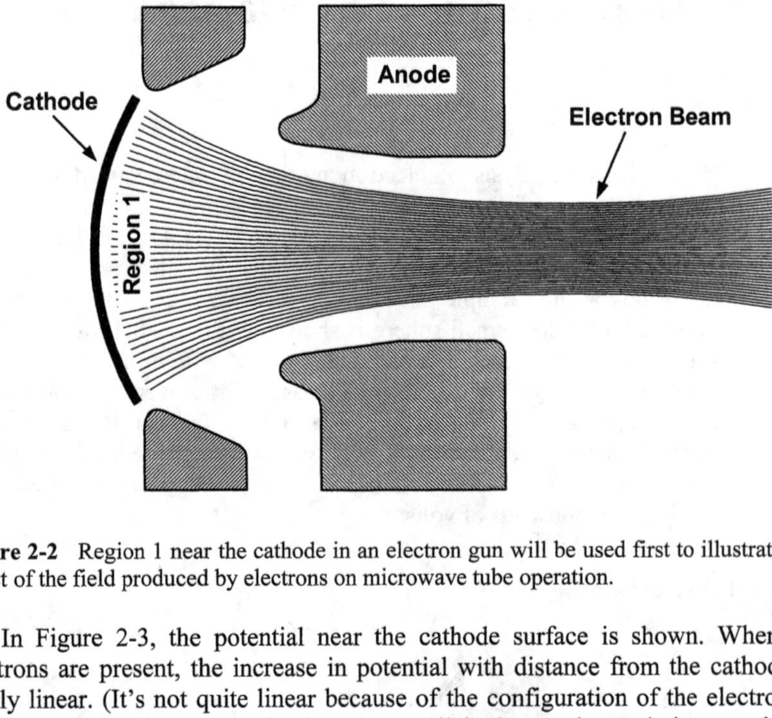

Figure 2-2 Region 1 near the cathode in an electron gun will be used first to illustrate the effect of the field produced by electrons on microwave tube operation.

In Figure 2-3, the potential near the cathode surface is shown. When no electrons are present, the increase in potential with distance from the cathode is nearly linear. (It's not quite linear because of the configuration of the electrodes. With a cathode and an anode that are parallel planes, the variation would be linear.)

When the cathode is heated so that a small current flows (this is called thermionic emission), each electron causes a slight depression in the potential. As the temperature of the cathode is increased, more current flows and the potential is depressed further. Eventually, when temperature is increased far enough, the density of electrons near the cathode surface becomes so high that the potential near the cathode surface decreases to zero. It is not possible to increase the current further by increasing temperature because the additional electrons would tend to cause the potential near the cathode to become negative. However, if the potential becomes negative, the additional electrons would be repelled and returned to the cathode surface. Thus, it turns out that emission is self-limiting and it is the potential depression produced by the emitted electrons that produces the limiting action. Operation of a cathode in this self-limiting mode is called *space charge limited operation.*

When emission is space charge limited, then cathode temperature, surface condition, and so forth have no effect on emission. This is indeed a very fortunate situation because it eliminates the necessity for extreme uniformity of temperature across the surface of a cathode. It also eliminates the necessity for precise control of the voltage and current to the heater in the cathode. This is why it can be said that, in this case, Mother Nature has been on our side.

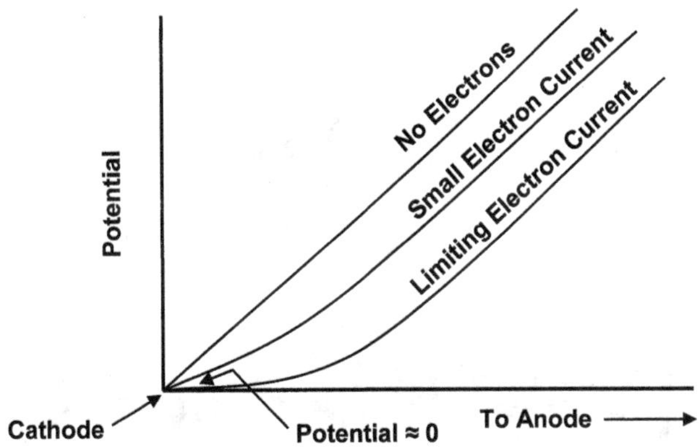

Figure 2-3 Effect of electrons on potential near the cathode in an electron gun.

In the electron gun shown again in Figure 2-4, another region (Region 3) is identified where the electric field produced by electrons plays a critical role in gun operation. The electrodes in an electron gun are designed so that, after electrons leave the cathode, they move toward the center of curvature of the cathode. As they enter the anode aperture, they pass through a region (Region 2, which is not shown here but is discussed in Chapter 3) where the action of the electric fields is analogous to that of an optical lens. The focal point of the electron flow is shifted from the center of curvature of the cathode to a point farther from the cathode.

Then, as the electrons pass through the anode, they enter Region 3 where they escape from the accelerating electric fields between the cathode and anode and the only forces affecting their trajectories are the forces between electrons. As the electrons move closer together, the increasing forces between electrons cause their trajectories to become nearly parallel in the center of Region 3. It is at this point that, to counteract the electron forces that cause the beam to expand, focusing forces must be applied to the beam if it is to be transported through the RF interaction structure of a klystron or TWT. Electron beam focusing and transport are discussed in Chapter 7.

The potential depression in the beam that accompanies the forces causing the beam to stop contracting and then to expand is dependent on the charge density in the beam. Also, as explained with the aid of Figures 2-5 and 2-6, the depression is dependent on the proximity of the beam to the metallic cylinder (tunnel) through which the beam is passing.

To find the potential in the beam, Gauss's law can be used. This law relates the electric field on a surface surrounding charged particles like electrons to the magnitude of the charge, Q, contained within that surface. Gauss's law is

$$\int_S E \cdot ds = \frac{Q}{\varepsilon_o} \tag{2-1}$$

where ε_o is the dielectric constant of free space.

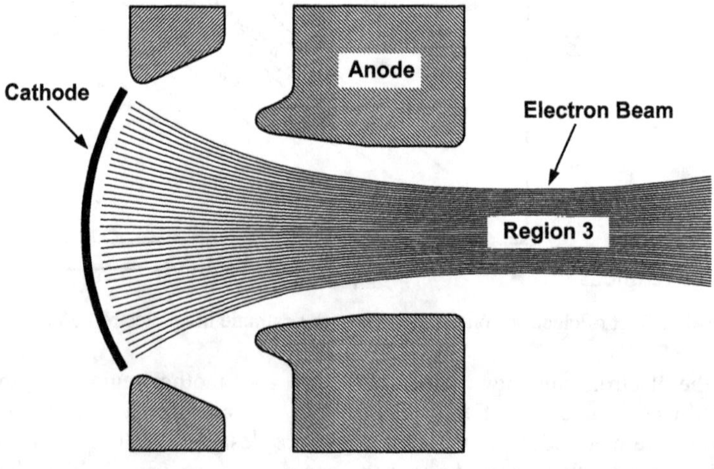

Figure 2-4 In Region 3, the forces between electrons are an essential part of the formation process of an electron beam.

Inside the electron beam for the segment shown in Figure 2-5, which is like that in the center of Region 3 where electrons have parallel trajectories, Gauss's law is

$$\int_S E_r \cdot ds = -\frac{\rho \pi r^2 \ell}{\varepsilon_o} \tag{2-2}$$

where ρ is the charge density, which is assumed to be uniform. The charge, Q, is the charge density multiplied by the volume, $\rho \pi r^2 \ell$.

Only the radial component, E_r, of the electric field exists and it is constant on the surface with radius r and zero at the ends of the beam segment so E_r is

$$E_r = -\frac{\rho r}{2\varepsilon_o} \tag{2-3}$$

The minus sign shows that the electric field is in the $-r$ direction, which means that it is pointed radially inward toward the center of the beam. (By definition, electric fields point from plus to minus.)

The potential, V, is the integral of the electric field and, inside the beam, is

$$V = \frac{\rho r^2}{4\varepsilon_o} \text{ volts} \tag{2-4}$$

The charge density, ρ, can be expressed in terms of current and velocity by referring to Figure 2-5. All of the charge in length ℓ passes position $z = \ell$ in time

$$t = \frac{\ell}{u_o} \tag{2-5}$$

where u_o is the beam velocity. The current, which is the charge per unit time, is

$$I = \frac{Q}{t} = \pi b^2 \rho u_o \tag{2-6}$$

and so the charge density is

$$\rho = \frac{I}{\pi b^2 u_o} \tag{2-7}$$

As a side note, the current density $J = I / \pi b^2$ is the current per unit area so

$$J = \rho u_o \tag{2-8}$$

This is, in fact, the definition of current in a beam, and it is a relation that will be used frequently.

Finally, the beam velocity is related to beam voltage by

$$u_o = \sqrt{2\eta V_b} \tag{2-9}$$

where η is the electron charge-to-mass ratio and V_b is the beam voltage.

When (2-4), (2-7), and (2-9) are combined and the constants are evaluated, the potential at the center of the beam with respect to the tunnel wall is found to be

$$V_o = 1.52 \times 10^4 \frac{I}{\sqrt{V_b}} \quad \text{volts} \tag{2-10}$$

For a 10,000-V, 1-A beam, the depression is 152V. For a 100,000-V, 30-A beam, the depression is 1,442V.

When the tunnel is larger than the beam, as shown in Figures 2-5 and 2-6, the electric field is

$$E_r = -\frac{\rho b^2}{2\varepsilon_o r} \tag{2-11}$$

and the potential depression in the center of the beam increases to

$$V = 2V_o \ln\frac{a}{b} + V_o \quad \text{volts} \tag{2-12}$$

If the tunnel radius is twice the beam radius, the depression is 363V for a 10,000-V, 1-A beam and 3,441V for a 100,000-V, 30-A beam.

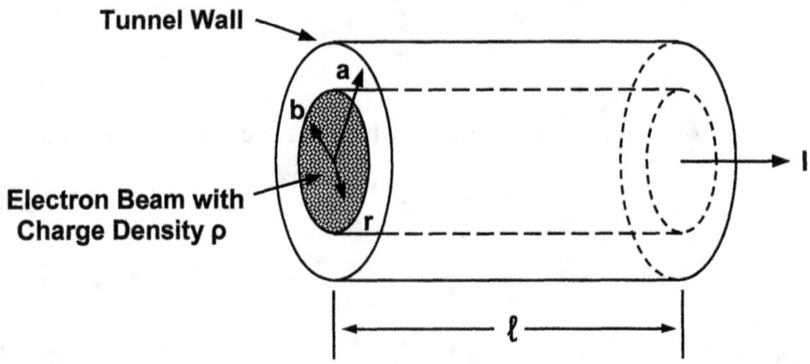

Figure 2-5 Segment of electron beam.

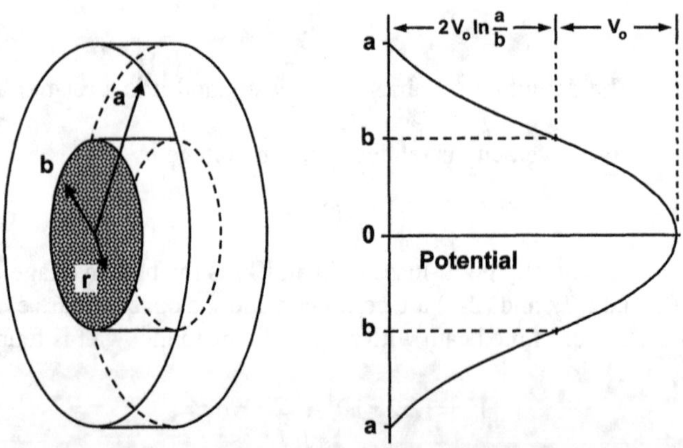

Figure 2-6 Slice of an electron beam and associated potential distribution.

These depressions are significant for at least two reasons. First, the electron beam is traveling more slowly than the beam voltage would indicate. For example, the electrons in the center of the 100,000-V beam with a 3,441-V depression have been accelerated through 96,559V rather than 100,000V.

The depression is also significant because positive ions, which can interfere with normal tube operation, can become trapped in the beam.

2.2 MAGNETIC FIELD

The electron beam in a microwave tube produces a magnetic field, but this field is small and is often neglected. The calculation of the field is straightforward. In any circular path concentric with the beam, the magnetic field is constant. At the outer edge of a beam with radius b and current I, the magnitude of the field is

$$H = \frac{I}{2\pi b} \quad \text{A/m} \tag{2-13}$$

The magnetic flux density is

$$B = \mu H = \frac{\mu I}{2\pi b} \quad \text{tesla} \tag{2-14}$$

where μ is the permeability of free space.

As an example, assume that a TWT contains a 1-A electron beam with a radius, b, of 1 mm as shown in Figure 2-7. The magnetic flux density is

$$B = 2\times10^{-4}\ \text{T} = 2\text{G} \tag{2-15}$$

This flux density is about three orders of magnitude smaller than the flux density used to focus the electron beam and so its effect is usually neglected.

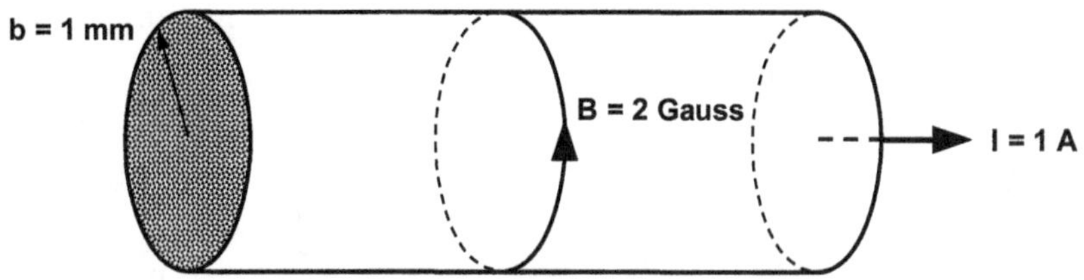

Figure 2-7 Magnetic flux generated by electron beam.

Chapter 3

Electron Motion in Static Electric Fields

In the electron guns and in some collectors of TWTs and klystrons, dc voltages are applied to accelerate or decelerate electrons. The ways in which those fields influence electron trajectories are described in this chapter. Included is a discussion of the relativistic velocity correction, which is important in very high-power devices.

The very simple case of electron deflection by transverse fields in a cathode ray tube is followed by an examination of electron motion in electric lenses. The focusing relation derived here will be of use in the analysis of electron guns.

Finally, the influence of the electric fields produced by the electrons on the electron trajectories in an axisymmetric configuration is examined. This will also be of use in the analysis of electron guns as well as in explanations of electron beam behavior.

3.1 MOTION PARALLEL TO FIELD

Assume that an electron is moving in the $+z$ direction and that the electric field, E_z, is in the opposite ($-z$) direction. Then the force on the electron is

$$F = m\,a = m\frac{d^2 z}{dt^2} = -\,eE_z \tag{3-1}$$

If E_z is constant and the initial electron velocity is zero, then

$$u_z = -\frac{e}{m}E_z t \tag{3-2}$$

The kinetic energy of the electron is

$$\frac{1}{2}mu^2 = -\,eE_z z \tag{3-3}$$

but, since E_z is constant, $-E_z\, z$ is the potential V through which the electron has been accelerated,

$$\frac{1}{2}mu^2 = eV \quad \text{joules} \tag{3-4}$$

The energy of the electron, eV, is often expressed in electron volts (eV) rather than in joules. The relation between electron volts and joules (J) is

$$1\,\text{eV} = 1.6\times10^{-19}\ \text{J} \tag{3-5}$$

An electron that has been accelerated through a potential of 1V has an energy of 1 eV. An electron that has been accelerated through 10,000V has an energy of 10,000 eV.

If the electron has an initial velocity of u_o resulting from being accelerated through a potential V_o, then

$$\frac{m}{2}\left(u^2 - u_o^2\right) = e\left(V - V_o\right) \tag{3-6}$$

The total energy of the electron is still eV and the increase in energy is $e(V - V_o)$.

3.2 RELATIVISTIC VELOCITY CORRECTIONS

The relation (3-4), showing that electron velocity is proportional to the square root of voltage, is valid only when velocities are low compared with the velocity of light. This is because the mass was assumed to be constant, whereas, by the theory of relativity, the mass of a particle changes with velocity. Also, from the theory of relativity, mass and energy are equivalent and are related by the square of the velocity of light, c^2; that is, the energy, w, is related to mass by

$$w = mc^2 \tag{3-7}$$

If a force, F, is applied over a distance, dz, to accelerate an electron, then the energy expended, dw, and the increase in electron mass, dm, are related by

$$F\,dz = dw = c^2 dm \tag{3-8}$$

Newton's second law is

$$F = \frac{d}{dt}(mu) \tag{3-9}$$

At high velocities, where m is variable, this equation must be written as

$$F = \frac{d}{dt}(m\,u) = m\frac{du}{dt} + u\frac{dm}{dt} \tag{3-10}$$

When this relation is combined with (3-9) and integrated with the constant evaluated at $u = 0$ so that the electron mass is the "rest" mass, m_o, the result is

$$m = \frac{m_o}{\left(1 - \frac{u^2}{c^2}\right)^{1/2}} \tag{3-11}$$

The relation between electron velocity and voltage can now be found by reconsidering the energy expended, $c^2\,dm$, in accelerating the electron over distance dz (see (3-8)). If the accelerating force is supplied by an electric field, E, then the energy supplied by that field is $-eEdx$ or edV where dV is the voltage increment through which the electron was accelerated. The energy supplied must equal the increase in electron energy so

$$e\,dV = c^2 dm \tag{3-12}$$

If the electron starts at rest, where it has "rest" mass, m_o, and is accelerated through a total voltage, V, then

$$e\,V = c^2\left(m - m_0\right) \tag{3-13}$$

With the substitution of the mass from (3-11)

$$u = c\left(1 - \frac{1}{\left(1 + \frac{eV}{c^2 m_o}\right)^2}\right)^{1/2} \tag{3-14}$$

Defining

$$V_n = \frac{m_o c^2}{e} = 5.11 \times 10^5 \quad \text{V} \quad (\approx 1/2\,\text{MV}) \tag{3-15}$$

where eV_n is the energy equivalent of the rest mass of the electron, then

$$u = c\left(1 - \frac{1}{\left(1 + \frac{V}{V_n}\right)^2}\right)^{1/2} \tag{3-16}$$

When this expression for velocity is substituted into (3-11) for mass,

$$m = m_o \left(1 + \frac{V}{V_n}\right) \tag{3-17}$$

Electron mass and velocity as functions of voltage are shown in Figure 3-1.

As is indicated in Figure 3-1, most microwave tubes operate at voltages between 5 kV and 250 kV. Some low-power devices operate below 5 kV and a very few extremely high-power devices operate above 250 kV.

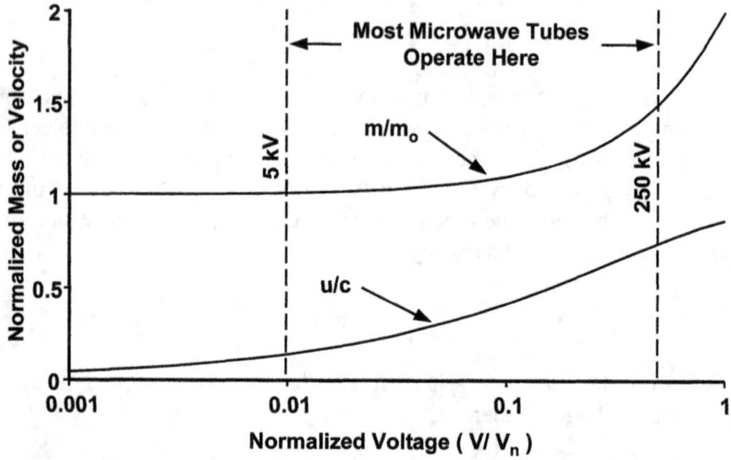

Figure 3-1 Normalized electron mass and velocity as functions of voltage.

3.3 ELECTRIC LENSES

In this section, the influence of an electric lens on electron trajectories is examined. The electron gun used in Chapter 2 and shown again here in Figure 3-2 will be of use in explaining one of the reasons why we want to understand the effects of electric lenses.

As electrons enter the anode aperture, they pass through Region 2 where the action of the electric fields is analogous to that of an optical lens. The focal point of the electron flow is shifted from the center of curvature of the cathode to a point farther from the cathode. To understand why this happens, it is very useful to examine the "equipotential" surfaces near the anode in the electron gun.

In Figure 3-2, the anode voltage is 10 kV above the cathode voltage. An equipotential surface at a potential of 5 kV is identified. If a voltage probe could be used that did not disturb the potential surfaces, and voltages were measured, this is the surface on which the 5-kV measurements would be plotted. Thus, an equipotential surface is an imaginary surface on which the potentials are equal.

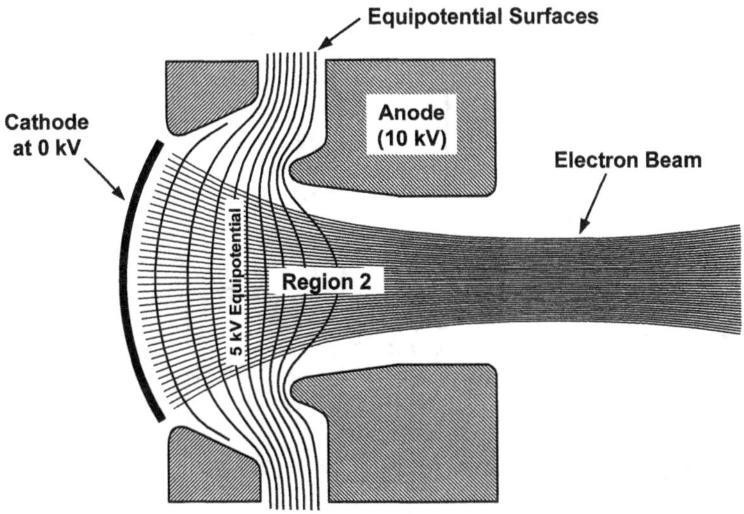

Figure 3-2 Equipotential surfaces that form a defocusing lens in an electron gun.

The equipotential surfaces to the right of the 5-kV contour in Figure 3-2 all bow toward the aperture in the anode. Electrons tend to have trajectories that are perpendicular to equipotential surfaces because that is the direction in which the electric field is strongest, so that is why Region 2 is a beam-defocusing region.

In Figure 3-3, single aperture lenses are shown and the focusing actions of these lenses are indicated. Each electrode set in Figure 3-3 consists of three parallel plates with voltages such that $V_3 > V_2 > V_1$. The center plate at potential V_2 is equally spaced from the other two and has a circular aperture. In this section, a simple relation is developed to determine the focal length, f, in terms of the applied voltages. This focal length can be used in the analysis of electron guns.

As shown in Figure 3-3(a), if the voltage difference between plates 1 and 2 is larger than that between plates 2 and 3, the equipotential contours near the aperture will bulge into the region between 2 and 3. An electron traveling parallel to the axis in regions 1–2 will have its trajectory deflected away from the axis in regions 2–3.

For the case shown in Figure 3-3(b), the voltage difference between plates 2 and 3 is larger than between 1 and 2 and so the equipotential contours near the aperture bulge into the region 1–2. An electron traveling parallel to the axis in region 1–2 will have its trajectory deflected toward the axis in region 2–1.

Assuming that $V_1 = 0$ and that an electron leaves electrode 1 with negligible velocity, then the velocity of the electron at electrode 2 at $z = 0$ is

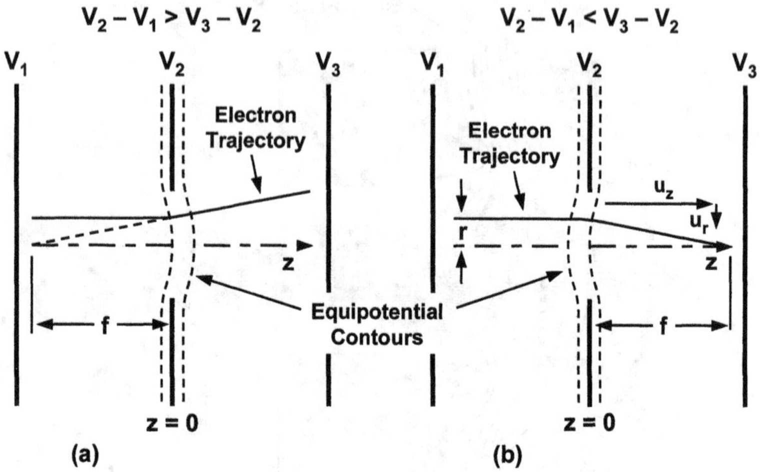

Figure 3-3 (a, b) Focusing action of single-aperture lens.

$$u_z = \left(2\frac{e}{m}V_2\right)^{1/2} \tag{3-18}$$

The approximation made here is that the voltage at $z = 0$ in the aperture is the same as that on electrode 2.

As is indicated in Figure 3-4(a), in the vicinity of the aperture in the center electrode, where the equipotential surfaces are curved, there is a radial component to the electric field, E_r. This radial field produces a radial force on the electron and a radial velocity component, u_r, which can be obtained from

$$\frac{d}{dt}m\,u_r = -e\,E_r \tag{3-19}$$

and is

$$u_r = -\frac{e}{m}\int_{-t_o}^{t_o} E_r dt \tag{3-20}$$

The time $-t_o$ to t_o is the time spent by the electron in the influence of the radial field. Assuming that u_z is constant during this time, then the time interval, $-t_o$ to t_o, corresponds to a distance interval, $-z_o$ to z_o, and since $u_z = dz/dt$, u_r may be written as

$$u_r = -\frac{e}{m u_z}\int_{-z_o}^{z_o} E_r dz \tag{3-21}$$

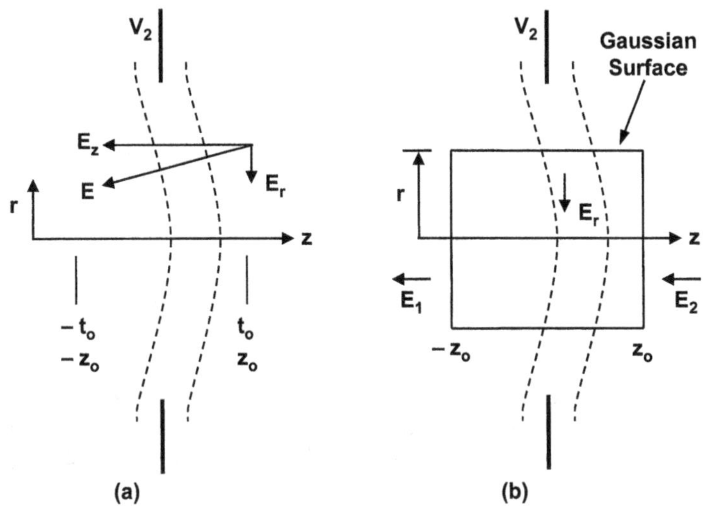

Figure 3-4 (a, b) Electric fields in lens.

An expression for the integral in (3-21) can be obtained by using Gauss's law. A surface to which Gauss's law will be applied (Gaussian surface) is shown in Figure 3-4(b). The surface has a radius r and a length $2z_o$ and is located at the center of one of the lenses in Figure 3-3. There is no charge within the surface and so the integral of the normal component of the electric field over the surface is zero. Assuming that E_1 and E_2 are constant, the result is

$$\int_{-z_o}^{z_o} E_r dz = \frac{r}{2}\left(E_1 - E_2\right) \tag{3-22}$$

so

$$u_r = -\frac{e}{m u_z}\frac{r}{2}\left(E_1 - E_2\right) \tag{3-23}$$

Dividing this equation by u_z and using the energy relation (3-18), then

$$\frac{u_r}{u_z} = -\frac{r}{4}\frac{\left(E_1 - E_2\right)}{V_2} \tag{3-24}$$

However, from Figure 3-3

$$\frac{u_r}{u_z} = \frac{r}{f} \tag{3-25}$$

so

$$f = -\frac{4\ V_2}{E_1 - E_2} \tag{3-26}$$

This is the single-aperture lens formula of Davisson and Calbick. This result is valid for $u_r << u_z$ in a thin lens ($z_o <<$ radius of the lens aperture).

3.4 UNIVERSAL BEAM SPREAD CURVE

Earlier in this chapter, the motions of individual electrons under the influence of applied fields were examined. In this section, as was done in the discussion of Region 3 in an electron gun in Chapter 2, it will be assumed that there are no applied fields. A beam of electrons is assumed and the only electric fields are those produced by the electrons. This analysis differs from that in Chapter 2 because the radial motions of the electrons and radial decelerating and accelerating forces are considered.

The situation that will be analyzed is like that shown in Figure 3-5. The electron beam is converging as it approaches $z = z_{m,}$ but the forces between electrons cause the beam to stop converging and then to diverge.

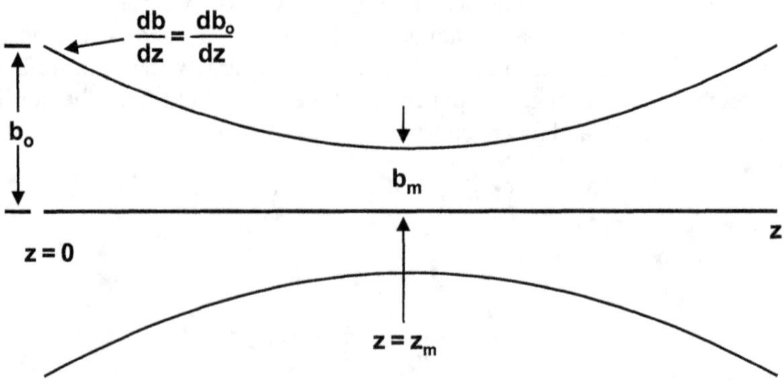

Figure 3-5 Assumed beam shape.

The radial acceleration of the electrons is

$$\frac{d^2 r}{dt^2} = -\eta E_r \tag{3-27}$$

where E_r is the electric field produced by the electrons. Using the relations given in Chapter 2 for the electric field, (2-3), and for the charge density, (2-7), the electric field at the beam boundary is

$$E_r(b) = -\frac{I}{2\pi b\,\varepsilon_o u_o} \tag{3-28}$$

so the equation of motion for electrons on the beam boundary as a function of time is

$$\frac{d^2 b}{dt^2} - \frac{\eta I}{2\pi b\,\varepsilon_o u_o} = 0 \tag{3-29}$$

or, as a function of z, is

$$\frac{d^2 b}{dz^2} - \frac{\eta I}{2\pi b\,\varepsilon_o u_o^3} = 0 \tag{3-30}$$

Next, let

$$A = \left[\frac{\eta I}{\pi\,\varepsilon_o u_o^3}\right]^{1/2} = 174\sqrt{P} \tag{3-31}$$

where

$$P = \frac{I}{V^{3/2}} \tag{3-32}$$

In Chapter 5, it will be shown that the relation between current and voltage in most thermionic diodes, including the electron guns on klystrons and TWTs is given by (3-32). The quantity P is called the *perveance* and depends only on the geometry of the diode or electron gun. Now, by substituting A into (3-30), the following expression for the beam radius is obtained:

$$\frac{d^2 b}{dz^2} - \frac{A^2}{2b} = 0 \tag{3-33}$$

This can be rewritten as

$$\frac{d^2 B}{dZ^2} - \frac{1}{2B} = 0 \tag{3-34}$$

where

$$B = \frac{b}{b_o} \quad \text{and} \quad Z = A\frac{z}{b_o} \tag{3-35}$$

Here, b_o is the initial radius of the beam at $z = 0$.

After multiplying by dB/dZ, integrating and evaluating the integration constant at $Z = 0$, (3-34) becomes

$$\left(\frac{dB}{dZ}\right)^2 = \ln B + \left(\frac{dB_o}{dZ}\right)^2 \tag{3-36}$$

or

$$B = e^{(dB/dZ)^2 - (dB_o/dZ)^2} \tag{3-37}$$

When the beam radius reaches the minimum value, b_m, and $B = B_m$, then $dB/dZ = 0$, so

$$B_m = e^{-(dB_o/dZ)^2} \tag{3-38}$$

To obtain Z, (3-36) may be rewritten as

$$Z = \int_1^B \frac{dB}{\left(\ln B + (dB_o/dZ)^2\right)^{1/2}} \tag{3-39}$$

This integral may be put in a recognizable form by changing variables and letting

$$u = \frac{dB}{dZ} \tag{3-40}$$

then

$$Z = 2e^{-(dB_o/dZ)^2} \int_{dB_o/dZ}^{dB/dZ} e^{u^2}\, du \tag{3-41}$$

Plots of B as a function of Z can be generated for various values of dB_o/dZ by selecting values of dB/dZ and then calculating corresponding values of B and Z. The result is in Figure 3-6. These curves are symmetrical about the position of the minimum beam radius. As dB_o/dZ is made more negative, the minimum first moves to the right ($+Z$ direction) and then to the left. Thus, for a given perveance, there is an optimum value of dB_o/dZ that maximizes the length of a tunnel through which a beam can travel without being intercepted by the tunnel walls.

In the design of an electron gun, the position of the minimum beam radius will be of interest. This position, Z_{min}, as a function of dB_o/dZ can be determined by evaluating (3-41) with $dB/dZ = 0$. The result is shown in Figure 3-7.

A version of this curve that can be obtained by changing the initial conditions to the position where the minimum beam radius occurs and by normalizing the radius and axial distance with respect to the minimum radius, b_m, is the universal beam spread curve shown in Figure 3-8. The curve is symmetrical about $z = z_o$. This curve describes the shape of an electron beam in a field-free space. One of its uses is in periodic focusing systems where this curve describes the shape of the beam between lenses.

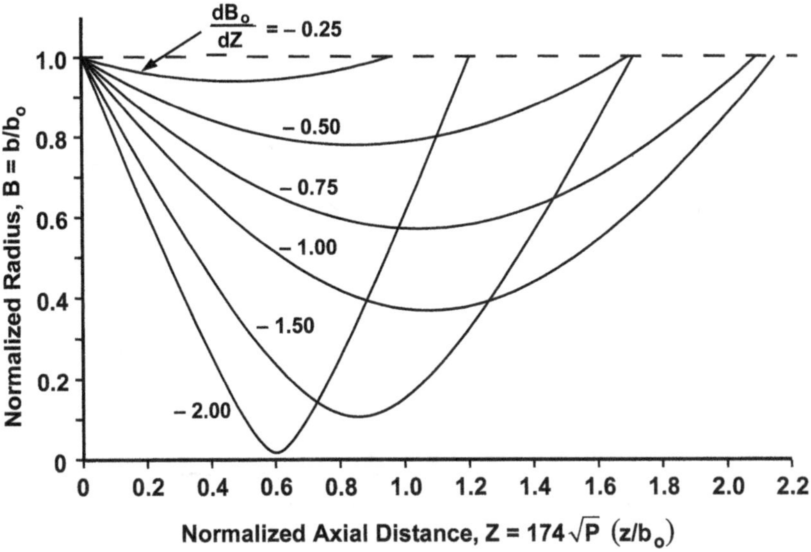

Figure 3-6 Electron trajectories under the influence of space charge.

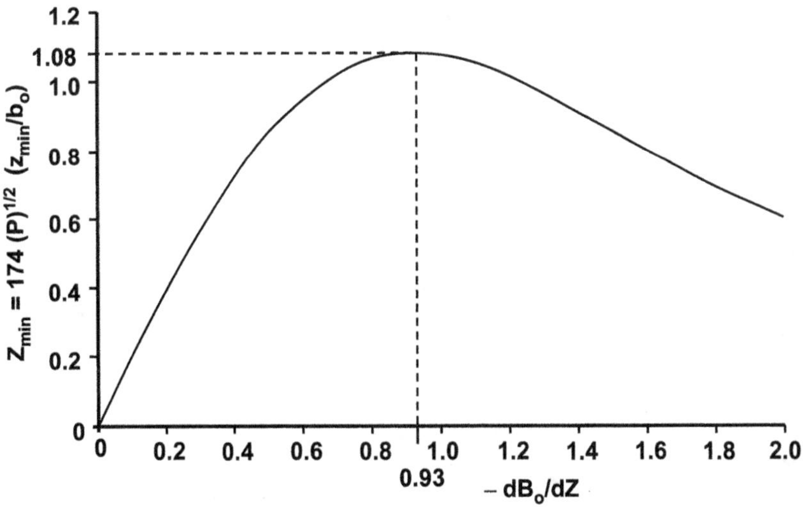

Figure 3-7 Axial position, Z_{min}, of minimum beam radius as a function of dB_o/dZ.

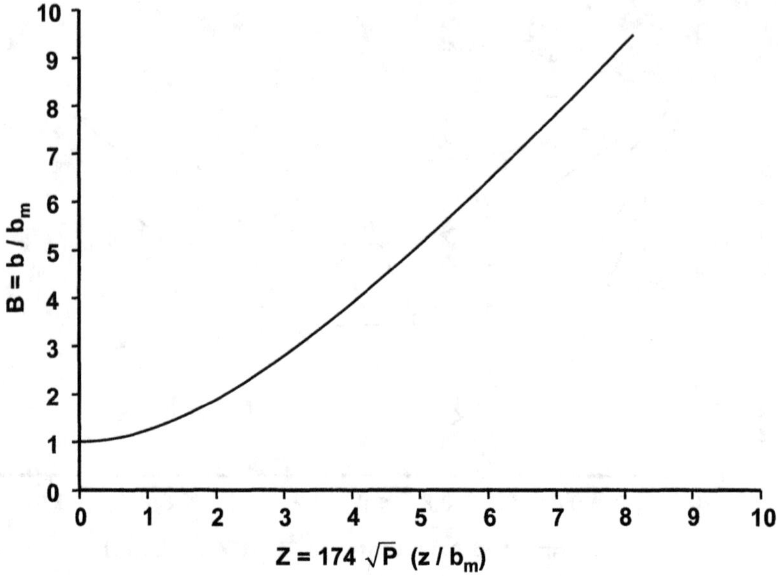

Figure 3-8 Universal beam spread curve.

Chapter 4

Influence of Magnetic Field on Electron Motion

Magnetic fields are used in nearly all microwave tubes for controlling the motions of the electrons. In this chapter, the motion of an electron in a static magnetic field is considered first. Then the motion in combined electric and magnetic fields is examined. After a general consideration of axially symmetric fields, Busch's theorem is derived. This theorem applies when axial and radial electric or magnetic fields are present, and is extremely important in the analysis of electron beams in all linear-beam tubes.

4.1 ELECTRON MOTION IN A STATIC MAGNETIC FIELD

The force, $\boldsymbol{F}$, produced by magnetic flux, $\boldsymbol{B}$, on an electron moving with velocity, $\boldsymbol{u}$, is

$$\boldsymbol{F} = -e\boldsymbol{u} \times \boldsymbol{B} \tag{4-1}$$

The quantities $\boldsymbol{F}$, $\boldsymbol{B}$, and $\boldsymbol{u}$ are bold to indicate that they are vectors. The vector product $\boldsymbol{u} \times \boldsymbol{B}$ is maximum when the vectors are perpendicular. For vectors that are not perpendicular, the vector product decreases by the sine of the angle between vectors.

Figure 4-1 shows the trajectory of an electron traveling with velocity u_o at a right angle to the direction of the flux density, B, which is constant. The force on the electron is eu_oB. An easy way to determine the direction of the force on the electron is by using the right-hand rule. As is indicated in Figure 4-1,

1. Point fingers toward direction of charge velocity, u_o.
2. Rotate fingers toward magnetic field, B.
3. Thumb points to force on positive charge, F+.
4. Force on electron is opposite.

The force deflects the trajectory of the electron as shown in Figure 4-1 and, because its direction is perpendicular to the trajectory of the electron, it does not change the speed of the electron. Since the speed is constant, the force is constant. The result is that the path of the electron is circular.

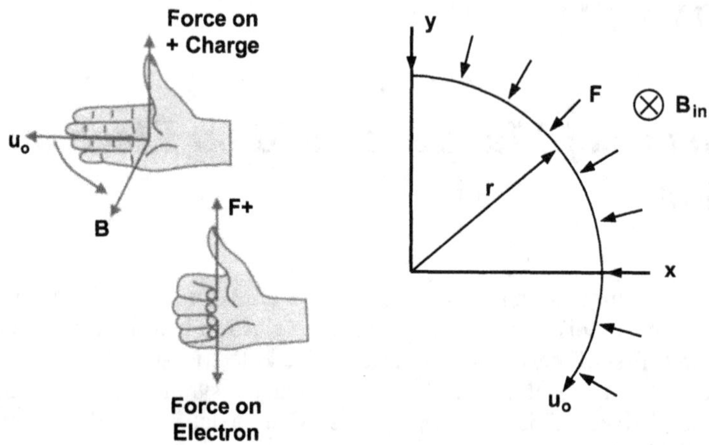

Figure 4-1 Deflection of electron by magnetic field.

The radius of the path can be determined by considering the outward centrifugal force on the electron as it follows the circular path. The magnitude of this force is mu_o^2/r, where r is the radius of the electron path. This outward force equals the inward force produced by the magnetic field so

$$e u_o B = \frac{m u_o^2}{r} \tag{4-2}$$

from which

$$r = \frac{u_o}{\eta B} \tag{4-3}$$

In terms of the voltage, V_o, through which an electron was accelerated to reach velocity u_o, r is

$$r = 3.37 \times 10^{-6} \frac{V_o^{1/2}}{B} \text{ m} \tag{4-4}$$

At $V_o = 10{,}000$V and $B = 0.1$T (1,000) G, $r = 3.37$ mm.

The frequency with which the electron rotates is easy to determine because the path length for one rotation is $2\pi r$ and the velocity is u_o. Thus, the time required for each revolution is

$$t = \frac{2\pi r}{u_o} = \frac{2\pi}{\eta B} \tag{4-5}$$

and the frequency of rotation is

$$f = \frac{1}{t} = \frac{1}{2\pi}\eta B \tag{4-6}$$

Note that this frequency depends only on the electron charge and mass and on the magnetic flux density but not on the electron velocity or the radius of the path. The angular frequency with which the electron rotates is called the *cyclotron frequency*, ω_c, and is given by $\omega_c = 2\pi f$ or

$$\omega_c = \eta B \tag{4-7}$$

Some values of the frequency of rotation, $f_c = \omega_c / 2\pi$, are:

f_c (MHz)	Flux Density (G)
2.8	1
28	10
280	100
2,800	1,000
28,000	10,000

4.2 ELECTRON MOTION IN COMBINED ELECTRIC AND MAGNETIC FIELDS

4.2.1 Perpendicular Fields in Rectangular Coordinates

When electric and magnetic fields are present, the force equation is

$$\boldsymbol{F} = -e(\boldsymbol{E} + \boldsymbol{u} \times \boldsymbol{B}) \tag{4-8}$$

Assume, as indicated in Figure 4-2, that the E field is in the $-y$ direction and that B is in the $-z$ direction. First, let the electric field be zero; then the force is $-e(u \times B_z)$ and this is the example considered earlier. As shown in Figure 4-2, with an initial velocity of $u_o = 0$ in the x direction, the electron follows a circular trajectory in the clockwise direction.

With u_o and B_z remaining constant, as the electric field is increased from zero, the electron trajectories change dramatically. For example, consider the trajectory when E_y has been increased so that $E_y / B_z = u_o / 3$. The force produced by the electric field is in the + y direction and, as the electron moves away from $x = 0$, this opposes the y-directed magnetic force. As a result, the electron trajectory gradually moves outside the $E_y = 0$ path. As E_y continues to decelerate the electron, the y component of velocity eventually becomes zero.

At the same time, as the velocity is decreased, the radius of curvature, which is u/ω_c, is reduced. Then, as the electric and magnetic forces accelerate the electron in the $+y$ direction, the trajectory takes on the shape of a teardrop.

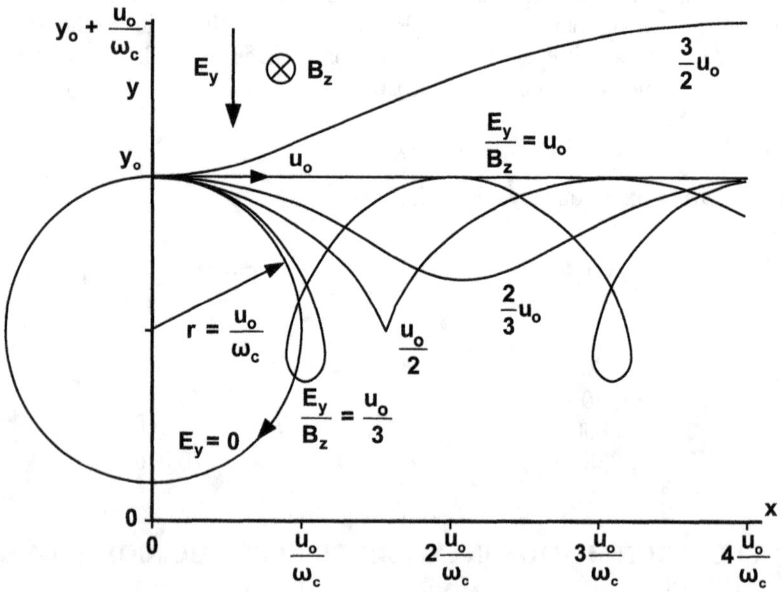

Figure 4-2 Electron trajectories for various electric field levels. The magnetic flux density, B_z, and initial electron velocity, u_o, are constant.

As the electric field is increased, the electron trajectories become cusp shaped and then somewhat sinusoidal. When the field is such that the y-directed force is zero, that is,

$$F_y = 0 = -e\left(-E_y + u_o B_z\right) \tag{4-9}$$

then

$$\frac{E_y}{B_z} = u_o \tag{4-10}$$

and the electron trajectory is a horizontal line.

Electron motions similar to these occur in a circular format in a magnetron where the electric field is radial and the magnetic field is parallel to the axis.

4.2.2 Axially Symmetric Fields

In Chapter 3, we examined the motions of electrons in a beam when only space charge forces are present. This led to the universal beam spread analysis. The effect of a magnetic field was not included in that analysis. In this section, we will derive Busch's theorem, which is a powerful tool used in the analysis of the motion of electrons when axial and radial electric and magnetic fields are present. In Chapter 7, Busch's theorem will be used with the universal beam spread analysis to define the behavior of magnetically focused electron beams. Forces resulting from axial electron motions, $\dot{z}$, and a radial component, B_r, of the magnetic flux density will be included so that the θ force equation is

$$F_\theta = m\left(2\dot{r}\dot{\theta} + r\ddot{\theta}\right) = e\dot{r}B_z - e\dot{z}B_r \tag{4-11}$$

which can be written as

$$\eta r\left(\dot{r}B_z - \dot{z}B_r\right) = \frac{d}{dt}\left(r^2\dot{\theta}\right) \tag{4-12}$$

Multiplying by *dt,* the following can be obtained:

$$\frac{\eta}{2\pi}\left(2\pi r dr B_z - 2\pi r dz B_r\right) = d\left(r^2\dot{\theta}\right) \tag{4-13}$$

To attach physical significance to this equation, consider a ring of electrons moving axially and expanding radially as shown in Figure 4-3. Assume that the ring is crossing magnetic flux lines. As the ring expands radially, it crosses axial components of the magnetic flux. The total axial flux crossed, $d\Psi_z$, in distance dr is

$$d\Psi_z = 2\pi r dr B_z \tag{4-14}$$

where B_z is the axial component of flux density. The resulting force on the electrons is in the clockwise ($+\theta$) direction.

As the ring moves axially, it crosses radial components of flux. The total radial flux crossed, $d\Psi_r$, in the distance dz is

$$d\Psi_r = 2\pi r dz B_r \tag{4-15}$$

where B_r is the radial component of flux density. The resulting force on the electrons is in the counterclockwise ($-\theta$) direction.

Now, (4-14) and (4-15) are the quantities on the left side of (4-13) so the net force producing rotation of the electron ring results from the flux change, $d\Psi = d\Psi_z - d\Psi_r$, and the rotation can be found from

$$\frac{\eta}{2\pi} d\Psi = d(r^2\dot{\theta}) \qquad (4\text{-}16)$$

When this is integrated, the result is

$$r^2\dot{\theta} - r_o^2\dot{\theta}_o = \frac{\eta}{2\pi}(\Psi - \Psi_o) \qquad (4\text{-}17)$$

where r_o, $\dot{\theta}_o$, and Ψ_o are the radius and rate of rotation of the ring and the flux contained within it at z_o. The quantities r, θ, and Ψ are the corresponding values at position z.

When $\dot{\theta}_o = 0$ at z_o (such as at the cathode), (4-17) becomes

$$\dot{\theta} = \frac{\eta}{2\pi r^2}(\Psi - \Psi_o) \qquad (4\text{-}18)$$

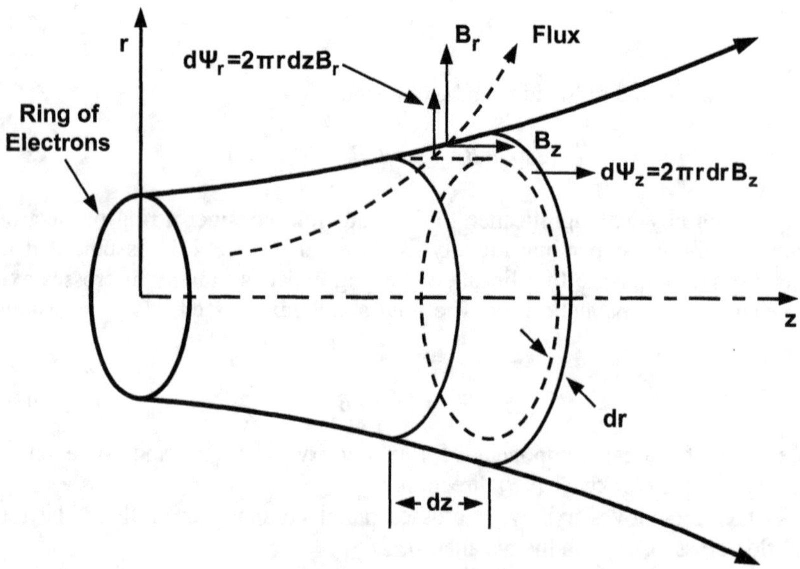

Figure 4-3 Ring of electrons crossing axially symmetric magnetic flux lines.

In most applications of interest in linear-beam tubes, Ψ changes slowly with z so

$$\Psi \approx \pi r^2 B_z \quad \text{and} \quad \Psi_o \approx \pi r_o^2 B_{zo} \tag{4-19}$$

where B_{zo} is the flux density at the point where $\dot{\theta}$ is zero. Now, Busch's theorem assumes the approximate form

$$\dot{\theta} = \frac{\eta}{2}\left(B_z - B_{zo}\frac{r_o^2}{r^2} \right) \tag{4-20}$$

As stated at the beginning of this section, this result is great value in understanding the behavior of magnetically focused electron beams.

Chapter 5

Thermionic Cathodes

The cathode is the source of electrons for the electron beam in every microwave tube. The current density of electron emission from the cathode ranges from milliamperes to tens of amperes per square centimeter of cathode area.

Two cathode emission mechanisms are used in conventional tubes:

- Thermionic emission;
- Secondary emission.

In cathodes for linear-beam tubes, only thermionic emission is used. Thermionic cathodes are the primary subject for discussion in this chapter. Secondary emission cathodes are used in crossed-field tubes such as magnetrons and crossed-field amplifiers. Those cathodes and their applications are discussed in Chapter 18.

J. R. Pierce [1] listed the primary characteristics that an ideal cathode should have:

1. Emits electrons freely, without any form of persuasion such as heating or bombardment (electrons would leak off from it into vacuum as easily as they pass from one metal to another);
2. Emits copiously, supplying an unlimited current density;
3. Lasts forever, its electron emission continuing unimpaired as long as it is needed;
4. Emits electrons uniformly, traveling at practically zero velocity.

Of course, real cathodes have none of these ideal characteristics. For example, real cathodes must be heated to a temperature near 1,000°C to produce appreciable electron emission. At that temperature, the current density is not unlimited, but instead is on the order of a few tens of amperes per square centimeter at most. Because of the necessity for a high operating temperature, some of the key constituents of a cathode evaporate, which leads to their depletion and, eventually, to the end of the life of the cathode.

The reason for the fourth characteristic of an ideal cathode is not as obvious as the other three. First, it should be noted that, in real cathodes, there are random microscopic fluctuations in the rate at which electrons are emitted. Second, electrons are emitted with finite random velocities in random directions relative to the cathode surface. There are at least two important reasons why an ideal cathode should "emit electrons uniformly, traveling at practically zero velocity":

1. Fluctuations in electron emission and variations in electron emission velocities lead to noise currents in the electron beam, which produce noise on the output from the tube.
2. Variations in emission velocities and in the directions with which electrons leave the cathode surface lead to problems in focusing the electrons into a well-defined beam. This problem is particularly severe in very small, high-density beams.

While real cathodes have none of the characteristics that Pierce's ideal cathode would have, there has been enormous progress in cathode capabilities since Pierce listed his ideal characteristics. This progress is indicated in Figure 5-1. Note, in particular, the large increase in emission capability that was achieved in the 1970s and 1980s. As pointed out by Thomas et al., this progress was due largely to the advent of modern surface analytical capabilities that made possible a better understanding of the physics and chemistry of emitter surfaces [2]. In this chapter, we attempt to provide sufficient background material so that, at least qualitatively, the operation of modern cathodes, with the exception of the scandate cathode, can be understood. The operation of the scandate cathode is the subject of active research at the present time.

The chapter starts with the basic theory of thermionic emission. This theory is straightforward and gives a reasonably accurate prediction of the maximum emission that can be expected from a cathode at a given temperature, in the absence of field effects, if the work function is known. Next, the enhancement of thermionic emission that results from the application of an electric field to the surface of the cathode (the Schottky effect) is discussed. The emission regulating effect of the electron cloud adjacent to the surface of the cathode (space charge limited emission) is of extreme importance in the use of thermionic cathodes in linear-beam tubes. For the person who must use a cathode in a linear-beam tube, the concept of space charge limited emission may be the most important material discussed in this chapter.

To maintain space charge limited emission, the cathode work function must remain low enough so that adequate emission is obtained at a given temperature during the life of a cathode. The work function of the cathode is critically dependent on the materials and condition of the surface of the cathode. Thus, a large portion of this chapter is devoted to a discussion of cathode materials and their impact on life of the cathode and the tube in which it is used.

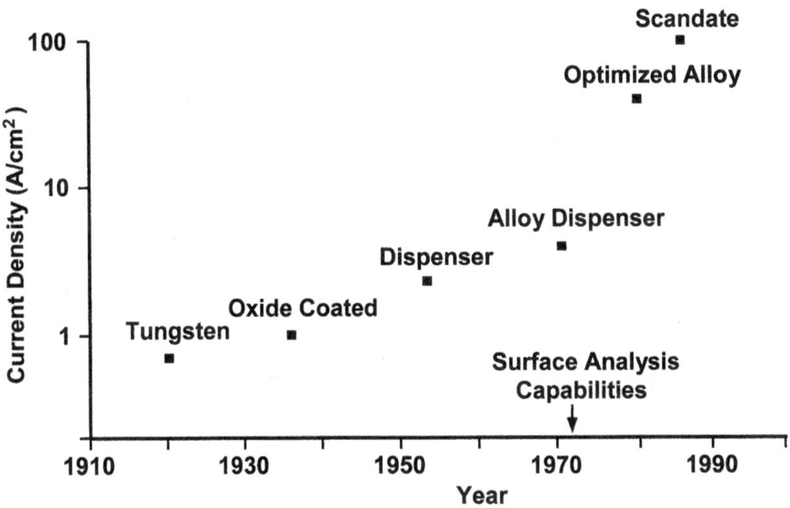

Figure 5-1 Historical perspective of thermionic cathode emission capabilities. (From: R. E. Thomas et al., *IEEE Trans. Electron Devices*, March 1990. © 1990 IEEE.)

The physical arrangement of the cathode and the elements that provide the power necessary to heat it to the required temperature in the time permitted by the application are described. Conventional heaters are discussed along with fast warm-up assemblies that are required in some special applications of tubes.

The magnetic field of the filament can have a significant impact on the operation of an electron gun. Electron trajectories near the cathode surface are perturbed by this field with the result that beam generation and focusing are degraded. If an ac filament current is used, the beam shape may be modulated at the ac frequency and this, in turn, modulates the power output of the tube. These effects are discussed at the end of the chapter.

5.1 EMISSION MECHANISMS

5.1.1 Thermionic Emission [3, 4]

At temperatures above absolute zero, some electrons have sufficient energy to escape from a cathode surface. As temperature is increased, the number of electrons with sufficient energy to escape increases. In addition to temperature, the nature of the surface of the cathode has an extremely strong effect on the rate at which electrons are emitted. Electron emission resulting from the heating of a surface is referred to as thermionic emission.

The fundamental aspects of thermionic emission may be understood with the aid of Figure 5-2, which is the classical energy level diagram for electrons near the interface between a cathode and vacuum. In the cathode, the parabolic curves represent electron energy levels adjacent to atoms. The energy levels merge to form the conduction band. At a temperature of absolute zero (0K), no electrons have energy greater than E_o, which is the top of the conduction band and is called the *Fermi level*. The difference in energy between the top of the conduction band in the cathode and the vacuum level adjacent to the cathode is known as the *work function*, commonly denoted by $e\varphi$.

At temperatures above absolute zero, some electrons have energy greater than E_o. Emission of the electrons can occur if their energy is $E_o + e\varphi$ or greater. It is necessary to keep in mind, however, that the electrons move in random directions within the cathode. Those moving toward the surface have the highest probability of being emitted.

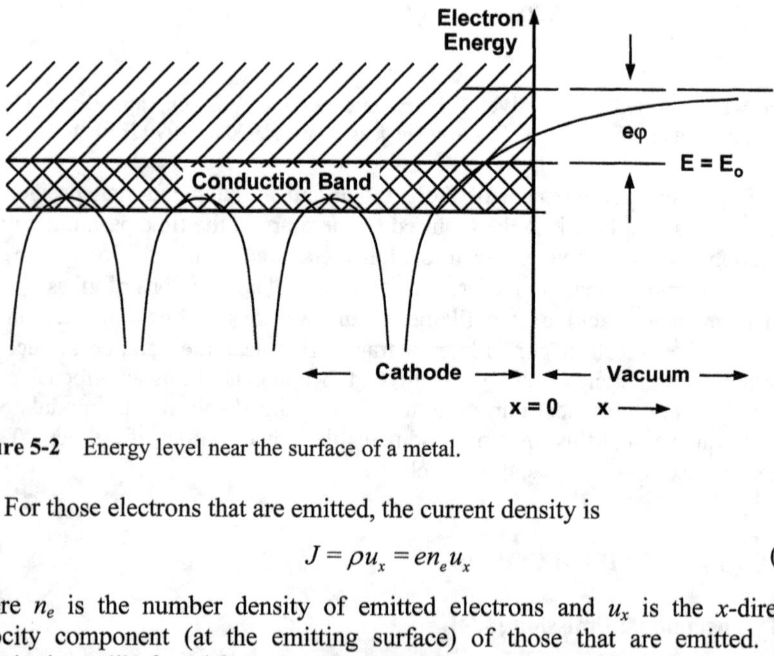

Figure 5-2 Energy level near the surface of a metal.

For those electrons that are emitted, the current density is

$$J = \rho u_x = e n_e u_x \tag{5-1}$$

where n_e is the number density of emitted electrons and u_x is the x-directed velocity component (at the emitting surface) of those that are emitted. This velocity is readily found from the energy relation

$$\frac{1}{2} m u_x^2 = E_o + e\phi \tag{5-2}$$

The number density being emitted, n_e, is not so easy to find. This is computed from the density of energy states in the cathode and the probability that they are occupied. When the analysis is carried out, it is found that

$$J = A_o T^2 e^{-e\phi/kT} \tag{5-3}$$

which is the Richardson-Dushman equation for thermionic emission. A_o is a universal constant and has the value

$$A_o = 1.20 \times 10^6 \text{ A/ m}^2\text{deg}^2 \tag{5-4}$$

The most significant aspect of the Richardson-Dushman equation is the exponential variation of current density with work function and temperature. The variation of current density with T^2 is negligible by comparison. This can be readily demonstrated by considering the relative effects of a small change in temperature. For example, for a typical operating temperature of 1,000°C, a 1% change in T produces roughly a 70% change in the value of the exponential term, but only a 2% change in T^2.

To compare theory with experiment, the Richardson-Dushman equation is first written in the following form:

$$\ln \frac{J}{T^2} = \ln A_o - \frac{e\phi}{kT} \tag{5-5}$$

This suggests that a plot of $\ln J/T^2$ versus $1/T$ should be a straight line. The work function, $e\varphi$, should be the slope of the line and $\ln A_o$ should be the intercept of the extrapolation of the line to infinite temperature ($1/T = 0$). The data are plotted in the form shown in Figure 5-3, which is known as a Richardson plot.

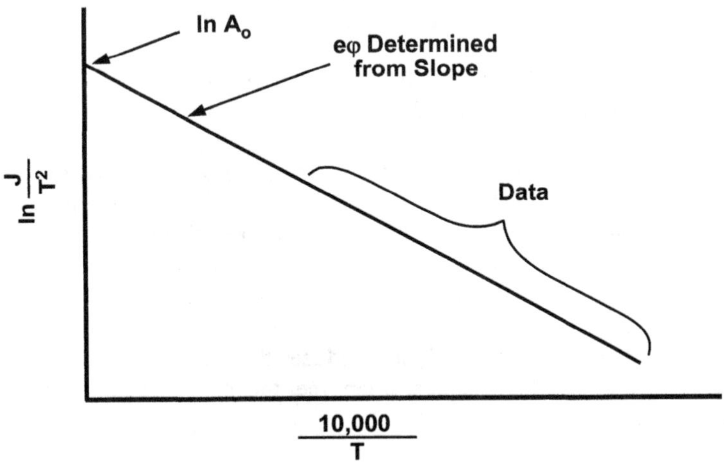

Figure 5-3 Richardson plot for determining work function and emission constant A_o.

In practice, the experimental value of A_o turns out to be on the order of one-quarter to one-half of the predicted value. Two reasons for this are:

1. The work function has been assumed to be independent of temperature. In fact, $e\varphi$ may vary with temperature.
2. The same value of $e\varphi$ has been assumed for all parts of the emitting surface. In fact, $e\varphi$ may be different for different crystal surfaces. Suppose, for example, that half of the emitter area has work function $e\varphi_1$ and the other half has work function $e\varphi_2$ where $e\varphi_2 > e\varphi_1$. Then, because of the exponential dependence of emission on work function, practically all of the emission will be from the $e\varphi_1$ areas. The emitting area is, therefore, only half of the nominal cathode area, and the measured value of A_o is only half of the theoretical value.

From the Richardson-Dushman equation, it is clear that to obtain high values of emission, the work function must be low and the temperature must be high. Unfortunately, as shown in Table 5-1, when pure metals are considered, materials with low work functions also have low melting temperatures. Cesium, with a work function of 2.1 eV and a melting temperature of 28°C, is an extreme example. Barium also has a reasonably low work function, but has a moderate melting temperature (725°C). It turns out that barium is commonly used in thermionic cathodes, but not because it has a low work function. Instead, barium is used because of its capability (accompanied by other elements including oxygen) to reduce the work function of the underlying metal surface. Thus, even though they have high work functions, it is possible to use high melting temperature metals such as tungsten and osmium by introducing barium and other elements to reduce the work function of the surface.

Figure 5-4 shows theoretical emission densities as functions of temperature and work function (dashed lines). Also shown are measured emission densities for several cathodes, most of which contain metals listed in Table 5-1. For W and ThW, the experimental curves are located where expected, based on the values of $e\varphi$ in Table 5-1. (For ThW, thorium diffuses to the surface of the tungsten and so the work function is approximately that of thorium.) Notice that, for the cathodes containing barium (BaW, BaNi, Type B, Alloy, Scandate, and Oxide), the locations of the experimental curves indicate work functions well below that of barium (~ 2.7 eV). In most cases it is the interaction of the barium and other elements with the underlying metal that produces the low work function. This will be discussed later in this chapter when the theory of cathode operation is examined.

Table 5-1
Room Temperature Work Functions for Various Metals

Metal	Work Function eV*	Melting Temperature °C
Barium	2.7	725
Calcium	2.9	839
Carbon	5.0	~3,550
Cesium	2.1	28
Hafnium	3.9	2,227
Iridium	5.2	2,410
Molybdenum	4.5	2,620
Nickel	5.2	1,455
Osmium	5.4	3,045
Platinum	5.3	1,773
Rhenium	5.1	3,180
Scandium	3.5	1,539
Sodium	2.7	97
Strontium	2.6	769
Tantalum	4.2	2,996
Thorium	3.4	1,750
Tungsten	4.6	3,410
Zirconium	4.1	1,852

*Work function is dependent on crystalline orientation. The values listed are estimates of average values.

5.1.2 Schottky Effect

In the derivation of the Richardson-Dushman equation, the effect of an electric field at the cathode surface was not taken into consideration and so neither voltage nor electric field appear in the equation. In fact, when an electric field is applied to the surface of a cathode, it is found that the emission increases slightly.

The reason for this effect may be explained with the aid of Figure 5-5. As was shown in Figure 5-2 and again in Figure 5-5(a), an electron must be emitted with energy $e\varphi$ or greater to overcome the work function barrier at the surface of the cathode. When a positive voltage exists on an anode near the cathode, then an electric field, E_a, is applied to the region between the cathode and anode. Near the cathode surface, as shown in Figure 5-5(b), this field reduces the barrier that electrons must overcome to be emitted so that more electrons are emitted. The resulting increase in emission current is called the *Schottky effect*. The amount by which the work function is reduced is $e\Delta\varphi$ where

$$\Delta\phi = \left(\frac{e E_a}{4 \pi \varepsilon_o} \right)^{1/2} \tag{5-6}$$

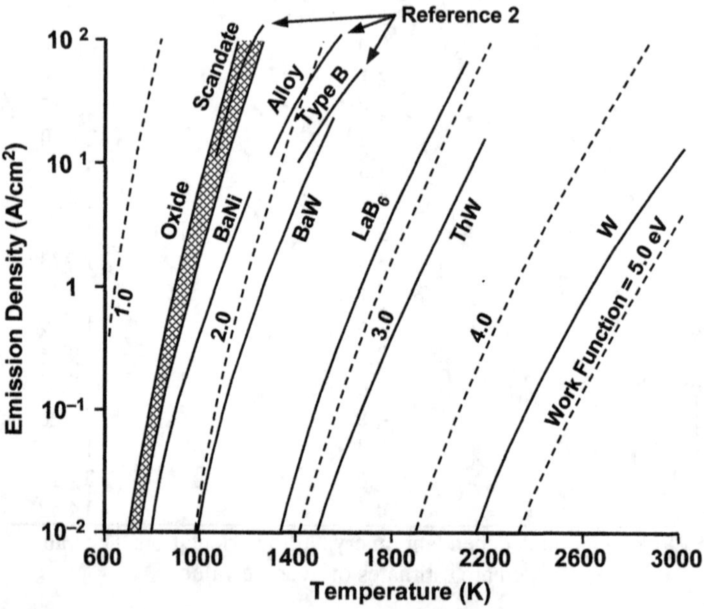

Figure 5-4 Emission characteristics of several cathodes. (Adapted from: [2] and from *Methods of Experimental Physics*, Vol. 4, Part A, Section 1.1.1 by G. A. Haas, copyright 1967 by Academic Press.)

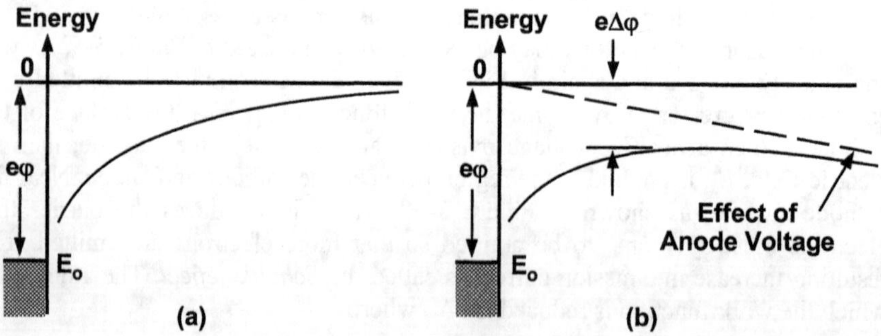

Figure 5-5 (a, b) Potential energy as a function of distance from the cathode surface.

Using this $\Delta\varphi$, the Richardson-Dushman equation can be modified as follows:

$$J = A_o T^2 e^{-\frac{e}{kT}\left[\phi - \left(\frac{e E_a}{4 \pi \varepsilon_o}\right)^{1/2}\right]} \qquad (5\text{-}7)$$

This can be written as

$$J = J_o e^{\frac{e}{kT}\left(\frac{e E_a}{4 \pi \varepsilon_o}\right)^{1/2}} \qquad (5\text{-}8)$$

where J_o is the Richardson-Dushman current density and is sometimes referred to as the zero-field current density [5]. J_o is a useful parameter for comparing performances of different cathodes.

In practice, to determine J_o experimentally, it is necessary to extrapolate from values measured at relatively high electric fields. This is because, at low electric field levels, electrons near the cathode surface alter the applied electric field. This is known as the electron space charge effect and will be discussed.

To determine J_o, proceed as follows:

$$\ln J = \ln J_o + \frac{e}{kT}\left(\frac{e E_a}{4 \pi \varepsilon_o}\right)^{1/2} \qquad (5\text{-}9)$$

This indicates that a plot of ln J versus $E^{1/2}$ should be a straight line with an intercept, when extrapolated to $E = 0$, of the value of ln J_o. This is illustrated in Figure 5-6. Note that, at low values of electric field, the measured value of current density falls below the extrapolated value because of the effect of the charge of the electrons in the region near the cathode surface. Some error in J_o may result from the change of the true work function by the applied electric field.

5.1.3 Field Emission

As the electric field applied to the surface of a cathode is increased to the $10^9 - 10^{10}$ V/m (0.1–1.0 V/Å) level, it is found that electron emission increases very rapidly. Furthermore, the increase in emission is almost independent of cathode temperature.

The energy profile is like that for the Schottky effect (Figure 5-5), except that the applied field is much larger. As a result, as indicated in Figure 5-7, the width of the potential barrier at the surface of the cathode is very narrow. Because of the wave nature of the electron, there is some probability that an electron can exist on the vacuum side of the barrier even though it does not have sufficient kinetic energy to overcome the barrier. This is known as the *tunneling effect.*

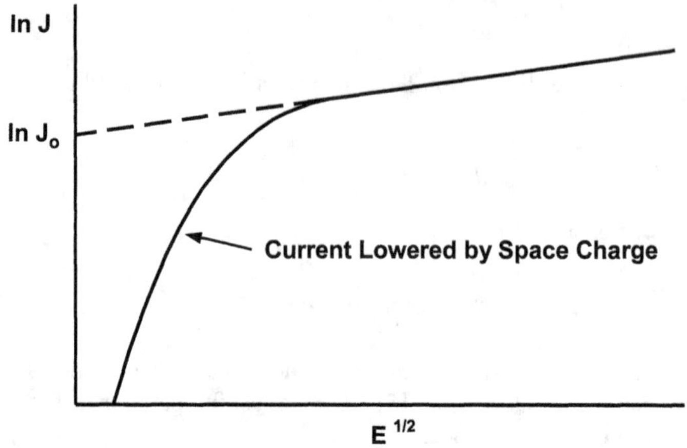

Figure 5-6 Illustration of extrapolation to find zero-field current density.

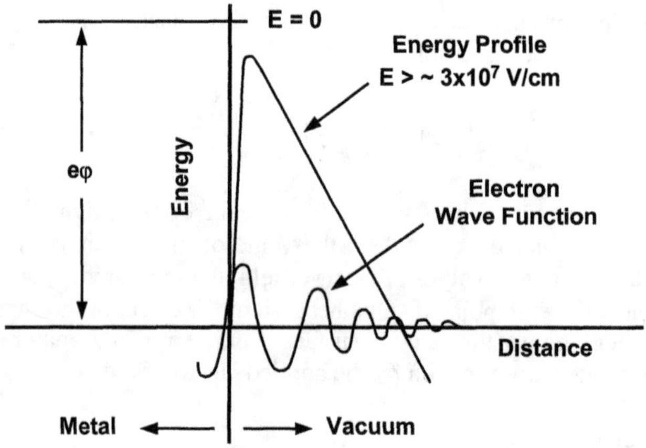

Figure 5-7 Energy profile with a very strong electric field.

The resulting electron is called *field emission.* Field emission increases exponentially with applied voltage. Because extremely large emission densities are possible, there have been many attempts to develop field emission cathodes for microwave tubes. To achieve the high electric field levels necessary for substantial emission, it has been necessary to fabricate the cathodes from arrays of very small needle-shaped emitters. The needle shape is used to enhance the electric field, so that emission occurs at the tip of each emitter, as illustrated in Figure 5-8. Unfortunately, emitted electrons ionize gas molecules to form positive ions, which damage the tips of the emitters and degrade emission.

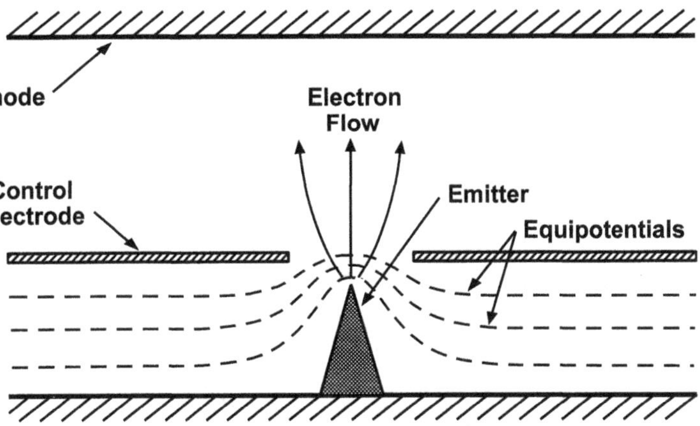

Figure 5-8 Field enhancement and electron emission from a needle-shaped emitter.

5.1.4 Space Charge Limitation [6, 7]

The effect of the negative charge of an electron is to reduce the potential that is present in the absence of the electron. Near an emitting cathode where many electrons are present, the reduction in potential can be appreciable. This is shown for a parallel-plane diode in Figure 5-9. With no electrons, the potential increases linearly from the cathode to the anode. With electrons, the potential profile is depressed as indicated by the dashed line.

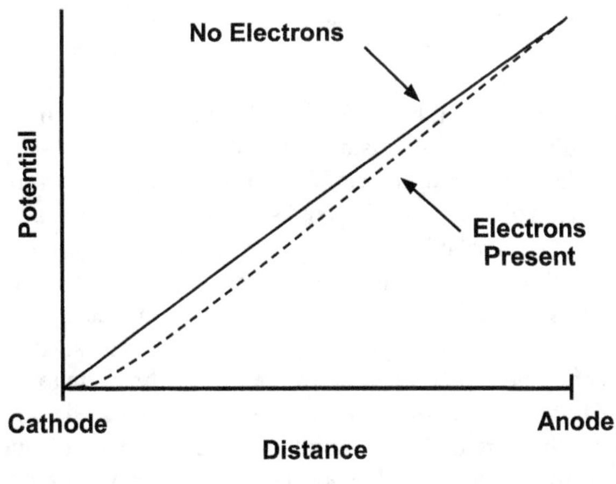

Figure 5-9 Potential distribution with and without electrons in a parallel-plane diode.

As the electron emission rate is increased (by increasing temperature, for example), the potential is further decreased. As indicated by Figure 5-10, there is a limit to how far the electron emission rate can be increased. This limit occurs when the density of electrons adjacent to the cathode surface becomes so large that it tends to depress the potential below zero. If the potential were to become negative, electrons would be forced back to the cathode.

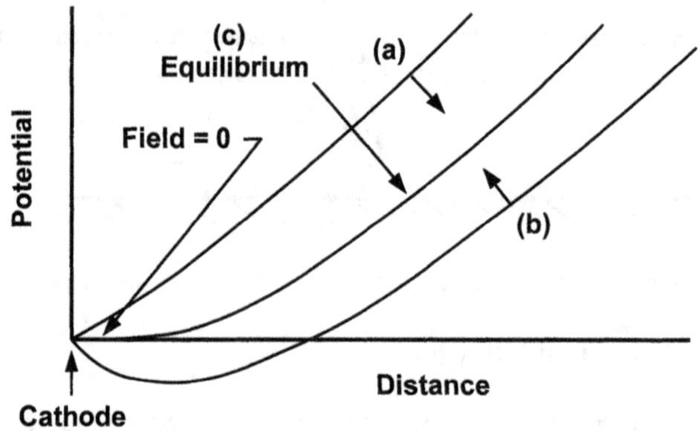

Figure 5-10 (a–c) Potential profiles near the surface of the cathode.

In Figure 5-10(a), the electric field at the cathode surface causes all emitted electrons to leave the cathode, thereby depressing the potential near the surface. In Figure 5-10(b) the electric field at the cathode surface forces electrons back to the cathode surface and increases the potential. An equilibrium condition occurs when, as shown in Figure 5-10(c), the potential adjacent to the cathode surface is zero, that is, when the electric field at the cathode surface is zero. If the potential tends to become positive, more electrons flow from the cathode, thereby depressing the potential. If the potential tends to become negative, fewer electrons flow from the cathode, thereby raising the potential.

When the electric field at the cathode surface is forced to zero by the electron cloud near the cathode surface, the emission is said to be space charge limited. When a sufficient number of electrons are present so that emission is space charge limited, then cathode temperature and surface condition have no effect on emission. This is indeed a very fortunate situation because it eliminates the necessity for extreme uniformity of temperature across the surface of a cathode. It also eliminates the necessity for precise control of the voltage and current to the heater in the cathode. When the voltage is increased in a diode with a space charge limited cathode, the potential at all points between the cathode and anode tends to increase. As a result, additional electrons flow from the cathode to cause the electric field at the cathode surface to remain at zero. It is evident, therefore, that diode current is dependent on anode voltage.

The relation between voltage and current in a parallel-plane, space charge limited diode can be determined in a straightforward manner. It is assumed that the electric field at the cathode surface is zero, as is indicated in Figure 5-10. Poisson's law applies and so the potential, V, can be found from

$$\Delta^2 V = \frac{\rho}{\varepsilon_o} \tag{5-10}$$

Assuming that the parallel plane diode extends to infinity in the transverse directions, then there are variations only in the x-direction, so

$$\frac{d^2 V}{dx^2} = \frac{\rho}{\varepsilon_o} \tag{5-11}$$

The charge density, ρ, is related to current density, J, and electron velocity, u, by

$$\rho = \frac{J}{u} \tag{5-12}$$

and, from the conservation of energy,

$$u = \sqrt{2\eta V} \tag{5-13}$$

so

$$\frac{d^2 V}{dx^2} = \frac{J}{\sqrt{2\eta V}\,\varepsilon_o} \tag{5-14}$$

This can be solved with the conditions that at $x = 0$, $V = 0$, and, for space charge limited emission with no initial electron velocity, $dV/dx = 0$. The result is

$$J = \frac{4}{9}\varepsilon_o (2\eta)^{1/2} \frac{V^{3/2}}{x^2} \tag{5-15}$$

When the values for η and ε_o are used,

$$J = 2.33 \times 10^{-6} \frac{V^{3/2}}{x^2} \tag{5-16}$$

This is the Child-Langmuir law for the flow of electrons in a parallel-plane, space charge limited diode. For a diode with a cathode-to-anode spacing $x = d$, this equation is

$$I = P V^{3/2} \tag{5-17}$$

where $I = JA$ with A being the cathode area. P is known as perveance and is

$$P = \frac{4}{9}\varepsilon_o (2\eta)^{1/2} \frac{A}{d^2} \tag{5-18}$$

or

$$P = 2.33\times10^{-6} \frac{A}{d^2} \tag{5-19}$$

for a parallel-plane diode. Perveance is a function only of geometry. Space charge limited diodes with geometries other than planar such as cylindrical and spherical all obey $I = PV^{3/2}$, where the expression for perveance is the one appropriate for the geometry used.

In this analysis, it has been assumed that the velocities of the electrons are zero when they are emitted. In fact, electrons are emitted with small initial velocities. If the electric field was zero at the cathode surface, all of the emitted electrons would be able to leave the surface and the saturation emission, J_o (total thermionic emission capability of the cathode), would flow. In practice, space charge limited emission is much less than saturated emission because the presence of large quantities of electrons near the surface causes a potential minimum to be formed just outside the cathode surface. The voltage at the minimum is a small fraction of a volt below cathode potential. Only the most energetic electrons can overcome this potential minimum and pass on to the anode. The less energetic electrons are returned to the cathode.

The effect of increasing the anode voltage so that more current flows is to decrease the depth of the potential minimum and move it closer to the cathode. As J approaches the saturation current density, J_o, dV/dx approaches 0 at the cathode surface. If the voltage is increased further, current will be limited to the saturation current density and space charge limited operation will cease.

In practice, space charge limited diodes behave as predicted. This is readily demonstrated by plotting current as a function of voltage using two-thirds power graph paper. As shown in Figure 5-11, the resulting plot is a straight line for space charge limited operation. The slope of the line is the perveance.

When the voltage is increased to a value high enough so that the cathode cannot supply sufficient current to maintain the space charge limited condition, the cathode is said to be *TL* (temperature limited). The TL region is readily apparent in the two-thirds power I-V graph because the plotted results depart from a straight line.

In Figure 5-12, current is shown plotted as a function of temperature for a thermionic cathode. In the region of TL operation, current increases with temperature. At high temperatures, current is independent of temperature, indicating space charge limited operation. As voltage is increased, the temperature at which the transition from TL to space charge limited operation occurs is increased.

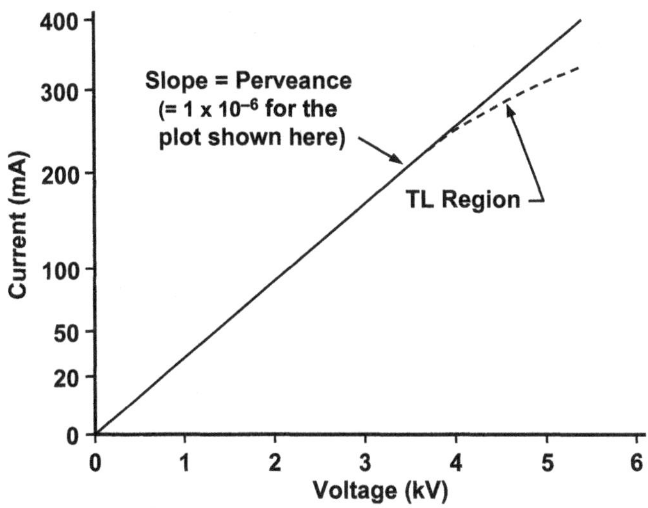

Figure 5-11 Current versus voltage plotted on two-thirds power graph paper.

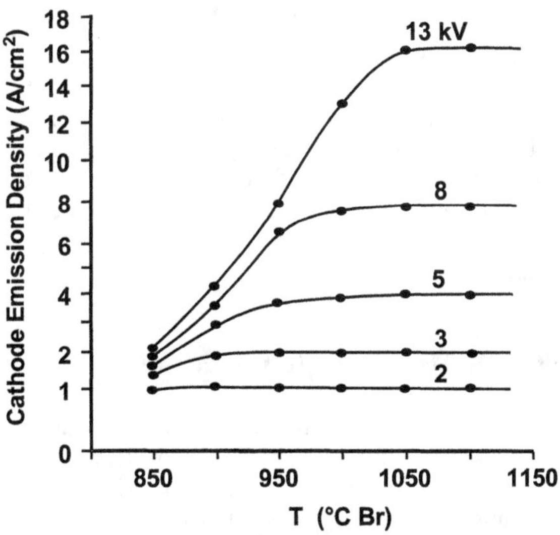

Figure 5-12 Emission as a function of temperature and anode voltage.

As indicated in Figure 5-13, the TL portion of a current versus temperature curve moves to higher temperatures as time increases. The reason for this will be discussed later in this chapter. It is necessary to select an operating temperature for a cathode that remains above the knee of the curve for the life of the cathode. The rate at which the TL curve moves increases with increasing temperature, so an excessive temperature will reduce cathode life.

In Figure 5-13, there is a slight upward slope with increasing temperature in the space charge limited region. This is caused by the movement of the cathode toward the anode as a result of thermal expansion of elements of the cathode support structure.

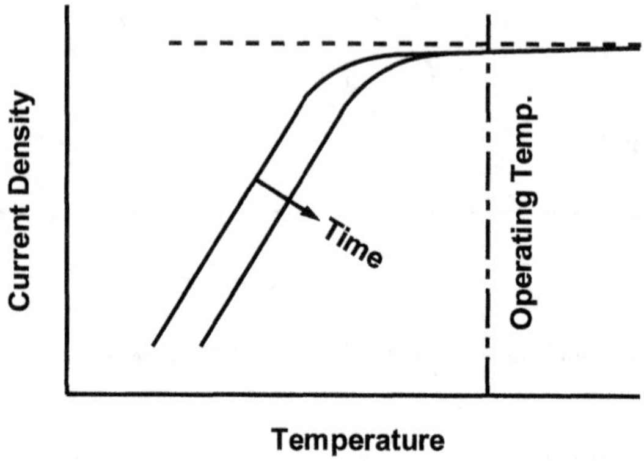

Figure 5-13 Movement of the TL curve with time.

The mechanisms dominating current flow in a thermionic diode are summarized in Figure 5-14. At low voltages, current is space charge limited. As voltage is increased, the emission limit of the cathode is reached and current becomes dependent on cathode temperature. Current continues to increase slowly with voltage because of the Schottky effect. At high voltages, field emission causes current to increase rapidly.

5.2 EVOLUTION OF THERMIONIC CATHODES

The first thermionic cathodes used in quantity were those in early radio tubes. They were made from pure tungsten, although other cathode materials were known that were capable of producing adequate emission at lower temperatures than tungsten. Because of the high work function (~ 4.6 eV), the operating temperature was high. For example, 2,200°C yields about 0.3 A/cm^2. Apparently, tungsten was used in early tubes rather than other cathodes because it was foolproof. A tungsten cathode always had plenty of emission. Other cathodes were temperamental and were sometimes "poisoned" by mysterious agents.

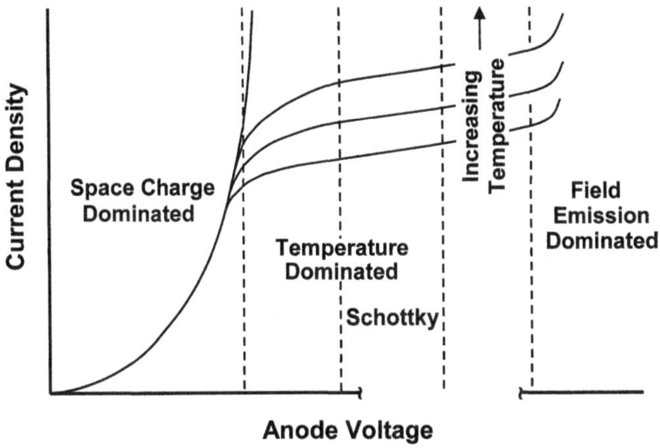

Figure 5-14 Summary of mechanisms dominating current flow in a thermionic diode.

For some applications, thorium oxide in a concentration of about 2% is added to tungsten in the process of making the tungsten wire for the cathode. By processing, thorium is brought to the surface of the tungsten. The work function of thorium is well below that of tungsten, so the emission is about 1,000 times higher than for tungsten.

The cathode surface is sometimes carburized, that is, heated in a vapor of a hydrocarbon such as alcohol, so that a layer of tungsten carbide is formed on the surface. This increases cathode life by reducing the rate of evaporation of thorium and protecting the cathode from ion bombardment.

The life expectancy of high-power switch tubes containing thoriated tungsten cathodes is shown in Figure 5-15. Thoriated tungsten cathodes are in common use in high-power switch tubes, transmitter tubes, and microwave oven magnetrons.

Early cathodes were all directly heated, that is, the filament, which was heated by passing a current through it, was also the cathode. Thus, as a result of the heater current, there was a voltage variation along the surface of the cathode. Because batteries were used as power sources, the filament voltage was dc, so there was no modulation of the emission current.

The development of ac radios and the need to heat the filaments with ac current brought about the need for a cathode, the surface of which was all at the same potential. The indirectly heated cathode resulted, whereby the heater was separate from the electron emitting surface. Directly and indirectly heated cathodes are illustrated in Figure 5-16.

In indirectly heated cathodes, tungsten is used for the filament. Because of heat losses and the relatively large cathode area, the cathode surface had to operate at a temperature substantially below the heater temperature. This provided the requirement for a low work function emitting surface

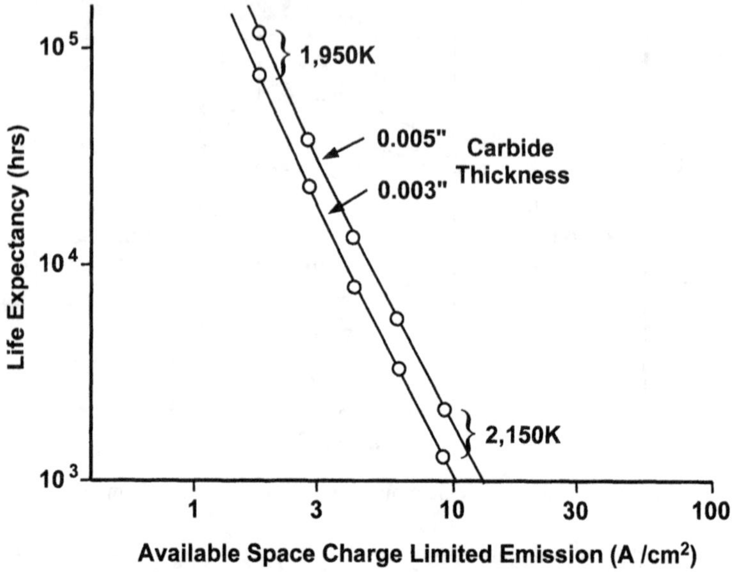

Figure 5-15 Life expectancy for a tube with a thoriated tungsten cathode (actual life ≈ 1/2 of life expectancy). (From: T. E. Yingst et al., *Proc. IEEE*, March 1973. © 1973 IEEE.)

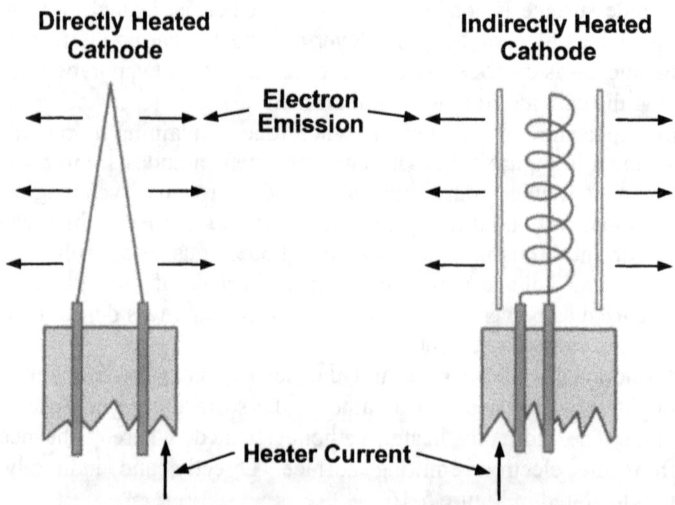

Figure 5-16 Comparison of directly and indirectly heated cathodes.

The oxide-coated cathode, which was discovered by Wehnelt in 1904, long before radio tubes, provided the solution to the problem because of its very low work function (≈1.5 eV) and resulting low operating temperature. Oxide-coated cathodes served the industry well from the 1920s through World War II and into the early 1950s. They are still prevalent today in applications requiring relatively modest cathode emission and long life. Early oxide-coated cathodes were described by J. R. Pierce [1] as follows:

> The early history of oxide coated cathodes seemed without rhyme or reason. It was found that if a wire intended to emit electrons was heated and rubbed with sealing wax and the sealing wax then burned off a truly phenomenal cathode was produced. Further work traced the emission to alkaline earth oxides, and the compound from which they came was humorously described as a barium resinate. The sealing wax technique died hard, and for several years the process of coating filaments consisted in stringing fine platinum ribbon out in long lengths, heating it with an electric current, and rubbing a resinous compound containing barium and strontium over it repeatedly. The narrow ribbon was then cut into lengths, put into a crude vacuum tube of the day, the tube was pumped and baked until a "hard vacuum" was achieved, and the filament was activated. This activation was a wishful process consisting of glowing the filament brightly, applying voltage, and drawing current to the plate until, in happy circumstances the plate got red-hot, giving off enough gas to produce a blue glow and further increase the activity and the current. After a lucky period of this treatment the filament became active; that is, it would emit electrons at a lower and reasonable temperature, at which it would last for a number of thousands of hours.

The art and science of oxide cathodes are now highly developed. A mixture of barium, strontium, and calcium carbonates is applied to a metal base (usually nickel) and then activated in vacuum. Activation consists of heating and reducing the carbonates to oxides.

As oxide-coated cathodes were improved, dc emission increased from mA/cm^2 to hundreds of mA/cm^2. Pulsed emission densities are dependent on pulse length and can be as high as tens of A/cm^2. Oxide-coated cathodes are not capable of high dc emission densities because of the high resistivity of the coating and the resistive heating that occurs at high currents. They are also easily poisoned.

The search for better cathodes than oxide-coated cathodes resulted from their low dc emission and susceptibility to damage. The L cathode, illustrated in Figure 5-17, was developed by Philips about 1950. In this cathode, the emissive material is sandwiched between the heater and a porous tungsten pellet. During operation, emissive material diffuses through the porous tungsten and forms a low work function emitting surface.

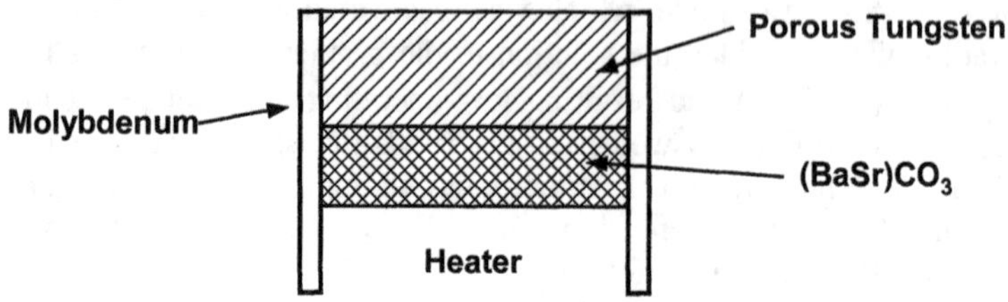

Figure 5-17 Essential features of the Philips L cathode.

Cathodes like the L cathode are referred to as *reservoir cathodes* for the obvious reason that a reservoir of emissive material is used. Early problems have been overcome and structures like the L cathode are now in use. Some of these are referred to as *MK cathodes*.

A reservoir cathode, developed by the Naval Research Laboratory, is known as the *controlled porosity dispenser* (CPD) cathode. The reservoir contains BaO, CaO, and Al_2O_3, as well as particles of tungsten, which act as a reducing agent that frees barium from the BaO. The CPD cathode uses a thin sheet of tungsten or alloy (to be discussed next) in which pores have been etched or drilled with a laser. The CPD approach to cathode construction offers advantages of improved emission uniformity and improved reproducibility and performance. At present, the difficulties of fabricating CPD cathodes, especially for concave cathodes, limit their usefulness.

From 1950 to 1955, the evolution of the impregnated dispenser cathode illustrated in Figure 5-18 occurred. The Philips B-type cathode developed by Levi is still in use by the microwave tube industry. The pores of a porous tungsten pellet are impregnated with a compound of BaO, CaO, and Al_2O_3. Barium is released when the impregnant reacts with the tungsten. The freed barium migrates to the surface of the porous tungsten to form the emitting surface. A common emissive mix used in the B-type cathode consists of five parts BaO, three parts CaO, and two parts Al_2O_3 and is frequently referred to simply as 5:3:2. Other mixes are 6:1:2 and 4:1:1. The 4:1:1 cathode is sometimes referred to as an *S-type cathode*. The choice of mix depends on the application of the cathode [8]. For example, the 5:3:2 cathode is often used in large tubes at low emission densities. The 6:1:2 cathode has very good poisoning resistance and a low barium evaporation rate and is recommended for most applications. The 4:1:1 cathode is very active and poison-resistant and can usually be operated 30°C cooler than other types.

The B-type cathode is rugged and relatively impervious to damage and can readily provide emission densities of several A/cm^2. At these high densities, the operating temperature must be 1,100°C or higher because of the relatively high work functions of these cathodes. As a result, life is limited to a few thousand hours.

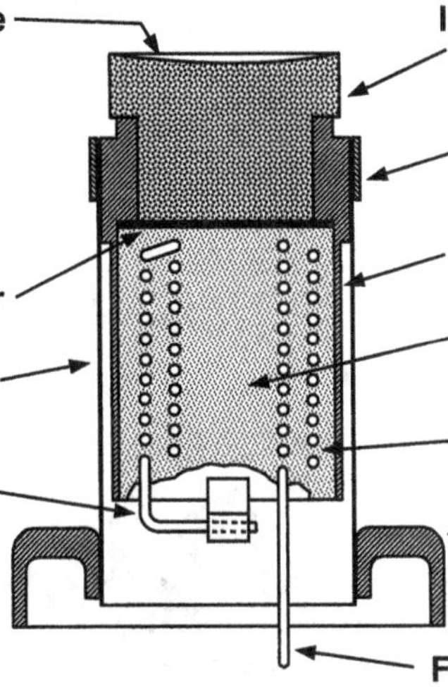

Figure 5-18 Typical B-type impregnated dispenser cathode.

The search for a cathode capable of an emission density comparable to the B-type cathode, but with a lower operating temperature and longer life, led to the development of the M-type cathode in the 1960s. In its simplest form, the M-type is a B-type with a film several thousand Angstroms thick of osmium, iridium, or rhenium applied to the surface. Compared to a B cathode, the effect of the film is to reduce the work function ~ 0.2 eV and the cathode operating temperature ~ 90°C (dependent on film metal).

A problem early in the development of M cathodes was degradation of the enhancing film. At normal operating temperatures, the tungsten substrate and coating metal would interdiffuse, and an alloy surface was formed with a composition that changed with time. The lowest work function occurs when the alloy is about 50/50 tungsten and coating metal. Thus, during the life of one a cathode, emission would actually increase at first, then degrade, and then eventually revert to that of a B cathode as the surface became tungsten.

Several fabrication techniques for controlling the properties of the alloy formed on the surface of an M cathode have been examined. In one cathode, the mixed metal matrix (MMM) cathode, the cathode pellet contains particles of the enhancing metal (usually iridium). Fabrication is difficult because the size distribution of the tungsten and iridium particles must be carefully controlled. In the controlled doping (CD) cathode developed by Varian, multiple layers of alloy material, having different compositions, are deposited. The composition of the layers is adjusted to minimize interdiffusion of the tungsten and the enhancing metal. A few percent of the enhancing metal can be added to the tungsten matrix prior to coating and this reduces the diffusion of tungsten into the coatings.

The performance characteristics of the cathodes just described are summarized in Figure 5-19. The increase in emission capability over B cathodes of cathodes containing osmium, iridium, or rhenium is clearly shown.

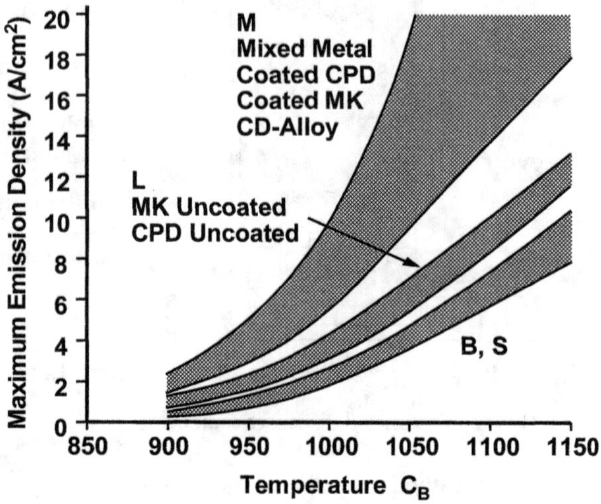

Figure 5-19 Performance comparison for various dispenser cathodes. (Adapted from: L. R. Falce, *Technical Record*, 1983 IEDM. © 1983 IEEE.)

For several years, there has been considerable interest in scandate cathodes. These cathodes have a very low work function (≈1.5 eV) and are capable of emission densities on the order of 100 A/cm^2 for thousands of hours. In one version developed by Philips, the emitting surface is composed of a layer of tungsten mixed with Sc_2O_3 (≈5% by weight). The entire matrix of the cathode is impregnated with BaO, CaO, and Al_2O_3. While these cathodes are being used in some applications (see, for example, [9, 10]), they are not yet in widespread use because of nonuniform emission, which leads to a very broad work function distribution (discussed in the following sections) and also because of a lack of robustness such as poor resistance to the effects of ion bombardment [11].

5.3 IMPREGNATED DISPENSER CATHODES

5.3.1 Fabrication

At present, the most commonly used cathodes in the microwave tube industry are impregnated dispenser cathodes. These are fabricated from small grains of tungsten that have been pressed together under high pressure and sintered at a temperature of over 2,000°C for 1 to 2 hours to form porous billets. The pores are then filled with a plastic material to facilitate rough machining to the desired shape. After machining, the plastic is removed and the porous matrix is filled with a compound containing barium oxide, calcium oxide, and aluminum oxide.

Usually, there is then an additional machining operation of the emitter surface to the final dimensions. This is a difficult procedure because the cathode is a porous structure. Care must be taken to avoid excess smearing of the tungsten over the pore openings. If the pore openings become plugged, then, during cathode operation, the emissive compound will be prevented from diffusing to the cathode surface. Figure 5-20 shows the surface and interior of a cathode after final machining. The differences between grain sizes and pore distributions between the surface and the interior show that significant smearing occurred during machining.

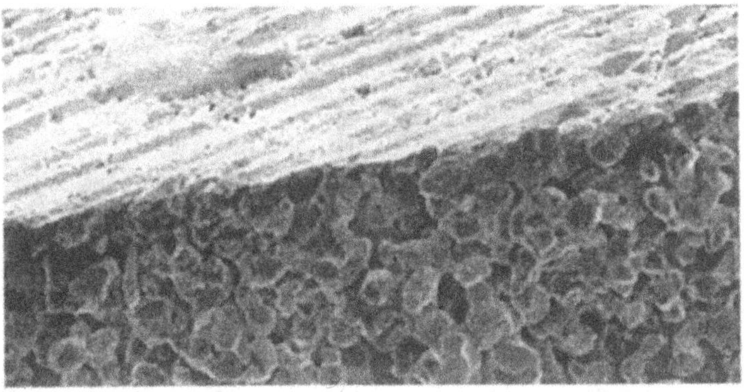

Figure 5-20 SEM image showing surface and interior view of a porous tungsten cathode. (From: R. L. Ives et al., *IEEE Trans. Electron Devices*, December 2005. © 2005 IEEE.)

Before a new dispenser cathode is installed in an electron gun, the cathode surface is usually processed to remove excess impregnant. One technique is to place the cathode in a vacuum chamber (not the tube in which it will eventually be installed) and heat it to a high temperature. This causes the excess impregnant to evaporate from the cathode surface. Another technique is a liquid cleaning process, referred to as SRLV or EDTA cleaning [12]. This process removes a shallow layer of impregnant from the tungsten surface.

Figure 5-21 shows the scale of the features of the surface (without smearing) of a typical B-type cathode [13–15]. A coated cathode (M type, for example) has a very thin film of enhancing metal on the surfaces of the tungsten grains. Note that all dimensions are typically a few microns. Note also that there is a wide distribution in the sizes of the tungsten grains and that they are irregularly shaped. This results in there being a lack of control over the porosity of the cathode surface and this contributes to the variability of the work function of the cathode (to be discussed next).

It should be pointed out that, over the years, there have been many attempts to control the porosity of a cathode surface. The work of the Naval Research Laboratory is mentioned earlier in this chapter. None of these attempts has found application for any one of several reasons [16]. Ives et al. [16] proposed a novel technique for controlling porosity that appears to overcome many of the shortcomings of previous techniques. Tungsten wires are sintered to form the uniform array of pores of uniform size shown in Figure 5-22.

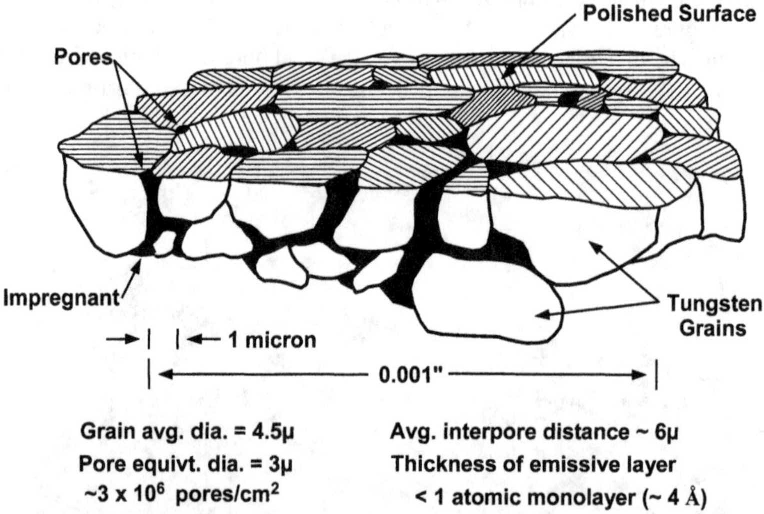

Figure 5-21 Scale of features of an impregnated cathode. (From: M. C. Green, *Technical Report,* RADC-TR-81-211, July 1981.)

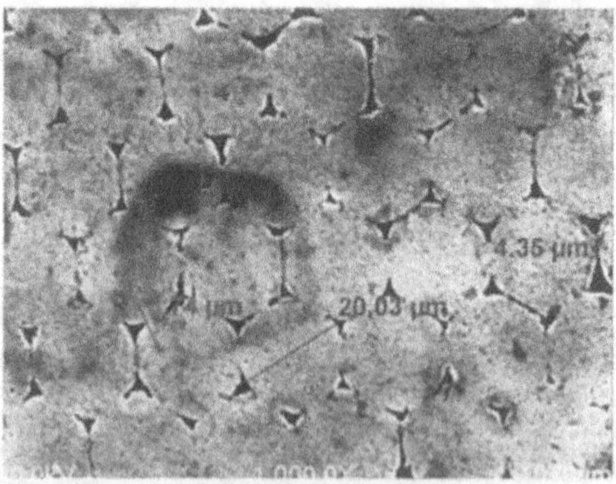

Figure 5-22 SEM photograph of sintered tungsten wires. (From: R. L. Ives et al., *IEEE Trans. Electron Devices*, December 2005. © 2005 IEEE.)

5.3.2 Operation

When a cathode is heated to the operating temperature, barium is released from the interior of the tungsten matrix by reactions of the impregnant with the tungsten. The freed barium migrates through the pores to the surface of the cathode. Once there, the barium migrates across the surface and eventually evaporates. The rate of diffusion of barium to the surface, along with the energy of desorption of barium atoms from the surface (which controls the evaporation rate), determines the surface coverage.

It is important to avoid operating a cathode at a temperature that is higher than necessary because the rate of evaporation of barium is highly temperature dependent. Clearly, it is also important for the surface to be heated to a uniform temperature. The hottest portions of the cathode become depleted of barium soonest and this leads to reduced emission and the end of cathode life. It is desirable that the temperature should not vary more than a few degrees centigrade over the entire surface of the cathode.

The work function of the cathode depends on the fraction of the cathode surface that is covered by barium along with the work function of the metal substrate. The reason for this will be discussed later in this chapter. An example of the variation of work function with surface coverage is shown in Figure 5-23. For optimum coverage, which turns out to be nearly a monolayer, the work function is a minimum. With time, as a cathode operates, the pores become depleted and the rate of diffusion of barium to the cathode surface decreases. As a result, the coverage of the surface decreases, the work function increases, the rate of emission decreases, and the space charge cloud may disappear. If that happens, emission becomes temperature limited. By increasing the cathode temperature, the rate of diffusion of barium to the surface can be increased, the rate of emission can be increased, and it may be possible to return to space charge limited operation. Of course, the increased temperature causes the rate of evaporation to increase and this hastens the return of temperature limited operation.

5.3.3 Miram Curves

Miram et al. developed a technique for presenting emission data that provides a powerful tool for comparing cathodes, for diagnosing their performance, and for helping to understand their operation [17, 18]. The basic concepts may be understood with the aid of Figure 5-24. Values of emission current are plotted as percentages of the FSCL current. Curves plotted in this manner are often referred to as *Miram curves*. Several curves are shown for TL operation. With the Miram technique, the curves are plotted for FSCL current densities that are separated by a factor of 2. From the Richardson-Dushman equation, the temperature difference required to produce an increase in current density by a factor of 2 is about 40°C, and, in fact, the TL curves are separated by about 40°C.

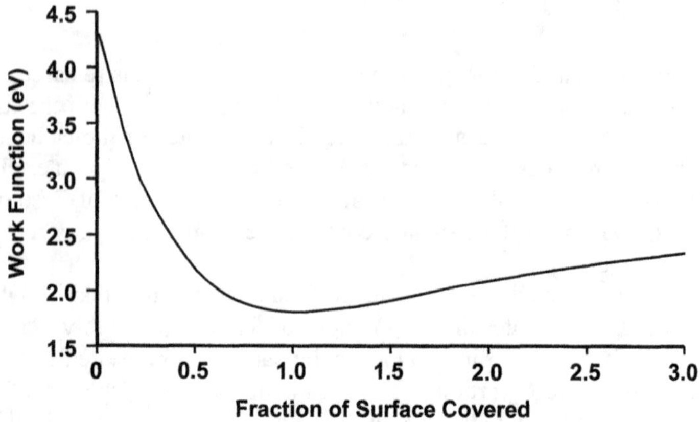

Figure 5-23 Variation of work function with the fraction of the surface covered by barium. (From: L. Falce and L. Garbini, Fifth IEEE IVEC. © IEEE 2004.)

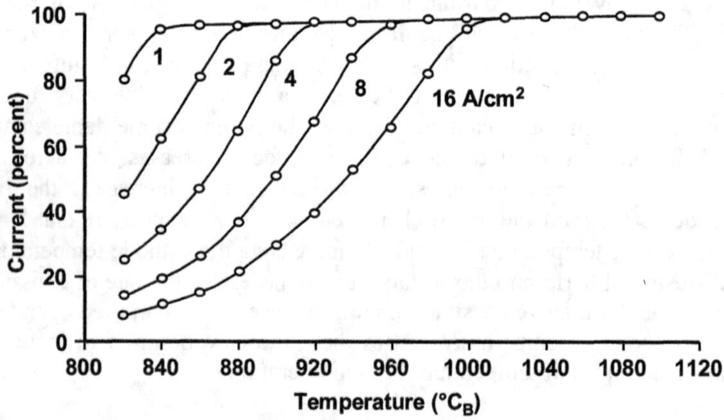

Figure 5-24 Miram curves for a good cathode. (From: T. J. Grant, *Technical Digest*, 1984 IEDM. © 1984 IEEE.)

The use of Miram curves and work function distribution curves provides a technique, not only for evaluating cathode performance as functions of temperature and loading, but also for revealing defects in cathode fabrication and operation. Examples of Miram curves for defective cathodes are given in Figure 5-25. The curves in Figure 5-25(a) were obtained from a cathode with chipped edges. The chipped regions had a higher effective work function than the central portion of the cathode. As a result, the emission plots exhibited two types of knees. The ones for the central portion were much like the knees for a good cathode. The other knees were very broad and occurred at higher temperatures.

Figure 5-25(b) shows results for a cathode for which the effective work function increased as the temperature was raised above 940°C. Note that, above 940°C, the separation between the 8 A/cm^2 curve and the 4 A/cm^2 curve is greater than the separations between the 4 and 2 A/cm^2 curves or the 2 and 1 A/cm^2 curves. This large change in separation indicates an increase in the work function.

In the plots shown in Figure 5-25(c), the knees of the curves have changed, as indicated by the dashed lines. The change may have occurred as a function of time, or the change may represent the difference between a cathode assembly that is known to be operating properly and another assembly that is supposed to be identical to the first. If only the knees of the curves have changed as shown in Figure 5-25(c), then it is likely that something has caused a nonuniformity in the electric field near the cathode surface. With a gridded electron gun, this could be movement or misalignment of one of the grids.

If both the spacing and shapes of the knees have changed as shown in Figure 5-25(d), then there is a cathode emission problem and possibly a misalignment of gun electrodes.

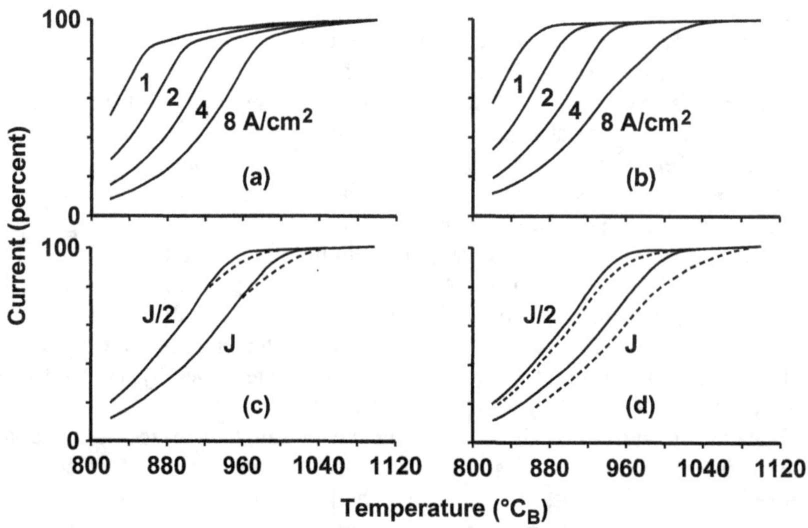

Figure 5-25 (a–d) Miram plots showing evidence of cathode and grid defects. (Adapted from: M. J. Cattelino et al. and T. J. Grant, *Technical Digests*, 1982 and 1984 IEDMs. © 1982 and 1984 IEEE.)

5.3.4 Work Function Distribution

Because the energy of desorption and the work function vary from grain to grain, the work function of a cathode is not a single valued quantity but instead has a distribution of values. For example, the B cathode can have twice as many crystalline surface orientations as a coated cathode, causing a broader work function distribution as shown in Figure 5-26.

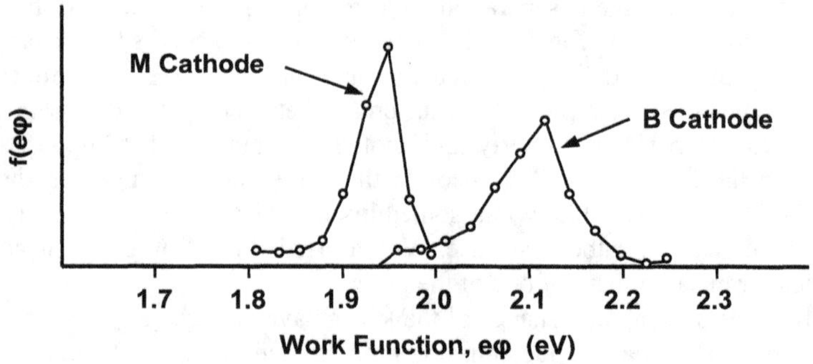

Figure 5-26 Work function distribution for a B cathode and an M cathode early in life. (From: T. J. Grant, *Technical Digest*, 1986 IEDM. © 1986 IEEE.)

To understand the overall emission characteristic of a cathode, it is necessary to understand the distributed nature of the work function. An important consideration is the thickness of the electron space charge cloud at the surface of the cathode during space charge limited operation. Because of the finite velocities of emission of electrons, there is a small depression in the potential (~0.2V) near the surface of the cathode. The location of the position where the potential is a minimum is a measure of the thickness of the space charge cloud. Figure 5-27 shows the distance between the potential minimum and the cathode surface. Note that, at emission densities of a few A/cm^2, which are typical for coated dispenser cathodes, the position of the potential minimum is comparable to the tungsten grain size (see Figure 5-21) and thus the emission patch size, so we can conclude that the various patches emit independently at rates determined by their individual work functions [19].

Because the various patches operate independently, the surface of the cathode behaves like an array of diodes with varying work functions that are connected together in parallel. Thus, the emission characteristic for the cathode can be modeled as a parallel combination of the emission characteristics of the diodes (patches). In the temperature-limited (TL) region of operation of each patch, the Richardson-Dushman equation applies. In the space charge limited region (indicated by *FSCL* to show that the calculation is based on an operation that is fully space charge limited, that is, no part of a patch is operating temperature limited), the Child-Langmuir law applies.

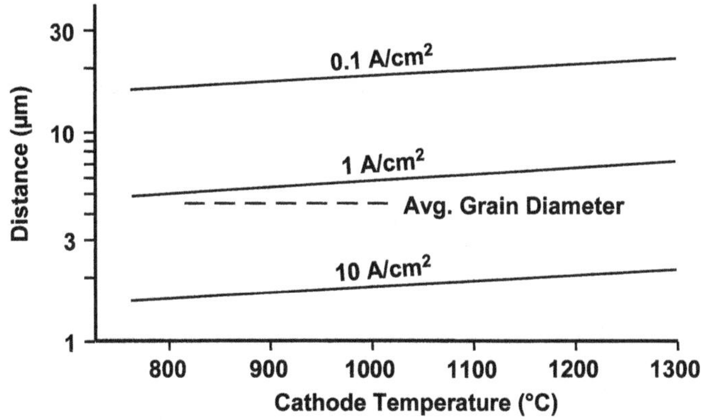

Figure 5-27 Distance between potential minimum and cathode surface.

Figure 5-28 shows, conceptually, how the TL and FSCL curves can be combined to obtain the overall emission characteristic (Miram curve), which is often called a *roll-off curve*. Notice that, below the FSCL current, emission consists of a combination of currents from patches that are operating TL and from patches that are operating SCL (space charge limited). In Figure 5-28, all currents are plotted as percentages of the FSCL values. It has been assumed that all emission patches operate at the same temperature and that the electric field applied by the anode is the same for all emission patches.

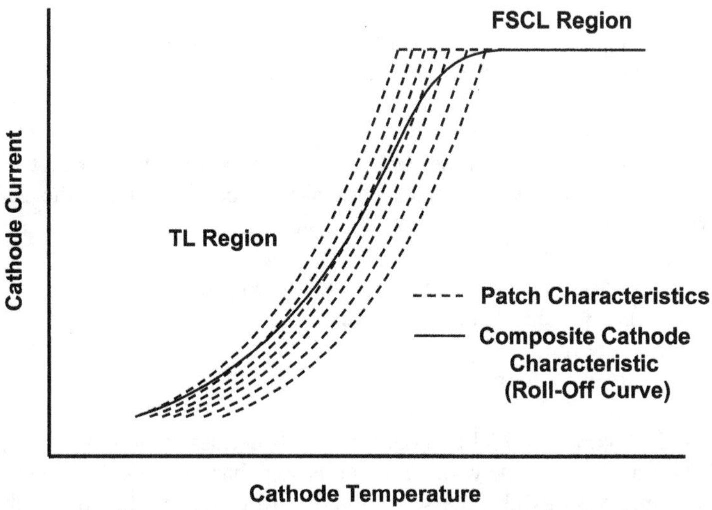

Figure 5-28 Combination of emission curves for several patches to form the overall emission characteristic (roll-off curve) for a cathode.

Cattelino and Miram have used the *practical work function distribution* (PWFD) as a tool for evaluating cathode quality and predicting life [20]. Figure 5-29 shows how the PWFD plot is constructed. First, the Miram curve is overlaid on a set of TL emission curves calculated using the Richardson-Dushman equation. Currents up to the 0.5 FSCL value are normalized with respect to the FSCL current and plotted against the work function. Currents above the 0.5 FSCL value are subtracted from the FSCL current, normalized, and plotted against the work function.

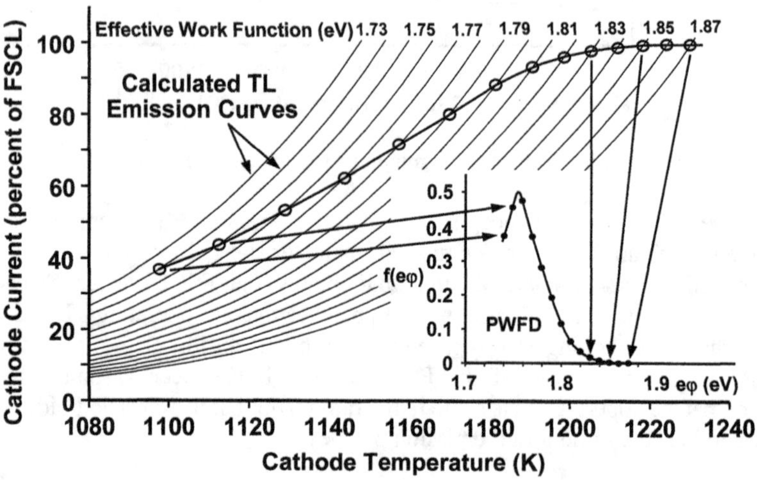

Figure 5-29 Determination of the practical work function distribution (PWFD). (Adapted from: Cattelino and Miram, *Applied Surface Science*, Vol. 111, 1997.)

Cattelino and Miram used PWFD to compare and cathodes. The best of class is the cathode of a particular type such as the M-type that has the best emission performance. The PWFD curves for best-of-class cathodes have the following characteristics:

1. The curves are nearly the same for all current densities.
2. The mean value and the width of the curves are nearly constant throughout life.
3. The curves are narrow.

Figure 5-30 contains PWFD plots for a best-of-class cathode and the plots in Figure 5-31 are for a cathode with a poor emission characteristic. The best-of-class curves in Figure 5-30 exhibit the second and third characteristics listed above. In addition, PWFD curves (not shown in Figure 5-30) at current densities of 2 and 4 A/cm^2 were nearly the same as those at 8 A/cm^2. By comparison, the PWFD curves in Figure 5-31 for a cathode with continual degradation in emission over its life show a continual shift in peak value toward higher work functions. Also, the widths of the curves increase with life indicating degradation in emission uniformity. Note that early in life, the best-of-class cathode had a lower peak PWFD work function (~1.78 eV) than the poor cathode (~1.81 eV).

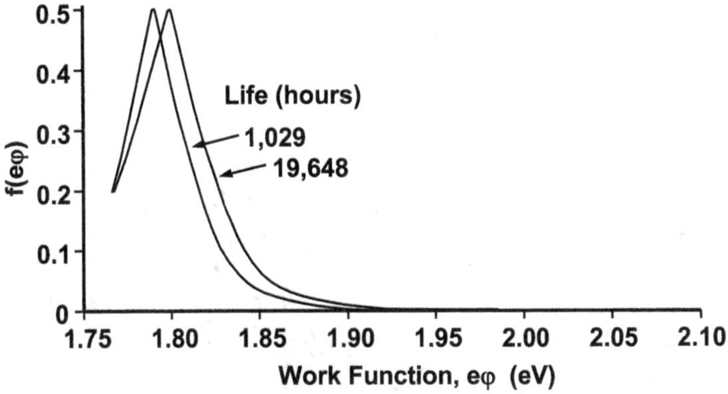

Figure 5-30 PWFD for best-of-class alloy-coated cathode at 8 A/cm^2. (Adapted from: Cattelino and Miram, *Applied Surface Science*, Vol. 111, 1997.)

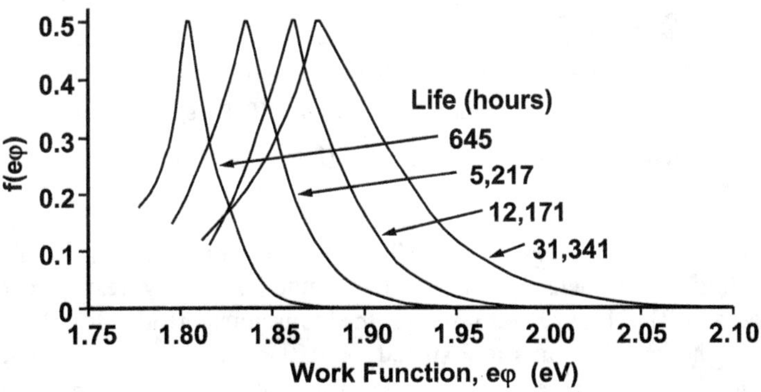

Figure 5-31 PWFD for an alloy-coated cathode at 8 A/cm^2 with poor emission and life characteristics. (Adapted from: Cattelino and Miram, *Applied Surface Science*, Vol. 111, 1997.)

Cattelino and Miram reported that the PWFD provides a means for predicting cathode performance and life expectancy. They noted that, in general, it has been observed that cathodes with similar PWFD characteristics tend to have similar life. Cathodes exhibiting best-of-class PWFD performance typically exhibit best-of-class life expectancy.

A quantity related to the PWFD is *emission spread.* Miram et al. reported that emission spread, along with PWFD, is useful for evaluating cathodes for applications like the gyrotron, which require TL emitter operation [21]. As shown in Figure 5-32, emission spread is determined from the roll-off curve. It is defined at a specific TL operating point (Miram uses 40% of the FSCL current) and is the difference between the calculated (Richardson-Dushman) work function that produces the emitted current and the lowest measurable work function for the emission characteristic (1.80 eV in Figure 5-32). For the example in Figure 5-32, the emission spread is 0.035 eV.

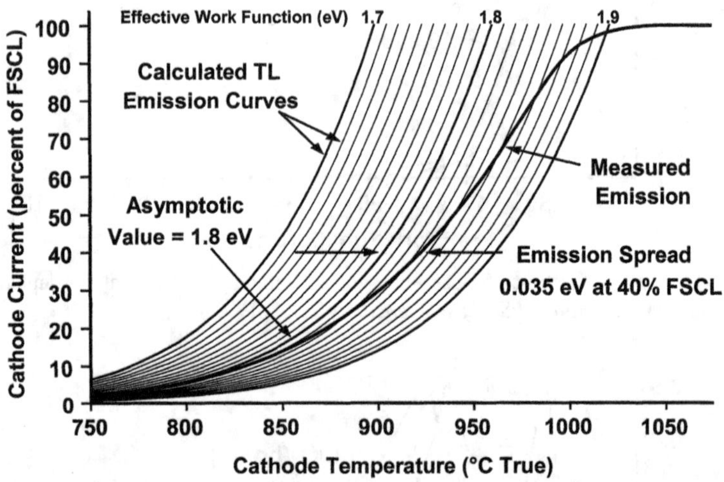

Figure 5-32 Determination of emission spread. (From: G. V. Miram et al., Fifth IEEE IVEC, © IEEE 2004.)

Before leaving the subjects of PWFD and emission spread, it is of value in understanding the characteristics of scandate cathodes to show results reported by Miram et al. [21]. Figure 5-33 shows the emission spread and PWFD for a scandate cathode. The emission spread is 0.2 eV and indicates that, at the temperature of 700°C, most of the surface of the cathode is operating temperature limited. However, over a range of ~300°C from ~750°C to ~1,100°C, the cathode becomes space charge limited. This is an indication of the emission nonuniformity of present state-of-the-art scandate cathodes.

5.4 LIFE CONSIDERATIONS

Long-life requirements often range from 10,000 hours to over 100,000 hours. While life testing is sometimes done, it is more common to predict cathode life on the basis of known cathode degradation mechanisms. Grant identified three distinguishable types of emission degradation mechanisms [19]:

- Type I gradual decrease in perveance, that is, FSCL emission decreases with time;
- Type II change in work function distribution with time and without dependence on changes in cathode temperature;
- Type III change in work function distribution with temperature as well as with time.

The type II mechanism results from a change in the base metal work function with time. This mechanism applies to coated cathodes and is attributed to the diffusion of tungsten through the coating. Grant pointed out that the work function distribution can move to lower as well as higher work functions early in life depending on coating composition, thickness, and interdiffusion rate.

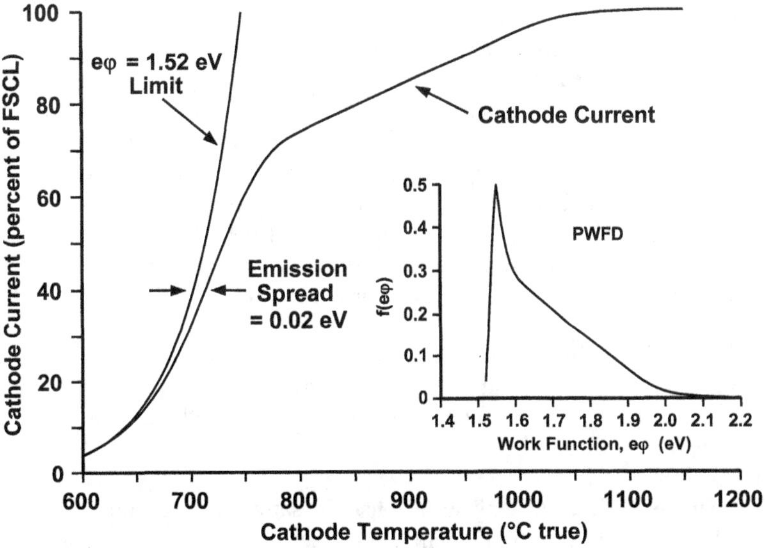

Figure 5-33 Emission spread and PWFD for a scandate cathode. (From: G. V. Miram et al., Fifth IEEE IVEC, © IEEE 2004.)

When the type III mechanism occurs, the work function distribution changes in a reversible manner as temperature changes. That is, if temperature is increased or decreased, the distribution increases or decreases and the measurements are repeatable. This can occur early in life and is attributed to an insufficient supply of barium, a change in sticking coefficient of the barium to the cathode surface or to external poisoning.

The reason given for the type I degradation is usually the gradual depletion of the barium supply in the impregnated tungsten pellet [2, 22–25]. Thus, the rate at which barium diffuses to the cathode surface decreases, so the barium coverage of the surface decreases and work function increases.

Grant noted that the Miram curves do not always support the conclusion that barium coverage of the entire cathode depletes with time [19]. In some cases, a reduction in current persists even if cathode temperature is increased 150°C above the point where the cathode is 99.5% FSCL. If depletion of the entire cathode had been the reason for the reduced current, the increase in temperature would have increased the current. Grant stated that "the preferred reason for the change in perveance is the complete depletion of barium from the edges of the cathode surface." Depletion from the cathode edge is rapid because, if there is no barrier to diffusion on the outer diameter of the cathode pellet, barium can diffuse and evaporate radially outward as well as from the front surface of the cathode.

A feature of Miram curves and of PWFD curves that results from barium depletion is the movement with time of the curves toward higher temperatures ([13] and Figure 5-31). Another example is given in Figure 5-34. By observing the rate at which the curves shift to higher temperatures, life projections can be made. End of life is sometimes predicted to occur when the knee of the roll-off curve of interest reaches the operating temperature and emission current starts to drop or drops by a prescribed amount such as 10%. For the curves shown in Figure 5-34, the movement of the curves over a 2,500-hour period is shown. Assuming that the curves continue to move at the same rate, then the time for the 1 A/cm^2 or the 2 A/cm^2 curves to reach the operating temperature (1,050°C) can be estimated. At 4 A/cm^2 the knee of the curve has already reached the operating temperature and current has dropped about 5%. For this current density, a 10% drop would soon occur.

Under ideal conditions, the selection of the temperature at which a cathode should be operated to maximize life would be straightforward. For example, consider the roll-off curve shown in Figure 5-35. A cathode could have this emission characteristic if:

1. All areas of the emitting surface had the same work function.
2. There was no variation in temperature across the emitting surface.
3. The electric field near the emitting surface was the same everywhere.

End of life would occur when the knee of the curve reached the operating temperature and suddenly there was a precipitous drop in emission current. The operating temperature could be selected by using the life prediction model of Grant and Falce [26]. Life predictions for B and M cathodes are shown in Figure 5-36 and in Figure 5-37.

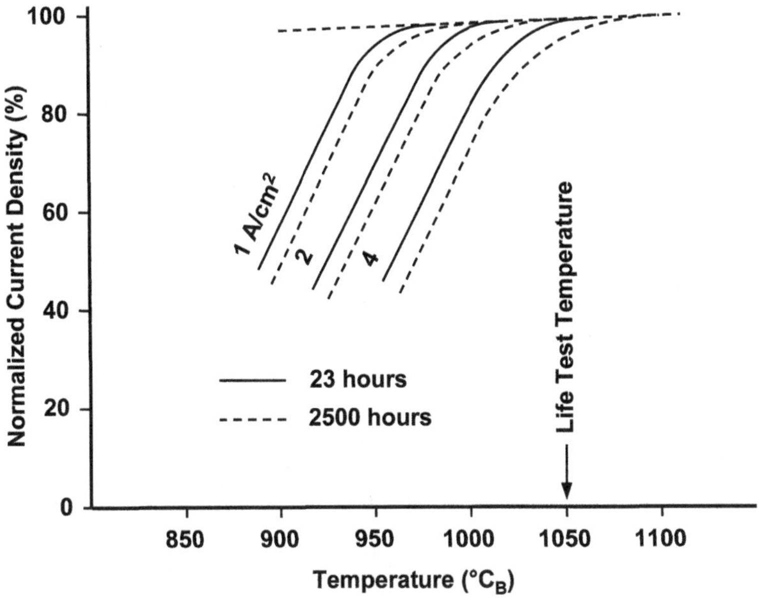

Figure 5-34 Miram plot for B-type cathode. (From: M. C. Green, *Technical Report* RADC-TR-81-211, July 1981.)

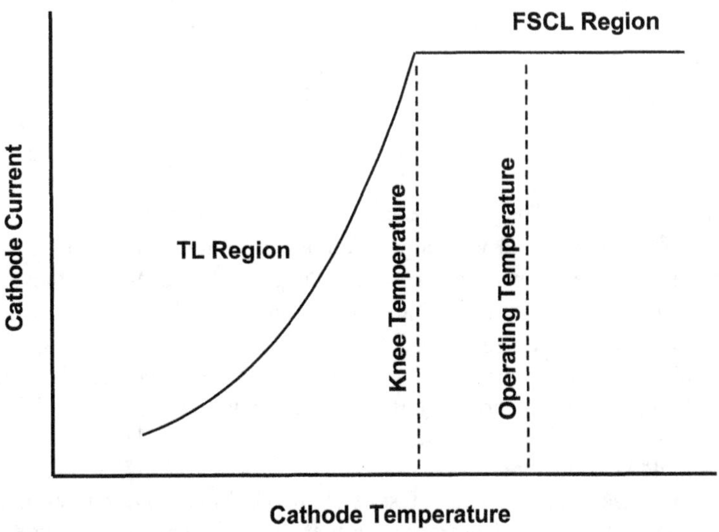

Figure 5-35 Idealized emission characteristic assuming no variations in work function or temperature across the cathode surface or electric field near the cathode surface.

5.4.1 Grant and Falce Life Prediction Model

The model used to generate the curves in Figures 5-36 and 5-37 is based on the type I and type II degradation mechanisms (gradual depletion and base metal work function change). The model is valid for both B- and M-type cathodes, the only difference being the variation with time of the base metal work function in the M-type alloy coatings. Life was defined as the first onset of temperature limited operation.

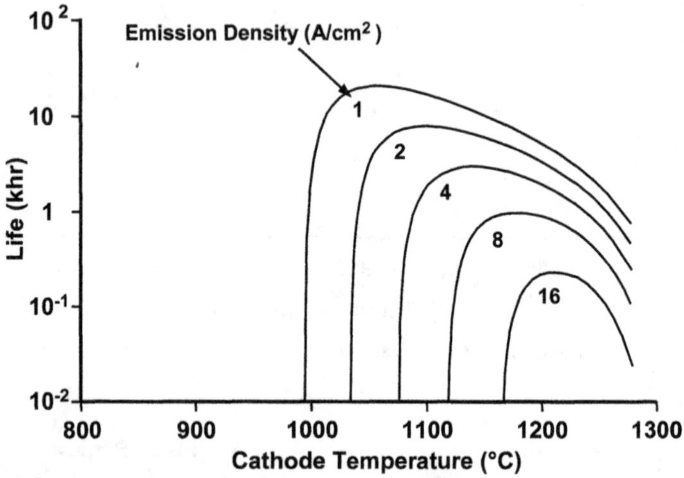

Figure 5-36 Life of a B-type cathode as a function of operating temperature. (From: T. J. Grant and L. R. Falce, Third IEEE IVEC, April 2002, © 2002 IEEE.)

As an example of the use of these curves for a B-type cathode having the emission characteristic in Figure 5-34, assume that life is to be maximized for operation at 1 A/cm^2. The knee temperature in Figure 5-36 is slightly below 1,000°C. Life is maximized with an operating temperature of ~1,050°C or ~50–60°C above the knee temperature. It should be noted, however, that there is very little sacrifice in life if the operating temperature is only ~40°C or as much as ~100°C above the knee temperature.

The life advantage (over an order of magnitude) of the M cathode, compared with the B cathode, is the most striking feature of Figures 5-36 and 5-37. Unless there is some compelling reason to use a B cathode, it is clear that an M-type cathode should be used.

An interesting characteristic of several of the M-cathode curves in Figure 5-37, for example, the 1 A/cm^2 curve, is that the cathodes have a longer life at temperatures below the initial knee temperature. This effect is due to a gradual change of the base-metal work function as tungsten diffuses into the M-type alloy coating.

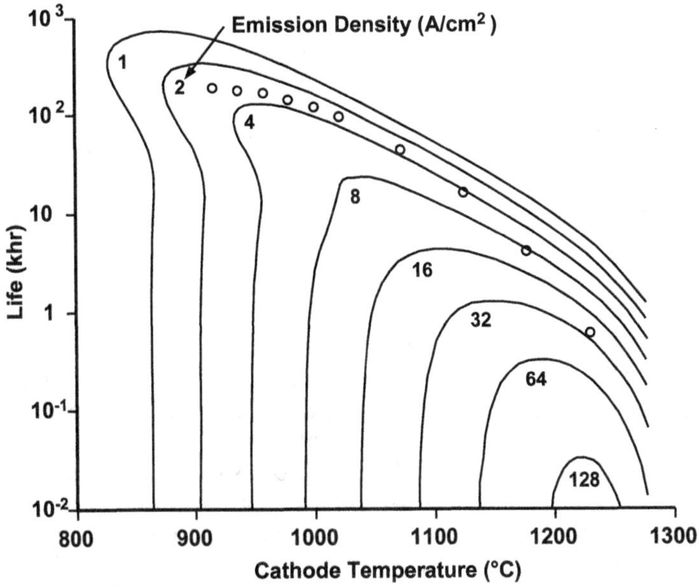

Figure 5-37 Life of an M-type cathode as a function of operating temperature. (From: T. J. Grant and L. R. Falce, Third IEEE IVEC, April 2002. © 2002 IEEE.)

Unfortunately, roll-off curves found for real cathodes don't have the sharp knee shown in Figure 5-35 but instead have a knee shape more like one of those shown in Figure 5-34. As a result, the selection of the temperature at which the cathode should be operated is not so straightforward. Sometimes a knee temperature is defined as shown in Figure 5-38. It is the intersection of linear extrapolations of the TL and FSCL lines that is used to determine the knee temperature. Then the operating temperature is specified to be a temperature increment, ΔT, above the knee. Unfortunately, this technique ignores the fact that there can be a great deal of variability in knee shapes for cathodes for a specific application. For example, for the cathode in Figure 5-38(a), FSCL operation would be possible for a time period of ~$\Delta T/2$, whereas for the cathode in Figure 5-38(b), the onset of TL operation and reduced current would occur much sooner.

Ideally, the operating temperature should be chosen so that end of life is determined when the cathode current in a tube has decreased by an amount that causes one of the operating parameters of the tube to fail to meet specification. In this case the "knee" temperature is the temperature that produces that cathode current. Then the operating temperature is specified to be a ΔT such as 50°C above that knee temperature.

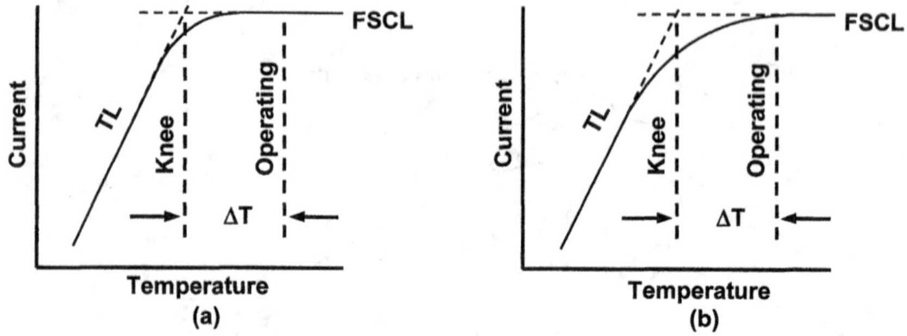

Figure 5-38 (a, b) A technique for defining the knee temperature for a real cathode.

Unfortunately, it is not as easy to determine end-of-life temperature in this manner as it would appear to be. This is because there are several factors that can cause irregularities in the shape of the roll-off curve in the FSCL region. Perhaps the most important factor is the thermal expansion characteristic of the cathode support structure. This causes the cathode-to-anode spacing and thus perveance to vary with temperature. This effect has to be compensated to find the true FSCL current. Other factors that complicate the determination of FSCL current include grid movement and scatter in measured data.

5.4.2 Longo Life Prediction Model

Another example of life prediction based on the depletion model is by Longo [22–24]. He observed that the following simple relation described the emission from an impregnated cathode over a wide range of voltages and temperatures.

$$\frac{1}{J_{OP}} = \frac{1}{J_{FSCL}} + \frac{1}{J_{TL}} \tag{5-20}$$

In this equation, J_{OP} is the observed current density, JF_{SCL} is the current density that would be observed if the entire cathode was operating space charge limited, and J_{TL} is the current density that would be observed if the entire cathode was operating temperature limited. Shown in Figure 5-39 is a comparison of Longo's equation with measured results.

Vaughan observed that the knee of the roll-off curve described by Longo's equation is much broader than that found for most cathodes. Vaughan proposed the relation

$$\frac{1}{J_{OP}^{n}} = \frac{1}{J_{FSCL}^{n}} + \frac{1}{J_{TL}^{n}} \tag{5-21}$$

where n can be adjusted to provide the best fit to data [27]. A value for n of 6 to 10 was given as being appropriate for good electron guns. The dashed curve in Figure 5-39 shows Vaughan's expression plotted for $n = 6$.

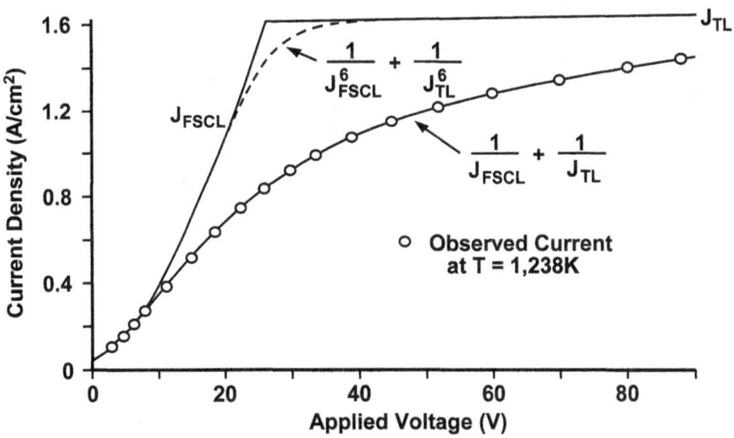

Figure 5-39 Empirical relations for emission current. (Adapted from: R. T. Longo, *Technical Digest*, 1978 IEDM. © 1978 IEEE.)

Longo used his relation to predict emission current density as a function of time by assuming that the work function in J_{TL} (the Richardson-Dushman current) varies with time. The time dependence of work function is assumed to result from the variation with time of the fraction of the cathode surface that is covered by barium and by how much barium is on the covered surface. Surface coverage is determined by considering losses by evaporation and increases due to the evolution of material in the porous tungsten and the migration of this material to the surface by gas flow under conditions known as *Knudsen flow*.

To use the prediction model for an M-type cathode, Longo took into account the diffusion of tungsten into the osmium coating and the resulting variation with time of the base metal work function and of the minimum cathode work function. He calibrated his model against the results of accelerated life tests as shown in Figure 5-40(a) and then used the model to predict the operating lifetime at normal operating temperatures as shown in Figure 5-40(b). In this figure, $J(0)$ is the initial cathode current density and $J(t)$ is the current density at a later time. Note that lifetimes over 100,000 hours are predicted with only small decreases in emission current.

Since the first use of cathodes many decades ago, there has been a continuing requirement for increased emission capability along with long life. At present, for microwave tubes, these requirements are driven primarily by millimeter-wave tubes where small, high-current density cathodes are required for acceptable beam formation. There are also requirements for high-current density cathodes for technologies other than microwave tubes. One of these is for free electron lasers where a high-current density cathode (commonly referred to as a *high brightness cathode*) is needed to help limit the divergence of the beam generated by the electron gun.

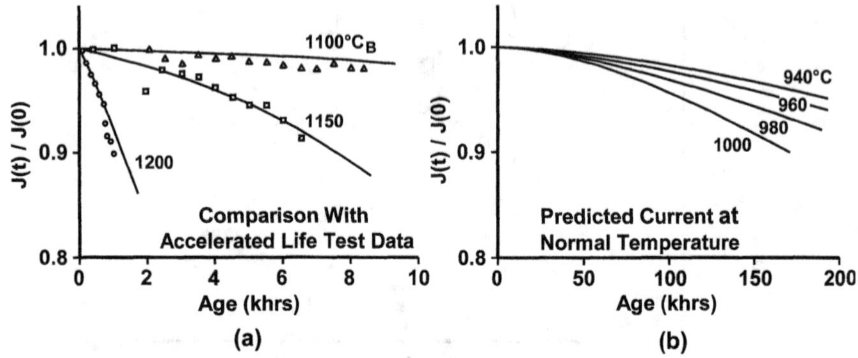

Figure 5-40 (a, b) Results of Longo's life prediction model. (From: R. T. Longo et al., *Technical Digest*, 1984 IEDM. © 1984 IEEE.)

There is a practical limit to the current density that can be drawn from a cathode, regardless of its emission capability. This limit results from the electrical breakdown characteristics of vacuum. For a 1-cm gap with a uniform field, the breakdown field is about 10^5 V/cm, so breakdown across the 1-cm gap would be expected for a voltage of about 100 kV. For a gap with hot electrodes (such as a thermionically emitting cathode) and filled with a high electron density, the breakdown voltage would be below 100 kV.

For a planar SCL diode with an anode-cathode separation of 1 cm, a voltage of about 120 kV is required for a current density of 100 A/cm^2. While the average field would be 120 kV/cm, the field at the SCL cathode would be zero and at the anode would be above the average field. Even if breakdown did not occur within the current flow region, it is probable that it would occur between the focus electrode and the anode where the spacing would be less than 1 cm. Thus, it appears that the practical upper limit to emission current densities is less than 100 A/cm^2. Higher-current densities are possible at very short pulse lengths because the breakdown field strength increases as pulse length becomes shorter.

By optimizing the alloy composition on coated cathodes, Varian (now CPI) has been able to demonstrate emission densities of over 100 A/cm^2 [28]. The Miram curves in Figure 5-41 show this emission capability and also show that the cathode was well behaved up to the highest temperatures and currents tested.

Thomas et al. used barium depletion calculations to predict cathode life at high emission densities [2]. The criterion used for end of life was when the temperature limited emission dropped to half the initial level. Evaporation rate data for B-type cathodes was used to predict life as a function of temperature for each cathode type. Thomas et al. noted that the approximation is crude because of differences in impregnant composition, pore structure, and surface composition, all of which affect the net rate of evaporation. Still, a B-type cathode tested at LLNL and an M type tested at NRL, both at about 40 A/cm^2, are in reasonable agreement with predicted life. Thus, the results as shown in Figure 5-42 appear to be valid within a factor of 2 or 3 uncertainty.

The life of the scandate cathode was also assumed to be limited by barium depletion. Thomas et al. noted that, at the time the paper was written, it was not clear what other mechanisms may limit the life of this cathode. Also, for all of the cathodes shown in Figure 5-42, it is often found that a higher temperature than expected (with resulting shortened life) is required to maintain emission because of the effects of poisoning gases.

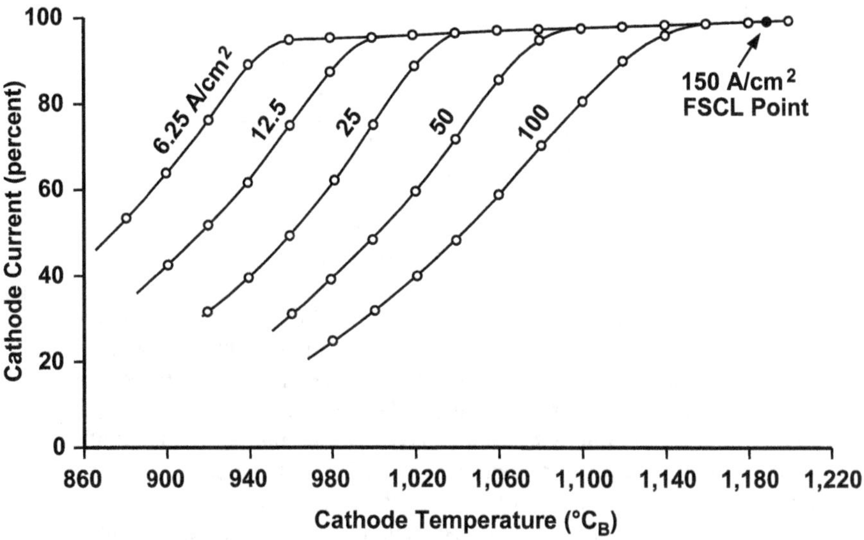

Figure 5-41 Test data for the 100 A/cm^2 dispenser cathode. (From: G. V. Miram et al., *Vacuum Electronics and Display Conference*, 1989.)

5.5 DISPENSER CATHODE SURFACE PHYSICS

The basic theory of thermionic emission discussed at the beginning of this chapter has been well established and understood for several decades. What has not been well understood until recently has been the work functions of surfaces with adsorbed atoms of various materials. As a result, cathodes have historically been developed more by trying various combinations of materials and techniques, rather than by design based on a thorough knowledge of surface physics. Fortunately, however, the modern state of cathode theories is significantly advanced from that given by Pierce in 1946 [1].

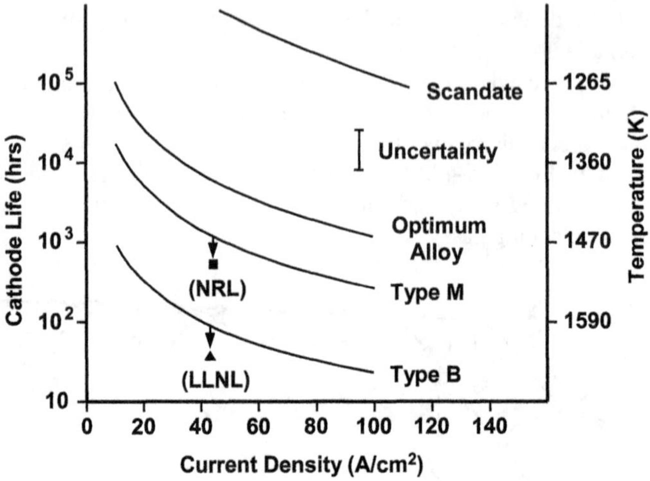

Figure 5-42 Predicted cathode life versus current density. (From: R. E. Thomas et al., *IEEE Trans. Electron Devices*, March 1990. © 1990 IEEE.)

> It is clear that as far as science goes, the state of the oxide coated cathode is confused indeed. Not only does theory fail to help us make good cathodes; it seems not to fit all the evidence. Maybe cathodes work sometimes one way and sometimes another, which isn't as ridiculous as it may sound. At any rate, while some physicists are uninterested and others confused, cathode making continues to be an art practiced, it seems, by artists with a good deal of witch doctor in their make-up. When something goes wrong, this or that demon is exorcised, and eventually things go right again. And, unlikely as it seems, cathodes continue to improve.

While it is not yet possible in theory to design a new type of cathode having a predetermined emission density, operating temperature, and life, present cathode theories do explain very well how most dispenser cathodes operate. The following section contains an overview of the present theories of dispenser cathode operation.

In the fundamental theory of thermionic emission, an energy level diagram for electrons in and near the metal surface, like that shown in Figure 5-43(a), was used. In dispenser cathodes, the emitting surface is far more complex than that shown in Figure 5-43(a). This is because of the diffusion of barium, oxygen, and perhaps other elements to the surface of the cathode. For the B-type cathode, for example, measurements show that the surface consists of monolayers of barium and oxygen on the tungsten base as indicated in Figure 5-43(b). The layers of barium and oxygen are each approximately a monolayer thick. The atoms in these layers are charged. The barium atoms have lost an electronic charge and are positively charged. The oxygen atoms have acquired an electronic charge and are negatively charged. Thus, the positive barium atoms and negative oxygen atoms, along with negative image charges on the surface of the metal (not indicated in Figure 5-43(b)), form electric dipoles. These dipoles affect the potential distribution at the surface of the metal, as indicated by the curved potential profile followed by a question mark. Thus, the work function that electrons must overcome to be emitted is changed.

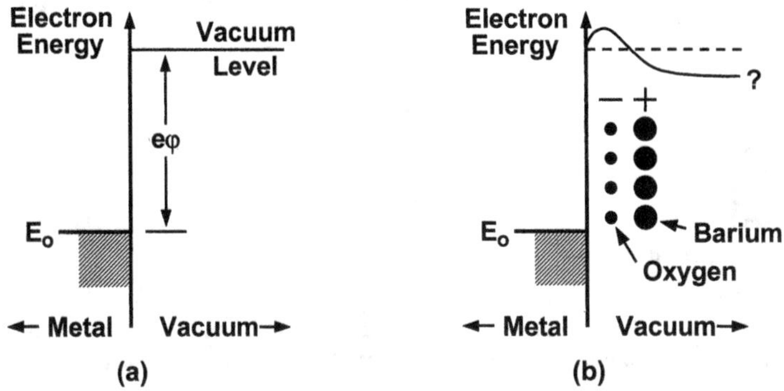

Figure 5-43 (a, b) Effect of oxygen and barium on the energy level diagram.

The positive charge of the barium atoms also results in electrostatic forces that attract the atoms to the surface of the metal. This causes the rate of evaporation of the barium atoms from the tungsten surface to be far below (by over an order of magnitude) the rate of evaporation of barium atoms from a barium surface.

Before attempting to explain the effect of two (or more) layers of charged atoms of different elements on the work function of a surface, it is helpful to consider a single layer of atoms of one element. Following the work of Zalm and van Stratum [29], who were early developers of the M-type cathode, consider again the energy level diagram for a metal surface in Figure 5-44(a). With no foreign atoms near the surface, the energy levels are as previously discussed. If a small number (far fewer than required to form a monolayer) of atoms that can easily give up electrons (have a low ionization potential) are brought into the electric field near the surface of the metal, it is probable that some of the atoms will become ionized and give up electrons to the metal surface (see Figure 5-44(b)). The presence of the resulting positive charge on the surface of the metal produces a reduction, $\Delta e\varphi$, in the work function, which, in turn, increases the probability of electron emission. The influence of the positive ions near the surface of the metal is similar to that of a microscopic grid at a positive potential. Such a grid would reduce the potential barrier at the surface of the metal and permit increased electron emission.

As additional atoms are brought near the surface, some become ionized and further reduce the work function (Figure 5-44(c)). For small numbers of atoms (much fewer than required for a monolayer), the amount by which the work function is reduced, $\Delta e\varphi$, increases linearly with the density of atoms on the metal surface.

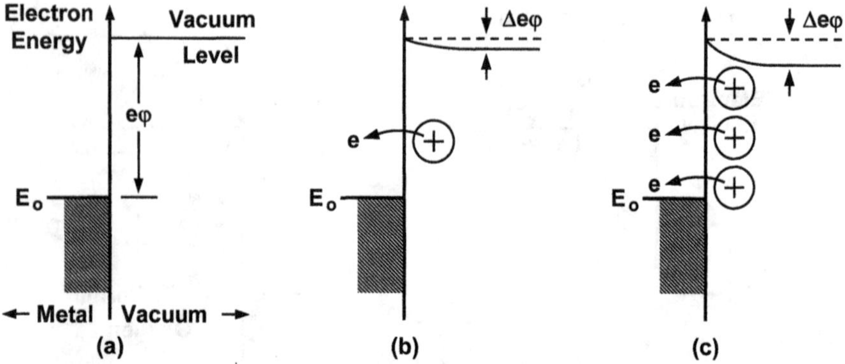

Figure 5-44 Energy level diagrams for a metal surface with: (a) no foreign ions on surface, (b) a few foreign ions on the surface, and (c) many foreign ions on the surface.

As more atoms are brought near the surface, and more ions are formed, two primary effects reduce the rate of ion formation:

1. The reduction in electric field near the surface that the ions produce results in a reduced probability of ionization. Eventually, the field is reduced enough that the probability of ionization approaches zero. The addition of more atoms will not produce additional reductions in the work function.
2. As the number of ions increases, the distance between ions decreases. As this distance decreases, the repulsion forces between ions increases, and this causes more ions to be ejected from the metal surface.

As still more atoms are brought near the surface, and the film thickness grows to over a monolayer, the work function gradually changes to that of the atoms on the surface.

Using the concepts outlined in the preceding discussion, it is possible to understand the reasoning that led Zalm and van Stratum to the use of osmium and other high work function metals in cathodes. Consider two metals, one having a higher work function than the other. For example, consider osmium with a work function of about 5.4 eV and tungsten with a work function of about 4.6 eV. The energy level diagram in Figure 5-45 indicates why, with an adsorbed layer of ions, Zalm and van Stratum expected the work function for osmium to be lower than for tungsten. In each case, as more atoms are brought near the surface, ionization will occur until the electric field has dropped to the point where the probability of ionization approaches zero ($P_i \to 0$). This field will be approximately the same for the two metals. Because the work function for osmium is higher than for tungsten, the field at the surface is higher, and more ions can be produced than for tungsten, so the resulting reduction in work function for osmium is higher than for tungsten. In fact, the reduction for osmium is high enough that the overall work function of the metal plus adsorbed ion layer is lower than for tungsten. This results in a higher electron emission from osmium (with an adsorbed layer) than for tungsten.

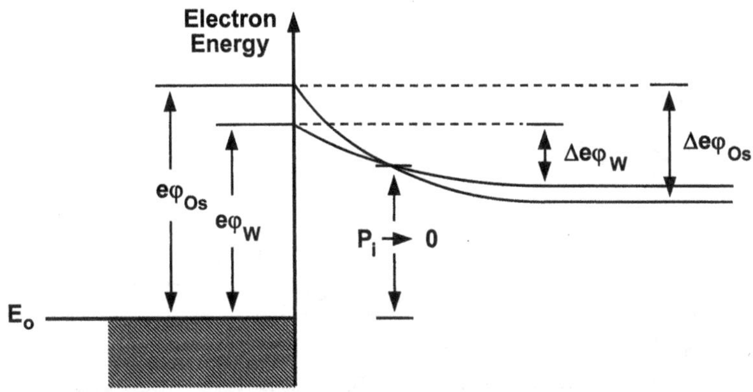

Figure 5-45 Effect of work function of metal on energy profiles.

The explanation given previously for the improved performance of an osmium cathode fails when some other metals are considered. For example, as indicated in Figure 5-46, the work function for platinum (which is a 5d transition metal in the same family as osmium) is comparable to that of osmium. However, the emission of a platinum coated cathode is significantly below that of osmium and is even below that of tungsten. As another example, the emission capability of a cathode coated with a tungsten-osmium alloy is superior to that of an osmium-coated cathode even though the work function of the tungsten-osmium alloy is below that of osmium.

To understand the differences in performance between cathodes coated with a tungsten-osmium alloy, osmium, or platinum, it is necessary to examine the atomic structure of the metal cathode surface and its relation to adsorbed oxygen and barium atoms. The work of Müller [30–32] has led to calculations of work functions for dispenser cathodes that are in excellent agreement with experimental results.

In Müller's work, a microscopic portion of the surface of a cathode is modeled as a cluster of atoms. The models are chosen on the basis of crystal structures found on actual cathodes. An example is shown in Figure 5-47. Here, the B-type cathode is modeled as a cluster consisting of 19 tungsten atoms in a body-centered cubic configuration and three groups of BaO molecules. The tungsten substrate in an actual B-type cathode is polycrystalline with many crystal faces present. The (100) orientation was chosen because of its simple structure. The clusters used for modeling the osmium and osmium-tungsten alloy substrates represented the hexagonal-close-packed lattice. For platinum, the face-centered cubic lattice was used.

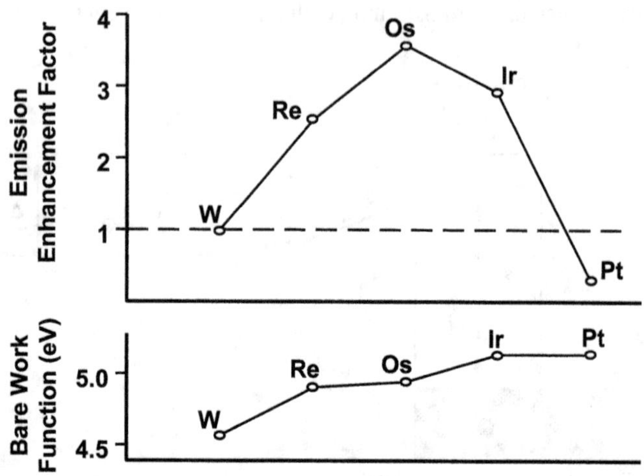

Figure 5-46 Emission enhancement and work function for dispenser cathode metals. (From: M. C. Green, *Technical Report,* RADC-TR-81-211, July 1981.)

The calculations were performed using large computer codes capable of analyzing the interactions between the relatively large numbers of atoms in the clusters. Calculated properties included electronic energies, densities of states, electron densities, charge distributions, surface electric dipoles, and barium binding energies on different substrates.

Only the work function change that occurs with the sorption of barium and oxygen on the metal surface is obtained from the calculated properties. Thus, the minimum work functions depend critically on the bare-metal work functions of the various crystal faces considered. Because accurate values of the bare surface work functions are not known in most cases, some assumptions and approximations had to be made. In addition, empirical temperature dependencies of polycrystalline work functions for pure metals were used to estimate work functions at 1,200K. The initial bare-metal work functions affect the results by their magnitude and temperature dependence. A high value leads to a large charge transfer from barium to oxygen and the metal substrate. Alloying with W, in general, reduces the work functions for Re, Os, Ir, and Pt.

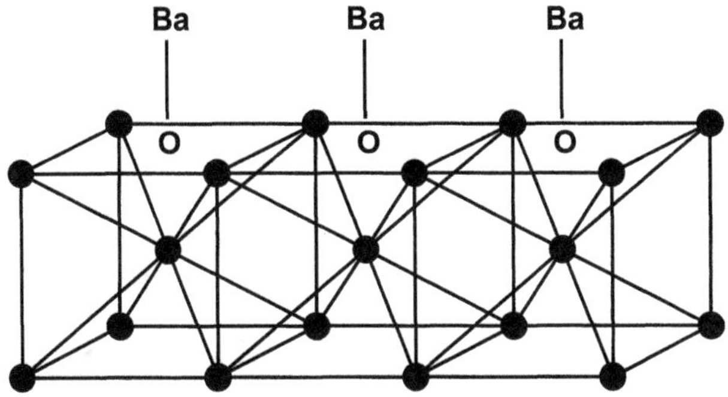

Figure 5-47 Body-centered cubic cluster (100 face) assumed for BaO/W computations. (From: W. Müller, *Technical Record*, 1991 IEDM. © 1991 IEEE.)

The work-function change that is induced by the sorption of BaO is found to be the largest for the hexagonal structures (Os, Ir, Os-W). The larger change more than offsets the initially higher bare-metal work functions of the alloy substrates and, as shown in Figure 5-48, leads to work functions that are up to 0.2 eV lower than for BaO on tungsten.

The calculated barium binding energies for high surface coverages are 5.4 eV for BaO/W and ~6.0 eV for the alloy substrates. This increase of 0.6 eV is in excellent agreement with the observed increase in the energy of desorption for Ba from B- and M-type cathodes. This, along with a reduced operating temperature for a given emission current, is the reason for the long life of the M-type cathode. For BaO on platinum, the calculated binding energy drops to 3.6 eV. Thus, barium evaporates at a higher rate from platinum than from the other metals considered, and the resulting reduced coverage of the surface leads to a higher minimum work function.

5.6 HEATERS

5.6.1 Conventional Heater Assemblies [33]

Typical oxide coated cathodes operate near 850°C and B-type cathodes operate near 1,100°C. To raise a cathode to this temperature, the heater must be very hot. The materials of choice for heaters are pure tungsten or tungsten with a small amount of rhenium added. The rhenium makes the tungsten easier to work.

After being formed, the heaters are coated with an insulating layer 0.002–0.003 inch thick of aluminum oxide. The oxide can be applied by spraying or by cataphoretic coating (an electrolytic process). The aluminum oxide insulation is sintered at high temperature to drive off organic binders and to form a hard, scuff-resistant coating.

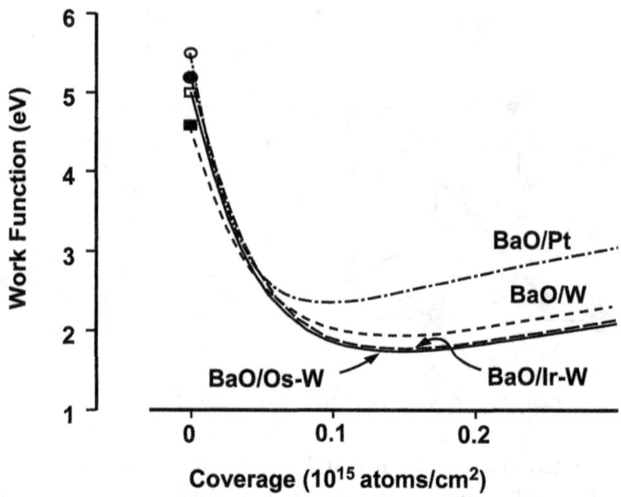

Figure 5-48 Calculated work functions for B- and M-type cathodes. (From: W. Müller, 1992 Tri-Service/NASA Cathode Workshop.)

Heaters for oxide-coated cathodes generally operate at a temperature near 1,100°C. This is not excessive for long life in tungsten or tungsten-rhenium heaters. Oxide cathode heaters are generally mounted in a fairly simple cathode structure with no special steps taken to improve the transfer of heat from the heater to the cathode. Two typical structures are shown in Figure 5-49.

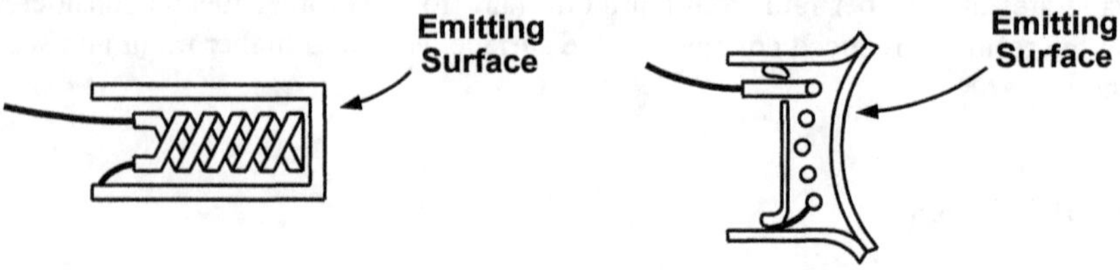

Figure 5-49 Typical heater assemblies for oxide-coated cathodes.

One side of the heater is often connected to the cathode so that no large voltage difference exists between the cathode and any part of the heater. The aluminum oxide insulation on the heater is thin and will not withstand much voltage without breakdown.

Heaters for B-type dispenser cathodes present a more severe problem because of the 1,100–1,150°C operating temperature of the cathode. The usual solution is to embed the heater in an aluminum oxide pellet at the back of the cathode as shown in Figure 5-50. Heat transfer between the heater and the cathode is much improved by this embedding or potting, so the heater operating temperature needs to be only 150–200°C higher than the cathode. Also, the potting gives the heater mechanical support, making it resistant to damage from vibration and shock. The heater potting material is added as a powder, possibly compacted under pressure and then sintered at high temperature.

Potted heaters have one disadvantage. The extra mass of the potting material increases the warm-up time over an unpotted version in a similar cathode. In most applications this warm-up time penalty (typically 2–3 minutes) is accepted as the price for improving heater reliability. Some applications, in missiles, for example, require fast warm-up. (Fast warm-up heaters are discussed in the following section.) Fortunately, such applications do not require long life, so the heater can be designed for low mass and high temperature operation.

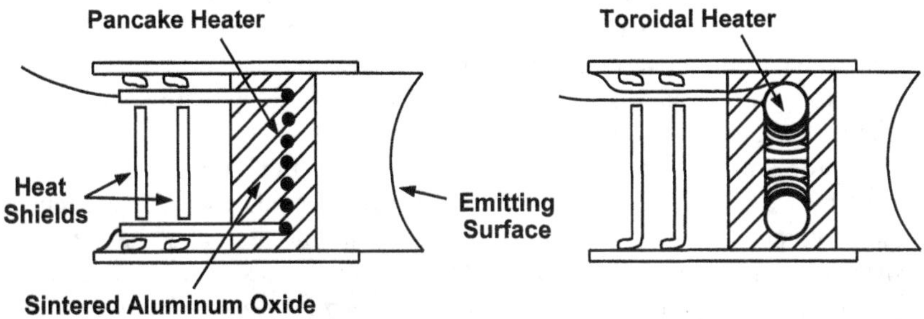

Figure 5-50 Heater assemblies for dispenser cathodes.

Dispenser cathodes require special steps to conserve heater power. Heat shields are used extensively to minimize radiation losses to the side and back. Support members are made with as thin a cross section as possible to minimize conduction losses. Finally, considerable care must be taken by the manufacturer to ensure consistency in the heater potting operation. Variations in contact between the heater, the potting material and the cathode can lead to dangerous heater "hot spots" and cathode "cold spots."

Heaters can be operated from either ac or dc supplies. When an ac supply is used, the heater may cause "hum" to appear as modulation of the signal. The source of this hum is often magnetic flux generated by the heater current. This flux extends to the cathode surface and perturbs the trajectories of electrons leaving the cathode. The toroidal heater on the right in Figure 5-50 is designed to minimize the generation of stray magnetic flux. The perturbation of the beam in a tube by magnetic flux from the cathode is discussed further in the following section.

The dc heater supplies require extra circuitry, but they are sometimes used. A dc heater will eliminate hum, but stray magnetic flux from the heater can still perturb the beam emerging from the gun and cause increased interception current. The rectifier circuitry for a dc heater supply adapts easily to voltage regulation, which is of value when ac line voltage variations are excessive.

When a dc supply is used, the polarity of the supply can be important. If one side of the heater is connected to the cathode, that side should go to the positive side of the heater supply. The heater is then negative with respect to the cathode. This precaution is taken to avoid emission current from the back of the cathode to the heater. If the heater were positive with respect to the cathode, it would attract any electrons unintentionally emitted from the back of the cathode. The voltage and current involved are small, but over a long period of time such a current can damage the heater.

5.6.2 Fast Warm-Up Heaters

Fast warm-up heaters are needed for cathodes in expendable microwave tubes for missiles and may be needed for some space defense applications. Full emission is required from these cathodes within a few seconds after heater power is applied. This emission requirement must be met even though the cathodes may have been dormant for years prior to turn on. Also, most tubes with this fast turn-on requirement must meet severe shock and vibration specifications.

The problems of rapid reactivation of a cathode after storage were discussed in detail by Haas et al. [34]. The factors involved include cathode type, surface composition, pore spacing, Ba impregnant concentration, Ba-O bonding, exposure to poisoning gases, and heating time and temperature. Haas et al. pointed out that it is the elimination of the poisoning products of ambient gases from the surface along with the arrival of new Ba surface atoms via the impregnant diffusing out of the pores that constitutes the reactivation process of the cathode surface. Some turn-on tests indicate that essentially no delay occurs. This could have been due to temperature overshoot as the cathode is very quickly brought to the operating temperature or to the passivating effect of other gases such as hydrogen. Other tests have correlated turn-on delays with measured pressures of oxidizing gases.

Shown in Figure 5-51 are the temperature characteristics of some fast warm-up heater-cathode assemblies [28]. The cathode diameter in each case was 0.2 inch (0.79 mm). In all cases, the fast warm-up times were achieved by using low thermal mass components. The 3-second design used an integral tungsten-3% rhenium heater cataphoretically coated with alumina. A voltage surge as high as twice the normal operating voltage was used.

The bombarder-type assembly, which is shown in Figure 5-52, achieved a warm-up time of 1 second [35]. The heater is made of thoriated tungsten and designed to be heated to about 1,700°C in a fraction of a second by a large overvoltage. At that temperature the heater is also an emitter. Several hundred volts are applied between the heater and the cathode. Bombardment of the back of the cathode by electrons from the heater raises the cathode to the operating temperature.

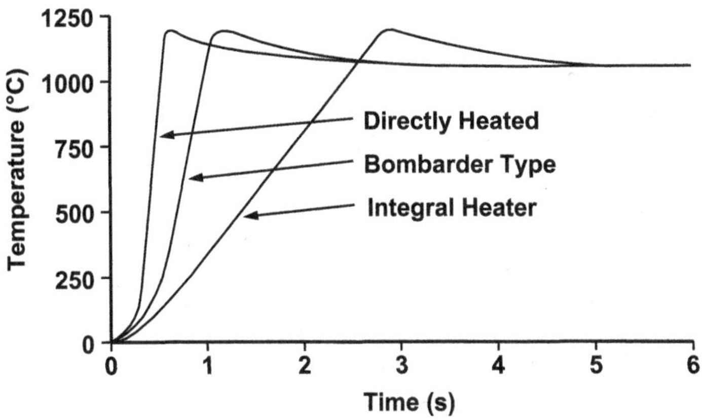

Figure 5-51 Temperature characteristics for fast warm-up heater-cathode assemblies. (From: G. V. Miram et al., *Vacuum Electronics and Display Conference*, 1989.)

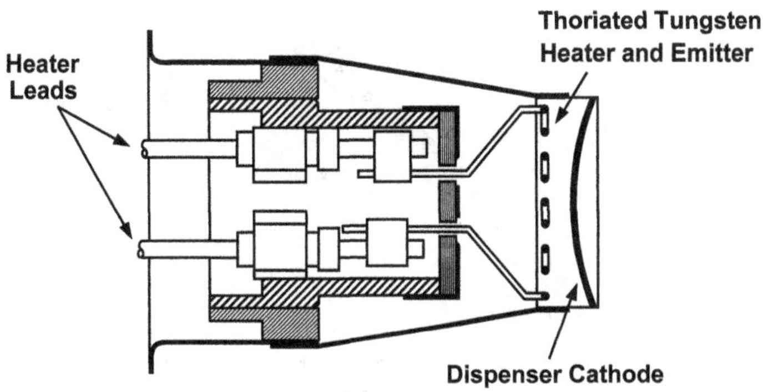

Figure 5-52 Bombarder-type fast warm-up heater. (From: R. C. Treseder et al., *Technical Digest*, 1983 IEDM. © 1983 IEEE.)

Cattelino et al. described fast warm-up heaters for large diameter cathodes [36]. Anisotropic pyrolytic graphite deposited onto anisotropic boron nitride is used to fabricate heater wafers. Both planar and spherical heaters can be made. Warm-up times of 1 to 2 seconds for 4–6.4-mm-diameter cathodes and less than 20 seconds for 60-mm-diameter cathodes have been demonstrated.

5.6.3 Heater Testing

By examining heater current and voltage, it is possible to determine if a heater is operating properly and, usually, if the cathode is at or near the proper operating temperature. It is important to realize that heater resistance varies rapidly with heater temperature. For the tungsten-3% rhenium wire often used for heaters, resistance increases by a factor of 6 as the wire is heated from room temperature to approximately 1,550°C. Thus, a plot of heater current versus heater resistance, as shown in Figure 5-53, can reveal malfunctions of the heater assembly.

To provide meaningful information, the region of acceptance must be determined for a particular class of heater assemblies that are known to operate at the desired temperature. Then Figure 5-53 becomes the benchmark against which other heater assemblies can be compared. Excessive heater current implies that the heater assembly is at a lower temperature than it should be (for that current, the heater resistance is too low so the temperature is too low). Possible problems are that the tube is gassy (gas cools the heater) or that a heat shield is touching the heater or cathode assembly and carrying heat away. Operation below the region of acceptance indicates excessive heater wire temperature and possible heater failure.

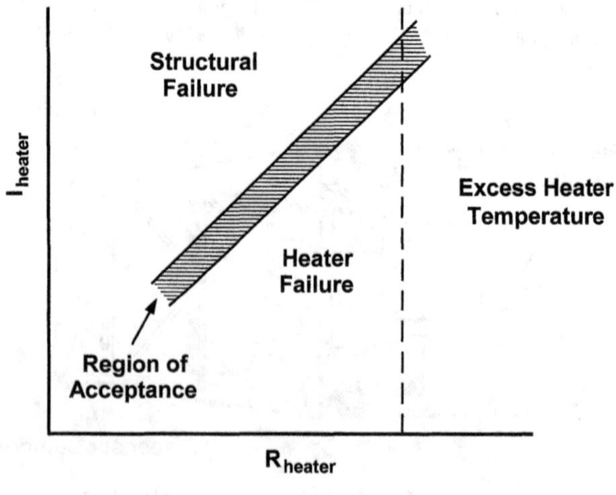

Figure 5-53 Heater performance. (Adapted from: T. J. Grant, *Technical Digest*, 1984 IEDM. © 1984 IEEE.)

5.6.4 Effect of Filament Magnetic Field

Detailed measurements made on the electron beam from a magnetically shielded Pierce gun (discussed in the following chapters) illustrate the effect that an ac heater can have on a beam. The current used in the cathode heater was ac, and the beam voltage was pulsed on at various times during the heater current cycle [37].

When the heater current was zero, the electron beam cross section of Figure 5-54 was obtained. When the heater current was at a maximum value, cross sections like that of Figure 5-55 were obtained. Under very specialized conditions, an image of the spiral-wound heater could be detected in the beam, as shown in Figure 5-56.

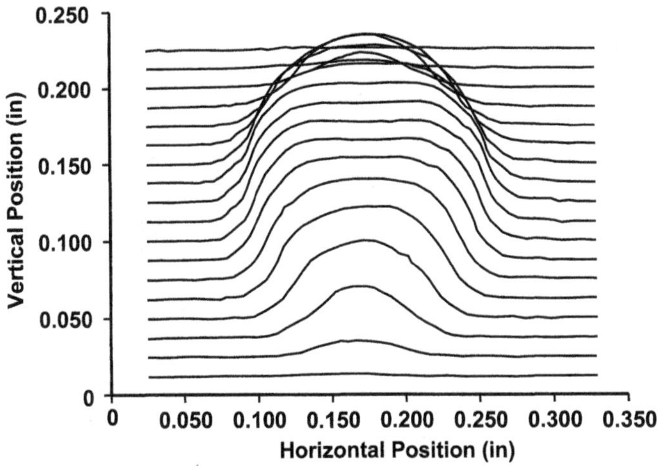

Figure 5-54 Electron beam cross section obtained for no current through the cathode heater. (From: A. S. Gilmour, Jr., *Proc. IRE*, Vol. 49, 1961. © 1961 IRE (now IEEE).)

The perturbations shown in Figures 5-55 and 5-56 resulted from the linking through the cathode surface of magnetic flux lines generated by the heater current. The variations in the shape of the beam as the filament current changes result in modulation of the RF output of the tube at the frequency of the filament current.

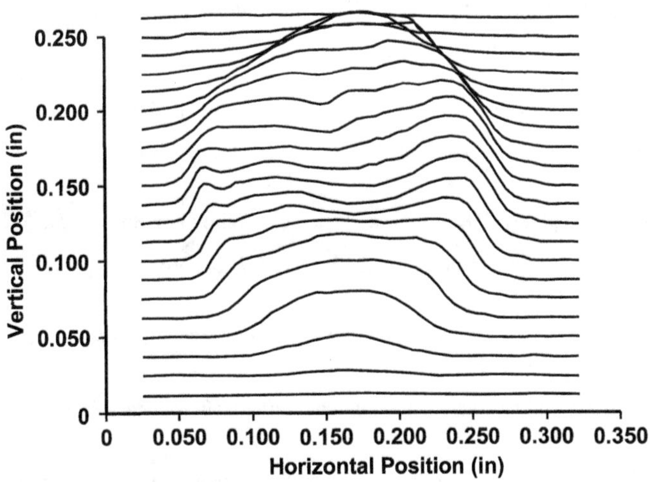

Figure 5-55 Electron beam cross section obtained with the cathode heater at a maximum value. (From: A. S. Gilmour, Jr., *Proc. IRE*, Vol. 49, 1961. © 1961 IRE (now IEEE).)

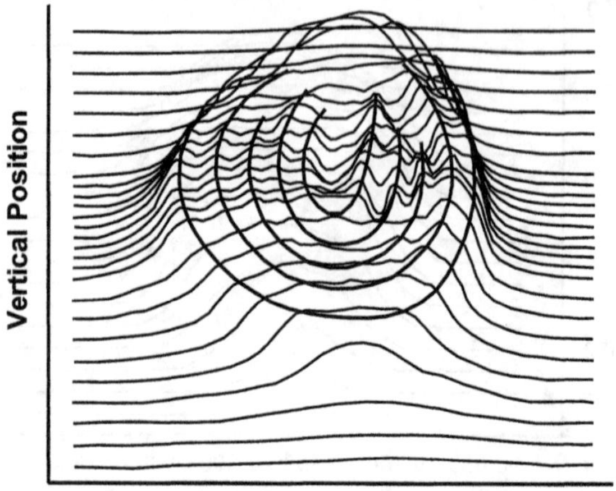

Figure 5-56 Image of spiral-wound cathode heater in electron beam cross section. (From: A. S. Gilmour, Jr., *Microwave Tubes*. Copyright 1986 by Artech House. Reproduced with permission.)

REFERENCES

[1] J. J. Coupling, a pseudonym of J. R. Pierce, *Astounding Science Fiction*, November 1946.

[2] R. E. Thomas et al., "Thermionic sources for high-brightness electron beams," *IEEE Trans. Electron Devices*, Vol. 37, No. 3, March 1990. pp. 850–861.

[3] G. Herrman, and P. S. Wagener, *The Oxide-Coated Cathode*, London: Chapman and Hall, 1951.

[4] R. L. Sproull, *Modern Physics*, New York: John Wiley and Sons, 1956.

[5] J. L. Cronin, "Modern dispenser cathodes," *IEE Proc*, Vol. 128, pt. 1, No. 1, February 1981, pp. 19–32.

[6] C. L. Hemenway, R. W. Henry, and M. Caulton, *Physical Electronics*, New York: John Wiley and Sons, 1967.

[7] M. Chodorow, and C. Susskind, *Fundamentals of Microwave Electronics*, New York: McGraw-Hill, 1964.

[8] J. O. Tarter, and J. J. Farrell, "Selection of dispenser cathode impregnant types," *Technical Digest*, Fifth IVEC, 2004, pp. 301–302.

[9] D. Sprehn, G. Caryotakis, and R. M. Phillips, "150-MW S-band klystron program at the Stanford Linear Accelerator Center," *SLAC Publication 7232*, July 1996.

[10] S. L. Gold, and G. V. Miram, "Assessing cathode quality in pulse and CW tubes," *Technical Digest*, Third IEEE IVEC, 2002, pp. 303–304.

[11] M. Zhang et al., "Investigation of recovery characteristics after ion bombardment on scandate cathode," *Technical Digest*, IVEC, 2004, pp. 153–155.

[12] J. J. Farrell, S. Conkwright, and J. O. Tarter, "The effects of chemical cleaning on impregnant removal as a function of impregnant type," *Technical Digest*, Fifth IVEC, 2004, pp. 297–298.

[13] M. C. Green, "Dispenser cathode physics," *Final Technical Report*, RADC TR 81 211, July 1981.

[14] M. C. Green, "Dispenser cathode design," *Final Technical Report*, RADC TR 83 99, May 1983.

[15] M. C. Green, "Modern thermionic cathodes," *Technical Digest*, IVED Meeting, Monterey, CA, 1987, pp. 925–928.

[16] R. L. Ives et al., "Controlled porosity cathodes from sintered tungsten wires," *IEEE Trans. Electron Devices*, Vol. 52, No. 12, December 2005, pp. 2800–2805.

[17] M. J. Cattelino, G. V. Miram, and W. R. Ayers, "A diagnostic technique for evaluation of cathode emission performance and defects in vehicle assembly," *Technical Digest*, IEDM, 1982, pp. 36–39.

[18] T. J. Grant, "A powerful quality assurance technique for dispenser cathodes and electron guns," *Technical Digest*, IEDM, 1984, pp. 334–337.

[19] T. J. Grant, "Emission degradation characteristics of coated dispenser cathodes," *Technical Digest,* IEDM, 1986, pp. 700–703.

[20] M. Cattelino, and G. Miram, "Predicting cathode life expectancy and emission quality from PWFD measurements," *Applied Surface Science*, Vol. 111, 1997, pp. 90–95.

[21] G. V. Miram et al., "Emission spread in thermionic cathodes," *Technical Digest*, Fifth IEEE IVEC, 2004, pp. 303–304.

[22] R. T. Longo, "Long life, high current density cathodes," *Technical Digest*, IEDM, 1978, pp. 152–158.

[23] R. T. Longo, "A study of thermionic emitters in the regime of practical operation," *Technical Digest*, IEDM, 1980, pp. 467–470.

[24] R. T. Longo, E. A. Adler, and L. R. Falce, "Dispenser cathode life prediction model," *Technical Digest*, IEDM, 1984, pp. 318–321.

[25] A. M. Shroff, and G. Firmain, "Long-life travelling-wave tubes. Use of M-type cathode. Life prediction model," *Technical Digest*, IEDM, 1985, pp. 346–349.

[26] T. J. Grant, and L. R. Falce, "Life prediction model for vacuum electron devices using barium calcium aluminate impregnated tungsten cathodes as electron sources," *Technical Digest*, Third IEEE IVEC, 2002, pp. 300–301.

[27] J. R. M. Vaughan, "A synthesis of the Longo and Eng cathode emission models," *IEEE Trans. Electron Devices*, Vol. ED-33, No. 11, November 1986, pp. 1925–1927.

[28] G. Miram, M. Green, and M. Cattelino, "Gridded gun design considerations for pulse and CW applications," *Vacuum Electronics and Display Conference*, Garmisch Partenkirchen, Austria, 1989.

[29] P. Zalm, and A. J. A. v. Stratum, "Osmium dispenser cathodes," *Philips Technical Review*, Vol. 27, No. 3/4, pp. 69–75, 1966.

[30] W. Müller, "Theoretical study of cathode surface," *Final Technical Report*, NASA CR-182166, July 1988.

[31] W. Müller, "Computational modeling of dispenser cathode emission properties," *Technical Digest*, International Electron Devices Meeting, 1991, pp. 399–402.

[32] W. Müller, "Mechanism of emission enhancement in barium dispenser cathodes," *Technical Digest*, Tri-Service/NASA Cathode Workshop, Greenbelt, MD, 1992.

[33] A. T. Raczinski, "TWT principles," TWT training course presented at NSWC, Crane, IN, 1982.

[34] G. A. Haas et al., "Rapid turn-on of shelf stored tubes: An update," *IEEE Trans. Electron Devices,* Vol. ED-38, No. 10, October 1991, pp. 2244–2251.

[35] R. C. Treseder, T. J. Grant, and G. V. Miram, "Design of quick start, high current density electron guns," *Technical Digest*, IEDM, 1983, pp. 453–455.

[36] M. J. Cattelino, G. V. Miram, and B. Smith, "Fast-and super-fast-warm-up cathodes using novel APG/APBN heaters," *IEEE Trans. Electron Devices*, Vol. ED-38, No. 10, October 1991, pp. 2239–2243.

[37] A. S. Gilmour, Jr., "Effect of filament magnetic field on the electron beam from a Pierce gun," *Proc. IRE,* Vol. 49, 1961, p. 976.

Chapter 6

Electron Guns

The electron guns for all linear-beam tubes have operating principles that are nearly the same. The only major differences are in size and operating voltage and current, which result from differences in power requirements. In this chapter, the operating principles of the electron guns used on nearly all linear-beam tubes are discussed. Most of these guns are designed using guidelines set forth by J. R. Pierce and are known as Pierce guns [1].

6.1 PIERCE GUNS

The electron gun in a microwave tube is used to form the electrons from the cathode into a beam suitable for interaction with a microwave circuit. Electron gun designers must cope with the following two basic problems:

1. Electrostatic repulsion forces between electrons tend to cause the beam to diverge.
2. The current density required in the electron beam is normally far greater than the emission density that the cathode can supply with an acceptable life expectancy.

The solution to these problems, described by Pierce, is summarized in Figure 6-1. It is assumed that there is no magnetic field in the gun. Thermal velocities of electrons are ignored and laminar electron flow is assumed. The gun is divided into three regions. In the first region, assuming that a beam with a circular cross section is desired, the shape of the cathode disk (referred to as the *cathode button*) is spherical. (This may not be the optimum shape as pointed out by Brewer [2].) The focus electrode is designed to produce equipotential surfaces in region 1 that are nearly spherical, with the same center of curvature as the cathode. As a result, electrons flow toward the center of curvature of the cathode.

In region 2, because the anode must contain a hole to let the electron beam pass through, equipotential surfaces bow into the anode aperture. As a result, a divergent electrostatic lens exists that produces a defocusing action on the electron beam.

Finally, in region 3, the electrons have escaped from the accelerating field of the cathode-to-anode regions and are drifting under the influence of space charge forces. Thus, the electrons in the beam follow universal beam-spread trajectories.

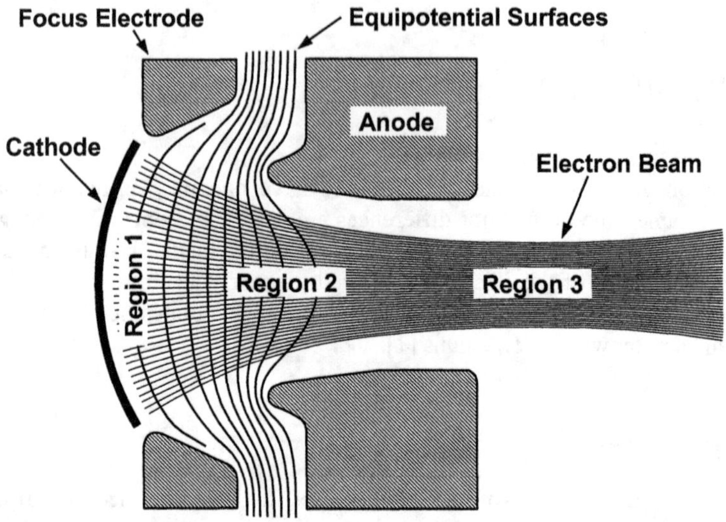

Figure 6-1 Overview of a Pierce gun.

6.1.1 Focus Electrodes for Parallel Flow

Dealing first with the repulsion force problem as electrons leave the cathode and move toward the anode, it is instructive to examine the problem of attempting to cause electrons to travel in parallel paths as they leave a cathode as indicated in Figure 6-2.

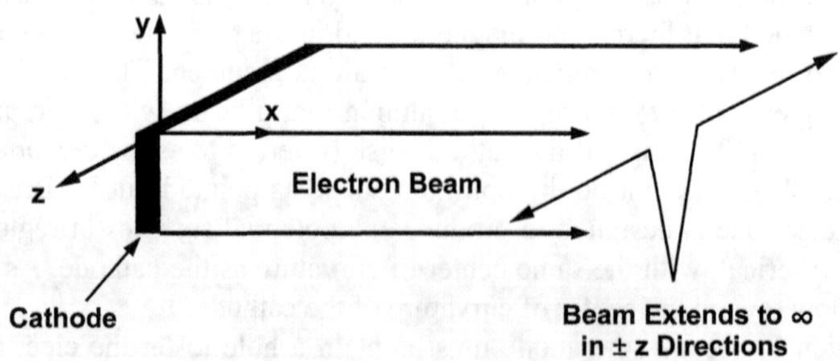

Figure 6-2 Parallel electron flow from a planar cathode.

When no electrons are present, the equipotential profiles in a parallel-plane diode are parallel and spaced equal distances apart for equal voltage increments. This is shown by the dashed lines in Figure 6-3. When electrons are present, the equipotential profiles are deflected to the right as indicated by the solid lines. Electrons tend to move perpendicularly to the equipotential profiles, so the electron trajectories diverge as the electrons move away from the cathode. This is simply another way of saying that the electrons are pushing on each other, and this causes their trajectories to diverge. If an electrode arrangement could be devised that would straighten the equipotential profiles even though electron space charge forces are present, then the electrons would travel in parallel paths.

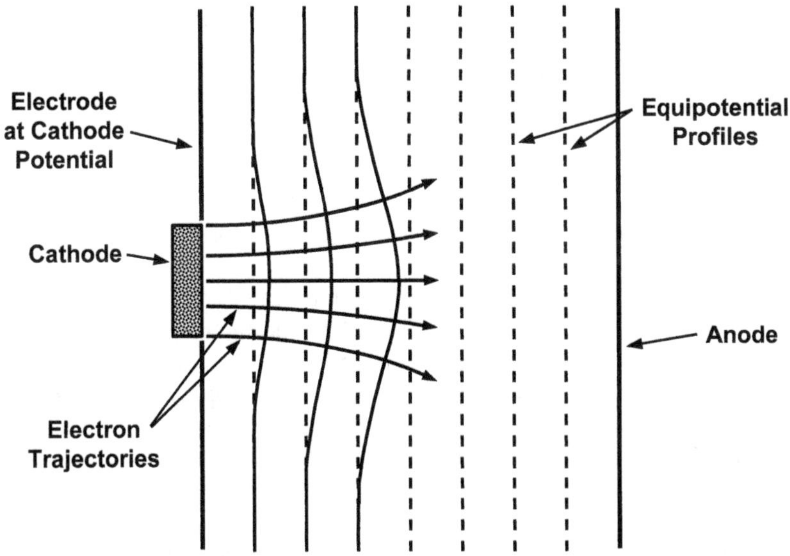

Figure 6-3 Equipotential profiles without (dashed lines) and with (solid lines) electrons.

Intuitively, it can be seen that, if the electrodes adjacent to the cathode are tipped toward the anode, as indicated in Figure 6-4, then the equipotential profiles will be straightened. In the absence of electrons, the profiles curve toward the cathode. The electron charge causes them to move toward the anode.

The mathematical solution that leads to a definition of the electrode configuration for parallel electron flow is similar to the solution for the problem of space charge limited flow discussed in Chapter 5. The main difference now is that the equations must be solved inside and outside the beam, and the solutions must match at the edge of the beam.

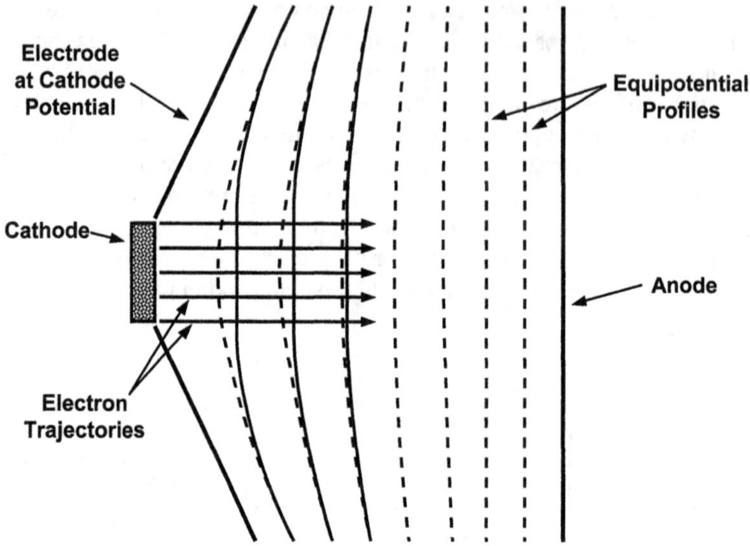

Figure 6-4 Straightening of equipotential profiles in electron beam.

Pierce showed that planar electrodes at the potential of the cathode placed at an angle of 67.5° with respect to the edge of the beam, as shown in Figure 6-5 (along with a curved electrode at the anode potential, V_a), will cause parallel electron flow. These electrodes at 67.5° are called Pierce electrodes and the 67.5° angle is referred to as the Pierce angle.

Figure 6-6 shows an electron gun containing 67.5° Pierce electrodes for producing parallel (rectilinear) electron flow. Note that grid wires are shown across the aperture in the positive electrode. This is to maintain the electrode potential within the beam and, thus, prevent divergence of the beam after it passes through the positive electrode. Beam divergence caused by an anode aperture will be discussed next.

6.1.2 Focus Electrodes for Convergent Flow

The current density required in the electron beam in a microwave tube is normally far greater than the emission density capability of the cathode. As a result, it is necessary to use a cathode having a relatively large area and then compress the flow of electrons as the beam is formed. The starting point in Pierce's analysis is a spherical diode. The inner surface of the outer sphere is the cathode and the inner sphere is the anode. A conical section of the diode is used as shown in Figure 6-7. Space charge limited emission is assumed.

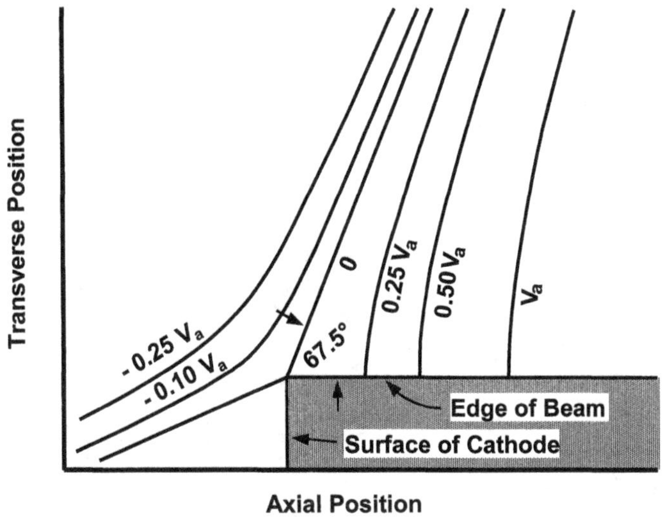

Figure 6-5 Equipotential profiles for parallel electron flow in a planar two-dimensional diode. (Adapted from: *Theory and Design of Electron Beams* by J. R. Pierce, copyright 1954 by D. Van Nostrand, Inc.)

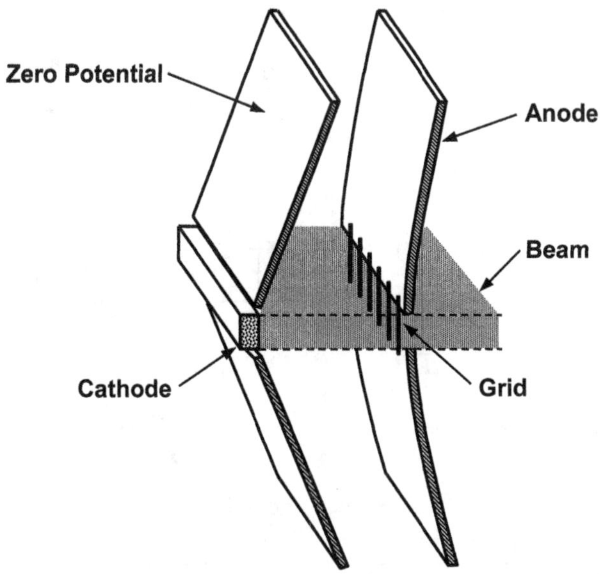

Figure 6-6 Electron gun for producing rectilinear (parallel) electron flow. (Adapted from: *Theory and Design of Electron Beams* by J. R. Pierce, copyright 1954 by D. Van Nostrand, Inc.)

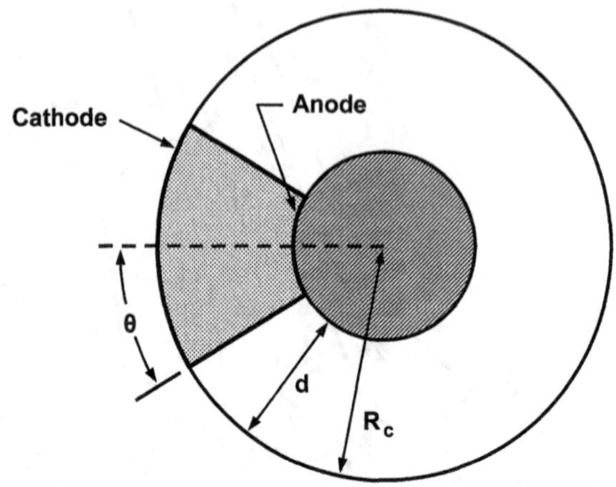

Figure 6-7 Conical diode used in a Pierce gun.

When the analysis is carried out, the perveance in the conical diode is found to be related to the half angle of the cone, θ, by

$$P = 14.67 \times 10^{-6} \frac{(1 - \cos\theta)}{(-\alpha)^2} \tag{6-1}$$

The quantity $(-\alpha)^2$ used here was first defined by Langmuir and Blodgett [3] and is a function of the dimensions of the diode as is shown in Figure 6-8.

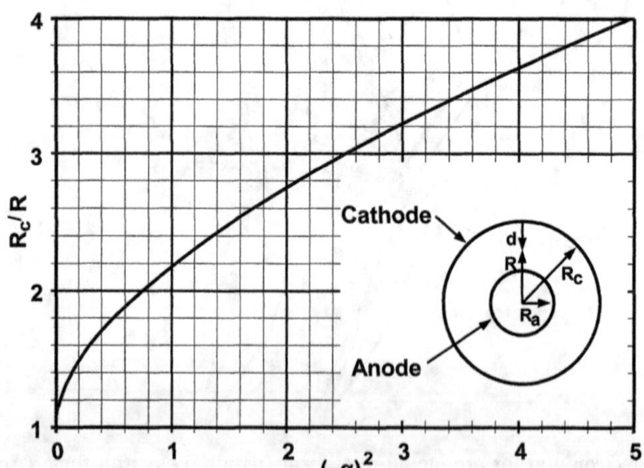

Figure 6-8 Langmuir and Blodgett's $(-\alpha)^2$.

Figure 6-9 shows perveance as a function of cone angle and normalized cathode-anode spacing. It will be shown that the extremely high values of perveance indicated in Figure 6-9 cannot be achieved in real electron guns when anode effects are considered.

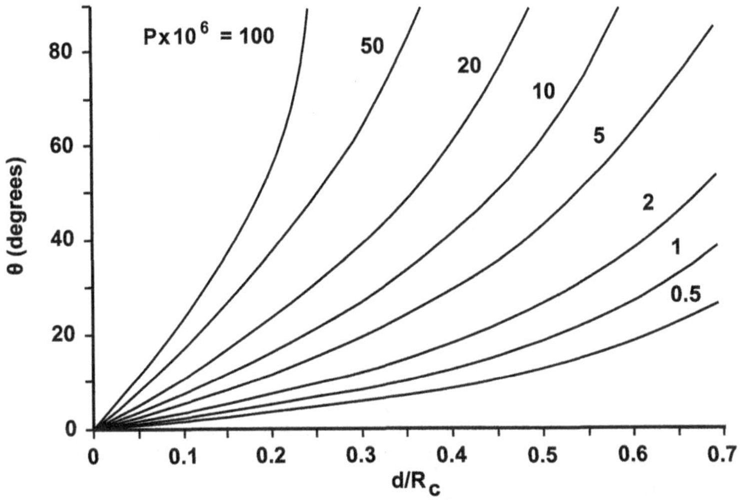

Figure 6-9 Perveance as a function of cone half angle and cathode-anode spacing.

The problem with the conical diode is that space charge forces will cause the beam to diverge just as they did for the planar diode. Equipotential contours, which should be spherical in shape, are deflected by the electron space charge potential as indicated in Figure 6-10. To retain radial electron flow when the remainder of the sphere is removed, it is necessary, as was the case for rectilinear flow in a parallel plane diode, to use electrodes that cause the equipotential profiles to have the shape which they had in the absence of the electrons (spherical in this case).

Conceptually, as for the planar diode, it can be seen that, by tipping the electrodes adjacent to the cathode forward, the equipotential contours can be made more nearly spherical. In fact, as indicated in Figure 6-11(a), close to the edge of the cathode, electron flow appears to originate from a planar portion of the cathode with electron flow perpendicular to that portion. Thus, an electrode placed at a 67.5° angle with respect to the edge of the beam will have the desired effect on the equipotential profiles near the edge of the cathode. For small cone half angles (small θ), this effect extends to the center of the cathode so no further consideration of the shape of the focus electrode is necessary.

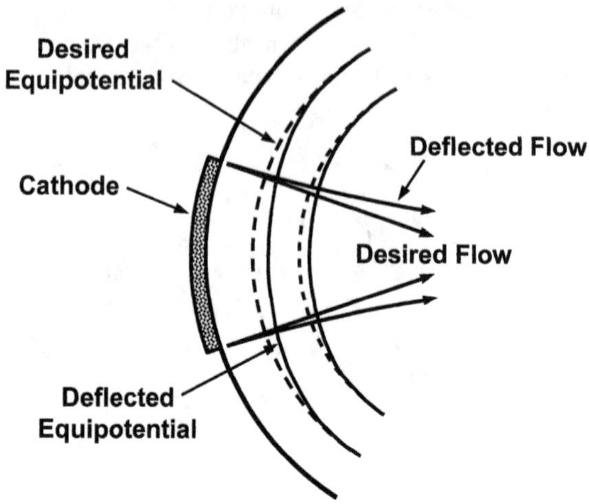

Figure 6-10 Deflection of equipotential contours in a conical diode by electron space charge potential.

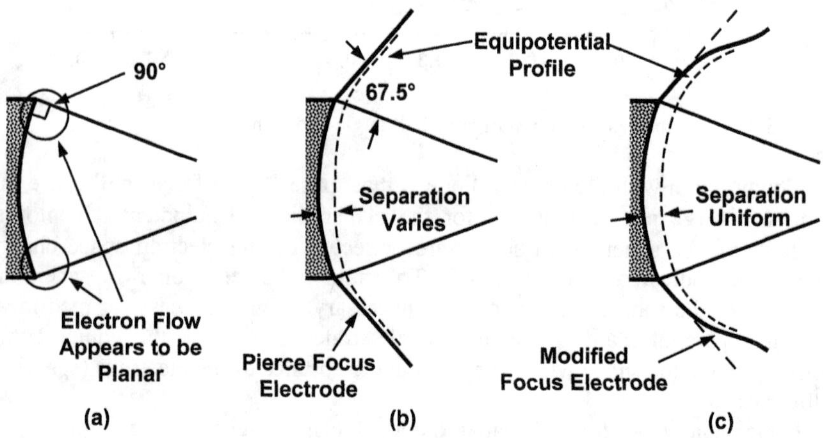

Figure 6-11 (a–c) Modification of Pierce electrodes to produce spherical equipotential profiles.

For large cone angles, as indicated in Figure 6-11(b), the potential profiles will still not be spherical. As a result, the electron trajectories, as the electrons leave the cathode region, will not be focused toward the center of curvature of the cathode.

A more important problem may be that nonuniform emission from the cathode will occur. For high perveance guns in which the cathodes are large in diameter, this problem can be severe. Assume, for example, that the separation between the cathode and one of the equipotentials near the cathode is twice as large at the cathode center as it is at the edge. Since the equipotential and the cathode are, in effect, a diode, the emission current density from the edge of the cathode will be four times that at the center. (Remember that current in a diode varies inversely with the square of cathode-anode spacing.) This will compound the focusing problem of the electron gun and will also shorten the cathode life because the entire cathode will have to be at a temperature high enough to provide the emission required at the outer edge.

To correct these problems, the shape of the focusing electrode must be modified so that the equipotential profiles near the cathode surface become more nearly spherical. Researchers have experimented with many possible shapes, one of which is shown in Figure 6-11(c) [4]. In effect, the modified focus electrodes lift the equipotential profiles away from the edges of the cathode. There is little effect on the shapes of the profiles in the center region of the cathode because the electron space charge fields dominate in controlling the potentials there.

6.1.3 Defocusing Effect of Anode Aperture

To remove electrons from the conical diode, it is necessary to cut a hole in the anode. The resulting distortion of the equipotential profiles near the anode aperture is illustrated in Figure 6-12 [5]. Notice that, at a low perveance as in Figure 6-12(a), the distortion of the profiles is limited to the region very close to the anode aperture. As perveance is increased, as in Figure 6-12(b, c), the size of the anode aperture increases and the separation between the anode and cathode decreases. At a perveance of 3×10^{-6} the distortion of the profiles extends to the cathode surface. As a result, the electron trajectories in the cathode-to-anode region are seriously perturbed and emission from the cathode is very nonuniform.

In [5], Brewer described the modification of the focus electrode as shown in Figure 6-12(d) to reduce distortion of the equipotentials and improve electron focusing and cathode emission uniformity. The modified focus electrode has the same effect as that shown in Figure 6-11(c). That is, the equipotentials are lifted away from the regions near the edge of the cathode. Although focusing and emission uniformity are improved, perveance decreases because of the reduced electric field near the cathode edges.

The distortion of the equipotential profiles near the anode aperture results in the formation of a divergent lens. This lens changes the focus of the electron flow to a point beyond the center of the sphere, as shown in Figure 6-13.

The lens formed by the distorted equipotential profiles is nowhere near a perfect lens because it suffers from spherical aberration. As is shown in Figure 6-14, this aberration results in a focal length that varies with the radial position in the lens. Notice that the trajectory of the electron that leaves the cathode at position 1 is focused to a point to the left of the trajectory of the electron that leaves the cathode at position 2.

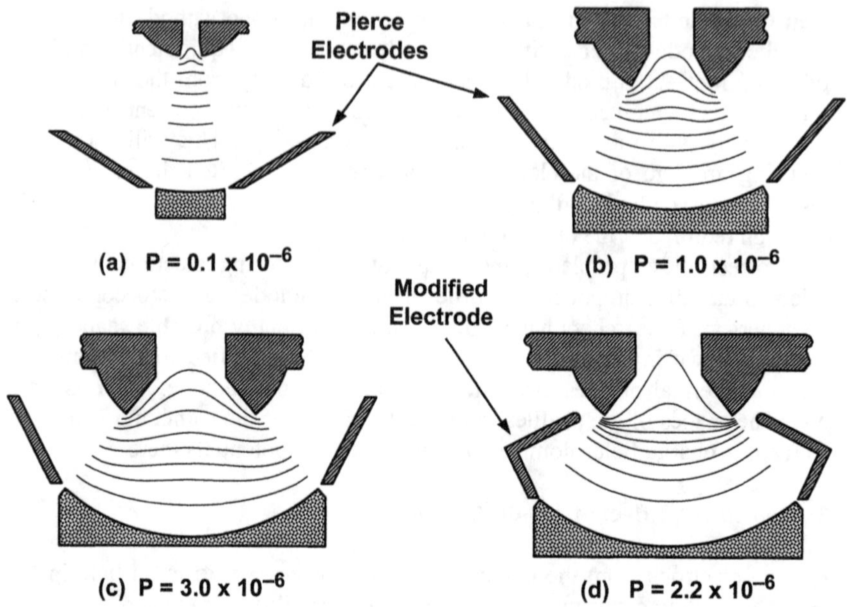

Figure 6-12 (a–d) Effect of anode aperture on equipotential profiles. (Adapted from: G. R. Brewer, *Jour. App. Phys.*, January 1957 and *Focusing of Charged Particles*, copyright 1967 by Academic Press.)

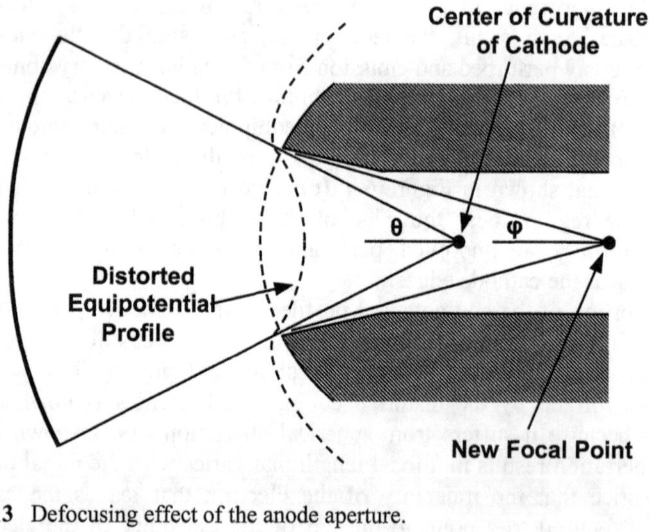

Figure 6-13 Defocusing effect of the anode aperture.

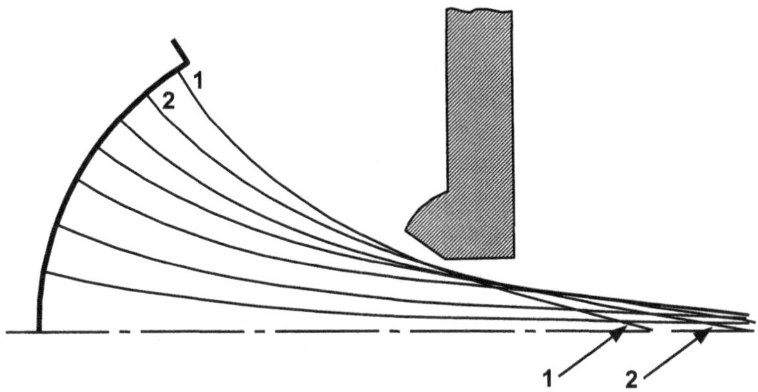

Figure 6-14 Effect on the focal length of spherical aberration. (Adapted from: G. R. Brewer, U.S. Patent 3,139,552, June 30, 1964.)

Frost, Purl, and Johnson [6] found that the measured beam from a high perveance gun was larger in diameter and less uniform in current density than expected from theory. By modifying the spherical radius of the cathode, as shown in Figure 6-15, they were able to correct for the lens aberration. The resulting improvements in beam characteristics are shown in Figure 6-16. The beam-current profiles at the top left of the figure were obtained before the correction was made. Note the severe nonuniformity in current as a function of radial position in the beam.

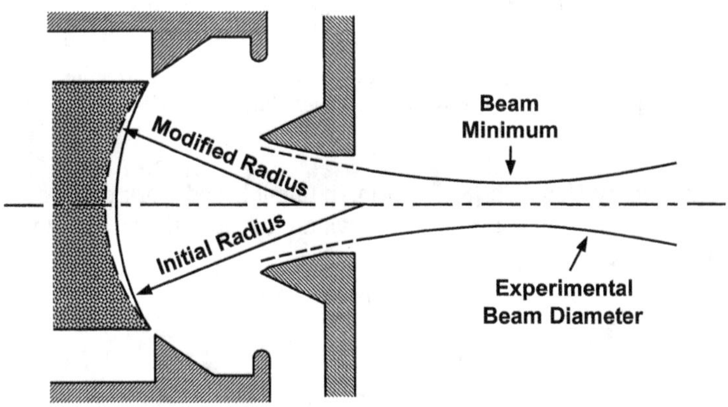

Figure 6-15 Modification to cathode radius used by Frost, Purl, and Johnson [6] to correct for spherical aberration.

After modifying the cathode radius, the profiles at the bottom left of Figure 6-16 were obtained. Not only was the uniformity of the current improved, but the radius of the beam at the position of the beam minimum was also reduced. After an additional small change in the focus electrode voltage, the profile to the right was obtained. As indicated by the ×s, the shape of this profile was very close to that predicted by theory (the theories of Pierce and of Cutler and Hines [7], who predicted thermal velocity effects).

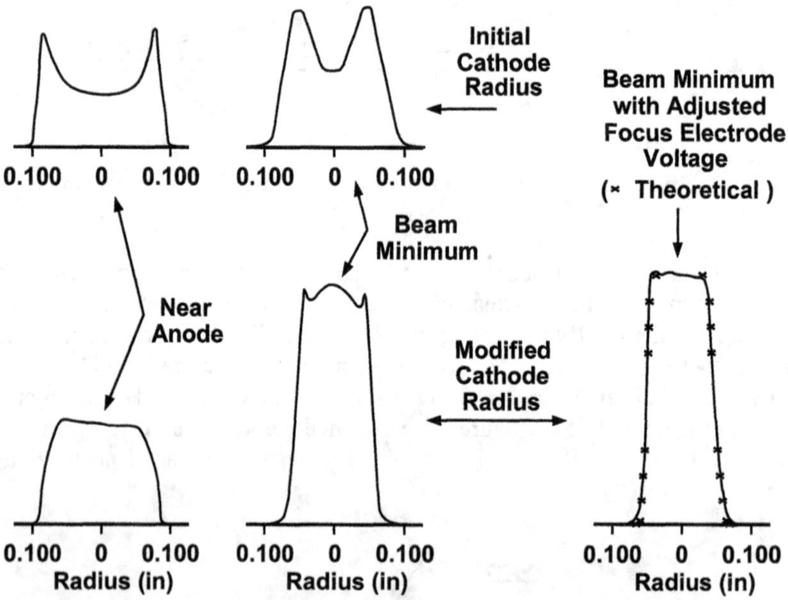

Figure 6-16 Beam-current profiles measured before and after correcting for spherical aberration. (Adapted from: R. D. Frost et al., *Proc. IRE*, August 1962. © 1962 IRE (now IEEE).)

As reported by Vaughan [8], True in an unpublished work defined the values of the gun angle, θ_T, and the radius of curvature of the cathode, R_{cT}, corrected for spherical aberration by

$$\frac{\sin\theta}{\sin\theta_T} = k \tag{6-2}$$

and

$$\frac{R_{cT}}{R_c} = k \tag{6-3}$$

where k is an empirical constant less than unity. For diode guns, $k = 0.905$ yields results that agree well with practice.

Before leaving the subject of spherical aberration, it should be pointed out that Brewer proposed the use of a nonspherical cathode for compensating for the effects of the aberration [2]. His parabolic-shaped cathode is shown in Figure 6-17.

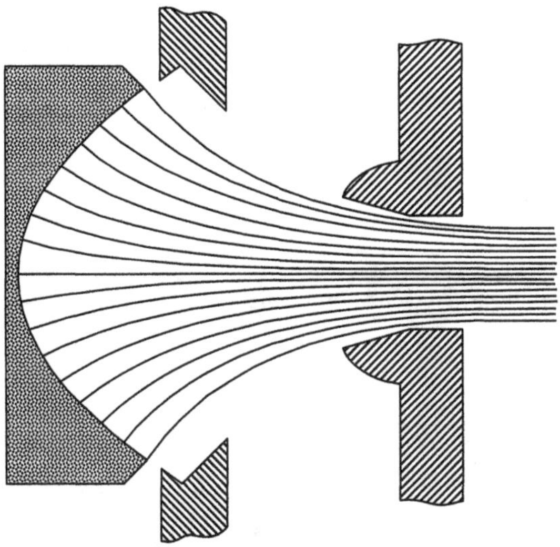

Figure 6-17 A nonspherical cathode that corrects for spherical aberration. (From: G. R. Brewer, U.S. Patent 3,139,552, June 30, 1964.)

6.1.4 Formation of Minimum Beam Diameter

After the beam leaves the anode aperture and enters region 3 (see Figure 6-1), the axial electric field is nearly zero and the only forces on the electrons are those resulting from the space charge. As a result, the shape of the beam is defined by the universal beam spread curve, shown in Figure 6-18. From the analysis for this curve (Chapter 3), the minimum beam radius, b_m, and the axial position of the minimum radius, z_m, can be found as functions of the entrance conditions into region 3. It is essential to have accurate knowledge of these quantities to properly inject the beam into the magnetic focusing structure.

It is important to realize that the values of b_m and z_m are very sensitive to R_c/R_a (the ratio of the cathode and anode radii of curvature). This is illustrated in Figure 6-19, where beam profiles are shown for values R_c/R_a of that differ by 25%. As R_c/R_a increases from 1.6 to 2.0, significant changes occur in both the minimum beam diameter and its axial position. As R_c/R_a increases to 2.5, the minimum beam diameter is again reduced by a significant amount. This figure clearly illustrates the importance of precisely locating the cathode with respect to the anode.

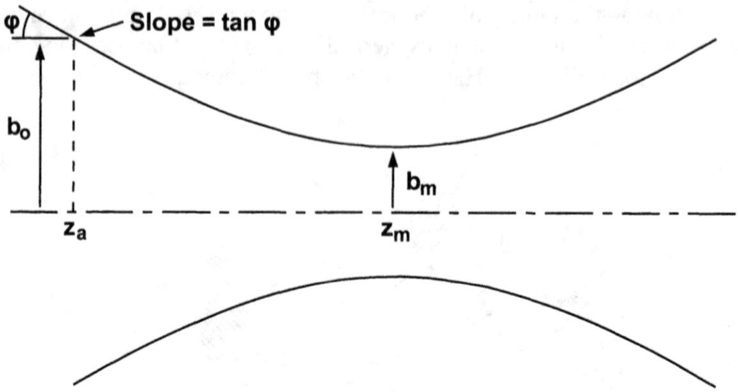

Figure 6-18 Beam shape in region 3 (drift region).

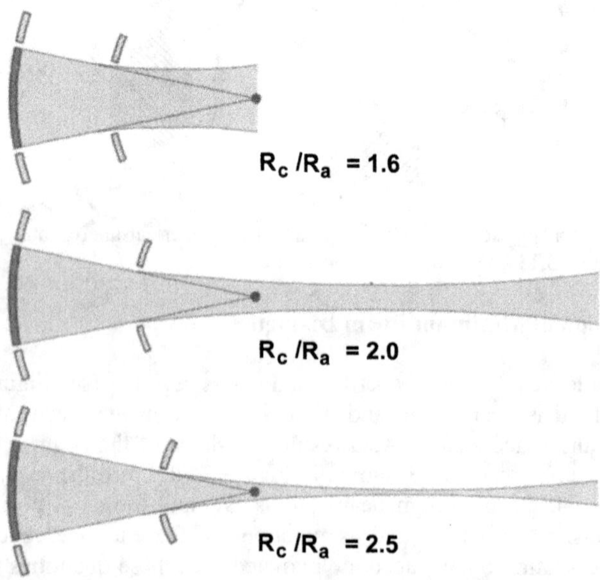

Figure 6-19 Emerging electron stream for various ratios of cathode radius to anode radius. (Adapted from: *Theory and Design of Electron Beams* by J. R. Pierce, copyright 1954 by D. Van Nostrand, Inc.)

Vaughan described a simple technique for synthesizing the design of a Pierce gun [8]. A design example is given in [9]. The technique provides a way to quickly check the design of an electron gun to see if dimensions and parameters are reasonable.

6.1.5 Thermal Velocity Effects

In this section we will consider the effects of the thermal velocities of electrons on the beam produced by an electron gun. It is assumed that the magnetic field is zero. The nature of electron emission from a thermionic cathode is discussed early in Chapter 5 and also in Chapter 23. The Maxwellian velocity distribution of the emitted electrons is of importance here.

Consider a small spot on the cathode of an electron gun. Electrons are emitted in all directions, and the beamlet of electrons from the spot grows in diameter and the current density decreases as the beamlet is accelerated toward the anode. The current has a Maxwellian distribution with a normal deviation, σ, as shown in Figure 6-20.

Cutler and Hines [7] analyzed the effects of thermal velocities on electron flow in a Pierce gun. They define the normal deviation by

$$\sigma = \frac{\Delta r}{(dr/dt)_c}\left(\frac{kT}{m}\right)^{1/2} \tag{6-4}$$

where Δr is the trajectory deviation of an electron with an initial transverse velocity at the cathode $(dr/dt)_c$. For a 10-kV, 1-A, 0.1-cm radius beam from a cathode at 1,000°C with a radius of 1 cm, the value of σ is ~0.01 cm.

Cutler and Hines summed the currents in beamlets from the entire cathode surface to find an expression for the total current distribution in the beam leaving the gun. The result is shown in Figure 6-21. Here, r is the radial position in the beam, b is the nominal beam radius (the radius in the absence of thermal effects), and J_o is the current density in a beam having the nominal radius.

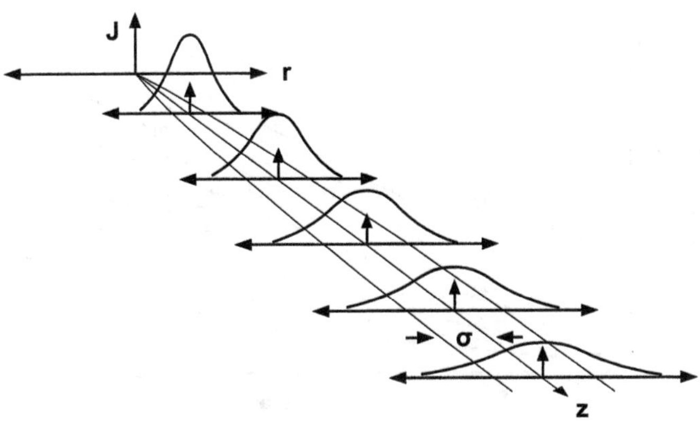

Figure 6-20 Effect of thermal velocities on a beamlet emitted by a spot on a cathode.

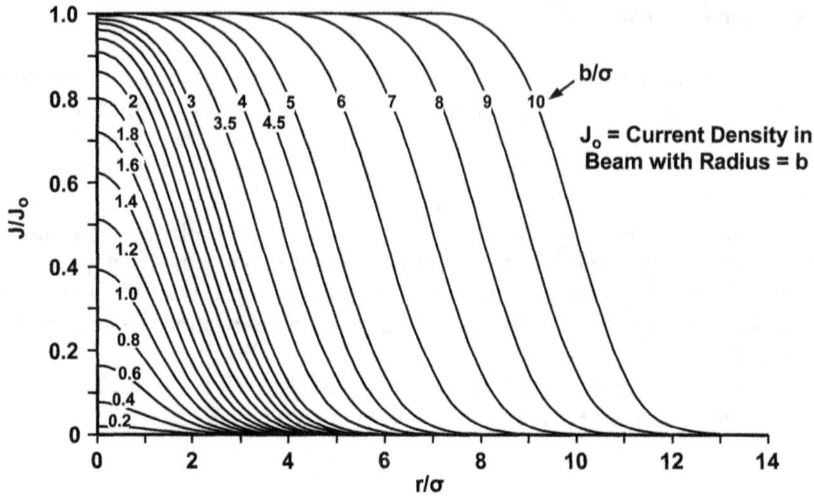

Figure 6-21 Current density versus radial position; b is the radius without thermal effects. (From: C. C. Cutler and M. E. Hines, *Proc. IRE*, March 1955. © 1955 IRE (now IEEE).)

For $\sigma = 0.01$ cm and a beam radius of 0.1 cm ($b/\sigma = 10$), the curve on the right in Figure 6-21 indicates the beam shape. Figure 6-22 shows the percentage of the total beam current to be found within any given radius of the beam. About 92% of the current would be contained within the 0.1-cm beam radius.

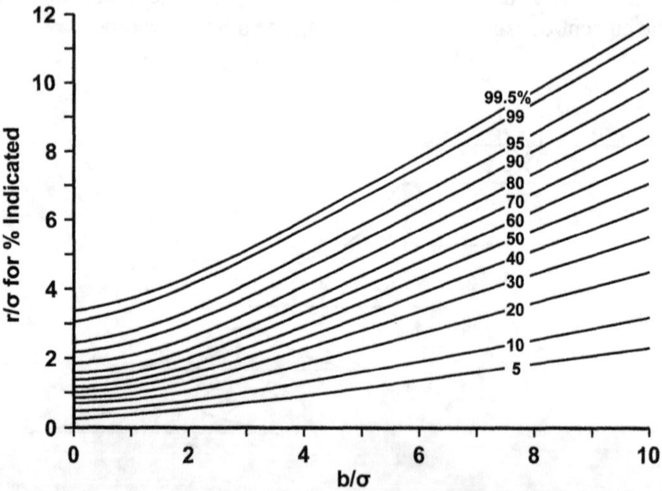

Figure 6-22 Percentage of total beam current to be found within any given radius. (From: Cutler and Hines, *Proc. IRE,* March 1955. © 1955 IRE (now IEEE).)

When very small beams are required, thermal effects are severe. An example from Danielson et al. [10] is given in Figure 6-23. Ignoring thermal effects, the minimum beam radius was calculated to be less than 2×10^{-3} in. The experimental radius enclosing 95% of the beam current (r_{95}) was over four times larger (about 8×10^{-3} inch). The minimum r_{95} radius calculated with the Cutler-Hines theory agreed well with experiment, but the variation of the radius with distance was quite different.

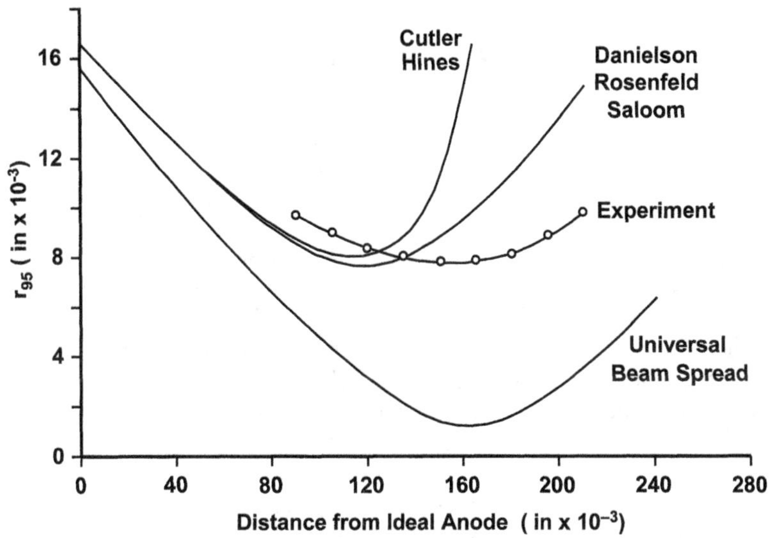

Figure 6-23 Computed and experimental beam profiles for a very small beam dispersed by thermal velocities. (Adapted from: Danielson et al., *Bell Sys. Tech. Jour.*, March 1956.)

Danielson et al. corrected the Cutler-Hines theory by including the correction to the Davisson-Calbick equation and by using the space charge field in the thermal beam rather than that in the nominal beam. The minimum radius that they predicted also agrees well with experiment, and the variation of radius with distance is closer to experiment than that predicted by Cutler and Hines.

Before leaving the discussion of thermal velocities, it is important to consider the effect of beam convergence in the electron gun on the transverse velocity distribution of the electrons. Figure 6-24 shows the trajectories of several nonthermal electrons, which are convergent on point F. The trajectory of one thermal electron is shown crossing the trajectories of the nonthermal electrons. Electric fields are assumed to be negligible. It is clear that the angle that the trajectory of the thermal electron makes with the trajectories of the nonthermal electrons increases as the beam converges. Thus, the component of thermal-electron velocity perpendicular to the nonthermal-electron trajectories increases as the beam converges.

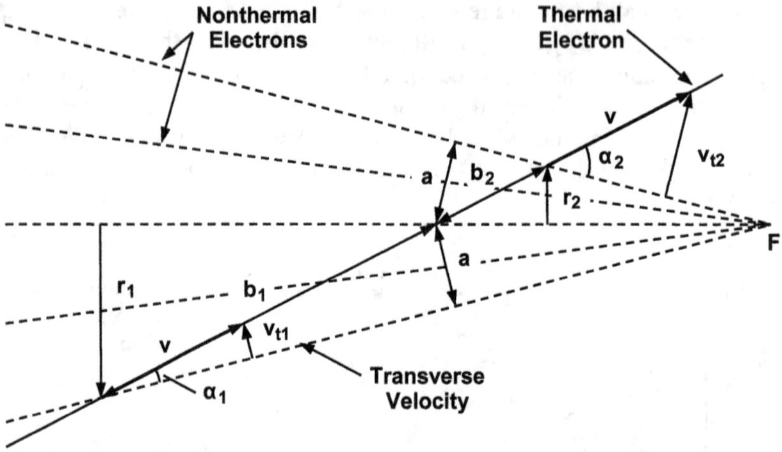

Figure 6-24 Illustration of increase in transverse velocity of thermal electron in convergent flow.

Using Figure 6-24, it is easy to calculate the increase in the transverse component of the velocity of the thermal electron. If v is the velocity of the thermal electron, then at the positions where the radius is r_1 and r_2, the transverse velocity components are

$$v_{t1} = v \sin \alpha_1 \quad \text{and} \quad v_{t2} = v \sin \alpha_2 \tag{6-5}$$

so

$$\frac{v_{t2}}{v_{t1}} = \frac{\sin \alpha_2}{\sin \alpha_1} \tag{6-6}$$

but

$$\frac{\sin \alpha_2}{\sin \alpha_1} = \frac{a / b_2}{a / b_1} = \frac{b_1}{b_2} \tag{6-7}$$

and

$$\frac{r_1}{b_1} = \frac{r_2}{b_2} \quad \text{so} \quad \frac{r_1}{r_2} = \frac{b_1}{b_2} \tag{6-8}$$

As a result,

$$\frac{v_{t2}}{v_{t1}} = \frac{r_1}{r_2} \tag{6-9}$$

That is, the transverse velocity increase is inversely proportional to the radius compression ratio. Thus, to limit the growth of transverse velocities in an electron gun, the amount by which the beam is compressed should be held to a minimum. A cathode with a high emission capability helps in this regard because, for a given beam size and current from a gun, the cathode can be smaller, so the radius compression ratio is smaller than for a cathode with a low emission capability.

6.1.6 Effects of Patchy Emission and Cathode Roughness

The thermal velocity effects on beam shape discussed in the previous section are frequently overshadowed by the relatively large transverse electron velocities produced by patchy emission and cathode roughness. The surface of a dispenser cathode is made up of grains of tungsten that have various crystalline orientations and that may have various work functions. Because the thickness of the space charge cloud at the cathode surface is comparable to the dimensions of the grains and the pores between grains, each grain emits independently of the others and the emission is said to be patchy. For those grains that are temperature limited, emission varies with grain work function and most of the emission comes from the grains with low work functions.

There are at least two ways by which the patchy nature of emission produces transverse velocity components to electrons leaving the cathode. These are indicated in Figure 6-25(a, b). In Figure 6-25(a) we note that space charge forces within the beamlet emitted from a patch cause spreading of the beamlet. In Figure 6-25(b) we show that equipotential surfaces near the cathode surface are rippled because of the varying space charge density near the patches. This produces a spreading of the beamlets.

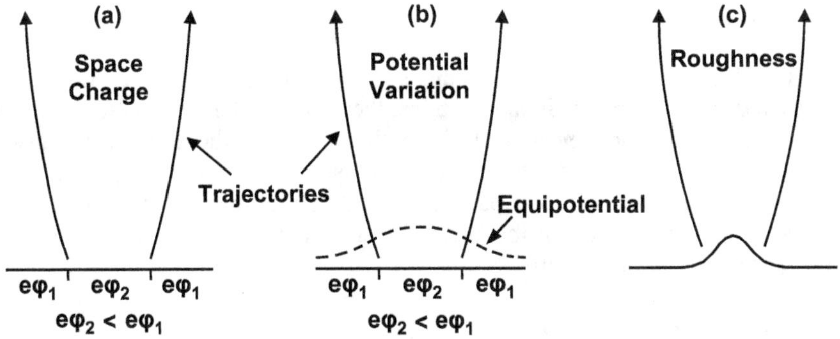

Figure 6-25 (a–c) Spreading effects of patchy emission and cathode roughness on electron trajectories.

As explained in Chapter 5, coated dispenser cathodes have a narrower distribution of work functions than uncoated cathodes, so patchy emission is less pronounced for coated than for uncoated cathodes. If the work function distribution of a cathode increases with time, then the effects of patchy emission become worse. This may cause the beam diameter to increase so that current intercepted by the circuit increases.

If the surface of the cathode is rough, as indicated in Figure 6-25(c), then local transverse electric fields are present, which produce transverse electron velocities. A problem sometimes encountered with coated dispenser cathodes is that the coating may develop microscopic cracks and peel away from the tungsten matrix. Then a cathode that was originally very smooth can become rougher as it is operated. If this happens, transverse velocities increase, beam size increases, and interception current may increase.

6.2 BEAM CONTROL TECHNIQUES

In all applications of microwave tubes, beam current must be controlled. Usually, the beam current must be switched on and off. Sometimes, continuous control is required. The "brute-force" way to control beam current is with the cathode-to-anode voltage as indicated in Figure 6-26(a). A second anode (called a *modulating anode*) may be placed between the cathode and final anode for current control as shown in Figure 6-26(b). In some guns current can be controlled with the focus electrode as shown in Figure 6-26(c). Finally, in the majority of applications, one or more grids is placed near the cathode surface, as indicated in Figure 6-26(d), for current control. In this section these control techniques are discussed. The primary emphasis is on grid control because of its wide use in TWTs.

6.2.1 Cathode Pulsing

Conceptually, cathode pulsing is the simplest technique for switching a microwave tube on and off. In virtually all applications, the anode of the gun and the body of the tube are at ground potential. To turn the beam on, the cathode is pulsed negative. To turn the beam off, the cathode is pulsed back to ground potential. The primary problem with cathode pulsing is that the power supply (the power modulator) must switch *all* of the beam power so the power modulator must be large, heavy, and, usually, inefficient.

6.2.2 Control Focus Electrodes

If a large enough negative voltage is applied to the control electrode, the current can be reduced to zero. The voltage needed to cut off beam current can be reduced by placing an additional electrode in the center of the cathode (Figure 6-27).

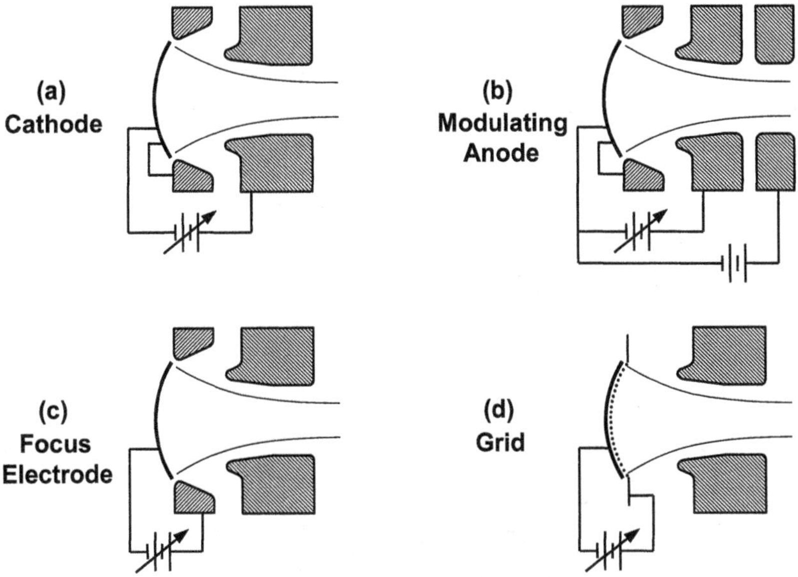

Figure 6-26 (a–d) Techniques for current control in an electron gun.

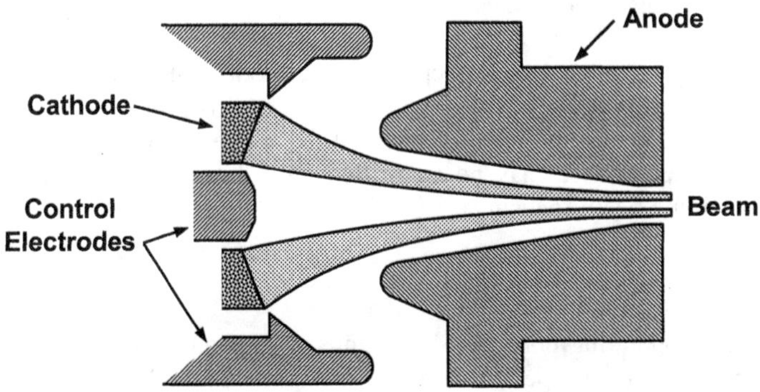

Figure 6-27 Control focus electrode gun. (From: Armand Staprans et al., *Proc. IEEE*, March 1973. © 1973 IEEE.)

The control focus electrodes can be used to turn on the beam by setting the voltage at or very close to the cathode potential. The beam is turned off by setting the voltage significantly below cathode potential. This system is not useful for the intermediate control of current because of the severe defocusing that occurs in the electron gun.

6.2.3 Modulating Anode

In the modulating anode gun, the anode of the electron gun (the modulating anode) is separated from the body of the tube, a part of which is a second (final) anode. The configuration of a modulating anode gun is shown in Figure 6-28.

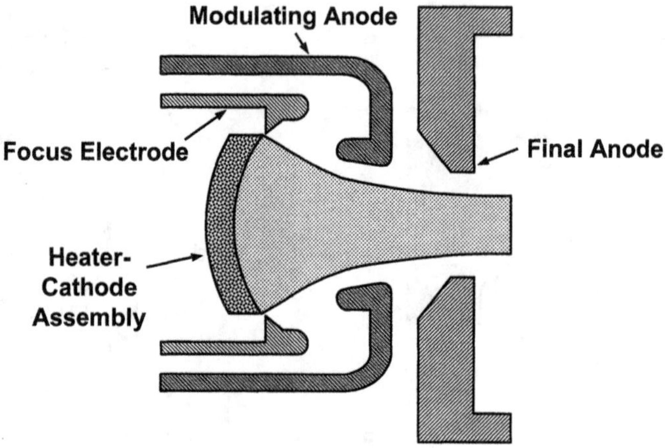

Figure 6-28 Modulating anode gun.

Although the voltage range over which the modulating anode must be varied to vary beam current is large, the power required to drive the modulating anode is small because the current intercepted by this anode is small. At intermediate voltage levels, perveance may be controlled with the modulating anode without severely affecting beam focusing.

6.2.4 Grids

Under normal operating conditions, the guns with modulating anodes and those with control electrodes operate as Pierce-designed guns. When grids are used, however, electron trajectories near the cathode are affected, and so beam focusing is degraded and beam dynamics are affected. Still, grid control is very widely used because of the small modulator size that it makes possible.

Several types of grid structures are illustrated in Figure 6-29. The simplest structure has a single grid element. A computer simulation of electron flow in a Pierce gun containing a single grid operating at a voltage positive with respect to the cathode is shown in Figure 6-30. Thermal velocities are ignored here, and, because the grid is positive, electron flow is very laminar. The problem with a single grid that is positive with respect to the cathode is that the cathode current intercepted by the grid can be very high. In some cases as much as 15% or more of the cathode current is intercepted. The results of the interception are excessive heating of the grid at high duty cycles and the need for a relatively powerful grid modulator.

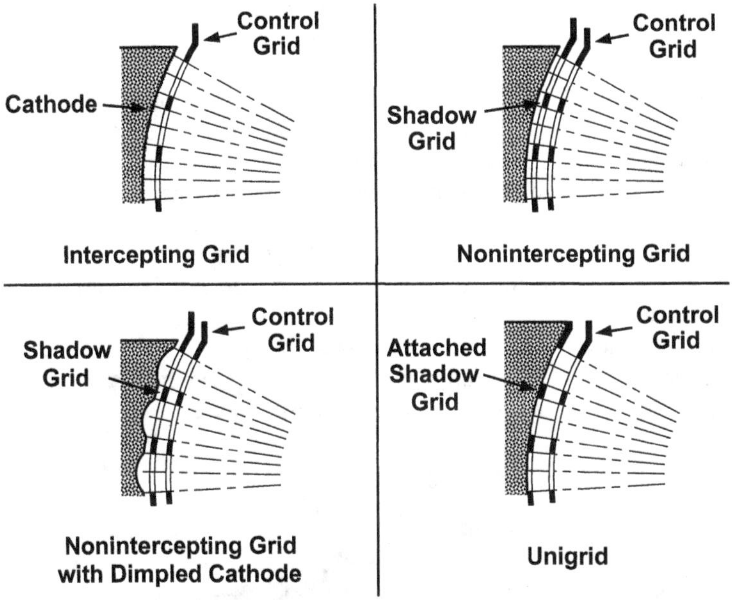

Figure 6-29 Various grid structures used in Pierce guns for current control. (From: *Microwave Tube Manual* by Varian, Air Force Publication No. T.O. 00-25-251, October 1979.)

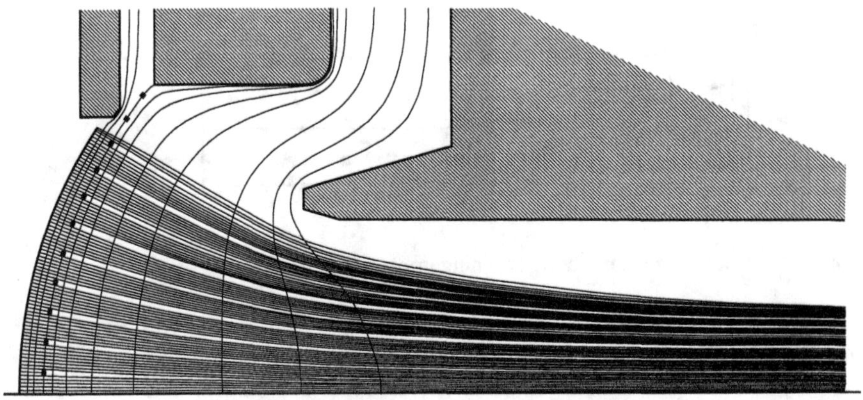

Figure 6-30 Computer simulation of electron flow in a Pierce gun with a single grid. (Adapted from: R. True, *Technical Digest*, 1985 IEDM. © 1985 IEEE.)

Many grid structures contain a shadow grid between the control grid and the cathode surface. The shadow grid operates at or near cathode potential so that heating by intercepted current is small. The elements of the shadow grid are aligned with the elements of the control grid so that the control grid is shielded and intercepts very little current. Several shadow-grid structures are shown in Figure 6-29.

Grid control of current in electron tubes has been used since the invention of the vacuum tube. In 1960, Wolkstein discussed methods of grid placement and associated design considerations for convergent-flow Pierce guns [11]. The basic concept is to form the grid so that its shape is identical to that of an equipotential contour near the cathode surface (usually spherical). Then, if the grid is at the potential of that contour and intercepts negligible current, electron flow will not be disturbed. To interrupt the current flow, the grid voltage is switched negative with respect to the cathode.

A simple relationship for grid voltage as a function of anode voltage and gun dimensions can be found by equating the current flow from the cathode to grid, I_c, to the current flow from grid to the anode, I (refer to Figure 6-31).

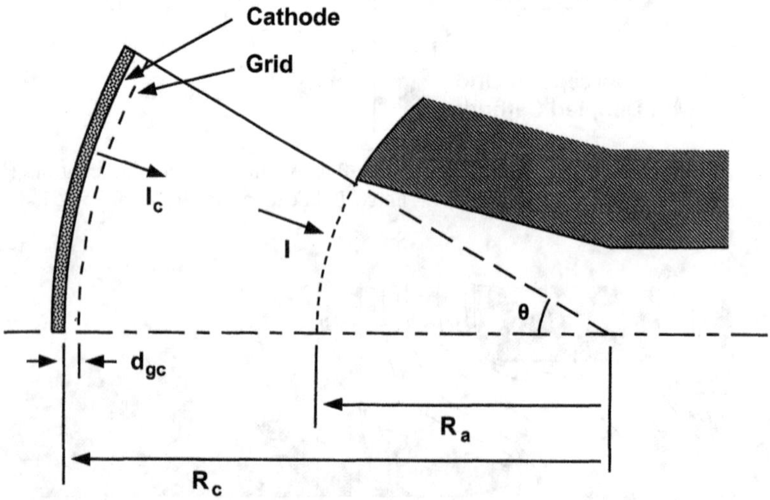

Figure 6-31 Gun parameters for determining grid-cathode spacing.

The current flow from the cathode to grid is given by the Child-Langmuir equation for parallel-plane diodes. The current flow from the grid to the anode is given by the Langmuir-Blodgett relation for spherical diodes. The result is in Figure 6-32 and is

$$\frac{V_a}{V_g} = \left(\frac{(-\alpha_a)}{d_{gc} / R_c} \right)^{4/3} \tag{6-10}$$

where V_g is the grid-to-cathode voltage, V_a is the anode-to-cathode voltage, and $(-\alpha_a)$ is Langmuir and Blodgett's $(-\alpha)$ evaluated at the anode. As an example of its use, consider a 10-kV, 1-A gun design with $R_c/R_a = 2.25$. For a grid voltage of 100V,

$$\frac{d_{gc}}{R_c} \approx 0.032 \tag{6-11}$$

If we assume that $R_c = 1$ cm, then

$$d_{gc} \approx 0.032 \text{ cm} \approx 0.013 \text{ inch} \tag{6-12}$$

So we see that the grid must be very close to the cathode if the grid voltage is to be low.

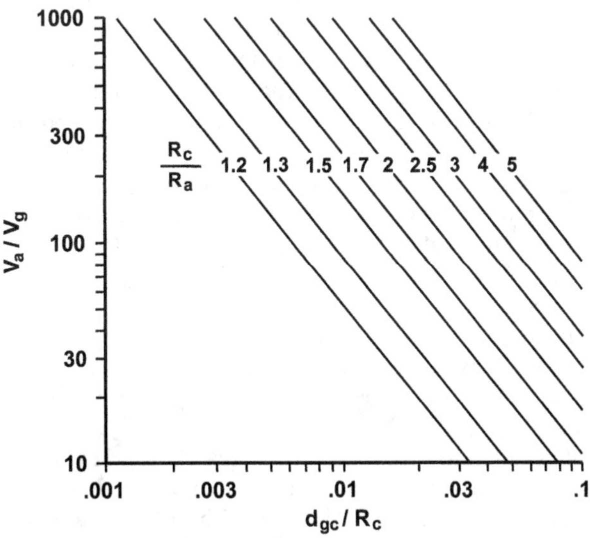

Figure 6-32 Relation of grid and anode voltages to grid, cathode, and anode positions. (Adapted from: H. J. Wolkstein, *RCA Review*, September 1960.)

Next, we will examine the configuration of the grid elements. Examples of control grids for Pierce guns are shown in Figure 6-33. The grid configuration on the left is often used in guns with intercepting grids and that on the right is used in shadow gridded guns.

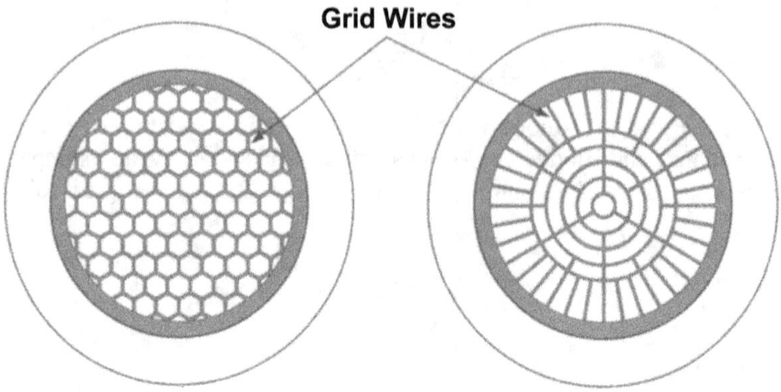

Figure 6-33 Examples of control grids for Pierce guns.

To stop (cut off) electron flow from the cathode, as shown in Figure 6-34, a negative voltage (with respect to the cathode) is applied to the grid. In Figure 6-34, the dashed curves show the potential variation between the cathode and anode for the 10-kV, 1-A electron gun with the grid 100V above the cathode potential. The solid curves show the potential variation when the grid is sufficiently negative with respect to the cathode to cut off electron flow. Note in the expanded drawing of the grid region that two solid curves are shown. One is the potential variation through a grid wire and the other is the potential variation midway between grid wires. The net effect of these two potential variations must be that the potential near the cathode is zero or negative if current flow is to be prevented.

To avoid electrical breakdown between the grid and the cathode, the grid voltage used for cutoff must not be excessive. The cutoff grid voltage is controlled largely by the spacing between grid wires. Clearly, for large spacings, the cutoff grid voltage must be large because of the strong tendency for the anode field to penetrate the regions between grid wires. Thus, for relatively small cutoff grid voltages (and low probability of grid-cathode breakdown), the grid wire spacing must be small.

In addition to small grid wire spacing, the area of the grid wires (and, thus, the width of the grid wires) must be small compared with the cathode area if the current intercepted by the grid is to be small. The ratio of the grid area to the cathode area is the screening fraction, which is the fraction of the cathode current intercepted by the grid. Thus, it is clear that, for small interception, the screening fraction must be small.

An example of grid dimensions was reported by True [12] for the control grid in a shadow gridded gun similar to that on the right in Figure 6-33. For a 9.6-kV gun with control grid voltages (with respect to the cathode) of 102V and –86V, d_{gc} = 0.015 inch, effective grid wire spacing = 0.023 inch and grid wire width = 0.002 inch.

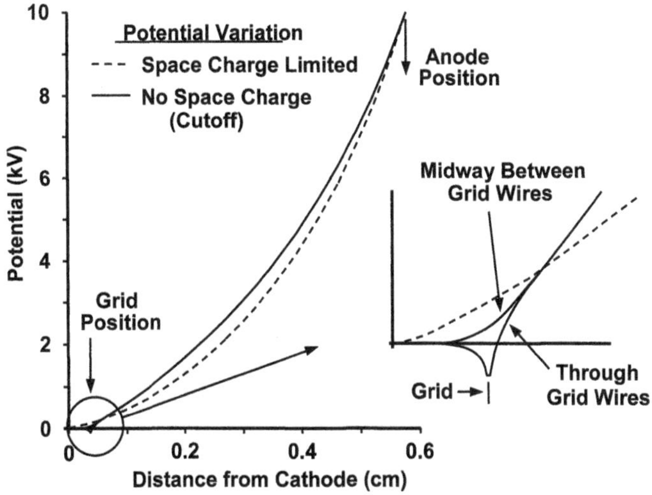

Figure 6-34 Potential variations in a gridded 10-kV Pierce gun.

At this point, we can start to appreciate the care that must be taken in designing and fabricating a gridded Pierce gun. Not only are dimensions small, but also proper grid-to-cathode spacing must be maintained as the cathode is heated to a temperature of 1,000°C or higher. The problem of maintaining spacings and alignments is even more critical in shadow gridded guns, which are discussed next.

The main problem with a single grid is that the cathode current intercepted by it can be very high. To alleviate this problem, many gridded Pierce guns use a shadow grid between the control grid and the cathode. The shadow grid was invented by Drees [13], who described its configuration and operation as follows: "In accordance with the present invention, the novel grid construction comprises a first control grid and a second grid spaced from and aligned with the first grid. The second grid is disposed between the control grid and the cathode. In this arrangement, with the second grid preferably being held at cathode potential, the second grid acts as a (shadow) grid in that it shadows electrons from the control grid so that they are not intercepted by the control grid."

The effects on electron trajectories of the shadow grid, its location, and its voltage are illustrated in Figure 6-35 [14]. Each of these plots shows computer-generated electron trajectories from a cathode for the region from midway between grid wires to the grid wires. In an ideal shadow gridded gun, both the shadow grid and the control grid would be at positions and voltages that coincide with the equipotential surfaces that exist in the absence of the grids. Thus, except for interception and the finite sizes of the grid wires, electron flow would not be disturbed. The plot in Figure 6-35(d) shows trajectories with grid voltages adjusted to approximate the ideal shadow gridded gun voltages.

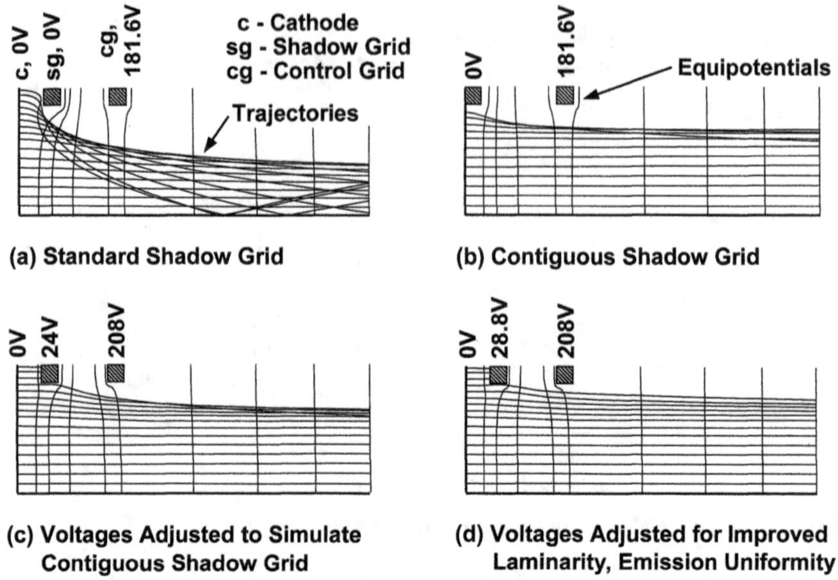

Figure 6-35 (a–d) Electron trajectories for shadow-gridded structures (simplified planar model). (Adapted from: R. True, *Technical Digest*, 1979 IEDM. © 1979 IEEE.)

For the standard shadow gridded gun (Figure 6-35(a)), the shadow grid is at cathode potential but is placed a short distance (usually a few thousandths of an inch) from the cathode surface. Thus, a lens action occurs near each wire of the shadow grid. This is evident from the bending of the equipotential surfaces near the shadow grid. As a result of the lens action, electron trajectories are severely distorted.

When the shadow grid is attached to or placed on the cathode surface, as in Figure 6-35(b), the distortion of the trajectories is significantly reduced. Unfortunately, of course, the shadow grid must operate at the cathode temperature, which can cause difficulties with grid emission and grid alignment.

An advantage of placing the shadow grid at cathode potential, as in Figure 6-35(a, b), is that only one grid power supply is required. If a separate power supply for the shadow grid can be used, then, as shown in Figure 6-35(c, d), the beam laminarity of the contiguous shadow grid can be equaled as in Figure 6-35(c) or improved upon as in Figure 6-35(d).

Shown in Figure 6-36 are plots of electron trajectories in a shadow-gridded gun with the standard shadow grid [15]. The beam radius containing 95% of the current, r_{95}, is indicated by dots. If the trajectories outside the r_{95} radius are followed back to the cathode, it is seen that they are the trajectories of electrons originating close to the grid wires.

If we examine Figure 6-36, we note that the grid-to-cathode spacing is somewhat smaller near the edge of the cathode than near the center. This was also evident in Figure 6-30. The reason for the reduced spacing at the cathode edge is that the equipotentials were spaced that way. As a result, the electric field and the emission density at the edge of the cathode are somewhat larger than at the center.

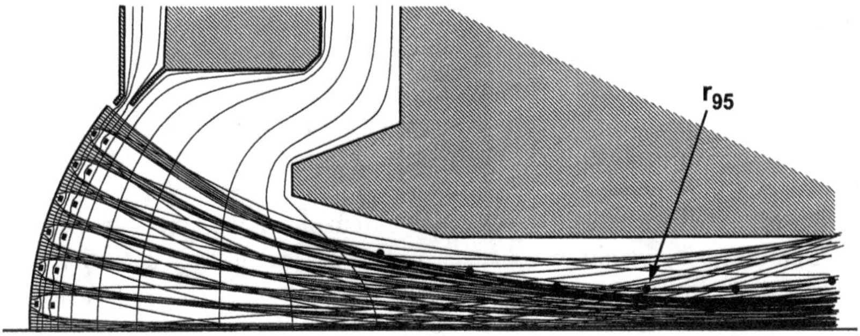

Figure 6-36 Electron trajectories in a shadow-gridded gun. (Adapted from: R. True, *IEEE Trans. Electron Devices*, March 1984. © 1984 IEEE.)

To cut off current flow, the control grid is pulsed negative with respect to the cathode. The resulting voltage distribution in the region of the grids is shown in Figure 6-37. Here, the grid voltage was –160V and the inner most potential contour from grid wire to grid wire was at –10V, so the grid voltage was slightly below that required for cutoff.

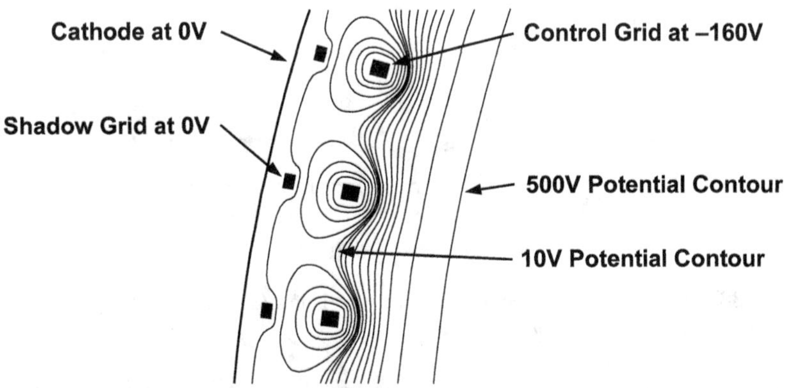

Figure 6-37 Voltage distribution near the grids at a control grid voltage slightly below cutoff. (Adapted from: R. True, *Technical Digest*, 1989 IEDM. © 1989 IEEE.)

We have noted that the electric field is somewhat higher at the edge of the cathode than at the center. Also, when grids like that shown on the right in Figure 6-33 are used, the spacing between grid wires increases with increasing radius. Thus, for negative control grid voltages, the anode field has a greater tendency to penetrate the regions between grid wires near the edge of the cathode than near the center. This means that, as control grid voltage is made negative with respect to the cathode, the last part of the cathode to be cut off is that near the outer edge. To compensate for this effect, True and Laycock tapered the widths of the grid wires so that they increased in width near the cathode edge [16]. As a result, cutoff voltage was reduced by a factor of almost 2.

As the control grid voltage is varied, cathode current varies as shown in Figure 6-38. In addition, the beam current intercepted (the body current), mostly by the RF structure, varies. This is because, as control grid voltage and cathode current are changed while cathode voltage remains fixed, the gun perveance changes and focusing conditions change. As a result, the entrance conditions of the beam from the gun into the magnetic focusing field change. This causes the beam shape and body current to vary.

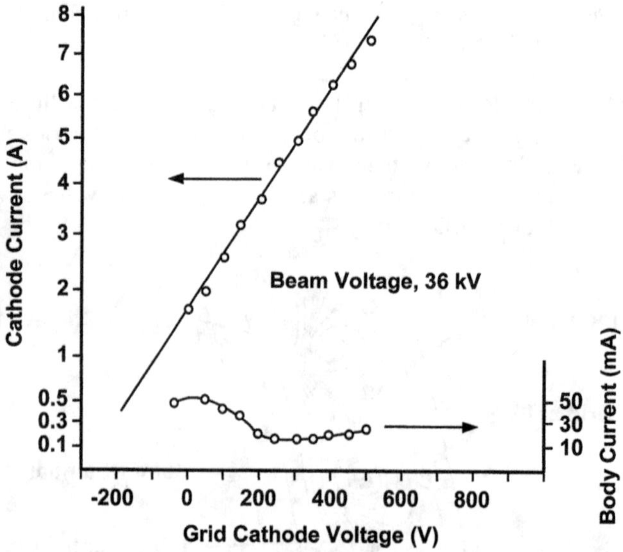

Figure 6-38 Shadow-gridded electron gun characteristics. (From: Armand Staprans et al., *Proc. IEEE*, March 1973. © 1973 IEEE.)

A particularly disturbing aspect of the body current characteristic shown in Figure 6-38 is that body current tends to become maximum when grid voltage is zero. This means that if grid bias is lost (perhaps because of a power supply failure), the RF circuit may be damaged because of excess body current. To prevent this damage, a beam shaver like that shown in Figure 6-39 is often placed between the gun and the RF circuit.

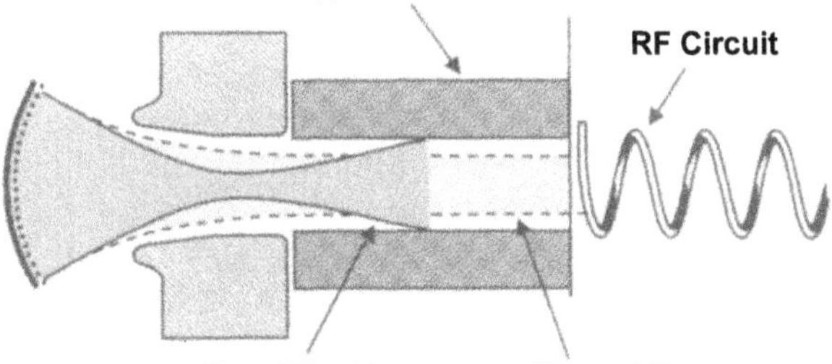

Figure 6-39 Placement of beam shaver to protect the RF circuit.

When the control grid is pulsed negative to cut off the beam current, complete cutoff may not occur if the grid is emitting electrons. The resulting small beam current can result in unwanted signal generation and interpulse noise.

Grids are often fabricated from molybdenum. Because of its high work function, grid emission from the molybdenum is insignificant even at high grid temperatures. However, when barium and barium oxide vapors from the cathode collect on the molybdenum, the work function can decrease to the point where grid emission occurs.

There are two obvious options for preventing grid emission:

1. Keep the grid cool enough so that emission is negligible regardless of the work function.
2. Keep the work function of the grid high enough so that emission will be negligible regardless of temperature.

The problem of controlling grid emission is complicated by the fact that there is a very large temperature variation from the center of a grid to the outer edge. An example calculated for a 3-mil-thick grid with a spiderweb configuration like that in Figure 6-33 is shown in Figure 6-40. The temperature drop from the center of the grid to the outer edge is nearly 400°C. Clearly, there can be no problem with grid emission near the outer edge of the grid, but near the center, where the temperature approaches 800°C, emission can occur if the work function is low enough.

A technique that is sometimes used in an attempt to keep the work function high is to coat the molybdenum with zirconium. Haas and Shih [17] reported that zirconium acts as a getter by reducing barium oxide molecules and absorbing the oxygen. The freed barium then immediately vaporizes from the zirconium surface so that the work function is not reduced.

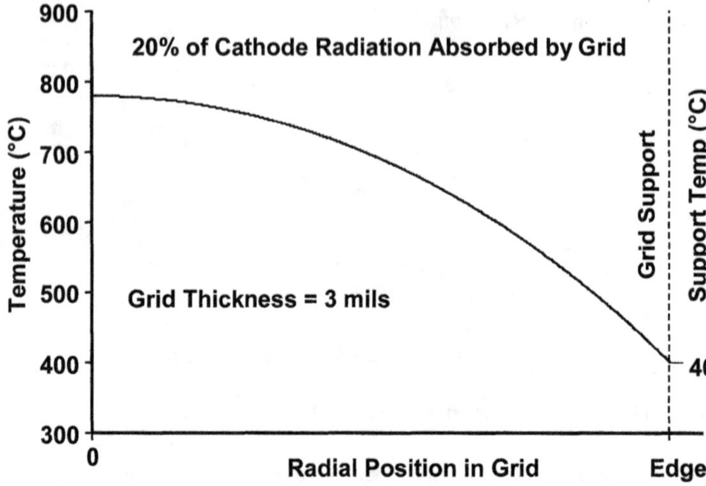

Figure 6-40 Grid temperature as a function of radial position.

Unfortunately, as the oxygen content of the zirconium increases, its effectiveness decreases and eventually excessive grid emission can occur. Green and Goeser [18] reported that a thin layer of tungsten deposited on the molybdenum prior to zirconium deposition serves to extend the life of the zirconium film. In fact, the zirconium/tungsten/molybdenum combination is found to outperform pure zirconium grids. The role of the tungsten layer is not known.

The best way to prevent excessive control grid emission is to ensure that the temperature of the center of the grid is less than about 600°C. Then, even if the grid work function is as low as 2.2 eV (highly unlikely), emission will be negligible. To do this, the grid support must be properly cooled and the radial thermal conduction of the grid must be as high as possible.

The fabrication of shadow grid structures is difficult, especially in high-frequency tubes, where dimensions are small. Alignment of the elements of the control and shadow grids is critical. Alignment must be maintained despite the fact that the shadow grid operates at a higher temperature than the control grid because it is closer to or perhaps in contact with the cathode.

An example of the grid motions that occur as a cathode warms up is given in Figure 6-41. These results were computed using the thermal expansion characteristics of the grids and their support structures [19]. Note that, at one point, the shadow grid moved to within about 0.001 inch of the cathode surface. This small spacing exceeded the total tolerance on the position of the grid. Thus, the shadow grid could touch the cathode, which, in turn, could lead to deformation of the shadow grid [20]. The author of [19] stated that this phenomena had also been noticed experimentally and corrective action was undertaken.

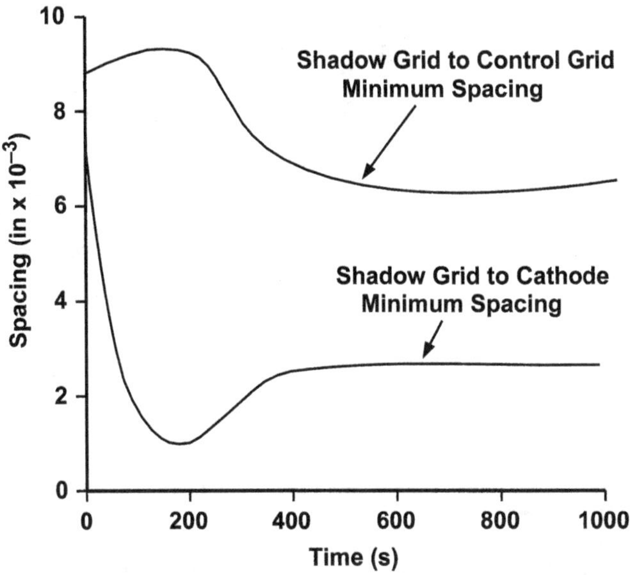

Figure 6-41 Relative spacing between shadow grid, control grid, and cathode during gun warm-up. (From: R. Sawicki, *Technical Digest*, 1978 IEDM. © 1978 IEEE.)

In an effort to overcome the problems of maintaining proper grid spacings in shadow gridded guns, Miram et al. developed the bonded grid structure shown in Figure 6-42 [21–23]. The grid elements were pyrolytic graphite, and these were deposited on a thin boron nitride spacer. To suppress grid emission, the control grid was overcoated with a thin boron nitride film. Technical feasibility was demonstrated in a 2,000-hour life test in which grid leakage and emission were shown to be very small and stable. The bonded grid has not yet found application because boron nitride with the required characteristics is not available and the cost of establishing a facility to produce it is, at present, prohibitive.

Very sophisticated grid structures have been developed for some microwave tube applications. In the example in Figure 6-43, two control grids are used along with a shadow grid. This grid structure was developed for a tube that could operate efficiently at two different power levels. This is referred to as dual-mode operation. The grid closest to the anode controls emission from the central part of the cathode. The next grid controls emission from the outer portion of the cathode. Thus, beam current and diameter can be switched to change the beam power level while maintaining constant beam voltage and proper focusing conditions.

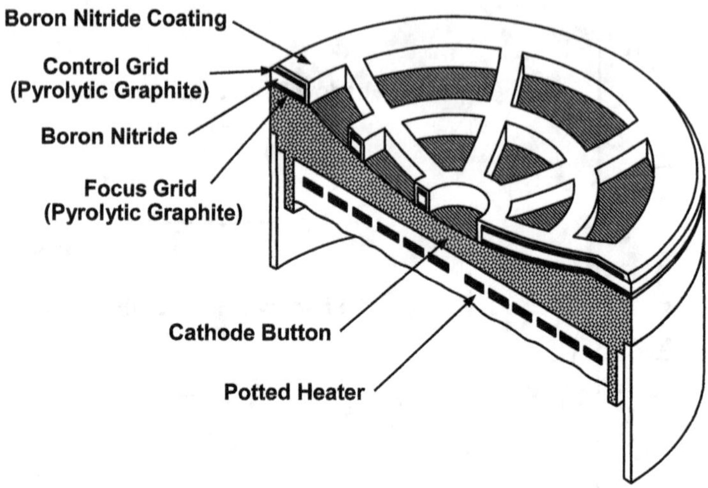

Figure 6-42 Bonded grid and cathode assembly. (From: T. J. Grant et al., *Technical Digest*, 1983 IEDM. © 1983 IEEE.)

In [24], True described a dual-mode structure using two grids rather than three. The grid farthest from the cathode is a control grid like that on the right in Figure 6-33. The grid next to the cathode is like a shadow grid except that the outer openings contain closely spaced wires. When this grid is pulsed positive with respect to the cathode (+36V in [24]), current flows from the entire cathode. When the grid is negative (–36V in [24]), the outer portion of the cathode is cut off and current flows only from the center portion.

The use of a shadow grid makes it possible to reduce interception on the control grid to about 0.1% of the cathode current. Unfortunately, at high frequencies where gun dimensions are very small, the fabrication of shadow grid structures becomes very difficult. To facilitate gun construction, negative grids are often used. With these grids, only a single grid is required and there is no grid interception. The beam optics are comparable to a shadow gridded gun. The current-voltage characteristic of a negative grid is compared with a unigrid (see Figure 6-29) in Figure 6-44.

6.2.5 Summary of Beam Control Electrode Characteristics

Table 6.1 contains a summary of the more important characteristics of beam control electrodes [25]. Two amplification factors are given. The total amplification factor, μ, is the ratio of beam to control electrode voltage required for beam control. The cutoff amplification factor, μ_c, is the ratio of beam voltage to the negative control electrode voltage (with respect to the cathode) required to reduce the beam current to zero.

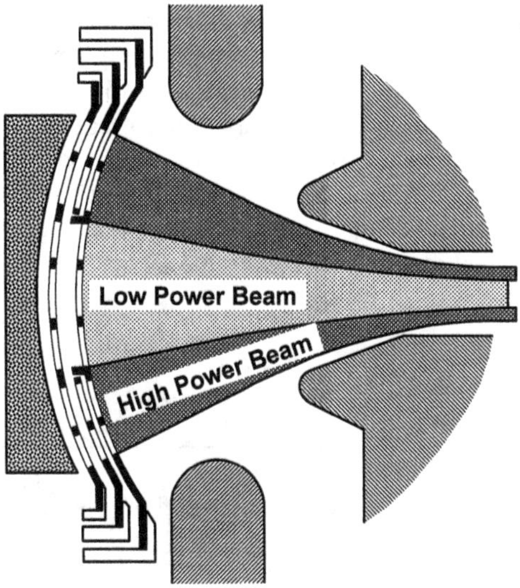

Figure 6-43 Dual-mode gun. (From: *Microwave Tube Manual* by Varian, Air Force Publication No. T.O. 00-25-251, October 1979.)

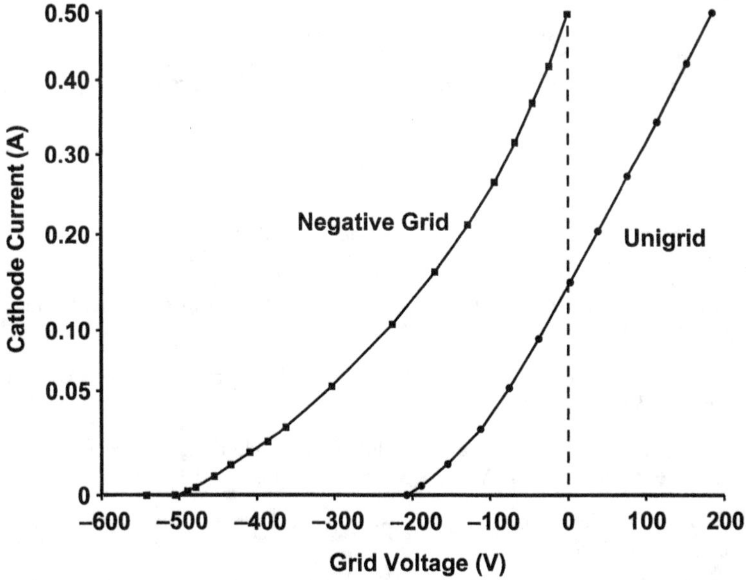

Figure 6-44 Control characteristics of a negative grid compared with a shadow grid. (From: B. James, Internal Varian Report.)

As would be expected, the gridded guns have by far the highest amplification factors and the modulating anode gun has the lowest amplification factor. The capacitances of all of the control electrode structures to the surrounding electrodes are essentially the same. Capacitance is important, of course, in the design of the modulator that must be used to drive the control electrode structure. Finally, as would be expected, the interception of the intercepting grid structure is much higher than the other structures. Only the modulating anode gun has good focusing because the design of that gun is essentially the design of an ideal Pierce gun.

Table 6.1
Typical Characteristics of Control Electrodes

Type	μ	μ_c	Capacitance	Grid Interception	Focusing at Low Voltage
Modulating Anode	1 to 3		50 pF	0	Good
Control Focus Electrode	2 to 10	2 to 10	100 pF	0	Poor
Intercepting Grid	50	100	50 pF	15%	Fair
Shadow Grid	30	300	50 pF	0.1%	Fair

Source: A. Staprans et al., *Proc. IEEE*, Vol. 61, No. 3, March 1973, pp. 299–330.

REFERENCES

[1] J. R. Pierce, *Theory and Design of Electron Guns*, 2nd ed., New York: Van Nostrand, 1954.

[2] G. R. Brewer, "Charged Particle Gun with Non Spherical Emissive Surface," U.S. Patent 3,139,552, 1964.

[3] I. L. Langmuir, and K. Blodgett, "Currents limited by space charge between concentric spheres," *Phys. Rev.,* Vol. 24, July 1924, p. 53.

[4] R. Helm, K. Spangenberg, and L. M. Field, "Cathode-design procedure for electron-beam tubes," *Elec. Comm.,* Vol. 24, March 1947, pp. 101–107.

[5] G. R. Brewer, "Formation of high-density electron beams," *Jour. App. Phys.,* Vol. 28, No. 1, January 1957, pp. 7–15.

[6] R. D. Frost, O. T. Purl, and H. R. Johnson, "Electron guns for forming solid beams of high perveance and high convergence," *Proc. IRE*, August 1962, pp. 1800–1807.

[7] C. C. Cutler, and M. E. Hines, "Thermal velocity effects in electron guns," *Proc. IRE*, March 1955, pp. 307–315.

[8] J. R. M. Vaughan, "Synthesis of the Pierce gun," *IEEE Trans. Electron Devices,* Vol. ED-28, January 1981, pp. 37–41.

[9] A. S. Gilmour, Jr., *Principles of Traveling Wave Tubes*, Norwood, MA: Artech House, 1994.

[10] W. E. Danielson, J. L. Rosenfeld, and J. A. Saloom, "A detailed analysis of beam formation with electron guns of the Pierce type," *Bell Sys. Tech. Jour.*, March 1956, pp. 375–420.

[11] H. J. Wolkstein, "Design considerations for grid-controlled electron guns for pulsed traveling wave tubes," *RCA Review*, September 1960, pp. 389–413.

[12] R. True, "Calculation and design of grids in Pierce guns," *Technical Digest*, IEDM, 1989, pp. 215–218.

[13] J. M. Drees, "Non-intercepting grid structure for an electron tube," U.S. Patent 3,484,645, December 16, 1969.

[14] R. True, "An ultra-laminar tetrode gun for high duty cycle applications," *Technical Digest*, IEDM, 1979, pp. 286–289.

[15] R. True, "A theory for coupling gridded gun design with PPM focusing," *IEEE Trans. Electron Devices,* Vol. ED-31, No. 3, March 1984, pp. 353–362.

[16] R. B. True, and D. E. Laycock, "Gridded electron gun," U.S. Patent 4,737,680, April 12, 1988.

[17] G. A. Haas, and A. Shih, "Study of high work function materials needed for close-spaced grid applications," *Applications of Surface Science,* Vol. 4, 1980, pp. 104–126.

[18] M. Green, and G. Goeser, "Deposition techniques for suppressed emission," *Final Report, Contract No. N00014-84-C-2270*, October 1988.

[19] R. Sawicki, "Analytical determination of the thermal/mechanical performance of traveling wave tube electron guns," *Technical Digest*, IEDM, 1978, pp. 160–163.

[20] H. Y. Yeh, "The role of structural stability in the reliability of microwave vacuum devices," *IEEE Trans. Electron Devices,* Vol. ED-34, No. 2, February 1987, pp. 494–497.

[21] G. V. Miram, and E. L. Lien, "Convergent electron gun with bonded nonintercepting control grid," *Technical Digest*, IEDM, 1978, pp. 164–167.

[22] G. V. Miram, and E. L. Lien, "Convergent electron gun with bonded nonintercepting control grid," *Technical Digest*, IEDM, 1979, pp. 290–292.

[23] T. J. Grant, R. Garcia, and G. V. Miram, "Bonded grid electron gun for 95 GHz extended interaction amplifier (EIA)," *Technical Digest*, IEDM, 1983, pp. 141–143.

[24] R. True, "Dual-mode electron gun," U.S. Patent 4,593,230, June 3, 1986.

[25] A. Staprans, E. W. McCune, and J. A. Ruetz, "High-power linear-beam tubes," *Proc. IEEE*, Vol. 61, No. 3, March 1973, pp. 299–330.

Chapter 7

Electron Beams

In Chapter 6, Pierce electron guns used in linear-beam tubes were discussed. In this chapter, the magnetic focusing of a beam from a Pierce gun is examined ([1–3] are recommended for further reading). Before proceeding with the discussion of magnetically focused beams, it is important to point out that numerous electrostatic focusing techniques have been examined in the more than 70 years that microwave tubes have been in existence. With the exception of very few special cases, only magnetic focusing is used currently.

The primary problem in focusing, as illustrated in Figure 7-1, is that space charge forces are so high in the beams in microwave tubes that, without focusing forces, rapid divergence of a beam occurs. With constraining (focusing) forces, beam divergence can be prevented.

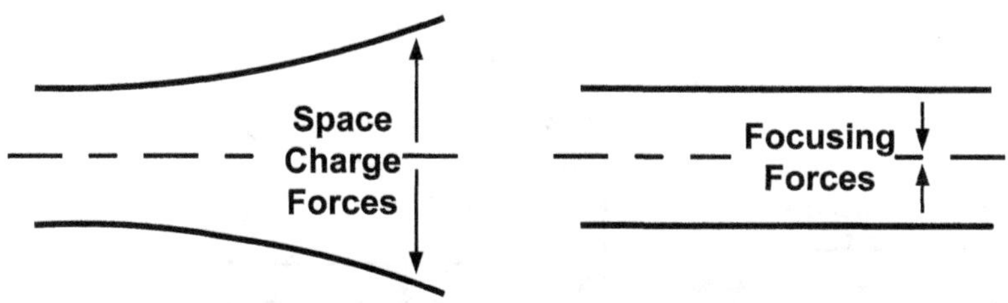

Figure 7-1 Illustration of the effects of space charge forces and focusing forces on an electron beam.

In linear-beam tubes, the focusing forces are provided by a magnetic field aligned with the axis of the electron beam. Often, a solenoid like that in Figure 7-2 is used to generate the field. Before developing the equations of beam motion in a magnetic focusing field, it is of value to discuss qualitatively how magnetic focusing works.

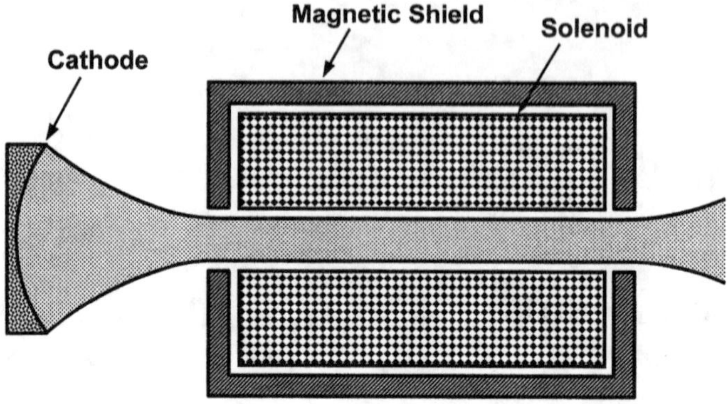

Figure 7-2 Illustration of the use of a solenoid to focus the beam in a TWT or klystron.

7.1 OVERVIEW OF UNIFORM-FIELD FOCUSING

To simplify the description of the focusing process, it is initially assumed that there is no magnetic flux through the cathode and that the beam first encounters magnetic flux lines as it enters the solenoid. It is also assumed that there are no transverse velocity components (from thermal or grid effects, nonuniform cathode emission, and so forth). Magnetic focusing of the beam is initiated as the electron beam crosses radially directed components of magnetic flux lines (illustrated in Figure 7-3).

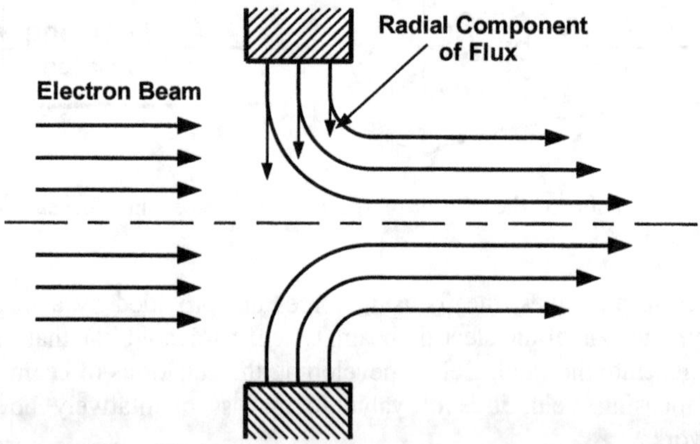

Figure 7-3 Configuration of magnetic flux lines at entrance of electron beam into solenoid.

Consider, for example, an electron above the axis of the beam (and solenoid). That electron encounters a component of magnetic flux that is radially directed downward. The magnetic force on the electron is out of the paper (the cross product of the velocity, *u*, and the flux density, *B*, is into the paper and the force on an electron is in the reverse direction from the $u \times B$ product).

Consider next an electron below the axis. That electron encounters a component of magnetic flux that is directed upward. The magnetic force on that electron is into the paper.

When the magnetic forces on all electrons are considered, the net effect is that the electron beam starts rotating in the clockwise direction as it enters the field with the flux lines in the directions shown in Figure 7-3. If the directions of the flux lines were reversed, the direction of rotation of the beam would be counterclockwise.

As the electron beam moves further into the magnetic field, it encounters stronger axial components of magnetic flux. Consider again an electron above the beam axis. As a result of crossing radial components of flux, this electron has a velocity component out of the paper. The cross product of that velocity component with the axial magnetic flux is a vector away from the axis so that the force on the electron is toward the axis. Similar considerations for other electrons show that they all experience a magnetic force toward the axis. Thus, it is the rotation of the beam (established as the beam crosses radially directed flux lines) interacting with the axial components of flux that produces the magnetic focusing force. As each electron in the beam is followed, it is found that a helical trajectory about the axis of the solenoid is traversed as indicated in Figure 7-4.

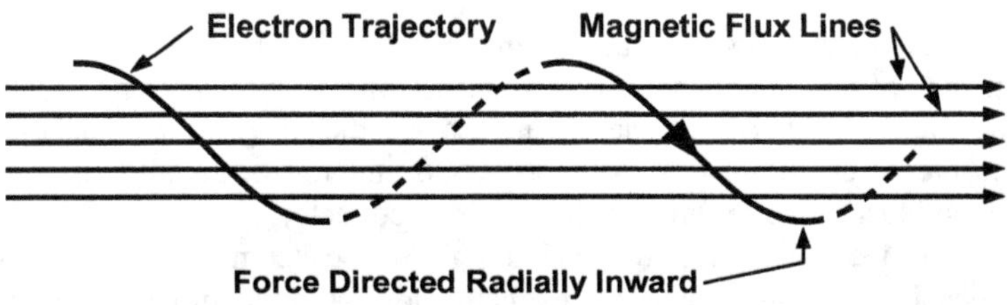

Figure 7-4 Electron trajectory in an axial field.

7.1.1 Brillouin Flow

If the space charge forces (plus centrifugal forces resulting from beam rotation) tending to cause the beam to diverge are just balanced by equal and opposite magnetic forces tending to compress the beam, and if the electrons enter the magnetic field with no radial velocity components, then the beam diameter will remain constant as the beam rotates and moves through the magnetic field.

For a beam with uniform charge density throughout and no magnetic flux through the cathode, the magnetic flux level that produces a magnetic force that exactly balances the space charge and centrifugal forces is called the *Brillouin flux level*, commonly denoted by B_B. The resulting electron motion in the form of a rotating beam with constant diameter is called *Brillouin flow*. The configuration of the electron gun and magnetic pole piece for launching Brillouin flow are shown in Figure 7-5.

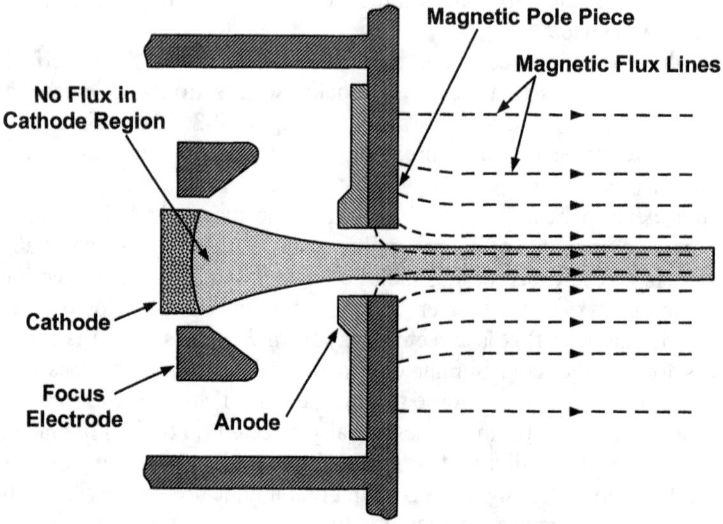

Figure 7-5 Gun and pole piece configuration for launching Brillouin flow.

When the beam exits from the magnetic field at the collector end of the tube, the sequence of interactions is just the opposite of that at the entrance of the beam into the field. That is, as the beam leaves the magnetic field, the radial compressive force disappears, rotation stops, and the beam expands under the influence of space charge forces as it enters the collector.

Now, as we have said, for Brillouin flow, the beam must have a uniform charge density and must enter the magnetic field on axis with no radial electron motion and with a radius, b_o, equal to the equilibrium radius, a (the radius at which space charge and centrifugal forces are balanced by magnetic forces). This situation is shown in Figure 7-6.

7.1.2 Scalloping

When the electron gun is shielded from magnetic flux, it is extremely difficult to achieve the perfectly balanced force condition that is required for Brillouin flow. For example, consider the case when the electron beam enters the magnetic field with a radius smaller than the equilibrium radius shown in Figure 7-7.

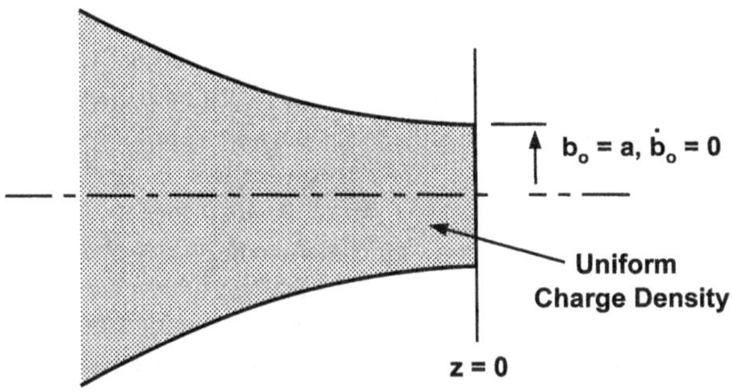

Figure 7-6 Entrance conditions for Brillouin flow. (The initial beam radius, b_o, is the equilibrium radius, a, and no radial velocities exist so $\dot{b}_o = 0$.)

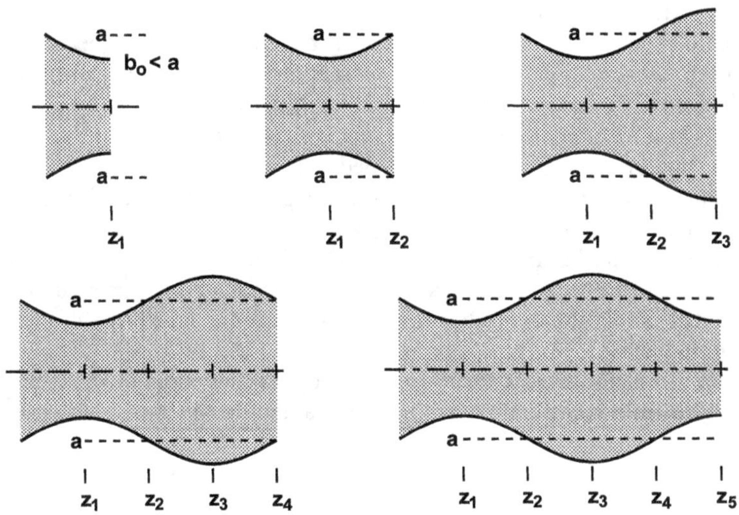

Figure 7-7 Beam dynamics resulting from the beam entrance into magnetic field at less than the equilibrium radius.

At plane z_1, where the beam enters the magnetic field, the space charge force is larger than the magnetic focusing force, so the beam starts to expand. When the beam reaches plane z_2, the radial forces are equal, but the electrons are moving radially outward. As a result, the electrons overshoot the equilibrium radius. By the time the electrons have reached plane z_3, the radial motion of the electrons has been stopped because the space charge and centrifugal forces have become smaller than the magnetic forces. In fact, this force imbalance causes the beam to become smaller again and, by plane z_4, the electrons are again crossing the equilibrium radius. Because the electrons are moving radially inward at plane z_4, they overshoot the equilibrium radius. By the time the beam reaches plane z_5, the inward radial motion has been stopped because the space charge and centrifugal forces exceed the magnetic forces.

In a uniform axial field, the beam radius at plane z_5 is the same as it was at plane z_1 and the entire process that took place between planes z_1 and z_5 starts over again. As a result, the beam radius continues to oscillate as the beam moves in the axial direction. This periodic variation of beam radius is referred to as *scalloping.*

For the preceding consideration (electrons moving parallel to axis as the beam enters the magnetic field), it is possible to adjust the magnetic field level so that the beam radius is the equilibrium radius. Then scalloping can be eliminated.

A similar situation occurs when a beam traveling parallel to the axis but having a radius larger that the equilibrium radius enters the magnetic focusing field. Then the beam scallops inward because the focusing forces are larger than the space charge and centrifugal forces. Again, it is possible to adjust the magnetic field level to eliminate scalloping.

If the electron gun is at the wrong axial position so that the beam is either diverging or converging as it enters the magnetic field, then no adjustment of the magnetic field can be made to eliminate the resulting scalloping.

At this point it is appropriate to comment on the beam behavior that occurs if the axis of the electron beam is not aligned with the axis of the magnetic field. This could happen if the electron gun was misaligned during tube fabrication. The gun axis could be either off center or tilted with respect to the axis of the tube. Other causes for misalignment include errors in cathode, grid, or control electrode positioning or nonuniformities in magnetic pole piece material. The result of beam misalignment is that the entire beam spirals about the magnetic field axis at the cyclotron frequency.

Figure 7-8 contains a summary of the problems that could be encountered in attempting to form a Brillouin beam flowing parallel to the axis. These problems are:

(a) Flux density is too small.
(b) Flux density is too large.
(c) Beam is converging at the entrance.
(d) Beam is diverging at the entrance.
(e) Beam axis is offset from the magnetic field axis.
(f) Beam axis is tilted with respect to the magnetic field axis.

Brillouin flow has the advantage that the magnetic field required is the lower than for any other focusing technique. The primary disadvantage is that the beam is extremely sensitive to misalignments and perturbations (such as those produced by RF modulation). The effect of a change in beam current, such as that resulting from RF bunching of electrons, is shown in Figure 7-9. Note that an increase in current of 50% results in an increase in beam diameter of over 40%.

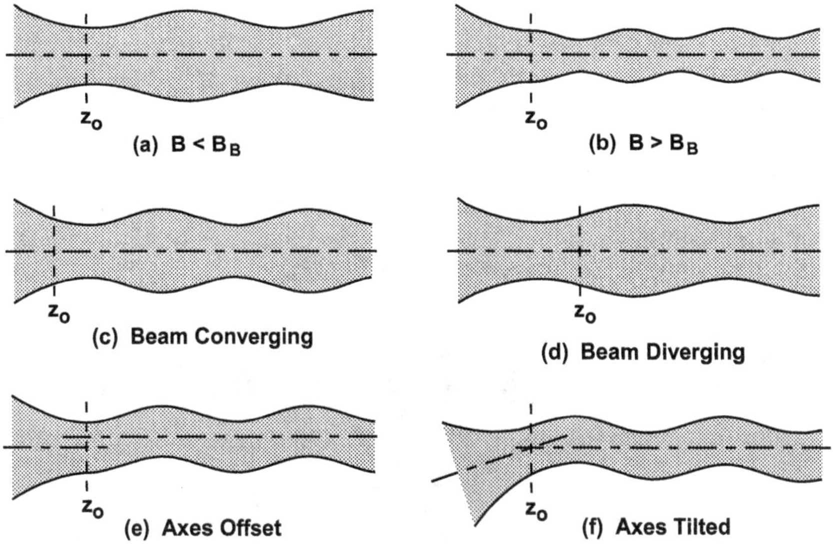

Figure 7-8 (a–f) Problems in forming a Brillouin beam flowing parallel to the axis.

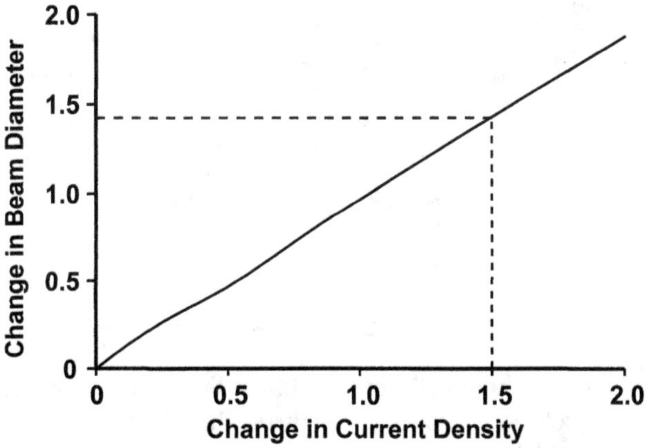

Figure 7-9 Change in beam diameter resulting from a change in current density for a Brillouin beam. (From: *Power Travelling Wave Tubes* by J. F. Gittins, published 1965 by American Elsevier Publishing.)

We will show in this chapter that Brillouin flow can be nearly achieved under laboratory conditions. In practice, the magnetic focusing field that is used, even when the cathode is shielded, is greater than the Brillouin value. Reasons for this include transverse electron velocities (from thermal and other effects in the gun) and the increase in beam size that results from RF modulation.

7.1.3 Confined (Immersed) Flow

In practice, the use of Brillouin focusing is avoided because of the high sensitivity of the beam to perturbations. Instead, as shown in Figure 7-10, the magnetic circuit providing the focusing field is purposely designed so that magnetic flux lines pass through the cathode. This makes possible the use of a much larger magnetic field than the Brillouin value for focusing the beam. In addition, this can be accomplished without causing scalloping of the beam. This type of focusing is referred to as confined flow or immersed flow focusing.

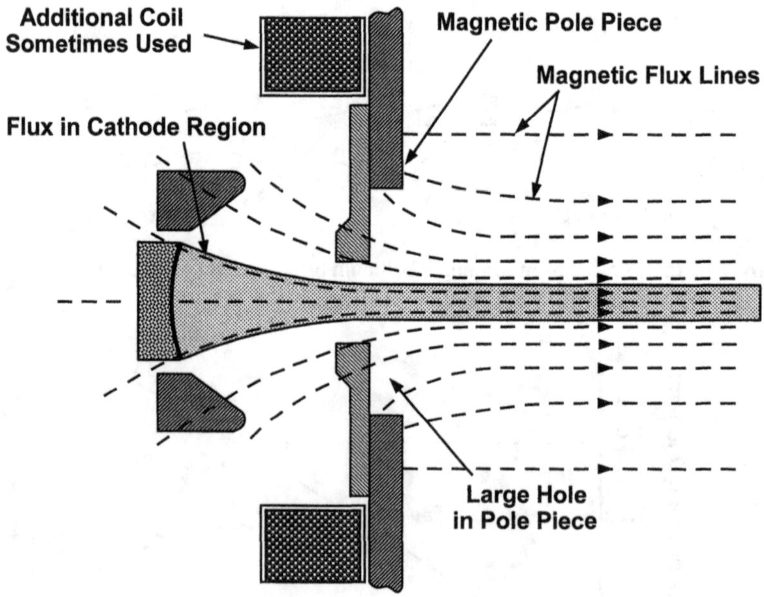

Figure 7-10 Magnetic circuit for launching confined (immersed) flow.

It is helpful, in understanding the effect of flux through the cathode, to assume that the electron trajectories follow the magnetic flux lines in the electron gun. Then the magnetic field has no effect on the electrons in the gun. It should be pointed out that electrons do, in fact, cross flux lines in the gun (as is indicated in Figure 7-10). The resulting behavior of the beam is essentially the same as it is if it is assumed that the electrons do not cross flux lines.

As the electron beam leaves the gun and enters the main part of the focusing field, the increase in flux density must be sufficient to produce a focusing force that counterbalances the space charge and centrifugal forces. The focusing force results, just as it did for Brillouin focusing, from the interaction of the beam rotation with the axial magnetic field. In this case, however, the axial field is larger than it was for Brillouin flow so the beam rotates more slowly than for the Brillouin case. Also, since the beam rotates more slowly, the centrifugal force is reduced and this further reduces the rotational rate required for equilibrium.

Next, consider a perturbation to the beam that causes the electrons to move radially outward. This could result from electron bunching caused by the interaction of the electrons with an RF field. As the outward moving electrons cross flux lines of the very large confined flow focusing field (often two to three times the Brillouin value), the rate of rotation of the beam increases rapidly. The interaction of this increased rotation with the large focusing field produces a focusing force far larger than that which occurs with Brillouin focusing. As a result, the radial expansion of the beam is much less than for Brillouin focusing.

The primary benefit of confined flow focusing is that beam control is significantly better than for Brillouin focusing. This is illustrated in Figure 7-11, which is like Figure 7-9 except that curves for field levels two times ($m = 2$) and three times ($m = 3$) the Brillouin value are shown. For an increase in beam current of 50%, the change in beam diameter is decreased from over 40% for Brillouin focusing to about 8% for $m = 2$ and less than 4% for $m = 3$. The primary disadvantage of confined flow focusing is that the magnet(s) must be larger and heavier than those for Brillouin focusing to produce the required field. Also, if electromagnets are used, the power and cooling requirements are larger than those for Brillouin flow.

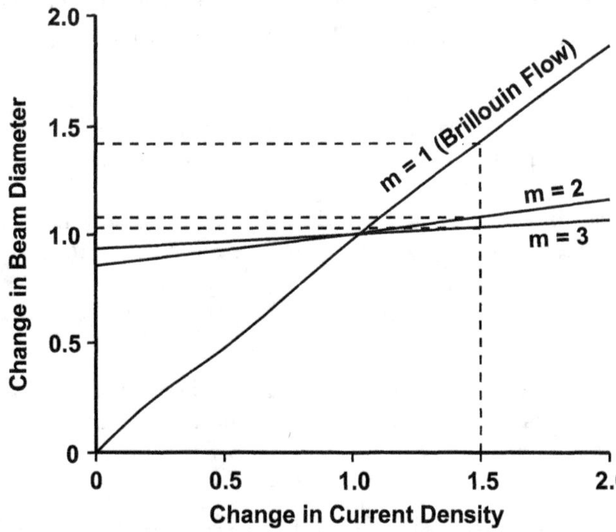

Figure 7-11 Effect of current density on beam diameter for fields greater than Brillouin. (From: *Power Travelling Wave Tubes* by J. F. Gittins, published 1965 by American Elsevier Publishing.)

7.2 UNIFORM-FIELD FOCUSING AND LAMINAR FLOW

7.2.1 The Beam Equation

In the electron beams to be analyzed, the fields are axially symmetric. The radial electric field, E_r, is produced by the electron beam and the magnetic flux density, B, used to focus the beam is in the axial direction. It is assumed that the charge density, ρ, and the axial velocity, u_o, are uniform in the radial direction (but not necessarily in the axial direction). From Chapter 4, the radial force equation, which will be used to find the motion of the electrons, is

$$\ddot{r} - r\,\dot{\theta}^2 = -\eta(E_r + B\,r\,\dot{\theta}) \tag{7-1}$$

where $\dot{\theta}$ is given by Busch's theorem:

$$\dot{\theta} = \frac{\eta}{2}\left(B - B_c\,\frac{r_c^2}{r^2}\right) \tag{7-2}$$

Here it has been assumed that the electrons originate at the cathode at a position, r_c, where the flux density is B_c. At a given axial position, it should be noted that, with the assumption of uniform charge density in the radial direction and an additional assumption that electron flow is laminar in the gun (electron trajectories do not cross), then r_c/r is a constant. Thus, at any axial position, $\dot{\theta}$ is constant throughout the beam.

Also, since the space charge force (and also the centrifugal force and the magnetic force) is proportional to radial position, the radial acceleration and radial motions of the electrons are proportional to r, so electron flow in the beam is laminar.

The electric field was found using Gauss's law in the universal beam spread analysis. At the beam boundary where $r = b$, the electric field, $E_r(b)$, is

$$E_r(b) = -\frac{I}{2\pi\,b\,\varepsilon_o\,u_o} \tag{7-3}$$

By combining the force equation with Busch's theorem and Gauss's law, the following relation may be obtained for the motion of the electrons on the outer edge of the beam:

$$\ddot{b} + b\,\omega_L^2\left[1 - \left(\frac{B_c\,b_c^2}{B\,b^2}\right)^2\right] - \frac{\eta\,I}{2\pi\,b\,\varepsilon_o\,u_o} = 0 \tag{7-4}$$

where $\omega_L = \eta B/2$ is the Larmor frequency, which is half of the cyclotron frequency. This equation is sometimes referred to as the beam equation.

While it is a simple matter to obtain numerical solutions to the beam equation, these solutions don't provide the insight into beam behavior that can be obtained from an approximate closed-form solution. It will be convenient to let

$$a = \frac{1}{B}\left(\frac{2I}{\eta \pi \varepsilon_o u_o}\right)^{1/2} \tag{7-5}$$

It will be shown that this quantity is the equilibrium radius for a Brillouin beam. As would be expected, as B increases, the magnetic forces increase, so the equilibrium radius must decrease. As the current increases, the charge density in the beam increases so the space charge forces and equilibrium radius must increase. As velocity increases, the space charge density and the space charge forces decrease so the equilibrium radius must decrease.

Using the equilibrium radius, the beam equation can be written as

$$\frac{\ddot{b}}{a} + \omega_L^2\left[\frac{b}{a}\left[1-\left(\frac{B_c}{B}\frac{b_c^2}{b^2}\right)^2\right]-\frac{a}{b}\right]=0 \tag{7-6}$$

The first term is proportional to the radial acceleration and, therefore, to the radial force on the electrons. When the quantity in square brackets is zero, the net radial force is zero. (This occurs because the space charge and centrifugal forces causing the beam to expand are equal and opposite to the magnetic focusing forces that compress the beam.) When the radial forces are zero, the beam is in equilibrium and there are no forces causing it to expand or contract. If this condition can be sustained, then the beam diameter remains constant and a cylindrical beam has been formed. This is normally the desired beam shape. If the radius at which the forces are zero (the equilibrium radius, b_e), then the quantity in square brackets is

$$\frac{b_e}{a}\left[1-\left(\frac{B_c}{B}\frac{b_c^2}{b_e^2}\right)^2\right]-\frac{a}{b_e}=0 \tag{7-7}$$

This can be simplified by examining the relationship between B and B_c. The value of B can be expressed in terms of B_c and the Brillouin flux density by examining the force equation. At equilibrium, the radial acceleration is zero and thus, for Brillouin flow,

$$\eta E_r = -\eta B_B a \dot{\theta}_B + a\dot{\theta}_B^2 \tag{7-8}$$

where $\dot{\theta}_B$ is the value of $\dot{\theta}$ for Brillouin flow.

For confined flow,

$$\eta E_r = -\eta B b_e \dot{\theta} + b_e \dot{\theta}^2 \tag{7-9}$$

However, the beam radii are the same for Brillouin and confined flow and E_r is the same because the current, velocity, and beam radii are equal, so these two equations can be combined with Busch's theorem to find

$$B^2 = B_B^2 + B_c^2 \frac{A_c^2}{A_e^2} \tag{7-10}$$

where $A_c \propto b_c^2$ is the area of the cathode and $A_e \propto b_e^2$ is the cross-sectional area of the equilibrium beam. The ratio A_c/A_e is referred to as the *area compression ratio* of the electron gun. This result is significant because A_c/A_e is often as high as 25 or more so a flux density of only a few tens of gauss at the cathode surface has an effect equivalent to a Brillouin flux density of 500 to 1,000 gauss or more.

As noted previously, it is often convenient to reference the actual flux density used for focusing, B, to the Brillouin value, B_B, by the confinement factor m, which is defined by

$$B = mB_B \tag{7-11}$$

Using this relation, the equilibrium beam radius becomes:

$$\frac{b_e}{a} = m \tag{7-12}$$

That is, for confined flow, the ratio of the equilibrium radius to the Brillouin radius is the confinement factor, which is the ratio of the focusing field to the Brillouin field.

Now, by assuming that the beam radius is close to the equilibrium radius, an approximate solution to the beam equation can be found. By letting

$$b = b_e(1+\delta) \tag{7-13}$$

where $\delta << 1$ and ignoring second-order terms, the beam equation reduces to

$$\ddot{\delta} + \omega_s^2 \delta = 0 \tag{7-14}$$

where

$$\omega_s = \sqrt{2}\left(2 - \frac{1}{m^2}\right)^{1/2} \omega_L \tag{7-15}$$

The general solution to this equation is

$$\delta = A \sin \omega_s t + B \cos \omega_s t \tag{7-16}$$

Since $b = b_e(1 + \delta)$ and $t = z/u_o$,

$$b = b_e\left(1 + A\,\sin\frac{\omega_s}{u_o}z + B\,\cos\frac{\omega_s}{u_o}z\right) \tag{7-17}$$

or

$$b = b_e\left[1 + C\sin\left(\frac{\omega_s}{u_o}z + \phi\right)\right] \tag{7-18}$$

This shows that the beam radius can oscillate (scallop) about the equilibrium radius, b_e, with a frequency ω_s. The wavelength of a scallop is

$$\lambda_s = \frac{u_o}{f_s} = \frac{0.030\times10^{-3}}{B}\left[\frac{V}{\left(2 - \frac{1}{m^2}\right)}\right]^{1/2}\ \text{m} \tag{7-19}$$

The values of C and φ can be determined by examining the conditions of entrance of the beam at $z = 0$ into the main magnetic focusing field.

7.2.2 Brillouin Flow

When Brillouin flow focusing is used, no magnetic flux exists in the cathode region, $m = 1$, so the configuration of the magnetic field at the entrance to the focusing structure is as shown in Figure 7-12. The equilibrium radius of the beam $b_e = a$ and $\omega_s = \sqrt{2}\,\omega_L$, so the beam radius is

$$b = a\left[1 + C\sin\left(\sqrt{2}\frac{\omega_L}{u_o}z + \phi\right)\right] \tag{7-20}$$

First, let the electron beam enter the focusing structure at the equilibrium radius traveling parallel to the axis. Then at $z = 0$, $b = a$, and $db/dz = 0$, so $C = 0$. As a result, the beam radius remains constant as the beam travels through the magnetic focusing field.

The magnetic flux density that produces a beam of constant radius when no magnetic flux links (passes through) the cathode is the Brillouin flux density, usually denoted by B_B. The resulting beam is a Brillouin beam, and the resulting electron motion is Brillouin flow. The magnetic flux required for Brillouin flow is smaller than for any other magnetic focusing technique. The Brillouin flux density can be determined from the equilibrium radius and is:

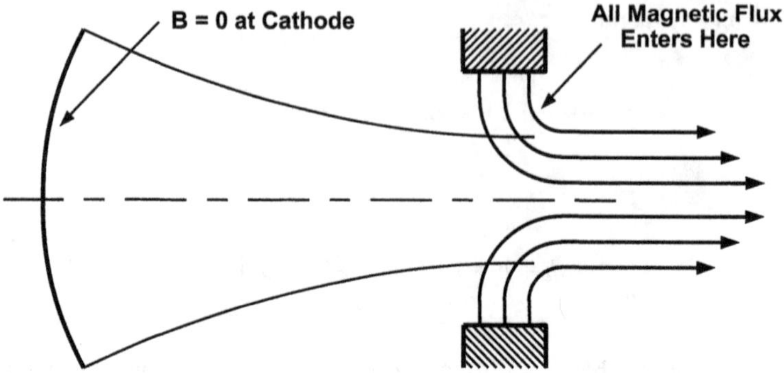

Figure 7-12 Configuration of the magnetic field at the entrance to the focusing structure for Brillouin flow ($m = 1$).

$$B_B = 0.83 \times 10^{-3} \frac{I^{1/2}}{a V^{1/4}} \quad \text{T} \tag{7-21}$$

Because the flux density is minimized, the size of the magnet structure (permanent magnet or solenoid) is minimized with Brillouin flow. The penalty for using Brillouin flow, as shown in the following discussion, is that the beam is very sensitive to misalignments, incorrect magnetic fields, RF drive, and so forth.

If the magnetic flux density is changed so that it is less than the Brillouin value, that is, $B < B_B$, then the value of the equilibrium radius, a, increases and so the beam diameter increases as the beam enters the focusing field. Subsequently, the beam diameter oscillates about the equilibrium diameter as shown in Figure 7-13.

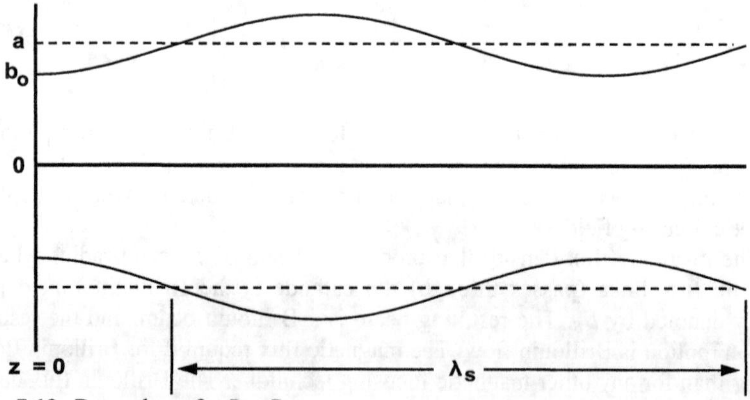

Figure 7-13 Beam shape for $B < B_B$.

As noted earlier in this section, this periodic fluctuation of the beam radius is referred to as scalloping. The wavelength, λ_s, of the scallop is

$$\lambda_s = 0.030 \times 10^{-3} \frac{V^{1/2}}{B} \text{ m} \tag{7-22}$$

The way in which scallop wavelength and beam shape vary with B is shown in Figure 7-14.

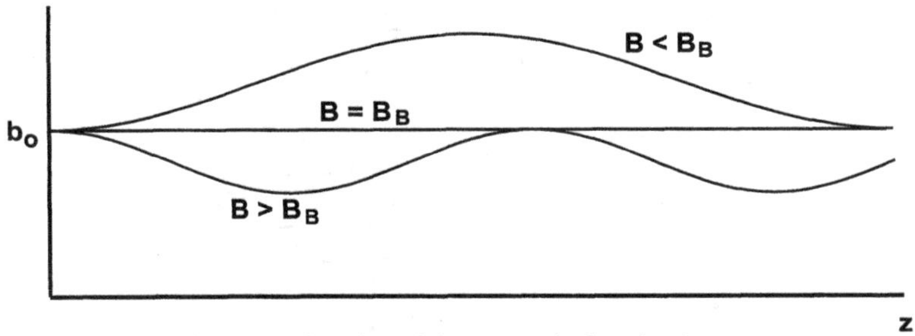

Figure 7-14 Beam shape as a function of the magnetic flux density.

Next consider the case when the beam enters the focusing field at $b = a$ but db/dz is not zero. That is, the beam is either expanding or contracting. When the beam is expanding, the beam radius varies with B as shown in Figure 7-15.

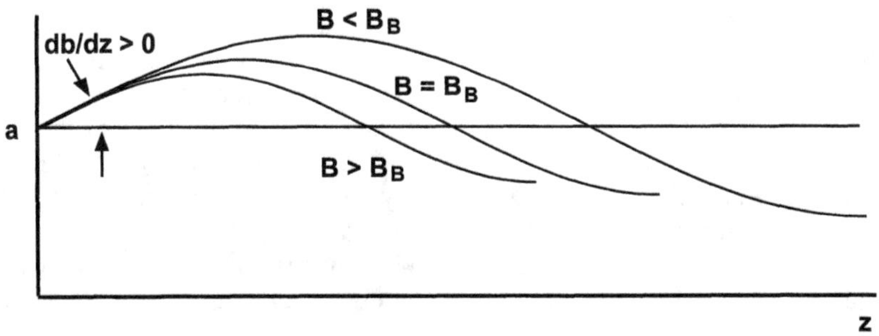

Figure 7-15 Beam shapes for $db/dz > 0$.

Experiments show that Brillouin-focused electron beams can behave very nearly as predicted. Results obtained by scanning a current probe across the center of a beam at equally spaced axial positions are shown in Figure 7-16 [4, 5]. The inward scalloping of the beam when $B > B_B$ is clearly shown. When $B = B_B$, the beam diameter was nearly constant.

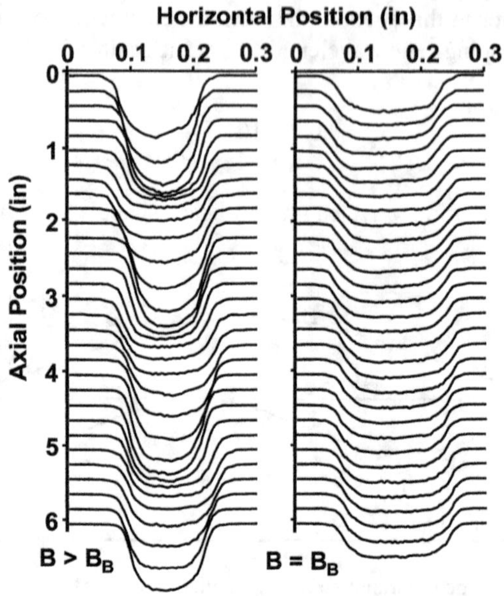

Figure 7-16 Current density profiles for a beam from a shielded gun. (From: A. S. Gilmour, Jr., *Proc. High Power Microwave Tube Symposium*, The Hexagon, 1962.)

Ideally, each of the current profiles in Figure 7-16 should be rectangular. At least three factors prevented the current from rising abruptly from zero to a constant value at the edges of the beam profiles. One factor was the finite diameter (0.010 inch (0.25 mm)) of the aperture in the current probe used for current sampling. The second factor was the thermal velocities of the electrons, and the third factor was the patchiness and roughness of the oxide coated cathode.

From data like that in Figure 7-16, the beam radius as functions of the axial position and the magnetic flux density were determined and are shown in Figure 7-17. At the predicted Brillouin flux density of 285G, the variation in beam diameter was less than 2%.

Brewer reported measurements on beams from several electron guns with a wide range of focusing conditions [6]. He observed translaminar streams of electrons that originated from aberrations in the gun but these constituted a small percentage of the current in the beam. He concluded: "The close agreement between the experimental data and the laminar-flow theory would appear to indicate that the bulk of the beam is in a very nearly laminar-flow situation. This does not mean that no electrons cross other electron paths in the beam, since many anomalies have certainly been observed which could result from electron-trajectory crossings. However, the author believes that most of the electrons in a well-formed beam do behave as in laminar flow for magnetic field values near the Brillouin-flow condition."

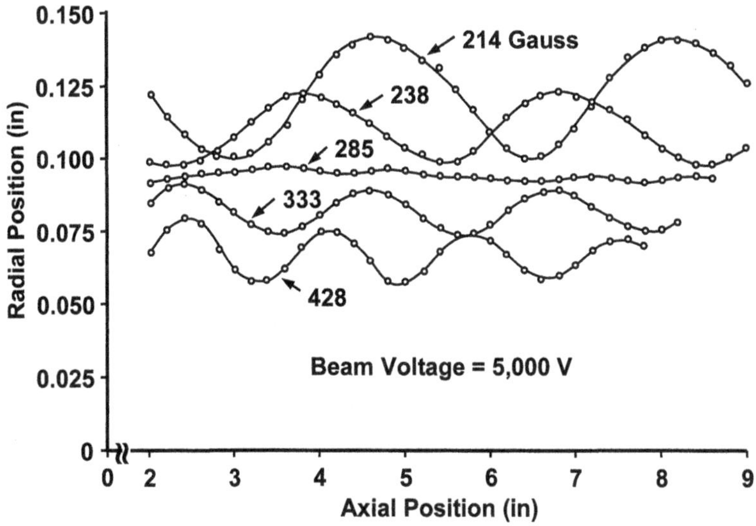

Figure 7-17 Beam radius versus axial position and magnetic flux density. (From: A. S. Gilmour, Jr., *Proc. High Power Microwave Tube Symposium*, The Hexagon, 1962.)

7.2.3 Confined (Immersed) Flow

When part of the magnetic flux used for focusing the beam passes through the cathode, the beam is said to be partially confined or confined (or sometimes partially immersed or immersed) and the behavior of the beam is quite different than that for Brillouin flow. First, the scalloping of the beam is reduced when the beam is expanding or contracting as it enters the focusing field. Second, when changes in beam conditions occur, such as bunching of the electrons by an RF field, the resulting perturbations of the beam radius are reduced significantly.

Using Busch's theorem, it is possible to more clearly understand the details of confined flow focusing. For example, the rate of rotation of the beam, $\dot{\theta}$, compared to the rate of rotation of the Brillouin focused beam, $\dot{\theta}_B$, can be found to be

$$\frac{\dot{\theta}}{\dot{\theta}_B} = m - \left(m^2 - 1 \right)^{1/2} \tag{7-23}$$

This relation is plotted in Figure 7-18. Note that, as m increases from unity (the Brillouin value) to the range of 2 to 3 commonly used for confined flow focusing, the rate of rotation of the beam drops by a factor of 4 to over 5 from the Brillouin value.

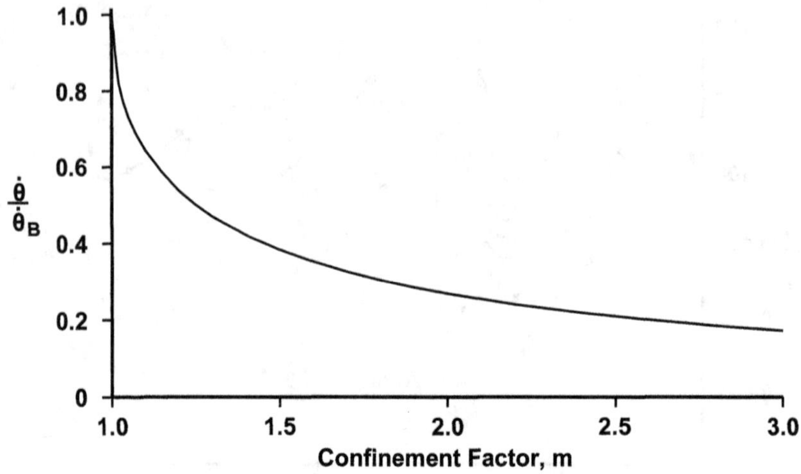

Figure 7-18 Ratio of rotation rates of confined and Brillouin beams as a function of the confinement factor.

To more clearly understand why the beam rotates so slowly for reasonable values of m, consider Figure 7-19. When flux links the cathode, the total flux through the cathode surface is $\pi\, b_c^2\, B_c$. This is the same as the total flux through the area defined by radius g (shown in Figure 7-19), which is $\pi\, g^2\, B$. Note that when $g = 0$, there is no flux through the cathode. When $g = b$, the total flux through the cathode is the same as the total flux contained in the electron beam at position z. As a result,

$$\pi\, b_c^2\, B_c = \pi\, g^2\, B \tag{7-24}$$

and the following relation between g and m can be found:

$$g = \left(1 - \frac{1}{m^2}\right)^{1/4} b_e \tag{7-25}$$

Notice that, for reasonable values of m, g is very close to b_e. For example, for $m = 2$, $g = 0.93b_e$. Thus, the flux that causes rotation of the beam is equivalent to that contained in the outer 7% of the beam.

Because the beam rotates so slowly with confined flow focusing, the centrifugal force, which is equal to the space charge force for Brillouin focusing, almost vanishes. Nearly all of the focusing force produced by the slowly rotating beam in the strong focusing field is used to offset space charge forces.

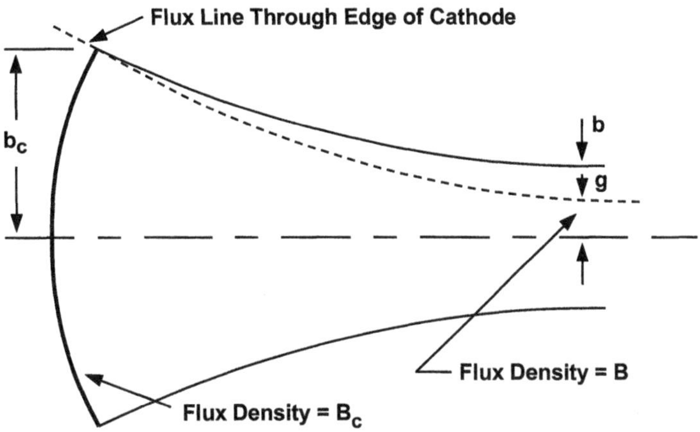

Figure 7-19 Definition of quantity g.

It is of value to compare the beam dynamics for confined focusing to those for Brillouin focusing. First, assume that the beam enters the focusing field at the equilibrium radius but with a radial component of velocity. Then the resulting scallop amplitude, C, compared with the amplitude of the scallop on a Brillouin beam, C_B, for the same entrance conditions ($a = b_e$), is [7]

$$\frac{C}{C_B} = \frac{1}{\left(2m^2 - 1\right)^{1/2}} \tag{7-26}$$

This result is shown in Figure 7-20. Note that a significant decrease in scallop amplitude occurs as the flux density is increased.

Next consider the effect on beam radius of a change in current in the beam. Here it will be assumed that the beam velocity remains constant. Current may change because of the use of a control electrode, such as a grid, in the electron gun to modulate the electron flow. Also, current changes in localized regions when the beam interacts with the RF field on a nearby circuit to produce bunching of the electrons.

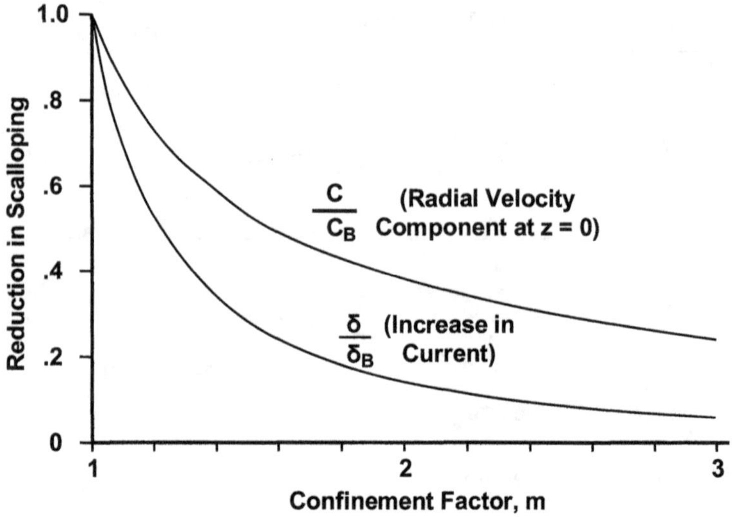

Figure 7-20 Scallop amplitudes for confined flow, C and δ, relative to those for Brillouin flow, C_B, δ_B, as functions of confinement factor.

When a change in current occurs, let the change in the beam radius for confined flow be δ and for Brillouin flow be δ_B. The ratio of δ to δ_B can be shown to be [7]

$$\frac{\delta}{\delta_B} = \frac{1}{2m^2 - 1} \tag{7-27}$$

The ratio, δ/δ_B, is plotted in Figure 7-20 as a function of the confinement factor m. It is clear that the effect of confined flow on beam perturbations resulting from current changes is significant and that the use of confined flow is very desirable to reduce these perturbations. The penalty is that the increase in flux density required for confined flow necessitates an increase in the size of the magnets used and, if solenoid focusing is employed, an increase in solenoid power.

In his paper on magnetically focused electron beams, Brewer clearly showed the confining effect of magnetic flux through the cathode [6]. As the cathode flux density was increased, beam scalloping decreased and variations in current density with distance were reduced. Brewer also noted that cathode flux prevents translaminar electrons from crossing or coming close to the beam axis, so cathode flux improves the laminarity of the beam.

7.3 UNIFORM-FIELD FOCUSING AND NONLAMINAR FLOW

In the chapters on cathodes and guns, several mechanisms have been identified that produce transverse velocity components to electron trajectories:

1. Thermal effects;
2. Patchy emission;
3. Cathode roughness;
4. Filament magnetic field;
5. Field distortion by grids;
6. Gun aberrations.

As a result of the transverse velocity components, electron trajectories cross and the electron flow is said to be nonlaminar.

There is no single theory for treating all forms of nonlaminar flow. In fact, the only way for determining, with reasonable accuracy, electron behavior in nonlaminar flow is by tracing trajectories using numerical techniques. Results of this approach will be shown in a following section, which is concerned with PPM-focused beams from gridded guns.

Several theories exist for special cases of nonlaminar flow. One of the first was that by Pierce and Walker for the effects of thermal velocities on Brillouin flow [8]. This theory was based on the assumption of a Maxwellian distribution of charge density within the beam. The beam was assumed to be in thermal equilibrium with a temperature that was suggested to be the cathode temperature multiplied by the area compression ratio of the gun. Pierce and Walker computed the percentage of the beam current to be found outside the nominal Brillouin radius.

Brewer compared his experimental beam profiles with the predictions of Pierce and Walker [8]. The temperature used was the cathode temperature multiplied by the area compression ratio of the gun. Brewer showed that this was reasonable by examining the radial spread of electrons passing through an aperture located on the axis of the beam. The observed temperature was within 20% of the calculated temperature. Brewer's measurement of the current profile for the beam edge compared favorably with the Pierce and Walker theory near the nominal beam radius. At larger radii, there was a low current-density "tail" on the current profile that was far from the prediction of the theory. This tail was thought to have resulted from the translaminar streams of electrons produced by gun aberrations.

Gilmour et al. [4, 5] and Brewer [6] showed that beams focused for near Brillouin flow conditions do behave very nearly as predicted by laminar flow theory. However, Ashkin [9] and Harker [10] reported that electron flow in beams, which had been thought to be laminar, was actually very nonlaminar. Following the works of Ashkin and Harker, Herrmann presented an optical theory for thermal velocity effects that discarded the assumption of laminarity [11].

Herrmann's theory treats beams as perfect optical systems. The thermal velocities of the electrons emitted in all directions by the cathode are assumed to have a Maxwellian distribution. Electron trajectories are assumed to obey the laws of geometrical optics of ideal optical systems. A comparison of electron trajectories in Herrmann's optical model with those for a laminar flow model is shown in Figure 7-21. For laminar flow, as discussed earlier in this chapter, the electrons move side by side without ever crossing each other's paths. For imbalanced space charge and magnetic forces, the beam radius oscillates with a scallop wavelength, λ_s.

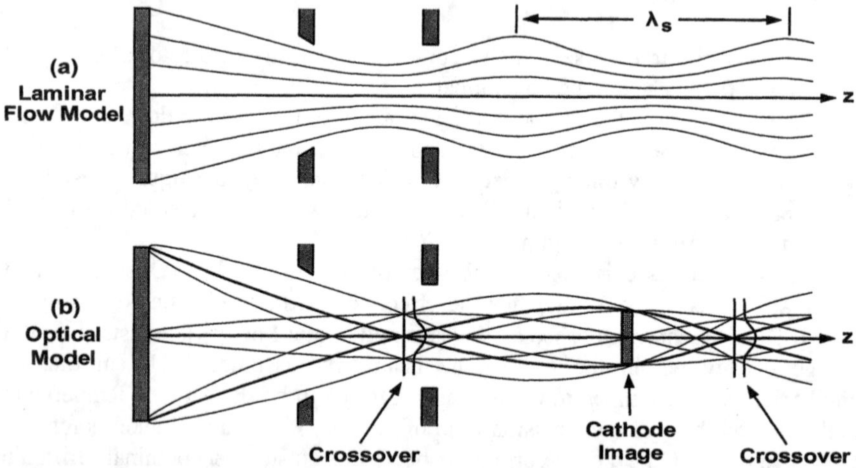

Figure 7-21 Comparison of (a) laminar and (b) optical models of electron flow. (Adapted from: Gabriel Herrmann, *Jour. App. Phys.*, February 1958.)

With the optical theory, electrons emitted from a point on the cathode focus periodically to form cathode images. As explained by Herrmann [11]: "One finds along the axis a succession of cathode images alternating with crossovers which impart to the beam a distinct structure which depends primarily on the internal motion of the electrons rather than on the shape of a boundary electron, as is the case in laminar theory."

Amboss compared the predictions of Herrmann with measurements on highly convergent, low-perveance (0.015×10^{-6}) beams focused by a flux density over an order of magnitude above the Brillouin value [12]. As shown in Figure 7-22, he found excellent agreement with the predictions of the optical theory. In this figure, λ_I is the distance between cathode images (the first and second images are shown). Note that the current profile changes from rectangular to Gaussian in the crossover region between images.

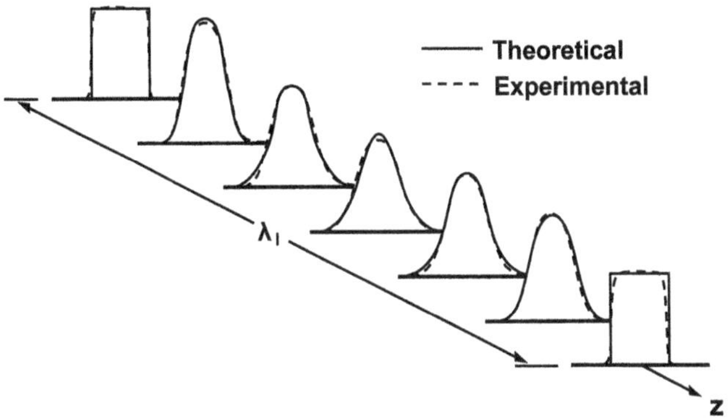

Figure 7-22 Comparison of optical theory predictions with experiment. (Adapted from: K. Amboss, *IEEE Trans. Electron Devices*, October 1964. © 1964 IEEE.)

The Amboss results show the validity of the optical theory under conditions far from normal beam focusing conditions. For fields closer to the Brillouin level, Brewer found that the behavior of the translaminar electrons, which were produced by gun aberrations, was similar to that predicted by optical theory. In Gilmour's beam studies, the only evidence of optical-theory behavior was when an image of the heater for the cathode appeared in the electron beam. Magnetic flux lines generated by the heater current perturbed the flow of electrons from the cathode and, apparently, led to the optical behavior of the subsequent electron trajectories

Other studies of nonlaminar beams include those of Kirstein [13, 14] and Hechtel [15]. Kirstein used essentially the same approach as Herrmann. Axially symmetric and two-dimensional strip beams were treated. Hechtel treated beams having a Gaussian current distribution in the radial direction at the exit from the gun. The reason for considering this type of beam is that, at millimeter-wave frequencies where beams are very small, thermal velocity effects in the gun result in a beam having a current distribution that is nearly Gaussian. Expressions are given for the magnetic field required to achieve various percentages of beam transmission through a tunnel.

7.4 FOCUSING WITH PERMANENT MAGNETS

7.4.1 Overview

A permanent magnet that produces a unidirectional field can be used in place of a solenoid to focus an electron beam. This eliminates the requirement for a power supply and ensures that the magnetic focusing field is always present. Disadvantages are size and weight. Also, in some applications, a disadvantage is that it is not possible to adjust the magnetic field to optimized tube performance.

As shown in Appendix C, the weight reduction of a periodic permanent magnet (PPM) focusing system, when compared with a solenoid or a unidirectional magnet, can be one to two orders of magnitude. A PPM focusing system is shown in Figure 7-23.

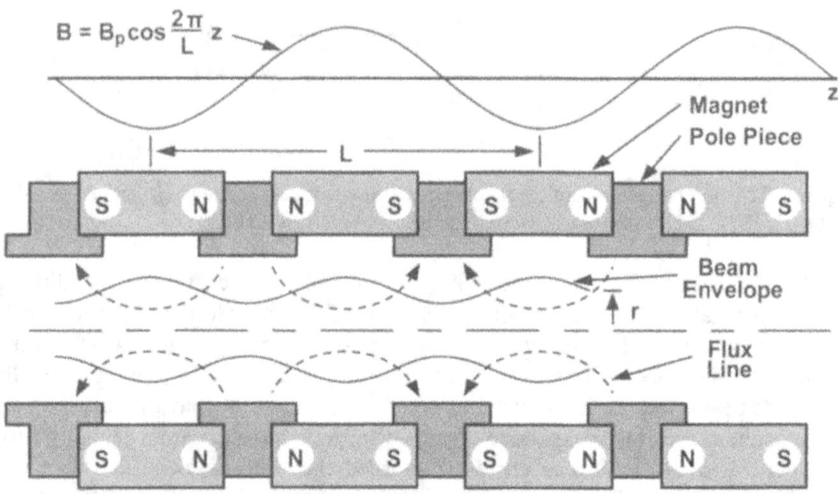

Figure 7-23 Periodic focusing of an electron beam with permanent magnets.

The basic principle of PPM focusing is straightforward. As a beam enters the field of a magnet section, the forces on the electrons act in exactly the same way as was described for Brillouin focusing. That is, the beam starts to rotate and the interaction of the rotational motion with the axial field produces a radial force that compresses the beam.

As the beam leaves the magnet section, rotation stops, focusing forces vanish, and the beam expands under the influence of space charge forces. The beam then enters another magnet section with the field in the opposite direction from the previous one. Then the beam rotates in the opposite direction, but is focused just as it was in the previous section.

The overall result, as the beam traverses the alternating field of the PPM structure, is that its rotation oscillates back and forth, producing alternating periods of magnetic focusing and beam expansion and beam ripple. The difference between beam ripple, which results from the periodicity of focusing, and beam scalloping, which is the oscillation of a beam that is not in equilibrium, is indicated in Figure 7-24.

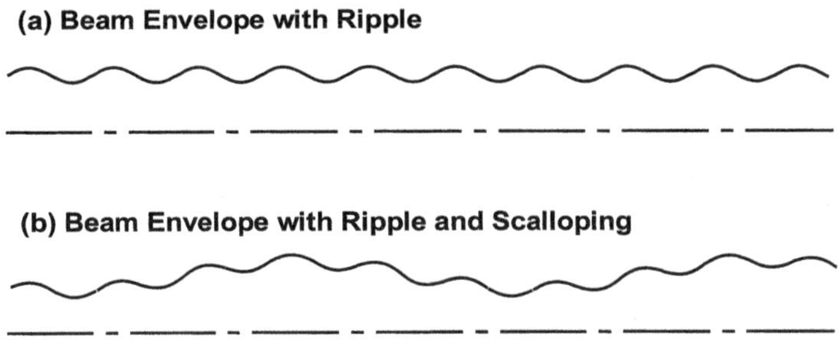

Figure 7-24 (a, b) Illustration of beam ripple and beam scalloping.

7.4.2 Laminar Flow, No Cathode Flux

While the qualitative explanation of PPM focusing is straightforward, the analysis is complex and not so straightforward. Still, much can be learned from an examination of the beam equation when PPM fields are used. In the analysis, it is assumed that the magnetic field is a function of axial position, z, only and that no radial variations occur. It can be shown that this assumption is realized along the axis when the magnetic field varies sinusoidally with distance. For the purposes of this analysis, it will be assumed that

$$B_z = B_p \cos \frac{2\pi z}{L} \tag{7-28}$$

where B_p is the peak field and L is the magnet period, or twice the distance between adjacent pole pieces as shown in Figure 7-23.

Next, assuming that no magnetic flux links the cathode, and assuming that the beam velocity is constant at u_o, the beam equation can be written as

$$\frac{d^2 b}{dz^2} + b\left(\frac{\eta}{2u_o}\right)^2 \left(B_p^2 \cos^2 \frac{2\pi z}{L} - B_B^2 \frac{a^2}{b^2}\right) = 0 \tag{7-29}$$

where it has been assumed that there is the equivalent of a Brillouin flux density, B_B, that produces an equilibrium radius, a, for the beam.

A relation between B_p and B_B can be found by equating the magnitudes of the restoring forces at the maximum and minimum excursions of the beam radius [7]. The result is

$$B_p = \sqrt{2}\, B_{rms} = \sqrt{2}\, B_B \tag{7-30}$$

That is, the rms value of the periodic field is the Brillouin field. As a result, the PPM focusing discussed here is similar to Brillouin focusing in that the space charge forces are balanced by the magnetic forces and no field is present at the cathode.

Next, the stability of PPM focusing will be examined. The stability question is important because there are ranges of values of flux density, magnet period, L, beam current, and beam voltage for which the beam diverges so that no beam current is transmitted through the PPM structure. Following the works of Mendel et al. [16] and Pierce [17], with the substitutions

$$T = \frac{2\pi z}{L} \tag{7-31}$$

and

$$\omega = \frac{2\pi u_o}{L} \tag{7-32}$$

which is the frequency at which the field appears to the moving electrons to alternate, the beam equation can be rewritten as

$$\frac{d^2 b}{dT^2} + b\left(\frac{\omega_{Lp}}{\omega}\right)^2 \cos^2 T - \frac{\omega_p^2 a^2}{2\omega^2 b} = 0 \tag{7-33}$$

where ω_{Lp} is the peak value of the Larmor frequency, $\eta B_p/2$.

Letting $b/a = \sigma$ be the normalized beam radius, α be the magnetic field coefficient,

$$\alpha = \frac{1}{2}\left(\frac{\omega_{Lp}}{\omega}\right)^2 \tag{7-34}$$

and β be the space charge coefficient

$$\beta = \frac{1}{2}\left(\frac{\omega_p}{\omega}\right)^2 \tag{7-35}$$

the normalized beam radius becomes

$$\frac{d^2\sigma}{dT^2} + \alpha(1+\cos 2T)\,\sigma - \frac{\beta}{\sigma} = 0 \tag{7-36}$$

Solutions to this equation were originally obtained using an analog computer [16]. It was assumed that the electrons were injected into the magnetic field with no radial velocity at $z = 0$, which was where the magnetic field was a maximum. Typical beam envelope curves for three values of the magnetic field are shown in Figure 7-25. Both scalloping and ripple are present in Figure 7-25(a, c) when the magnetic field is not adjusted for optimum focusing.

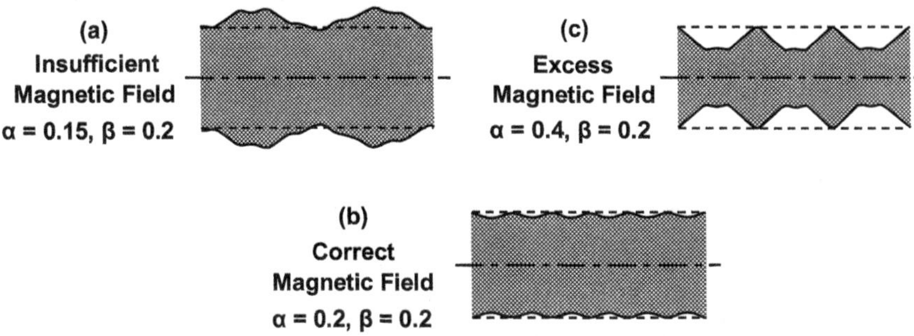

Figure 7-25 (a–c) Analog computer plots of beam contours. (From: J. T. Mendel et al., *Proc. IRE*, May 1954. © 1954 IRE (now IEEE).)

Curves similar to those in Figure 7-25 were obtained by Mendel et al. for a wide range of values of β and from these curves the universal plot for optimum focusing given in Figure 7-26 was constructed. Note that optimum focusing occurs when

$$\alpha = \beta \tag{7-37}$$

or when

$$\omega_{Lp} = \omega_p \tag{7-38}$$

When substitutions are made for rms flux density (B_B), charge density, and velocity, this relation is seen to be that for the equilibrium radius for Brillouin flow. This is not surprising since, on average, the space charge forces are the same as for the Brillouin case and the rms field is the Brillouin field so the centrifugal and focusing forces are the same as for the Brillouin case.

The minimum ripple for optimum focusing is also shown in Figure 7-26. For values of β below about 0.35, the ripple amplitude is reasonably small. As β increases above 0.35, the ripple amplitude becomes very large and, in fact, for $\beta >$ 0.66, the beam diverges.

We can learn more about the onset of the unstable region by further examination of the equation for the normalized beam radius. The term containing

β is the space charge term and represents a diverging force. If this term is neglected and the remainder of the equation diverges, then the solutions including space charge forces must diverge.

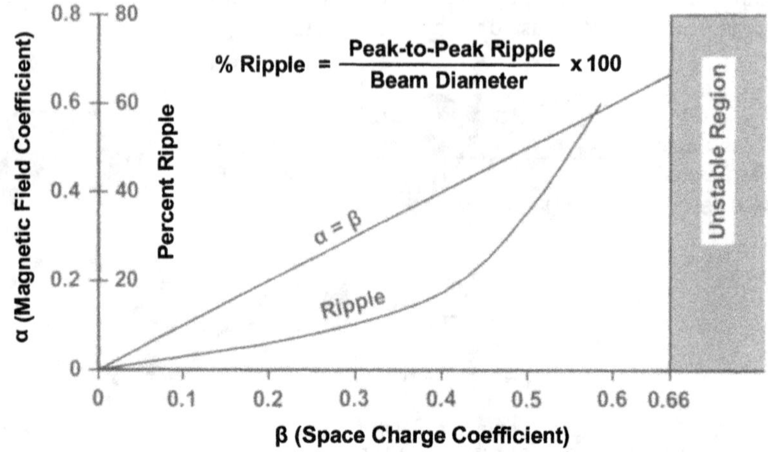

Figure 7-26 Optimum focusing conditions and minimum ripple amplitude. (Adapted from: J. T. Mendel et al., *Proc. IRE*, May 1954. © 1954 IRE (now IEEE).)

The equation for the normalized beam radius is a form of the Mathieu equation, which is characterized by solutions for α that are periodic in T for some ranges of α and unstable for other ranges of α. Figure 7-27 shows the ranges of α for which the solutions are unstable. Note that, as α increases from zero, the first unstable region starts at

$$\alpha = 0.66 \tag{7-39}$$

For $\alpha = \beta$, the unstable region for β would be expected for $\beta > 0.66$ as shown in Figure 7-27.

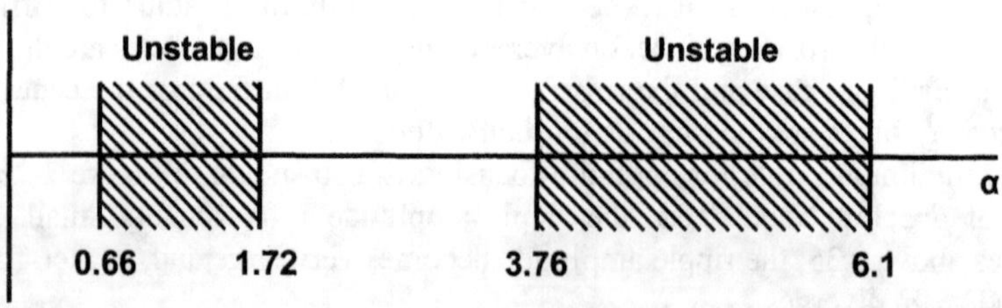

Figure 7-27 Regions of α for which (7-36) is unstable.

Using the equations for the space charge coefficient, the magnetic field coefficient, and the fact that the unstable region starts at $\alpha = 0.66$, the relations between ω_p, ω_{Lp}, and ω at which the unstable region starts are

$$\omega_p = 1.15\ \omega \tag{7-40}$$

and

$$\omega_{Lp} = 1.15\ \omega \tag{7-41}$$

The reason for the first of these relations is straightforward. If the periodic impulses provided by the focusing field at frequency ω are near synchronism with the natural frequency of oscillation of the electrons, ω_p, the beam diverges. The reason for the second relation can be understood by remembering that PPM magnets form a lens system that periodically focuses the electron beam as is indicated in Figure 7-28. The strength of each lens is proportional to B^2 and, therefore, to ω_{Lp}^2 and the focal length is inversely proportional to ω_{Lp}^2. Thus, if the lenses are too strong or are too far apart as indicated qualitatively by analogy to optical lenses in Figure 7-29, the beam diverges. In Figure 7-29(a), the lens strength and separation are such that the focal length, *f*, is one half of the lens separation. The ray, which is analogous to an electron trajectory, remains focused throughout the lens system. In Figure 7-29(b), the focal length is much shorter than in Figure 7-29(a) but the ray remains focused. In Figure 7-29(c), the focal length has been reduced slightly compared with Figure 7-29(b) and the ray diverges as it travels through the lens system.

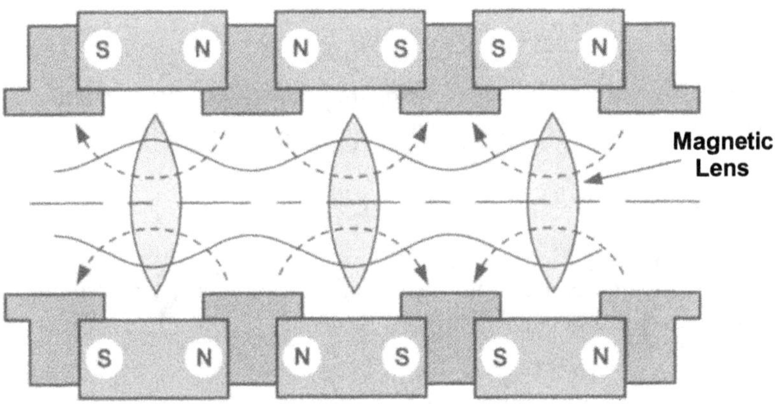

Figure 7-28 PPM focusing indicated by a series of convergent lenses.

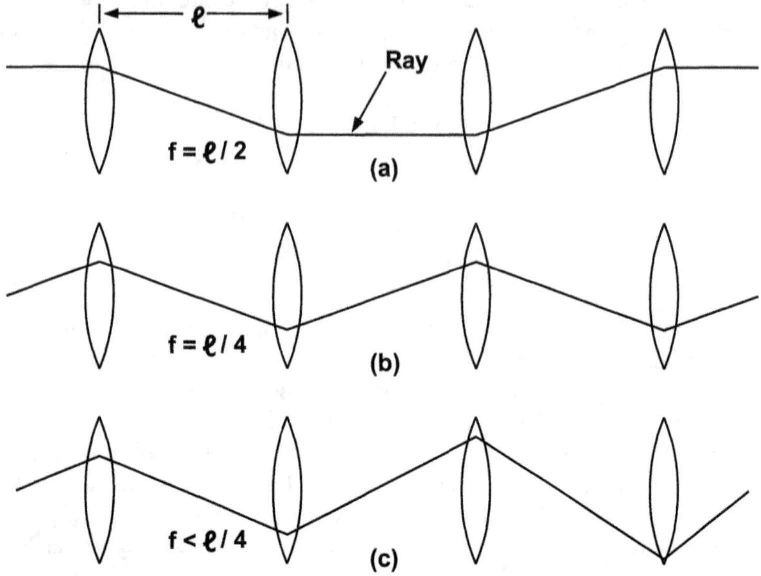

Figure 7-29 (a–c) Examples of optical rays (analogous to electron trajectories) in a series of lenses. Focusing instability occurs for $f < \ell / 4$.

In practice, PPM structures are designed to operate well below the unstable region. As a result, when the parameters contained in ω_{Lp} and ω of the magnetic field coefficient are used, it is found that

$$\frac{L^2 B_p^2}{V_o} < 2.37 \times 10^{-9} \tag{7-42}$$

where V_o is the dc beam voltage. Also, from the space charge coefficient the following, which is known as the *beam stiffness factor,* can be found

$$\frac{\lambda_p}{L} > 0.87 \tag{7-43}$$

When V_o is adjusted so that beam transmission stops, the resulting beam voltage is known as the *stopband voltage*, V_s, and is

$$V_s = 0.92\, L^2 B_{rms}^2 \tag{7-44}$$

where V_s is in kV, L is in cm, and B_{rms} is in kGauss.

True [18] reported the use of the stopband voltage as a tool for determining the magnetic field and as a diagnostic tool. An example of the determination of the stopband voltage is shown in Figure 7-30. Beam transmission is plotted as a function of beam voltage with perveance held constant. True reported that the magnetic field strength measured using a gaussmeter was in excellent agreement with the field calculated from the stopband voltage. Uses reported by True for the measurement of stopband voltage are the following:

- Optics development (hot tube magnetic field, field modification due to shunting, and magnetic entrance conditions);
- Manufacturing (stack correctness, backwards magnets, and so forth);
- Automated test (stopband voltage for record of magnetic field);
- Failure analysis (helpful in sorting out cause of tube failure);
- Environmental testing (magnetic field reduction with temperature);
- Remote sensing (electrical interrogation of tube in field).

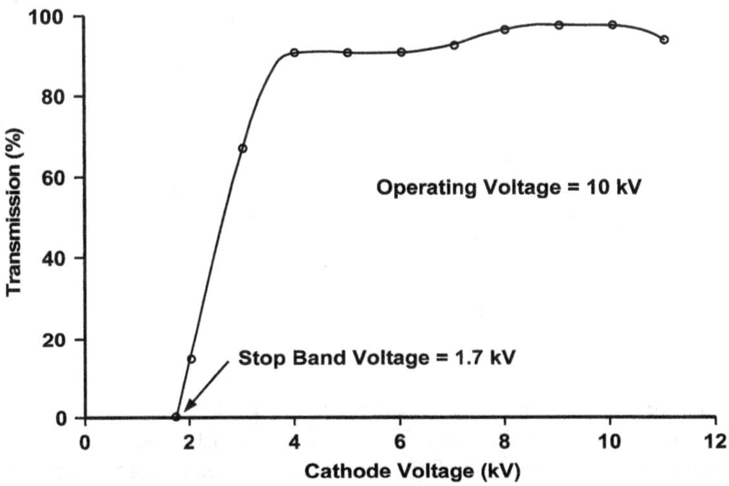

Figure 7-30 Beam transmission as a function of cathode voltage in a PPM-focused TWT with a shadow-gridded gun. (From: R. B. True, 1988 Microwave Power Tube Conference.)

7.4.3 Laminar Flow with Cathode Flux

In the previous section we examined PPM focusing of beams from electron guns with cathodes that are shielded from magnetic flux. It is also possible to achieve the equivalent of confined flow focusing with PPM fields and with flux through the cathode. Several researchers, including Chang [19], Mendel [20], Harker [21], and Harmon [22], have treated periodic focusing with cathode flux.

Confined flow PPM focusing achieves many of the advantages of confined flow focusing with uniform fields. For example, resistance to ion and RF defocusing is increased and the control of beams from gridded guns is improved. A deficiency in the PPM case that is not present in uniform-field focusing is that the rotational energy of the beam increases as the beam becomes more confined. As a result, the energy in the axial motion of the beam and the efficiency decrease.

The reason for the increase of rotational energy with the cathode magnetic field can be understood with the aid of Figure 7-31. Consider first the rotation of the beam within the first magnet. From Busch's theorem, we know that the rate of rotation is determined by the *change* in flux density encountered by the beam as it enters the first magnet. However, with cathode flux as shown in Figure 7-31 in the opposite direction to that in the first magnet, the *change* in flux density is larger than it is with no flux through the cathode. Thus, the rate of rotation of the beam in the first magnet is larger than it is with no flux through the cathode.

As an example of the effect of cathode flux, assume a cathode flux density of 25 gauss, a peak flux density in the first magnet of 1,000 gauss, and an area convergence ratio of 20:1. Then the peak rate of beam rotation in the first magnet is

$$\dot{\theta} \propto B_p + 20 \times B_c = 1{,}500 \text{ gauss} \tag{7-45}$$

As the beam moves from the first magnet to the second, the direction of the magnetic field reverses and is aligned with the cathode field. Then the peak rotation of the beam is

$$\dot{\theta} \propto 1{,}000 - 20 \times 25 = 500 \text{ gauss} \tag{7-46}$$

As the beam moves on to the third magnet, rotation again reverses and the peak value of the rate of rotation is again proportional to 1,500 gauss.

The ratio of magnetic focusing forces from magnet to magnet is equal to the ratio of rotation rates, which is 3. This large ratio of focusing forces is offset by an equally large ratio of centrifugal forces (the large rate of rotation in magnet 1 produces a centrifugal force that is three times that in magnet 2). As long as the beam velocity remains constant, the ripple amplitude from magnet to magnet remains constant.

It should be observed that, because there is more energy in rotation in magnet 1 than in magnet 2, the axial velocity of the beam in magnet 1 is lower than in magnet 2. As a result, the beam density in magnet 1 is larger than in magnet 2 and this produces a fluctuation in the radius of the beam that is similar to beam scalloping.

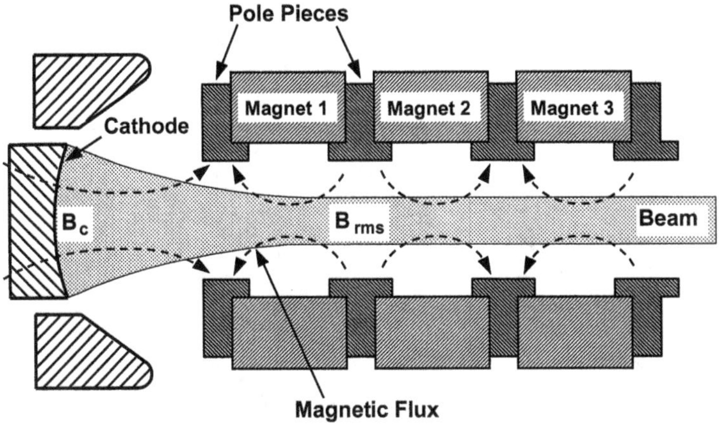

Figure 7-31 Magnetic field configuration for confined flow PPM focusing.

Figure 7-32 shows the variation of peak rotational beam energy with magnetic flux density and with microperveance (perveance $\times 10^6$) [23]. Note that, for reasonable values of confinement, a large fraction of the beam energy is in rotation for microperveances above ~0.5. As a result, efficiency is degraded. Thus, confined flow PPM focusing is limited to low-perveance designs. Hence, it is particularly suitable for millimeter-wave tubes for which perveance is usually limited by cathode loading considerations and in which beam interception problems are particularly critical.

A primary reason for using confined flow focusing is to limit the expansion of the beam during RF operation. In [23], Ayers et al. presented beam expansion factors (the ratio of the maximum beam radius with RF drive to the dc radius) for 0.45 and 0.036 microperveance TWTs. The results for 0.036 microperveance are shown in Figure 7-33. Notice that:

1. Beam expansion is dramatically decreased as the magnetic field is increased.
2. There is little change in beam expansion as λ_p/L is increased.
3. The ripple decreases rapidly as λ_p/L is increased. Ripple is defined as

$$\%\text{ripple} = \frac{r_{\max} - r_{\min}}{r_{\max} + r_{\min}} \times 100 \tag{7-47}$$

4. The maximum flux density and, to some extent, the energy product of the magnets limit operation at high magnetic fields and high values of λ_p/L.

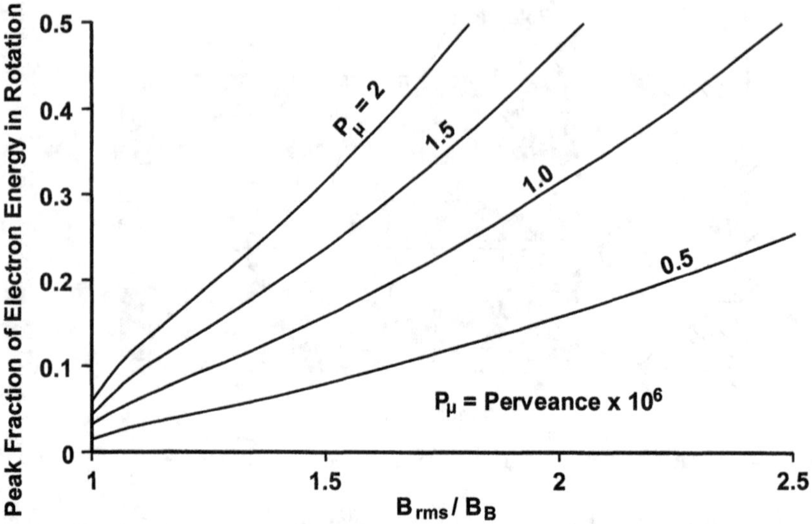

Figure 7-32 Dependence of peak transverse energy on the magnetic field for various microperveances. (From: W. R. Ayers et al., *Technical Digest*, 1985 IEDM. © 1985 IEEE.)

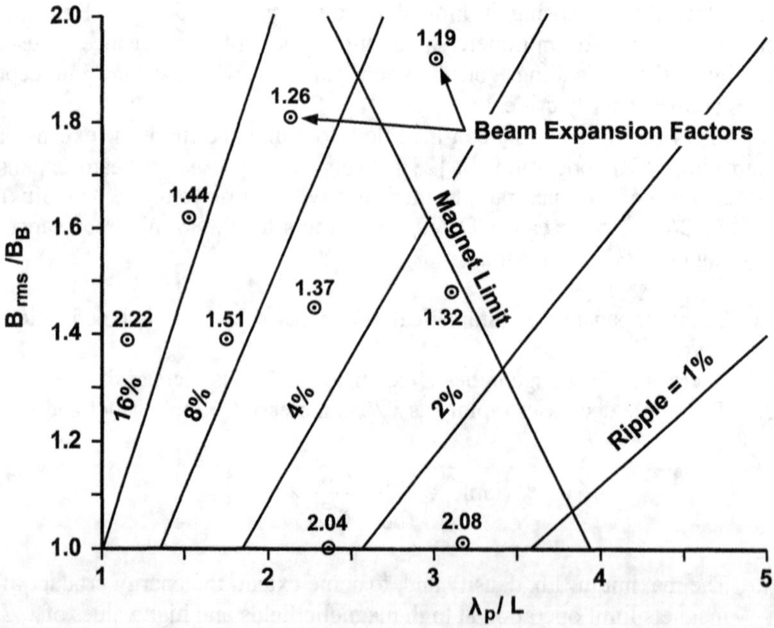

Figure 7-33 Beam expansion factors for 0.036 microperveance TWT. (From: W. R. Ayers et al., *Technical Digest*, 1985 IEDM. © 1985 IEEE.)

7.4.4 Nonlaminar Flow

In the chapters on cathodes and guns, many sources of beam nonlaminarities were discussed. The most important were the grids in electron guns. Most PPM-focused TWTs use gridded electron guns and the nonlaminarities produced by these guns are usually far more significant than other nonlaminarities. To gain an appreciation for the nonlaminar nature of the beam from a gridded gun, consider Figure 7-34. This figure contains computer-generated electron trajectories for a PPM-focused beam from a shadow gridded gun [24]. The radial scale has been expanded to 10 times the axial scale to help illustrate the apparently random nature of the electron trajectories.

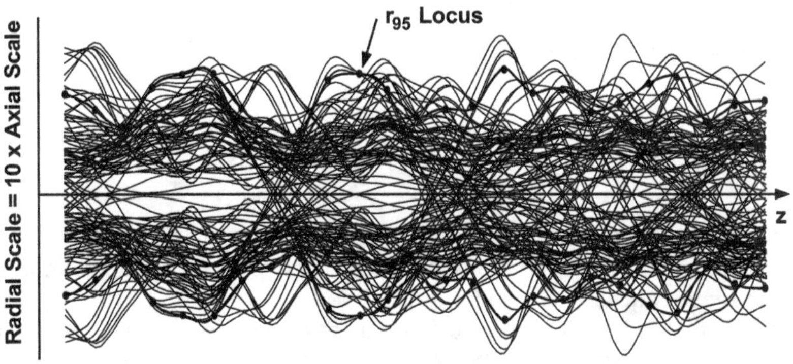

Figure 7-34 PPM-focused beam from shadow gridded gun. (Adapted from: Richard True, *IEEE Trans. Electron Devices*, February 1987. © 1987 IEEE.)

The most definitive procedure that has been reported for dealing with PPM-focused beams and gridded guns is that by True [24–26]. In this section, we will describe the basic elements of True's theory. There are basically three steps to True's procedure:

1. Most importantly, the transverse velocity content of the beam at its entrance into the magnetic focusing field is found. This is a measure of the transverse velocities of the electrons and is expressed in normalized form as

$$\sigma = \frac{\overline{v}_r}{u_o} \tag{7-48}$$

where u_o is the dc velocity of the beam.

2. The transverse velocities, along with space charge forces, must be controlled by the magnetic focusing structure. In effect, True treated the

transverse velocity content as an additional force (in addition to the space charge force) that must be focused. This is done by finding an effective microperveance, P_μ^+, for the beam (microperveance equals perveance multiplied by 10^6), which is the sum of the regular beam microperveance, P_μ, and a scattering microperveance, $P_{\mu s}$, that results from the transverse velocity content. That is,

$$P_\mu^+ = P_\mu + P_{\mu s} \tag{7-49}$$

3. In the absence of magnetic flux through the cathode, the field level required to focus an ideal beam is the Brillouin field, and this is proportional to the square root of perveance. True assumed that the increased focusing field, B^+, made necessary by transverse electron velocities should be at least

$$\frac{B^+}{B_B} = \left(\frac{P_\mu^+}{P_\mu}\right)^{1/2} \tag{7-50}$$

The ratio B^+/B_B is referred to as the magnetic field intensification ratio. True finds that the perveance ratio, P_μ^+/P_μ, must be multiplied by a factor of 1.65 to minimize the beam ripple in actual magnet designs.

The normalized transverse velocity content of the beam, σ, is composed of two parts. The first part, σ_T, results from the thermal velocity content of the beam, given by

$$\sigma_T = \frac{r_c}{r_{95}}\left(\frac{kT}{2eV_o}\right)^{1/2} \tag{7-51}$$

where r_c is the disc radius of the cathode and r_{95} is the beam radius containing 95% of the beam current. The other part of σ is the nonthermal part, σ_{NT}, which results from gun aberrations and scattering by grid wires. This part can be found by a computer simulation that includes grid wire details. True found that a good approximation for the combined thermal and nonthermal parts is

$$\sigma = \left(\sigma_{NT}^2 + \sigma_T^2\right)^{1/2} \tag{7-52}$$

For gridded guns, $\sigma_{NT} >> \sigma_T$ so $\sigma \approx \sigma_{NT}$. In the remainder of this section, we will treat only gridded guns and we will ignore thermal effects. We will also substitute σ for σ_{NT}.

The value for σ is found at or near the beam minimum from the weighted standard deviation of the slopes of the electron trajectories. Although σ is defined in terms of slopes, it is essentially equal to v_r/u_o. True found, as shown in Figure 7-35, that σ can be approximated by

$$\sigma = 0.0015\ \theta \tag{7-53}$$

where θ is the half angle of the electron gun. Note that guns which have small half angles also have low values of σ. The reason was given in Chapter 6 when an electron crossing a convergent electron stream was considered. The transverse velocity increase was found to be inversely proportional to the radius compression ratio.

The next step in the analysis is to incorporate the effect of transverse velocities on the spreading of the electron beam. As pointed out earlier in this section, True treated the transverse velocity content as a force that, along with the space charge force, must be focused by the magnetic field. This is done by finding an effective microperveance, which is $P_\mu^+ = P_\mu + P_{\mu s}$. The scattering microperveance, $P_{\mu s}$, is found to be related to σ by

$$P_{\mu s} = 197.6\,\sigma^2 \tag{7-54}$$

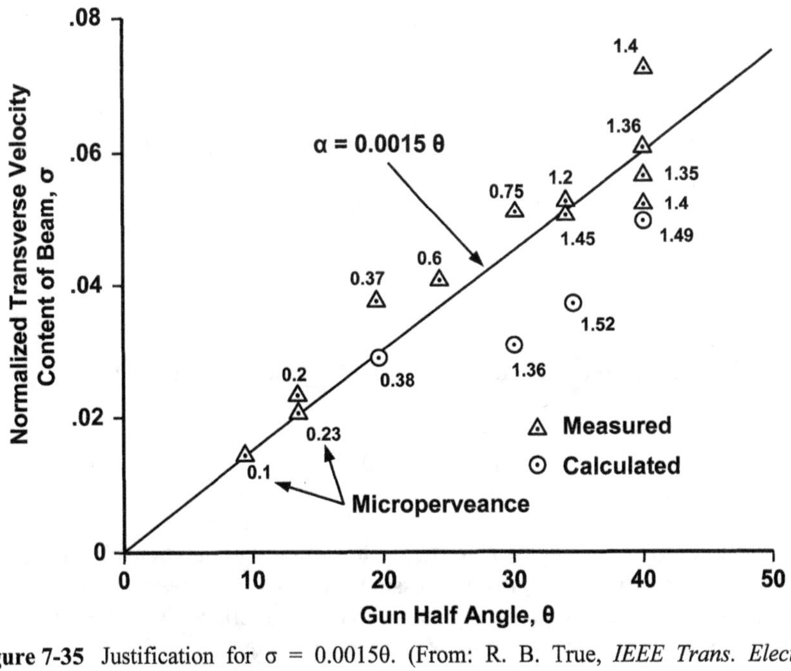

Figure 7-35 Justification for σ = 0.0015θ. (From: R. B. True, *IEEE Trans. Electron Devices*, March 1984. © 1984 IEEE.)

This relation is derived by considering electron behavior and the rate of beam spread near the beam minimum. Thus, the effective microperveance is

$$P_\mu^+ = P_\mu + 197.6\,\sigma^2 \tag{7-55}$$

The effective microperveance is next used to find the magnetic field intensification factor, B^+/B_B. The magnetic field is proportional to the square root of perveance. Also, allowance must be made for effects such as spherical aberration, nonideal entrance conditions and gun-design imperfections, so True assumed that the intensification factor should take the form

$$\frac{B^+}{B_B} = \left(F\,\frac{P_\mu^+}{P_\mu} \right)^{1/2} \tag{7-56}$$

where a value for F of 1.65 is chosen on the basis of numerical and experimental data. Now, substituting the relation for effective microperveance, the intensification factor becomes

$$\frac{B^+}{B_B} = \left[F \left(1 + \frac{197.6\,\sigma^2}{P_\mu} \right) \right]^{1/2} \tag{7-57}$$

To make use of this equation, the relation between σ and P_μ is needed. Equation (7-53) was an empirical relation between σ and the gun half angle, θ. The exact relation between perveance and θ was discussed in Chapter 6. For our purposes here, a much simpler equation developed by True can be used. That is,

$$P_\mu = 0.006\,\theta^2\,\frac{r_{95}}{r_c} \tag{7-58}$$

By substituting this equation and that for σ into the intensification factor,

$$\frac{B^+}{B_B} = \left[1.65 \left(1 + 0.0741\,\frac{r_c}{r_{95}} \right) \right]^{1/2} \tag{7-59}$$

This result is plotted in Figure 7-36. It is interesting that the intensification factor is a function of only the radius compression ratio. We now see why beams from high convergence guns require field levels that are significantly higher above the Brillouin level than do beams from low convergence guns.

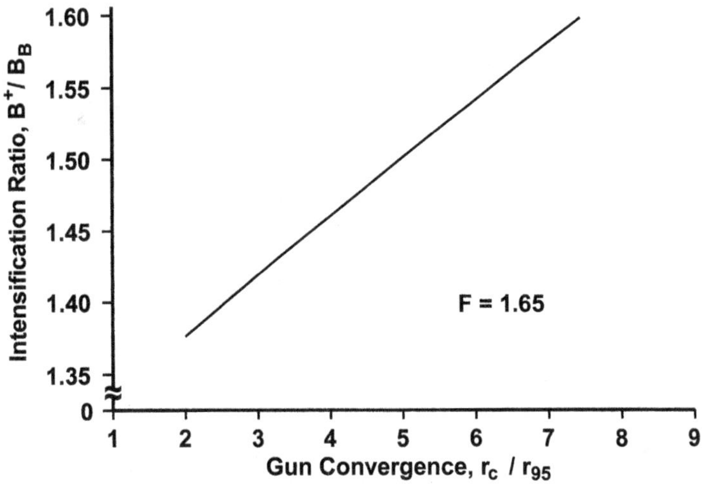

Figure 7-36 PPM field intensification ratio versus gun convergence. (From: R. B. True, *IEEE Trans. Electron Devices*, March 1984. © 1984 IEEE.)

The magnetic field intensification factor may be combined with r_{95}/a, where a is the tunnel radius (so that r_{95}/a is the fraction of the tunnel filled by the beam) to obtain a normalized focusing factor, denoted by True as FOC F_1 and given by

$$\text{FOC } F_1 = \frac{B^+ / B_B}{r_{95} / a} \tag{7-60}$$

Acceptable limits for this parameter for good tube design are shown in Figure 7-37.

In the investigation of the stability of PPM-focused beams, we noted that the beam stiffness factor, λ_p/L, should be greater than 0.87, but we gave no indication of how much greater it should be. True stated that for pulse tubes, λ_p/L should exceed 2.7 and for CW tubes, it should be greater than 3.2. These values are large enough so that the effects of the individual lenses of the PPM field blend together so that the beam approximates a beam focused by a uniform magnetic field.

Before leaving the subject of nonlaminar PPM-focused beams, it is important to point out that factors other than RF drive level may cause a beam to increase in diameter as it drifts through a tube. This, in turn, would lead to higher interception than is expected in the absence of increased beam diameter. These factors may also affect the rate at which a beam expands as it enters the collector and this, in turn, affects the collector design. To deal with these phenomena, True [26, 27] defined the term *tunnel emittance*, which is given by

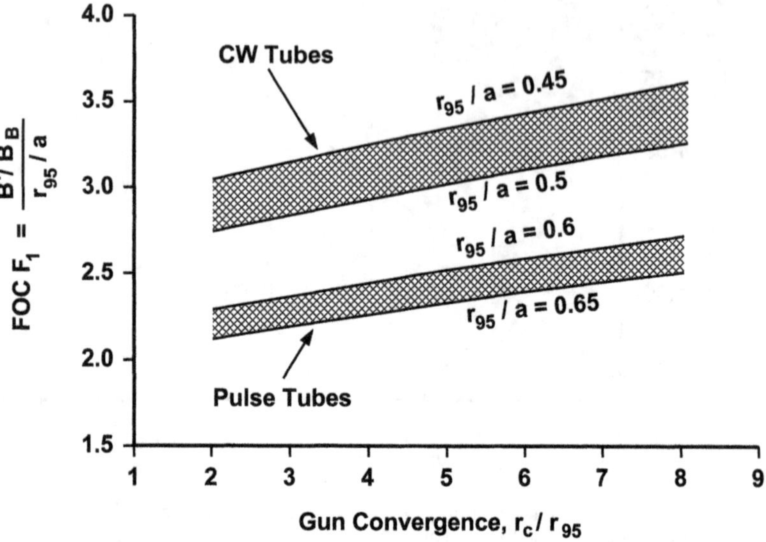

Figure 7-37 Relation of field intensification and beam-filling factors to gun convergence. (From: R. B. True, *IEEE Trans. Electron Devices*, March 1984. © 1984 IEEE.)

$$\text{Tunnel emittance} = \sigma\,\frac{r_{95}}{a} \tag{7-61}$$

Figure 7-38 shows the growth of tunnel emittance with distance. This growth is due partly to an increase in normalized transverse velocity (σ) and partly to a small increase in beam size.

One important factor that affects tunnel emittance growth is the current profile of the beam leaving the electron gun. The ideal beam shape is shown in Figure 7-39(a). This profile is shown slightly dished to match the radial variation in the magnetic field. The profiles shown in Figure 7-39(b, c) are less desirable shapes because of the concentration of charge in the center of the Gaussian distribution or in the ears of the "rabbit ears" of the hollow distribution. These charge concentrations contain potential energy that is converted to increased transverse electron velocities as the beam flows through the tube.

Another factor that affects tunnel emittance growth is the entrance point of the beam into the magnetic field. True considers the beam minimum position (the beam waist) to be the point where tunnel emittance is at a minimum. The axial position of the waist should be at the point where the first magnetic field peak has risen to $1/\sqrt{2}$ of its maximum value.

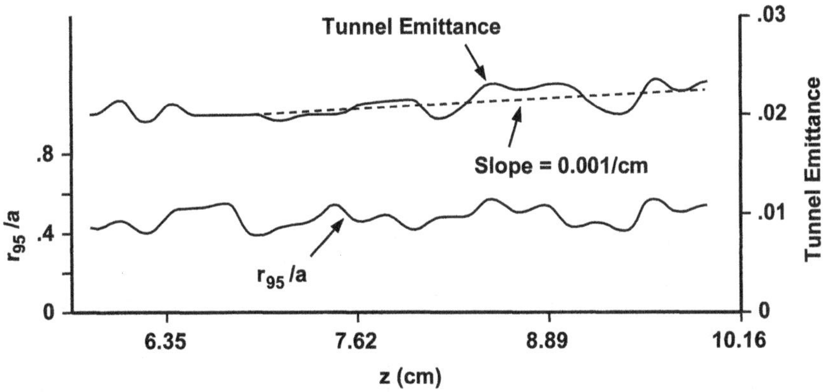

Figure 7-38 Illustration of growth of tunnel emittance with distance. (From: R. B. True, *Technical Digest*, 1985 IEDM. © 1985 IEEE.)

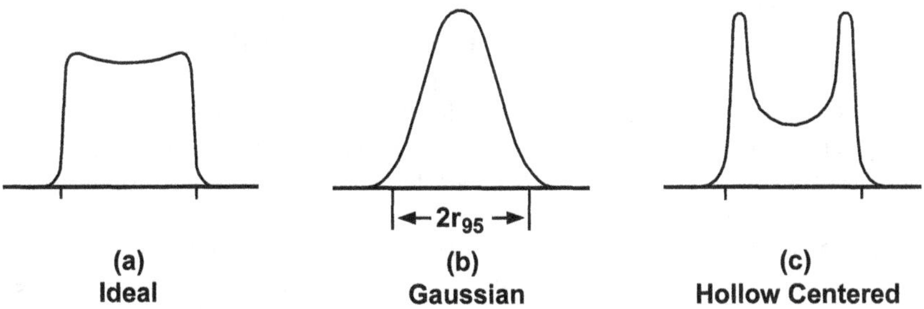

Figure 7-39 (a–c) Current density versus radius at beam minimum. (From: R. B. True, *IEEE Trans. Electron Devices*, February 1987. © 1987 IEEE.)

7.5 ION EFFECTS IN ELECTRON BEAMS

Instabilities caused by positive ions have been observed in microwave tubes since their invention. The source of the ions is gas, which is ionized by the electron beam. Linear-beam tubes are far more susceptible to ion effects than crossed-field tubes because potential wells (depressions), which trap ions, can exist. Ion density increases with time so that ion effects are most severe during long-pulse or CW operation. The use of confined flow focusing is effective in reducing or, in some cases, eliminating instabilities resulting from ions.

7.5.1 Examples of Ion Effects

Ions may affect tube performance in at least the following ways:

1. Positive ions neutralize electron space charge potentials within a beam. Beam diameter becomes smaller (ion focusing is said to occur), and so power output may be affected. Also, interception current decreases with time. In extreme cases, the electron beam remains focused in the collector and melts a hole through the collector wall which, of course, destroys the tube.
2. Because the potential depression in an electron beam is reduced by ions, the beam velocity is increased. This reduces the phase length of a tube. The change in beam potential can be on the order of 0.01% to 1%. Since beam velocity is proportional to the square root of potential, the fractional change in velocity is one half of the change in potential. Phase change is one half of velocity change. Thus, for a tube phase length of 10,000°, the resulting decrease in phase is about 0.25° to 25° and varies with tube operating conditions. Under some conditions, the ion density in the beam may vary at a frequency of hertz to kilohertz and produce periodic changes in power and phase.
3. Gas released from a cathode is ionized in an electron gun and can affect electron focusing. If this gas release varies with time (there are reports of microbursts of gas from a cathode), the RF power and phase of a tube can be affected.
4. Ions can produce high-frequency (megahertz to tens of megahertz) oscillations or noiselike fluctuations in the electron beam of a tube. These fluctuations result in the modulation of the power output of the tube (*ion sidebands* are said to occur). The frequency of the fluctuations is dependent on electron beam density and ion mass.

Figures 7-40 through 7-43 show effects of ions in an electron beam. In Figure 7-40, an electron beam cross section is shown with a significant peak in electron density in the center of the beam. It is thought that the peak resulted from positive ions in the center of the beam near the electron gun being accelerated toward the cathode. Near the center of the cathode, these ions, along with sputtered cathode material that becomes ionized, cause sufficient neutralization of the electron space charge cloud so that additional electrons leave the cathode. These results were obtained at a pressure of about 10^{-6} torr by the author in the experiments described in [27].

At low pressures ($< 1\times10^{-8}$ torr) and long-pulse or CW operation, a low-frequency (tens of hertz to kilohertz) instability is observed in some linear-beam tubes. An example is shown in Figure 7-41 [28]. These results were obtained with a Brillouin-focused klystron designed for CW operation. The tube was equipped with a shadow-gridded electron gun. The modulator provided pulse lengths from 1 μs to 100 ms. The tube was not instrumented to permit an investigation of beam behavior, so the details of tube behavior are not known.

At the beginning of each pulse, after an initial turn-on transient, the tube operated stably for about 15 ms. Then the tube exhibited an ion induced instability with a period of about 5 ms. Transients in the output power were the evidence of the instability. If the beam was pulsed off for a period as short as 1 μs, the instability would not recur until about 15 ms after the beam had been pulsed on. Apparently, the ions dissipated during the 1-μs off period. An important conclusion of McCune's work is that linear-beam amplifiers can be pulsed off periodically for a very short time and be inherently free of ion oscillation problems, assuming that the tube has been adequately evacuated.

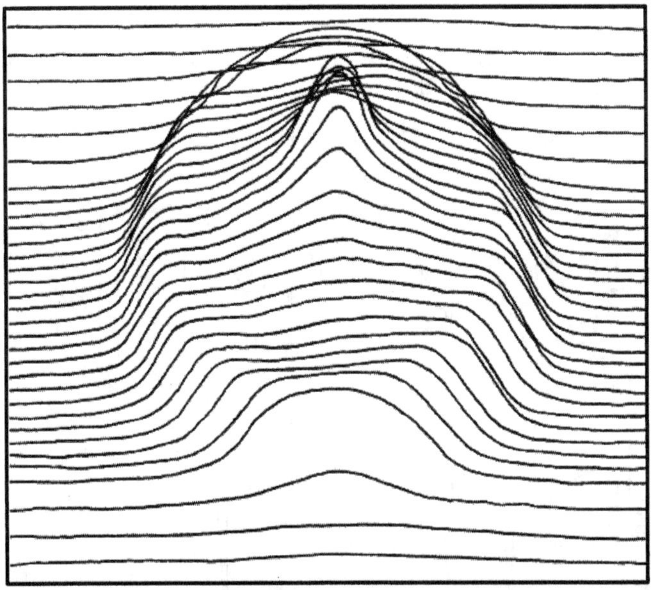

Figure 7-40 Electron beam cross section showing the effect of ions.

Additional examples of low-frequency instabilities caused by ions are shown in Figure 7-42 [29]. In this case, variations with time in the phase length of CW, confined flow, coupled-cavity traveling wave tubes are shown. Even though these phase fluctuations are small (a fraction of a degree for a TWT with a total phase length of nearly 10,000°), they result in phase noise that interferes with system operation.

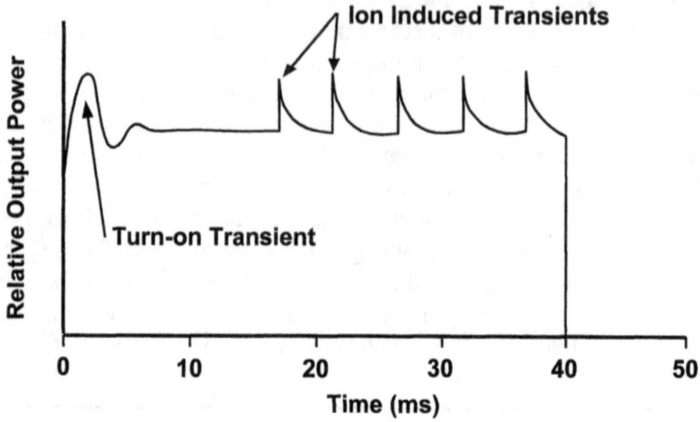

Figure 7-41 RF output with 40-ms beam pulse width. (From: E. W. McCune, *Technical Digest*, 1983 IEDM. © 1983 IEEE.)

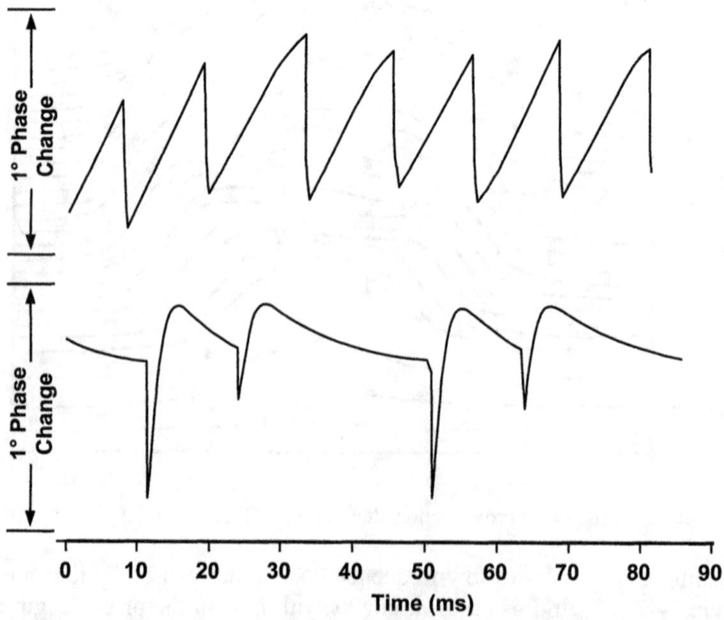

Figure 7-42 Phase fluctuations induced by ions in CW, confined-flow, coupled-cavity TWTs. (From: A. S. Gilmour, Jr., et al., *IEEE Trans. Plasma Science*, June 2004. © 2004 IEEE.)

When the ion concentration becomes sufficient, oscillations may occur as indicated by Figure 7-43(a). This figure shows the amplitude of the current to a probe in a pulsed Brillouin-focused beam in the presence of hydrogen at a pressure of about 10^{-6} torr. When ion effects were absent, the probe current was as shown in Figure 7-43(b).

At the beginning of the pulse with gas present, several cycles of a relatively high frequency of oscillation occurred (~8 MHz). After about 2 μs, a lower frequency of ~4.5 MHz occurred. Then the higher frequency returned. Finally, at the end of the pulse, the probe current become very noisy. These oscillations were attributed to the radial motion of ions in the electron beam.

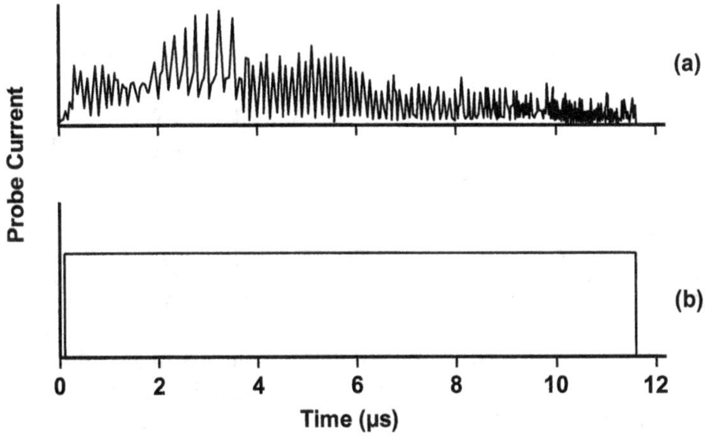

Figure 7-43 Probe current in electron beam analyzer (a) with and (b) without ion oscillations. (From: A. S. Gilmour, Jr., RADC-TDR-63-465, December 1963.)

To understand why ions produce instabilities, it is necessary to understand ion generation and ion behavior. In addition, the gas pressure in the gun that results from evaporation of cathode material, ion sputtering of the cathode and the neutralizing of ions that enter the gun must be understood. In the examples considered in this chapter, it is not thought that electrons released from the ions have an effect on instabilities because they very quickly escape from the beam. Ion, electron, and gas behaviors (because of its dependence on ion motion) all depend on the potential distribution in the electron beam, so we start with the analysis of the potential depression in a beam without ions. Then the effects of ionization are introduced. It is shown that, if the potential distribution is such that ion traps exist, then low-frequency and high-frequency instabilities may occur. Some of these instabilities and possible reasons for them are discussed in detail. Frequent reference is made to 20-kV, 3-A coupled-cavity TWTs for which detailed measurements were made and operating conditions were carefully documented [29].

7.5.2 Gas Sources

Before discussing ions and their effects, it is important to understand the source of the ion, which is, of course, the gas in the tube and its interaction with the electron beam. Even though great care is taken in materials selection, construction, and processing for modern microwave tubes (see Appendix B), the gas pressure in these tubes (except for the electron gun) is usually in the 10^{-8} to 10^{-9} torr range. At these pressures, there are ~10^7 to 10^8 gas molecules/cm^3, so there is no shortage of gas from which ions can be generated.

There are many sources of this gas. Sometimes, as reported in [29], most of the structure of the tube is copper, which has been brazed in hydrogen, so it is thought that hydrogen is the primary gas species. Because argon is used as the shield gas in welding processes, argon is sometimes the primary gas species. Even though cleanliness is of utmost importance in the assembly of a tube, carbon-based particles may remain after assembly and processing, so there may be a carbon-based gas in the tube. A virtual leak, which results in the gradual release of gas from any one of various sources, can result in the primary gas species being any one of many gases. Still, the ultimate pressure in a tube (except for the electron gun) is usually in the 10^{-8} to 10^{-9} torr range.

The gas pressure in the gun can be about two orders of magnitude above that in the remainder of the tube and depends on three phenomena.

1. The pressure depends on the constituents of the cathode and the temperature at which the cathode is operating. For the dispenser cathode used in virtually all linear-beam tubes, it is the barium on the surface of the cathode that is evaporated at the cathode temperature of ~1,000°C that is thought to be the primary gas constituent in the gun. The pressure of that gas is estimated to be on the order of 10^{-6} torr at 1,025°C (estimated from the pressure of 7.5×10^{-6} torr given in [30] for 1,100°C).
2. Ions formed by the interaction of the electron beam with gas tend to congregate along the center of the beam. If the potential along the axis of the beam is such that the ions can flow into the gun, they form a beam that impinges on the center of the cathode (see Figure 7-44). At the cathode, they recombine with electrons and produce an increase in gas pressure in the gun. As the pressure increases, the gas flows out as shown in Figure 7-45 through the anode to the remainder of the tube. The gas is then reionized and reinjected into the gun as an ion beam. The gas flow from the gun is restricted by the sizes of the anode tunnel and the apertures in the tunnel leading to the interaction region. The increase in pressure in the gun can be estimated from the gas flow rate (which is assumed to be the same as the rate of flow of ions into the gun) and from the conductances of the tunnel and the apertures. In the study reported in [29], hydrogen at a pressure of about 10^{-8} torr was the predominant gas

and the increase in pressure in the gun was 10^{-8} torr or less. For gases heavier than hydrogen, the pressure in the gun would be much higher.

3. A very important result of ion impingement on the cathode is the sputtering of the cathode surface and the pressure increase that results from the sputtered material. This pressure increase depends on ion species, gun voltage, and sputtering coefficient of the cathode surface and is thought to be significant.

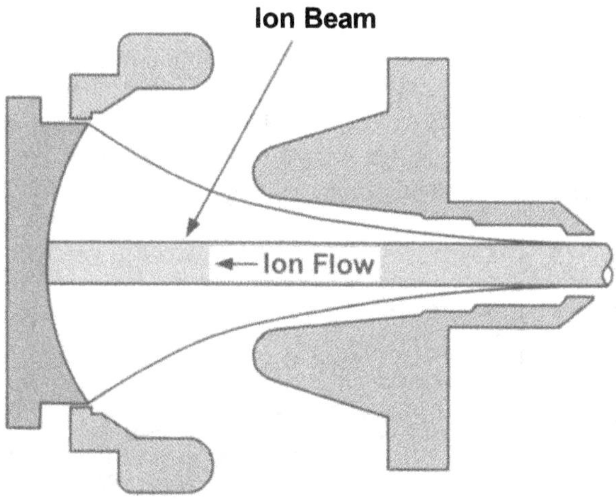

Figure 7-44 Ion flow into an electron gun.

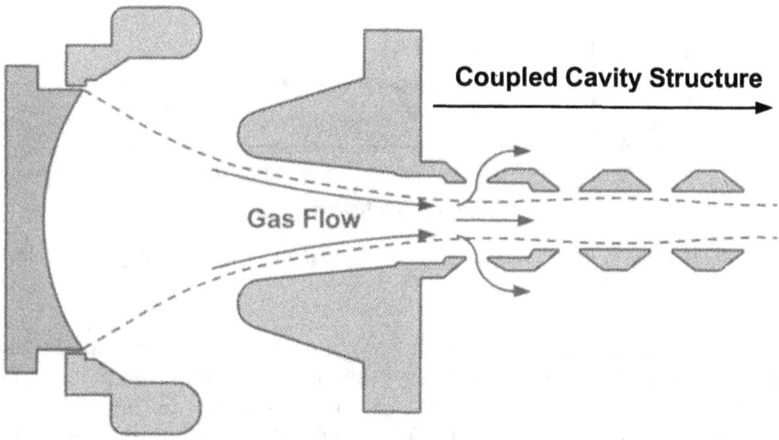

Figure 7-45 Gas flow from the gun resulting from ion flow into the gun.

7.5.3 Ionization

The rate at which ions are formed in an electron beam depends on the gas density (and thus the pressure, P), the ionization cross section of the gas molecules, σ_i, the electron beam current, I, and the length of the beam, d, and is given by

$$\dot{q} = 2.19\times10^{35}\, I\, \sigma_i\, P\, d \text{ ions/s} \tag{7-62}$$

where I is in amperes, σ_i is in cm^2, P is in torr, and d is in cm. Figure 7-46 contains plots of the ionization cross section for some common gases as a function of primary electron energy [31]. Values of σ_i for H, N, and O for the high-energy collisions shown in Figure 7-46 were not available in [32], so these were estimated by scaling from lower energies (0 to 4,000 eV) in the same way that known ionization cross sections scale with primary energy.

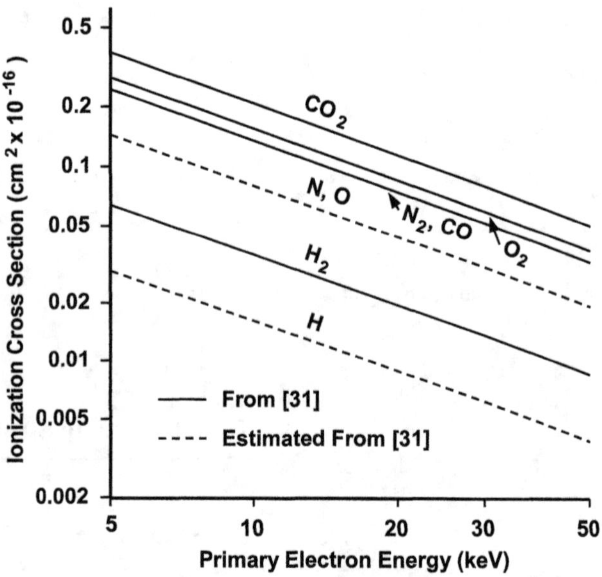

Figure 7-46 Ionization cross sections for some common gases at high primary electron energies.

As pointed out previously, the primary gas constituent in the electron gun is thought to be barium. The ionization cross section for barium is not given in [31]; however, the cross section for indium is given and is shown in Figure 7-47. The atomic weight of barium is about 20% greater than that of indium and the ionization potential of barium is about 10% less than that of indium so the ionization cross section of barium should be slightly larger than that of indium.

Because of this large cross section and the high barium pressure in the gun, the possibility that electron focusing in the gun is affected cannot be overlooked.

From the equation for the rate at which ions are formed, the following more useful equation can be derived and gives the rate at which the ion density increases, $\dot{\rho}_i$, relative to the electron density in the beam, ρ_e.

$$\frac{\dot{\rho}_i}{\rho_e} = 6.58 \times 10^{22}\, \sigma_i\, P \sqrt{V} \text{ ions/ electron/ millisecond} \tag{7-63}$$

where V is in kilovolts. This relation is plotted in Figure 7-48 for a pressure of 1×10^{-8} torr. To find the ion generation rate at a pressure other than 1×10^{-8} torr, simply multiply the generation rate found from Figure 7-48 by the factor by which the pressure differs from 1×10^{-8} torr. Notice that, if the gas in the beam is nitrogen or oxygen and the beam voltage is 10 kV, enough ions will be generated to completely neutralize the beam in about 50 ms. Of course, for complete neutralization to take place, the ions must be trapped in the beam.

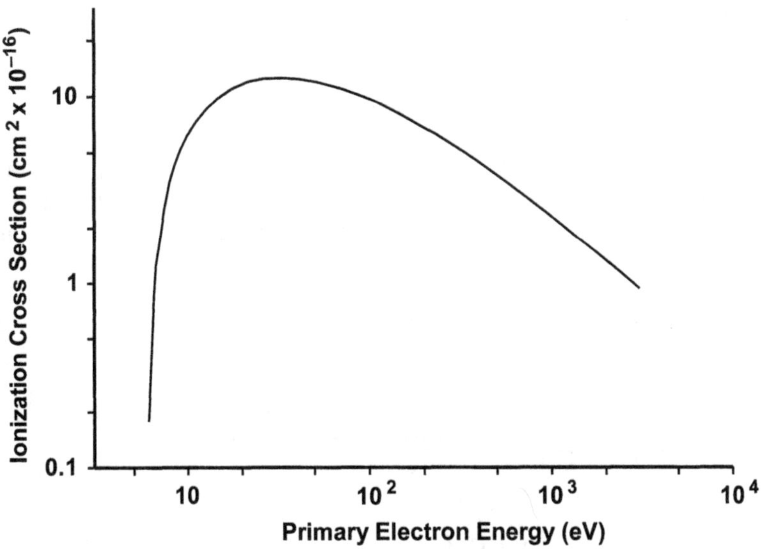

Figure 7-47 Ionization cross section for indium (thought to be similar to barium).

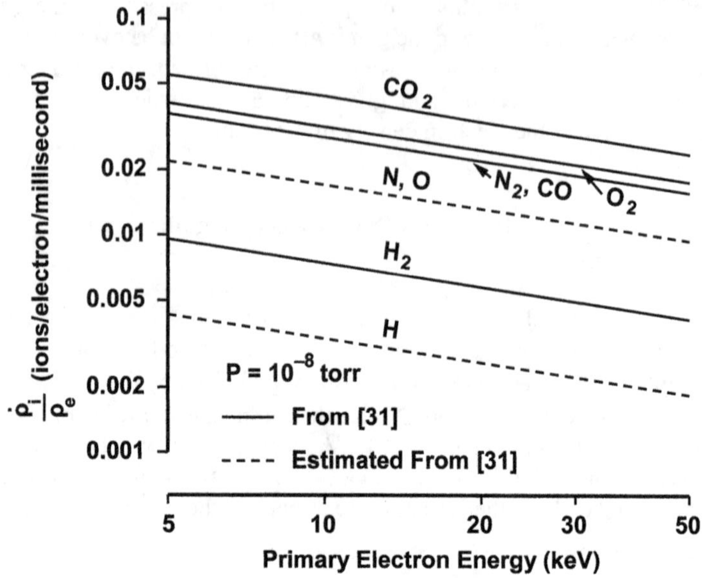

Figure 7-48 Rate of ion accumulation in a high energy electron beam at a pressure of $1{\times}10^{-8}$ torr.

7.5.4 Potential Depression in an Electron Beam

The key to understanding the reasons for the effects of ions on a beam is to understand the potential depression in the beam. Ions can affect this depression and this can lead to the deleterious effects that ions have on tube operation. The potential distribution in a beam of electrons was examined in Chapter 2. As a reminder, the cross section of an electron beam that was considered is shown in Figure 7-49 and the potential distribution in a metal tunnel containing the electron beam is shown in Figure 7-50.

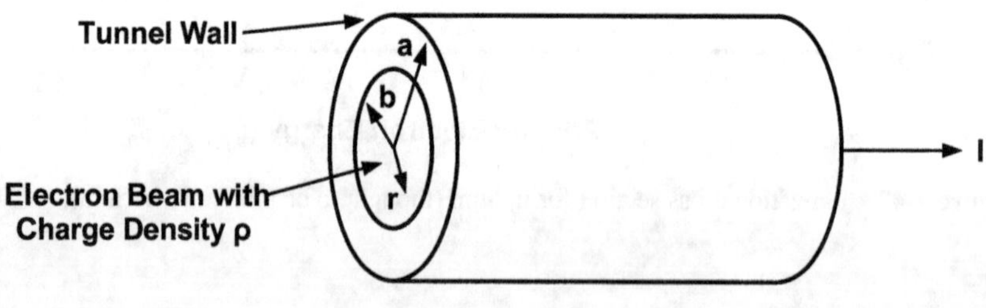

Figure 7-49 Section of electron beam.

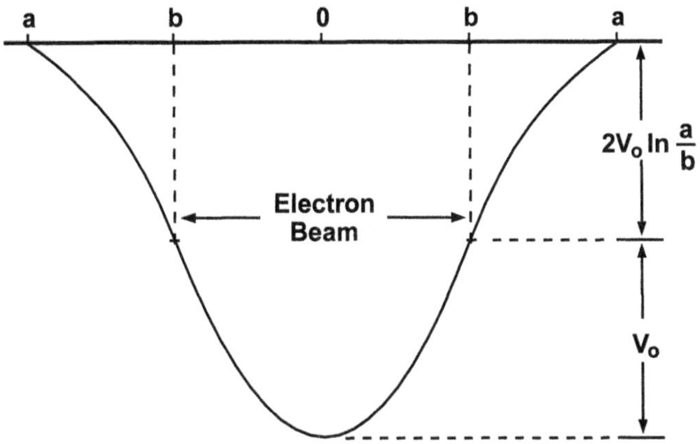

Figure 7-50 Potential distribution in a metal tunnel of radius *a* containing an electron beam.

The potential drop from the wall to the center of the beam (the potential depression in the beam) increases with an increasing ratio of the tunnel radius, *a*, to the beam radius, *b*, as shown in Figure 7-51. Notice that the potential depression increases rapidly as the beam radius becomes smaller. For a typical beam-filling factor (inverse of a/b) of 2/3, $V = 1.8\ V_o$, where V_o is the potential drop within the beam given by

$$V_o = \frac{I}{4\pi\varepsilon_o u_o} = 479 \frac{I}{\sqrt{V_b}} \text{ volts} \tag{7-64}$$

and where V_b is the beam voltage in kilovolts. For a 20-kV, 3-A beam, the potential depression is 583V.

An example of the potential distribution that may exist along the center of a 20-kV, 3-A beam is shown in Figure 7-52. Here it has been assumed that the filling factor (the fraction of the tunnel radius that is filled by the electron beam) for the average beam radius is 2/3. At the gun end of the tube, the potential drops very rapidly to the cathode voltage of –20 kV. At the collector end, the potential drops to the depressed collector voltage. It has been assumed here that there is no scalloping so that the beam radius is constant from the gun to the collector.

Now another example of the usefulness of Figure 7-48 can be seen by considering the following example. Assume a TWT like that in [29] with a 20-kV electron beam and that contains hydrogen. The ion accumulation rate at 10^{-8} torr is a little over 0.002 ion/electron/ms. In other words, in a millisecond, the beam is about 0.2% neutralized.

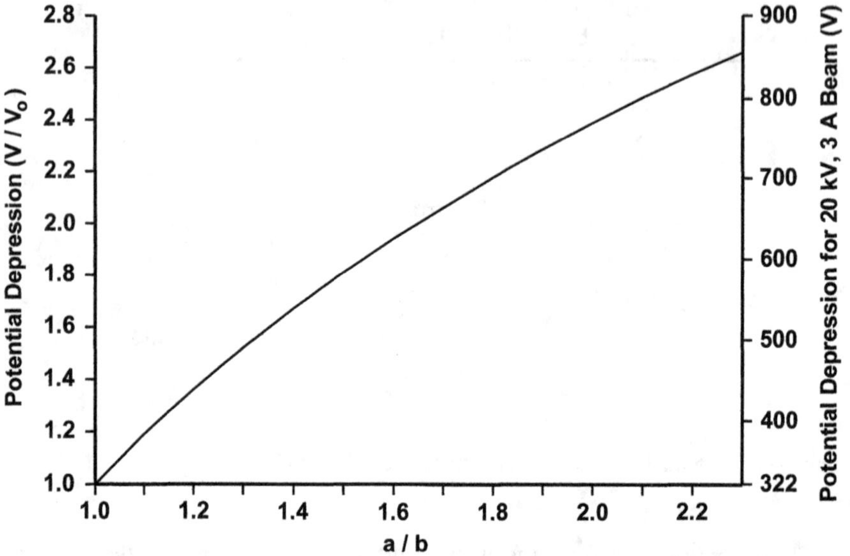

Figure 7-51 Potential depression in an electron beam as a function of the ratio of the tunnel radius to the beam radius.

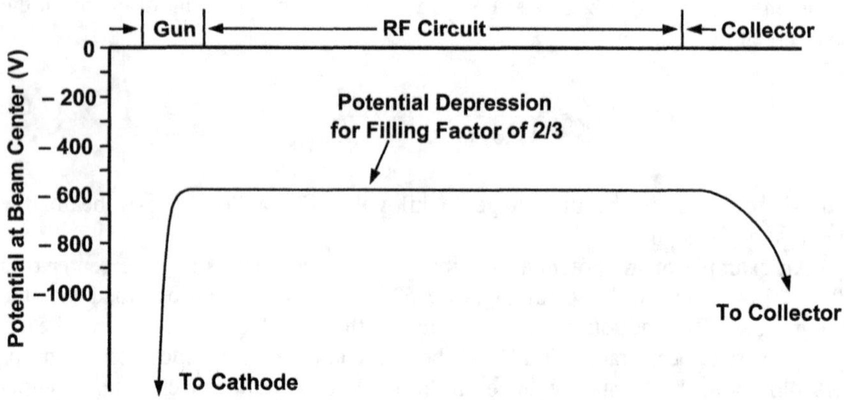

Figure 7-52 Potential depression at beam center as a function of axial position.

The potential depression in a 20-kV beam with a perveance of 1×10^{-6} is about 600V. In a millisecond, this depression would be reduced by 1.2V so the beam voltage would be increased by 1.2V or 0.006%. The beam velocity varies as the square root of beam voltage so the beam velocity would increase by 0.003%. In a TWT, as the beam goes faster, the phase length of the tube decreases. The incremental change in phase is one half of the change in velocity or, in the case being considered, 0.0015%. The total phase length of a TWT is on the order of 10,000°, so the change in phase length would be about 0.15°/ms. Notice in the upper trace in Figure 7-42 that the phase fluctuations are about 1°/10 ms or 0.1°/ms, which is surprisingly close to the estimate of 0.15°/ms.

7.5.5 Steady State Effects of Ionization

Next consider the effects of ionization on the potential profiles shown in the previous section. For each ionization event that results from the collision of a beam electron with a gas molecule, an electron is released. The first question to be answered is: What happens to these ionization electrons?

7.5.5.1 Ionization Electron Flow

In the regions where the ionization electrons are generated, the potential is several hundred volts below the tunnel wall, so these electrons are strongly attracted toward the tunnel wall. However, where the axial magnetic field used to focus the electron beam is high, the ionization electrons are prevented from reaching the tunnel wall. This can be shown by examining the magnetic flux distribution and using the Hull cutoff condition to investigate the radial electron flow.

The Hull cutoff condition is derived in Chapter 17 for axially symmetric electric and magnetic fields and is usually applied to the smooth-bore magnetron. To use that analysis here, as shown in Figure 7-53, we replace the cathode of the magnetron with the electron beam containing ionization electrons. The anode is the tunnel wall.

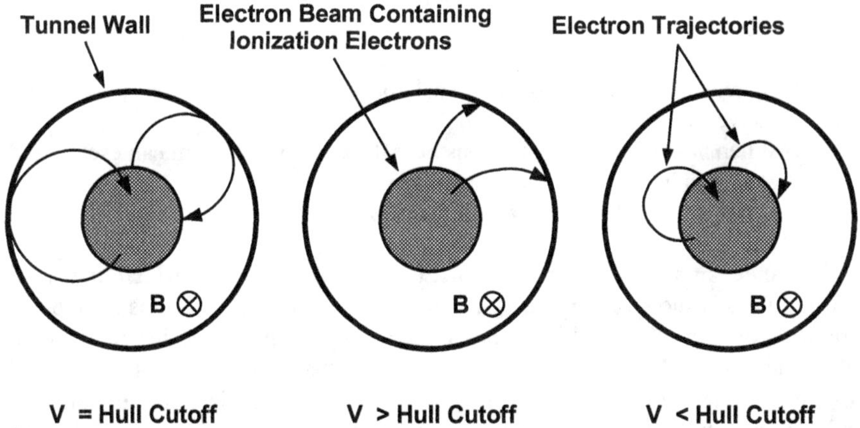

Figure 7-53 Ionization electron trajectories for various potential depressions.

If the potential depression in the beam equals the Hull cutoff potential, the electron trajectories just graze the tunnel wall. (This applies throughout the beam and assumes that the initial radial velocity of the electrons is zero.) If the depression is greater than the Hull cutoff voltage, the ionization electrons escape to the wall. If the depression is too small, the electrons return to the beam.

The Hull cutoff voltage can be written as

$$V_H = m^2 V_o \frac{a^2}{b^2}\left(1-\frac{r^2}{a^2}\right)^2 \tag{7-65}$$

where r is the radial position in the beam and m is the beam confinement factor. Now the potential from the beam edge to the tunnel wall,

$$V_{e-t} = 2V_o \ln \frac{a}{b} \tag{7-66}$$

and from the beam center to the tunnel wall,

$$V_{c-t} = V_o + 2\, V_o \ln \frac{a}{b} \tag{7-67}$$

can be compared with the Hull cutoff potential. Assuming that ionization electrons are initially at rest, then when $V_{e\text{-}t} / V_H$ or $V_{c\text{-}t} / V_H > 1$, ionization electrons can escape from the beam. When $V_{e\text{-}t} / V_H$ or $V_{c\text{-}t} / V_H < 1$, ionization electrons cannot escape.

Figure 7-54 shows the confinement factors, m, and the ratio of a to b for which ionization electrons can and cannot escape from the beam. For a filling factor of 2/3,

1. No electrons escape for the usual focusing condition of B substantially above B_B.
2. When $m = 1$ ($B = B_B$), electrons near the edge of the beam can escape and electrons in the beam center are trapped.
3. When B drops to 0.8 B_B, all of the ionization electrons can escape.

As an example, consider again a 20-kV, 3-A beam for which the axial focusing field is shown in Figure 7-55. The confinement factor for this example is somewhat over $m = 2$, so all ionization electrons in the main focusing field would be confined to the beam. However, near the collector pole piece where the axial flux density rapidly drops through $B = B_B$ to $B \approx 0$, all ionization electrons can easily escape from the beam. (In this case the electric field produced by the collector voltage has no effect on ionization electrons near the pole piece.)

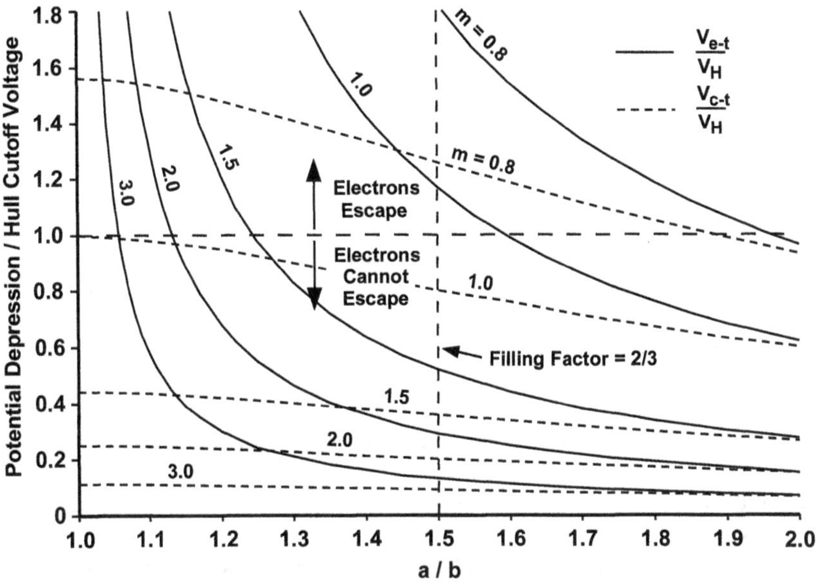

Figure 7-54 Conditions under which ionization electrons can and cannot escape from a beam.

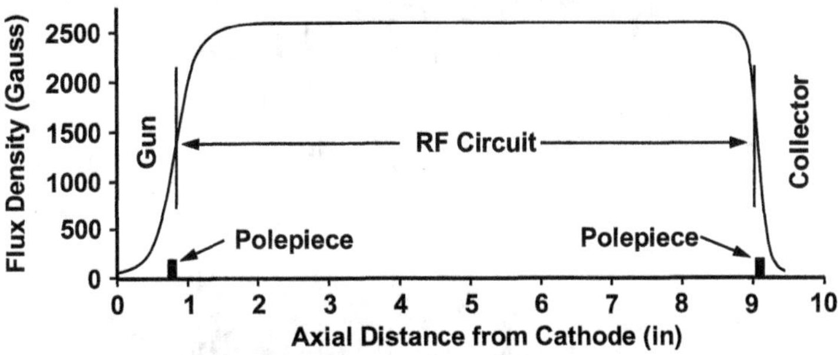

Figure 7-55 Axial magnetic flux density for a confined flow, 20-kV, 3-A beam.

Near the gun pole piece, the situation is more complicated because the electric field from the gun does penetrate into this region and repel ionization electrons. Thus, even though the flux density drops to a low value, the gun electric field prevents the electrons from reaching this region and escaping. As a result, we conclude that, for this confined flow example, the ionization electrons can escape from the beam only at the collector end of the tube.

Another example of importance is the PPM-focused beam. Theoretically, the peak flux density is $\sqrt{2}\,B_B$ and this is increased by a factor of about 1.5 in practical applications [24–26]. Thus, the peak axial flux density is more than strong enough to confine the ionization electrons. However, periodically, the flux density drops through B_B to zero as the focusing field reverses so all ionization electrons can escape.

7.5.5.2 Ion Flow

With a constant axial potential depression, it would be expected that ions would flow to the gun and the collector. However, as the beam leaves the region of the magnetic pole piece at the collector end of the tube and enters the collector, it may expand and pass into close proximity to the tube body. This causes the potential depression in the beam to decrease, which, in turn, produces a barrier to ion flow that prevents all but the most energetic ions from entering the collector. (Ions are generated at various potentials because of the radial variation of potential in the beam, so they have a distribution of velocities.) Similarly, when the RF drive is applied to the TWT and the tube is operating near saturation, the RF bunching causes the beam to be large in diameter near the output cavity. This causes the potential depression to decrease, which impedes the flow of ions to the collector.

Another factor to consider is beam scalloping. Variations in the magnetic focusing field affect the barrier amplitude at the collector end of the tube and the ion level in the tube. The reason for this effect is illustrated in Figure 7-56.

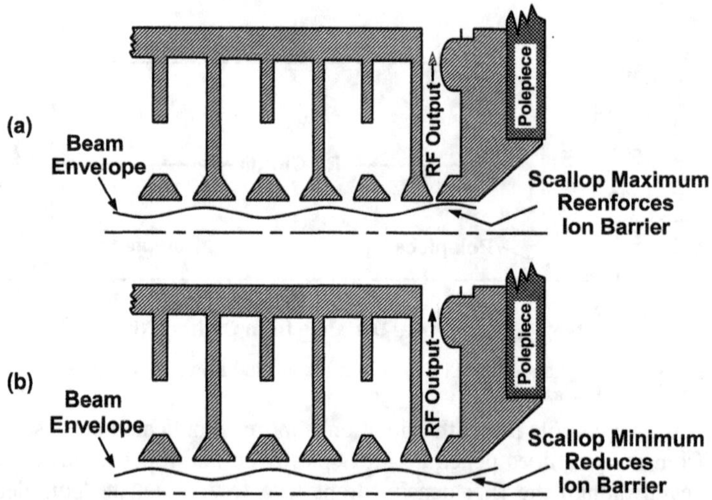

Figure 7-56 Illustration of how scalloping can (a) reinforce or (b) reduce the ion barrier at the collector end of a TWT.

Normally the RF interaction region is many wavelengths long. As the magnetic focusing field is varied, large-diameter (and low depression) regions of the beam sometimes coincide with the barrier at the collector end and reinforce it

(Figure 7-56(a)), which prevents all but the most energetic ions from entering the collector. At other magnetic field levels, small diameter (and high depression) regions of the beam coincide with the barrier and reduce its amplitude allowing more ions to flow to the collector (Figure 7-56(b)).

If the beam produced at the electron gun is imperfect (which is always the case to some extent), then there may be a large-diameter region in the beam near the gun as shown in Figure 7-57. This region produces a decrease in the potential depression and a barrier to the ion flow to the gun. As a result, as shown in Figure 7-58, there may be barriers to ion flow at the collector and the gun ends of the tube. The relative heights of the barriers at the gun and the collector control the relative flows of ions to the gun and collector ends of the tube.

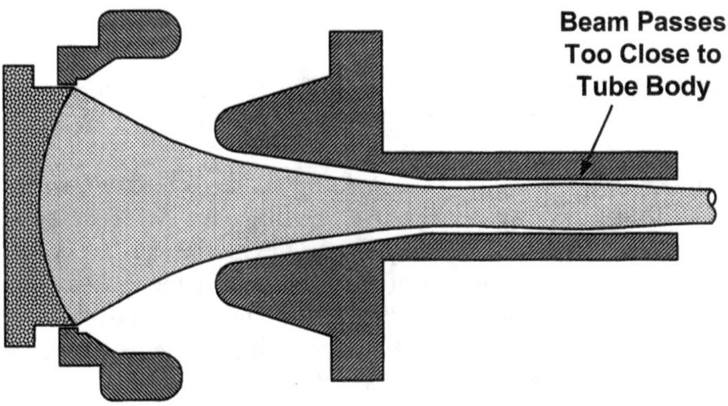

Figure 7-57 Possible cause of decrease in the potential depression near the gun.

In some TWTs, ions are prevented from flowing to the gun by a positive anode. There is speculation for some magnetic field configurations that the beam may fill with ions so that the potential depression disappears. Then ions can flow radially to the tunnel wall.

7.5.6 Low-Frequency Instabilities

With barriers to ion flow at the gun and collector ends of the tube, the ion concentration in the beam increases with time. Often, in long pulse and CW operation of linear-beam tubes, low-frequency (hertz to kilohertz) fluctuations are observed in the power and phase of the output signal (see Figure 7-42). There have been many attempts to attribute these fluctuations to interactions between the electron beam and the ions in the RF interaction region of the tube. To the best of the author's knowledge, none of these attempts has produced results that come even close to explaining the experimental results. The work that does explain the experimental results is that reported in [29]. That work is summarized here.

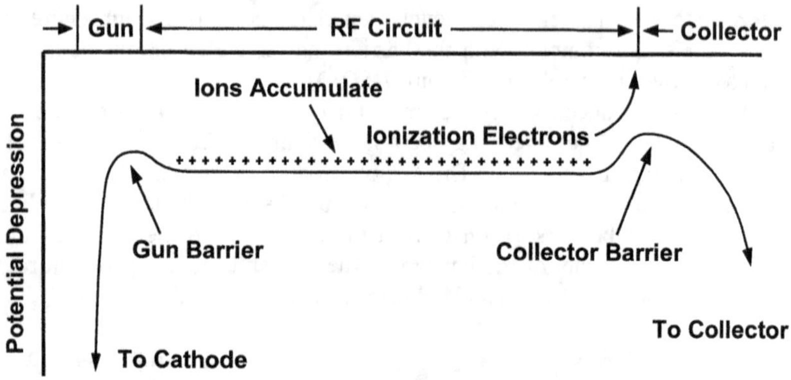

Figure 7-58 Potential depression on the axis of a TWT with barriers to the ion flow near the gun and near the collector.

Before proceeding, it should be noted that, if the collector barrier is lower than the gun barrier, then, as the ion level in the beam increases, more ions of increasingly lower energy are able to spill over the barrier into the collector. This appears to be an ideal way to get rid of the ions because there does not appear to be any way for ions in the collector to cause instabilities. (The exception to this is when the ion density is so high in the collector that the electron beam doesn't expand properly. Then the beam can damage the collector.)

If the gun barrier is lower than the collector barrier, then, as the ion level in the beam increases, ions spill into the gun. Directly and/or indirectly, these ions neutralize part of the electron space charge forces in the beam waist (see Figure 7-59) in the gun and this causes the waist to shrink. As the beam waist shrinks, this, in turn, causes the gun barrier to shrink, which allows more ions to spill into the gun. This process continues until the beam is drained of ions that can spill over the barrier. At this time, the beam waist returns to its normal diameter and the gun barrier returns to its normal height. The beam then fills with ions and eventually reaches the point where ions again spill into the gun causing the relaxation-type oscillation to repeat.

The way in which ions entering a gun affect gun operation is complicated by several factors.

1. As was pointed out previously, the ions sputter the cathode, and this produces a barium/oxygen gas in the gun. The high density of this gas coupled with the large ionization cross section of the sputtered atoms results in a high rate of generation of ions by the beam in the gun.

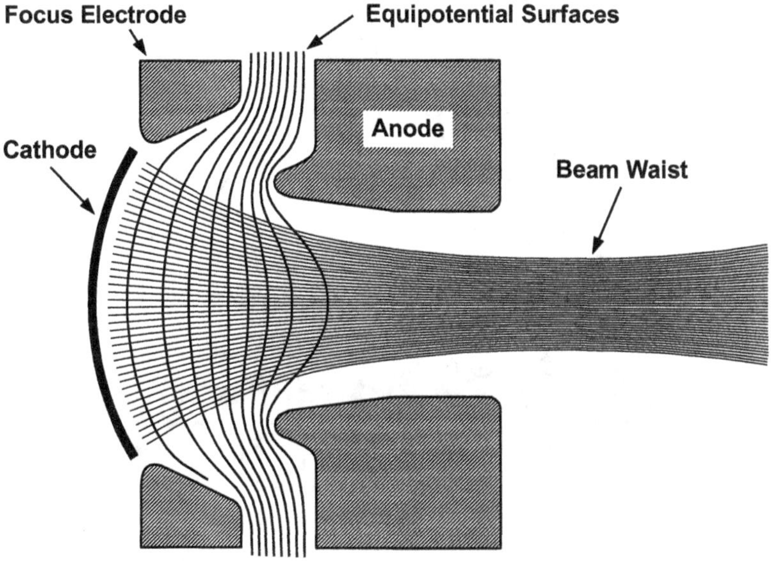

Figure 7-59 Cross section of electron gun showing beam waist region.

2. Usually, as ions accumulate in the beam, the potential depression in the beam decreases. However, in the beam waist in the gun, ions can cause the potential depression to increase. The reason for this is that increase in depression caused by beam shrinkage more than offsets the decrease in depression resulting from ion accumulation. The increase in depression causes more ions to be trapped.
3. The effect of ions in the beam waist is sometimes exacerbated by an ion trap in the waist. The reasons for this trap and the potential depression caused by it are shown by the example in Figure 7-60 [29]. Notice that, as the beam approaches the beam waist, it passes a nearby ridge on the anode and this causes the potential depression on the axis to decrease. Then the beam moves away from the anode for a short distance and so the depression increases. The beam then approaches a smaller drift tube (and, at the same time, the beam diameter increases) so the potential depression decreases again.
4. The size of the waist trap also depends on the pressure of cathode evaporants in the gun. This pressure, in turn, is dependent on cathode temperature and on the total time that the cathode has been operating. The reason for the pressure effect is that ions produced by electron collisions with these gas atoms reduce the size of the beam entering the beam waist. This increases the potential depression (decreases the waist barrier) to the left of the waist trap, which reduces the size of the trap.

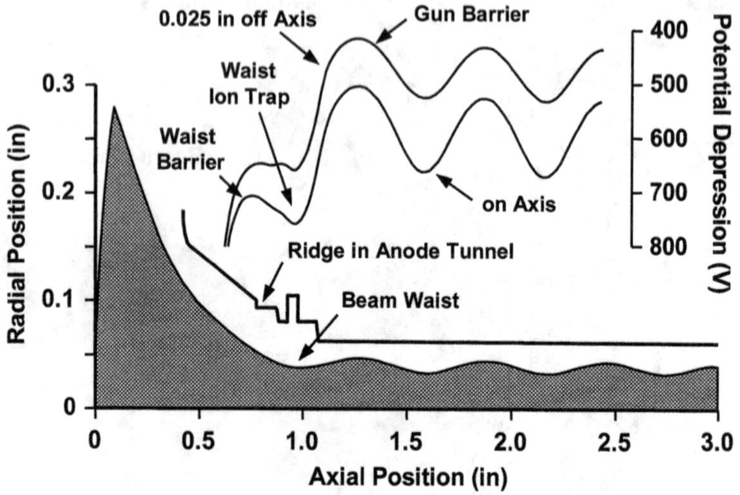

Figure 7-60 Effect of the anode-tunnel shape on the potential depression near the beam waist.

7.5.7 High-Frequency Instabilities

In Figure 7-43, two high-frequency instabilities are shown by the fluctuations in the probe current in beam analyzer experiments. During the lower-frequency portions of the pulse, a transverse oscillation of the electron beam was found. That is, the peaks and valleys in current in the 2–3-μs time frame of Figure 7-43 were found to correspond to alternating transverse positions of the electron beam. The higher-frequency oscillation in the 4–5-μs time frame was found to result from a fluctuation of the beam radius.

The transverse mode of oscillation, shown in Figure 7-61, may be explained with the aid of Figure 7-62, which shows the potential distribution in a drift tube containing an electron beam. The motion of an ion is given by

$$\delta = \delta_o \sin \frac{\omega_i}{\sqrt{2}} t \tag{7-68}$$

where ω_i is given by

$$\omega_i = \left(\frac{e\rho_e}{m_i \varepsilon_o} \right)^{1/2} \tag{7-69}$$

and is the frequency with which the ion oscillates back and forth in the potential well formed by the electron beam.

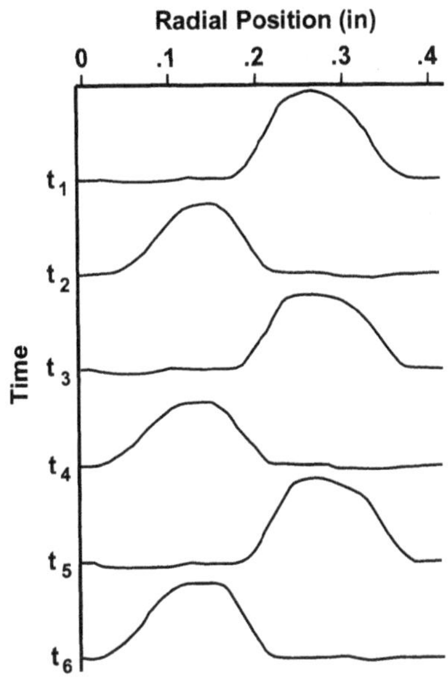

Figure 7-61 Probe current as a function of radial position for transverse mode of oscillation. (From: A. S. Gilmour, Jr., RADC-TDR-63-465, Vol. 1, December 1963.)

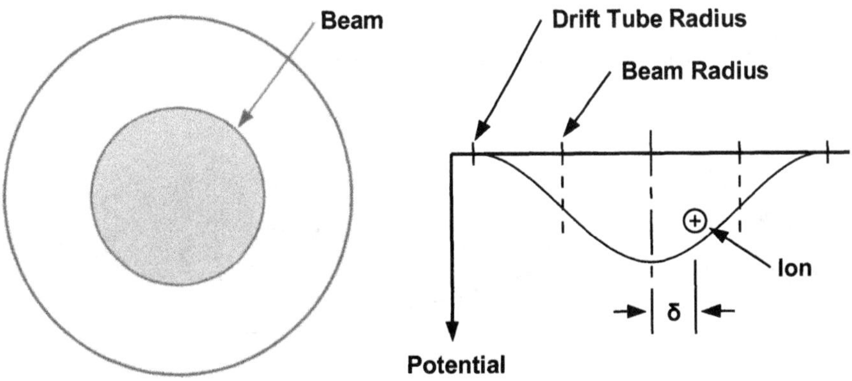

Figure 7-62 Potential distribution in a drift tube containing an electron beam.

When an ion beam of radius c (as indicated in Figure 7-63) with a charge density equal to the electron density is considered, then the analysis becomes more complex, but the equation for small displacements of the ion beam is still straightforward and is

$$\delta = \delta_o \sin \frac{\omega_i}{\sqrt{2}} \left(1 - \frac{c^2}{d^2}\right)^{1/2} t \tag{7-70}$$

where d is the radius of the drift tube. Thus, the frequency of ion oscillation is reduced. If the diameter of the ion beam increases with time as ion generation takes place, it would be expected that the ion oscillation frequency would decrease. In fact, this effect has been observed as indicated by Figure 7-64.

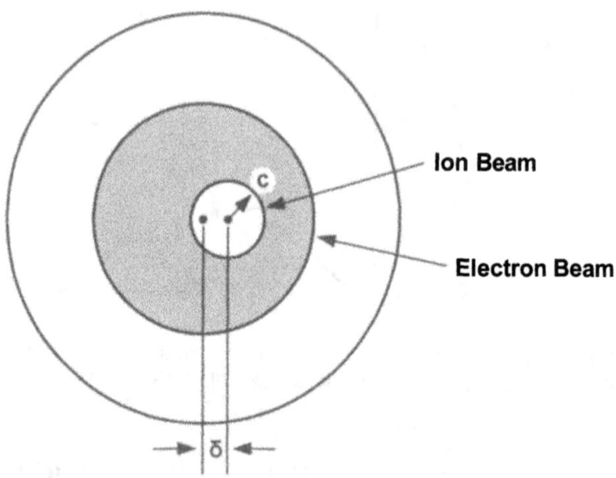

Figure 7-63 An ion beam contained within an electron beam.

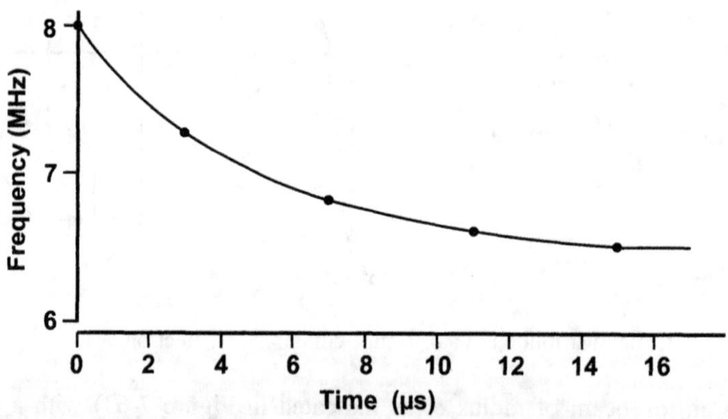

Figure 7-64 Frequency of transverse oscillations as a function of time. (From: A. S. Gilmour, Jr., *Proc. 5th International Congress on Microwave Tubes.*)

During high-frequency portions of the pulse shown in Figure 7-43, a radial oscillation of the electron beam was found. That is, the beam diameter changed periodically from a relatively large value to a smaller value, as shown in Figure 7-65. This figure shows probe current as a function of radial probe position in the beam at times of maximum and minimum beam diameter.

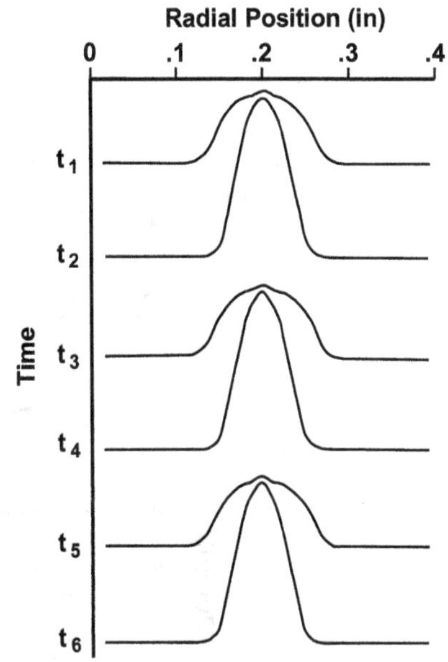

Figure 7-65 Probe current as a function of radial position for radial mode of oscillation. (From: A. S. Gilmour, Jr., RADC-TDR-63-465, Vol. 1, December 1963.)

The radial mode of oscillation may be explained with the aid of Figure 7-66. Assume that, initially, the positive ions are uniformly distributed throughout the potential well of the electron beam. The ions are attracted and flow to the center of the beam. The resulting concentration of ions produces a positive hump in the potential well (see Figure 7-67) and a repulsive force, which causes the ions to reverse direction and flow radially outward. As the ions flow outward, the potential well deepens, causing a strong enough electrostatic force to pull them inward again. The radial electrostatic forces in the electron beam depend on the ion concentration and so the beam oscillates radially in synchronism with the ions.

When an analysis of the radial mode of oscillation is carried out, it is found that the frequency at which the radial oscillations are expected to occur is $\sqrt{2}$ greater than the frequency of transverse oscillations (neglecting ion beam diameter effects).

The radial oscillation frequency, like the transverse oscillation frequency, is a function of ion mass as well as electron beam density and density profile.

Oscillation frequencies from the range of megahertz to several tens of megahertz are common.

The presence of ion oscillations is indicated by ion sidebands on the output frequency of a microwave tube. Also, the oscillation frequency is a function of the square root of beam voltage (see Figure 7-68). This provides a means for checking to determine if gas is the cause of spurious tube output signals.

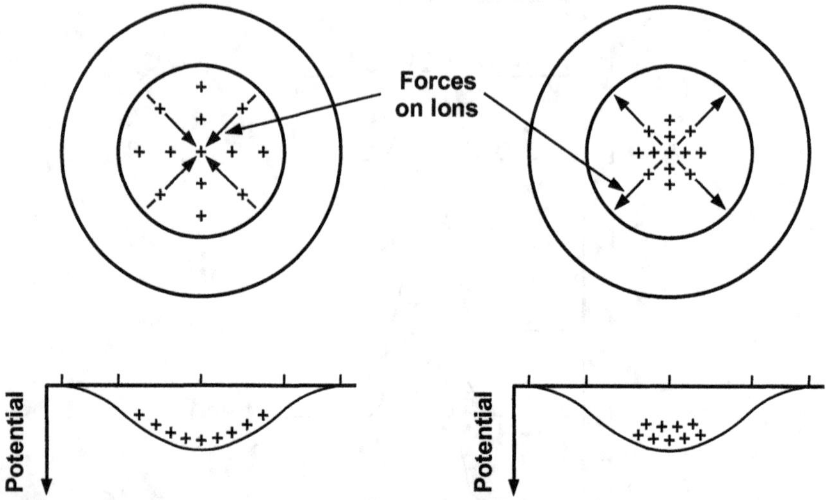

Figure 7-66 Ion distribution in the radial mode of oscillation.

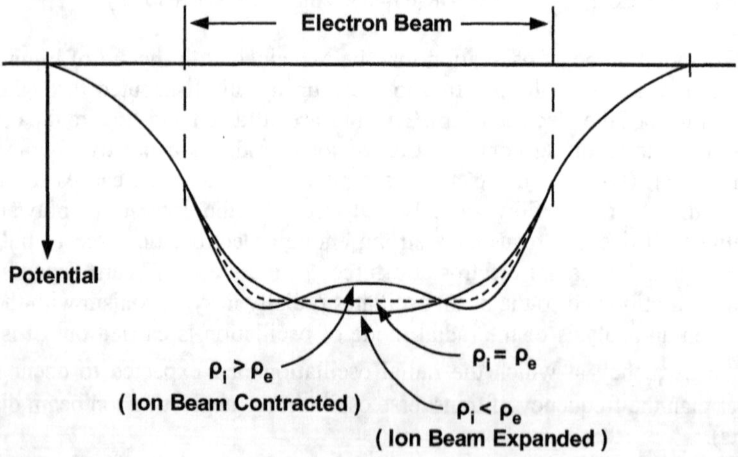

Figure 7-67 Potential profiles for radial ion oscillations.

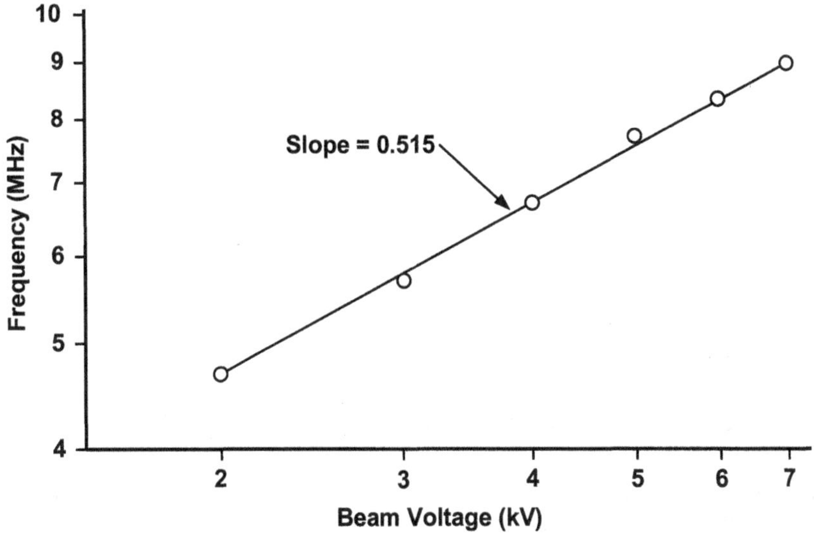

Figure 7-68 Ion-induced beam-oscillation frequency as a function of beam voltage. (From: A. S. Gilmour, Jr., RADC-TDR-63-465, Vol. 1, December 1963.)

REFERENCES

[1] M. Chodorow, and C. Susskind, *Fundamentals of Microwave Electronics*, New York: McGraw-Hill, 1964.

[2] J. F. Gittins, *Power Travelling Wave Tubes*, New York: American Elsevier, 1965.

[3] W. J. Kleen, *Electronics of Microwave Tubes*, New York: Academic Press, 1958.

[4] A. S. Gilmour, Jr., G. C. Dalman, and L. F. Eastman, "The current and velocity distributions in a velocity modulated Brillouin focused electron beam," *Proc. 4th International Congress on Microwave Tubes*, Eindhoven, the Netherlands: Centrex, 1963.

[5] A. S. Gilmour, Jr., and D. Hallock, "A demountable beam analyzer for studying magnetically confined electron beams," *Advances in Electron Tube Techniques*, New York: Pergammon Press, 1963, pp. 153–157.

[6] G. R. Brewer, "Some characteristics of a magnetically focused electron beam," *Jour. App. Phys.*, Vol. 30, No. 7, July, 1959, pp. 1022–1038.

[7] A. S. Gilmour, Jr., *Principles of Traveling Wave Tubes*, Norwood, MA: Artech House, 1994.

[8] J. R. Pierce, and L. R. Walker, "Brillouin flow with thermal velocities," *Jour. App. Phys.*, Vol. 24, No. 10, October 1953, pp. 1328–1330.

[9] A. Ashkin, "Electron beam analyzer," *Jour. App. Phys.*, Vol. 28, No. 5, May 1957, pp. 564–569.

[10] K. J. Harker, "Nonlaminar flow in cylindrical electron beams," *Jour. App. Phys.*, Vol. 28, No. 6, June 1957, pp. 645–650.

[11] G. Herrmann, "Optical theory of thermal velocity effects in cylindrical electron beams," *Jour. App. Phys.*, Vol. 29, No. 2, February 1958, pp. 127–136.

[12] K. Amboss, "Verification and use of Herrmann's optical theory of thermal velocity effects in electron beams in the low perveance regime," *IEEE Trans. Electron Devices*, Vol. 11, No. 10, October 1964, pp. 479–485.

[13] P. T. Kirstein, "On the effects of thermal velocities in two-dimensional and axially symmetric beams," *IEEE Trans. Electron Devices*, Vol. 10, No. 2, March 1963, pp. 69–80.

[14] P. T. Kirstein, "Thermal velocity effects in axially symmetric solid beams," *Jour. App. Phys.*, Vol. 34, No. 12, December 1963, pp. 3479–3490.

[15] J. R. Hechtel, "Magnetic focusing of electron beams in the presence of transverse velocity components," *IEEE Trans. Electron Devices,* Vol. ED-28, No. 5, May 1981, pp. 473–482.

[16] J. T. Mendel, C. F. Quate, and W. H. Yocom, "Electron beam focusing with periodic permanent magnet fields," *Proc. IRE*, Vol. 42, No. 5, May 1954 pp. 800–810.

[17] J. R. Pierce, "Magnetic focusing system," U.S. Patent 2,847,607, 1958.

[18] R. B. True, "The role of the stop band voltage in PPM focused TWTs," *Microwave Power Tube Conference*, 1988.

[19] K. K. N. Chang, "Beam focusing by periodic and complementary fields," *Proc. IRE*, Vol. 43, No. 1, January 1955, pp. 62–71.

[20] J. T. Mendel, "Magnetic focusing of electron beams," *Proc. IRE*, Vol. 43, No. 3, March 1955, pp. 327–331.

[21] K. J. Harker, "Periodic focusing of beams from partially shielded cathodes," *IEEE Trans. Electron Devices*, Vol. 2, No. 4, October 1955, pp. 13–19.

[22] W. A. Harmon, *PPM Focusing Concepts*, Hughes Electron Dynamics Division Technical Report No. 70, 1987.

[23] W. R. Ayers, et al., "RF beam spread in PPM confined-flow TWTs," *Technical Digest,* IEDM, 1985, pp. 357–360.

[24] R. True, "A theory for coupling gridded gun design with PPM focusing," *IEEE Trans. Electron Devices*, Vol. ED-31, No. 3, March 1984, pp. 353–362.

[25] R. True, "Emittance and the design of beam formation, transport, and collection systems in periodically focused TWTs," *IEEE Trans. Electron Devices*, Vol. ED-34, No. 2, February 1987, pp. 473–485.

[26] R. True, "Tunnel emittance growth in PPM focused TWTs," *Technical Digest,* IEDM, 1985, pp. 192–195.

[27] A. S. Gilmour, Jr., "Transverse and radial oscillations induced in electron beams by ions," *Proc. 5th International Congress on Microwave Tubes*, Paris: Dunod, 1965, pp. 453–456.

[28] E. W. McCune, "Ion oscillations in pulsed klystron amplifiers," *Technical Digest,* IEDM, 1983, pp. 148–150.

[29] A. S. Gilmour, Jr., et al., "Ion instabilities initiated in the gun of a linear-beam microwave tube," *IEEE Trans. Plasma Science*, Vol. 32, No. 3, June 2004, pp. 1251–1261.

[30] K. L. Jensen, Y. Y. Lau, and B. Levush, "Migration and escape of barium atoms in a thermionic cathode," *IEEE Trans. Plasma Science*, Vol. 28, No. 3, June 2000, pp. 772–783.

[31] Y. K. Kim, *Electron-Impact Ionization Cross Section Database,* PhysRefData/Ionization/intro.html: http://physics.nist.gov/, 2005.

Chapter 8

Beam-Gap Interactions

In all microwave tubes, two fundamental interactions of the electron beam and the RF circuit are responsible for the operation of the tube. These are indicated on a klystron in Figure 8-1. First, in amplifiers, the RF input signal produces a voltage on an RF circuit (gap in cavity in Figure 8-1) that initiates bunching (modulation) of an electron beam. Second, the RF current in the beam induces current in the nearby RF circuit. This current, in turn, interacts with the impedance of the circuit to produce a voltage that, until RF power is removed from the tube, enhances the modulation of the beam. These interactions, beam modulation and current induction, are the subjects of this chapter.

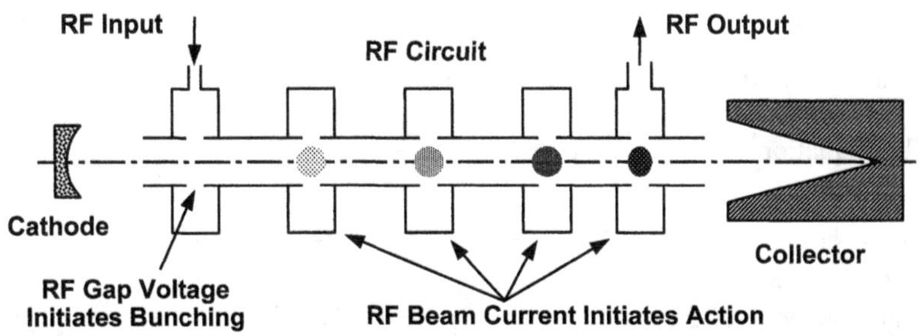

Figure 8-1 Illustration of where gap interactions take place in a microwave tube (klystron shown here).

8.1 BEAM MODULATION

In the first klystrons, the beam was modulated by passing it through a gap between two parallel grids across which RF voltage was applied. Beam power was very limited because of grid heating by electron impact. With the use of gridless gaps, beam power can be orders of magnitude above that possible with gridded gaps. Modulation by gridded and by gridless gaps is described in the following sections.

8.1.1 Gridded (Planar) Gaps

Consider the interaction of an electron beam with the electric field in a gridded (planar) gap as shown in Figure 8-2. The velocity of the beam entering the gap is u_o and is modified by an amount Δu as the beam passes through the gap. The time and position at the entrance to the gap are t_o and $z = 0$. The width of the gap is ℓ and the voltage applied to the gap is $V_1 \sin\omega t$.

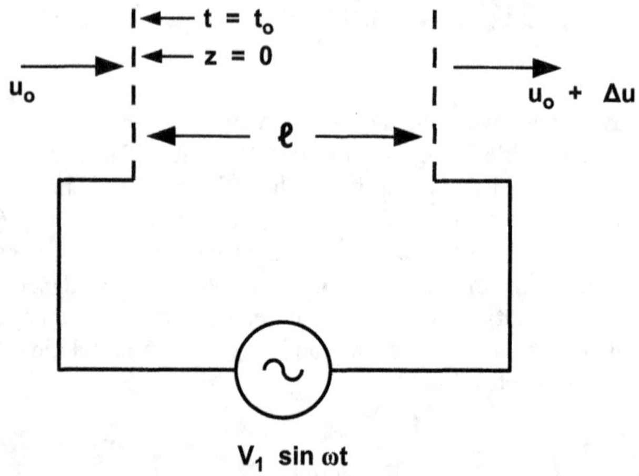

Figure 8-2 Velocity modulation of an electron beam.

The equation of motion for electrons in the gap is

$$\frac{d^2 z}{dt^2} = -\eta E = \eta \frac{V_1}{\ell} \sin \omega t \tag{8-1}$$

and so the velocity of the electrons leaving the gap is

$$u = u_o \left(1 + \frac{\alpha M_p}{2} \sin \omega t \right) \tag{8-2}$$

where α is called the *depth of modulation* and is defined as follows:

$$\alpha = \frac{V_1}{V_o} \tag{8-3}$$

The dc acceleration voltage for the electrons in the electron gun is V_o. The quantity M_p, which is plotted in Figure 8-3, is the *modulation coefficient* or the *gap coupling coefficient* and is

$$M_p = \frac{\sin\frac{\varphi_o}{2}}{\frac{\varphi_o}{2}} \tag{8-4}$$

Here, φ_o is the dc transit angle through the gap and is defined as follows:

$$\varphi_o = \omega\frac{\ell}{u_o} \tag{8-5}$$

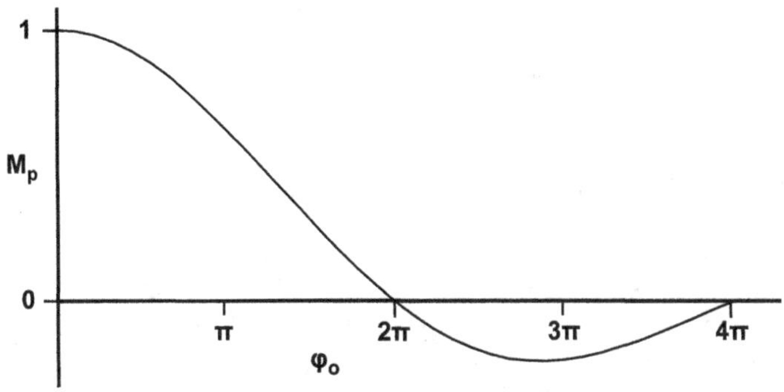

Figure 8-3 Gap coupling coefficient as a function of gap transit angle.

As is indicated in Figure 8-4, for a very narrow gap, φ_o is small so an electron "sees" a nearly constant voltage as it traverses the gap. For this case, the gap coupling coefficient is maximum and is nearly unity. As the gap is widened, the voltage varies as an electron traverses the gap and so the effective gap voltage (V_{eff} in Figure 8-4) is less than the maximum gap voltage (V_{max}). The effectiveness of the gap in accelerating the electron decreases (i.e., the gap coupling coefficient goes down). If the gap width is increased to the point where $\varphi_o = 2\pi$, the applied voltage varies through one complete cycle as an electron traverses the gap and the net acceleration of the electron is zero. In this case, $M_p = 0$.

The value of the gap coupling coefficient is that, in the analysis, a gap with a finite width ℓ can be replaced with a gap having an infinitesimal width and voltage reduced by M_p without affecting the velocity. In addition, the reference plane where $z = 0$ can be shifted from the first grid of the gap to the center of the gap where $z = \ell/2$. This has been done in (8-2).

The modulation process described here is referred to as *velocity modulation.* The reason, which is clear from (8-2), is that the velocities of electrons are "modulated" as they traverse a gap to which an alternating voltage is applied. This is in contrast to the situation in a triode or a tetrode where the density of the electrons is modulated.

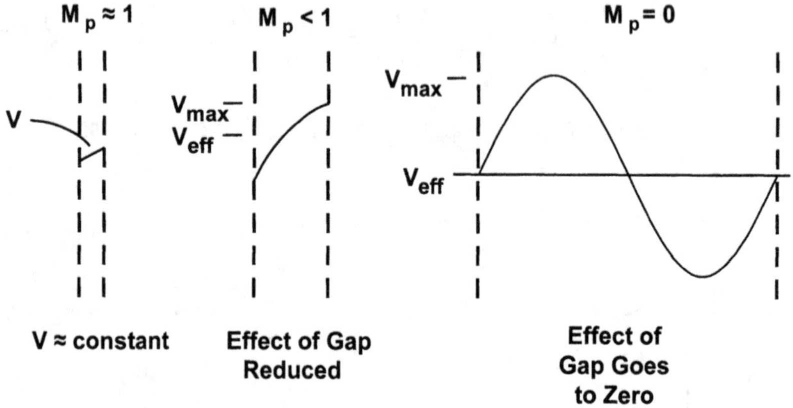

Figure 8-4 Effect of the gap width on the gap coupling coefficient.

With velocity modulation, the electron beam is first accelerated to high velocity and then subjected to the RF modulating field rather than being modulated at low (nearly zero) velocity and then accelerated. The result is that the transit time limitations of the triode and tetrode are overcome with the velocity modulation technique and operation at much higher frequencies and power levels is made possible.

8.1.2 Gridless (Nonplanar) Gaps

Gridless gaps are used for velocity modulation in virtually all microwave tubes. These gaps are usually formed by two facing electrodes. For cylindrical gaps, the fields produced are somewhat as shown in Figure 8-5.

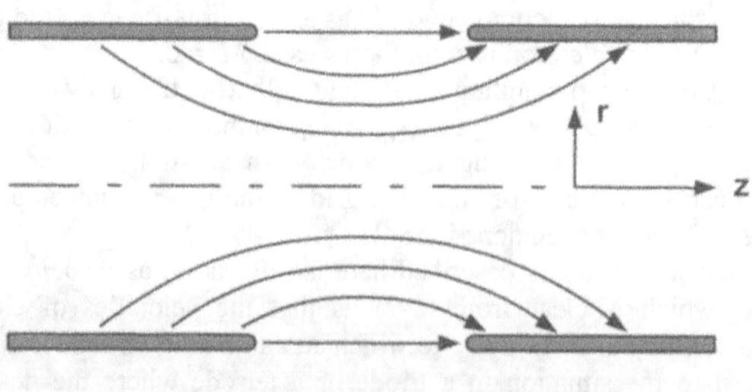

Figure 8-5 Electric field lines in a gridless gap.

The electric field extends over a large region inside the gap and has variations in both the axial and radial directions. When a beam passes through the gap, electrons near the axis must travel further to "see" the full gap voltage than electrons near the gap. The field on axis may go through a large portion of a cycle while the field near the gap varies through a small portion as the electrons pass through, so electrons on axis receive a smaller net velocity change than those near the gap. Thus, the gap coupling coefficient varies with radius.

The analysis of the interaction of the beam in the gridless gap is significantly more complicated than in the gridded gap. Still, under small signal conditions, the beam velocity can be expressed quite simply as

$$u = u_o \left(1 + \frac{\alpha M_{np}}{2} \sin \omega t\right) \tag{8-6}$$

where M_{np}, which is a function of radius, is the gap coupling coefficient for the nonplanar gap. The magnitude of M_{np} depends on the electric field distribution within the gap. The field, in turn, depends on several factors including the gap width, the cylinder radius, and the shapes of the cylinder tips.

For example, as shown in Figure 8-6, if the gap is narrow compared with the cylinder radius and the cylinder tips are blunt, then the field within the gap will be nearly constant. On the other hand, if the gap is relatively wide and the cylinder tips are very sharp, the field will vary in a parabolic-like manner. Kosmahl and Branch [1] used a hyperbolic-cosine approximation to the field shape and found rather complex relationships for the gap coupling coefficient. A more straightforward, but less elegant, approach is that described by Vaughan [2]. He used the gap coupling coefficient found for a uniform gap field and then devised an empirical correction factor.

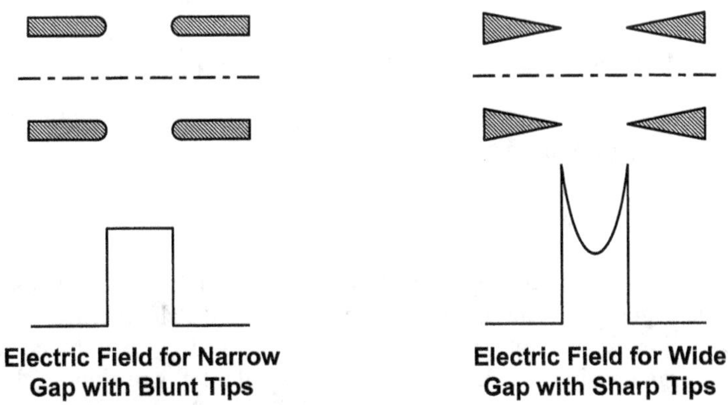

Figure 8-6 Electric field shapes in gaps.

With a uniform field in the gap, the field experienced by the electron traveling close to the walls of the cylinders is the same as the field in the gridded gap, that is, constant, and so the electron velocity leaving the gap (at $z = \ell/2$) is given by (8-2). The expression found for M_{np} as a function of radius is shown in Figure 8-7 and is

$$M_{np} = M_p \frac{\mathrm{I}_0(\beta_e r)}{\mathrm{I}_0(\beta_e a)} \tag{8-7}$$

where M_p is the modulation coefficient for the planar (gridded) gap (8-4), $\beta_e = \omega/u_o$ is the propagation constant for the electron beam, and a is the cylinder radius. Thus, the modulation coefficient for the gridless gap M_p is modified by I_0 Bessel functions. This shows what is expected for a gridless gap, that the modulation is weaker on the axis than at the wall. It is usual practice to limit $\beta_e a$ to ≈ 1 to 1.5 to ensure adequate interaction with the electron beam.

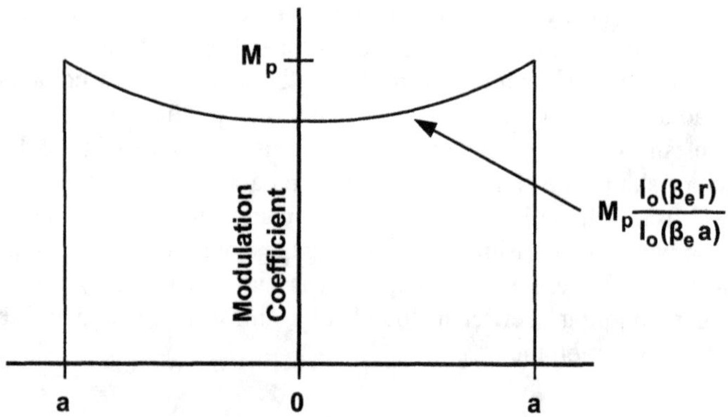

Figure 8-7 Modulation coefficient for a gridless gap.

8.2 CURRENT INDUCTION

8.2.1 Gridded (Planar) Gaps

Consider first a sheet of electron charge passing between two planar parallel grids as shown in Figure 8-8. The surface charge density on the charge sheet is assumed to be σ_e coulombs per unit area. Electric fields E_L and E_R are produced by σ_e. Positive charge densities σ_L and σ_R are induced on the left and right grids, and as the charge sheet moves, electron current I_i flows in the circuit connecting the grids.

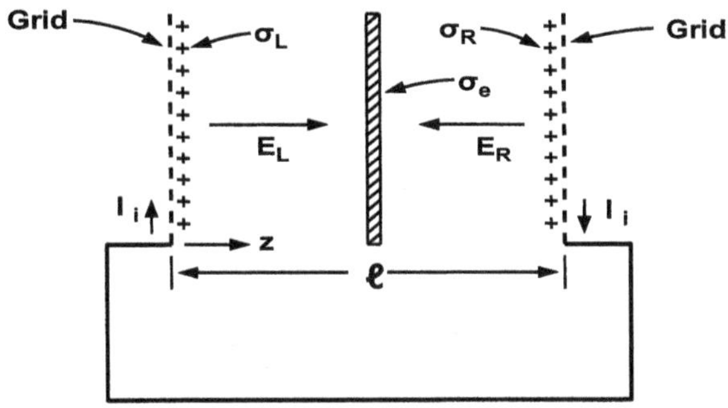

Figure 8-8 Electrode and charge-sheet configuration for the current induction analysis.

Neglecting fringing fields at the edges of the grids and charge sheet, the charge density induced on the left grid is

$$\sigma_L = \sigma_e\left(1 - \frac{z}{\ell}\right) \tag{8-8}$$

and that induced on the right grid is

$$\sigma_R = \sigma_e \frac{z}{\ell} \tag{8-9}$$

As the position of the charge sheet changes, σ_L and σ_R vary as shown in Figure 8-9. The electron current density, J_i, induced in the grids is

$$J_i = \frac{\sigma_e}{\ell} u \tag{8-10}$$

where u is the velocity of the charge sheet. The circuit current is $I_i = A\,J_i$, where A is the area of a grid.

The preceding result is a form of Ramo's theorem. The importance is that it shows that, immediately upon the entrance of a charge into a region, current starts to flow in nearby electrodes. The charge can be emitted from one of the electrodes and collected by the other or can enter and leave the region through grids. This analysis can be applied to RF gaps by assuming that RF magnetic fields are negligible (as in conventional microwave tubes) and that field propagation effects are negligible (valid if dimensions are small compared with a wavelength).

There is much to be learned from this simple relationship (8-10). First and most importantly, the current waveform induced in the grids and circuit may be very different than the current waveform of the charge sheet. Since the thickness of the charge sheet has been assumed to be infinitesimal, the volumetric charge density, ρ_e, must be infinite. The current density of the sheet, $J_e = \rho_e u$, is therefore infinite.

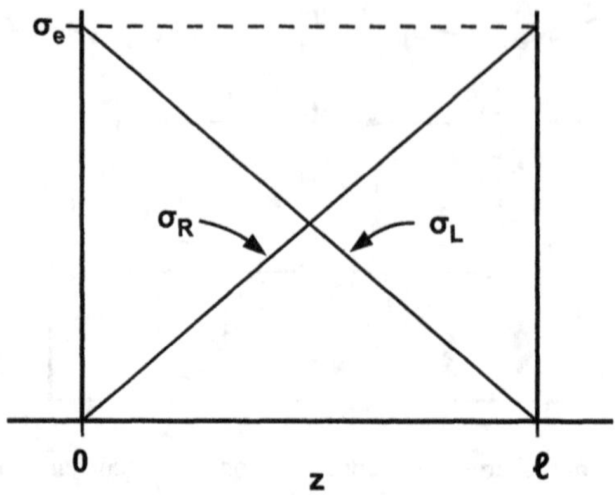

Figure 8-9 Variation of charge densities induced on the left and right grids by charge sheet at position *z*.

On the other hand, for constant velocity, the current density in the grids is constant as the sheet of charge traverses the gap. Thus, J_i has a rectangular waveform with a duration $\tau = d/u$, which is the transit time of the charge sheet across the gap. The waveforms for J_i and J_e are illustrated in Figure 8-10 for the charge sheet located at position *z* in the gap.

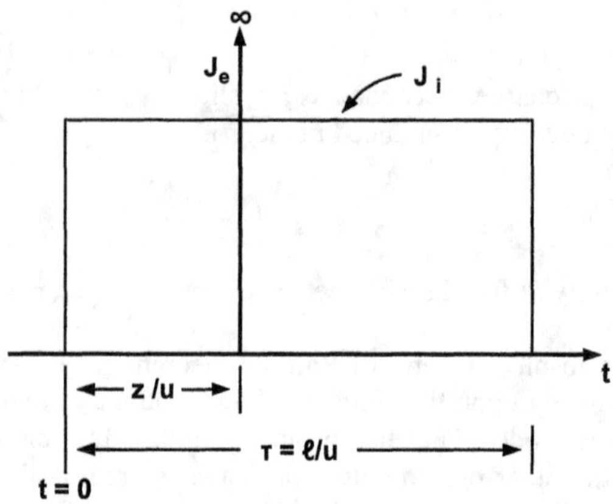

Figure 8-10 Current waveform induced in the circuit (J_i) and current waveform of the charge sheet (J_e).

Next, consider the situation when the voltage between the grids is large enough to change the electron velocity. Then the induced current waveform is changed. For example, assume, as indicated in Figure 8-11, that the left grid is the cathode of a diode that emits a sheet of charge at zero velocity and that the right grid is an anode at potential V. When the charge in the sheet is small enough that the electric field is approximately V/ℓ, the velocity of the electrons is

$$u = \frac{dz}{dt} = \eta \frac{V}{\ell} t \tag{8-11}$$

The velocity of the charge sheet increases linearly with time until the sheet arrives at the anode. The induced current is $I_i = A\,J_i$ where A is the area of the sheet and

$$J_i = \frac{\sigma_e}{\ell} \eta \frac{V}{\ell} t \tag{8-12}$$

The induced current has a triangular waveform, as shown in Figure 8-12.

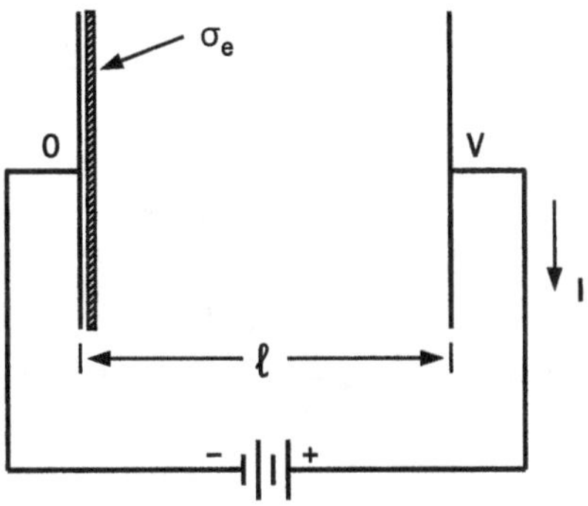

Figure 8-11 Charge sheet emitted by a cathode with an accelerating voltage on the anode.

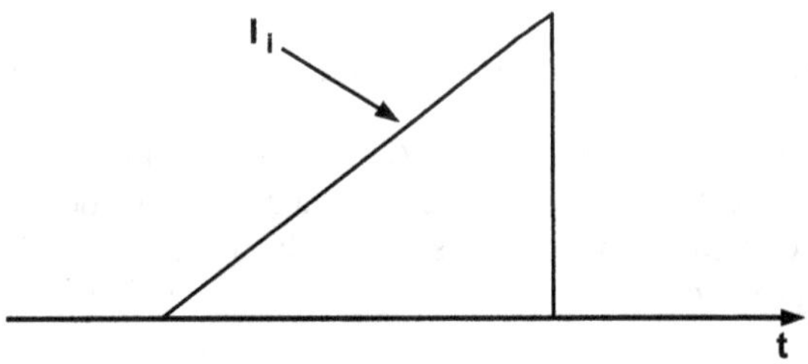

Figure 8-12 Current waveform induced by a charge sheet accelerated from rest to the anode potential in a diode.

If several sheets of charge are considered, then, as long as interactions between them are ignored, the current waveforms produced are superimposed. For the case of several sheets of electrons emitted at random times and leaving the cathode with zero velocity and being accelerated across the gap to an anode, the current pulses might appear as shown in Figure 8-13. This situation is fundamental to the analysis of shot noise in the current emitted by a cathode.

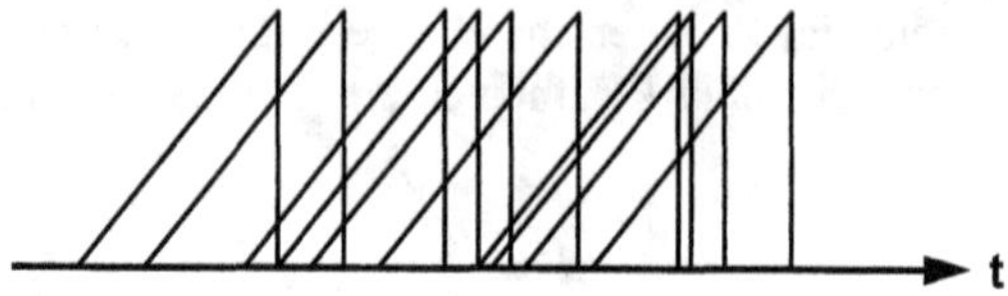

Figure 8-13 Superimposed current pulses induced by several charge sheets having the same charge.

To determine the current induced by an electron beam, the beam can be divided into slices with a thickness Δz, each one of which induces the current density, ΔJ_i, into the circuit. This is indicated in Figure 8-14. Then all of the increments of current can be added to obtain $I_i = A\,J_i$ where

$$J_i = \frac{1}{\ell}\int_0^{\ell} J_e\,dz \tag{8-13}$$

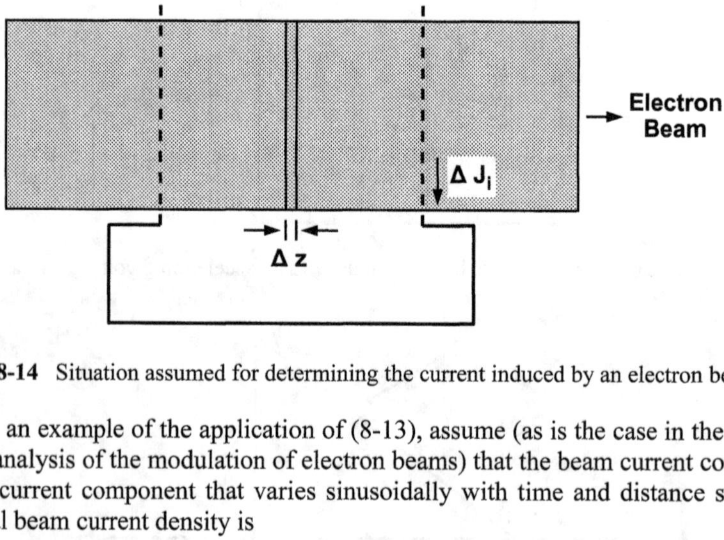

Figure 8-14 Situation assumed for determining the current induced by an electron beam.

As an example of the application of (8-13), assume (as is the case in the small signal analysis of the modulation of electron beams) that the beam current contains an RF current component that varies sinusoidally with time and distance so that the total beam current density is

$$J_e = J_o + J_1 \sin\left(\omega t - \frac{\omega}{u_o} z\right) \tag{8-14}$$

When this current is used in (8-9), the RF electron current induced in the circuit and referenced to the center of the gap is found to be

$$I_i = J_1\,A\,M_p\,\sin\omega t \tag{8-15}$$

where M_p is the gap coupling coefficient for the planar gap (8-4) and A is the cross-sectional area of the electron beam.

The process by which the sinusoidal component of beam current is induced in a circuit is shown graphically in Figure 8-15. At time t_o, the position of the beam is such that the increase in beam charge density (electron bunch) is equal and opposite to the adjacent decrease in density (sometimes called an antibunch and denoted here by plus signs), so the electron current, I_i, induced in the circuit at t_o is zero. At time, t_1, only an electron bunch is between the grids and the induced current is maximum. The situation at t_2 is similar to that at t_o in that the net beam charge between the grids is zero and the induced current is zero. At t_3, only an antibunch is between the grids and the circuit current is minimum.

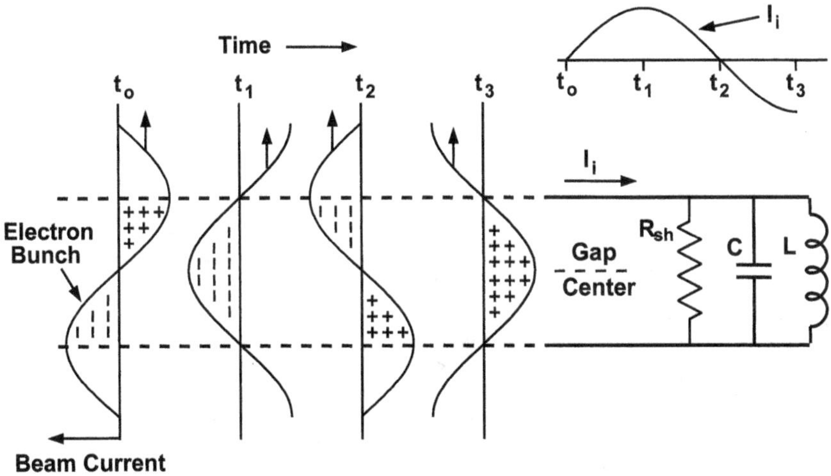

Figure 8-15 Current induction by a beam passing through a cavity.

In these discussions of current induction, as long as the gap voltage is small compared to the beam voltage, the induced current depends only on the geometry of the gap. However, the interaction of the induced current with the impedance of the circuit attached to the gap does produce a voltage across the gap. That voltage, in turn, does produce effects on the beam.

First, it is important to remember that, as a function of frequency, the impedance of a parallel RLC circuit is as shown in Figure 8-16. When an electron beam modulated at frequency f_o passes through the gap, the induced current flows through the resistive portion of the circuit and produces the voltage shown in Figure 8-17. The phase of the induced voltage is such that it slows the bunch (slightly at small signal levels). At the same time, a new signal is launched on the beam, which is 90° out of phase with I_i. Part of the energy extracted from the bunch is dissipated in R_{sh} and part is used to launch the new signal.

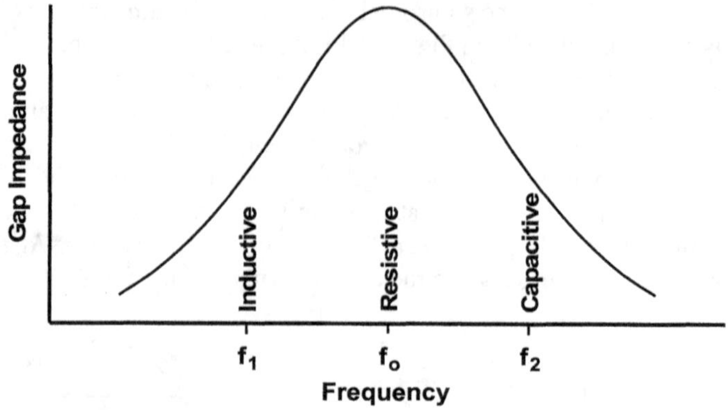

Figure 8-16 Impedance presented to an electron beam by a resonant circuit as a function of frequency.

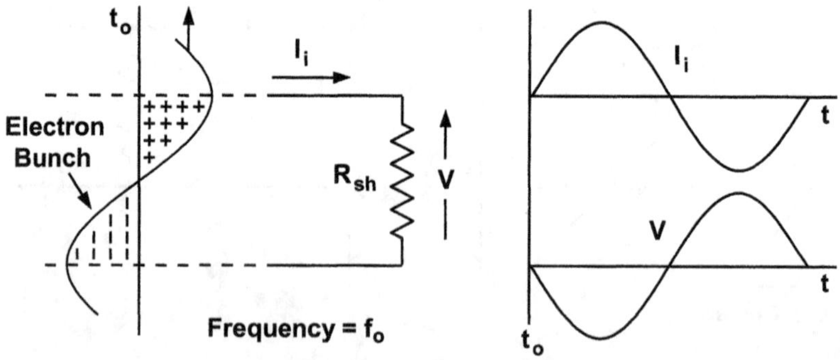

Figure 8-17 When the circuit is resistive (because the electron beam is modulated at f_o), the gap voltage generated is phased 180° with respect to the induced current. Energy is extracted from the bunch.

When the electron beam is modulated at $f_1 < f_o$, the impedance presented to the beam is primarily inductive. The phase of the induced voltage with respect to I_i is as shown in Figure 8-18. (Remember, with conventional current in an inductive circuit, voltage leads current. With electron current, voltage lags current.) The voltage across the gap at t_o decelerates electrons in and near the leading edge of the bunch and enhances bunching. When the voltage has reversed 180° later, it accelerates electrons in and near the trailing edge of the bunch and again enhances bunching. Thus, at small signal levels, the signal launched on a beam by an inductive circuit voltage is in phase with the signal on the beam. At large signal levels near the output circuit of a klystron, an inductively tuned resonator is frequently used to enhance bunching by sweeping electrons from regions between bunches (decelerating electrons ahead of a bunch and accelerating those behind).

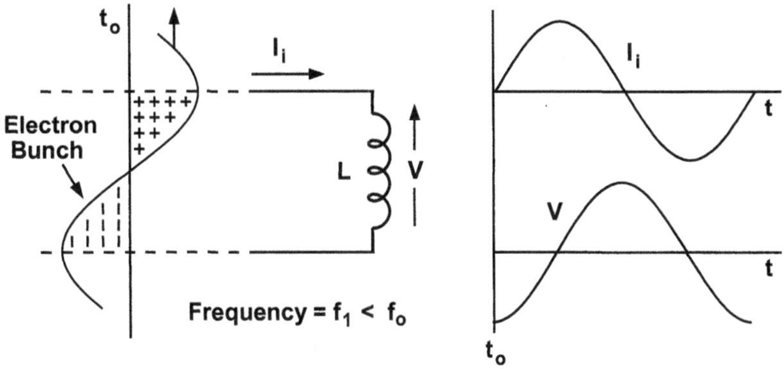

Figure 8-18 When the circuit is inductive, the induced voltage lags the electron current by 90°.

When the electron beam is modulated at $f_2 > f_o$, the impedance of the resonant circuit is primarily capacitive as shown in Figure 8-19. In this case the induced electron current lags the voltage by 90°. The resulting effect on the electron bunches in the beam is the opposite of that which takes place with an inductive circuit. Thus, the voltage accelerates electrons away from the leading and trailing edges of a bunch, which tends to destroy the bunch. At small signal levels, the signal launched on the beam by a capacitive circuit is 180° out of phase with the signal on the beam.

In the following chapter, it will be shown that the RF beam current density produced by a modulating cavity, J_1, is proportional to the velocity modulation, that is,

$$J_1 \propto u_1 \tag{8-16}$$

Since $u_1 \propto M_p$ (8-2), then

$$J_1 \propto M_p \tag{8-17}$$

This means that the induced current must contain M_p^2. (This assumes that the modulating gap and the current induction gap are the same. If the two gaps are different, then the induced current contains the product of the modulation coefficients.) Thus,

$$I_1 \propto M_p^2 \sin \omega t \tag{8-18}$$

For wide gaps, therefore, the induced current is much smaller than for narrow gaps. This effect is even more pronounced in gridless gaps, where M is a function of the radial as well as the axial dimensions of the gaps.

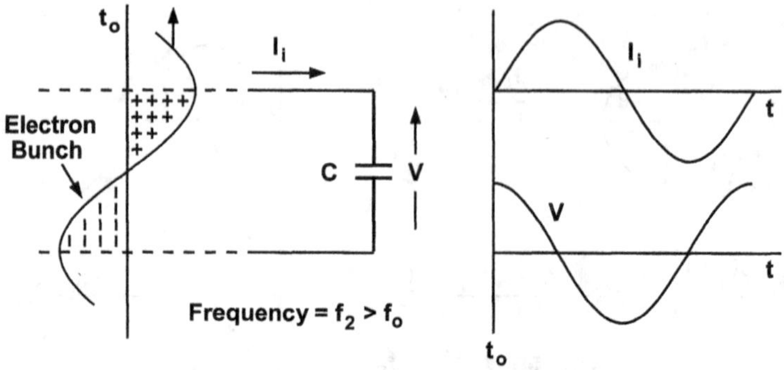

Figure 8-19 When the circuit is capacitive, the electron current lags the voltage by 90°.

8.2.2 Gridless (Nonplanar) Gaps

The principles described for current induction in a gridded gap apply equally well in a gridless gap. Here, the appropriate gap coupling coefficient must be used. At a very large signal level, such as that in the output cavity of an efficient klystron, it is important to accurately model radial and axial fields in the interaction region. This was shown by Kosmahl and Albers in an accurate three-dimensional analysis of the performance of output gaps in klystron amplifiers [3].

The electron beam was divided into axially and radially deformable rings. Figure 8-20 shows the trajectory of an outer ring that was slowed down substantially and finally intercepted. In this figure, only the right half of the gap is shown. The left half is a mirror image of the right half. The phase of the RF is indicated along the trajectory. At ωt near 300° and close to the tunnel tip, the ring rapidly decelerated, stopped moving in the axial direction, and then reversed its axial motion. The convergent action of the field pushed the ring inward and then, in the accelerating phase of the field, the axial motion again reversed and the ring was accelerated almost radially outward and was intercepted by the tunnel wall.

8.3 BEAM LOADING

The bunching action in a cavity of a klystron speeds up electrons over half the RF cycle and slows them down over the other half. When the gap transit angle is small, the energy of electrons leaving the gap, averaged over a cycle, is nearly equal to the energy with which they enter the gap. As the gap transit angle is increased, the average energy of electrons leaving the gap becomes greater than the entering energy. As a result, power must be supplied to bunch the electrons and there is an equivalent resistance that can be assigned to the beam to represent the load in which the power is dissipated. This resistance is called the beam loading resistance.

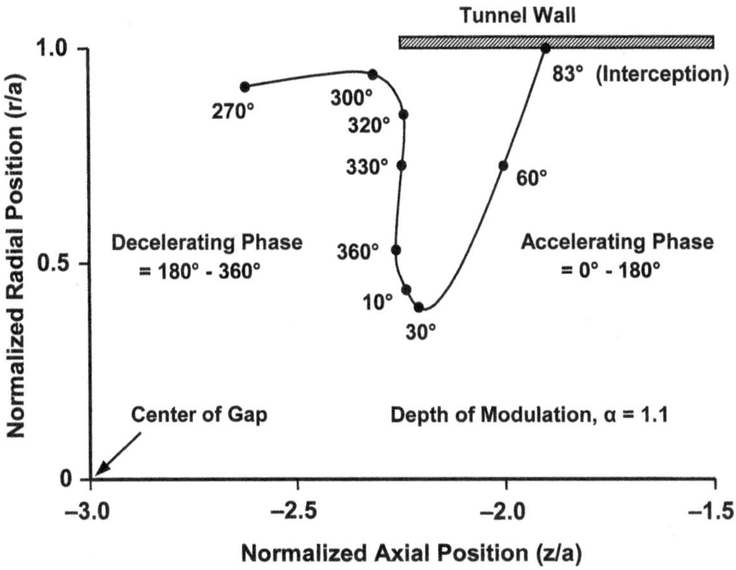

Figure 8-20 Trajectory of the outer ring of charge in the output cavity of a klystron. (Adapted from: H. G. Kosmahl and L. U. Albers, *IEEE Trans. Electron Devices*, October 1973. ©1973 IEEE.)

In examining equivalent circuits of klystron cavities, we will use admittances rather than impedances, so the beam loading conductance will be of interest. The beam also provides a reactive component of admittance, which changes the resonant frequency of the cavity by a small amount.

The beam loading admittance can be determined by using Ramo's theorem (8-13) for induced current in a circuit. The result is

$$I_i = G_B V_1 \sin \omega t + B_B V_1 \cos \omega t \tag{8-19}$$

where G_B is the beam loading conductance given by

$$G_B = G_o \frac{2(1-\cos\varphi_o) - \varphi_o \sin\varphi_o}{2\varphi_o^2} \tag{8-20}$$

and B_B is the beam loading susceptance given by

$$B_B = G_o \frac{2\sin\varphi_o - \varphi_o(1+\cos\varphi_o)}{2\varphi_o^2} \tag{8-21}$$

and G_o is the dc beam conductance, I_o/V_o. Plots of beam loading conductance and susceptance for a gridded gap are given in Figure 8-21 [4].

The beam loading conductance varies from small values for short gaps to a maximum of about $0.2G_o$ for a gap somewhat longer than π radians. For a 10,000-V, 1-A beam, $G_o = 1\times10^{-4}$ mhos ($R_o = 1\times10^4$ ohms) and so $G_B = 2\times10^{-5}$ mhos ($R_B = 5\times10^4$ ohms). At small signal levels, this beam loading resistance is comparable to the cavity resistance.

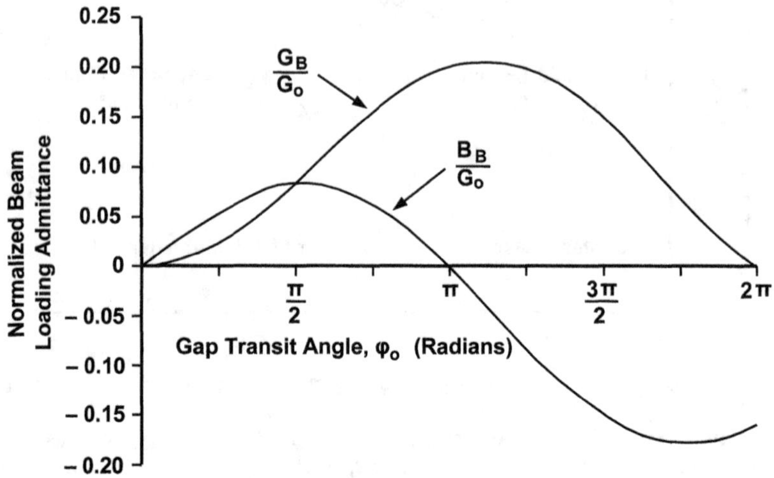

Figure 8-21 Beam loading admittance for a gridded gap. (Adapted from: *Vacuum Tubes*, by K. R. Spangenberg, copyright 1948 by McGraw-Hill.)

REFERENCES

[1] H. G. Kosmahl, and G. M. Branch, "Generalized representation of electric fields in interaction gaps of klystrons and traveling-wave tubes," *IEEE Trans. Electron Devices*, Vol. ED-20, No. 7, July 1973, pp. 621–629.

[2] J. R. M. Vaughan, "A model for the klystron cavity gap," *IEEE Trans. Electron Devices*, Vol. ED-32, No. 11, November 1985, pp. 2482–2484.

[3] H. G. Kosmahl, and L. U. Albers, "Three-dimensional evaluation of energy extraction in output cavities of klystron amplifiers," *IEEE Trans. Electron Devices*, Vol. ED-20, No. 10, October 1973, pp. 883–890.

[4] K. R. Spangenberg, *Vacuum Tubes*, New York. McGraw-Hill, 1948.

Chapter 9

Electron Bunching Produced by a Gap

In Chapter 8, the modulation of an electron beam was examined. It was shown that the velocities of the electrons in the beam are modulated during the interaction with the circuit. In a klystron, this interaction occurs at discrete locations along the length of the device. In a traveling wave device (a TWT and a crossed-field device), the interaction is continuous because the electric field on the circuit that modulates the beam travels along the circuit in synchronism with the electron bunches in the beam. There are significant differences between the bunching action in a klystron and that in a traveling wave device. In this chapter, we describe klystron bunching. The bunching that occurs in the presence of a traveling wave in a TWT is discussed in Chapter 12. Bunching in a crossed-field device is covered in Chapters 19 and 20.

9.1 BALLISTIC BUNCHING

The term *ballistic bunching* is used to describe the bunching of electrons when space charge forces are neglected. Consider again the equation that was developed in Chapter 8 for the velocity of electrons that have interacted with a modulating electric field,

$$u = u_o\left(1+\frac{\alpha M}{2}\sin\omega t\right) \tag{9-1}$$

This relation shows, as is expected, that when an electron beam is passed through a modulating electric field, some electrons have their velocities increased, some have their velocities decreased, and some have their velocities unchanged. After the beam leaves the field, those electrons that have had their velocity increased gradually overtake the slower electrons. This process is illustrated in Figure 9-1, which is known as an *Applegate diagram*. Axial positions of electrons are plotted as a function of time, and so the slopes of lines represent electron velocities. The trajectories of electrons leaving the modulating field when it is decelerating (indicated at the bottom of the figure) are represented by lines of reduced slope.

Similarly, the trajectories of accelerated electrons are represented by lines of increased slope.

In Figure 9-1, space charge forces between electrons have been ignored, so the electron trajectories are shown to cross. In practice, at small and medium drive signal levels, space charge forces prevent electron trajectories from crossing. That situation is discussed later in this chapter. At large signal levels, electron trajectories do cross, so the following discussion is relevant to that situation.

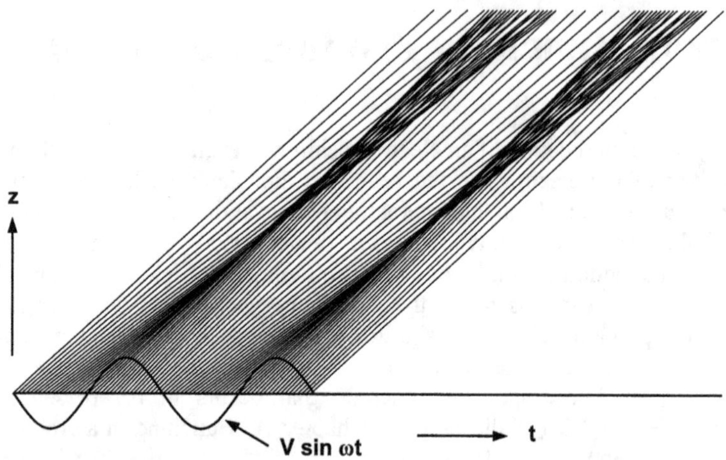

Figure 9-1 Distance-time curves for the electrons in a velocity-modulated beam in a field-free drift space (Applegate diagram).

As a result of the convergence of electron trajectories illustrated by the Applegate diagram, electron bunching is said to occur. The traveling electron bunches constitute an RF current in the electron beam. For this ballistic case, the RF current in the beam is

$$i(z) = \frac{i_1}{|1 - X \cos \omega t|} \tag{9-2}$$

where

$$X = \frac{\omega z}{u_o} \frac{\alpha M}{2} \tag{9-3}$$

is called the *bunching parameter*. If α is small, then $i_1 \approx I_o$, that is, i_1 is approximately the dc beam current.

The current waveforms as a function of axial position are sketched in Figure 9-2 (see also [1]). Note that, because of the cosine function in the denominator with an arbitrarily large multiplier, infinite current peaks are possible with this analysis. In practice, very large peaks may occur but, as will be shown later, space charge forces limit the amplitudes of the peaks.

The current waveforms shown in Figure 9-2 are rich in harmonics, but usually the only component of interest is the fundamental component. This may be obtained [2] by doing a Fourier analysis of the waveforms. When this is done, the fundamental component of RF current, i_f, is found to be

$$i_f = 2I_o J_1(X) \tag{9-4}$$

where J_1 is a Bessel function of the first kind. This relation is as plotted in Figure 9-3. Note that the peak value of i_f is larger than the dc current. The value of X for which i_f is maximum is 1.84, and so the axial position, z, at which the peak value occurs is

$$z = 1.84 \frac{2u_o}{\omega \alpha M} \tag{9-5}$$

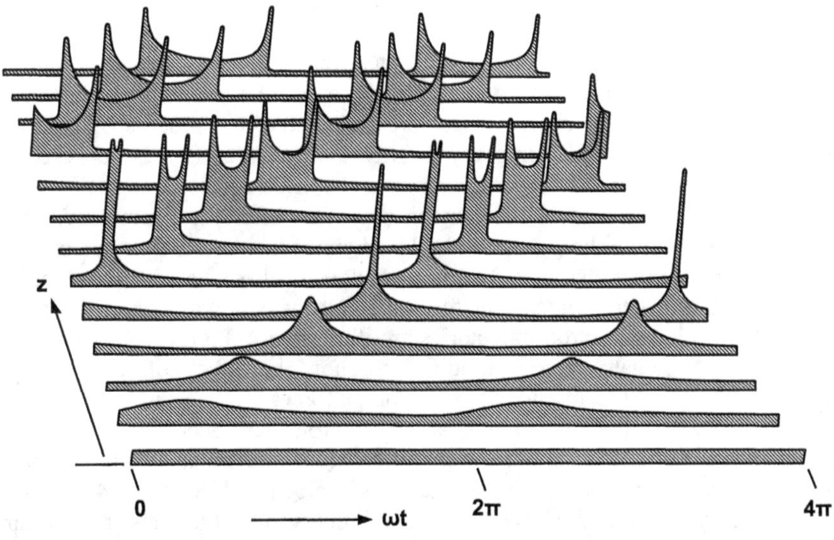

Figure 9-2 Current waveforms as a function of distance. (Adapted from: *Vacuum Tubes* by Karl R. Spangenberg, copyright 1948 by McGraw-Hill Book Company, Inc.)

In a device like a two-cavity klystron where it is desirable to locate a second gap at the point of maximum fundamental current, a problem arises because of (9-5). The position of maximum current is inversely proportional to the depth of modulation, α, and therefore to the RF drive level. Once a position has been chosen, then for larger or smaller drive levels, the peak current in the beam remains the same but the axial position of the peak changes. Also, once a drive level and position have been chosen, the relationship between drive level and RF current is not linear because of the J_1 function. Fortunately, this problem is alleviated when space charge effects are taken into consideration.

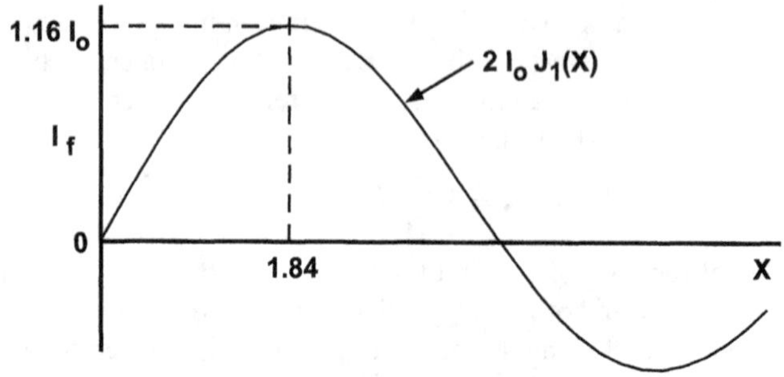

Figure 9-3 Amplitude of fundamental component of RF current.

9.2 BUNCHING WITH SPACE CHARGE FORCES

The Applegate diagram in Figure 9-1 illustrates electron bunching after a beam is passed through a modulating gap when space charge forces are ignored. For small signal levels, when space charge forces are considered, the process of velocity modulation within the modulating gap is altered very little. This is because the electrons barely move from their dc positions as the beam passes through the gap.

As electrons converge in the bunching process after the beam leaves the modulating gap, space charge forces build up, impeding the rate at which fast electrons overtake slow electrons. In addition, slow electrons have their velocities increased because of the bunch that forms behind them (toward the modulating cavity). The resulting electron trajectories are illustrated in Figure 9-4, which is a portion of an Applegate diagram modified to show the effects of space charge forces. Trajectories in the absence of space charge forces are also shown.

At the axial position denoted by $\lambda_p/4$, fast electrons have been slowed to the dc velocity of the beam and slow electrons have been accelerated to the dc beam velocity. At $\lambda_p/4$, all electrons have the same velocity. Also, at $\lambda_p/4$, the RF electron density and the RF current reach maximum values. For small to medium RF signals, the RF current is nearly sinusoidal.

A very important characteristic of the bunching process with space charge forces is that *all* electrons are either speeded up or slowed down to the same velocity (the dc beam velocity) at the same axial position ($\lambda_p/4$). In addition, even if the amplitude of the modulating field is changed so that initial electron velocities are changed, the axial position of the bunch remains the same. This result is extremely important to the klystron engineer because, unlike the situation when space charge forces are ignored, the cavity location for maximum RF beam current is not a function of signal level, of gap width, or of frequency of operation. The distance, λ_p, is the electron plasma wavelength. The origin of this wavelength is the natural frequency of oscillation (the electron plasma frequency) of a cloud of electrons and the velocity with which the cloud of electrons is moving.

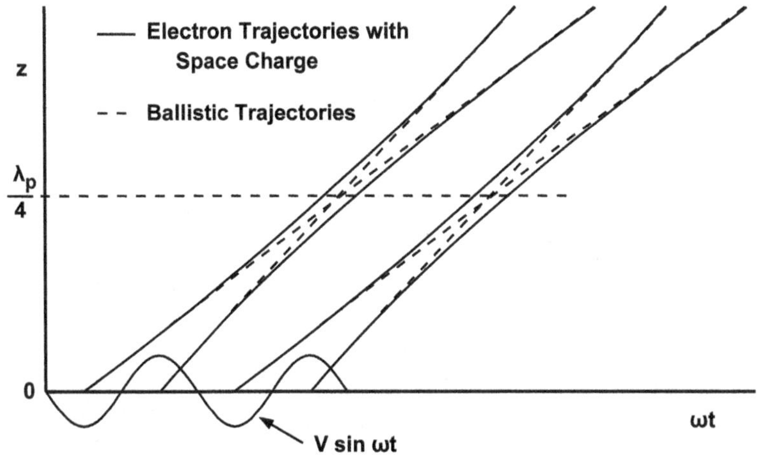

Figure 9-4 Modified Applegate diagram indicating space charge effects.

To understand electron plasma oscillations, it is useful to start by considering an electron cloud that, as a whole, is not moving. The transverse dimensions of the cloud are assumed to extend to infinity and stationary ions are assumed to be uniformly distributed throughout the cloud to neutralize the average charge of the electrons. (These ions will not affect the electron plasma oscillations.)

Initially, assume that the electrons are uniformly distributed as shown in Figure 9-5(a). Assume further that some electrons have been accelerated to the left and some have been accelerated to the right as indicated by the arrows. (This could be accomplished by moving a pair of parallel grids, between which an RF voltage is applied, through the cloud. This is similar to what happens in modulating gaps in a klystron except that, in those cases, the electrons move and the gaps are stationary.)

After acceleration, electrons move to the right and to the left as indicated in Figure 9-5(b). Regions of high density (bunches) form and the resulting space charge forces in these bunches cause the electron movement to stop and then reverse directions. Next, as indicated in Figure 9-5(c), the electrons become uniformly distributed again but are moving in directions opposite to those shown in Figure 9-5(a). As the electrons continue to move, new bunches form as indicated in Figure 9-5(d). Again the space charge forces that build up in the bunches cause the electron movement to stop and reverse directions. After that, the distribution shown in Figure 9-5(a) returns and the process starts over again. The electrons continue to move back and forth at a frequency known as the *electron plasma frequency*, which is given by

$$\omega_p = 2\pi f_p = \left(\eta \frac{\rho_o}{\varepsilon_o}\right)^{1/2} \tag{9-6}$$

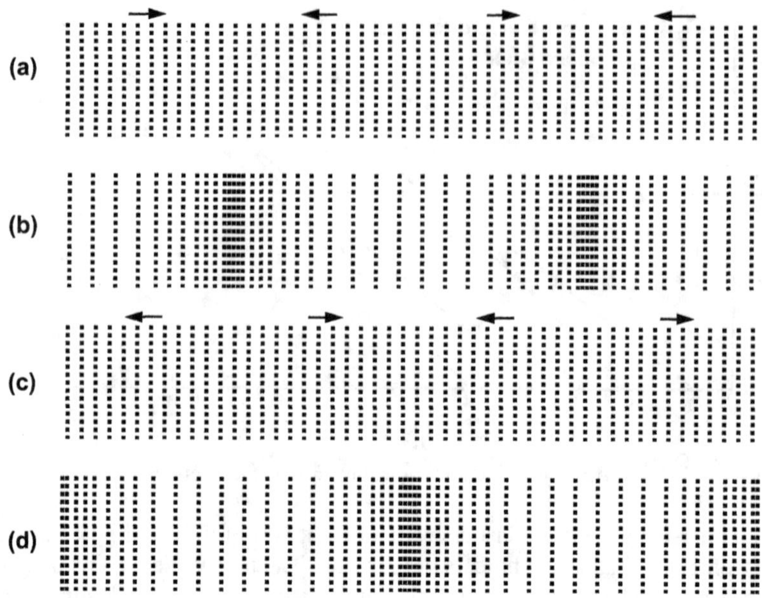

Figure 9-5 (a–d) Electron distributions during plasma oscillations.

The plasma frequency is dependent *only* on the electron density, ρ_o. We will see later that ω_p is an extremely important quantity in the analysis and understanding of wave motion in a TWT.

To appreciate the magnitude of the plasma frequency, consider a 10,000-V, 1-A electron beam with a diameter of 2 mm. The electron plasma frequency, f_p, is

$$f_p = 1.64 \text{ GHz} \tag{9-7}$$

which is well below the operating frequency of the TWT containing that beam.

The preceding discussion of electron plasma oscillations was for a stationary cloud of electrons. When the cloud is moving so that an electron beam is formed, the plasma oscillations are not affected (unless the beam velocity is high enough that relativistic effects are important). Then, referring back to Figure 9-4, the situation at $z = 0$ is like that in Figure 9-5(a). At $\lambda_p/4$, the electrons are compressed as shown in Figure 9-5(b). This is one-quarter of a cycle of oscillation after $z = 0$ so

$$\frac{\lambda_p}{4} = \frac{u_o}{f_p / 4} \tag{9-8}$$

or

$$\lambda_p = \frac{u_o}{f_p} = \frac{2\pi u_o}{\omega_p} \tag{9-9}$$

For the 10,000-V, 1-A beam considered previously,

$$\lambda_p = 3.62 \text{ cm} \tag{9-10}$$

This length is much greater than the beam diameter, so boundary effects (the relatively small diameter of the beam and the presence of a nearby metallic structure) have a considerable impact on the magnitude of the electric fields between the bunches. The fields are reduced, so ω_p is reduced and λ_p is increased.

For a confined beam (no transverse electron velocities because of large B_z), as shown in Figure 9-6, the effect of finite beam radius on plasma frequency is shown in Figure 9-7. Here ω_q is the reduced plasma frequency and F is called the *plasma frequency reduction factor*. Note that, if $\omega b/u_o = 1$ and $n = 1.5$, the value of F is about 0.5, so the plasma frequency is one-half of the value calculated for an infinite beam. Correspondingly, the plasma wavelength for the confined beam, λ_q, is twice the value calculated for an infinite beam. Thus, the maximum RF current in the beam occurs at a distance from the modulating cavity that is twice that for the infinite beam.

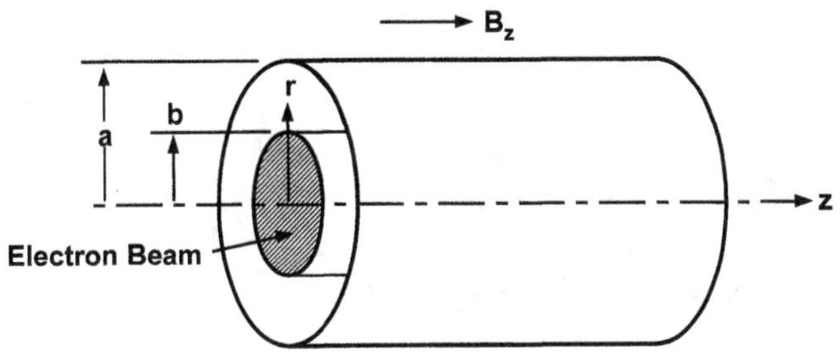

Figure 9-6 Cylindrical beam in a cylindrical tunnel.

When transverse electron velocities are present, as in a Brillouin focused beam, the plasma frequency is reduced even more than it is in a confined beam. The reason for this can be explained with the aid of Figures 9-8 and 9-9. For cavity modulation of a Brillouin beam at small signal levels, the electron bunching exists as a modulation of the radius of the beam rather than as a modulation of the beam density. In Figure 9-9, the electric field pattern for the bunches in a Brillouin beam (depicted as plus and minus charges on the surface of the beam) is compared to that for the confined beam. Note that the field lines can bow inward as well as outward from the surface bunches on the Brillouin beam. As a result, the axial component of the field and the space charge forces are smaller than for the confined beam. This results in a lower plasma frequency and a larger plasma wavelength for the Brillouin beam than for the confined beam.

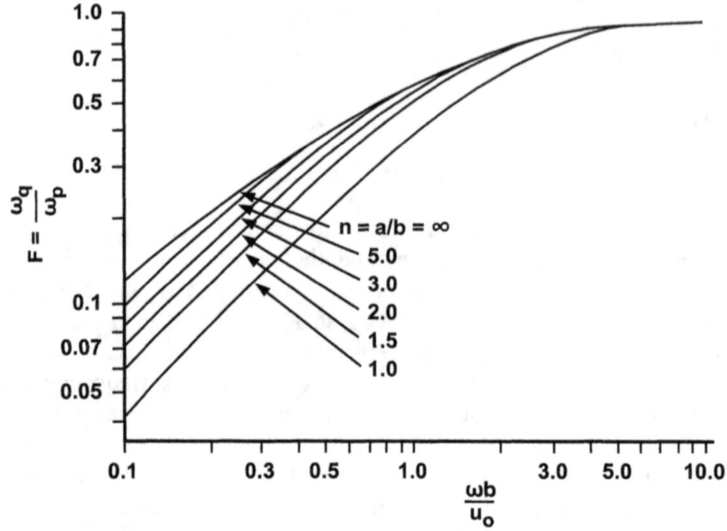

Figure 9-7 Plasma frequency reduction factor for a confined beam.

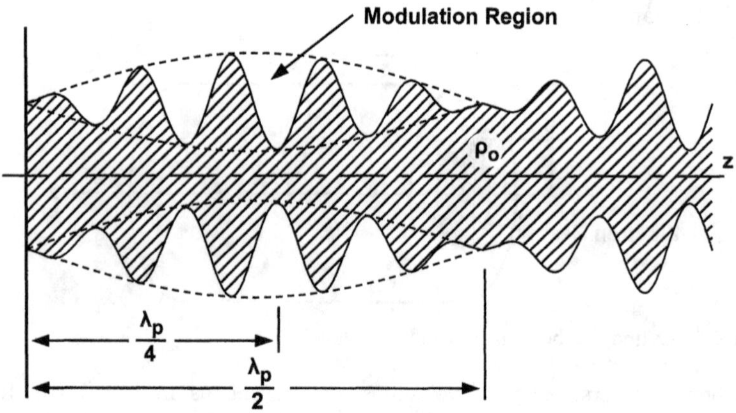

Figure 9-8 RF modulation of beam radius for a Brillouin beam.

The plasma frequency reduction factors for a Brillouin beam are shown in Figure 9-10. It should be noted that the values of F for the Brillouin beam are somewhat smaller than for confined flow. For example, for $n = 1.5$ and $\omega b/u_o = 1$, the value of F is about 0.4, whereas, for the confined beam, the value is 0.5.

In the ballistic analysis there is no need to consider propagating electric and magnetic fields. However, when space charge effects are considered, these fields must be included and the analysis is far more complex. When the analysis is carried out, the results are straightforward as would be expected from the previous qualitative discussion.

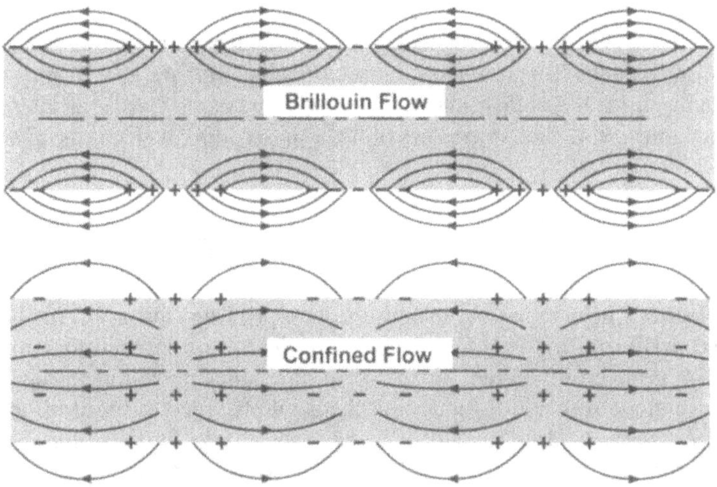

Figure 9-9 Comparison of RF electric field patterns near bunches in a Brillouin beam and a confined beam.

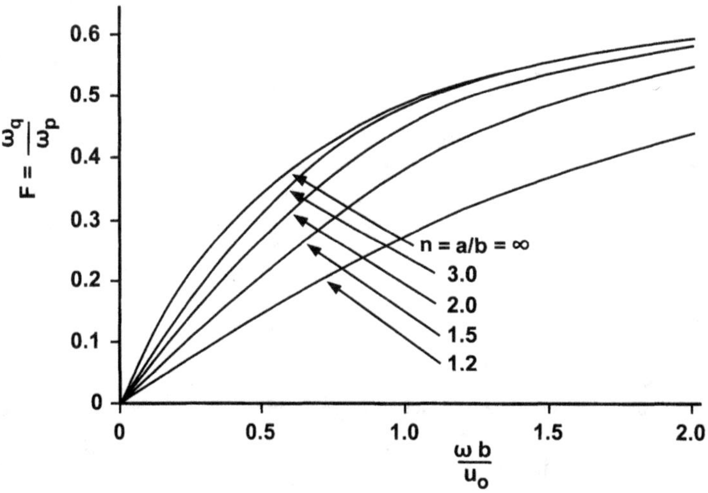

Figure 9-10 Plasma frequency reduction factor for a Brillouin beam.

The axial component of the electron beam velocity found from the analysis is

$$u_z = u_o \frac{\alpha M}{2} \cos\beta_q z \cos(\omega t - \beta_e z) \tag{9-11}$$

where $\beta_q = \omega_q / u_o$ and $\beta_e = \omega / u_o$ are propagation constants.

If we move along the beam at a velocity such that $\omega t = \beta_e z$ or at $z/t = u_o$ (the dc beam velocity), then the ac velocity, u_z, of the electrons appears as shown by curve 1 of Figure 9-11. If we choose a time such that $\omega t = \beta_e z - \pi$, then u_z appears as shown by curve 2. At other times, u_z lies between curves 1 and 2 as indicated by the dashed lines. Note that, regardless of ωt, u_z is zero for all electrons at $\beta_q z = \pi/2$ or at

$$z = \frac{\pi}{2\beta_q} = \frac{\pi}{2}\frac{u_o}{\omega_q} = \frac{u_o}{4f_q} = \frac{\lambda_q}{4} \tag{9-12}$$

and all electrons have the same velocity (u_o) at $\lambda_q/4$. The reason for this has been discussed earlier in this chapter. As the fast electrons approach the slow ones at $z = \lambda_q/4$, a bunch is formed. Furthermore, the bunching action causes the fast electrons to slow to the dc velocity and the slow electrons to speed up to the dc velocity. As electrons drift beyond $\lambda_q/4$, the space charge forces within the bunch cause the electrons in the front of the bunch to speed up and those in the rear of the bunch to slow down.

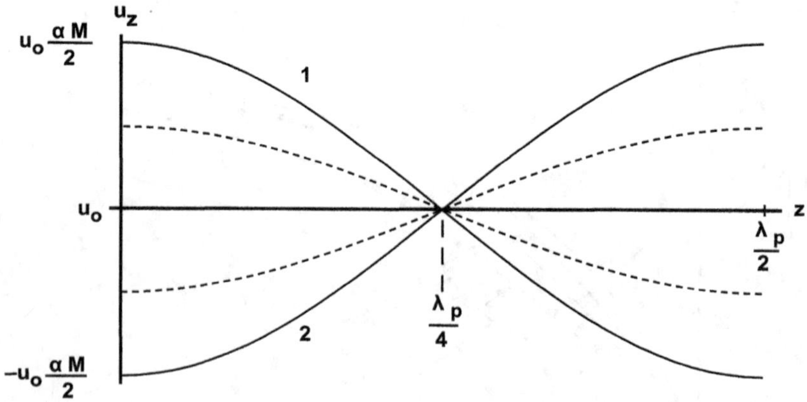

Figure 9-11 Velocities of electrons as a function of position.

The RF current density in the beam is

$$J_z = J_o \frac{\omega}{\omega_q}\frac{\alpha M}{2} \sin \beta_q z \sin\left(\omega t - \beta_e z\right) \tag{9-13}$$

Now, $J_z = 0$ at $z = 0$ and varies sinusoidally with distance as shown in Figure 9-12. The maximum current occurs when $z = \lambda_q/4$, which corresponds to the position where the ac velocity goes to zero. If the electrons drift beyond $\lambda_q/4$, then the bunch disperses and the current density decreases. The bunch and current density disappear at $\lambda_q/2$ and then reappear as the beam passes $\lambda_q/2$.

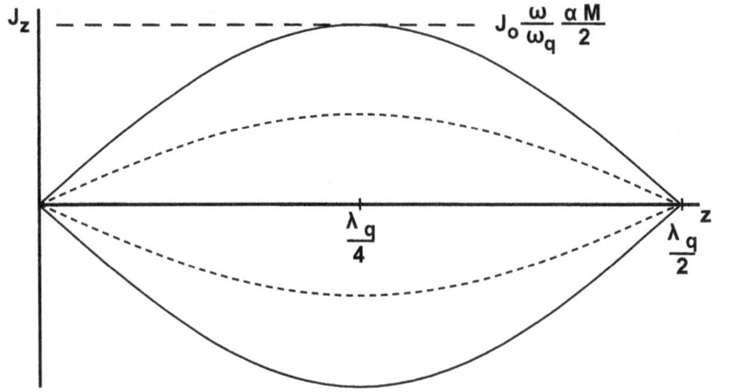

Figure 9-12 Current density as a function of position (as seen by an observer moving at the beam velocity).

The variations of current density shown in Figure 9-12 are those that an observer moving at the beam velocity would see. If, instead, at an instant in time, a "snapshot" of current density is considered, it would appear as shown in Figure 9-13. (ωt = 0, 2π, 4π, and so on for this figure.) The current peaks labeled 1, 2, and 3 result from bunches of electrons that are in the process of forming. Beyond $\lambda_q/4$, the bunches disperse and the current peaks (4 and 5) decrease.

For the plot in Figure 9-13, $\omega = 10\omega_q$ and within a distance λ_q there would be 10 cycles of RF current and 10 bunches of electrons in various stages of formation (five bunches are indicated in $\lambda_q/2$ in Figure 9-13).

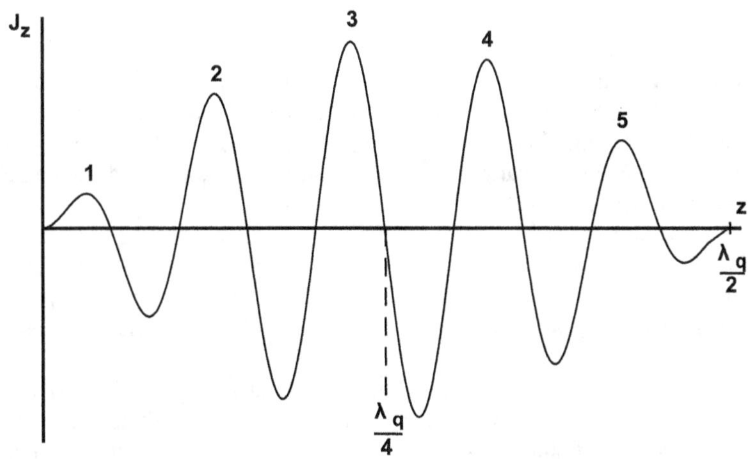

Figure 9-13 Instantaneous current density as a function of position.

9.3 LARGE SIGNAL EFFECTS

At the beginning of this chapter, we examined bunching with no space charge forces. This ballistic bunching produced an RF current that depended only on the bunching parameter, X, which was proportional to αz (α = depth of modulation and z = distance along the electron beam). The maximum RF current was 1.16 times the dc beam current. This same maximum RF current could be obtained at any drive level (any α) simply by choosing the correct axial position in the beam. Thus, current as a function of distance and drive level is as shown in Figure 9-14.

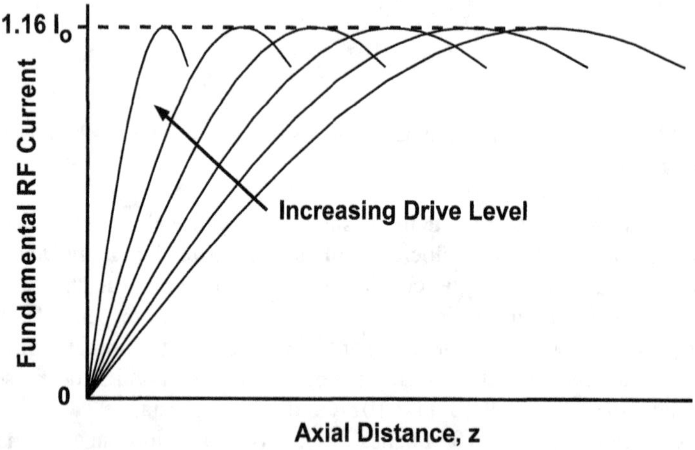

Figure 9-14 RF current as functions of axial position and drive level for ballistic bunching.

When the effects of space charge on bunching were considered, the variation of RF current with α and z was found to be very different than for the ballistic case. At small signal levels, the maximum RF current always occurred at the same distance from the modulating cavity, $\lambda_q/4$. As the drive level was varied, the amplitude of the current varied and the current as a function of distance and drive level is as shown in Figure 9-15.

The transition from space charge wave to ballistic behavior was perhaps best illustrated by Mihran [3]. He used the experimental apparatus in Figure 9-16 to measure RF currents in a beam modulated by a klystron cavity. Mihran's apparatus was a two-cavity klystron with the following additional features:

1. The output cavity was movable allowing drift length to be variable.
2. A control grid allowed variation of beam current at fixed beam voltage.
3. A series of apertures could be placed over the control grid allowing variation of the beam diameter.

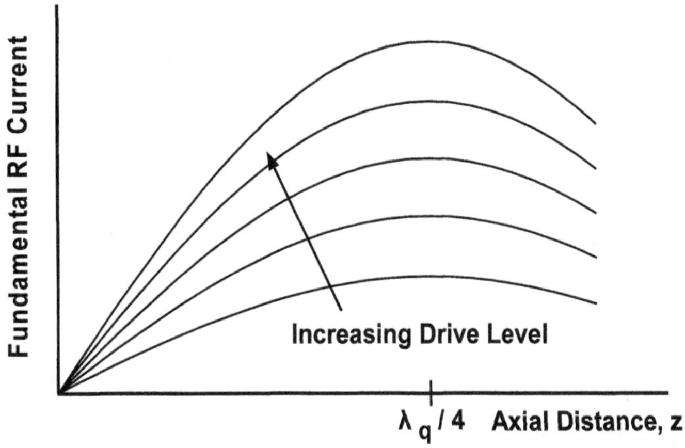

Figure 9-15 RF current as functions of distance and drive level when space charge forces are considered.

4. The output cavity was tuned to a frequency, f_2, which was sufficiently far from the input frequency to ensure that the RF output voltage would always be much lower than the beam voltage. This guaranteed that small signal conditions would prevail in the output gap, thus giving direct information on the electron bunching in the beam.

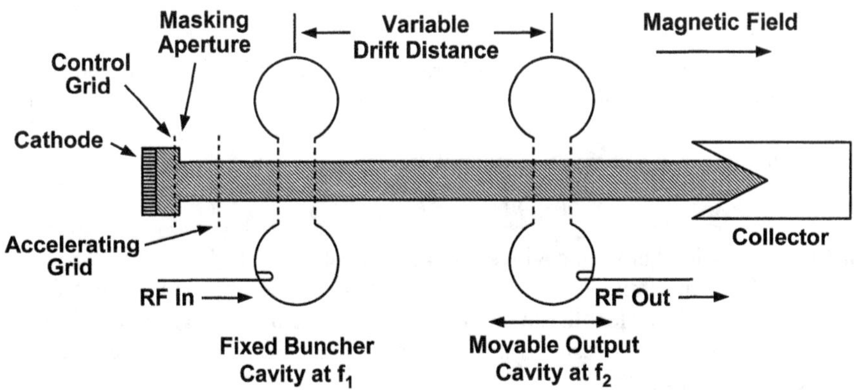

Figure 9-16 Mihran's two-cavity klystron experimental setup.

A set of experimental results is shown in Figure 9-17. These curves show the transition from space charge wave to ballistic behavior as signal level was increased. The dashed lines show the current maxima predicted by large signal theories.

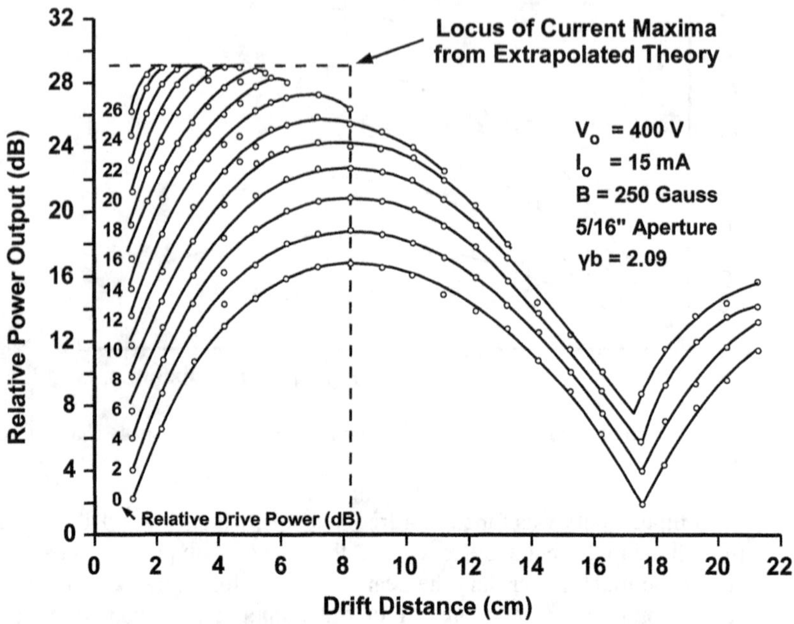

Figure 9-17 Mihran's results showing the transition from space charge wave to ballistic behavior. (From: T. G. Mihran, *IRE Trans. Electron Devices*, January 1959. © 1959 IRE (now IEEE).)

The most significant result of Mihran's work was the demonstration of the amount by which the distance between cavities should be below $\lambda_q/4$ at large signal levels. Mihran gave the optimum distances between cavities of a multicavity klystron as being approximately:

$$90°: 90°: \cdots : 65° : 45°: 30°$$

that is, the spacing between cavities should vary from $\lambda_q/4$ to 1/3 $\lambda_q/4$ as the signal increases from the input level to saturation. Mihran was the first to specify the amount by which drift lengths should be reduced in the output region of a klystron. These reduced spacings resulted in improved efficiencies. This will be discussed in detail in following chapters.

Large signal theories failed to predict the drift tube shortening that Mihran demonstrated was necessary to optimize efficiency. However, Webber [4] was able to use a disk electron model like that shown in Figure 9-18 and compute curves for current versus distance. In Webber's model, 16 disks per wavelength were used for most calculations. Charge was uniformly distributed throughout the disks and the disks were invariant throughout the calculation. The space charge force was determined by first calculating the electric field due to a single disk and then obtaining the average axial force on another disk. The total force on a disk was obtained by integrating this force expression over all other disks in the beam.

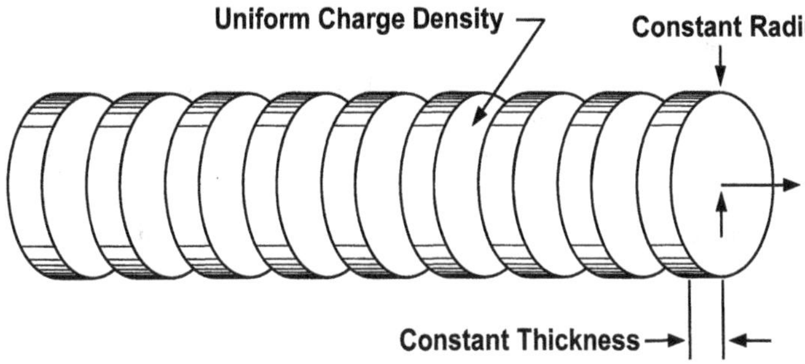

Figure 9-18 Disk model of an electron beam.

The disks were assumed to behave ballistically, like individual electrons, in that each was given an initial velocity and/or phase position and then was allowed to bunch accordingly. Disks were allowed to cross over or pass through one another. Some shortcomings of Webber's model are:

1. Only axial motions of electrons or disks were considered.
2. The disk approximation did not allow any variation of velocity or density modulation across the beam.

By assuming the beam to consist of concentric rings of charge as shown in Figure 9-19, these shortcomings could be overcome. Webber did not do this because of the large computer storage capacity required and because he thought the effort to be of questionable value in view of other factors which limit the performance of actual devices.

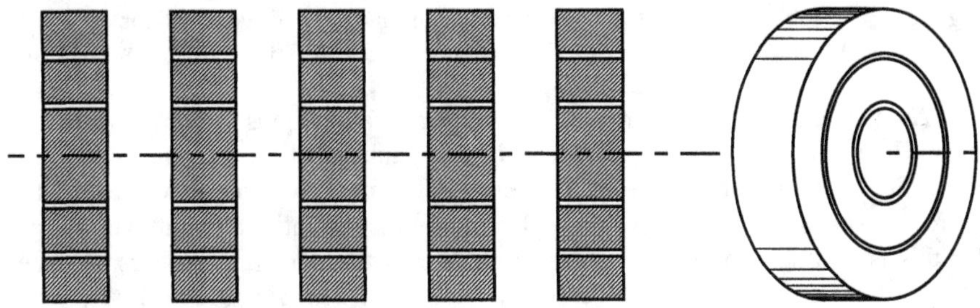

Figure 9-19 Ring model of an electron beam.

A set of Webber's curves showing current versus distance is shown in Figure 9-20. The quantity α contained in $\omega_q/\omega\alpha$ is the depth of modulation so decreasing values of $\omega_q/\omega\alpha$ correspond to increasing drive levels. In general, there is good agreement with Mihran's experimental results. The primary difference is in the transition from space-charge limited behavior to ballistic behavior. Mihran's results show a more gradual transition than do Webber's calculations. Mihran points out that the disagreement is probably attributable to the inaccuracy of the disk electron assumption for large γb where $\gamma = \omega/u_o$. For $\gamma b = 2$ (Mihran's results are for $\gamma b = 2.09$ and Webber's calculations are for $\gamma b = 1.6$), there is a 2 to 1 variation of fields from the beam edge to the beam center. This would warp the electron disks assumed in Webber's calculations.

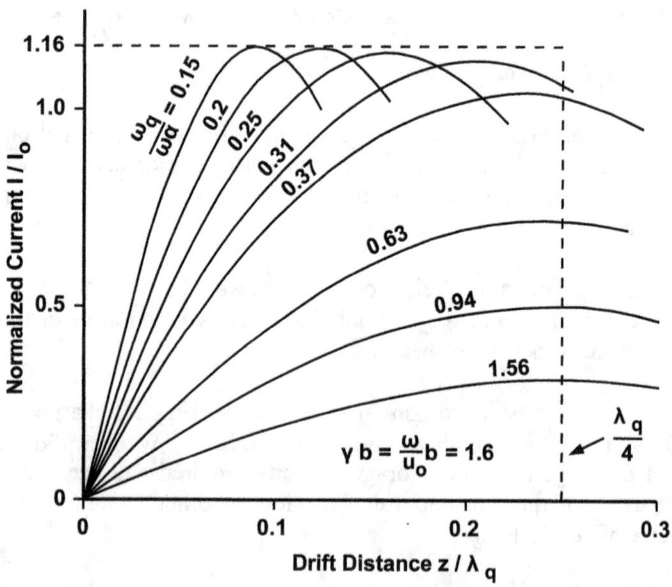

Figure 9-20 Webber's calculated current versus distance and drive level for $\gamma b = 1.6$. (From: S. E. Webber, *IRE Trans. Electron Devices*, April 1958. © 1958 IRE (now IEEE).)

Another example of the results of Webber's calculations is shown in Figure 9-21. Here $\gamma b = 0.4$ and the beam was considerably smaller in diameter than for the example shown in Figure 9-20. Notice that, for very large drive levels, the current exceeds the ballistic value of 1.16 I_o by a substantial amount. The reason for this is that space charge is actually enhancing the bunching process. Webber showed that electrons near the center of the bunch, which would normally cross over (fast electrons would pass slow ones) if moving ballistically, are stopped and reversed relative to the center of the bunch before crossing takes place. The result is that the fundamental component of current in the beam is enhanced. Webber showed that the charge densities at which this effect is optimized correspond to relatively low beam perveances. We will see in a later chapter when high-efficiency klystrons are discussed that a relatively low perveance (substantially less than 1×10^{-6}) is necessary for high efficiency.

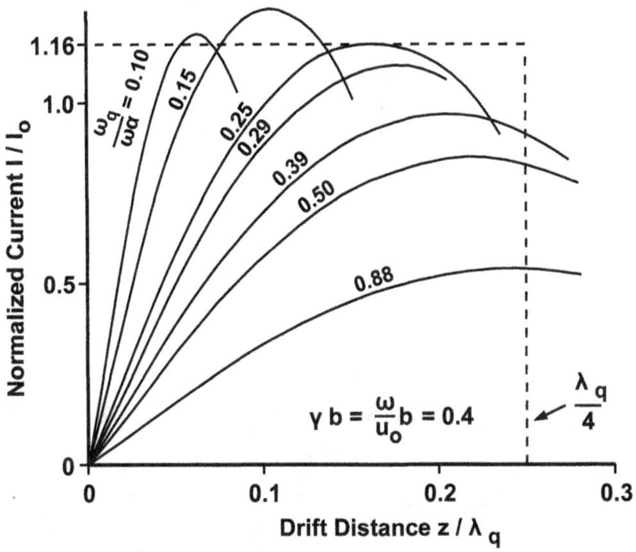

Figure 9-21 Webber's calculated current versus distance and drive level for γb = 0.4. (From: S.E. Webber, *IRE Trans. Electron Devices*, April 1958. © 1958 IRE (now IEEE).)

Experiments have been performed to verify the prediction that the RF current in a Brillouin-focused beam exists as a modulation of the radius of the beam [5]. The RF beam analyzer used for the measurements is shown in Figure 9-22. This was the analyzer used for the dc beam measurements described in Chapter 7. The analyzer was modified for RF current measurements. The beam from a shielded Pierce gun was focused with a solenoid. A vacuum ball valve was used to permit changes to be made in the measurement apparatus without exposing the gun and cathode to air (which could have changed the beam characteristics). The beam was velocity-modulated at approximately 2,000 MHz. A beam-scanning mechanism with a 0.010-inch aperture was used for sampling currents at various transverse and axial positions.

Plots of the fundamental component (harmonics could also be observed) of RF beam current as functions of the transverse and axial position in the beam are shown in Figure 9-23 [6]. The RF drive powers of 10W, 50W, and 200W corresponded to approximate values of α of 0.1, 0.25, and 0.5, respectively. At the smallest drive level (10W) the RF current was entirely on the surface of the beam.

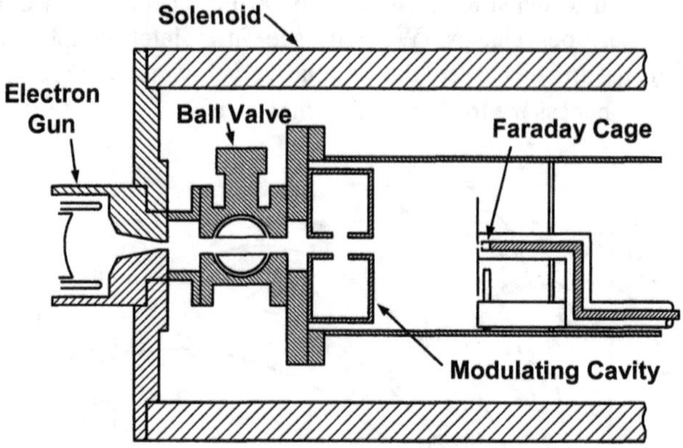

Figure 9-22 RF beam analyzer for measuring surface currents in a Brillouin beam. (Adapted from: A. S. Gilmour, Jr., *IEEE Trans. Electron Devices*, July 1972. © 1972 IEEE.)

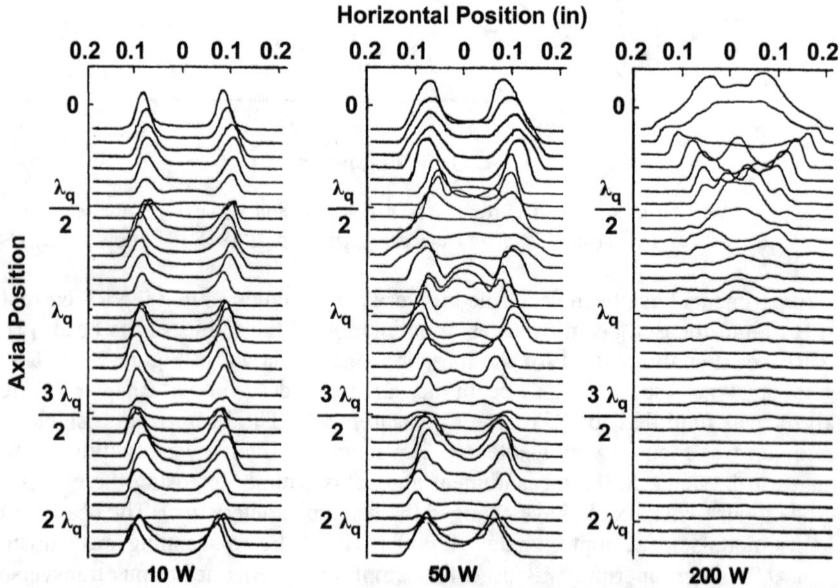

Figure 9-23 Distributions of fundamental RF current for various modulation powers. (From: A. S. Gilmour, Jr., *IEEE Trans. Electron Devices*, July 1972. © 1972 IEEE.)

The predicted space-charge wave behavior of the electrons is clearly illustrated by four half plasma wavelengths of variation of the beam current. At the medium drive level the space-charge wave behavior became severely distorted. At the highest drive level, the space-charge wave behavior disappeared.

Electron energy measurements were made to obtain a measure of the RF velocities of electrons in the beam. First, the peak values of the electron energies were determined at relatively low values of α. The Faraday cage used for collecting the electrons passing through the 0.010-inch aperture in the beam-scanning mechanism was biased to a negative voltage (4,800V) comparable to the dc beam voltage. In addition, a 10,000-MΩ resistor was placed in series with the bias batteries. With this large resistance, only a fraction of a microampere of current to the Faraday cage caused an additional bias of several hundred to several thousands of volts. A fraction of a microampere was less than 0.1% of the normal peak Faraday cage current and consisted therefore of only those electrons having the maximum energy.

Shown in Figure 9-24 are the peak energies (expressed in terms of the bias voltage on the Faraday cage) up to an α of approximately 0.1. At small values of α, the variation of energy was a series of half sinusoids, as expected. When α exceeded 0.05, the energy profiles became distorted. At values of α much above 0.1 (not shown), the profiles became so severely distorted as to be essentially meaningless.

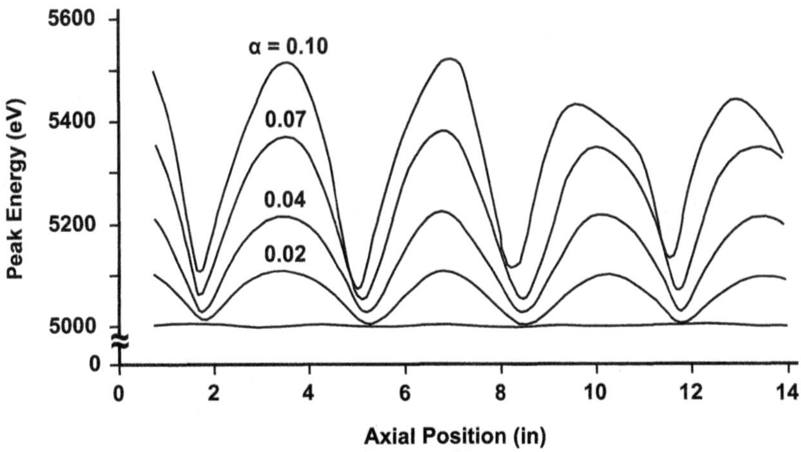

Figure 9-24 Peak beam energy as a function of axial position and RF drive level for Brillouin flow conditions. (From: A. S. Gilmour, Jr., *Proc. High Power Microwave Tube Symposium*, The Hexagon, 1962.)

REFERENCES

[1] W. J. Kleen, *Electronics of Microwave Tubes*, New York: Academic Press, 1958.

[2] D. L. Webster, "Cathode-ray bunching," *Jour. App. Phys.*, Vol. 10, No. 7, July 1939, pp. 501–508.

[3] T. G. Mihran, "The effect of space charge on bunching in a two-cavity klystron," *IRE Trans. Electron Devices*, Vol. ED-6, No. 1, January 1959, pp. 54–64.

[4] S. E. Webber, "Ballistic analysis of a two-cavity finite beam klystron," *IRE Trans. Electron Devices*, Vol. 5, No. 2, April 1958, pp. 98–108.

[5] A. S. Gilmour, Jr., G. C. Dalman, and L. F. Eastman, "The current and velocity distributions in a velocity modulated Brillouin focused electron beam," *Proc. 4th International Congress on Microwave Tubes,* Eindhoven, the Netherlands: Centrex, 1963.

[6] A. S. Gilmour, Jr., "Radial and axial current and velocity distributions in large-signal velocity-modulated electron beams," *IEEE Trans. Electron Devices,* Vol. ED-19, No. 7, July 1972, pp. 886–890.

Chapter 10

Basic Klystrons and Their Operation

The klystron is one of two major classes of devices categorized as linear-beam (O-type) tubes. The other is the traveling wave tube (TWT). The family tree of linear-beam tubes given in Figure 10-1 shows some types of klystrons and traveling wave tubes, and also indicates that there are hybrid devices that combine the klystron and TWT technologies.

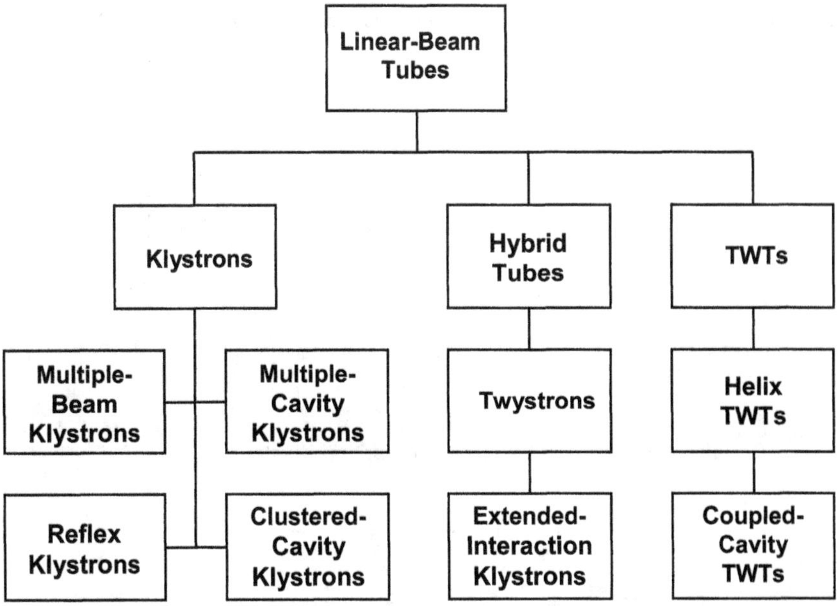

Figure 10-1 Family tree of linear-beam tubes.

Figures 10-2 through 10-5 indicate the differences between the klystron, TWT, and hybrid device technologies [1]. In the klystron, interactions take place in discrete locations along the beam and there is no coupling of the signal between parts of the interaction structure. The RF signal is carried from cavity to cavity by the electron beam.

In the coupled-cavity version of the TWT, cavities similar in appearance to klystron cavities are used. There are coupling apertures in the walls of the cavities so that the RF energy can propagate along the interaction structure in addition to being carried along by the electron beam. The propagation of the signal along the interaction structure is a primary distinguishing feature between TWTs and klystrons.

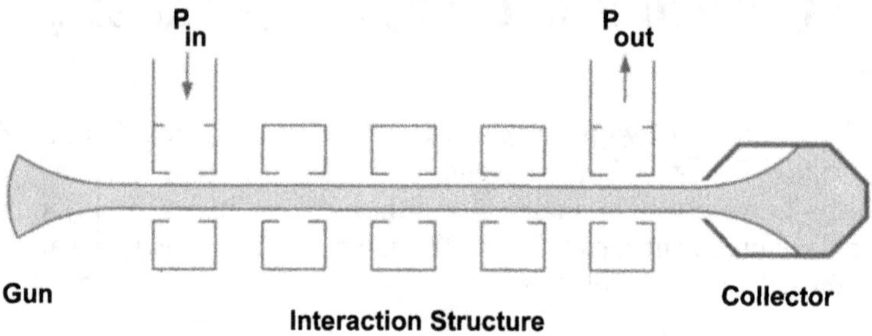

Figure 10-2 Klystron. (Adapted from: Armand Staprans, et al., *Proc. IEEE*, March 1973. © 1973 IEEE.)

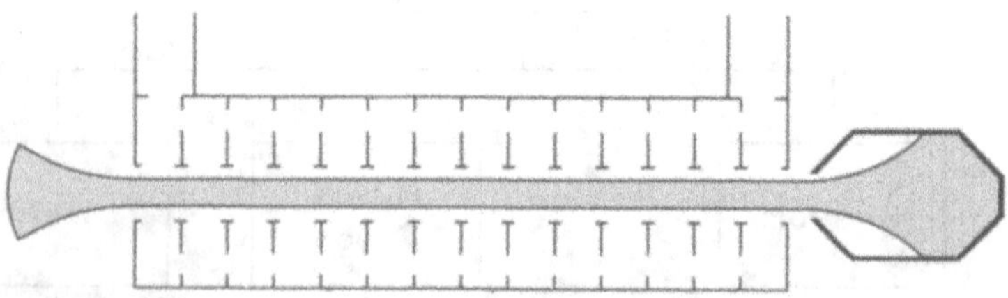

Figure 10-3 Coupled-cavity TWT. (Adapted from: Armand Staprans, et al., *Proc. IEEE*, March 1973. © 1973 IEEE.)

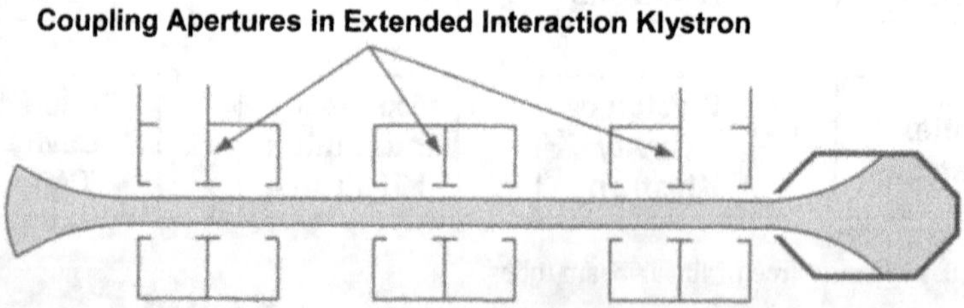

Figure 10-4 Clustered-cavity klystron and extended interaction klystron.

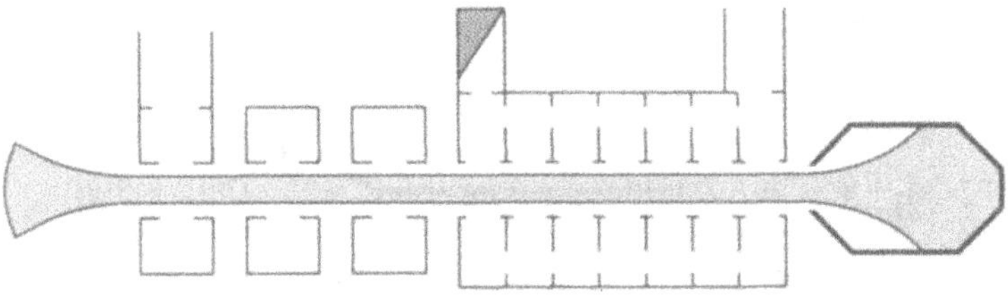

Figure 10-5 Twystron amplifier. (Adapted from: Armand Staprans, et al., *Proc. IEEE*, March 1973. © 1973 IEEE.)

In the extended interaction klystron (EIK) one or more of the single cavities of the conventional klystron is replaced by a structure containing two or more interaction gaps. Bandwidth is increased over that of the conventional klystron. In addition, at the output end of the tube, the voltage at each interaction gap is reduced, permitting operation at very high-power levels.

The twystron consists of a klystron input section and a coupled-cavity output section. Twystrons are capable of several megawatts at bandwidths of 10–15%.

10.1 THE INVENTION AND BASIC OPERATION OF THE KLYSTRON

Research leading to the development of the klystron had its start in the mid- and late 1930s. The impetus for this research was the need for higher RF powers at higher frequencies than could be generated by triodes and tetrodes, which were limited in frequency by the transit time of electrons in the cathode-grid region. When the transit time approaches the period of the RF signal, gain decreases rapidly. The first published account of a major advance that would eventually lead to the klystron (magnetron-like devices had appeared earlier) was given by Heil and Heil [2]. This device, called the Heil tube, is shown in Figure 10-6.

In the Heil tube, an electron beam passed through a floating drift tube, which the Heils called an oscillating Faraday cage, to a collector. The drift tube was connected to a tuned circuit. A small RF voltage across the gap at the entrance into the drift tube caused velocity modulation of the electrons. That is, the voltage across the gap accelerated some electrons and decelerated others. The faster electrons from the accelerating portion of a cycle caught up with the slower ones from the decelerating portion and so bunches of electrons were formed. When the bunches emerged from the drift tube, they induced a signal in the tuned circuit and, at the same time, enforced the velocity modulation of the beam entering the drift tube. The way in which this device operated made use of the transit time of the electrons, which had been a problem in triodes and tetrodes. The Heil tube was an experimental oscillator that apparently was not used for any practical application; however, it is credited as being the first linear-beam device to use the principle of velocity modulation of electrons.

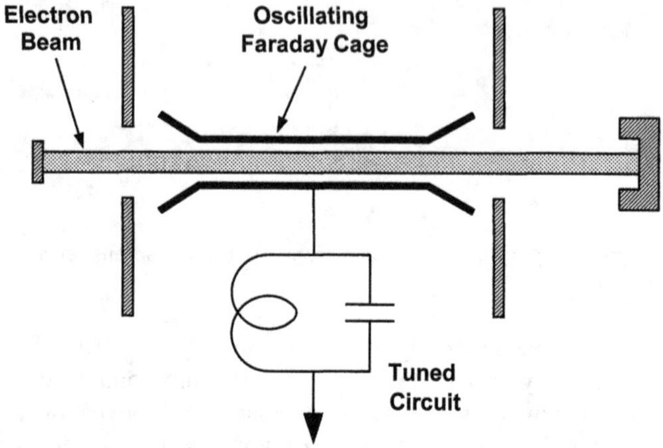

Figure 10-6 The Heil tube.

In 1937, the Varian brothers first described their work on the klystron, which also used the principle of velocity modulation, but also replaced the lumped element tuned circuits with resonant cavities. Figure 10-7 is an entry from Russell Varian's notebook and shows the basic two-cavity configuration of the klystron.

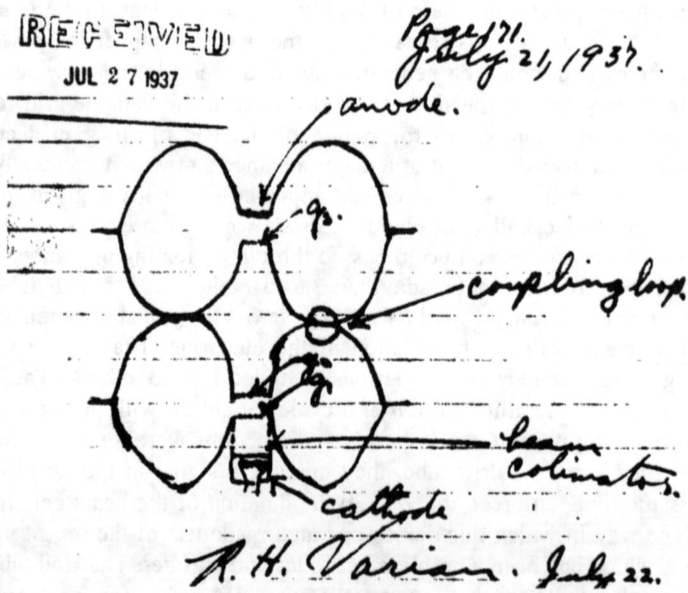

Figure 10-7 An entry from Russell Varian's notebook.

The Varians described the klystron in the *Journal of Applied Physics* in 1939 [3]. The following abstract from the Varians' paper contains a concise and very clear description of the basic operating principles of the klystron.

"A High Frequency Oscillator and Amplifier"
Russell H. Varian and Sigurd F. Varian

Abstract

> A dc stream of cathode rays of constant current and speed is sent through a pair of grids between which is an oscillating electric field, parallel to the stream and of such strength as to change the speeds of the cathode rays by appreciable but not too large fractions of their initial speed. After passing these grids the electrons with increased speeds begin to overtake those with decreased speeds ahead of them. This motion groups the electrons into bunches separated by relatively empty spaces. At any point beyond the grids, therefore, the cathode ray current can be resolved into the original dc plus a nonsinusoidal ac. A considerable fraction of its power can then be converted into power of high frequency oscillations by running the stream through a second pair of grids between which is an ac electric field such as to take energy away from the electrons in bunches. These two ac fields are best obtained by making the grids form parts of the surfaces of resonators of the type described in [this Journal].

Figure 10-8 shows the two-cavity amplifier described by the Varians in the *Journal of Applied Physics*. An electron beam (stream of cathode rays) from the cathode at the bottom of the figure is velocity modulated by the lower resonator (called a "rhumbatron" by the Varians). Electron bunching occurs as the beam drifts toward the upper cavity. The resulting ac current in the beam induces current in the second cavity and results in RF power out of the klystron which is substantially larger than the RF input power.

It was not long after the invention of the two-cavity amplifier that a three-cavity klystron was developed. In this device, the third cavity (sometimes called an idler cavity) was inserted between the input and output cavities and was not coupled to any external circuit. As the electron bunches produced by the first cavity passed through the idler cavity, RF current was induced in the cavity. This current produced an RF electric field across the gap which enhanced the velocity modulation of the beam. As a result, the gain of the device was increased.

It was next found that, by adding a fourth cavity between the idler and the output cavity, the efficiency could be increased. This fourth cavity, which was called the penultimate (next to last) cavity, was tuned to a higher frequency than the center frequency of operation of the tube. The effect, as is explained in Chapter 9, was to enhance the bunching of the beam in the output cavity.

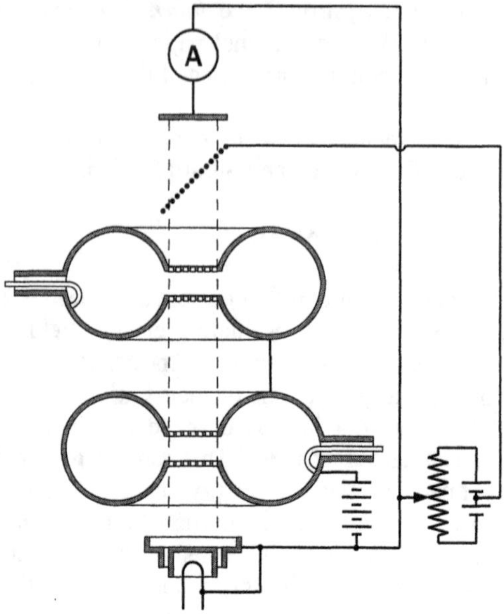

Figure 10-8 Basic klystron configuration shown by the Varian brothers. (From: R. H. Varian and S. F. Varian, *Jour. App. Phys..*, May 1939.)

Two-, three- and four-cavity klystrons are relatively narrowband devices. The possibility of increasing the bandwidth by adding idler cavities increase bandwidth was postulated in the 1940s. However, increasing the bandwidth in this way is not as easy as it first appears to be because the cavities of a klystron are all coupled to the same electron beam and so their tuning characteristics are not independent. If not tuned properly, some of the idler cavities will demodulate the beam and produce holes (regions of reduced gain) in the frequency response. It was not until 1957 that Kreuchen et al. [4] satisfactorily analyzed the multicavity klystron and described a method for increasing bandwidth with reasonably flat power output across the band.

In addition to properly tuning the idler cavities, it is necessary to have an output structure that can cover the required bandwidth. It was the development of double-tuned and triple-tuned cavities along with the merging of TWT and klystron technologies that resulted in devices with bandwidths exceeding 10%. The configurations of some of these devices are shown in Chapter 11.

In principle, the gain of the klystron amplifier can be increased indefinitely by adding cavities along the beam. In practice, with a six-cavity amplifier, a stable gain in excess of 110 dB has been observed. Normally, the gain is limited to 50 to 60 dB. At higher levels, the shot noise of the beam is increasingly amplified, and the signal-to-noise ratio of the output is reduced. In addition, to obtain stability, there must be no feedback from output to input, either because of leakage of transmission line connections, RF modulation of collector leads, and feedback to the electron gun, or reflected electrons within the beam.

The elements of a five-cavity klystron are shown in Figure 10-9. The intermediate cavities are used to increase the gain and usually the bandwidth. Often, the spacings between cavities vary in the way reported by Mihran and discussed in Chapter 8. The intermediate cavities are often one-quarter of a plasma wavelength ($\lambda_q/4$) apart. The spacing between the penultimate cavity (next to last cavity) and the last cavity is less than $\lambda_q/4$ because of the ballistic nature of the bunching process at the large signal levels in the beam near the output end of the tube.

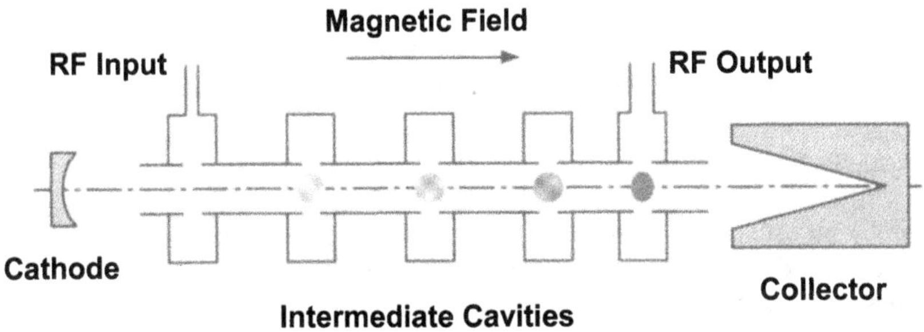

Figure 10-9 Basic elements of a five-cavity klystron.

In many modern klystrons, the spacings between intermediate cavities are not $\lambda_q/4$. The spacings are adjusted to optimize bandwidth and/or efficiency. Spacings for high efficiency and extremely broad bandwidths are discussed in Chapter 11. Optimization is often done using computational techniques and a disk model of the electron beam. An example of a calculation for a seven-cavity klystron is shown in Figure 10-10. This figure is basically an Applegate diagram in which dc velocities of the electrons have been removed. Each trace represents the phase of a disk of electrons relative to the undisturbed phase. Thirty electron disks per RF cycle were used in Figure 10-10 and two RF cycles are shown. The pairs of dashed vertical lines show positions of cavities. In this case the last cavity was an extended interaction cavity (to be discussed in Chapter 11) with two gaps.

In the regions of cavities 1 and 2, where the signal levels were very small, the change in velocities of the disks as they pass through the gaps, shown by the leading or lagging phase, was small. At the third gap, the signal was large enough to produce large velocity modulation, the magnitude of which is indicated by the slopes of the phase lines. The resulting bunching of the beam is indicated by the convergence of the phase lines just before the fourth cavity. At the fourth and subsequent cavities, the nonlinear nature of the interaction is readily apparent because of the crossing of phase lines, which means that some disks were passing others.

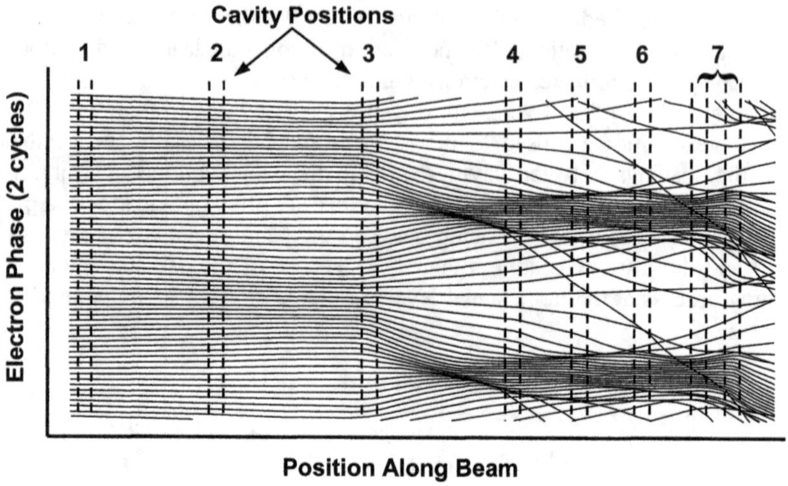

Figure 10-10 Electron phase diagram for a seven-cavity klystron.

From the calculation of phase positions, velocities, and densities, RF currents were calculated and power output could be determined. By adjusting cavity locations, resonant frequencies, and *Q*s, the bandwidth and efficiency were optimized.

10.2 KLYSTRON CAVITIES

10.2.1 Cavity Operation

Cavity resonators are used in klystrons to apply a signal to the electron beam and then to remove the amplified signal from the beam. From a circuit point of view, each resonator can be considered to be a resonant parallel LC circuit that has been modified so that it will operate at microwave frequencies. This is illustrated in Figure 10-11. As frequency is increased, the multiturn-inductor, multiplate-capacitor circuit in Figure 10-11(a) is eventually replaced by the circuit in Figure 10-11(b) in which the inductor has a single half turn and the capacitor has a single pair of closely spaced parallel plates. To further increase frequency, the capacitance is reduced by increasing plate spacing and inductance is reduced by adding inductors in parallel as indicated in Figure 10-11(c).

Finally, with the addition of more inductors, the inductive portion of the circuit becomes a hollow toroid as shown in Figure 10-11(d). The magnetic field is concentrated in the toroidal portion of the circuit and the electric field is concentrated in the central (capacitive) portion.

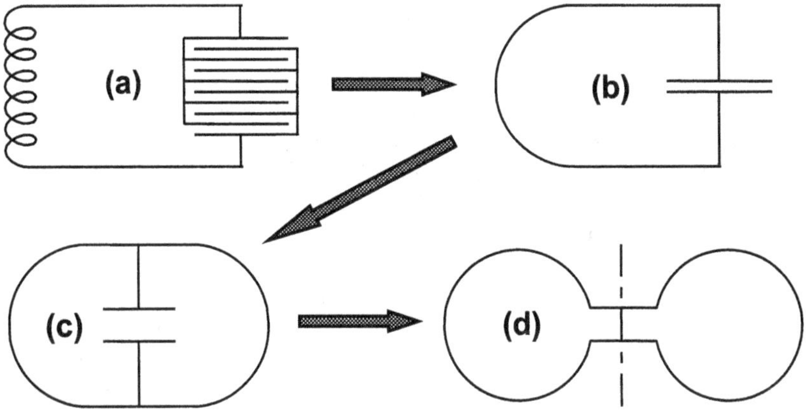

Figure 10-11 (a–d) Modification of parallel resonant circuit for operation at microwave frequencies.

At resonance, energy oscillates back and forth between the toroidal and parallel-plate portions of the cavity, just as it does in a parallel LC circuit. By replacing the parallel-plate portion of the cavity in Figure 10-11(d) with parallel grids, it is possible to pass a low-power electron beam through the center portion.

At high-power levels, the grids cannot be used because they would be destroyed by the electron beam. The natural evolution of Figure 10-11(d) is Figure 10-12(a). To simplify fabrication, the cavity usually takes the shape in Figure 10-12(b). This type of cavity is referred to as a doubly reentrant cavity. The ends of the tubular electron beam tunnels (referred to as the drift tube tips) are the reentrant portions of the cavity. The capacitance of the cavity results from the energy stored in the electric field between the drift tube tips.

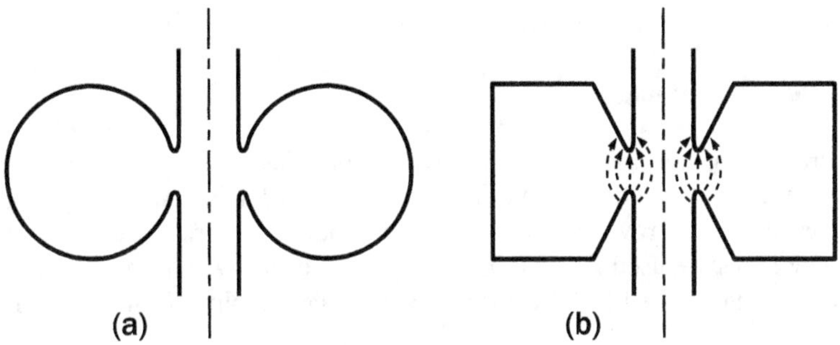

Figure 10-12 (a, b) Gridless cavities for high-power klystrons.

The electric field pattern (dashed lines in Figure 10-12) is not as well defined in the gridless cavity as it is in the gridded cavity. Still, the field exists in the central region of the cavity and is available for interaction with an electron beam.

The cavities of most klystrons are entirely within the vacuum envelope and tubes of this type are referred to as internal cavity klystrons. However, in some low-frequency tubes (below S-band), the RF cavities are outside the vacuum envelope as indicated in Figure 10-13. These tubes are called external cavity klystrons. The cavities are placed onto the tube by the user and may be mechanically tuned over a frequency range of 10% to 15%. Advantages of this type of construction are that the vacuum structure may be kept small while maintaining good cavity-to-beam coupling, and also the tubes can be less expensive to replace because the RF cavities are reusable. The disadvantages are problems with multipactor (to be discussed in Chapter 24) on the ceramics and RF arcing, along with the need to tune the tubes in the field.

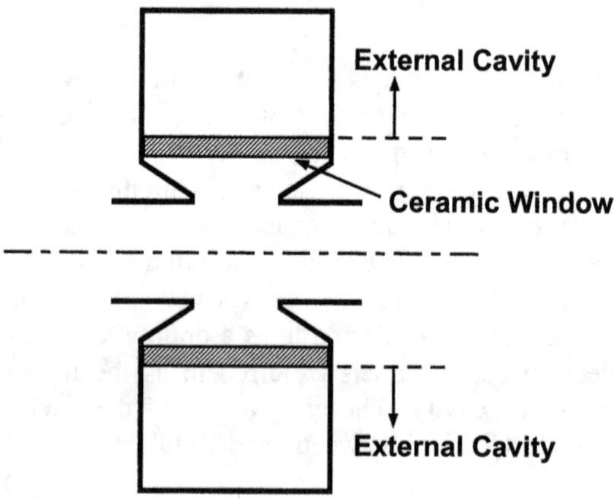

Figure 10-13 External klystron cavity.

10.2.2 Power Coupling

In the drawing of the original Varian klystron (see Section 10.1), small coupling loops attached to the ends of coaxial lines are shown for injecting a signal into a cavity or taking energy from a cavity. In modern klystrons, either loops or waveguide irises are used for coupling energy to a cavity. A typical coupling loop is shown in Figure 10-14 and cavity-to-waveguide coupling is shown in Figure 10-15.

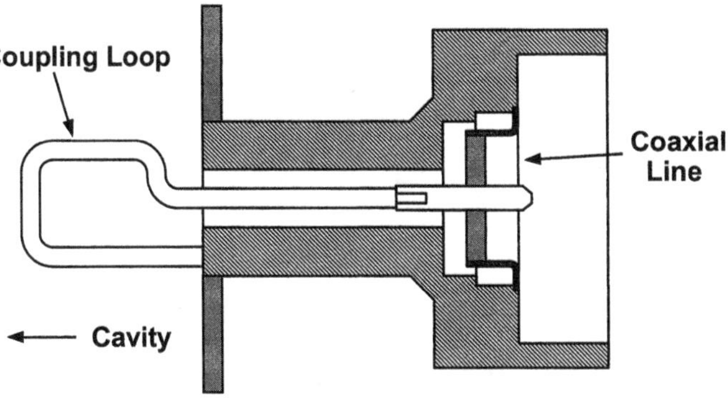

Figure 10-14 Loop for coupling energy into or out of a cavity.

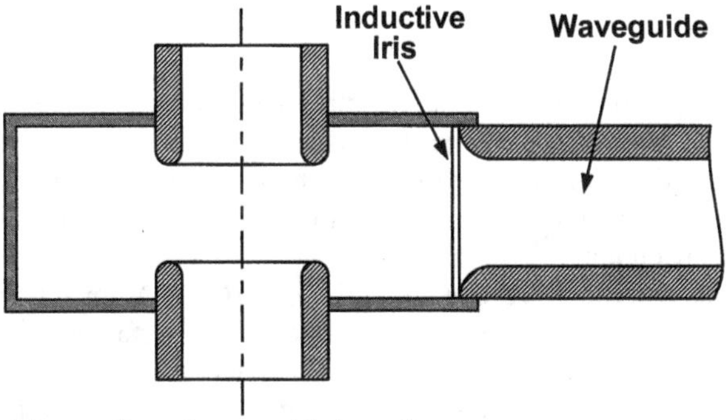

Figure 10-15 Iris coupling of waveguide to cavity.

The principle of operation of coupling loops is straightforward. To inject a signal into a cavity, an RF current, I_1, is injected into the coupling loop as is indicated in Figure 10-16. As a result, some magnetic field lines are established within and around the coupling loop and some circle through the toroidal portion of the cavity. These field lines induce a current flow, I_2 in the walls of the cavity. The resulting flow of charge produces an electric field, E, across the capacitive portion of the cavity. As the RF input current, I_1, oscillates, the magnetic field, the current, I_2, and the electric field, E, are all caused to oscillate.

When energy is extracted from a cavity by a coupling loop, the situation described in the previous paragraph is reversed. A modulated electron beam passing through the capacitive portion of the cavity induces current, I_2, in the walls of the cavity. The way in which current is induced is described in Chapter 8. Associated with the current, I_2, there is a magnetic field within the toroid. A portion of the magnetic field links the coupling loop and induces a current, I_1, in the output coaxial line and, thus, in the load. The direction of the induced current is such that it opposes the flux producing the current.

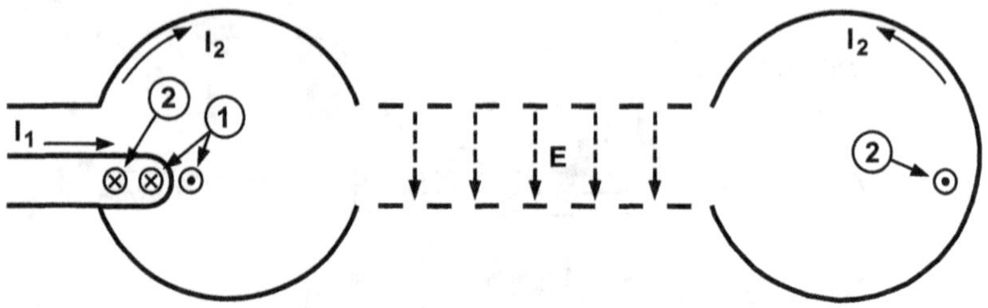

Figure 10-16 Use of coupling loop to inject signal into cavity.

When iris coupling of a waveguide is used, the orientation of the waveguide is such that the electromagnetic field propagates directly into the cavity and establishes an electric field across the interaction gap.

10.2.3 Tuners

It is essential to be able to adjust the resonant frequencies (tuning) of the cavities of a klystron. As was shown in the discussion associated with Figure 10-16, the outer donut-shaped region of a cavity is the inductive portion and the region near the interaction gap is the capacitive portion. Tuners are used to vary either inductance or capacitance.

Two inductive tuners are shown in Figure 10-17. With either the plunger shown in the left side of the figure or the diaphragm shown to the right, the volume of the cavity can be adjusted. Inductance varies directly with cavity volume and thus, as the plunger or diaphragm is pushed into the cavity, the inductance is reduced and the resonant frequency is increased. In some plunger-type tuners, chokes are used to reduce current flow in the tuner. In others, the plunger makes electrical contact with a channel surrounding it through spring-loaded fingers.

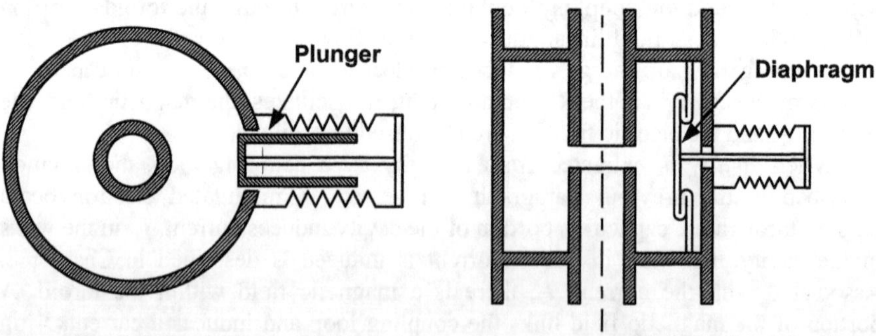

Figure 10-17 Inductive tuners. (Adapted from: *Power Klystrons Today* by M. J. Smith and G. Phillips, copyright 1995 by Research Studies Press, Ltd.)

Capacitive tuners are shown in Figure 10-18. With these tuners, a paddle-shaped electrode near the interaction gap is used to vary capacitance. As the paddle is moved closer to the gap, capacitance is increased and resonant frequency is decreased. For high-power applications, it is sometimes necessary to introduce a coolant into the central spindle.

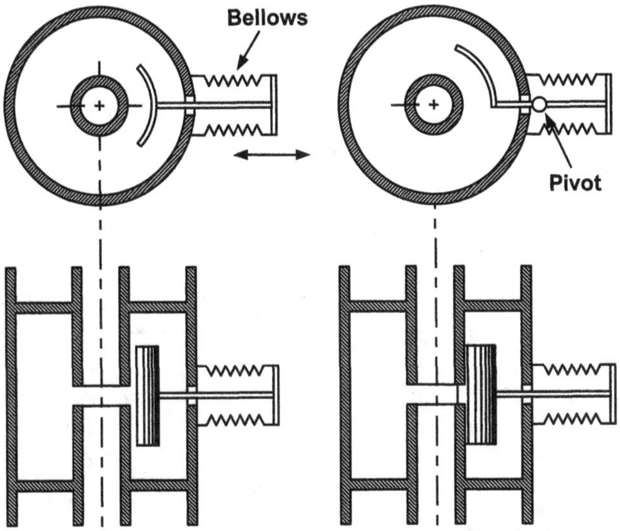

Figure 10-18 Capacitive tuners. (Adapted from: *Power Klystrons Today* by M. J. Smith and G. Phillips, copyright 1995 by Research Studies Press, Ltd.)

10.2.4 Equivalent Circuits and Circuit Parameters [5]

We showed in Chapter 8 that an electron beam passing through a cavity induces a current in the cavity and the beam can be represented by an equivalent current generator. In addition, the electron beam has impedance that has a resistive and a reactive component. This impedance must be included to accurately model the operation of a cavity. First, the equivalent circuit for the cavity will be considered and then beam loading can be introduced.

The primary problem is to find parameters that describe the impedance characteristics of a cavity at and near resonance. Parameters are needed that will determine the voltage and power generated by the induced beam current as a function of frequency. Even though a klystron cavity has regions containing primarily E fields and those containing primarily H fields, it is difficult to try to define capacitances or inductances associated with these regions. Instead, parameters can be used, which are defined for lumped constant circuits. These parameters can then be defined in terms of fields and energies to describe the interaction of the electron beam with the cavity.

Consider the circuit shown in Figure 10-19. The resistance, R, not only represents the cavity wall losses but also can include power delivered to an external load coupled to the circuit by an inductive coupling loop or through an aperture joining the cavity to a waveguide.

By transforming the series R to an equivalent shunt resistance, R_{SH}, as follows

$$R_{SH} = \frac{\omega^2 L^2}{R} = \omega L Q \tag{10-1}$$

where

$$Q = \frac{\omega L}{R} \tag{10-2}$$

the circuit in Figure 10-19 can be converted to the form shown in Figure 10-20.

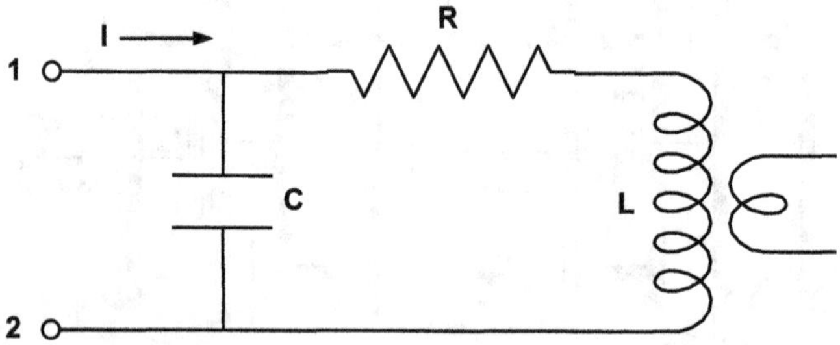

Figure 10-19 Equivalent circuit for a klystron cavity.

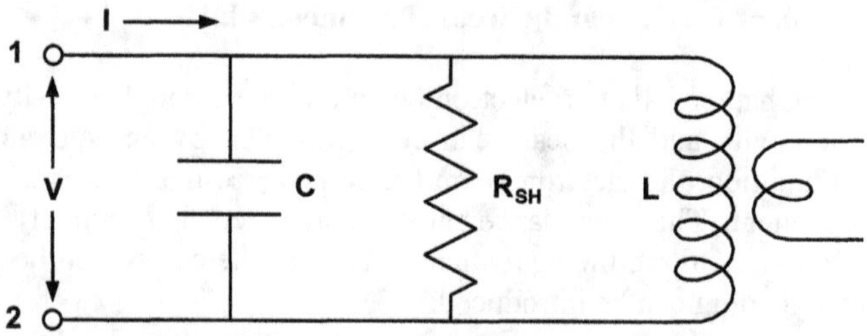

Figure 10-20 Circuit with series resistance transformed to shunt resistance.

These relations for R_{SH} and Q (10-1) and (10-2) have been derived for the equivalent circuit of the cavity with lumped circuit elements. Before discussing the significance of these relations, it is of value to redefine R_{SH} and Q so that they can be used for cavities. The definition of Q ($= \omega L/R$) can be rewritten as

$$Q = \frac{\omega L\left(i^2/2\right)}{R\left(i^2/2\right)} = \frac{\omega U}{W} \tag{10-3}$$

where i is the peak current in the series circuit and so $L\ i^2/2$ is the peak energy stored, U, and $R\ i^2/2$ is the average power dissipated, W. Similarly, R_{SH} can be written as

$$R_{SH} = \frac{\omega^2 L^2}{R} = \frac{\omega^2 L^2\left(i^2/2\right)}{R\left(i^2/2\right)} = \frac{V^2}{2W} \tag{10-4}$$

where $V = \omega L\ i$ is the peak voltage across L. This is also very nearly the voltage across C and terminals 1 and 2 of the equivalent circuit because the voltage across R in a resonant cavity is small compared to the voltage across the reactive elements. Finally, the ratio of R_{SH} to Q is

$$\frac{R_{SH}}{Q} = \frac{V^2}{2\omega U} \tag{10-5}$$

Now, R_{SH}, Q, and ω are the three quantities which can be used to describe any cavity. Also, their definitions can be given in terms of cavity quantities. (As indicated in Figure 10-21, U is the peak energy stored and W is the power lost.) These quantities are for an unloaded cavity. With a load, the power dissipated in the load must be added to W. Then, both R_{SH} and Q are reduced, but their ratio, R_{SH}/Q, is a geometrical property of the cavity and remains unchanged.

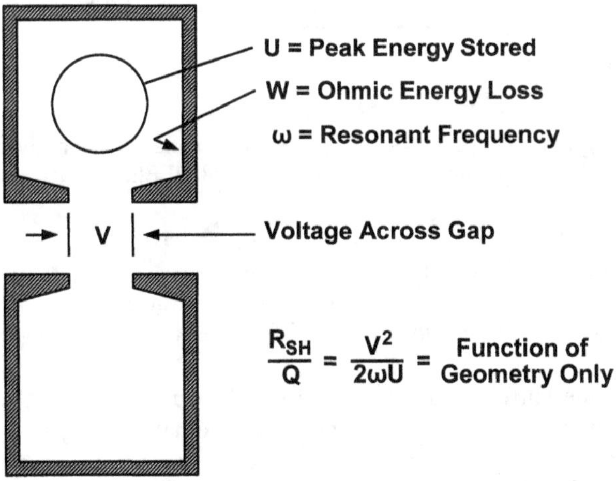

Figure 10-21 Quantities used to define cavity performance.

The value of R_{SH}/Q is a figure of merit for a cavity and is independent of Q for any particular cavity geometry. Thus, the shunt impedance, R_{SH}, may be set to the value required for matching the external load to the cavity by adjusting the loaded Q. This then fixes the Q and, since Q is inversely proportional to the bandwidth, the expected frequency response of the cavity (and the klystron) is determined. The lower the beam impedance is and the higher the R_{SH}/Q is, the wider the bandwidth is.

The R_{SH}/Q of a cavity of the type shown in Figure 10-21 (referred to as a reentrant cavity) varies with the dimensions of the cavity as shown in Figure 10-22. It is possible to obtain values of R_{SH}/Q of over 150. Often, practical considerations limit the degree of design freedom for the cavity shape, but an R_{SH}/Q of 120 is not normally difficult to obtain [6].

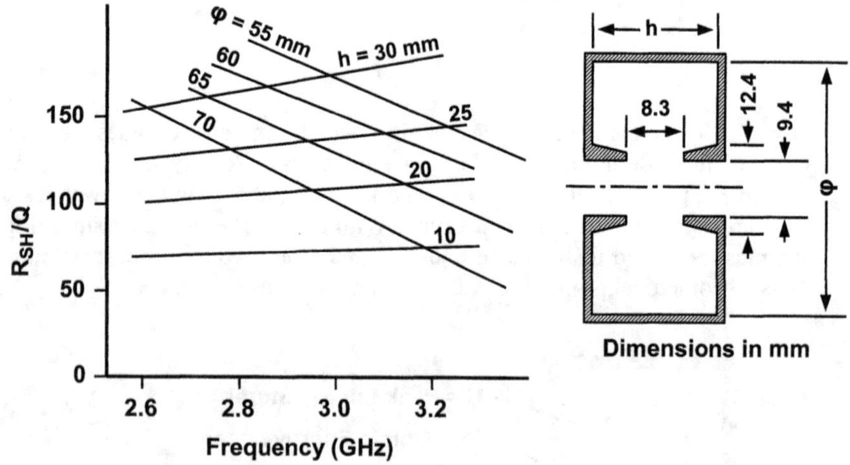

Figure 10-22 R_{SH}/Q as a function of dimensions for reentrant S-band klystron cavities. (Adapted from: G. Faillon, *Technical Digest,* IEDM. ©1985 IEEE.)

Returning to the effect of the electron beam, we have already mentioned that R_{SH} can be controlled to optimize power to the load. There are two cases of importance, but in either case, for any given R_{SH}, a large R_{SH}/Q is an advantage for bandwidth. Concerning the adjustment of R_{SH}, the two cases of importance are:

1. Where the current generator has a small value as is indicated in Figure 10-23 it can be shown that optimum external loading (and maximum power delivery) is for the case where the power dissipated inside the cavity is just equal to the power delivered to the external load. In such a case, the external loading is equivalent to adding another resistance, R_L, having the value $R_L = R_{SH}$ in parallel with R_{SH}.

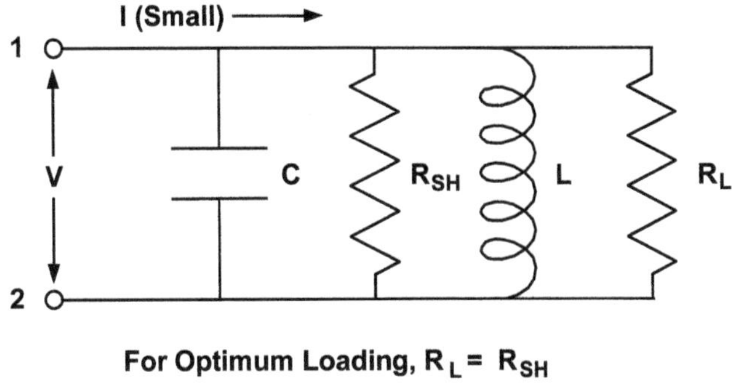

Figure 10-23 Optimum loading for small current.

2. Under the circumstance indicated in Figure 10-24 when the voltage produced across the circuit, V, is greater than the dc beam voltage, V_o, electrons passing through the gap may be stopped and turned back, and then it is necessary to increase the loading (decrease R_L). This will result in more power being delivered to R_L than to R_{SH} and an optimum adjustment of R_L can be done empirically by maximizing power to R_L. Optimum design and performance of klystrons is normally obtained under this condition. Under any circumstance, the bandwidth is still optimum when R_{SH}/Q is maximum.

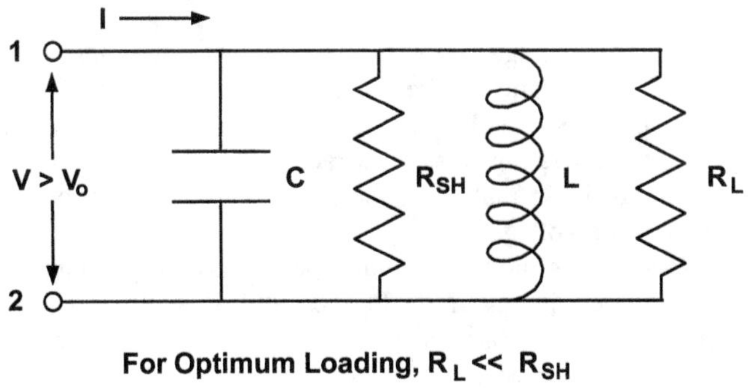

Figure 10-24 Optimum loading for circuit voltage > dc beam voltage.

10.2.5 RF Cavity Losses

RF losses occur in cavities because of microwave currents flowing in the cavity walls. It is important to be able to estimate these losses for at least two reasons:

1. Adequate cooling for the cavity walls can be provided.
2. The operating power level of the cavity can be limited so that losses do not exceed the cooling that has been provided.

One approach to estimating cavity losses has been described by Symons and Jory [7, 8]. They have defined a figure of merit, F_m, which relates the square of the peak cavity gap voltage (which is a measure of energy stored in the cavity) to the power dissipated in the cavity walls. When energy and power are expressed in terms of R_{SH}/Q and Q_o, it is seen that F_m is invariant as all cavity dimensions are scaled with wavelength. This figure of merit is

$$F_m = \frac{V^2}{P_L/A}\frac{\delta}{\lambda^3} = \left(\frac{R_{SH}}{Q}\right)\left(Q_o\frac{\delta}{\lambda}\right)\left(\frac{A}{\lambda^2}\right) \tag{10-6}$$

in which V is peak voltage, P_L/A is the loss per unit area, δ is the skin depth, λ is the operating wavelength, and Q_o is the Q of the cavity with no load attached.

Symons pointed out [8] that not only are the three quantities on the right side of this equation invariants for wavelength scaling of a particular cavity geometry, but, for klystrons, they vary over extremely limited ranges. In most klystrons, $R_{SH}/Q \approx 125$, $Q_o\delta/\lambda \approx 0.8$, and $A/\lambda^2 \approx 1$, and so $F_m \approx 100$. Now, using this value of F_m and the value of δ for copper, (10-6) can be rewritten as

$$V^2 = 0.83\times 10^6\,\lambda^{5/2}\,\frac{P_L}{A} \tag{10-7}$$

For pulsed devices, the average power loss is the peak power loss reduced by the duty cycle, D_u. For this relation to be useful in estimating power losses, a value must be assigned to the peak voltage, V. This can easily be done in the case of a klystron output cavity operating so as to extract maximum power from the electron beam passing through it. In that case, the peak voltage across the cavity gap is approximately the beam voltage.

Figure 10-25 contains plots of average power loss for various voltages and operating wavelengths. The usual cooling capability in klystrons is limited to about 1 kW/cm^2; however, dissipations in excess of 10 kW/cm^2 have been demonstrated for use in crossed-field devices.

10.3 SMALL SIGNAL OPERATION [5]

In discussing the power output and gain of a klystron and their dependence on cavity parameters, it is useful to consider the load to be connected to the cavity through a coupling loop as shown in Figure 10-26. The results of the analysis do not depend on the particular circuit representation of the external loading, coupling loop, and so forth. The external load can be coupled to the cavity in any one of many ways, for example, with an aperture in the cavity wall, and the load can be placed somewhere out in an external waveguide. Also, the load can have a reactive component that appears in the equivalent circuit as additional *L* or *C* and that changes the resonant frequency. All of these require suitable parameters for their description, but ultimately they can be reduced to something equivalent to the description given here.

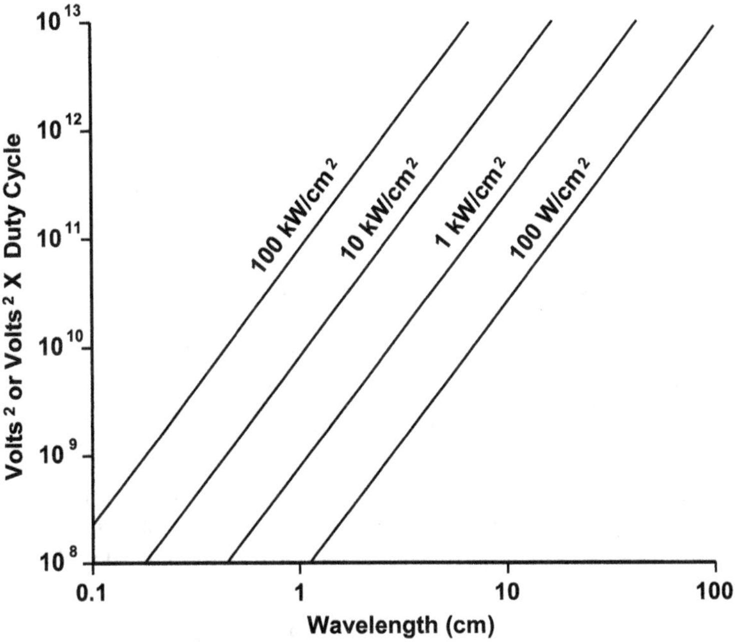

Figure 10-25 RF cavity losses as functions of cavity gap voltage and wavelength. (Adapted from: R. S. Symons, 1986 IEDM. © 1986 IEEE.)

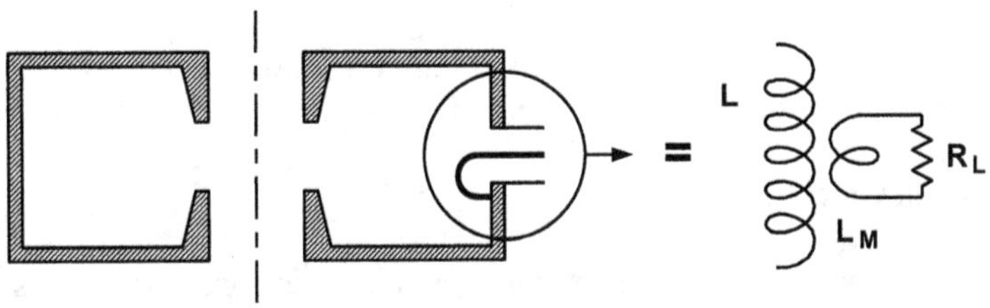

Figure 10-26 Circuit representation of coupling loop.

10.3.1 Load Representation

In Figure 10-26, L_M is the mutual inductance of the cavity and coupling loop and R_L is a resistive load terminating a transmission line attached to the coupling loop. The effect of this external load is to add another resistance, R_L', in series with the R in the equivalent circuit for the cavity as shown in Figure 10-27. The loss in this resistor represents power delivered to the load.

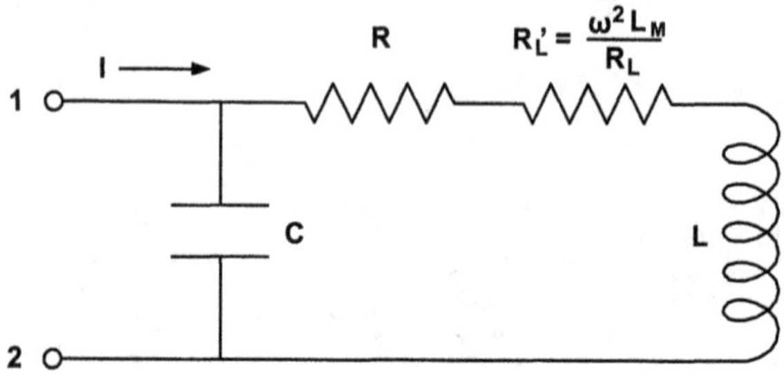

Figure 10-27 Equivalent series circuit containing load resistance.

When R and R_L' are transformed to parallel resistances, the total shunt resistance, R_{SHT}, is

$$R_{SHT} = \frac{\omega^2 L^2}{R + R_L'} \tag{10-8}$$

It will be useful to introduce conductances to represent the cavity and the external load. The total conductance, G_T, which is the inverse of the total shunt resistance, R_{SHT}, is the sum of cavity conductance, G_C, and load conductance, G_L, that is

$$G_T = G_C + G_L = \frac{1}{R_{SHT}} = \frac{1}{R_{SH}} + \frac{1}{R_{SHL}} \tag{10-9}$$

or

$$G_T = \frac{R}{\omega^2 L^2} + \frac{R_L'}{\omega^2 L^2} \tag{10-10}$$

The equivalent parallel circuit is shown in Figure 10-28. Now the voltage gain and the conditions for optimum power output of a klystron can be considered.

10.3.2 Gain Calculation

When small-signal conditions exist, the gap voltage is considerably less than the beam voltage. This was referred to as Case 1 earlier in this chapter. The voltage produced at the cavity gap, V, can be written to take into account the effects of the load on the shunt resistance, R_{SHT}, (or the total conductance, G_T) and the Q_T (loaded) of the cavity as

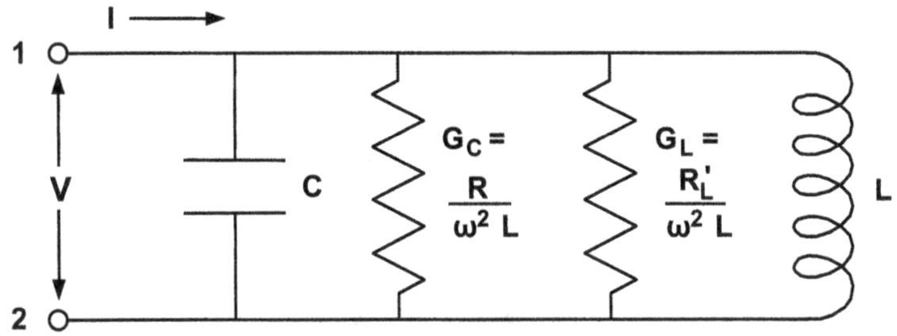

Figure 10-28 Equivalent parallel circuit containing the load resistance.

$$V = \frac{I\,R_{SHT}}{1+2\,j\,\delta Q_T} = \frac{I}{G_T\left(1+2\,j\,\delta Q_T\right)} \tag{10-11}$$

where I is the current induced in the cavity. As usual, $\delta = f - f_o$ is the difference between the operating frequency and the cavity resonant frequency. For ballistic bunching, the induced current was discussed in Chapter 8 and is

$$I = M_2 2 I_o \,\mathrm{J}_1\left(\frac{\theta}{2}\frac{V_1}{V_o}\,M_1\right) \tag{10-12}$$

As shown in Figure 10-29, θ is the drift angle from the modulating cavity to the cavity in which current is being induced, M_1 as the gap coupling coefficient of the modulating cavity and V_1 as the voltage at the first cavity. Here beam loading and any additional terms due to velocity modulation on the beam as it enters the output gap are being ignored. The velocity modulation term is usually small and the beam loading can easily be included as an equivalent admittance in parallel with the other elements of the equivalent circuit.

The power to the load can be shown to be

$$P_L = \frac{\left|I\right|^2 G_L}{G_T^2\left(1+4\delta^2 Q_T^2\right)} \tag{10-13}$$

Maximum power transfer occurs at resonance so $\delta = 0$ and the power becomes

$$P_L = \frac{\left|I\right|^2 G_L}{G_T^2} = \frac{\left|I\right|^2 G_L}{\left(G_L + G_C\right)^2} \tag{10-14}$$

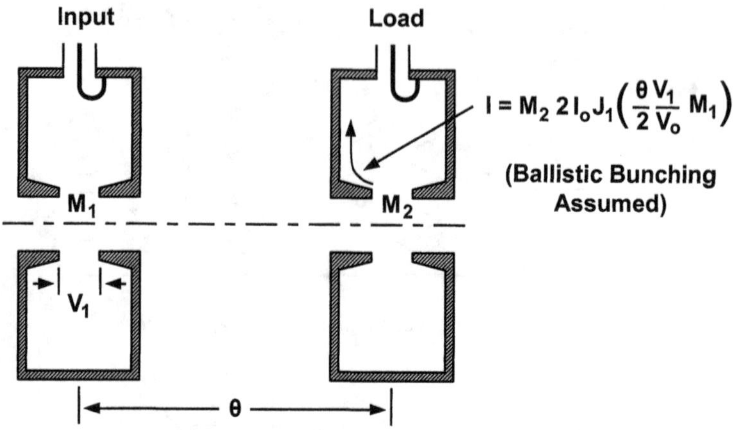

Figure 10-29 Illustration of current coupling in a two-cavity klystron.

Now, power is maximized when $G_L = G_C$, that is, when the load resistance (considered as a shunt resistance in the equivalent circuit) equals the shunt resistance of the cavity. This condition means, of course, that the total power is divided equally between internal losses and the external load. We should note here that beam loading changes these results only by adding the beam loading conductance to the internal losses.

It is also possible to calculate voltage gain using considerations similar to those used for power. In most cases, a klystron amplifier has more than two cavities. The cavities between the input and output are intended to provide greater gain than is possible with just two cavities. The current arriving at gap n due to velocity modulation from previous gaps produces a voltage across n which is greater than the previous gap voltages. By using several such cavities, we can then produce the required maximum current at the output cavity with much smaller voltage (and power) input at the first cavity than in the simple two-cavity klystron.

The single-stage voltage gain from cavity m to cavity n can be calculated from (10-11) and (10-12). First, we note that, for small values of X, $J_1(X) \approx X/2$ so that the equation for induced current in gap n becomes

$$I_n = M_n M_m \frac{I_o}{V_o} \frac{\theta}{2} V_m \tag{10-15}$$

Now substituting this into (10-11) and, assuming that there is no external load (as is the case in most intermediate cavities in a klystron), then the voltage gain becomes

$$\frac{V_n}{V_m} = M_n M_m \frac{I_o}{V_o} \frac{\theta}{2} \frac{R_{SH}}{(1+2j\delta Q)} \tag{10-16}$$

This relation is the single stage gain. As expected, the gain and gain-bandwidth product depend on R_{SH}/Q.

The typical klystron has several stages. The intermediate cavities between the first and the last stages produce extra stages of gain as just described. That calculation was simplified because we assumed that the only velocity modulation and current that existed at any particular cavity was that due to the velocity modulation from the previous cavity. Then we calculated a new voltage produced by this current. This is exactly true for the second cavity in a chain. At the third cavity, however, the modulation and current will be the sum of two terms:

1. A current due to velocity modulation at the first cavity, which then produced a current at the third cavity;
2. A current due to the voltage of the second cavity which, in turn, was due to the current arriving from the first cavity.

There will still be some persistent velocity modulation from two earlier cavities, and the current terms will produce voltage on the third cavity, which remodulates the beam and continues on to the next.

As a result, as we go up the chain of cavities, we see that the complexity of the calculation increases. At each cavity there are currents that depend directly on the voltage at some previous cavity through velocity modulation converted into current by bunching. In addition, there are terms that contain currents produced by voltages at intermediate cavities by currents arriving at earlier cavities. These are so-called "feed-forward" terms that make the calculation of klystron performance very complex.

The components required for determining gain in a multicavity klystron have been illustrated by Symons and Vaughan as shown in Figure 10-30 [9]. Here, gap impedances are indicated by Z_1, Z_2 ... where the subscripts indicate cavity numbers. The quantities g_{21}, g_{32}, ... are transconductances defined by, for example,

$$g_{21} = \frac{I_2}{V_1} \tag{10-17}$$

and I_2 is the current induced in cavity two by the voltage, V_1, at gap one. The total current gain for the five-cavity klystron is given in (10-18).

Thus, the klystron is unlike an ordinary amplifier where the input voltage (or current) to one stage is proportional to the output voltage (or current) of the previous stage. The klystron behaves like a multistage amplifier gain where there are voltages from each stage fed into the stage ahead of them and also fed around into the following stages. Another factor that must be considered is that it is often desirable to tune the cavities in various ways so that the resonant frequency of different cavities is different. In addition, some cavities may be loaded to lower the Q and thereby increase the bandwidth.

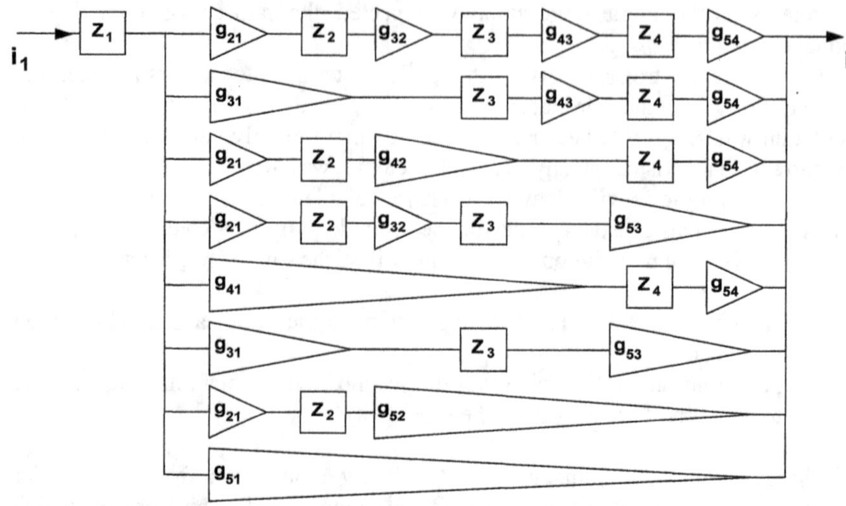

Figure 10-30 The equivalent circuit of a five-cavity klystron. (From: R. S. Symons and J. R. M. Vaughan, *IEEE Trans. Electron Devices*, October 1994. © 1994 IEEE.)

$$\begin{aligned}\frac{I_5}{I_1} = Z_1(&g_{21}\, g_{32}\, g_{43}\, g_{54}\, Z_2\, Z_3\, Z_4 \\ &+ g_{31}\, g_{43}\, g_{54}\, Z_3\, Z_4 \\ &+ g_{21}\, g_{42}\, g_{54}\, Z_2\, Z_4 \\ &+ g_{21}\, g_{32}\, g_{53}\, Z_2\, Z_3 \\ &+ g_{41}\, g_{54}\, Z_4 \\ &+ g_{31}\, g_{53}\, Z_3 \\ &+ g_{21}\, g_{52}\, Z_2 \\ &+ g_{51})\end{aligned} \tag{10-18}$$

10.4 POWER OUTPUT CHARACTERISTICS

The discussions up to this point have been for gap voltages very much less than the beam voltage ($V_n < V_o$) so that small signal theory could be applied in calculating transit effects, power transfer, and so forth. However, when $V_n \approx V_o$, or more strictly, when $M_n V_n \approx V_o$, then small-signal calculations, which have assumed that the modulated velocity of the electrons in a gap is a small fraction of their average velocity, are inaccurate and the calculations for gain or optimum power transfer are not valid.

When $M_nV_n \approx V_o$, the average electron velocity is substantially reduced and, in fact, some electrons may even have their motion reversed. Empirically, the output load is varied until optimum power is delivered. This corresponds to $G_L > G_C$ so that more of the power is delivered to the load than to the output cavity. This additional loading will tend to reduce the voltage across the output gap enough so that electron motion is not reversed. Instead, many of the electrons will be just slowed down to zero velocity as they pass through the gap and will have given up all of their energy to the cavity plus loads. Then as much power as possible will have been extracted from the electrons.

These conditions apply in all high efficiency-klystrons. High efficiency is obtained because most of the energy taken from the electrons is delivered to the external load rather than being dissipated in losses in the cavity. To calculate optimum power transfer under these conditions, it is necessary to go through a numerical computation of electron trajectories through the output gap and find the conditions of optimum energy loss by the electrons. Under these circumstances, electrons will be passing each other and, in some cases, will be going backwards. Highly efficient klystrons operate with these conditions in the output cavity.

10.4.1 Tuning of Conventional Klystrons

Conventional klystrons (not extremely efficient or broadband klystrons) can be synchronously tuned, efficiency tuned, or broadband tuned. In all cases, the tubes are usually tuned for maximum power output. With synchronous tuning, all cavities are tuned to the same frequency and gain is maximized. For high-efficiency tuning, a klystron is first synchronously tuned and then the penultimate cavity is tuned upward to maximize power. By being tuned upward, the penultimate cavity is inductive at the operating frequency and this enhances the bunching process at the output cavity (see Chapter 8).

Tuning characteristics of a penultimate cavity are shown in Figure 10-31. With a klystron operating at, or near, full output power, it is important not to tune the penultimate cavity down in frequency into the shaded region indicated in Figure 10-31. This is because there is no load on the penultimate cavity, and the voltage induced could be high enough to cause breakdown across the cavity gap.

To broadband tune a klystron, it is first synchronously tune at a low power level. Then, as indicated in Figure 10-32, the intermediate cavities are detuned downward and upward to achieve the desired bandwidth and gain flatness. The bandwidth of the output cavity determines the ultimate bandwidth of the tube. For broad bandwidth, a high perveance beam is normally used to achieve low beam impedance, which, in turn, results in a low beam-loading impedance in the output cavity. Also, bandwidth can be broadened by using an extended interaction (multiple-coupled-cavity) output circuit (discussed in Chapter 11).

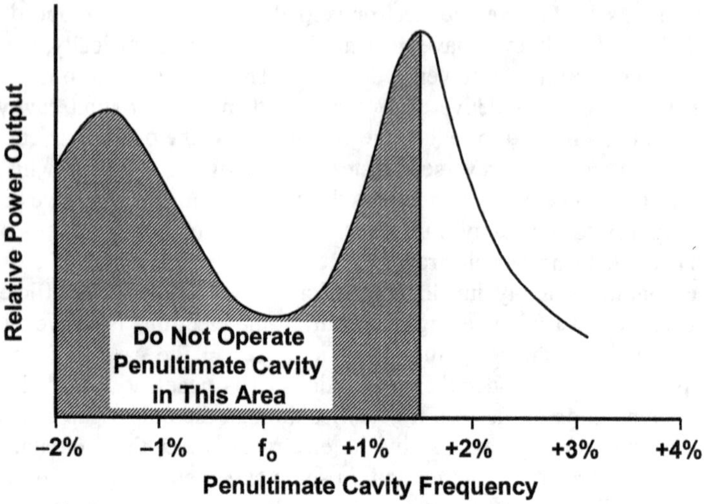

Figure 10-31 Penultimate cavity tuning characteristics. (From: *Microwave Tube Manual*, by Varian Associates, Air Force Publication Number T.O.00-25-251, October 1979.)

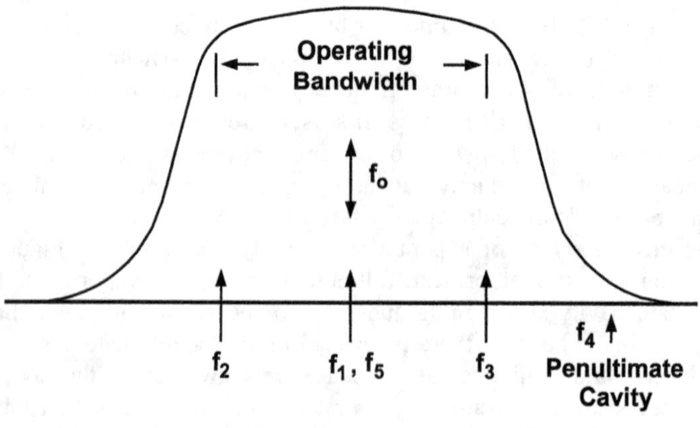

Figure 10-32 Typical cavity resonant frequencies for a broadband-tuned klystron. (From: A. Staprans, et al., *Proc. IEEE*, March 1973. © 1973 IEEE.)

Figures 10-33 and 10-34 show output power as a function of drive frequency for an efficiency-tuned and a broadband-tuned klystron. Finally, Figure 10-35 shows the output characteristics for the VA-884C and VA-884D klystrons when they are broadband tuned.

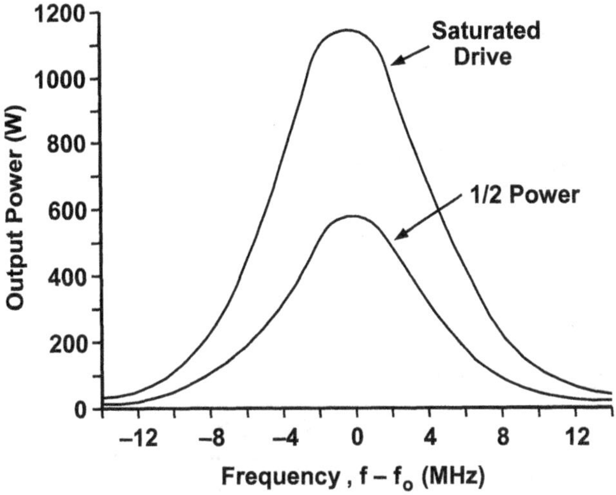

Figure 10-33 Power versus frequency for an efficiency-tuned klystron. (From: *Microwave Tube Manual*, by Varian Associates, Air Force Publication Number T.O.00-25-251, October 1979.)

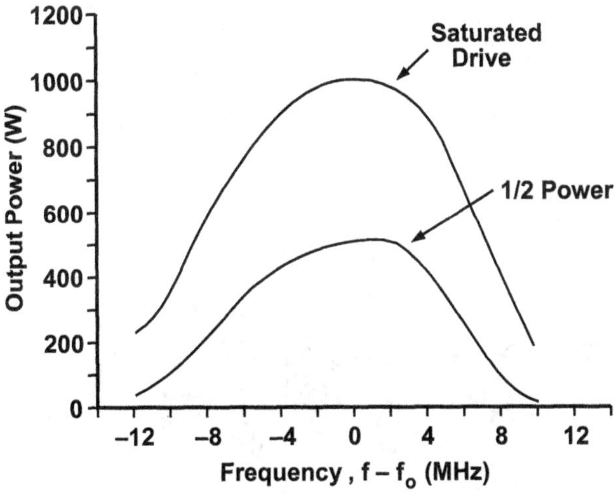

Figure 10-34 Power versus frequency for a broadband-tuned klystron. (From: *Microwave Tube Manual*, by Varian Associates, Air Force Publication Number T.O.00-25-251, October 1979.)

The center frequency of operation of a klystron can be varied while maintaining the desired broadband characteristics by simultaneously tuning all cavities. This can be done mechanically with a channel tuner mechanism like that illustrated in Figure 10-36. With this mechanism, which is sometimes manually operated and sometimes motor driven, a gear-coupled drive is used to simultaneously change the position of the tuning plunger in all cavities. An example of the resulting shift in operating frequency is given in Figure 10-37.

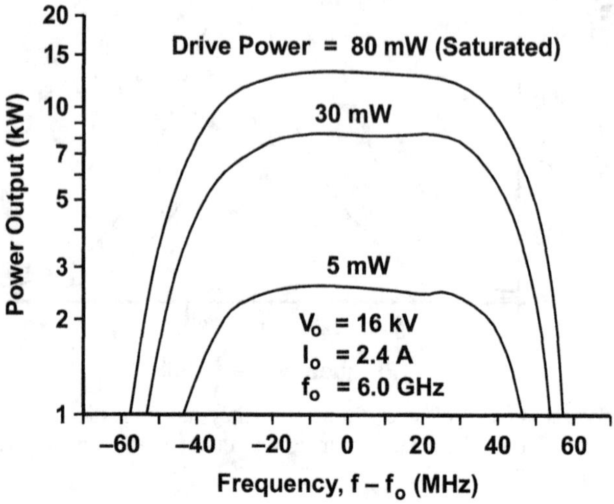

Figure 10-35 Output characteristics for the VA-884C and VA-884-D klystrons. (From: A. Staprans, et al., *Proc. IEEE*, March 1973. © 1973 IEEE.)

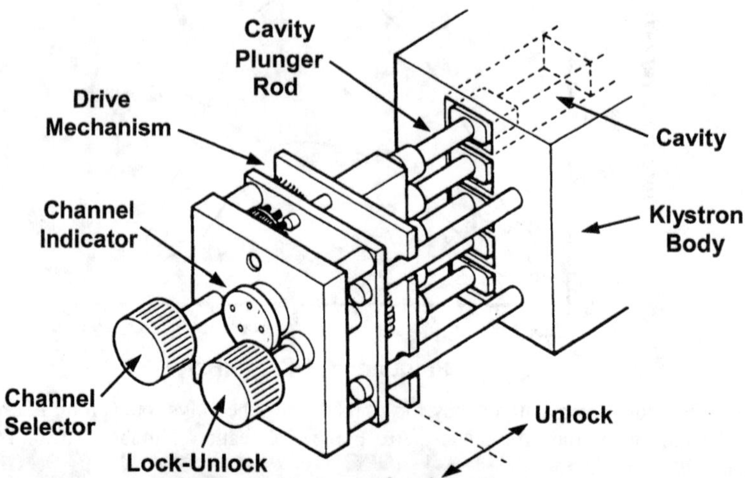

Figure 10-36 Klystron channel tuner mechanism. (From: A. Staprans, et al., *Proc. IEEE*, March 1973. © 1973 IEEE.)

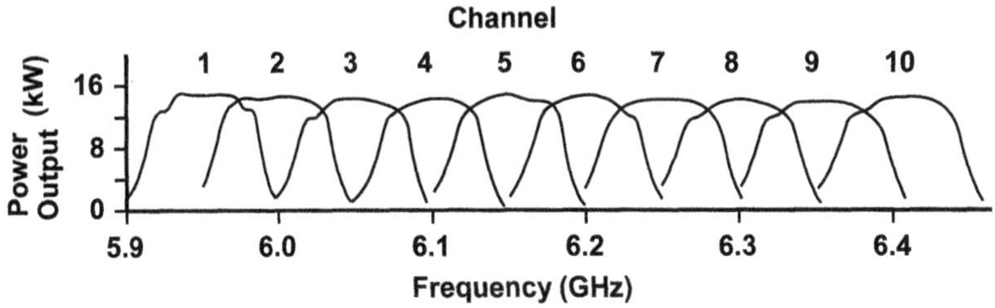

Figure 10-37 Channel tunings characteristics of a VA-884D klystron. (From: A. Staprans, et al., *Proc. IEEE*, March 1973. © 1973 IEEE.)

10.4.2 Transfer Characteristics

Transfer characteristics for a klystron are shown in Figure 10-38. As is typical for all linear-beam tubes, output power usually varies linearly with drive power at low power levels. At high drive levels, there is a point where optimum bunching occurs in the output cavity and output power is maximized. Beyond that drive level, output power decreases.

When a klystron is broadband tuned, there is significant decrease in gain compared with synchronous or efficiency tuning. This is clear from Figure 10-38, which shows an increase in drive power of nearly two orders of magnitude required for broadband-tuned operation as compared with efficiency-tuned operation.

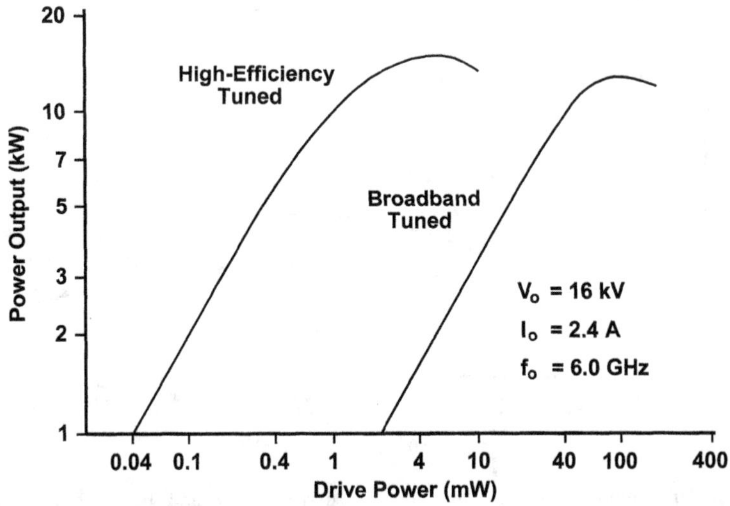

Figure 10-38 Gain characteristics for the VA-884C and VA-884D klystrons. (From: A. Staprans, et al., *Proc. IEEE*, March 1973. © 1973 IEEE.)

The output power of a klystron (as well as for other microwave tubes) varies dramatically with beam voltage as shown in Figure 10-39. This is as expected for at least two main reasons. In the first place, beam power, P_o, varies as the 5/2 power of beam voltage, V_o, because

$$P_o = I_o \; V_o \tag{10-19}$$

but

$$I_o = p V_o^{3/2} \tag{10-20}$$

so

$$P_o = p V_o^{3/2} \tag{10-21}$$

where I_o is the beam current and p is perveance. Thus, an increase in beam voltage by a factor of 1.5 (as shown in Figure 10-39) is accompanied by an increase in beam power of a factor of 2.8. For constant tube efficiency, RF power output would go up by the same amount. Efficiency is not constant, however, but decreases with decreasing voltage because of changes in beam bunches and diameter at voltages below the design voltage.

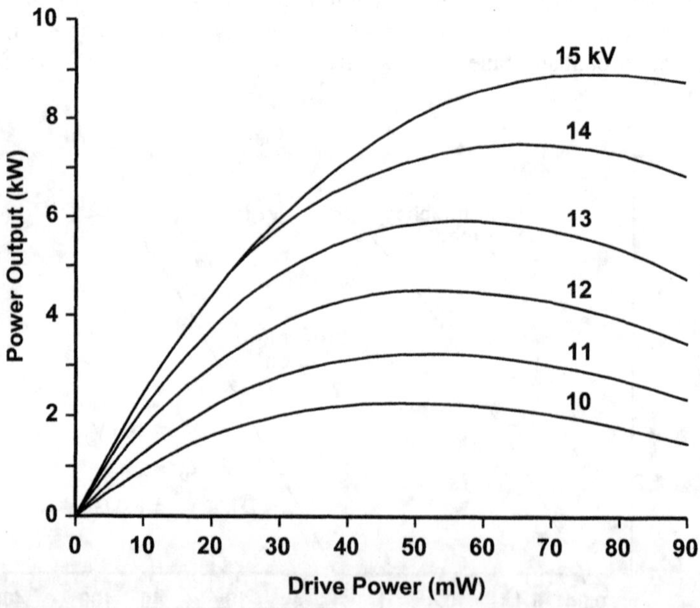

Figure 10-39 Output power as functions of drive power and beam voltage. (From: *Microwave Tube Manual*, Varian Associates, Air Force Publication Number T.O.00-25-251, October 1979.)

REFERENCES

[1] A. Staprans, E. W. McCune, and J. A. Ruetz, "High-power linear-beam tubes," *Proc. IEEE*, Vol. 61, No. 3, March 1973, pp. 299–330.

[2] A. Arsenjewa-Heil, and O. Heil, "Eine neue methode zur erzeugung kurzer ungeämpfter elektromagnetishen welle von großer intensität," *Zeit. für Physik*, Vol. 95, 1935, pp. 752–773.

[3] R. H. Varian, and S. F. Varian, "A high frequency oscillator and amplifier," *Jour. App. Phys.*, Vol. 10, May 1939, pp. 321–327.

[4] K. H. Kreuchen, B. A. Auld, and N. E. Dixon, "A study of the broadband frequency response of the multicavity klystron amplifier," *Jour. App. Phys.*, May 1957, pp. 529–567.

[5] M. Chodorow, Lecture notes.

[6] M. J. Smith, and G. Phillips, *Power Klystrons Today*, Taunton, Somerset, England: Research Studies Ltd., 1995.

[7] R. S. Symons, and H. R. Jory, "Cyclotron Resonance Devices," *Advances in Electronics and Electron Devices*, Vol. 55, New York: Academic Press, 1981.

[8] R. S. Symons, "Scaling laws and power limits for klystrons," *Technical Digest*, IEDM, 1986, pp. 156–159.

[9] R. S. Symons, and J. R. M. Vaughan, "The linear theory of the clustered-cavity klystron," *IEEE Trans. Plasma Science*, Vol. 22, No. 5, October 1994, pp. 713–718.

Chapter 11

Special-Purpose Klystrons

11.1 HIGH-EFFICIENCY KLYSTRONS

Many calculations of the maximum efficiency for klystrons have been made. Mihran et al. [1] predicted an efficiency of 80%. In several Soviet publications, predictions in the 80% to 90% range have been made [2–9]. There is no evidence that these high efficiencies have been achieved.

Bastien et al. [10] showed the predicted variations of efficiency for low- and high-perveance klystrons (see Figure 11-1). Notice that the effects of load impedance on output gap voltage and output cavity current are shown. Maximum efficiencies occur at the point where electrons are reflected in the output cavity. For low perveance, Faillon predicted 70% efficiency, and for high perveance, he predicted about 47%.

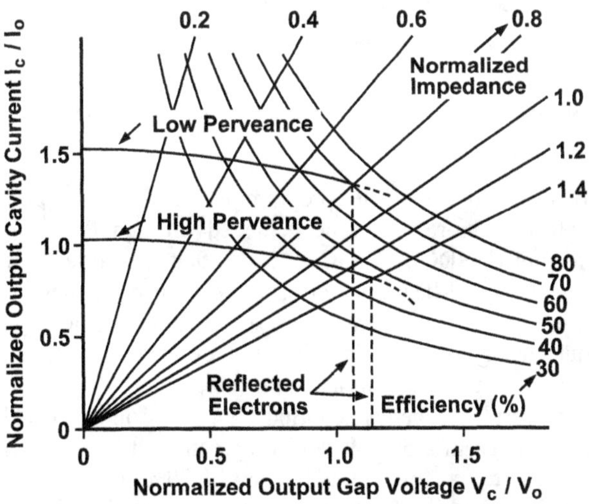

Figure 11-1 Predicted output cavity characteristics for a high-efficiency klystron. (From: C. Bastien, G. Faillon, and M. Simon, 1982 IEDM. © 1982 IEEE.)

Kosmahl and Albers [11] computed an efficiency of 80.3% with 6% interception. This computation was carried out using a three-dimensional analysis of the performance of output gaps in klystrons. Their work (see Figure 11-2) showed the variation in efficiency with perveance. Note that efficiency falls almost linearly with increasing perveance. Also, interception increases with increasing perveance. Kosmahl and Albers also showed the strong dependence of efficiency on drift tube diameter in the output cavity. These results, which were computed for Mihran's [1] and Lien's [12] designs, are shown in Figure 11-3.

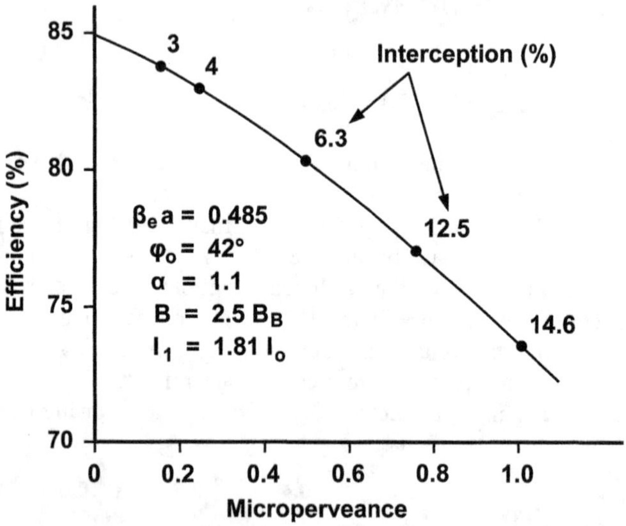

Figure 11-2 Efficiency versus perveance assuming a constant bunching level. (Adapted from: H.G. Kosmahl and L. U. Albers, *IEEE Trans. Electron Devices*, October 1973. ©1973 IEEE.)

In the solar-powered satellite program, Kosmahl [13] predicted that an efficiency of 80% or more might be achieved through the use of multistage depressed collectors (MSDC). It was predicted that the collector would recover from 40–50% of the spent-beam power. The collector efficiency is limited to the 40–50% range because the electrons in the spent beam have a very large velocity spread. Some electrons are barely moving as they leave the output cavity, while others have as much as twice the initial beam energy.

The highest overall efficiency reported for a klystron is 74% [12]. This high efficiency was achieved in a 50-kW device operating at 2,450 MHz. In the same reference, Lien reported an overall efficiency of 70% for an 800-MHz klystron with a power of about 70 kW. These efficiencies were achieved without the use of depressed collectors. Some of the more significant aspects of this high-efficiency work are summarized here.

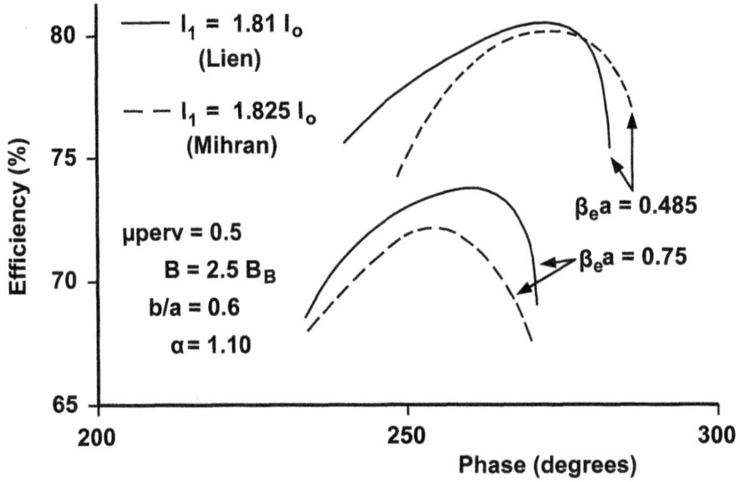

Figure 11-3 Efficiencies computed for two values of β_ea. (Adapted from: H. G. Kosmahl and L. U. Albers, *IEEE Trans. Electron Devices*, October 1973. ©1973 IEEE.)

Perhaps the most important consideration in achieving the efficiencies in [12] was the use of second-harmonic bunching to increase the RF beam current. Some of the cavity configurations considered for second harmonic bunching are shown in Figure 11-4. In narrowband tubes, the combinations of fundamental frequency and second harmonic cavities were used. In an alternate approach, applicable to broadband devices, an abnormally long drift length was used. In this case, the harmonic bunching was produced by second-harmonic space-charge forces within the beam rather than with a second-harmonic field.

The electron phase plot for the case of the single second harmonic prebuncher shown in Figure 11-5 shows computed trajectories for 32 reference electrons in one RF period. In this plot, electrons with dc velocity have trajectories parallel to the abscissa. Electrons with a velocity higher than the dc velocity have a negative slope and those with velocities lower than the dc velocity have a positive slope. It should be noted that the majority of the reference electrons is contained within the bunch at the location of the output cavity and that the interbunch region is essentially free of electrons. In addition, the velocity spread among the electrons is small.

The fundamental and second-harmonic frequency components of beam current are shown as functions of normalized distance along the tube in Figure 11-6. The maximum fundamental current produced by the buncher at the output resonator is 1.8 times the dc beam current. Lien noted that higher currents could be produced, but the velocity spread among electrons is increased and efficiency is not increased.

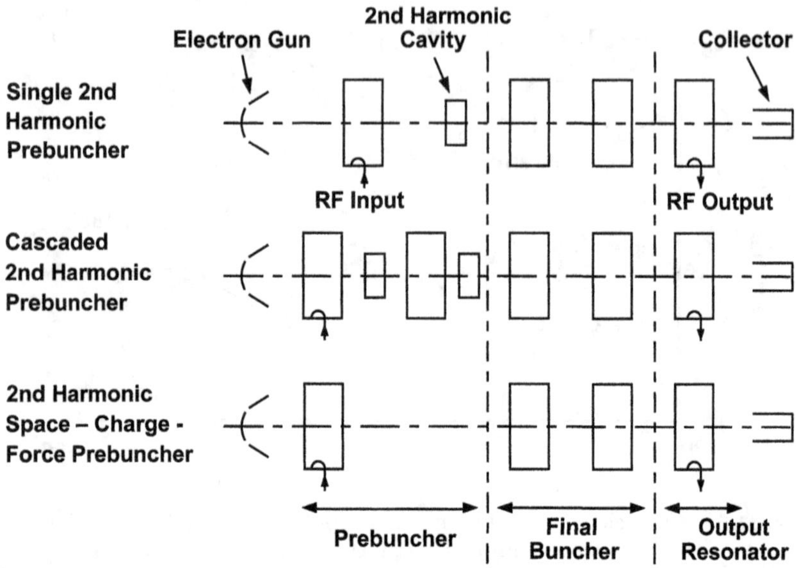

Figure 11-4 High-efficiency klystron circuits using second-harmonic bunching. (From: E. L. Lien, *Proc. 8th MOGA*, September 1970.)

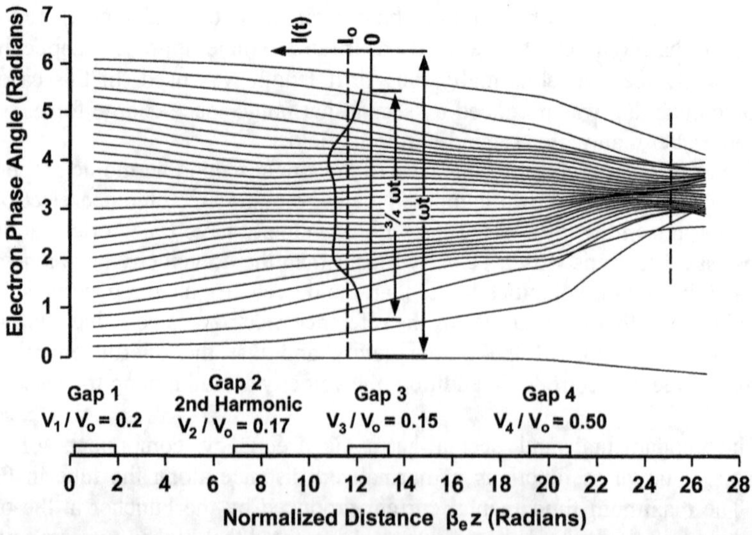

Figure 11-5 Electron phase angle versus distance for a high-efficiency klystron. (From: E. L. Lien, *Proc. 8th MOGA*, September 1970.)

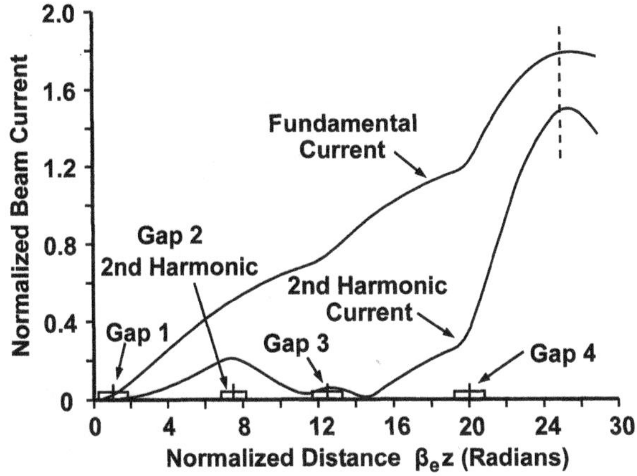

Figure 11-6 Normalized RF beam current for a high-efficiency klystron. (From: E. L. Lien, *Proc. 8th MOGA*, September 1970.)

Output power and efficiency curves for two klystrons using second-harmonic cavities are shown in Figures 11-7 and 11-8. The X-3074B operated at about 800 MHz. This tube had six fundamental-frequency cavities and one second-harmonic cavity. The results in Figure 11-8 are for the 2,450 MHz tube, which had two second-harmonic cavities in addition to six fundamental-frequency cavities. It should be reiterated that these high efficiencies were attained without the use of depressed collectors.

Experience with multistage depressed collectors for high-power operation is limited to tubes operating at about 100 kW. For example, Varian (now CPI) [14] developed klystrons for television service that used multistage depressed collectors. Figure 11-9 shows the improvement in efficiency that can be achieved with an MSDC on a TV klystron. With a standard collector, these tubes have an efficiency up to about 50%. With an MSDC recovering 57% of the spent beam power, the overall efficiency becomes 71%.

11.2 HIGH-POWER KLYSTRONS

Klystrons are capable of generating microwave power from over a megawatt CW to tens or even hundreds of megawatts pulsed. Symons [15] provided a technique for estimating the power that should be obtainable. Of course, the RF power is primarily dependent on the beam power, which is the product of beam voltage and beam current, so it is necessary to determine limits on beam voltage and current in order to estimate power capability.

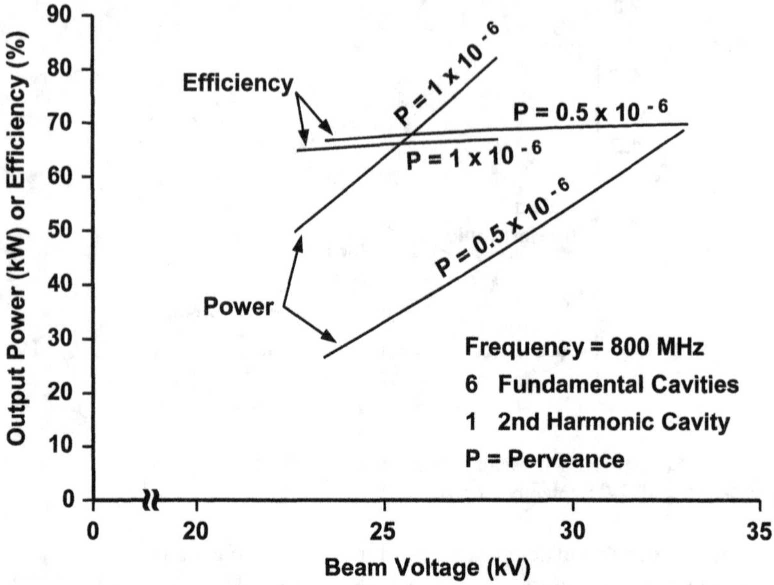

Figure 11-7 Power and efficiency for X-3074B, 800-MHz klystron. (From: E. L. Lien, *Proc. 8th MOGA*, September 1970.)

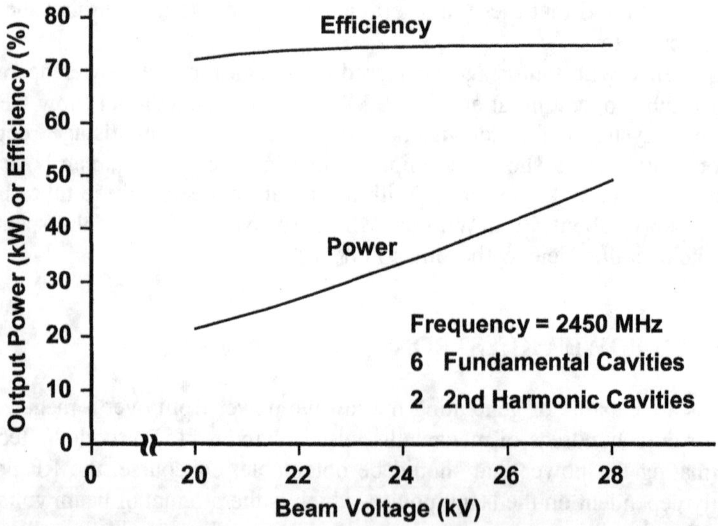

Figure 11-8 Power and efficiency for VKS-7773, 2,450-MHz klystron. (From: E. L. Lien, *Proc. 8th MOGA*, September 1970.)

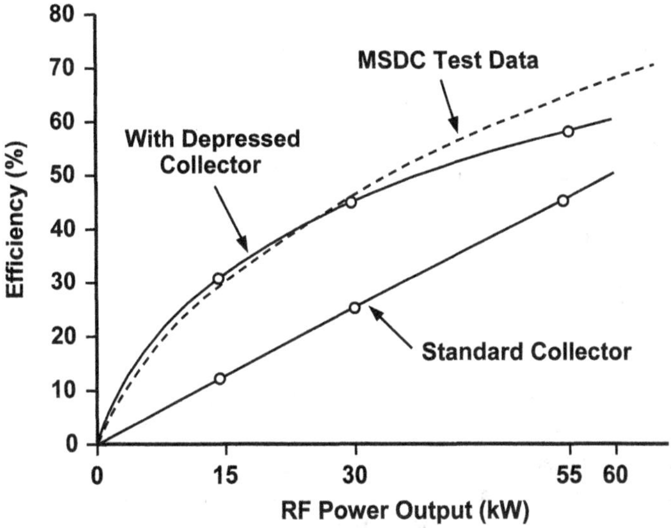

Figure 11-9 Overall efficiency for a Varian TV klystron with and without a multistage depressed collector (MSDC). (From: E. W. McCune, 1986 IEDM. © 1986 IEEE.)

11.2.1 Limits on Beam Voltage

The most obvious limit on beam voltage is the onset of arcing in the electron gun. This limit depends on pulse length, as Table 11.1 indicates. Note that hold-off voltage decreases with increasing pulse length. Symons pointed out that the use of intermediate potential grading electrodes within the gun could make higher operating voltages possible.

Table 11.1

Voltage Limits Achieved in Practice

Pulse Length	Voltage	Manufacturer	Band
0.1 μs	700 kV	Varian	S
2.0 μs	475 kV	SLAC	S
6.5 μs	315 kV	SLAC	S
10 μs	300 kV	Litton	L
continuous	200 kV	Varian	S

Source: [15].

Other factors, beside gun arcing, limit the operating voltage of a klystron. For efficient operation, the voltage across the output gap must be sufficient to remove most of the energy from the electron beam as it passes through the gap. This happens when the peak voltage across the output gap is nearly equal to the beam voltage. However, arcing can place a limit on gap voltage and, consequently, on beam voltage.

There is also a limit imposed on cavity gap voltage by the RF losses in the output cavity. These losses are proportional to the square of the cavity gap voltage. Also pulsed heating of the cavity surface can lead to cracking and fatigue failure. These heating effects were discussed in Chapter 10. Heating as a function of cavity voltage and operating wavelength were shown in Figure 10-20.

Kilpatrick [16] proposed a criterion for the onset of RF arcing across gaps and Peter et al. [17] proposed the modification

$$f = 1.6\ E^2\ e^{-8.5/E} \tag{11-1}$$

where f is the frequency in megahertz and E is the electric field in megavolts per meter.

It should be noted that Loew and Wang [18] demonstrated an RF breakdown limit far above that given by (11-1). Their results were obtained for electron linac structures. However, Phillips and Sprehn [19] found that the RF breakdown in their high-power klystron work matched well with (11-1). They pointed out that a major difference between the klystron and the linac is the presence in the klystron of a nearby intense electron beam. They hypothesize that stray electrons or x-ray photons strike the cavity walls and initiate breakdown at a relatively low field gradient more like that given by (11-1).

The use of (11-1) is simplified by noting that the fields used in high-power klystrons are high enough that the exponential term is ~ unity. An approximation for the magnitude of the electric field can be found by examining the gap transit angle. First of all, the field is related to gap voltage by

$$E = \beta \frac{V}{\ell} \tag{11-2}$$

where ℓ is the gap length and β is the field enhancement factor resulting from the field concentration at the gap edges (typically 1.9 for klystron gaps).

Because the gap transit angle is nearly 1 radian in all klystrons, the gap length can be determined to be

$$\ell = \frac{\lambda}{2\,\pi}\left[1 - \gamma^{-2}\right]^{1/2} \tag{11-3}$$

where, with V in volts,

$$\gamma = 1 + \frac{V}{511{,}000} \tag{11-4}$$

The quantity γ is the increase in electron mass that occurs as beam voltage (and electron velocity) increase. By combining (11-3), (11-2), and (11-1), the following relation between voltage limit and wavelength can be obtained. This is the limit on gap voltage, which is also the limit on beam voltage for a klystron with a single-gap output. The units are volts and centimeters.

$$V^2 = 1.32 \times 10^{10}\left(1-\gamma^{-2}\right)\lambda \tag{11-5}$$

Figure 11-10 shows this voltage limit plotted as a dashed line on Figure 10-25 (which showed heating as functions of voltage and operating wavelength).

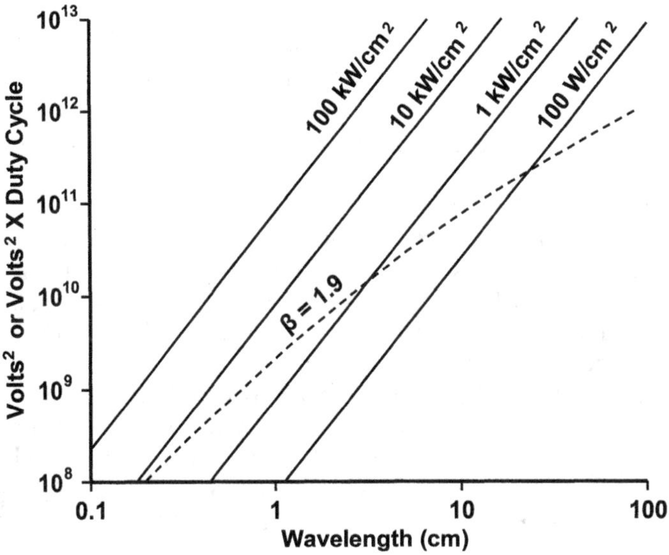

Figure 11-10 Limits on beam voltage due to cavity losses and RF arcing. (From: R. S. Symons, 1986 IEDM. © 1986 IEEE.)

By using multiple output cavities, such as those in extended interaction klystrons, the voltage across each gap of the output circuit is lower than for a single output cavity and both cavity losses and output gap arcing can be reduced. As a result, the voltage limits for multiple output klystrons can be much higher than for klystrons with single output cavities.

11.2.2 Limits on Beam Current

There are several limits to beam current in a linear-beam tube. The most obvious limits are evident in the relation for RF power output, which is

$$P = \eta I V = \eta P_\mu \times 10^{-6} V^{5/2} \tag{11-6}$$

where η is efficiency and P_μ is the microperveance of the electron gun. It is very difficult to design guns with $P_\mu > \sim 2$ because of space charge effects on potential contours within the guns. Since microperveance is limited and beam voltage is limited, beam current is limited.

Symons [15] observed that efficiency is fairly well described by the relation

$$\eta = 0.9 - 0.2\ P_\mu \tag{11-7}$$

Figure 11-2 very nearly supports this relation. (In their work at SLAC, Phillips and Sprehn [19] found a somewhat different relation, however, we will use Symons' relation here.)

The reason for the variation of efficiency with microperveance is that, as perveance is increased, it becomes more difficult to form electron bunches that are optimally configured for maximum power extraction from the electron beam. As a result, for maximum efficiency, perveance must be limited, which limits beam current.

As the frequency of operation of a klystron is increased, other factors come into play in limiting beam current. Beam current is JA, that is, the product of current density and the cross-sectional area of the electron beam. The area of the beam is limited by the drift tunnel radius, which scales with wavelength. Symons pointed out that the electron beams in most klystrons have a radius that is approximately one-half of a radian of electron transit angle.

The beam current density is the cathode emission density multiplied by the area convergence ratio of the gun. For impregnated cathodes and long life, a current density of 5 A/cm^2 can be obtained with little dependence on pulse length. For short pulses, Symons assumed that an oxide coated cathode could be used. The emission density, J_c in A/cm^2, is given by the empirical relation

$$J_c^{2.61}\tau = 814 \tag{11-8}$$

where τ is the pulse length in microseconds.

If we assume a maximum area convergence ratio of 200:1, then the obtainable beam current density is 1,000 A/cm^2 for $\tau \geq 10$ μs, 2,600 A/cm^2 for $\tau = 1$ μs, and 6,300 A/cm^2 for $\tau = 0.1$ μs.

11.2.3 Estimate of Obtainable Power

Regardless of frequency, space charge effects limit perveance and therefore beam current and power. By combining (11-6) and (11-7), we find that the dependence of RF power on perveance and on voltage is

$$P = P_\mu\left(0.9\ -\ 0.2\,P_\mu\right)10^{-6}V^{5/2} \tag{11-9}$$

As pointed out in the previous section, as frequency increases, beam area decreases and so maximum beam current decreases. The limit on beam current is

$$I = J \frac{\lambda^2}{16\pi}\left(1 - \gamma^{-2}\right) \tag{11-10}$$

The limit on J is obtained from the maximum cathode current density and the area compression ratio.

Now, by combining (11-10) with (11-7) and (11-6), the following expression for power as functions of wavelength, J, V, and P_μ, can be obtained.

$$P = \left(0.9 - 0.2\ P_\mu\right) \frac{\lambda^2 J V}{16\pi}\left(1 - \gamma^{-2}\right) \tag{11-11}$$

We now have two limiting relations, (11-9) and (11-11), for obtainable power. Figure 11-11 shows these power limits plotted as a function of $\lambda^2 J$ with voltage as a parameter.

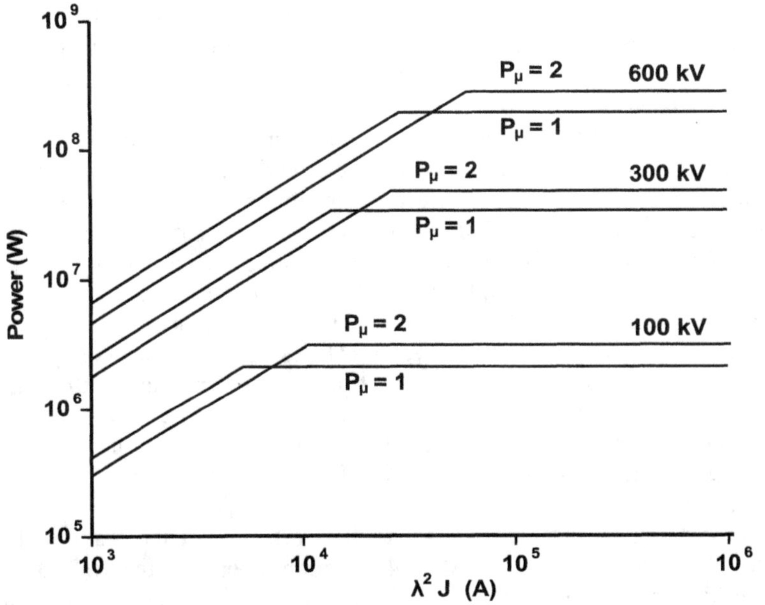

Figure 11-11 Estimate of obtainable power as functions of wavelength, beam current density, and voltage for microperveances of 1 and 2. (From: R. S. Symons, 1986 IEDM. © 1986 IEEE.)

The procedure outlined by Symons for establishing the reasonableness of a power requirement for a klystron is as follows:

1. Find the maximum allowable beam voltage from Table 11.1 and Figure 11-10.

2. Using Figure 11-11 or (11-11) and (11-9), see if the desired power can be achieved.

As an example, the 10-cm, 50-MW, 6.5 μs klystron used at the Stanford Linear Accelerator Center operates at approximately 300 kV and microperveance 2. From Figure 11-10, it can be seen that it is operating close to Kilpatrick's criterion limit for RF output gap arcing of about 10^{11} square volts. The product, $\lambda^2 J$, for this device is about 10^5 and thus, from Figure 11-11, it is reasonable to expect an output power of 50 MW at a microperveance of 2.

There are several possible techniques for developing klystrons with power capabilities above those predicted by Symons' procedure. As pointed out earlier in this section, by using multiple output cavities, output gap voltage and output gap arcing can be reduced so there is the possibility of using higher beam voltages.

The use of voltage grading electrodes in the electron gun could make possible higher gun voltages. True discussed the use of grading electrodes in the design of a 1-MV, 1,825-A Pierce gun [20]. Three electrodes were used to limit the voltage gradients to less that 200 kV/cm and True said that the gun should operate reliably for pulse lengths up to 5 microseconds.

There is considerable effort at the present time to develop multiple beam klystrons (discussed later in this chapter). These hold the promise of being able to generate extremely high powers because many beams, each operating at a microperveance near unity, can be used with the result that the klystron has an overall microperveance ~10 or higher. It is possible that a similar increase in microperveance can be achieved by using a beam in the form of a flat sheet.

Two outstanding examples of existing high-power klystrons are the 1-MW, CW, 352-MHz klystrons manufactured by Thales [21] and the 65-MW, 3.5-μs pulse length, 180-pps, S-band klystrons manufactured by the Stanford Linear Accelerator Center (SLAC) [21]. The Thales tube, the TH 2089, was built for CERN, the European Nuclear Research Center. The weight of the klystron, electromagnet, and support structure is 2,400 kg and the dimensions are 4.80 × 1.85 × 1.20 meters.

The SLAC tube, the SLAC 5045, is the workhorse of the Stanford Linear Collider (SLC). A total of 245 SLAC 5045 klystrons operate in the SLC. Over 600 of these klystrons have been built at SLAC. The mean time between failures (MTBF) is higher than 50,000 hours.

The Stanford Linear Accelerator Center klystron group designed, fabricated, and tested high-power klystrons, which range from 477 MHz to 11.424 GHz and from 1 MW under continuous operation to 150-MW at a 3-μs pulsewidth. Particularly noteworthy are the 150-MW S-band tube and a 75-MW X-band tube.

The operating parameters for the 150 MW S-band tube are listed in Table 11.2 [22]. The tube has operated at powers greater than 150 MW. The tube uses a two-cell output cavity to increase efficiency, a scandate cathode, and lossy drift tubes to improve stability.

Table 11.2

Operating Parameters for 150-MW S-Band SLAC Klystron

Beam voltage	535 kV
Beam current	700A
RF pulse width at rep rate	3 μs at 60 Hz
RF output power	150 MW
Saturated gain	> 50 dB
Efficiency	≥ 40%
Operating frequency	2,998 MHz
Solenoid focusing field	2,100 gauss (3×B_B)

The 75-MW X-band program was established to provide klystrons for SLAC version of the Next Linear Collider. Phillips and Sprehn described the development efforts for 75-MW solenoid-focused and PPM-focused klystrons [19]. The solenoid-focused tube operated at 75 MW and 450 kV with pulse lengths up to 1.2 μs and an efficiency of 50%. The PPM-focused klystron operated at 71 MW and 463 kV with an efficiency between 53% and 63%.

11.3 BROADBAND KLYSTRONS

Traveling wave tubes are usually considered to be broadband amplifiers, while klystrons are thought to have narrow bandwidths. To make a correct comparison, however, output power must be specified. At power levels up to about a kilowatt, TWT bandwidths can exceed an octave because broadband helix circuits are used. When the average power level becomes high enough so the coupled-cavity circuits must be used, the bandwidth drops considerably and at the 100- to 200-kW level, the bandwidth is no wider than about 15% [23].

In the case of the klystron, the lower the cavity Q is, the wider the bandwidth is. The cavity Q, in turn, varies with beam impedance, R_o, which may be related to tube operating parameters as follows:

$$R_o = \left(\frac{\eta}{P_{RF}}\right)^{1/5} \frac{1}{p^{4/5}} \tag{11-12}$$

where η is efficiency and p is perveance. As a result, R_o and Q decrease and bandwidth increases as PRF and p increase. Within reasonable limits of perveance and efficiency (two opposing characteristics), the above relation indicates that it is two to three times easier to achieve the same bandwidth at 20 MW than it is at 200 kW. In fact, there are now klystrons available that deliver 3–10 MW and have a bandwidth of 10% or more, while at 100–200 kW, bandwidth is usually limited to 3% or 4%.

An example of a high-power klystron capable of 10% bandwidth is the VA-812C, for which the operating characteristics are given in Table 11.3 [24]. The efficiency as a function of frequency is shown in Figure 11-12. Because this tube operated at a low frequency, it was very large and this, in turn, made the use of a very high perveance beam possible. Also, to achieve the flat broadband response, each of the four cavities in the driver section had a different gap length and a different loading. Broadbanding of the output circuit was achieved by using a triple-tuned circuit. The characteristics of driver and output sections required for broad bandwidths are described in the following sections.

Table 11.3
VA-812C Operating Characteristics

Frequency	400–450 MHz
Peak Beam Voltage	145 kV
Perveance	2.5×10^{-6}
Peak Power Output	8 MW
Average Power Output	30 kW
Pulse Length	6 μs
Efficiency	40%
Electronic Bandwidth	12%
Gain	30 dB

Source: [24].

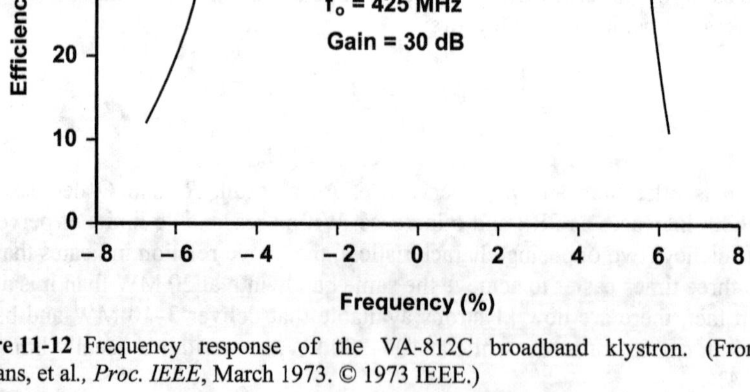

Figure 11-12 Frequency response of the VA-812C broadband klystron. (From: A. Staprans, et al., *Proc. IEEE*, March 1973. © 1973 IEEE.)

At microwave frequencies in the power range from about 200 kW to a few megawatts, bandwidths are required that conventional klystrons and TWTs cannot provide. As a result, special designs have been developed that make klystrons operating at bandwidths of 10% or more possible. The driver sections and the output sections of these tubes are discussed in the following sections.

11.3.1 Driver Sections

The driver section of a klystron consists of all of the cavities up to the output structure. In conventional narrowband tubes, the driver section usually has only three or four cavities and the output structure is a single cavity. In broadband klystrons, the driver section may consist of eight to ten or more cavities. In fact, for some clustered-cavity klystrons, the use of over 20 cavities in the driver section has been suggested.

As pointed out in Chapter 10, when large numbers of cavities are used, the interactions between cavities become extremely complex. This is because the modulation produced by each cavity is carried to all subsequent cavities by the electron beam. For wide bandwidths, the total frequency response of a cavity must be considered, not just the response near resonance. The total response depends not only on the cavity itself but also on the lengths of the drift tubes on either side. This is because the gain depends on the ratio of the RF current produced by a cavity to the RF current entering the cavity. However, these RF currents depend on the velocity modulation on the beam and on the distances that the beam must travel for the velocity modulation to become density modulation. Figure 11-13 shows the gain pole that occurs near resonance along with a gain dip at a frequency that becomes higher as the drift tubes are made longer [25].

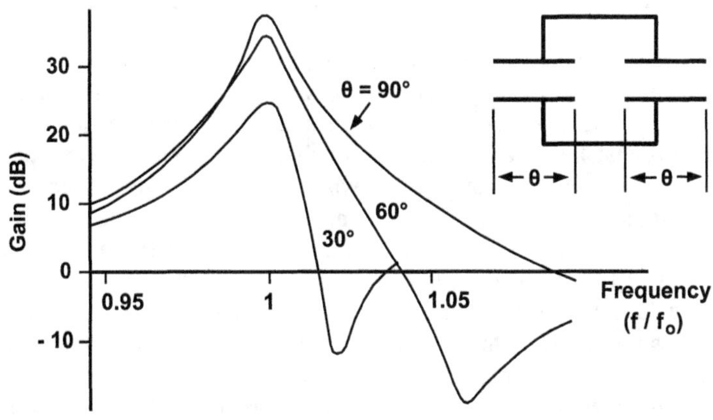

Figure 11-13 Example of the pole and zero of a single cavity as a function drift tube lengths. (Adapted from: G. Faillon, 1985 IEDM. © 1985 IEEE.)

In trying to achieve wide bandwidths for driver sections, it is not sufficient to simply spread the tuning frequencies of the individual cavities over the desired bandwidth. The gain dips must be taken into account, and especially those of the cavities tuned to the low end of the band. Those gain dips can be moved to the high end of the band by lengthening the associated drift tubes. Also, the cavity *Q*s may be adjusted to help control the overall shape of the composite gain curve.

An example showing the gain curves of seven cavities in a broadband driver section is shown in Figure 11-14. Here it is clear that long drift tubes have been used on the cavities at the low end of the band. The low-frequency (left-hand) gain curves are very broad and resemble, except for the broad resonances produced by low *Q*s, the curve for 90° drift tubes in Figure 11-13. The gain dips of these low frequency cavities are compensated for by the gain poles of the cavities tuned to the high end of the band.

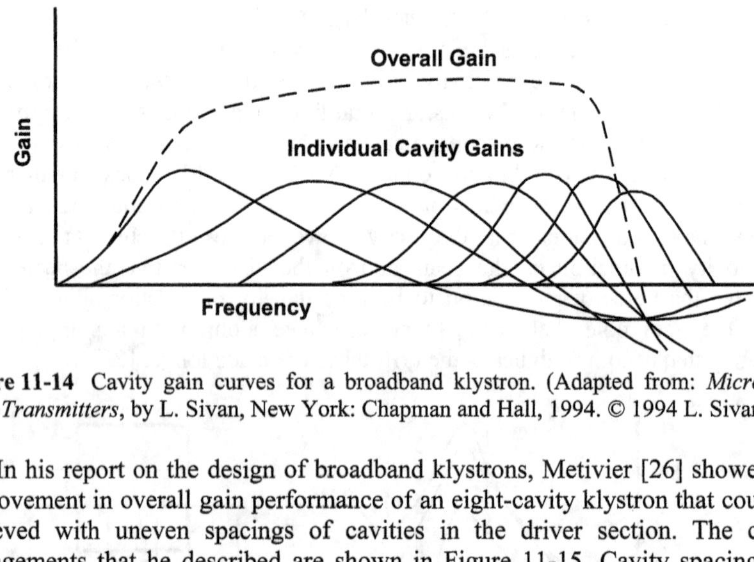

Figure 11-14 Cavity gain curves for a broadband klystron. (Adapted from: *Microwave Tube Transmitters*, by L. Sivan, New York: Chapman and Hall, 1994. © 1994 L. Sivan.)

In his report on the design of broadband klystrons, Metivier [26] showed the improvement in overall gain performance of an eight-cavity klystron that could be achieved with uneven spacings of cavities in the driver section. The cavity arrangements that he described are shown in Figure 11-15. Cavity spacings are shown along with *Q*s and frequencies to which the cavities were tuned.

Small-signal gain curves corresponding to the cavity spacings in Figure 11-15 are shown in Figure 11-16. With evenly space cavities, a large gain dip occurred at the low end of the frequency band. By increasing the spacings between cavities one and two and between 1 and 3, which were tuned to the low end of the band, the gain dips produced by these cavities were moved to the high end of the band. There these gain dips compensated for a large gain peak that had previously occurred. The overall small-signal gain ripple was reduced to 7 dB, which, with the gain compression occurring at saturation, resulted in an overall gain ripple of less than 1 dB at constant drive.

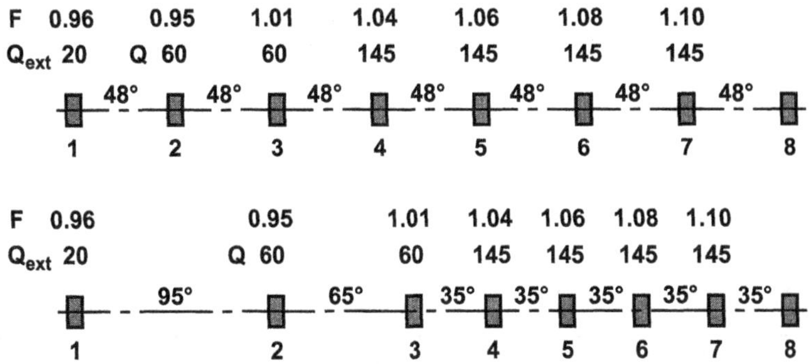

Figure 11-15 Cavity spacing used by Metivier. (From: R.L. Metivier, *Microwave Jour.*, April 1971.)

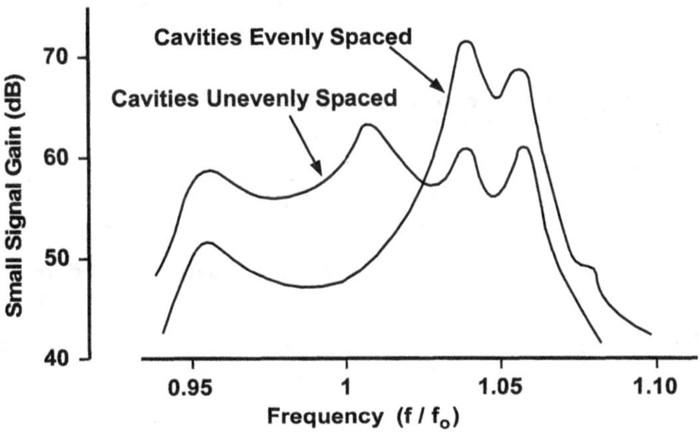

Figure 11-16 Gain of eight-cavity klystrons with cavity spacings used by Metivier. (From: R. L. Metivier, *Microwave Jour.*, April 1971.)

In his patent of a high-efficiency broadband klystron [27], Friedlander described a klystron in which "the gain-bandwidth product and amplitude response is made flatter by successive intermediate floating cavities downstream of the input cavity tuned to successively higher frequencies, and the drift lengths between them successively shorter." The frequencies to which the cavities are tuned in Friedlander's klystron (called the *tapered-drift-length klystron*) are indicated in Figure 11-17. The drift-length tapering is indicated in Figure 11-18.

The power output of a tapered-drift-length klystron for a radar system is shown in Figure 11-19. This tube has nine buncher cavities and one filter loaded output cavity (discussed in the following section) and has a bandwidth of 10%. This tube was completely designed using a variety of computer codes. The tube met specification "right out of the box" (when it was first operated). Minor adjustments were made to increase the operating margin.

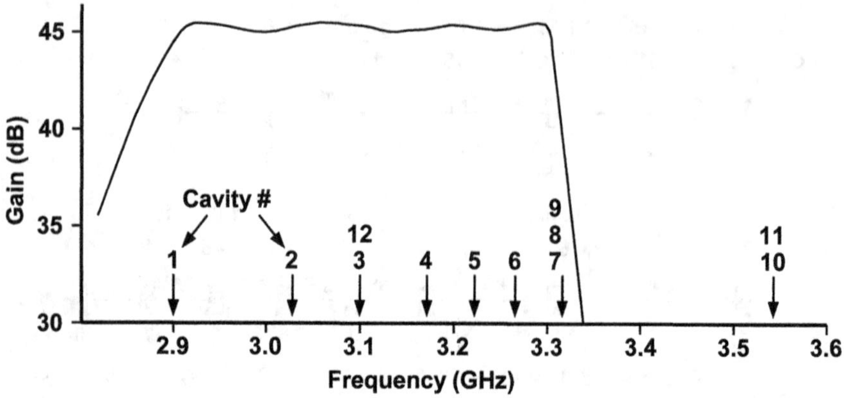

Figure 11-17 Cavity tuning of a tapered-drift-length klystron. (Adapted from: F. I. Friedlander, Patent No. 4,764,710, August 16, 1988.)

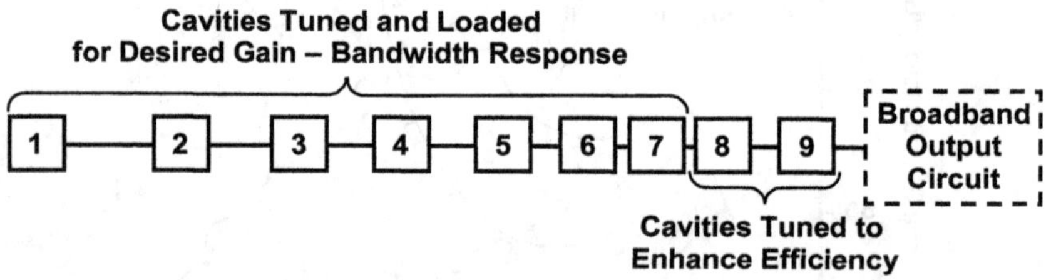

Figure 11-18 Cavity configuration in the driver section of a tapered-drift-length klystron. (Adapted from: F. I. Friedlander, 1988 Monterey Microwave Power Tube Conference.)

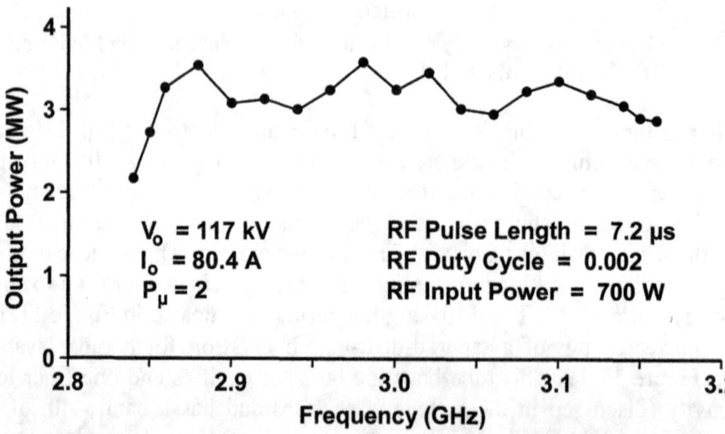

Figure 11-19 Peak output power of a VKS-8345A tapered-drift-length klystron. (Adapted from: B. Stockwell, et al., 1996 Monterey Microwave Power Tube Conference.)

Other efforts to increase bandwidth are based on the fact that, if a single cavity is replaced by a pair of cavities and the gap voltage of each cavity is half the gap voltage of the single cavity, then the stored energy in the pair of cavities would be half that of the single cavity. The voltage applied to the beam would be the same in either case. Because the stored energy has been decreased, Q has been decreased and bandwidth has been increased.

In an early effort to increase the bandwidth of a klystron, Chodorow and Wessel-Berg used structures resembling sections of TWT circuits in place of the cavities normally used in klystrons [28]. These structures are referred to as extended interaction structures. The klystron built by Chodorow and Wessel-Berg contained three extended interaction structures, each of which was formed by a short section of a ring-bar circuit that acted as a half-wave-long resonant circuit. The interaction impedance is generally proportional to the length of the circuit, resulting in a higher gain-bandwidth product in the tube. This tube demonstrated an improvement in bandwidth over conventional cavity-type tubes that existed at that time without any sacrifice in efficiency.

The most common form of extended interaction structure contains klystron-like cavities coupled together by slots in the adjacent cavity walls. Symons et al. showed, in the clustered-cavity klystron, that the coupling slots between cavities are not necessary [29, 30]. In fact, Symons pointed out that the coupling between cavities in an extended interaction structure introduces a zero in the response between the poles associated with the individual cavities. This zero places constraints on tuning and loading that are not present when the zero is eliminated by eliminating the coupling. Figure 11-20 gives a comparison of a conventional stagger-tuned klystron with a clustered-cavity klystron.

The cavity clusters can take on any one of a variety of configurations as indicated in Figure 11-21. In each of the clusters, the gaps of the cavities are close enough together so that there is essentially no gain between them. They are each driven by essentially the same current that was generated by the bunching at the previous cavities. Not only are the gap voltages essentially equal, but also because they are excited by the same bunch, they are in exactly the right phase relationship to cooperate in exciting a new bunch on the beam. Therefore, the cavity cluster acts as if it was a single cavity with half the capacitance and Q (for two cavities) or one-third the capacitance and Q (for three cavities) of a single cavity and a total voltage equal to the voltage for a single cavity.

Figure 11-22 shows the calculated small-signal gain of a prototype 4-MW cluster-cavity klystron. In this plot, Symons and Vaughan illustrated the necessity of including velocity currents in the calculation of performance. The total current density in the electron beam is the product of charge density and velocity,

$$J = \rho u = \rho u_o + \rho_o u \tag{11-13}$$

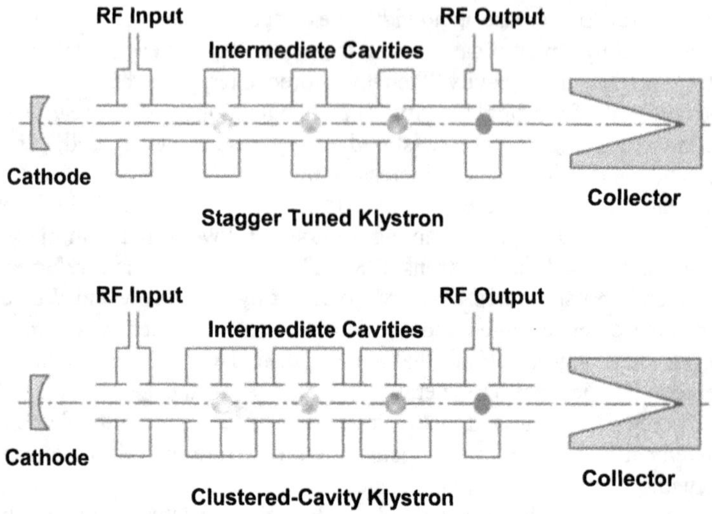

Figure 11-20 Comparison of a stagger-tuned klystron with a clustered-cavity klystron. (From: R. S. Symons and J. R. M. Vaughan, *IEEE Trans. Plasma Science*, October 1994. © 1994 IEEE.)

Usually, as we did in Chapter 8, only the ρu_o term is used in calculating induced currents because the other term (the velocity current term) is considered to be negligible. Symons and Vaughan showed that, for their designs, the velocity term is not negligible. Notice that the bandwidth in Figure 11-22 exceeds 10% and that the gain ripple is very small.

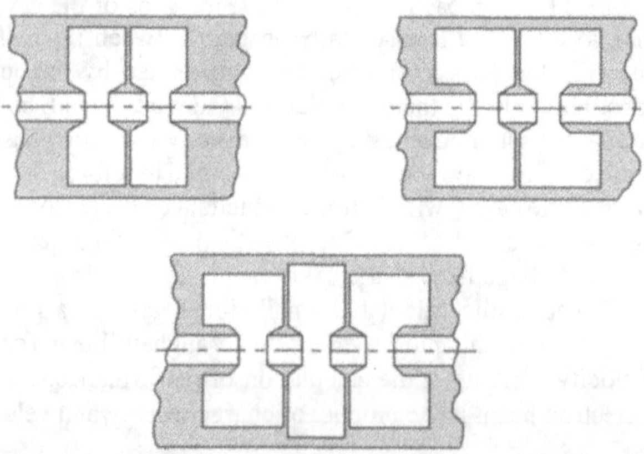

Figure 11-21 Cavities for a clustered-cavity klystron. (From: R. S. Symons, internal Litton report.)

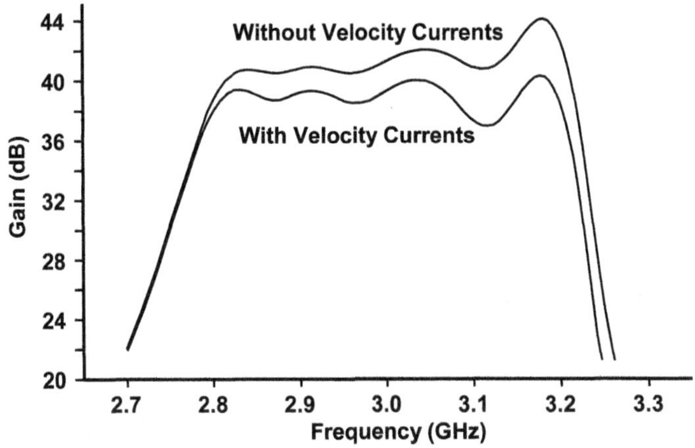

Figure 11-22 Calculated small-signal gain of a prototype clustered-cavity klystron. (From: R. S. Symons and J. R. M. Vaughan, *IEEE Trans. Plasma Science*, October 1994. © 1994 IEEE.)

11.3.2 Output Sections

The output section is normally designed to provide optimum coupling from the beam to the load for best efficiency. Impedance characteristics of some output sections are shown in Figure 11-23. The simplest output circuit consists of a standard reentrant cavity with an iris or loop coupling into waveguide or coaxial line.

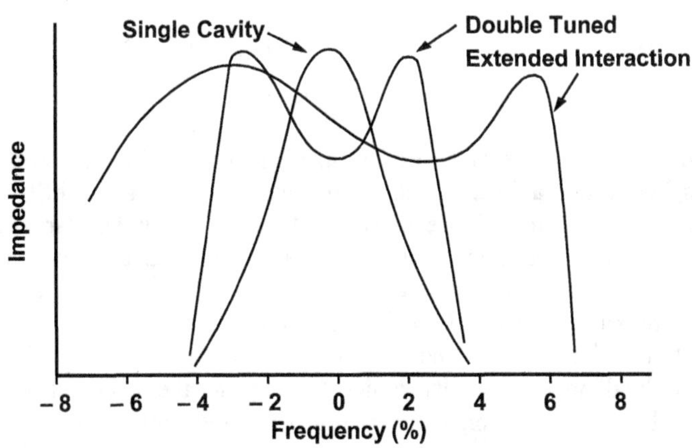

Figure 11-23 Impedance characteristics of single-tuned, double-tuned, and extended interaction circuits. (Adapted from: *Power Klystrons Today* by M. J. Smith and G. Phillips, © 1995 by Research Studies Press, Ltd.)

The double-tuned cavity consists of a conventional reentrant cavity coupled to the beam along with a filter cavity between this cavity and the load as shown in Figure 11-24. The equivalent circuit for the double-tuned cavity is given in Figure 11-25. Both cavities are tuned to the same frequency and the combination makes it possible to load the structure so that optimum impedance occurs at the gap over a greater bandwidth than with a single cavity.

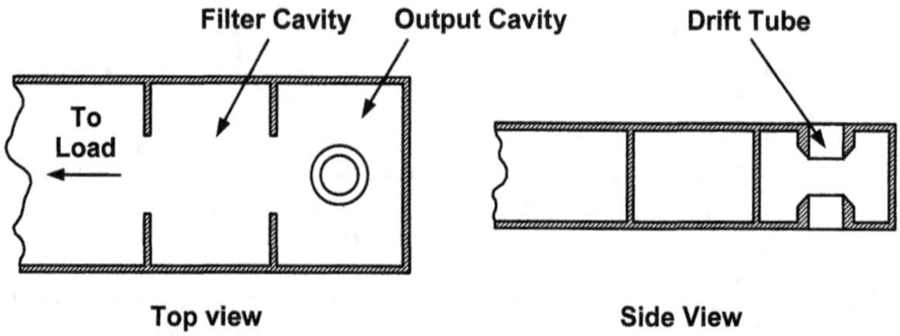

Figure 11-24 Double-tuned output circuit.

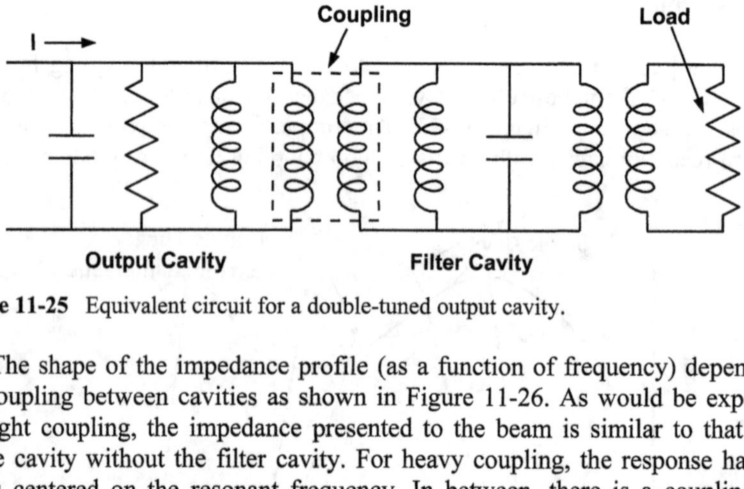

Figure 11-25 Equivalent circuit for a double-tuned output cavity.

The shape of the impedance profile (as a function of frequency) depends on the coupling between cavities as shown in Figure 11-26. As would be expected, for light coupling, the impedance presented to the beam is similar to that for a single cavity without the filter cavity. For heavy coupling, the response has two peaks centered on the resonant frequency. In between, there is a coupling that produces the maximum bandwidth-efficiency product.

Triple-tuned circuits are sometimes used. These consist of two filter cavities between the load and the cavity which couples to the beam. The VA-812C discussed early in this chapter, which had a bandwidth in excess of 10%, used a triple-tuned output circuit. A comparison of the impedance profile for a triple-tuned circuit with those for a single cavity and for a double-tuned circuit is shown in Figure 11-27. Compared with the double-tuned circuit, the impedance of the triple-tuned circuit is flatter over a wider bandwidth.

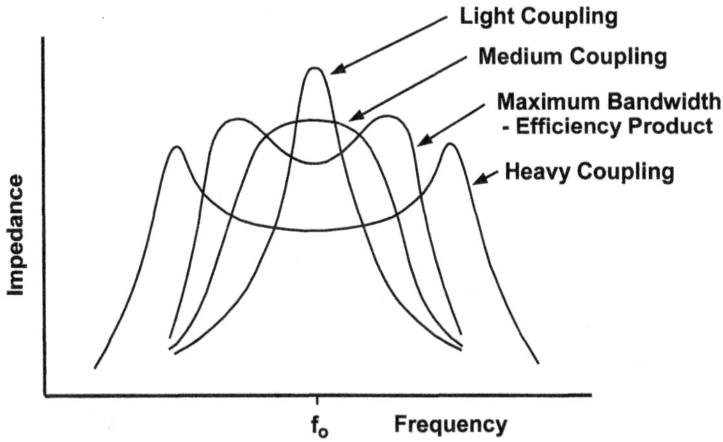

Figure 11-26 Effect of coupling on impedance characteristic of a double-tuned circuit. (Adapted from: *Power Klystrons Today* by M. J. Smith and G. Phillips, © 1995 by Research Studies Press, Ltd.)

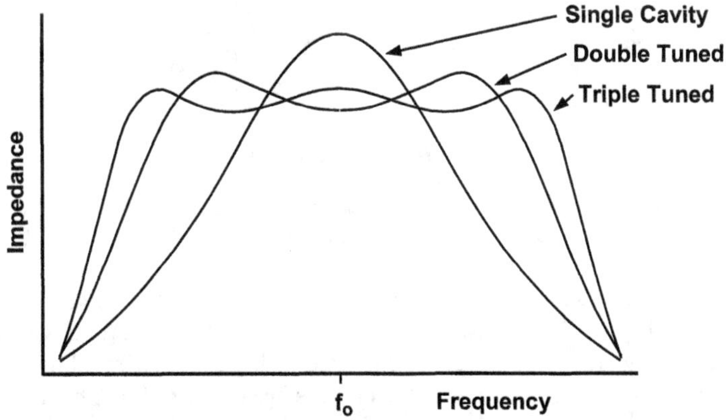

Figure 11-27 Impedance characteristic of a triple-tuned circuit.

Extended interaction circuits, which were discussed briefly in the previous section, are often used as output sections on klystrons. The impedance characteristic of an extended interaction structure is shown in Figure 11-23. An extended interaction structure consists of heavily coupled cavities as shown in Figure 11-28. In one mode of operation, the fields in all cavities are in phase and the electron transit angle between gaps is 2π radians. The other common mode of operation is similar to the fundamental backward wave mode of coupled-cavity TWT operation. That is, the fields in adjacent cavities are approximately 180° out of phase.

In comparing an extended interaction resonator with a single-gap output cavity, the bandwidth is, of course, increased. Perhaps more important is the fact that the power-handling capability is increased. This is because the total output voltage, which must be applied across a single gap in a conventional klystron, is divided among the output cavities of the extended interaction structure. The high-power capability of the extended interaction structure was perhaps best illustrated by a 1-MW average-power X-band device that used a five-gap coupled-cavity resonator like that shown in Figure 11-28 [31].

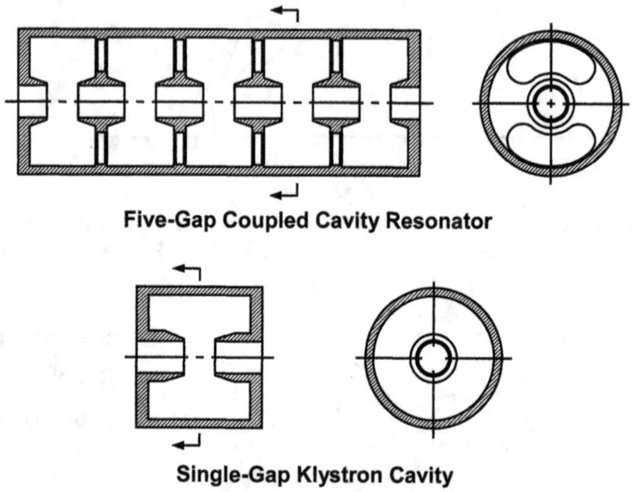

Figure 11-28 Comparison of a five-gap extended interaction structure with a single-gap klystron cavity. (From: A. Staprans, et al., *Proc. IEEE*, March 1973. © 1973 IEEE.)

The cloverleaf structure in Figure 11-29 is sometimes used as the output section for a klystron as shown in Figure 11-30. The cloverleaf structure is a fundamental forward wave circuit that is used in high-power traveling wave tubes [32]. Fundamental forward wave TWT circuits are discussed in Chapter 15. When the cloverleaf structure is used as the output section of a klystron, the tube is called a twystron.

To help increase the bandwidth of a twystron, a high-perveance beam is used to provide strong beam loading. With this circuit and beam combination, the low gain of the TWT section at band edges is offset by the relatively high gain of the stagger-tuned-klystron driver section. The overall result, as Figure 11-31 indicates, can be a very flat gain over a bandwidth of up to 15%. The resulting power output as a function of frequency is also very flat, as shown in Figure 11-32. Twystrons with peak-power outputs up to at least 10 MW have been built. Average powers over 30 kW are available and can be increased about an order of magnitude before circuit losses and interception present fundamental limitations.

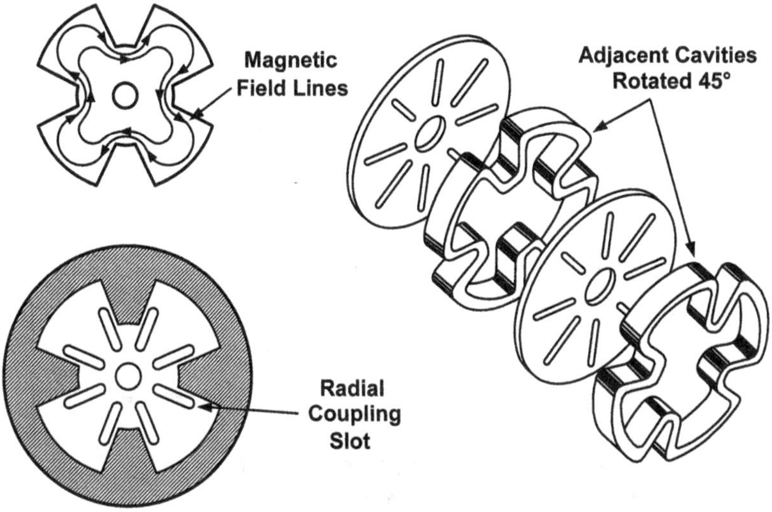

Figure 11-29 Cloverleaf structure. (Adapted from: M. Chodorow and R. A. Craig, *Proc. IRE*, August 1957. © 1957 IRE (now IEEE).)

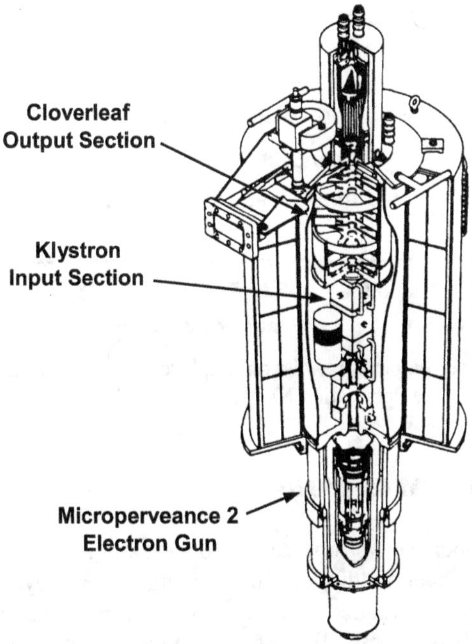

Figure 11-30 A multimegawatt twystron. (From: *Microwave Tube Manual* by Varian, Air Force Publication No. T.O.00-25-251, 1979.)

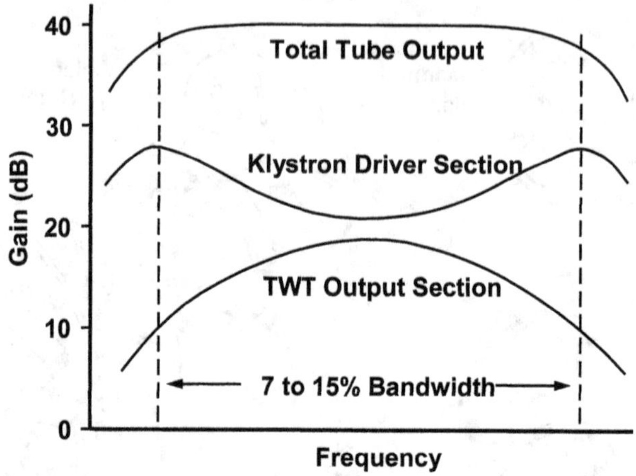

Figure 11-31 The gain of a twystron amplifier. (From: *Microwave Tube Manual* by Varian, Air Force Publication No. T.O.00-25-251, 1979.)

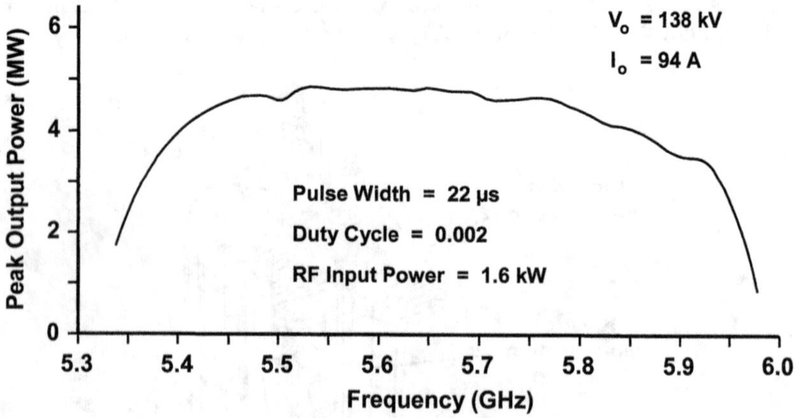

Figure 11-32 The power output for twystron amplifier VA-146. (From: *Microwave Tube Manual* by Varian, Air Force Publication No. T.O.00-25-251, 1979.)

11.4 MULTIPLE BEAM KLYSTRONS

Conventional single-beam klystrons can produce extremely high powers at high efficiency and with bandwidths of 10% or more. However, they have the disadvantage that microperveance is limited to a maximum value of ~2 and must be ~0.5 for high efficiency. Because of these relatively low values of microperveance, beam currents are relatively low and very high voltages are required if high power and high efficiency are to be obtained.

By using a device containing multiple electron beams, each one of which has a microperveance of ~1 or less, the total microperveance can be increased by as much as an order of magnitude or more, while, at the same time, efficiency can be optimized. For a given power level, as perveance is increased, beam current is increased, and beam voltage is decreased. Thus, the primary advantages of using multiple beam klystrons (MBKs) are the reduction in voltage that occurs and the ability to optimize efficiency.

In addition, because voltage is reduced, the length of the device is reduced. Because the individual beam currents are reduced, the magnetic field required for focusing is reduced. As a result of these two factors, for a given power level, the weight of the MBK is much less that the weight of a single beam klystron. Other advantages resulting from the use of MBKs are that power supplies are smaller, arcing and breakdown are reduced, it may be possible to use air insulation rather than oil, and x-ray shielding can be reduced or eliminated.

Apparently, the concept of the MBK was first proposed in the U.S.S.R. (now Russia) by Kovalenko [33] and in France by Bernier [34]. Then, in the 1960s, several versions of multiple beam klystrons (MBKs) were reported [35, 36]. Some results of the feasibility study in [35] are shown in Figures 11-33 and 11-34. For flexibility and convenience, Boyd et al. used conventional external cavity klystrons and inserted them into multiwavelength waveguides to test the basic principles of the MBK.

Four-beam and 10-beam configurations of MBKs were tested. The four-beam MBK is shown in Figure 11-33. The waveguides were operated in the $\pi/2$ mode, which ensured that the voltages at the interaction gaps of the individual tubes were the same. Loading capacitors were placed at the positions of the voltage nulls to help prevent interference from adjacent modes in the waveguides.

Test results for the 10-beam MBK are shown in Figure 11-34. The upper curve represents the 10-beam MBK performance and the lower curve represents the average performance of the single-beam klystrons. The authors concluded from this study that the MBK has the gain, efficiency, bandwidth, and stability characteristics of conventional klystrons. However, the power output of the MBK can be made to exceed that of the prototype klystron by a factor equal to the number of beams used.

While the work of Boyd et al. showed the feasibility of the MBK, it was not pursued in the United States. However, in the early 1960s, MBK designs developed by Zusmanovsky [37] and Korolyov [38] led the way for the rapid growth of MBK work in Russia. Now MBKs are used in a majority of new, prospective mobile ground-based airborne- and space-based radar and communication systems in Russia [39].

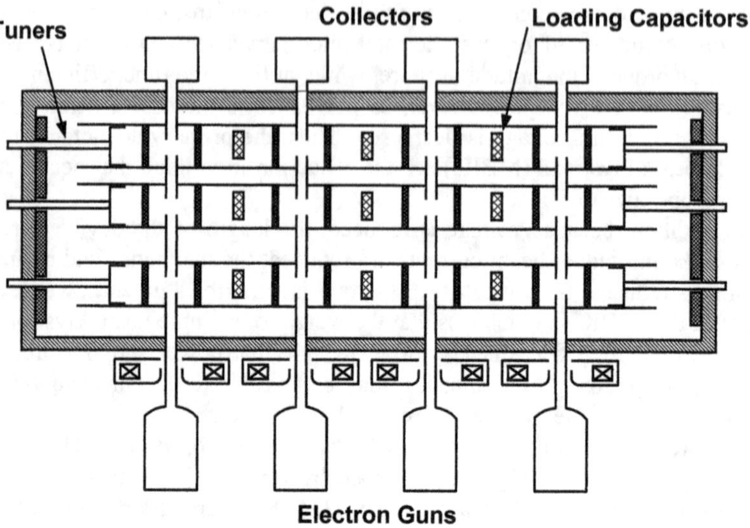

Figure 11-33 Configuration of four-beam MBK. External cavity klystrons without cavities were combined in a common RF structure. (Adapted from: Boyd et al., *IRE Trans. Electron Devices*, May 1962. © 1962 IRE (now IEEE).)

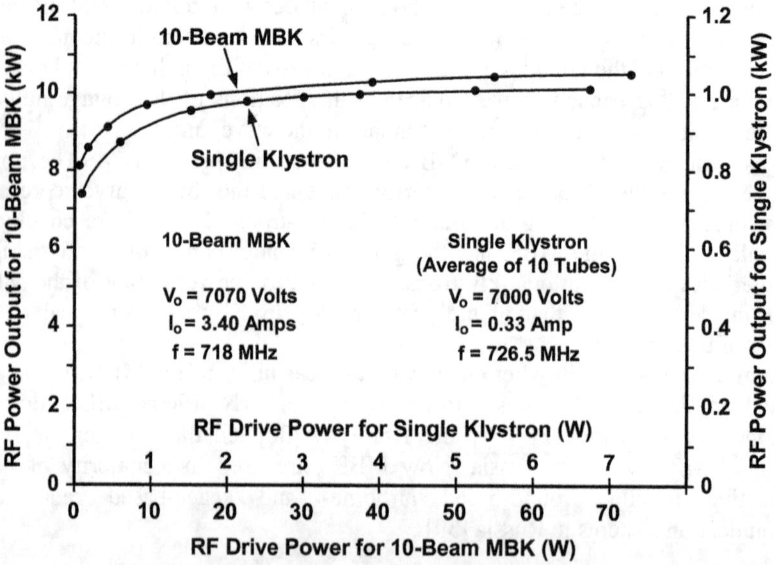

Figure 11-34 RF power characteristics of the single klystron and the 10-beam MBK. (Adapted from: Boyd et al., *IRE Trans. Electron Devices*, May 1962. © 1962 IRE (now IEEE).)

MBKs can be classified by the geometry of their resonators [39]. Fundamental mode (FM) resonators are shown in Figure 11-35. These may have toroidal or ring configurations. A higher-order mode (HM) resonator is shown in Figure 11-36. Boyd et al. used a linear version of this resonator [35].

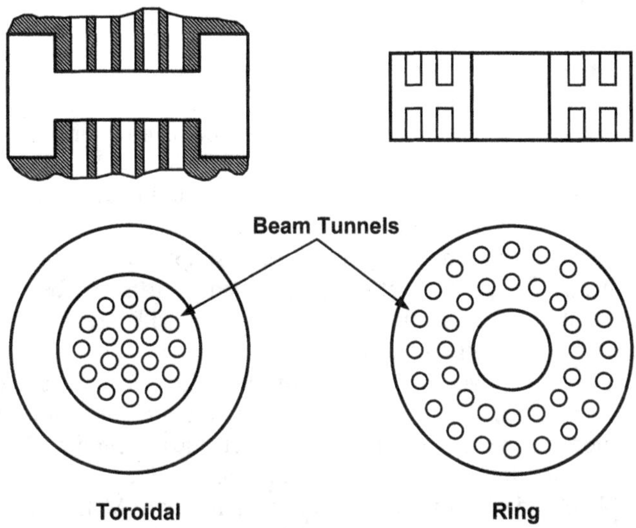

Figure 11-35 Fundamental mode resonators. (Adapted from: A. N. Korolyov et al., *IEEE Trans. Plasma Science*, June 2004. © 2004 IEEE.)

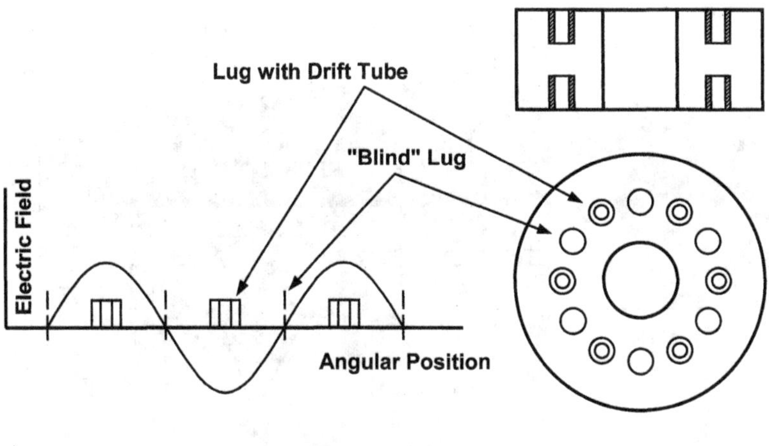

Figure 11-36 Harmonic mode resonator. (Adapted from: A. N. Korolyov et al., *IEEE Trans. Plasma Science*, June 2004. © 2004 IEEE.)

Advantages of the fundamental mode (FM) resonator are:

1. Large instantaneous bandwidth (TM_{01} mode);
2. Very low voltage because of the large number of beams that can be used.

Disadvantages of the FM resonator are:

1. Cathode loading may be high because beams are close together and so the cathode diameter is limited.
2. Because of high cathode loading, life may be severely limited.

The basic configuration of an FM MBK is shown in Figure 11-37 [40]. In this case toroidal resonators are shown. The control electrode, which is used to modulate the beam currents, may in practice be as simple as a flat plate containing holes for the beams to pass through. No convergence of the electron beams near the cathode assembly is shown, although electron guns with a relatively small area convergence ratio are sometimes used.

In [39] an excellent overview is given of the range of capabilities of Russian FM MBKs. Devices capable of RF powers from watts (in miniature MBKs) to hundreds of kilowatts are described. Emission densities as high as 30 A/cm^2 are mentioned along with lifetimes of thousands of hours.

Figures 11-38 and 11-39 show predicted parameters of MBKs operating at 10 GHz with a cathode emission density of 15 A/cm^2 [40]. Notice that a 100-kW device containing 19 beams would operate at 10 kV. By comparison, a single-beam klystron that produced 100 kW would have to operate at 30 kV or more.

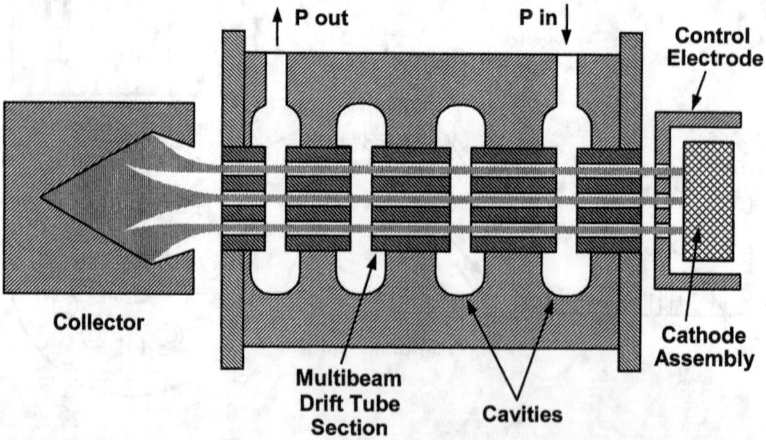

Figure 11-37 Basic configuration of a fundamental mode MBK. (Adapted from: A. S. Pobedonostsev et al., *Microwave Theory and Techniques Conf.*, Atlanta, Georgia, 1993.)

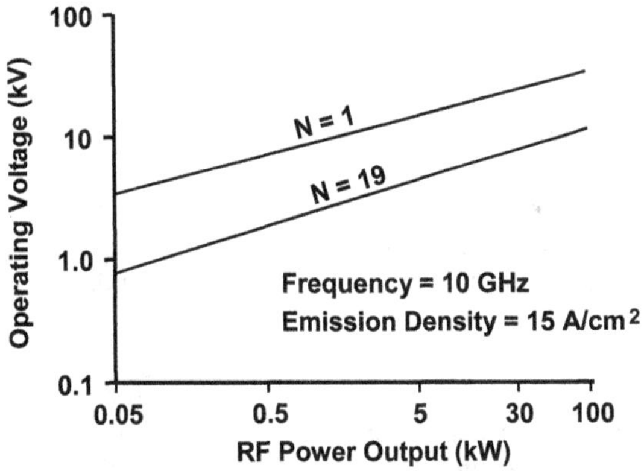

Figure 11-38 Predicted operating voltage as a function of power output for single beam and 19-beam MBKs. (Adapted from: A. S. Pobedonostsev et al., *Microwave Theory and Techniques Conf.*, Atlanta, Georgia, 1993.)

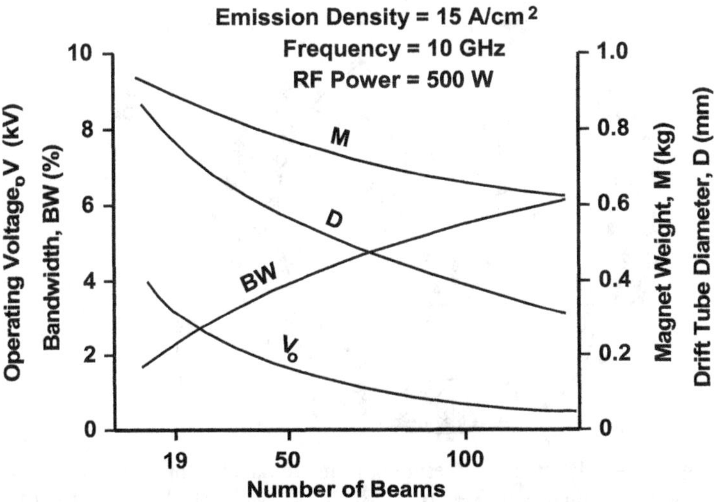

Figure 11-39 Predictions of the dependence of several MBK parameters on the number of beams used. (Adapted from: A. S. Pobedonostsev et al., *Microwave Theory and Techniques Conf.*, Atlanta, Georgia, 1993.)

The higher-order mode (HM) resonator MBK overcomes some of the limitations of the FM MBK but has other restrictions. The advantages of the HM resonator are:

1. Cathode loading is low because beam separation can be large so cathode size can be much larger than for FM devices.

2. Operation at very high average power levels is possible because of the large separation between beams and the improved cooling that is possible.

The disadvantages of the HM resonator are:

1. Bandwidth is narrow because of the RF circuit that is used.
2. It is increasingly difficult to focus beams as the distance from the center of the magnetic field increases. This increases the difficulty in fabricating electron guns.

The difficulty in beam focusing arises because of the radial components of magnetic flux that exist near and in the electron guns as well as in the drift space along the length of the klystron. As shown in Figure 11-40, this field component causes theta-directed forces on the electron beams. The resulting effect on the beams is shown in Figure 11-41.

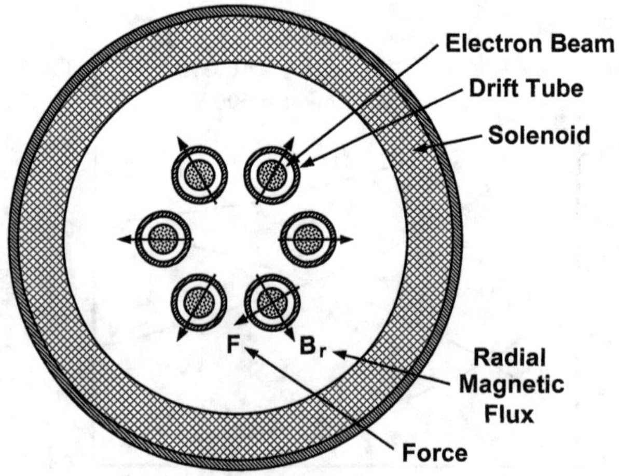

Figure 11-40 Theta-directed force resulting from the radial components of magnetic flux. (Adapted from: C. Bastien et al., Report No. SLCET-TR-87-0713-F, March 1989.)

Bastien et al. [41] analyzed the radial flux distribution in an electromagnet without pole pieces and found that B_r is proportional to the distance from the magnet axis and to the square of the magnet radius. For this reason, cavities with drift tubes packed as closely as possible to the magnet axis should be used and the radius of the magnet should be as small as possible.

For the electromagnet to be used on an MBK, Bastien et al. used pole pieces on the ends of the magnet and also, to reduce B_r to an acceptable level, divided the electromagnet into separate coils and used pole pieces between the coils.

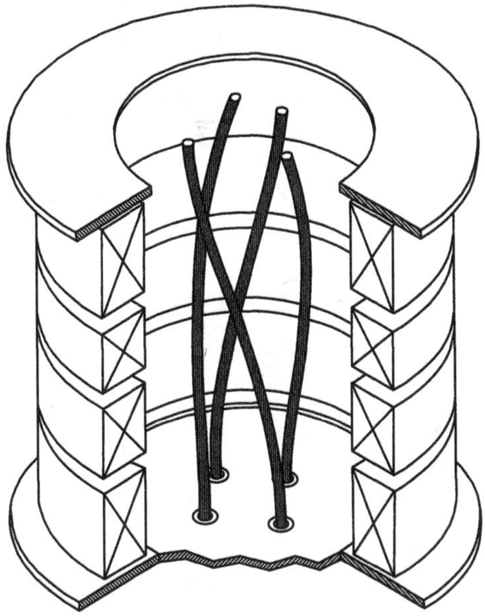

Figure 11-41 Effect on electron beams of radial magnetic flux components near and in the electron gun. (Adapted from: C. Bastien et al., Report No. SLCET-TR-87-0713-F, March 1989.)

When magnetic flux is required in the guns of an MBK to permit confined (immersed) flow focusing, the problem of radial flux components in the guns and within the apertures in the pole pieces becomes particularly severe. This problem is illustrated in Figure 11-42 [42]. In Figure 11-42(a), the field pattern that would exist in a gun of an MBK is shown if no attempt was made to eliminate the radial components of flux. In Figure 11-42(b), the desired flux distribution is shown. A technique for obtaining the desired flux distribution is shown in Figure 11-43. The asymmetry of the pole piece provides a magnetic reluctance configuration that eliminates B_r.

An application for MBKs that will require very high-power levels is the TESLA Linear Collider at Deutsches Elektronen-Synchrotron (DESY) [43]. The requirement to use MBKs came about when the specifications for the RF sources were established. The power requirement per tube was set at 10 MW at an efficiency of 70%. The pulse length requirement was 1.5 ms. Reaching this goal with a single-beam klystron would require a beam voltage of 241 kV. Quoting from [43]: "The construction of a klystron meeting these parameters seems to be impossible. The solution to this problem is the use of several low perveance beams in parallel at lower voltage in one vacuum vessel. This technology is utilized in the multi beam klystron."

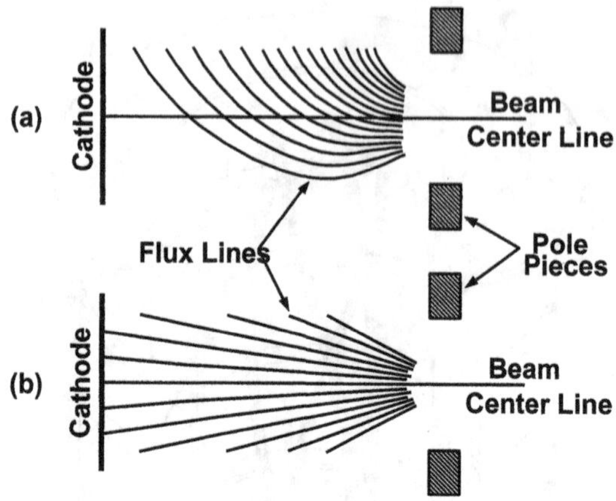

Figure 11-42 (a) Uncorrected and (b) corrected magnetic flux lines for off-axis electron guns. (Adapted from: R. L. Ives et al., *Proc. 6th Workshop on High Energy Density and High Power RF*, June 2003.)

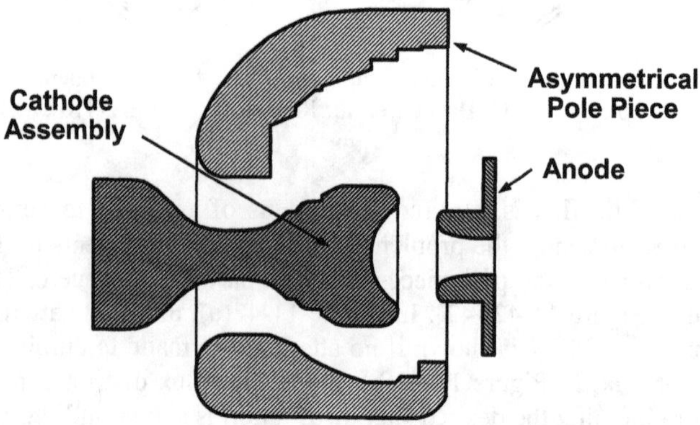

Figure 11-43 Asymmetric pole piece for correcting flux distribution in a gun. (Adapted from: R. L. Ives et al., *Proc. 6th Workshop on High Energy Density and High Power RF*, June 2003.)

Because the frequency of operation of TESLA is relatively low (1.3 GHz), device dimensions are large enough that either FM or HM MBKs could be considered to fulfill the RF source requirements. Figure 11-44 from Communications and Power Industries (CPI) indicated that the cathode emission density for the FM configuration would be excessive if long life was to be achieved. Because the CPI goal was a design for which emission density could be limited to 2 A/cm^2 to ensure long life, the HM configuration was chosen for its development effort.

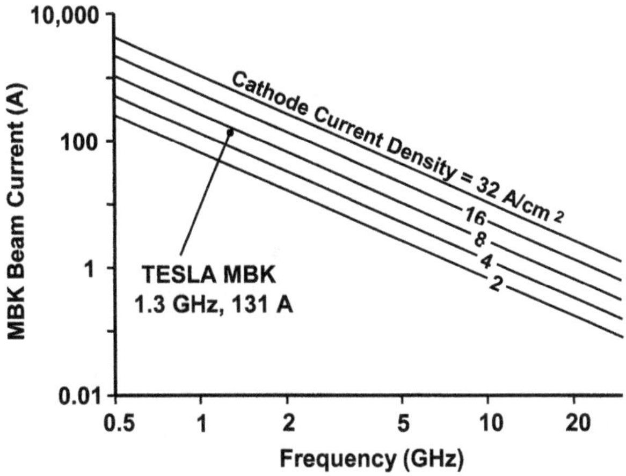

Figure 11-44 Cathode loading predicted by Communications and Power Industries (CPI) for the FM configuration of the TESLA MBK [44].

Thales chose the FM configuration for their device and have delivered several tubes to DESY. The cathode emission density is 5.5 A/cm^2 and the device has been operating at nominal performance level for 6 years (for an overview, see [44]). The power output of the Thales tube (the TH 1801) is shown in Figure 11-45 and the measured performance characteristics are listed in Table 11.4.

The Thales tube illustrates very well the benefits of the MBK. With a beam voltage only a little over 100 kV, a power of 10 MW is generated at an efficiency of 65%. An additional benefit that the multiple beam and low voltage parameters provide is reduced device size. Still, as shown in the outline drawing in Figure 11-46 of the Toshiba E3736 MBK, which is also being developed for TESLA, the device size is substantial. This is because of the relatively low operating frequency and the high-power requirement.

The U.S. Naval Research Laboratory is developing a 3.27-GHz MBK for beam transport and beam-wave interaction studies (Figure 11-47) [45]. This is an eight-beam four-cavity device operating in the fundamental mode. The gun is magnetically shielded and operates at 32A and 45 kV. Detailed results of calculations of electron energies, RF current, and power are given in [45].

A PPM-focused S-band MBK has been developed to produce 120–160 kW with a bandwidth of 10% [46]. The device uses 18 beams at 16 kV in a fundamental mode RF system. M-type cathodes operating at an emission density of 13–15 A/cm^2 are used to obtain a life of more than 1,000 hours. A control electrode at a bias voltage of –5.5 to –6 kV is used to modulate the electron beams. It is thought that the dc beam transmission has been limited to about 90% because of: (1) misalignment between the electron gun and the RF interaction region; and (2) nonuniform magnetic field at different beam locations.

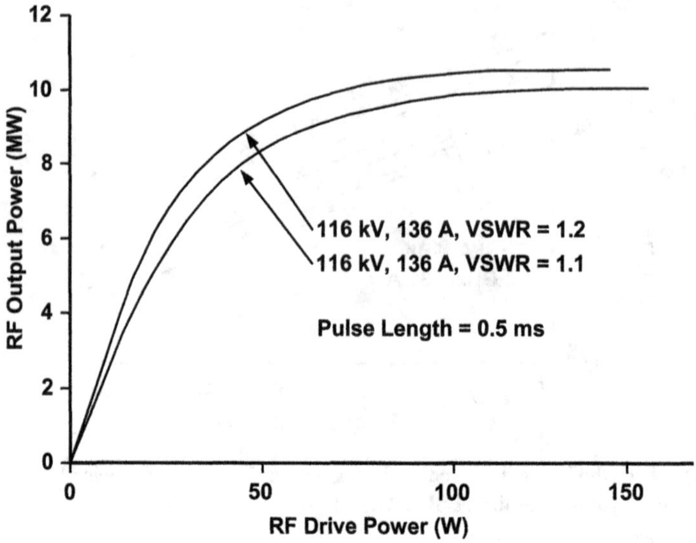

Figure 11-45 Power output characteristics of the Thales TH 1801 MBK [44].

Table 11.4
TH 1801 Operating Characteristics

Cathode Voltage	114–116 kV
Beam Current	134–136A
Perveance	3.3–3.5 μPerv
Output RF Peak Power	10 MW
Output RF Average Power	150 kW (500 μs at 30-Hz Factory)
Output RF Average Power	150 kW (1,500 μs at 10-Hz DESY)
Efficiency	65%
Drive RF Peak Power	140W
Gain	48.5 dB
Electromagnet Power	5.5 kW

11.5 EXTENDED INTERACTION KLYSTRONS

Extended interaction klystrons (EIKs) may be amplifiers (EIAs) or oscillators (EIOs). These devices are used at millimeter-wave frequencies to help overcome the difficulties of reducing the dimensions of conventional klystrons so that they can operate at these frequencies. A cross-sectional view of an EIA is shown in Figure 11-48.

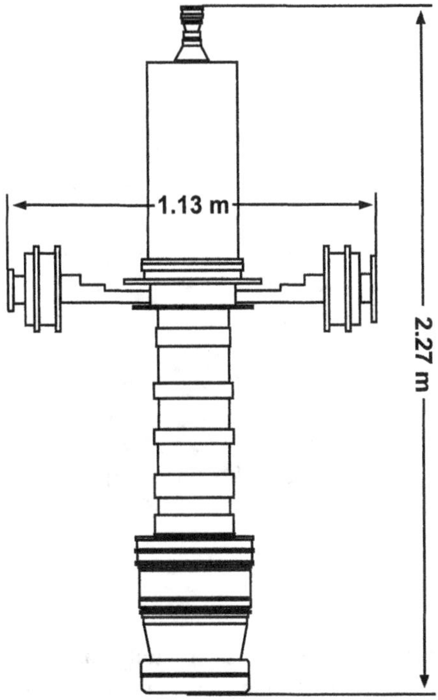

Figure 11-46 Toshiba E3736 MBK being developed for TESLA [44].

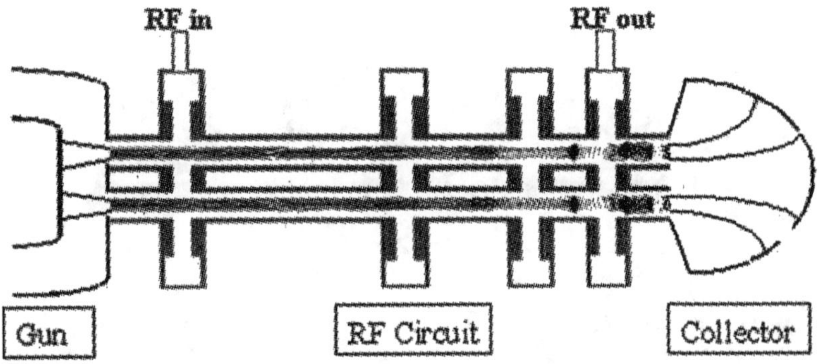

Figure 11-47 Schematic of the NRL MBK. Beam trajectories and cavities are from a MAGIC-3D simulation of an eight-beam, four-cavity MBK. (From: K. T. Nguyen, et al., *IEEE Trans. Plasma Science*, June 2004. © 2004 IEEE.)

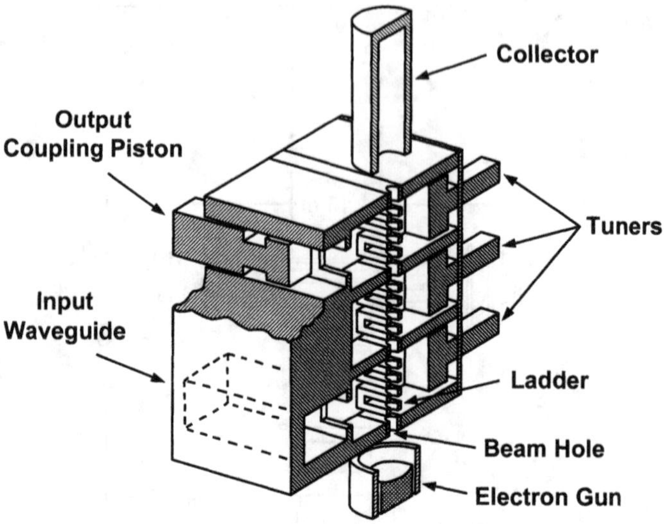

Figure 11-48 Cross-sectional view of an extended interaction amplifier (EIA). (Courtesy of CPI Canada.)

In extended interaction devices, each cavity of the conventional klystron is replaced by a section of resonant slow wave circuit. This circuit, which is referred to as a ladder structure, often contains long and short slots as shown in Figure 11-49. The electron beam passes through a hole in the center of the circuit and interacts with the RF wave on the structure as shown in Figure 11-50.

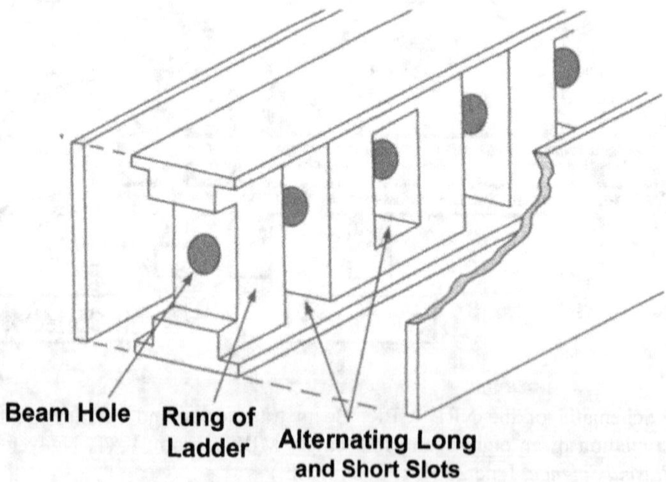

Figure 11-49 Ladder structure used in an EIK. (Courtesy of CPI Canada.)

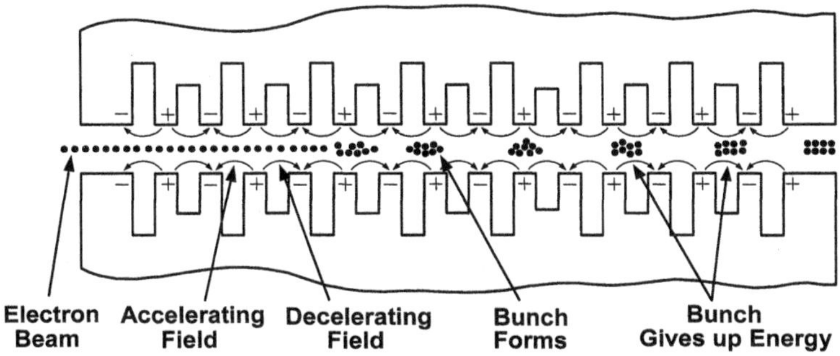

Figure 11-50 Beam interaction with the circuit in an EIK. (Courtesy of CPI Canada.)

The interaction of the electron beam with the ladder structure is identical to that occurring in a coupled-cavity traveling wave tube and is described in Chapter 15. Upon entering the ladder structure, some electrons experience an accelerating field and others experience a decelerating field. The accelerated electrons catch up with the decelerated electrons to form bunches. If the bunch velocity is somewhat higher than the phase velocity of the circuit wave, energy transfer to the circuit occurs and the RF signal is amplified.

Because the beam voltage is high and the beam current is low (compared with lower-frequency devices), the beam impedance is relatively high. As a result, to achieve reasonably high gain and high power, it is necessary to use a circuit with high interaction impedance and that is a characteristic of the ladder structure. The bandwidth, however, is fairly narrow as indicated in Figure 11-51 for a 140-GHz device. Here, at 1 dB below maximum power, the bandwidth is 0.16%.

The dimensions of the circuit in an EIA such as the 94-GHz device described in [47] can be readily estimated. The RF fields in adjacent slots are about 180° out of phase. The beam voltage is 16 kV and the electrons travel at one-fourth of the speed of light. The electrons must reach adjacent slots every half cycle, so, at 95 GHz, the slot spacing is

$$\text{slot spacing} = 0.25\,\frac{c}{2f} = 0.395\,\text{mm} = 0.0156\,\text{inch} \tag{11-14}$$

The average slot length is about a half wavelength long or 1.6 mm (0.064 inch). The beam radius is 0.125 mm (0.005 inch) and the beam focusing flux density is 10,000 Gauss.

The EIA for which the RF power characteristic is shown in Figure 11-52 was developed for use in the 94-GHz cloud profiling radar on NASA's CloudSat Program [48]. This EIA is the first space-qualified high-power W-band amplifier. The device uses a single-stage depressed collector to help achieve an overall efficiency of 32%. The high area convergence ratio of the gun, which is 100:1, makes possible low cathode loading and an operational life of over 2 years.

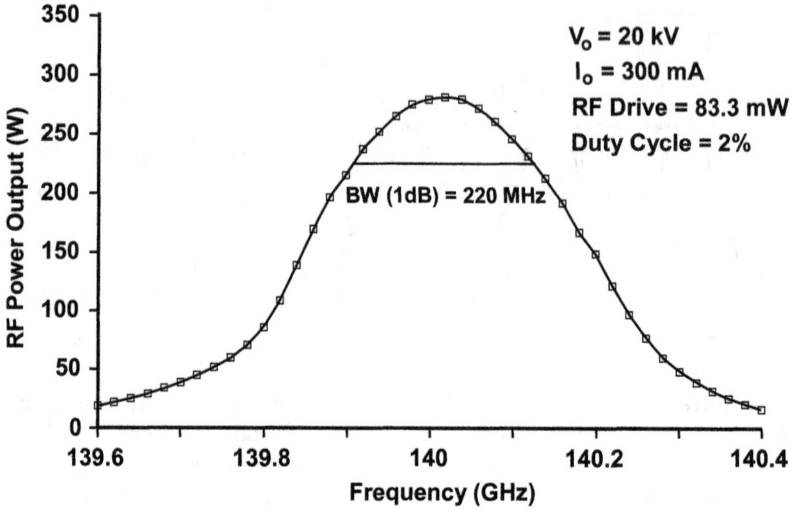

Figure 11-51 Bandwidth characteristic of a 140-GHz EIA. (Courtesy of CPI Canada.)

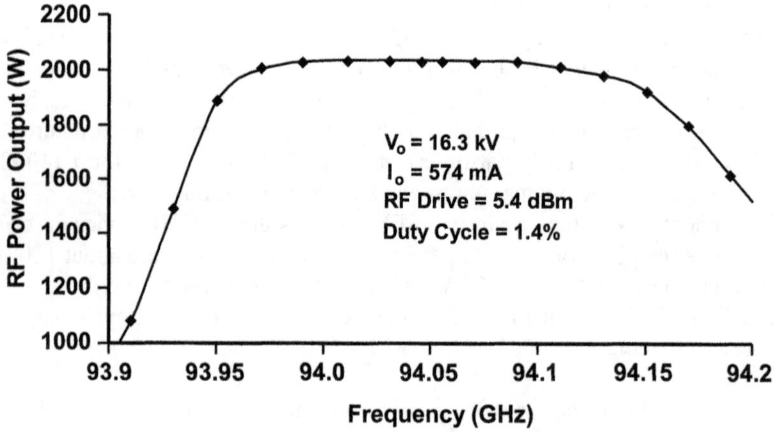

Figure 11-52 RF power characteristic of EIA used for the NASA CloudSat Program. (Adapted from: A. Roitman et al., *IEEE Trans. Electron Devices*, May 2005. © 2005 IEEE.)

The extended interaction oscillator is a single-cavity device as shown in Figure 11-53. The cavity can be mechanically tuned over a bandwidth of about 4%. A schematic of an EIO is shown in Figure 11-54. The EIO can be CW or pulsed using a cutoff grid. With proper design and low gas content, both the EIO and the EIA have very good phase noise performances [49]. Figure 11-55 contains plots of noise characteristics of several EIKs.

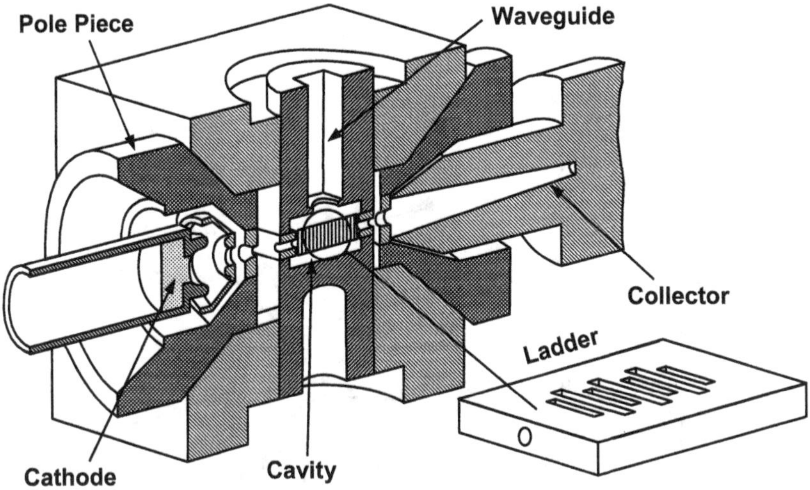

Figure 11-53 Cross-sectional view of EIO. (Courtesy of CPI Canada.)

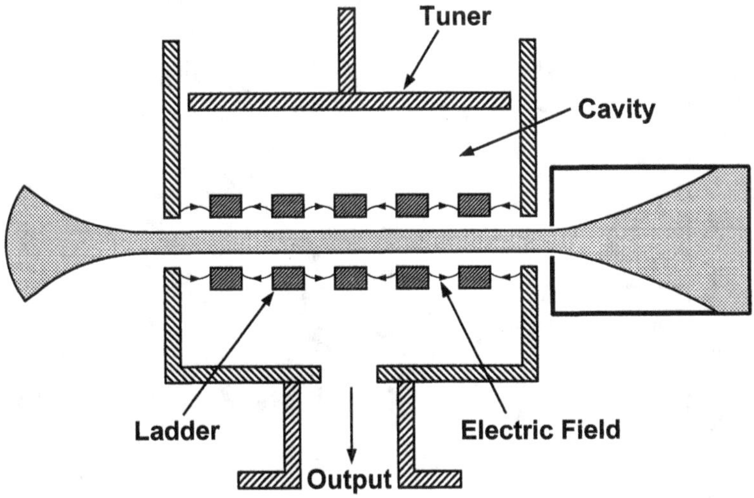

Figure 11-54 Schematic of extended interaction oscillator. (Courtesy of CPI Canada.)

The chart in Figure 11-56 displays the RF power capabilities of many of the EIKs manufactured by CPI Canada. Power levels range from 3 kW at 30 GHz to 30W at 280 GHz. The state of the art is clearly represented by the CloudSat device with 2 kW at 95 GHz.

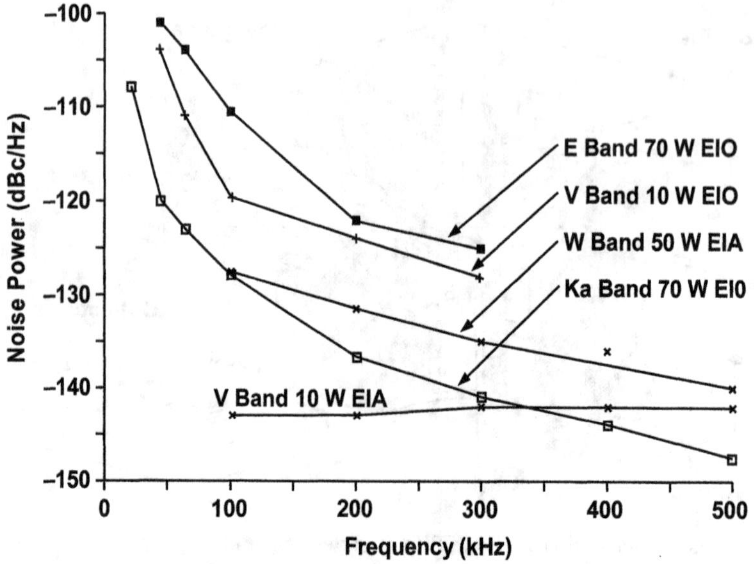

Figure 11-55 Noise performance of several EIKs. (Courtesy of CPI Canada.)

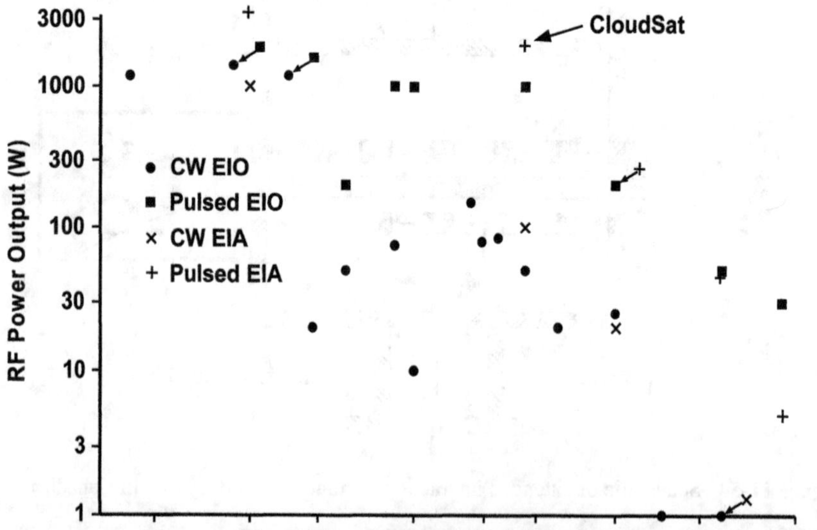

Figure 11-56 RF power capabilities of EIKs. (Courtesy of CPI Canada.)

11.6 REFLEX KLYSTRONS

A reflex klystron is a single-cavity oscillator. As shown in Figure 11-57, an electron beam passes through a cavity gap (normally gridded) into a region where the negative voltage on a repeller (reflector) electrode causes the electrons to reverse direction.

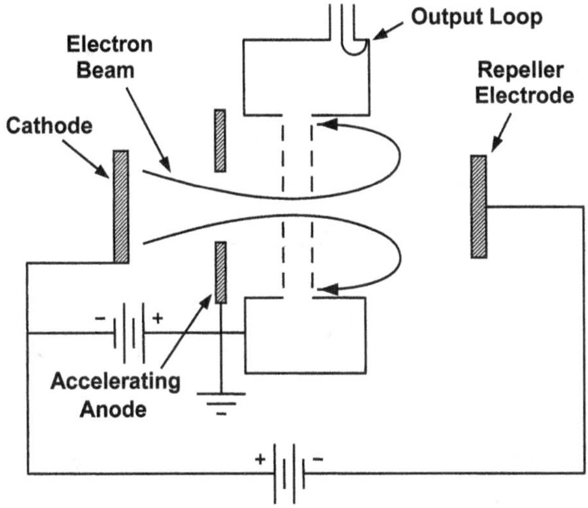

Figure 11-57 Principles of operation of a reflex klystron.

Assuming that a signal exists across the cavity grids, then the electron beam is velocity modulated as described in previous chapters. As the electron beam drifts into the repeller region, bunching occurs as indicated in Figure 11-58. If the bunched electrons return to the modulating gap at the correct time, they induce a current that reinforces the current in the cavity. Thus, the electron beam provides the feedback loop that causes the device to oscillate.

In Figure 11-58, the phase, α, at which the electron bunch returns to cause oscillations is 2.75 cycles after the beam initially passed through the bunching cavity. By changing the repeller voltage, the phasing of the return of the electrons to the modulating cavity can be changed. As a result, the electron bunch can be caused to return to the modulating gap 2.75, 3.75, 4.75 cycles, and so forth after initially passing through it. As indicated in Figure 11-59, oscillations can occur in any one of several modes that correspond to the phase angle through which the signal changes as the electrons drift through the repeller region.

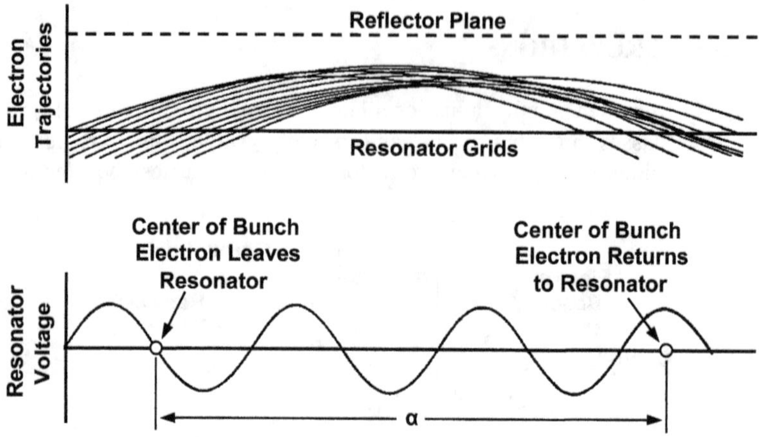

Figure 11-58 Electron bunching in a reflex klystron.

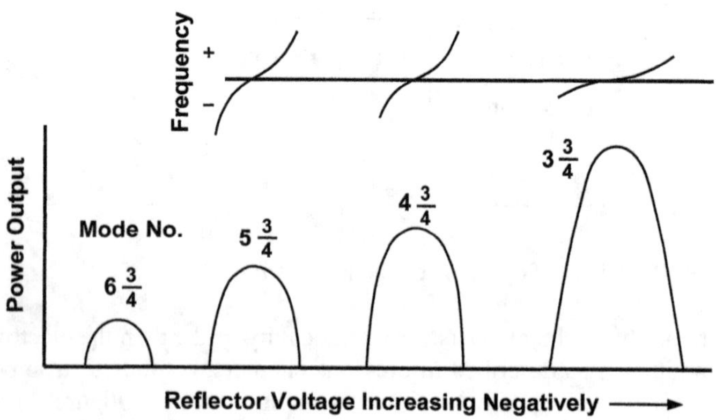

Figure 11-59 Power output versus reflector voltage for a reflex klystron.

As also indicated in Figure 11-59, the oscillation frequency in any mode of operation is a function of the time that it takes the electron bunch to return in that mode. For example, if the repeller voltage is above the mode-center value, then the electron transit time is less than the mode-center value, and the frequency is above the mode-center value. As a result, the RF current induced in the cavity is at a frequency above the resonant frequency of the cavity (mode-center frequency), so the frequency of operation is above the resonant frequency. Power output is reduced because operation off resonance reduces the cavity impedance and gap voltage.

Reflex klystrons are low-power (usually less than a watt) devices that have historically been used as local oscillators. They are low noise devices that are not difficult to phase lock.

REFERENCES

[1] T. G. Mihran, G. M. Branch, Jr., and G. J. Griffin, Jr., "Electron bunching and output gap interaction in broadband klystrons," *IEEE Trans. Electron Devices*, Vol. ED-19, No. 9, September 1972, pp. 1011–1017.

[2] V. M. Pikunov, V. Y. Prokop'yev, and A. N. Sandalov, "A fast method of calculating nonlinear processes in devices with longitudinal interaction," *Radiotekhnika i Elektronika*, No. 4, 1985, pp. 774–780.

[3] A. N. Sandalov, and A. V. Terebilov, "Special features of bunching and energy exchange in a relativistic multiresonator klystron," *Radio Engineering and Electron Phys.*, Vol. 28, No. 9, September 1983.

[4] Y. I. Vasil'yev, V. I. Kanavets, and A. V. Terebilov, "Effect of the velocity distribution in a bunch on the kinetic energy extraction efficiency in a klystron," *Radio Engineering and Electron Phys.*, Vol. 28, No. 4, April 1983.

[5] A. V. Aksenchik et al., "Computer modeling and investigation of most efficient interaction processes in multicavity klystrons," *Radio Engineering and Electron Phys.*, Vol. 28, No. 2, February 1983.

[6] A. V. Aksenchik et al., "Results of optimization of the efficiency of multiresonator klystrons," *Radio Engineering and Electron Phys.*, Vol. 27, No. 12, December 1982.

[7] V. A. Kochetova, A. V. Malykhin, and D. M. Petrov, "Synthesis of an electron bunch and the conditions for realizing it in a klystron," *Radio Engineering and Electron Phys.*, Vol. 26, No. 1, January 1981.

[8] V. A. Kochetova et al., "High efficiency drift klystron. Some theoretical and experimental questions (ordered bunching, gathered bunch, and convection current harmonics)," *Radio Engineering and Electron Phys.*, Vol. 26, No. 1, January 1981.

[9] V. A. Kochetova et al., "High efficiency drift klystron. Some theoretical and experimental questions (optimization of the AFC, comparison of calculated and experimental characteristics)," *Radio Engineering and Electron Phys.*, Vol. 26, No. 1, January 1981.

[10] C. Bastien, G. Faillon, and M. Simon, "Extremely high-power klystrons for particle accelerators," *Technical Digest,* IEDM, 1982, pp. 190–194.

[11] H. G. Kosmahl, and L. U. Albers, "Three-dimensional evaluation of energy extraction in output cavities of klystron amplifiers," *IEEE Trans. Electron Devices*, Vol. ED-20, No. 10, October 1973, pp. 883–890.

[12] E. L. Lien, "High efficiency klystron amplifiers," *Proc. 8th International Conference on Microwave and Optical Generation and Amplification,* Amsterdam, the Netherlands, 1970.

[13] H. G. Kosmahl, "Analytic investigation of efficiency and performance limits in klystron amplifiers using multidimensional computer programs; multistage depressed collectors; and thermionic cathode life studies," *Solar Power Satellite Microwave Power Transmission and Reception*, NASA CP 2141, 1980, p. 206.

[14] E. W. McCune, "A UHF TV klystron using multistage depressed collector technology," *Technical Digest,* IEDM, 1986, pp. 160–163.

[15] R. S. Symons, "Scaling laws and power limits for klystrons," *Technical Digest,* IEDM, 1986, pp. 156–159.

[16] W. D. Kilpatrick, "Criterion for vacuum sparking designed to include both RF and dc," *Rev. Sci. Inst.*, Vol. 28, No. 10, 1957, pp. 824–826.

[17] W. Peter et al., "Criteria for Vacuum Breakdown in RF Cavities," *Particle Accelerator Conference*, Santa Fe, NM, 1983.

[18] G. A. Loew, and J. W. Wang, "RF breakdown studies in room temperature electron linac structures" SLAC PUB 4647: Stanford Linear Accelerator Center, 1988.

[19] R. M. Phillips, and D. W. Sprehn, "High-power klystrons for the Next Linear Collider," *Proc. IEEE*, Vol. 87, No. 5, May 1999, pp. 738–751.

[20] R. True, "Beam optics calculations in very high-power microwave tubes," *Technical Digest,* IEDM, 1991, pp. 403–406.

[21] G. Caryotakis, "High-power microwave tubes: in the laboratory and on-line," *IEEE Trans. Plasma Science*, Vol. 22, No. 5, October 1994, pp. 683–691.

[22] D. Sprehn, R. M. Phillips, and G. Caryotakis, "The design and performance of 150-MW S-band klystrons," SLAC-PUB-6677, Stanford Linear Accelerator Center, September 1994.

[23] G. Faillon, and B. Villette, "Wide band high-power tubes," *Proc. of Military Microwaves,* MM90, London, 1990, pp. 401–406.

[24] A. Staprans, E. W. McCune, and J. A. Ruetz, "High-power linear-beam tubes," *Proc. IEEE,* Vol. 61, No. 3, March, 1973, pp. 299–330.

[25] G. Faillon, "A 200 kW S band klystron with TWT bandwidth capability," *Technical Digest,* IEDM, 1985.

[26] R. L. Metivier, "Broadband klystrons for multimegawatt radars," *Microwave Jour.*, April 1971, pp. 29–32.

[27] F. I. Friedlander, "High-efficiency broad-band klystron," U.S. Patent 4,764,710, August 16, 1988.

[28] M. Chodorow, and T. Wessel-Berg, "A high-efficiency klystron with distributed interaction," *IRE Trans. Electron Devices*, Vol. ED-8, No. 1, January 1961, pp. 44–55.

[29] R. S. Symons et al., "An experimental clustered-cavity klystron," *Technical Digest,* IEDM, 1987, pp. 153–156.

[30] R. S. Symons, and J. R. M. Vaughan, "The linear theory of the clustered-cavity klystron," *IEEE Trans. Plasma Science*, Vol. 22, No. 5, October 1994, pp. 713–718.

[31] W. Luebke, and G. Caryotakis, "Development of a one megawatt CW klystron," *Microwave Jour.*, Vol. 9, No. 8, August 1966.

[32] M. Chodorow, and R. A. Craig, "Some new circuits for high-power traveling-wave tubes," *Proc. IEEE*, Vol. 45, No. 8, August 1957, pp. 1106–1118.

[33] V. F. Kovalenko, "The electron tube of klystron type," Author's certificate, No. 72 756, class 21.13, 1940.

[34] J. Bernier, "Multiple beam device," French patent 992,853, September 15, 1944.

[35] M. R. Boyd et al., "The multiple beam klystron," *IRE Trans. Electron Devices,* Vol. ED-9, No. 3, May 1962, pp. 247–252.

[36] W. J. Pohl, "The design and demonstration of a wideband multiple-beam traveling-wave klystron," *IEEE Trans. Electron Devices*, Vol. ED-12, No. 6, June 1965, pp. 351–368.

[37] S. A. Zusmanovsky, "The multiresonator klystron," Author's certificate No. 15 556, January 5, 1955.

[38] S. V. Korolyov, "About the possibility of reducing the transit klystron weight and dimensions," *Elektron Tekh*, Vol. ser. 1, No. 9, 1968, pp. 176–184.

[39] A. N. Korolyov et al., "Multiple-beam klystron amplifiers: performance parameters and development trends," *IEEE Trans. Plasma Science*, Vol. 32, No. 3, June 2004, pp. 1109–1118.

[40] A. S. Pobedonostsev et al., "Multiple-beam microwave tubes," *Microwave Theory and Techniques Conference*, Atlanta, GA, June 1993.

[41] C. Bastien, M. Bres, and G. Faillon, "Compact multibeam klystron study," *Final Report no. SLCET-TR-87-0713-F*, March 1989.

[42] R. L. Ives et al., "Development of multiple beam guns for high-power RF sources for accelerators and colliders," *Proc. 6th Workshop on High Energy Density and High Power RF,* Berkeley Springs, WV, 2003.

[43] A. Beunas et al., "A high efficiency long pulse multi beam klystron for the TESLA linear collider," mdk2001.web.cern.ch/mdk2001/Proceedings/Session12/beunas.pdf, 2001.

[44] tesla.desy.de/new pages/hamburg meeting 4 2004/itrp public program.htm," 2004.

[45] K. T. Nguyen et al., "High-power four-cavity S-band multiple-beam klystron design," *IEEE Trans. Plasma Science*, Vol. 32, No. 3, June 2004, pp. 1119–1135.

[46] Y. Ding et al., "S-band multibeam klystron with bandwidth of 10%," *IEEE Trans. Electron Devices*, Vol. 52, No. 5, May 2005, pp. 889–894.

[47] C. Nilsen, "95 GHz 1 kW peak high performance radar transmitter," *Proc. Military Electronics 90*, London, 1990, pp. 115–120.

[48] A. Roitman, D. Berry, and B. Steer, "State-of-the-art W-band extended interaction klystron for the CloudSat Program," *IEEE Trans. Electron Devices*, Vol. 52, No. 5, May 2005, pp. 895–898.

[49] C. Nilsen, "Low noise MM wave transmitted tube oscillators," *Microwave Jour.*, July 1988, pp. 91–97.

Chapter 12

Traveling Wave Tubes

12.1 INTRODUCTION

The traveling wave tube (TWT) is one of two major categories of microwave devices known as linear-beam or O-type tubes. The other is the klystron. While many different RF circuits have been devised for use in TWTs, two predominate. These are:

1. Helix, for broadband applications;
2. Coupled-cavity, for high-power applications.

Other TWT circuits are mentioned in this book, but the majority of the discussions center on helix and coupled-cavity TWTs.

Traveling wave tubes are used from frequencies below 1 GHz to over 100 GHz. Power generation capabilities range from watts to megawatts. For helix TWTs, bandwidths may be as high as 2 octaves or more. For coupled-cavity TWTs, bandwidths in the 10–20% range are common.

Over 50% of the sales volume of all microwave tubes are for TWTs. Applications for TWTs are varied. They serve as the final amplifiers in most electronic warfare systems. In many radar systems, one or more TWTs are used as the high-power amplifier that generates the transmitted RF pulse. In others, a TWT may be used as the driver for some other device such as a crossed-field amplifier. They are the final amplifiers of choice in a large percentage of communications satellites and systems. Of particular note is the performance of the X-band TWT on the Voyager I satellite, which was still in operation after 28 years in space. This TWT was the backup for the solid-state amplifier that failed early in the mission.

12.1.1 Early History of the TWT [1]

The possibility of a traveling wave interaction between an electron beam and a nearby RF circuit was recognized in 1933 by Haeff [2, 3]. In patents filed in that year, Haeff described electron beam deflection tubes to be used as detectors or oscillographs that contained some of the features of helix traveling wave tubes. In the Haeff devices, an RF signal propagating on a helical structure was used to deflect a nearby hollow electron beam transverse to the direction of beam propagation. The velocity of the electron beam was equal to the velocity of propagation of the traveling wave on the helical structure. The possibility of amplification was pointed out.

In 1935, Posthumus [4] first described and built a cavity-type magnetron oscillator. He explained its operation as being due to an interaction between the tangential component of a traveling RF wave rotating with a velocity equal to that of the average velocity of the electrons. As a result of the interaction, electron energy is converted to amplification of the RF wave.

In May 1940, Lindenblad [5] was the first to describe helix traveling wave amplifiers that were similar to a helix traveling wave tube. He was the first to explain that a synchronous axial interaction between an electron beam and the RF wave on a helix could produce amplification of a signal on the helix. In a form of the tube described in the patent and shown schematically in Figure 12-1, Lindenblad produced useful amplification over a 30-MHz band at a carrier frequency of 390 MHz. In this tube the signal was applied to the beam with a grid in the gun. In other experiments, the signal was applied to the helix.

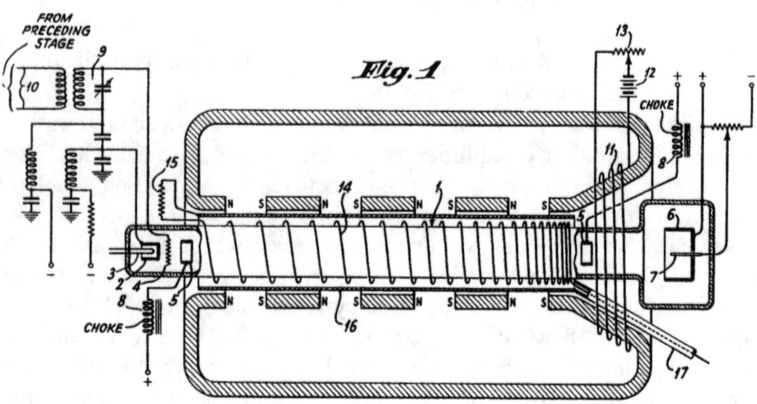

Figure 12-1 N. E. Lindenblad's traveling wave amplifier; patent filed May 1940.

The first Lindenblad tube was apparently a modified version of Haeff's inductive output tube. Lindenblad extended the glass vacuum envelope of the Haeff tube and replaced the cavity resonator with a helix. The helix was wound around the outside of the glass envelope and had a pitch such that the axial velocity of a wave on the helix was equal to the velocity of an electron beam inside the envelope. It is interesting that Lindenblad recognized the value of tapering the pitch of the helix to maintain synchronism with the electron beam, the velocity of which is reduced as energy is extracted to produce amplification.

In his patent, in addition to the helix, Lindenblad described the use of a helical waveguide as the slow wave circuit. He also stated that the helical conductors can, if desired, be placed within the envelope of the tube and surround the electron beam.

Kompfner [6–10], working in England, was apparently unaware of Lindenblad's earlier work. He reported that he realized in late 1942 (shortly after Lindenblad's patent was issued) that the basic growing wave principle of the magnetron could be used for amplification of RF signals. His objective was to develop an amplifier capable of a sensitivity and noise figure comparable to the best of the crystal-mixer receivers available at that time. Shown in Figure 12-2 is Kompfner's first sketch of a traveling wave tube.

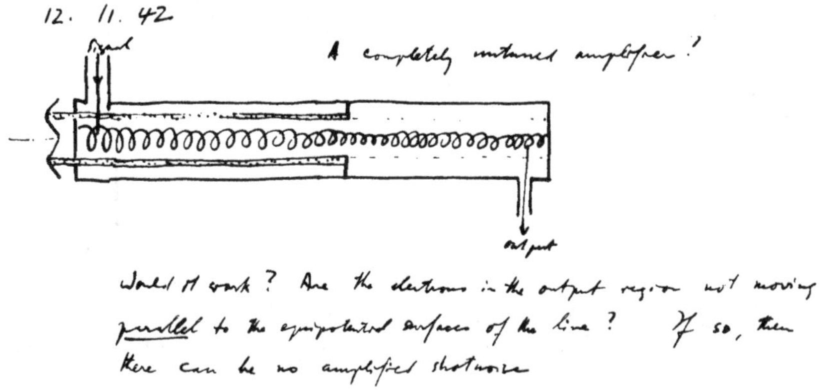

Figure 12-2 Sketch of a traveling wave tube from Kompfner's notebook. (From: *The Invention of the Traveling Wave Tube* by R. Kompfner, copyright 1963 by San Francisco Press, reproduced with permission.)

Kompfner built his first traveling wave tube in early 1943. He summarized the operation of the tube as follows:

> "When the radio frequency power emerging from the helix with the beam switched on was compared with the radio frequency power without the beam, it was found that, at a beam voltage of 2440 volts, there was an increase of 49%, while at a beam voltage of 2200 volts, there was a decrease of 40%."

After various measurements and experiments, Kompfner built and tested the tube shown in Figure 12-3. With an electron beam current of 110 μA at 1,830V a power amplification of 6 was achieved at a frequency of 3.3 GHz. The noise factor was 14 dB and was observed to depend on the fraction of the total current that arrived at the collector electrode. A further improvement in design increased power amplification to 14 and reduced noise factor to 11 dB. In theoretical studies completed before 1944, Kompfner developed first-order theories for power amplification and noise factor. The important result was that the wave tends to grow exponentially with distance along the transmission line.

The first public announcement of the British wartime work on helix traveling wave tubes was made at the Fourth Institute of Radio Engineer's Electron Tube Conference at Yale on June 27 and 28, 1946. At the same conference, work in the United States by J. R. Pierce and L. M. Field at Bell Telephone Laboratories was described. Subsequently [11], Pierce and Field described helix traveling wave tubes like that shown in Figure 12-4 (figure taken from Patent 2,602,148 by J. R. Pierce [12]). One unique feature of this device, described by Pierce, was the use of longitudinal insulating rods for supporting and accurately positioning the helix. Another was the use of a system of solenoids to produce a uniform magnetic field for focusing the electron beam. Pierce also described techniques for providing loss to suppress backward traveling waves and oscillations.

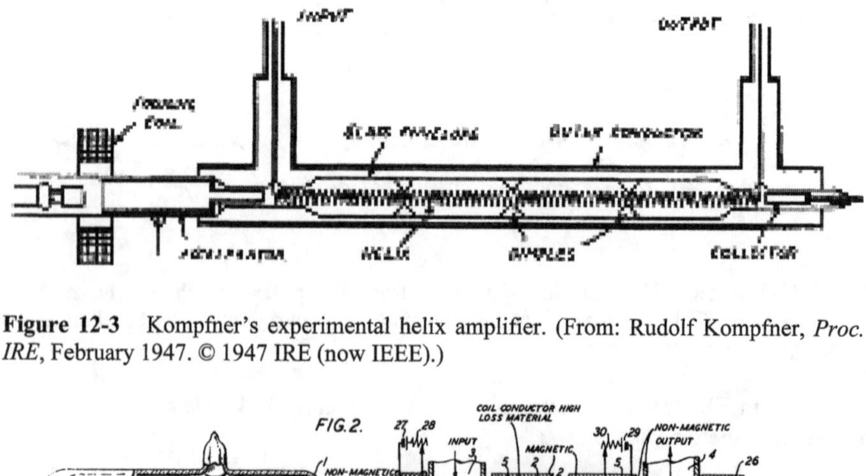

Figure 12-3 Kompfner's experimental helix amplifier. (From: Rudolf Kompfner, *Proc. IRE*, February 1947. © 1947 IRE (now IEEE).)

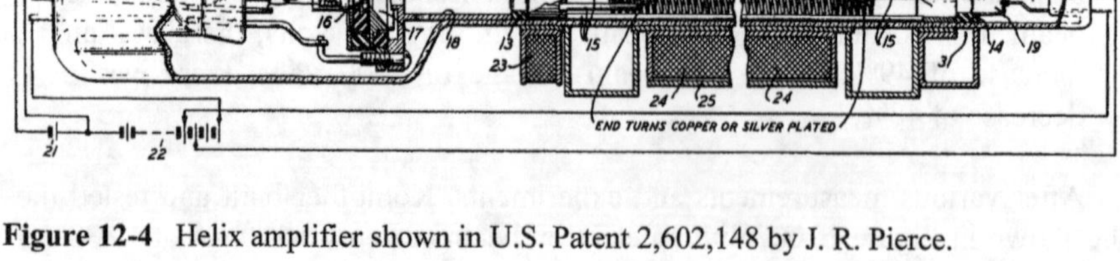

Figure 12-4 Helix amplifier shown in U.S. Patent 2,602,148 by J. R. Pierce.

In the patent by Field [13] the suppression of oscillations through the use of thin layers of colloidal graphite on the ceramic support rods for the helix is described. Near the midpoint of the helix, the conductivity of the coating was increased to provide the proper dissipation of reflected energy with a minimum of reduction in the gain of the forward growing wave.

During the postwar years from 1946 to 1950, there was intense activity in the development of adequate theories to explain the operation of the TWT. In addition to Pierce's theoretical work [14], noteworthy contributions [15, 16] were made by A. Blanc-Lapierre and P. Lapostolle in France in December 1946 that contained an analysis of traveling wave interaction using a structure consisting of resonant cavities coupled by axially aligned apertures.

It seems appropriate to conclude this brief early history by citing [17]. This book provided a unified coordination and exposition of the theory of the traveling wave tube and has been used as the standard reference for the theory of traveling wave tubes ever since.

12.1.2 Basic Operation of the TWT

There are two basic types of TWTs. The helix TWT, the main elements of which are shown in Figure 12-5, is a relatively low-power (usually tens to hundreds of watts) broadband (over 2 octaves is possible) device. The coupled-cavity TWT is capable of powers of megawatts, but bandwidth is limited to about 10–20%. Conceptually, the electron gun, electron beam, and collector are the same for all TWTs. However, there are substantial differences in the RF circuits. In spite of the differences, the operating principles are the same. We will discuss the operation of the helix TWT here.

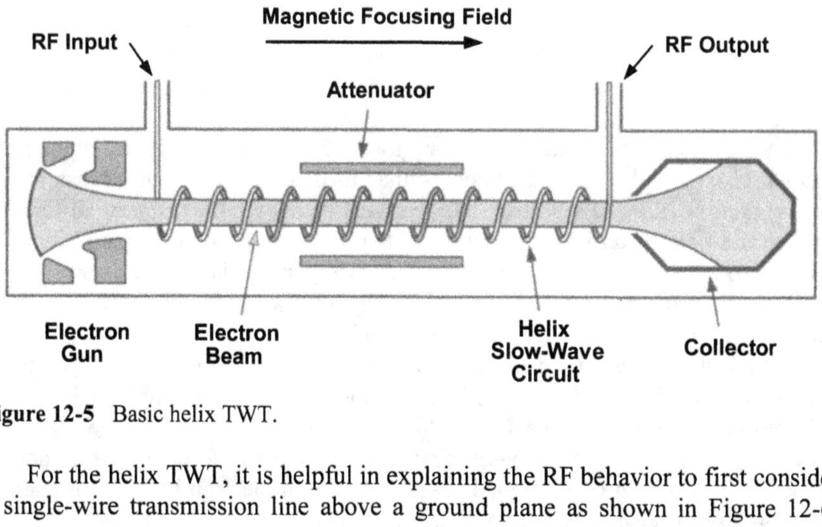

Figure 12-5 Basic helix TWT.

For the helix TWT, it is helpful in explaining the RF behavior to first consider a single-wire transmission line above a ground plane as shown in Figure 12-6.

Charges on the transmission line and the RF electric field pattern are shown at an instant in time. The RF magnetic field pattern is not shown and will not be considered because the RF magnetic forces on the electron beam in a microwave tube are insignificant compared with the forces produced by the electric field. If the RF generator is on the left end of the transmission line in Figure 12-6 and if a matched load is on the right end, then the charge and field patterns move to the right with constant amplitude. The velocity of propagation is that of light and is independent of frequency.

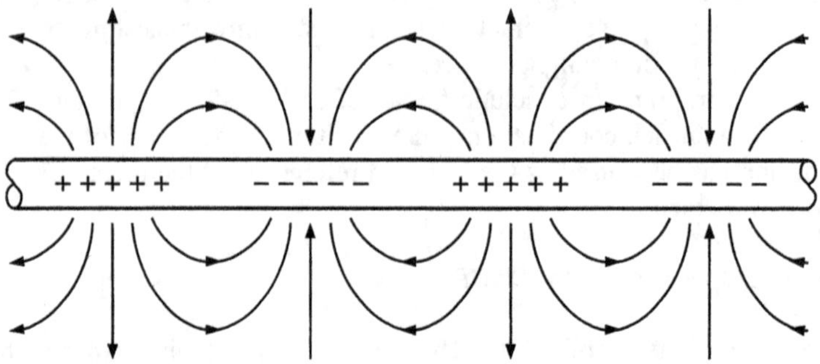

Figure 12-6 RF charge and electric field patterns for a single-wire transmission line above a ground plane.

Next, assume that the single-wire transmission line is formed into a helix as shown in Figure 12-7. An RF signal applied to the left end of the helix will travel at a velocity near that of light in a helical path along the conductor. The average velocity in the axial direction, which is called the *phase velocity* (usually denoted by v_p) of the helix, will be the velocity in the helical direction reduced by the pitch of the helix. The polarity of the signal will alternate every half-wavelength along the helical conductor. In Figure 12-7, two complete turns are required for each half-wavelength.

Electric field lines extend from regions of positive charge to regions of negative charge with a configuration similar to that shown for the transmission line in Figure 12-6. Other than the reduced velocity, the primary difference between the field pattern for the helix and that for a single-wire transmission line is that there is an electric field inside the helix with large axial components

When an electron beam is injected along the axis of the helix, the axial electric field components accelerate some electrons and decelerate others. In Figure 12-7, the forces on the electrons would be toward regions A and away from regions B. The field pattern would vary sinusoidally in the axial direction, as indicated in Figure 12-8. If the axial velocity of the electric field and the beam are nearly the same, then electrons will experience a continual force toward region A as the beam travels through the helix. As a result, an electron bunch will start to form in region A (also indicated in Figure 12-8(a)).

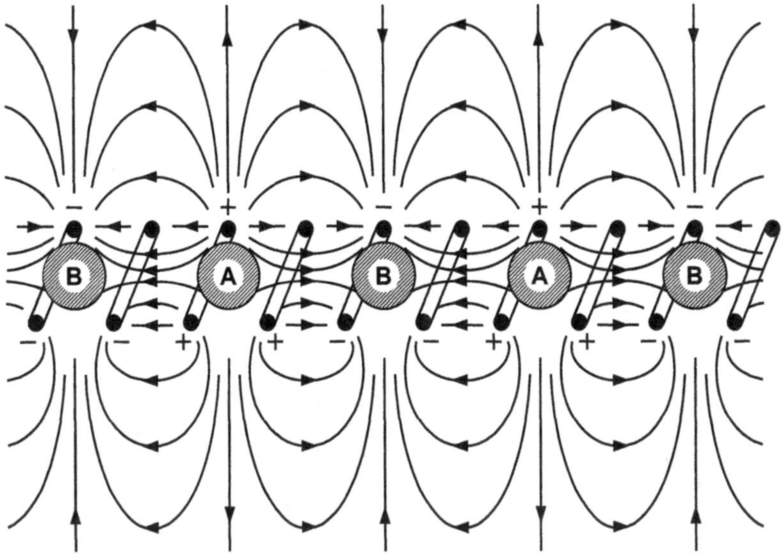

Figure 12-7 RF charge and electric field patterns for a helix.

The bunches that form in the beam repel electrons on the helix causing those electrons to travel to the right and to the left. The electrons on the helix travel at the phase velocity, v_p, of the helix which is nearly the same velocity as the electron bunches in the beam. Relative to a bunch, electrons on the left move left at $2v_p$. This produces a deficiency of electrons, which makes the circuit positive.

Relative to the bunch, electrons on the right, which are moving at approximately the bunch velocity, are nearly standing still. As a result, electrons accumulate ahead of the bunch and make the circuit negative.

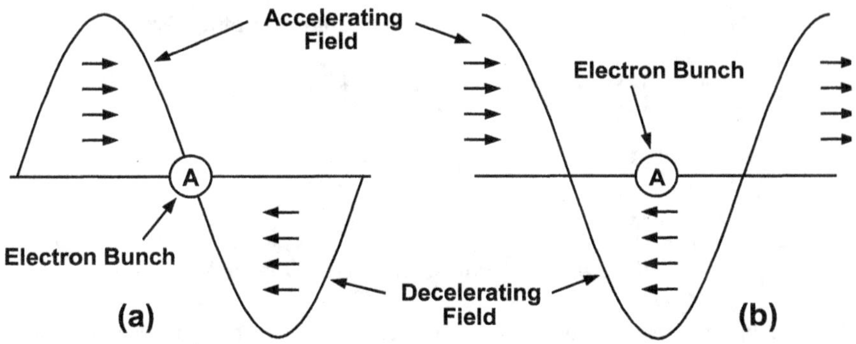

Figure 12-8 Axial field that bunches and extracts energy from beam (a) as beam enters circuit and (b) after interaction has occurred.

The increase in electron charge ahead of the bunch and decrease behind the bunch produces a voltage waveform on the helix as is shown in Figure 12-8(b) that is phase shifted 90° to the left of the initial waveform. As the voltage waveform on the helix shifts to the left, decelerating field regions move into the regions where the bunches are located. As more and more electrons are forced into the bunch, the induced waveform rapidly becomes larger than the initial waveform so the RF power on the circuit increases.

The initial bunching action in a TWT is like that produced by a gap (fast electrons catch up with slow electrons to produce a bunch). However, when the field shift shown in Figure 12-8(b) occurs, only decelerating forces are present, so electrons are slowed down but not accelerated. As a result, bunches are formed that move more slowly than the electron beam.

The fundamental aspects of the propagation of these bunches can be understood by considering an analogy with acoustic-wave propagation. In Figure 12-9(a), back-to-back acoustic transducers launch acoustic pressure (and density) waves in the air. The waves result from motion of the diaphragms in the transducers, which periodically compress the air. The waves propagate to the right and to the left at a velocity that is dependent on air pressure. With increased pressure, wave velocity increases.

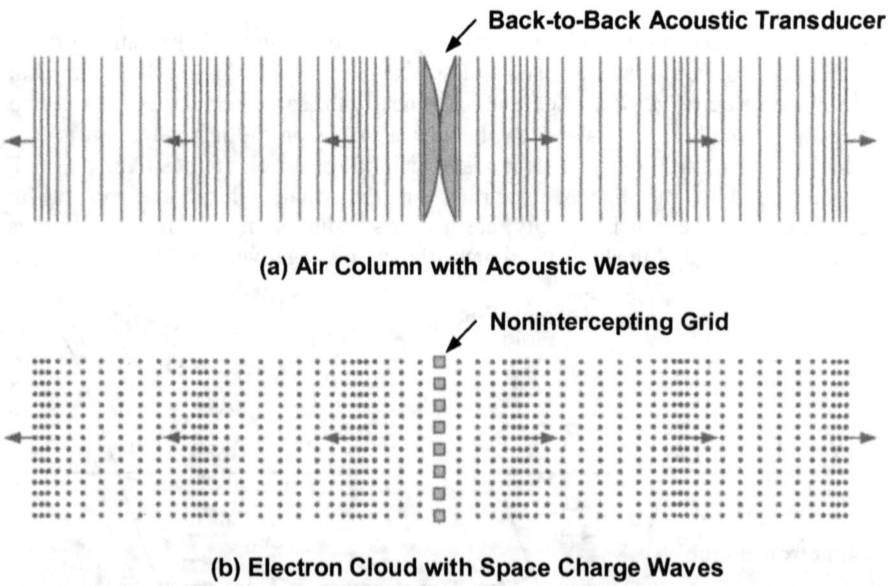

Figure 12-9 Arrangements for launching (a) acoustic waves and (b) electron density (space charge) waves.

In Figure 12-9(b), a grid is shown inserted in an electron cloud that is not moving. We will assume that the grid does not collect electrons. As the voltage on the grid oscillates from positive to negative and back to positive again, electrons are attracted, then repelled and then attracted again. As a result, the electron density near the grid is alternately reduced and increased and electron density waves are launched to the right and to the left. The velocities of these waves are dependent on the electron density. With increased density, wave velocities increase. These electron density waves are called space charge waves.

Next, let the electron column move to the right at a velocity that is much higher than the velocities of the space charge waves. The moving electron column is, of course, an electron beam. Both space charge waves are moving to the right with the beam. One is moving faster than the beam and is called a fast space charge wave. The other is moving slower than the beam and is called a slow space charge wave. The bunches produced by the decelerating field in Figure 12-8(b) constitute a slow space charge wave.

The energy extracted from the electron beam in slowing electrons to the bunch velocity is transferred to the circuit field, thereby producing amplification of that field. The mutual interaction of the beam and circuit result in an exponential growth of the circuit voltage as shown by the left half of the waveforms in Figure 12-10.

As the interaction process continues, more electrons are slowed to the bunch velocity and space charge forces within the bunches continue to increase. Eventually, these forces become large so that a portion of each bunch is retarded in phase enough so that it leaves the decelerating field region and enters an accelerating field (see electron density plots in the right third of Figure 12-10). Electrons in the acceleration portion of each bunch extract energy from the circuit field. Eventually, as the bunches continue to fall back in phase, energy extracted from the circuit wave becomes equal to the energy supplied and the wave on the circuit stops growing. At this point, "saturation" is said to occur and the signal amplitude is maximum.

12.2 TRAVELING WAVE INTERACTION

There are many approaches to the analysis of the interaction of an electron beam with a traveling wave on a circuit. The classical small-signal theory by J. R. Pierce [17] is a logical extension of the qualitative description of traveling wave interaction given here. Also, many of the parameters defined by Pierce are now part of the accepted vocabulary of the TWT industry. For these reasons and because the Pierce theory provides considerable physical insight into the mechanisms of TWT operation, the basic elements of the Pierce theory will be described here.

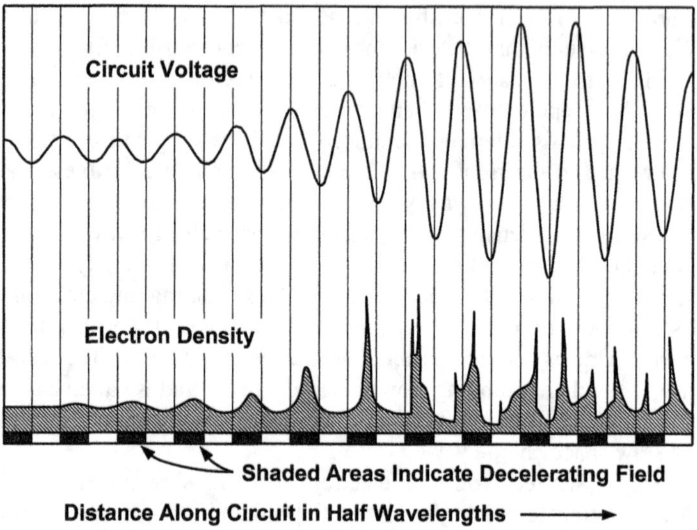

Figure 12-10 Voltage and charge density buildup in a TWT. (Adapted from: P. Hess, Ph.D. dissertation, University of California at Berkeley, 1960.)

The analysis is divided into several distinct parts. First, the RF current in the beam is given. (The relation used is the electronic equation, which can be derived in an analysis of space charge effects on bunching.) Then the slow wave circuit is analyzed. The circuit equation is then combined with the electronic equation to determine the propagation constants of the waves on the circuit and on the beam. Operation under synchronous (beam velocity equal to cold circuit phase velocity) and nonsynchronous (beam velocity not equal to cold circuit phase velocity) conditions are examined. Finally, the effects of circuit loss and of space charge are considered. The analysis is a small signal analysis, and it is assumed that quantities vary sinusoidally.

12.2.1 RF Current in a Beam

The current in the beam, i_z, as a function of the electric field, E_z, can be found from an analysis of the effect of space charge on electron bunching. The result is the electronic equation

$$i_z = \frac{j\beta_e I_o}{2V_o(\beta_e - \beta)^2} E_z \tag{12-1}$$

where I_o is the dc beam current, V_o is the dc beam voltage, $\beta_e = \omega/u_o$ is the propagation constant for the electron beam, and β, which will be found to have several values, is the propagation constant for the waves on the beam and circuit.

12.2.2 Circuit Equation

In the Pierce theory, the RF circuit of a TWT is modeled as a transmission line as shown in Figure 12-11. This model serves as a starting point, but then the analysis is generalized so that the exact form of the model is not important. The line has distributed inductance and capacitance per unit length of L_ℓ and C_ℓ. The beam with current i_z is assumed to pass very close to the circuit and, from Ramo's theorem, the current induced in the circuit is $I = i_z$.

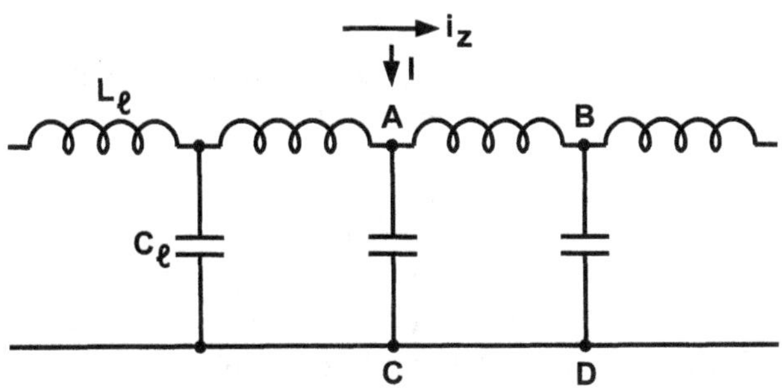

Figure 12-11 Transmission line model for a slow wave circuit.

The quantities L_ℓ and C_ℓ can be eliminated by considering the properties of the circuit. With no electron beam ($i_z = 0$), the phase velocity and impedance of the circuit are

$$v_p = \left(\frac{1}{L_\ell C_\ell}\right)^{1/2} \quad \text{and} \quad Z_o = \left(\frac{L_\ell}{C_\ell}\right)^{1/2} \tag{12-2}$$

and so

$$\beta^2 = \beta_c^2 = \omega^2 L_\ell C_\ell \tag{12-3}$$

and

$$\omega L_\ell = \beta_c Z_o \tag{12-4}$$

With these circuit quantities (Z_o and β_c), the equation for the electric field of the circuit (without the effects of space charge) that is acting on the electrons is found to be

$$E_z = j\,\frac{\beta^2 \beta_c Z_o}{\left(\beta^2 - \beta_c^2\right)} i_z \tag{12-5}$$

12.2.3 The Determinantal Equation

The electronic equation (12-1) and the circuit equation (12-5) provide two relationships between i_z and E_z. These may be combined in the absence of space charge forces to obtain the determinantal equation from which the propagation constants can be obtained:

$$\frac{\beta_e}{(\beta-\beta_e)^2}\frac{\beta^2\beta_c}{(\beta^2-\beta_c^2)}2C^3+1=0 \tag{12-6}$$

For reasons that will become clear later, Pierce defines a quantity C, used in this equation and called the *gain parameter*, by

$$C^3=\frac{Z_o}{4V_o/I_o} \tag{12-7}$$

The beam impedance is V_o/I_o, so C^3 is one-quarter of the ratio of the circuit impedance to the beam impedance. Because Z_o is usually on the order of tens of ohms and V_o/I_o is on the order of thousands to tens of thousands of ohms, C^3 is a very small quantity so C is usually in the 0.01 to 0.1 range.

The determinantal equation is a fourth-degree equation, which means that there will be four possible solutions for β, and these represent four natural modes of propagation along the electron beam and the circuit. The circuit alone has two modes of propagation, and this is consistent with the fact that the voltages at the two ends can be specified independently, so two boundary conditions must be satisfied. Four boundary conditions must be satisfied with the combination of circuit and electron stream. These can be taken as the voltages at the two ends of the circuit and the RF velocity and current of the electron beam at the point where the beam enters the circuit. The four waves resulting from the determinantal equation make it possible to satisfy these boundary conditions.

12.2.4 Synchronous Operation

Synchronous operation occurs when the velocity of the electron beam is made equal to the velocity of the wave in the absence of electrons. By assuming synchronous operation, the problem of solving the determinantal equation is simplified. After the solution for synchronous operation has been obtained, nonsynchronous operation will be considered.

Three of the propagation constants found from the determinantal equation, β_1, β_2, and β_3, contain the propagation constant of the electron beam, β_e, along with a small quantity (small because it is multiplied by C). As a result, each wave is a forward wave (traveling in the same direction as the electron beam) propagating at a velocity close to the beam velocity.

In β_1 and β_2 there is a quantity $+\beta_e C/2$, which indicates that the waves travel slower than the electron beam. In β_3, the quantity $-\beta_e C$ occurs, which indicates that the wave travels faster than the electron beam. The dominant quantity in the fourth wave is $-\beta_e$, which shows that the wave travels in the direction opposite to that of the electron beam.

In β_1 and β_2, the imaginary part of the propagation constant indicates that the wave is either a growing wave (+ in β_1) or a decaying wave (–in β_2). Finally, the propagation constants β_1 through β_4 represent the four waves illustrated in Figure 12-12.

It is of particular interest to determine the rate at which the growing wave increases. This may be determined from the equation for β_1, which describes the propagation of the electric field. Power is proportional to the square of the electric field, so the power gain is

$$10 \log_{10}\left(e^{\frac{\sqrt{3}}{2}C2\pi N}\right)^2 = 47.3\, CN \text{ dB} \tag{12-8}$$

where N is distance along the circuit in wavelengths. The reason that C is called the gain parameter is now clear. Note that gain is directly proportional to C.

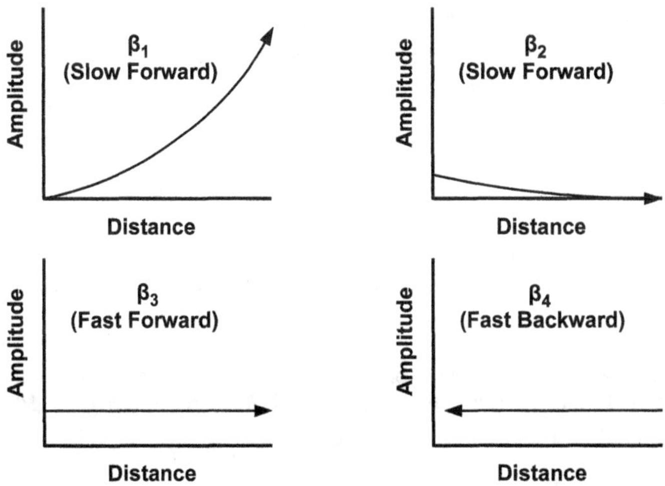

Figure 12-12 Waves represented by β_1 through β_4.

Equation (12-8) gives the gain for the lossless case. In addition to being the lossless case, this does not include losses that occur when a signal is launched on an electron beam. Since three forward waves are launched on the beam, a signal applied to the input of a TWT must divide in some way among these three forward waves and so only a portion of the input signal is used to launch the growing wave. To determine the relative amplitudes of the launched waves, the boundary conditions at the input must be matched. Here, the three waves must add to give the correct RF electric field, velocity, and current. The result is that the input signal divides evenly among the three waves. (This is only true at the input where there is no initial velocity or current modulation on the beam.)

Since only one-third of the input signal is used to launch the growing wave, the power gain, G, for the growing wave becomes

$$G = 10\log_{10}\left(\frac{1}{3}e^{\frac{\sqrt{3}}{2}C2\pi N}\right)^2 \tag{12-9}$$

or

$$G = -9.54 + 47.3\,C\,N \;\; \text{dB} \tag{12-10}$$

Power gain as a function of CN is shown in Figure 12-13. The solid curve includes the effects of all three input waves.

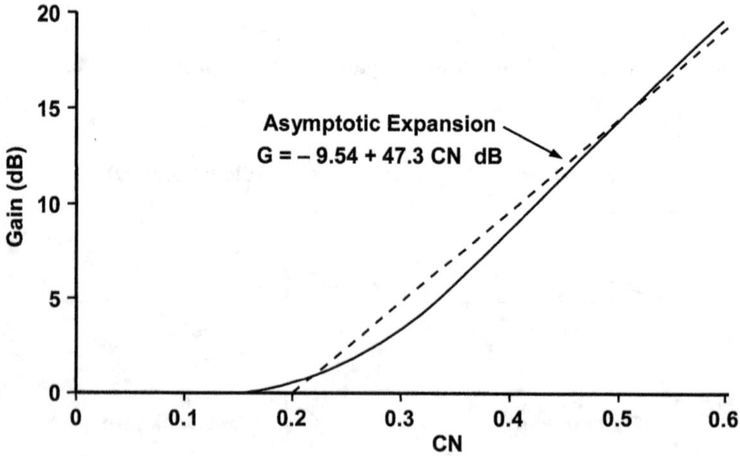

Figure 12-13 Power gain in a TWT for the synchronous case with zero loss and no space charge. (Adapted from: *Traveling Wave Tubes*, by J. R. Pierce, copyright 1950 by D. Van Nostrand Inc.)

Initially, the voltage does not change with distance and there is no gain. This is because the electron beam has no RF current and cannot induce a signal into the circuit until it becomes bunched. Eventually, the increasing wave predominates over the other two and the slope of the line must correspond to that for the growing wave, that is, 47.3. The dashed line represents the increasing wave, which starts at –9.54 dB and has the slope 47.3. Pierce wrote an asymptotic expression for the gain as

$$G = A + B\,C\,N \tag{12-11}$$

where, for this synchronous case with no loss, no space charge, and no initial RF velocity or current on the beam,

$$A = 10\log_{10}\left(\frac{1}{3}\right)^2 = -9.54 \text{ dB} \tag{12-12}$$

and

$$B = 10\log_{10}\left(e^{\sqrt{3}\pi}\right)^2 = 47.3 \tag{12-13}$$

This result is reasonably good for $CN > 0.2$.

12.2.5 Nonsynchronous Operation

When the electron beam travels at a velocity other than the velocity of the wave on the circuit in the absence of the beam, solutions to the quartic equation for the propagation constants are obtained numerically.

Pierce's velocity parameter, b, is useful in relating the dc electron velocity, u_o, to the phase velocity of the circuit, v_p, in the absence of electrons.

$$u_o = (1 + Cb)v_p \quad \text{or} \quad b = \frac{u_o - v_p}{Cv_p} \tag{12-14}$$

For $b > 0$, the electrons go faster than the undisturbed wave, and for $b < 0$, the electrons go slower than the undisturbed wave. For $b = 0$, the synchronous case occurs.

Characteristics of the growing wave are shown in Figure 12-14. The quantities x_1 and y_1 are components of the propagation constant. Because x_1 appears in the exponent in B in the gain equation as follows, it has a strong effect on gain.

$$B = 10\log_{10}\left(e^{2\pi x_1}\right)^2 \tag{12-15}$$

Clearly, if x_1 is positive, the wave is growing. If y_1 is positive, it is traveling faster than the electron beam. Notice in Figure 12-14 that, regardless of beam velocity, the wave is growing. Also, since $y_1 < 0$, the wave travels more slowly than the electrons, even when the electrons travel more slowly than the undisturbed wave ($b < 0$). In fact, the electrons must travel faster than the increasing wave to give energy to it.

Pierce also noted that, in a tube in which the total gain is large, a change in b of ± 1 about $b = 0$ can make a change of several decibels in gain. This change produces a difference between the phase velocity of the undisturbed wave, v_p, and the electron velocity, u_o, by approximately $\pm C$. Thus, the allowable difference between v_p, which may be a function of frequency, and u_o, which, of course, is not, is of the order of C.

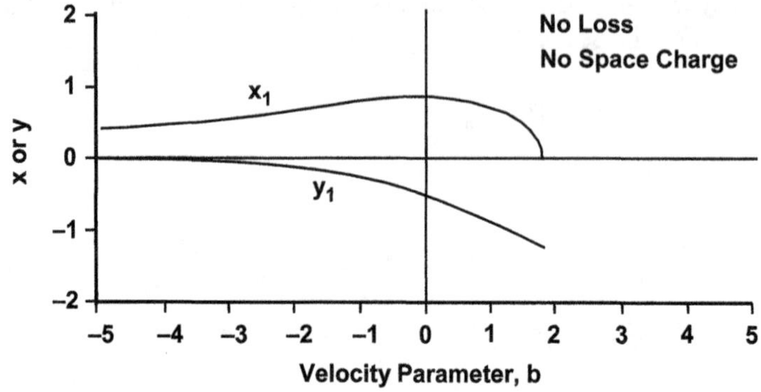

Figure 12-14 Components of incremental propagation constant for growing wave. (Adapted from: *Traveling Wave Tubes*, by J. R. Pierce, copyright 1950 by D. Van Nostrand Company, Inc.)

12.2.6 Effect of Circuit Loss

In addition to the losses that occur naturally in RF circuits, additional losses may be introduced to attenuate the backward wave. The backward wave normally originates from reflections either inside or outside the tube and can serve as a feedback mechanism to cause oscillations or large variations in gain (because of beating effects with the input signal). Pierce took loss into account by introducing the loss parameter, d, into the circuit propagation constant. The loss per wavelength, L, is

$$L = 54.5\, Cd \ \text{dB} \tag{12-16}$$

As an example, when $C = 0.025$ and $d = 1$, the loss is 1.36 dB/wavelength.

The most significant effect of attenuation is, of course, that the gain of the growing wave is decreased. This may be shown by examining the very large effect of the reduction in x_1 on the value of B in the gain equation (12-15). Figure 12-15 illustrates the effect of d on B. Note that the maximum value of B occurs very close to $b = 0$.

Pierce showed that, for small losses, the reduction in gain for the growing wave is about one-third of the circuit loss. The effect on the net gain of a TWT can be calculated only by taking into account the initial conditions in the presence of loss.

12.2.7 Effect of Space Charge

At the beginning of this chapter the electronic equation for the beam current as a function of electric field was combined with the circuit equation, which gave the circuit field acting on the electron beam. There was no consideration of space charge forces and their contribution to the electric field. When the space charge field is included in the electric field, the effects of space charge appear in the determinantal equation as an additional term containing the quantity Q, which is defined as follows

$$4QC = \left(\frac{\beta_q}{C\beta_e} \right)^2 \tag{12-17}$$

and which Pierce called Q the space charge parameter. The meanings of Q and of QC are not straightforward.

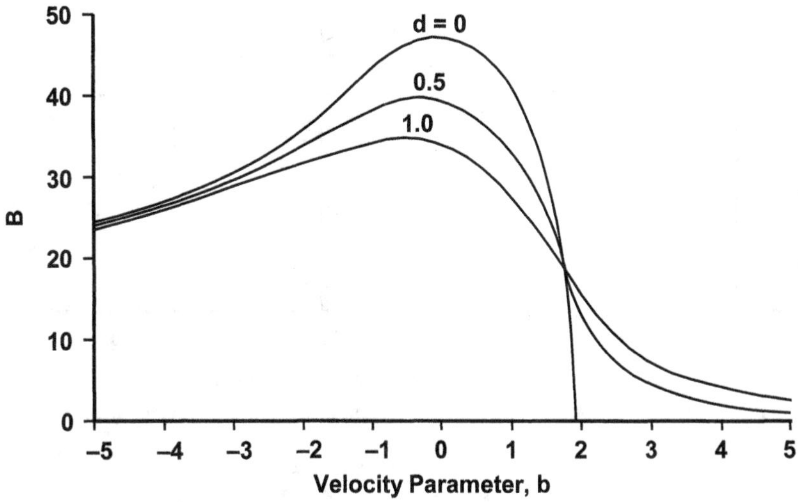

Figure 12-15 Effect of loss on the B factor in the gain of a growing wave. (Adapted from: *Traveling Wave Tubes*, by J. R. Pierce, copyright 1950 by D. Van Nostrand Company, Inc.)

However, by rewriting $4QC$ as follows, the significance can be inferred.

$$4QC = \frac{(\beta_q / \beta_e)^2}{C^2} = \frac{(\omega_q / \omega)^2}{C^2} \tag{12-18}$$

Now the ratio of plasma frequency to RF frequency is a measure of space charge debunching forces. Pierce's gain parameter, C, is a measure of the intensity of the interaction between the slow wave circuit and the electron beam. As a result, QC is a measure of the relative strength of the debunching forces to the bunching forces. Thus, QC has been adopted to describe the seriousness of space charge effects in traveling wave tubes.

Figure 12-16 shows the values of x and y for the growing wave only for several values of QC. Several effects of space charge are apparent. The most important is that the range of beam velocities (indicated by b) over which gain occurs (x_1 positive) is severely restricted. Furthermore, the velocity at which there is maximum gain in the increasing wave is now higher than in the absence of space charge. Also, the maximum gain is reduced.

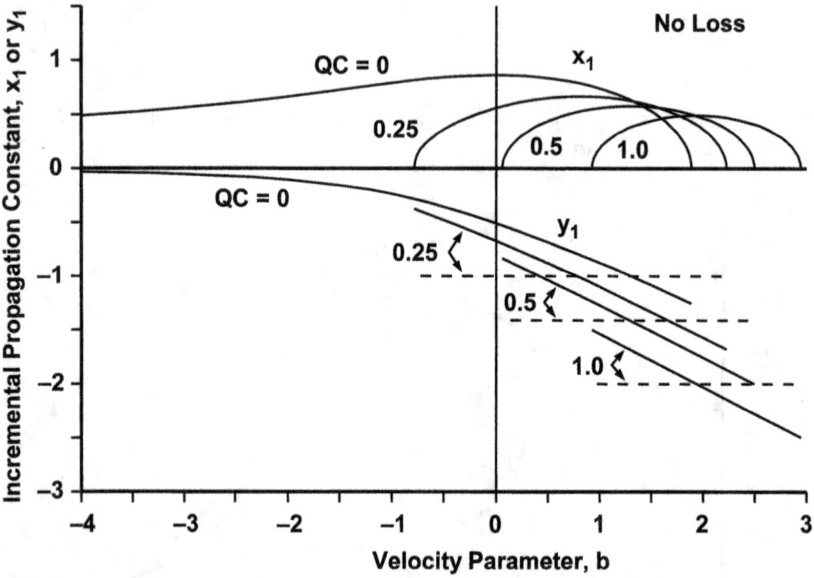

Figure 12-16 Effect of space charge on incremental propagation constants for a growing wave.

Slow space charge waves are indicated by the dashed lines in Figure 12-16. The value of y for slow space charge waves is

$$y = -\frac{\omega_q}{\omega C} = -\sqrt{4QC} \tag{12-19}$$

The significance of the slow space charge waves in TWT operation can be clearly seen by referring y and b to the incremental propagation constant for the slow space charge waves rather than β_e as is done in Figure 12-17. Now it is very clear that maximum gain occurs when the slow space charge wave velocity is the phase velocity of the circuit. As the velocity of the electron bunches moves away from the circuit velocity, gain decreases.

In Figures 12-16 and 12-17, it is also shown that the maximum gain decreases as the beam density (and QC) increases. This is to be expected because, as the beam density increases, space charge forces increase and the intensity of the bunching process decreases, which, in turn, means that the RF current in the beam decreases.

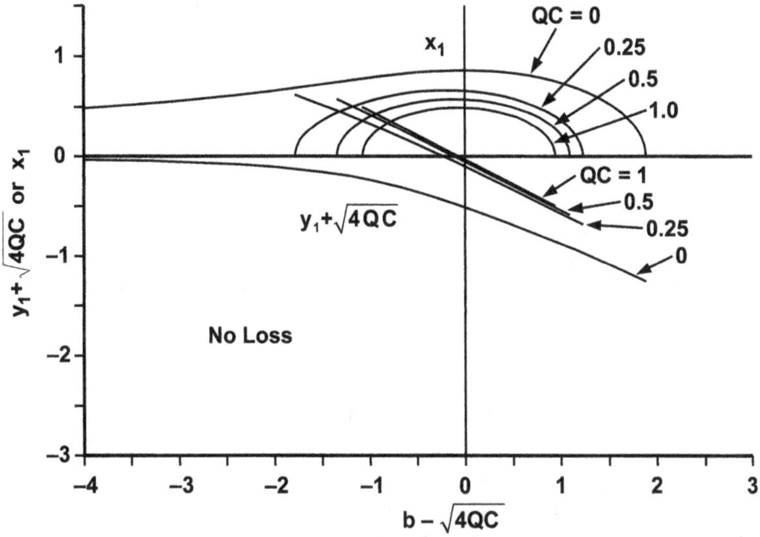

Figure 12-17 Effect of referring the growing wave to the slow space charge wave.

12.3 HIGH-LEVEL INTERACTION

12.3.1 Discussion of Interactions

Before considering high-level (nonlinear) interactions in a TWT, it is of value to briefly review the bunching action that occurs at small signal levels. As the beam enters the RF circuit of a TWT, the axial component of the electric field from the circuit accelerates some electrons and decelerates others to start the formation of bunches as indicated in Figure 12-18(a).

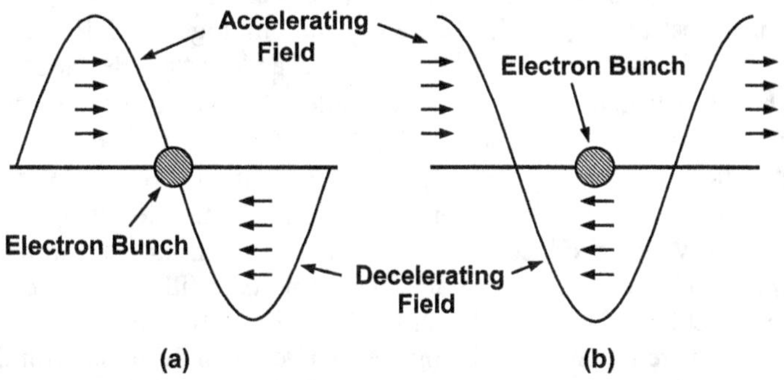

Figure 12-18 The axial field bunches and extracts energy from the beam (a) as beam the enters circuit and (b) after interaction has occurred.

The resulting RF current in the beam induces a current and voltage on the circuit that lags behind the initial voltage by 90°. Thus, the induced wave travels more slowly than the initial wave. As the bunches and the induced wave grow, decelerating field regions develop where the bunches are located as shown in Figure 12-18(b). The energy extracted from decelerated electrons is transferred to the circuit field, thereby producing amplification of that field. As long as small signal conditions exist, Pierce's theory applies and the various amplified waves on the beam and circuit grow exponentially with distance.

As the waves grow, the interaction between the beam and circuit becomes nonlinear. Perhaps the clearest paper ever published describing nonlinear phenomena in a TWT was that by Cutler [18]. A large-scale model of a TWT was constructed. Using an electron deflection system, the electron velocities and currents in the beam leaving the TWT interaction region were measured as a function of phase position relative to the circuit wave. Some of Cutler's results showing the events leading up to saturation are shown in Figure 12-19. The RF drive levels are in decibels below the level that produced saturation. The circuit voltage is shown at the top of each of the frames of Figure 12-19. RF current is indicated by the width of the shaded region and RF velocity is indicated by the deflection of the shaded region, from the center horizontal line in each frame. Thus, in each frame there is a complete story of fundamental circuit voltage, electron velocity, and current as a function of phase, for a particular drive level.

In Figure 12-19(a) the drive level was well below saturation and the current and velocity patterns were representative of small-signal conditions. Cutler noted that the velocity and current modulations at small-signal levels check calculated values well. The phase of the voltage relative to the RF electron velocity was computed from small-signal theory, but everything else in Figure 12-19, including subsequent variations in phase, was measured.

The first evidence of nonlinear behavior occurred at 17 dB below saturation (Figure 12-19(b)). The velocity and current (in Cutler's words) "show the beginnings of a cusp in the velocity curve and a definite non-sinusoidal bunching of the electrons in the retarding field region (between +30° and 210°)."

The next pattern at 12 dB below saturation, Figure 12-19(c), shows the development of a spur on the velocity curve and further development of the electron bunch between +60° and +180°. Cutler observed that the velocity within the bunch did not change significantly from lower drive levels. For this to be true, since the bunch is in a decelerating portion of the circuit field, the space charge field must compensate for the circuit field. Near 60°, the decelerating field must increase rapidly to produce the sharp deceleration of electrons and the development of the velocity spur. Furthermore, Cutler observed that the decelerating field must be much larger than the circuit field because electrons just behind the cusp are much more sharply decelerated than those preceding the cusp. He concluded that there are very sharply defined space charge fields much stronger than the circuit field.

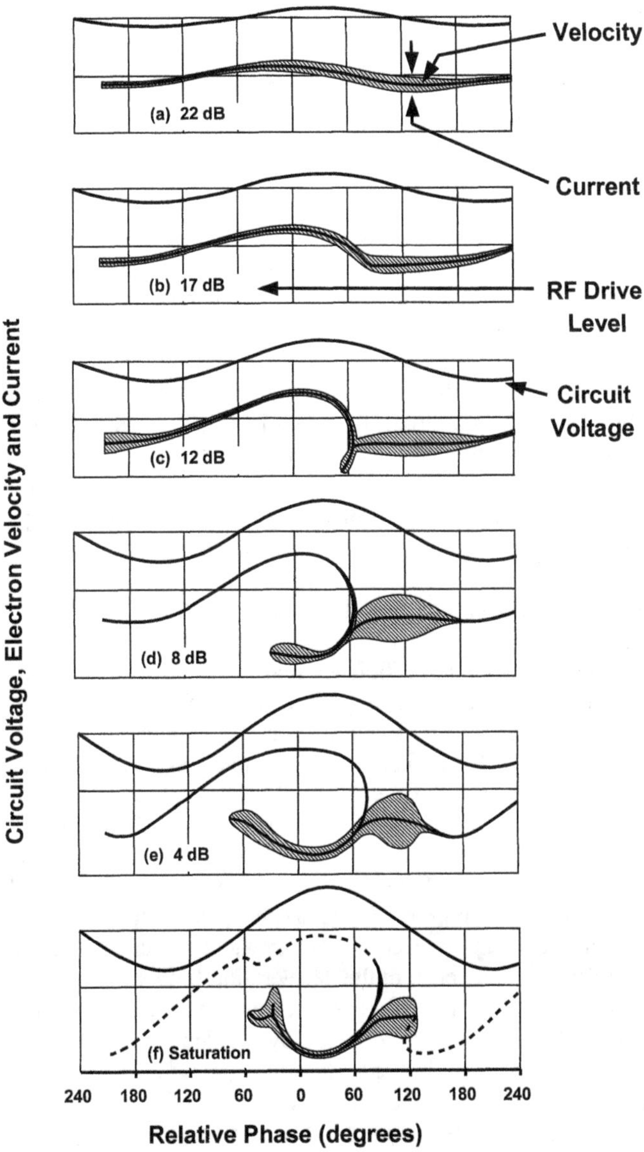

Figure 12-19 (a–f) Electron bunch behavior as a function of drive level relative to saturation drive. (Adapted from: C. C. Cutler, *Bell Sys. Tech. Jour.*, July 1956.)

The next two patterns (Figure 12-19(d, e)) show continuing development of the spur, continued movement of electrons from the high velocity regions, and continued consolidation of the main bunch just ahead of the spur. Apparently, because the electron velocities within the bunch are still nearly the same as at lower drive levels, the space charge and its associated field in the bunch has increased enough to compensate for the circuit field so that the net field remains near zero.

At 4 dB below the saturation drive level, the velocities of a large fraction of the electrons in the spur have increased because the spur has moved well into the accelerating region. The main bunch of electrons to the right of the spur has been concentrated into the 60°–120° phase interval. Apparently, the space charge field in advance of this region no longer compensates for the decelerating circuit field so electron velocity is decreasing.

At saturation, the tip of the spur has formed into a bunch that is nearly equal to the main bunch and is in a circuit field that is nearly equal and opposite to that in the main bunch. The few electrons (indicated by the dashed line) ahead of the main bunch are being decelerated into a second low-velocity loop. The reason for saturation is that the power being extracted from the circuit by the spur bunch is nearly the same as the power being added by the main bunch.

12.3.2 Estimates of Maximum Efficiency

Early estimates of the maximum efficiency that might be achieved in a TWT were made without the aid of Cutler's experimental results and without the aid of computer models. Slater [19] suggested that energy extraction from the beam in a TWT could be analyzed by observing the motions of electrons while traveling along at the speed of the circuit wave. We will use one of the frames from Cutler's results (Figure 12-19(d)) to illustrate Slater's suggestion. That frame is shown in Figure 12-20. The initial beam velocity is u_o. The velocity of the circuit wave is Δu_o below the beam velocity. Slater suggested that electrons eventually become trapped and oscillate in the region of the maximum decelerating circuit field. This trapping is clearly evident in Figure 12-20, where a bunch is shown to have formed ahead of the velocity cusp in the decelerating field.

If the electrons initially have an average velocity, u_o, greater than that of the circuit wave by Δu_o, and they are reflected by the decelerating field, they cannot acquire a velocity backwards relative to the wave greater than Δu_o. Those electrons in the small bunch to the left of the velocity cusp in Figure 12-20 fall into this category. These have been decelerated below the initial beam velocity by $2\Delta u_o$. The maximum efficiency would occur if all of the electrons (rather than the small fraction shown in Figure 12-20) were reflected and decelerated by the velocity increment $2\Delta u_o$.

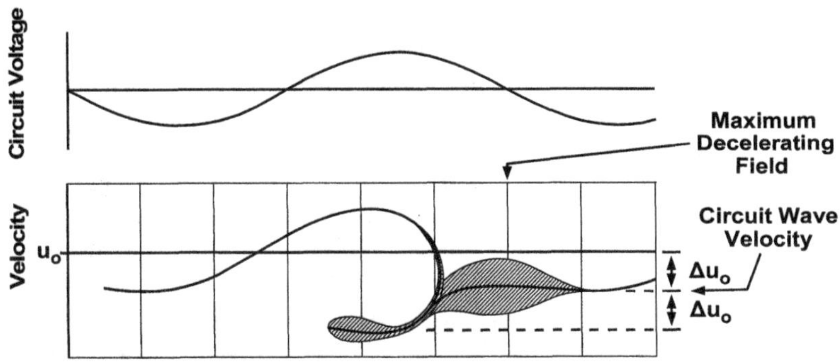

Figure 12-20 Considerations in Slater's estimate of maximum TWT efficiency.

Pierce [17] followed Slater's line of reasoning and used the Pierce small-signal expression for the velocity of the growing circuit wave to determine Δu_o. Pierce noted that "such considerations are complicated by the fact that the phase velocity of the wave in the large-signal region will not be the same as its phase velocity in the small-signal region. It is interesting, however, to see what limiting efficiencies this leads to."

Pierce found that, for $Cy_1 << 1$,

$$\eta \approx -4\,C\,y_1 \tag{12-20}$$

As an example, for a value of y_1 of -1 and C of 0.05, the maximum efficiency is predicted to be about 20%, which is high but not totally unrealistic.

It should be noted in Cutler's work (Figure 12-19(f)) that the maximum efficiency (saturation) actually occurred when the electron bunch to the left of the velocity cusp was nearly equal to the main bunch. The velocities of the two bunches were nearly the same and were about the velocity of the circuit wave (Δu_o below u_o). As a result, the efficiency in Cutler's experiment is predicted to be about half that in (12-20) or

$$\eta = -2\,C\,y_1 \tag{12-21}$$

12.3.3 Comment on Computer Modeling

Cutler's results give a clear picture of events leading up to saturation. An important impact on theoretical considerations is that the velocity becomes multivalued, and it does so at a relatively low drive level (see Figure 12-19(c), for example). Thus, the theoretical techniques (usually referred to as the Eulerian approach) that treat the electron beam as a fluid cannot be used. For example, our technique for analyzing space charge wave behavior in Chapter 9 and the Pierce theory of traveling wave interaction in this chapter both treat the beam as a fluid. These approaches are not valid when velocity becomes multivalued.

Instead of the fluid approach, the theory must be able to follow individual particles or groups of particles. This is referred to as the Lagrangian approach to the analysis. The ballistic theory of velocity modulation in Chapter 8 and of ballistic bunching in Chapter 9 used this approach. Space charge forces, which were ignored in those calculations, must be included.

No algebraic theory is available for use with the large-signal particle approach. Instead, numerical techniques must be used. It is beyond the scope of this book to examine the computer modeling of interactions in microwave tubes. This is not to degrade the importance of computer modeling. All modern TWTs are designed with the aid of computer codes. Most of the very important TWT characteristics such as power, electronic and overall efficiencies, and bandwidth can be accurately predicted with these codes. In the early 1990s, there was even a code available (the IBC code [20]) that ran on a personal computer and provided a great deal of insight into the performance of a TWT as various operating and design parameters were varied.

From the early numerical computations of Nordsieck [21], who ignored space charge effects, there was a steady development of computational techniques in the late 1950s and 1960s. Some of the more important of these works are listed as [22–32]. Some results from Hess's analysis [29] are given in Figure 12-21. Various quantities are plotted as a function of distance along the circuit in half wavelengths. The shaded areas of the distance scale indicate regions of decelerating circuit field. Note the similarity of the overall appearance of the electron velocity plots with those given by Cutler (Figure 12-19).

The earliest indication of nonlinear interaction in Hess's results appears in the electron density plot, when density becomes nonsinusoidal. A few half wavelengths later, a spur, like the Cutler spur, develops on the velocity plot and the velocity becomes multivalued. Also, the electron density bunch splits and, in subsequent half wavelengths, a portion falls back into an accelerating field. The resulting acceleration of electrons is shown by the increase in velocity of electrons in the spur. Finally, saturation occurs (the efficiency and circuit voltage reach maximum values) when a sufficient number of electrons have drifted back into the accelerating field.

12.3.4 Velocity Tapering

A major reason for saturation of the power output in TWTs is the reduction in average beam velocity caused by the transfer of electron energy to the circuit fields. An example of this reduction in average beam velocity is shown in Hess's results in Figure 12-21. The dashed line in the electron velocity plot shows average beam velocity. As we have discussed in a previous section, partly as a result of the reduced velocity near saturation, electrons fall back out of phase with the circuit field, move into accelerating regions, and extract energy from the circuit.

It has been recognized since the original invention of the TWT (see the varying pitch of the helix in Figure 12-1 from Lindenblad's patent) that efficiency might be improved if the circuit is modified so that the velocity of the circuit wave decreases along with the beam velocity. Then, even though the bunches are slowed down as they lose energy to the circuit, they will remain in phase with the circuit wave and continue to deliver energy.

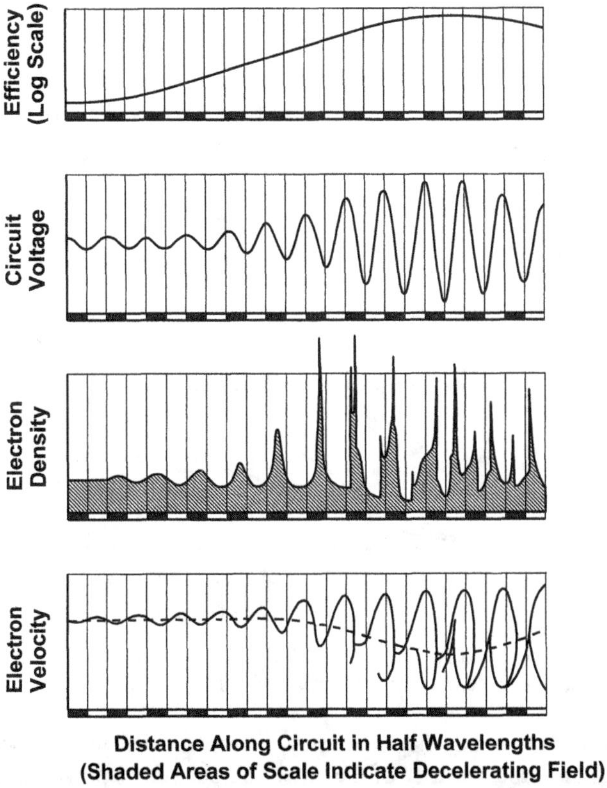

Figure 12-21 Characteristics of interactions in a TWT determined by computer analysis. (Adapted from: P. Hess, Ph.D. dissertation, University of California at Berkeley, 1960.)

This technique of modifying the circuit to enhance efficiency (and often other TWT characteristics) is known as velocity tapering. Early attempts to use velocity tapering produced little or no improvement in efficiency over untapered circuits. The advent of computer analyses (see [30], for example) has resulted in the design of tapers that produce appreciable improvements in efficiency. In [31], Meeker and Rowe reported an efficiency improvement from 18% to 35% through the use of a variable-pitch helix.

Another example of a computer-designed taper is reported in [33]. In a NASA development of a 20-GHz coupled-cavity TWT, a phase-adjusted taper was used to achieve an RF efficiency of 22.6% and a peak power of 1,000W. Previously, the interaction structure for this particular tube had a piecewise linear taper that consisted of sequences of cavities of equal length and cavity sequences of linearly tapering length. The RF efficiency of that tube was 9.6% and the peak power output was 420W. In the phase-adjusted taper, the phase of the electron bunches with respect to the phase of the circuit field changes gradually from a value (about 0°) conducive to the formation of strong bunches at the input end of the taper to a value (about 90°) conducive to strong transfer of beam power to the circuit at the output end of the taper. A comparison of the efficiencies of the TWTs with the piecewise linear taper and with the phase-adjusted taper is shown in Figure 12-22.

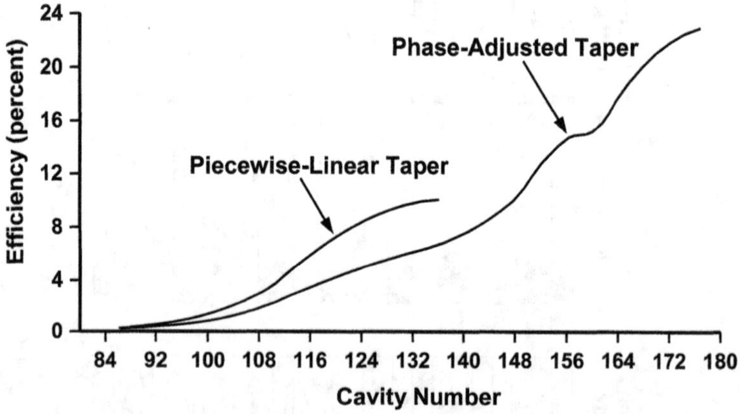

Figure 12-22 Computed RF efficiencies for a 20-GHz TWT with a phase-adjusted taper and one with a piecewise linear taper. (From: *NASA Tech Briefs*, September 1992.)

In addition to enhancing efficiency, a properly designed velocity taper can be used to enhance the linearity of the power output versus power input characteristic of a TWT. This is particularly important in communications applications of TWTs, where low distortion is required. An example of a linearizing taper is the dynamic velocity taper (DVT) developed by Kosmahl and reported in [34]. The basic philosophy of the DVT may be understood by examining the relation for Pierce's velocity parameter (12-14), which is

$$b = \frac{u_o - v_q}{C v_q} \tag{12-22}$$

In the small signal region, the gain of a TWT is constant and corresponds directly to the value of b. As the signal level grows and the TWT approaches nonlinear operation the beam gives up energy to the circuit. The average velocity of the beam decreases and so the velocity parameter decreases. To prevent the gain from decreasing, it is necessary to modify the circuit and v_q so that the velocity parameter remains constant.

Results are presented in the referenced report [34] for TWTs with and without DVTs. For one of the TWTs with the DVT (the 12-GHz, 50-W Hughes 8802 experimental TWT), the pitch variation is shown in Figure 12-23 and the resulting improvement in linearity and power output is shown in Figure 12-24.

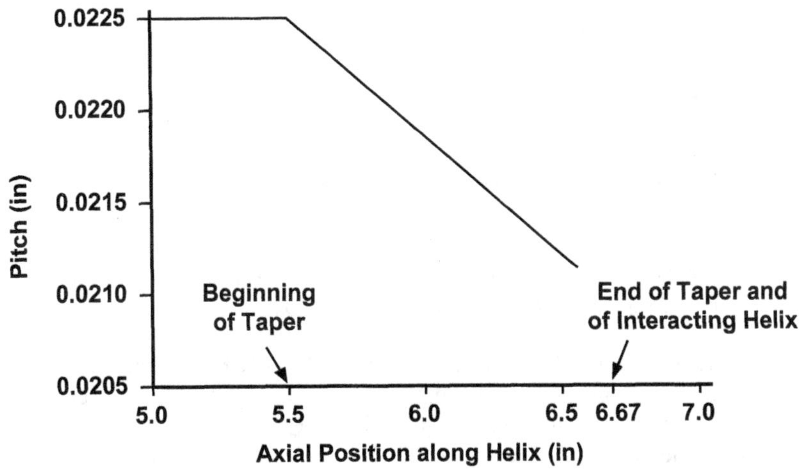

Figure 12-23 Pitch variation in the dynamic velocity taper (DVT) for the Hughes 8802 TWT. (Adapted from: H. G. Kosmahl and J. C. Peterson, *NASA Tech. Memo. 83590*, March 1984.)

Note in Figure 12-23 that the pitch decreases linearly (this decrease is actually the first part of an exponential decrease) by about 7% over the final 18% of the length of the helix. The results of this change in pitch are impressive as shown in Figure 12-24. The linearity near saturation is significantly improved, and the power characteristic approaches that of a hard limiter at saturation. (The discrepancy between computed and measured powers resulted from the use of cold rather than hot helix loss values in the computation.) The power output is increased about 1.5 dB and the RF efficiency (efficiency of conversion of beam power to RF output power) is about 16%, which is 5% better than without the DVT. The relative phase delay near saturation is about 30° and the AM to PM conversion does not exceed 3°/dB, which is nearly the same as that in an unmodified TWT.

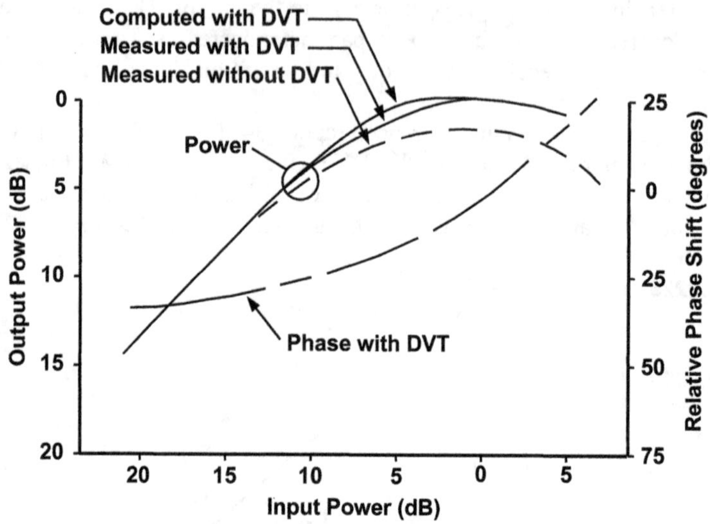

Figure 12-24 Effect of DVT on power and phase for the Hughes 8802. (From: H. G. Kosmahl and J. C. Peterson, *NASA Tech. Memo. 83590*, March 1984.)

Finally, Kosmahl and Peterson noted that, with an RF efficiency of 16%, a well-designed multistage depressed collector should raise the overall efficiency above 50%. By replacing the beryllium oxide (BeO) support rods for the helix with diamond, losses could be reduced, the interaction impedance would be improved, and the overall efficiency would increase to about 65%.

REFERENCES

[1] R. L. Wathen, "A history of the traveling wave tube," *Report No. 5202-5020, Sperry Gyroscope Company*, May 18, 1953.

[2] V. Haeff Andrew, "Device for and method of controlling high frequency currents," U.S. Patent 2064469, 1936.

[3] V. Haeff Andrew, "Device for and method of controlling high frequency currents," U.S. Patent 2233126, 1941.

[4] K. Posthumus, "Oscillations in split anode magnetron," *Wireless Engineer,* Vol. 12, No. 138, 1935, pp. 126–132.

[5] E. Lindenblad Nils, "Electron discharge device system," U.S. Patent 2300052, 1942.

[6] R. Kompfner, "Travelling wave valve," *Wireless World,* Vol. 52, No. 11, 1946, pp. 369–372.

[7] R. Kompfner, "The traveling-wave tube as amplifier of microwaves," *Proc. IRE,* Vol. 35, No. 2, 1947, pp. 124–127.

[8] R. Kompfner, "Traveling-wave tube as amplifier of microwaves," *Wireless Engineer,* Vol. 24, No. 288, September 1947, pp. 255–266.

[9] R. Kompfner, "On the operation of the traveling wave tube at low level," *British Institution of Radio Engineers*, August–September 1950, pp. 283–289.

[10] R. Kompfner, *The Invention of the Traveling Wave Tube*, San Francisco, CA: San Francisco Press, 1963.

[11] J. R. Pierce, and L. M. Field, "Traveling-wave tubes," *Proc. IRE,* Vol. 35, No. 2, 1947, pp. 108–111.

[12] J. R. Pierce, "High-frequency amplifier," U.S. Patent 2602148, 1952.

[13] L. M. Field, "High-frequency amplifying device," U.S. Patent 2575383, 1951.

[14] J. R. Pierce, "Theory of the beam-type traveling-wave tube," *Proc. IRE,* Vol. 35, No. 2, 1947, pp. 111–123.

[15] A. Blanc-Lapierre, and P. Lapostolle, "Contribution a l'etude des amplificateurs a ondes progressives," *Annales des Telecommunications*, No. 1, December 1946, pp. 283–302.

[16] A. Blanc-Lapierre, and P. Lapostolle, "Sur l'interaction entre une onde progressive et un faisceau d'electrons de vitesse voisine de celle de l'onde," *Comptes Rendue Rebdomadaires des Seances de l'Academie des Sciences*, No. 224, January 13, 1947, pp. 104–105.

[17] J. R. Pierce, *Traveling-Wave Tubes*, New York: Van Nostrand, 1950.

[18] C. C. Cutler, "The nature of power saturation in traveling wave tubes," *Bell Sys. Tech. Jour.,* Vol. 35, July 1956, pp. 841–876.

[19] J. C. Slater, *Microwave Electronics*, Princeton, NJ: Van Nostrand, 1950.

[20] I. J. Morey, and C. K. Birdsall, "Traveling-wave-tube simulation; the IBC code," Electronics Research Laboratory, University of California at Berkeley, September 26, 1989.

[21] A. Nordsieck, "Theory of the Large Signal Behavior of Traveling-Wave Amplifiers," *Proc. IRE,* Vol. 41, No. 5, 1953, pp. 630–637.

[22] H. C. Poulter, *Large Signal Theory of the Traveling Wave Tube*, Tech. Report No. 73, Electronics Research Laboratory, Stanford University, January 1954.

[23] P. K. Tien, L. R. Walker, and V. M. Wolontis, "A large signal theory or traveling-wave amplifiers," *Proc. IRE,* Vol. 43, No. 3, 1955, pp. 260–277.

[24] J. E. Rowe, *A Large Signal Analysis of the Traveling Wave Amplifier*, Tech. Report No. 19, Electron Tube Laboratory, University of Michigan.

[25] J. E. Rowe, "Design information on large-signal traveling-wave amplifiers," *Proc. IRE,* Vol. 44, No. 2, 1956, pp. 200–210.

[26] J. E. Rowe, "Correction," *Proceedings of the IRE,* Vol. 44, No. 6, 1956, pp. 818–819.

[27] P. K. Tien, "Large signal theory of traveling wave amplifiers including effects of space charge and finite coupling between beam and circuit," *Bell System Technical Jour.,* Vol. 35, No. 2, 1956, pp. 349–374.

[28] J. E. Rowe, "One-dimensional traveling-wave tube analyses and the effect of radial electric field variations," *IRE Trans. Electron Devices,* Vol. 7, No. 1, 1960, pp. 16–21.

[29] P. Hess, "Traveling wave tube large signal theory with application to amplifiers having dc voltage tapered with distance," Ph.D. dissertation, University of California at Berkeley, 1960.

[30] H. Sobol and J. E. Rowe, "Theoretical power output and bandwidth of traveling-wave amplifiers," *IRE Trans. Electron Devices,* Vol. 7, No. 2, 1960, pp. 84–94.

[31] J. G. Meeker and J. E. Rowe, "Phase focusing in linear-beam devices," *IRE Trans. Electron Devices,* Vol. 9, No. 3, 1962, pp. 257–266.

[32] H. K. Detweiler, "Characteristics of magnetically focused large signal traveling-wave amplifiers," Ph.D. dissertation, University of Michigan, Ann Arbor, October 1968, also RADC Tech. Rep. TR-68-433; ASTIA Doc. AD-842733.

[33] NASA Lewis Research Center, "Coupled-cavity traveling-wave tube has phase-adjusted taper," *NASA Tech Briefs*, September 1992, pp. 26–30.

[34] H. G. Kosmahl and J. C. Peterson, "A TWT amplifier with a linear power transfer characteristic and improved efficiency," NASA Technical Memorandum 83590, prepared for the Tenth Communications Satellite Systems Conference, Orlando, FL, March 18–22, 1984.

Chapter 13

Wave Velocities and Dispersion

For amplification to occur in a traveling wave tube, the axial component of the velocity of the wave on the RF circuit must be close to the velocity of the bunches of electrons in the beam. If, as frequency is varied, the velocity of the wave on the circuit moves away from the electron bunch velocity, gain will decrease. As a result, the variation of circuit wave velocity with frequency is an important consideration in the design of a TWT.

A circuit in which the wave velocity varies with frequency is said to have dispersion or to be dispersive. Before discussing dispersion in detail, it is important to examine wave velocities and, in particular, the concepts of the group and phase velocities of a wave.

13.1 GROUP AND PHASE VELOCITY

Figure 13-1 may help clarify group and phase velocity. Waves are breaking at an angle θ on a beach. The velocity, v, with which the wave is moving is indicated by a vector perpendicular to the wave crest. The velocity with which the wave crest sweeps along the beach is indicated by a vector parallel to the beach. This is the velocity at which any point on the wave (or any particular phase of the wave) sweeps along the beach. This velocity is the phase velocity and is denoted by v_p. A person walking along the beach and desiring to keep up with the wave crest (or with any other point on the wave) would have to walk at the phase velocity. Similarly, a person on a surfboard desiring to ride the crest of the wave in a path parallel to the beach would have to move at this same phase velocity.

The magnitude of the phase velocity depends on the direction chosen for phase velocity relative to the direction of motion of the wave. In Figure 13-1(b) the angle between the crest of the wave and the beach (and the direction chosen for the phase velocity) is smaller than in Figure 13-1(a). As a result, the wave crest in Figure 13-1(b) sweeps along the beach more rapidly than in Figure 13-1(a). Thus, the phase velocity has increased.

The group velocity, v_g, is the velocity of the energy of the wave in the direction of interest. For example, in Figure 13-1(a), as the energy in the wave at point A moves to point B, the distance that energy moves in parallel to the beach (and to the phase velocity) is from A to C. Thus, in this case, the group velocity is the component of the wave velocity in the direction of the phase velocity. Like the phase velocity, the magnitude of the group velocity depends on direction. This can be seen by comparing Figure 13-1(a) to Figure 13-1(b). The group velocity in Figure 13-1(b) is smaller than that in Figure 13-1(a).

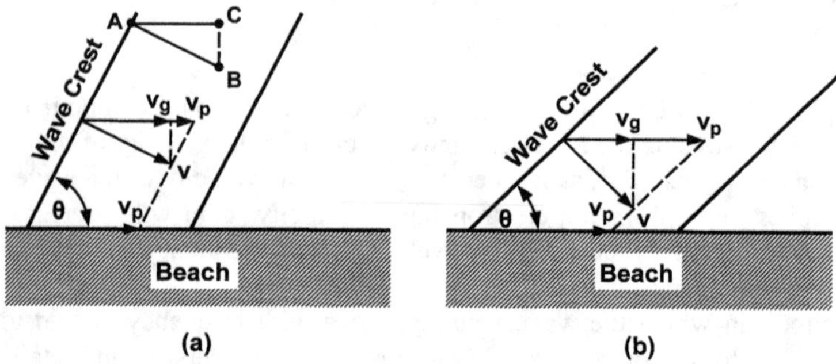

Figure 13-1 Illustration of the difference between (a) phase velocity and (b) group velocity.

Group velocity and phase velocity can have opposite directions. Consider, for example, Figure 13-2. The waves are moving to the left and the group velocity is to the left. It is assumed that there is a seawall with equally spaced openings along the shoreline. It is also assumed that a person walking along the beach would be able to see the waves only when passing one of the openings in the seawall. The person could walk from right to left and pass successive openings at the same time that one of the wave crests did. Assuming that the angle θ is the same as in Figure 13-1(a), then the magnitudes of velocities would all be the same as in Figure 13-1(a), but the directions would be reversed.

Alternatively, the person could walk from left to right (from 1 to 2, for example) and pass the openings in synchronism with successive wave crests moving from right to left (from 3 to 2, for example). In this case, the phase velocity is to the right while the group velocity is to the left. For a person on a surfboard to move to the right and continue to be accelerated by the wave crests (which are moving to the left), that person would have to jump from wave crest to wave crest.

When the group velocity and the phase velocity are in opposite directions, interaction with the wave is referred to as a *backward wave interaction*. An important aspect of this interaction, which was evident in the previous discussion, is that the interaction is periodic rather than continuous as it is when the group velocity and the phase velocity are in the same directions.

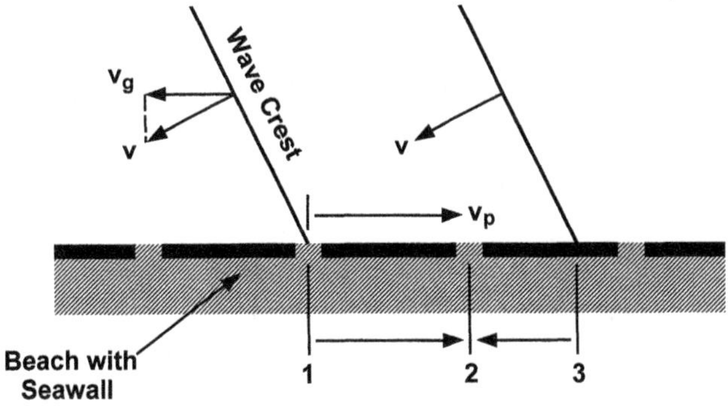

Figure 13-2 Illustration of backward wave interaction.

13.2 DISPERSION

A circuit in which the wave velocity varies with frequency is said to have dispersion or to be dispersive. Two ways are commonly used to illustrate the dispersion characteristics of a circuit. One, shown in Figure 13-3, is straightforward. The vertical axis represents phase velocity and the horizontal axis represents frequency. The plot for a circuit with no dispersion is simply a horizontal line. For a circuit with dispersion, the line is not horizontal.

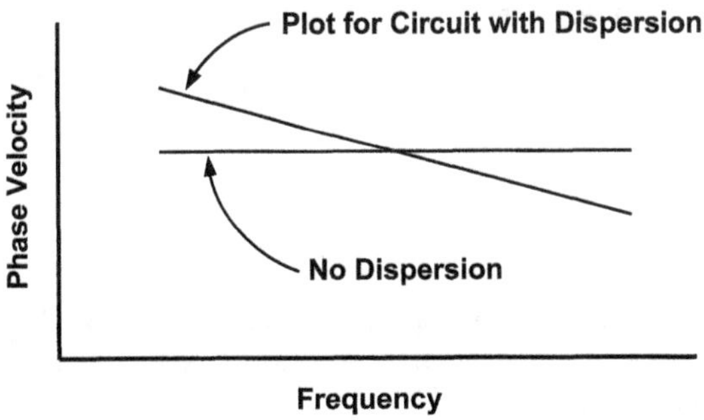

Figure 13-3 Straightforward representation of dispersion.

The second way to show dispersion characteristics is with a Brillouin diagram, which is often referred to as an ω-β diagram. Here, ω is angular frequency and β is the wave propagation constant. This diagram is not as straightforward as the first (phase velocity versus frequency). However, its use permits the display of considerably more information than the first. Because of the importance and widespread use of ω-β diagrams in the microwave component industry, their development is considered here.

A convenient way to develop an understanding of ω-β diagrams is to consider, first, simple microwave circuits and how their characteristics are represented. For this purpose, we will use a coaxial transmission line, which is not dispersive, and a waveguide, which is dispersive. Following these two examples, the more complicated example of a periodically loaded waveguide is examined.

13.2.1 Coaxial Transmission Line

In the fundamental TEM mode in a vacuum insulated coaxial line shown in Figure 13-4, the wave velocity and the electric and magnetic field lines are in planes perpendicular to the axis of the transmission line. The group velocity and the phase velocity are the same and are the velocity of light, c, for all frequencies. Thus, the coaxial line has no dispersion.

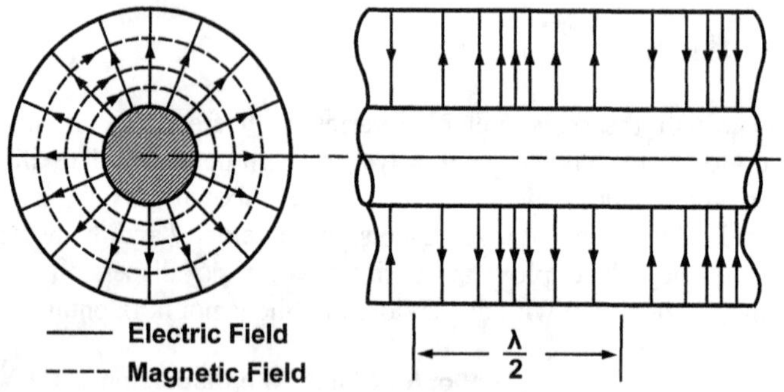

Figure 13-4 The TEM mode in a coaxial transmission line.

The propagation constant, β, is

$$\beta = k = \frac{\omega}{c} \tag{13-1}$$

The plot of ω *versus* β for the coaxial line is given in Figure 13-5. The plot is a straight line with slope c. In general, those portions of an ω-β diagram with constant slope represent frequency regions of operation of a circuit in which phase velocity is constant, so there is no dispersion.

13.2.2 Rectangular Waveguide

The phase velocity of a wave in rectangular waveguide varies with frequency so the waveguide is dispersive. To understand why phase velocity varies with frequency and how phase and group velocity are shown on an ω-β diagram, it is useful to examine the configuration of waves in a rectangular waveguide.

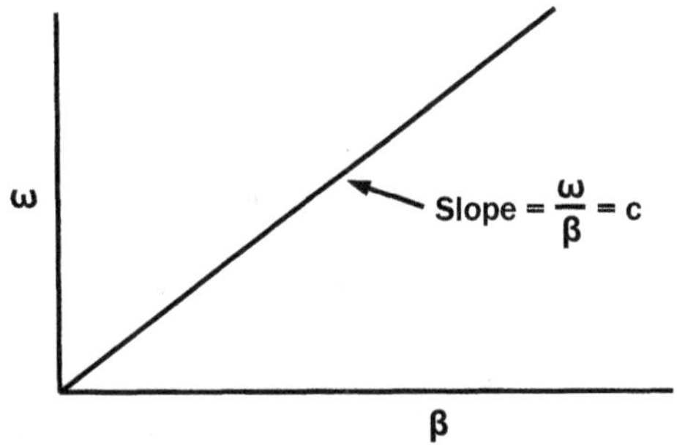

Figure 13-5 The ω-β diagram for the TEM mode in a coaxial transmission line.

A convenient place to start is with the consideration of two waves of equal amplitude traveling in free space, as indicated in Figure 13-6. The direction of propagation of one is at an angle $+\alpha$ with respect to the z axis and the other is at an angle $-\alpha$. The solid lines represent positions of wave crests (maximum value of E field out of the figure) and the dashed lines represent positions of wave troughs (minimum value of E field into the figure).

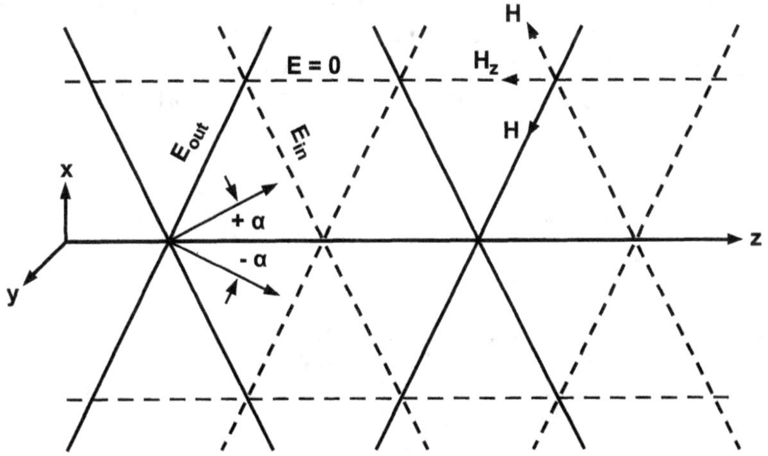

Figure 13-6 Plane waves of equal amplitude propagating at angles of $\pm\alpha$ with respect to the z direction.

Wherever wave crests and troughs (solid and dashed lines) intersect, the electric field is zero and only a magnetic field exists. In fact, all along a plane through zero-field points and parallel to the y-z plane, the electric field is zero. The E field is zero not only for the instant of time implied in Figure 13-6, but also at other times for waves propagating at the angles shown. Thus, conductors can be inserted along these planes without disturbing the E fields as long as the currents in the conductors correspond to those that would be induced by the magnetic fields.

The magnetic fields are also parallel to any *x-z* plane, and the electric fields are perpendicular to these planes. Thus, conductors with currents and charges corresponding to those that would be induced by the fields can be placed in the *x-z* planes.

Finally, two conductors in the *y-z* plane can be joined to the conductors in the *x-z* plane to form a rectangular pipe (waveguide) containing portions of the two waves. These two waves can propagate as though they were in free space, except that, in reality, they are continually reflected back and forth from one side of the guide to the other. In fact, there is only one wave because each wave is a reflection of the other.

The velocity with which the wave propagates along the waveguide (the group velocity) can easily be determined with the aid of Figure 13-7. Each component of the wave travels at the speed of light, c, at an angle α with respect to the axis of the guide. As a result, the group velocity is smaller than c and is

$$v_g = \mathrm{c}\cos\alpha \tag{13-2}$$

The phase velocity of the wave (the velocity of a particular phase or amplitude of the wave) is indicated by v_p. Thus, as the crest of the wave propagates at angle α a distance proportional to the velocity of light, c, the crest sweeps along the wall of the waveguide a distance proportional to higher velocity, v_p, which is

$$v_p = \frac{\mathrm{c}}{\cos\alpha} \tag{13-3}$$

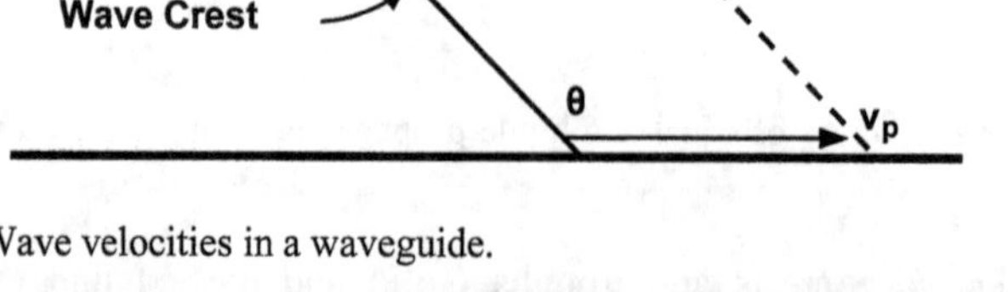

Figure 13-7 Wave velocities in a waveguide.

If an electron traveling parallel to the wall of the waveguide was to "ride" the crest of the wave and continually be deflected into the figure by the electric field, it would have to move at velocity v_p. Of course, this is not possible because, in this example, $v_p > \mathrm{c}$. TWT circuits are designed to be "slow-wave" circuits so that $v_p < \mathrm{c}$ and so electrons can move at a velocity equal to the phase velocity.

To examine the dispersion of waveguide, it is necessary to consider the change in wave configuration that takes place in a waveguide when frequency is varied. Shown in Figure 13-8 are wave configurations for three different frequencies $f_1 > f_2 > f_3$. Group velocities and phase velocities are also indicated.

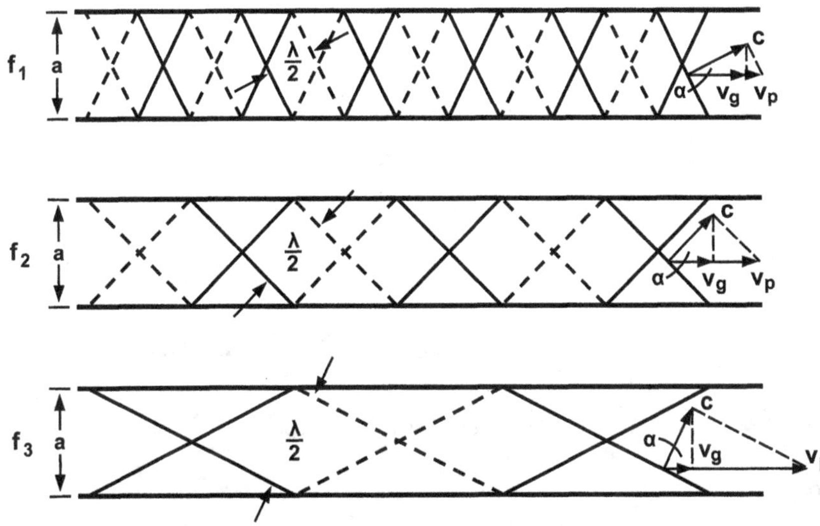

Figure 13-8 Wave configurations for frequencies $f_1 > f_2 > f_3$.

Note that, as frequency decreases, wavelength increases and the angle α must increase for the wave to fit in the waveguide. Note also that, as $\lambda/2$ approaches the width of the guide, a, α approaches 90° and most of the wave motion is back and forth across the waveguide rather than along the waveguide. When $\alpha = 90°$, wave propagation down the guide stops and the condition known as cutoff occurs. The wavelength and frequency at which cutoff occurs are often denoted by λ_c and f_c. As frequency decreases from f_1 through f_2 to f_3 and then to f_c, the phase velocity steadily increases and, in fact, becomes infinite at cutoff. At the same time, the group velocity steadily decreases and becomes zero at cutoff.

The ω-β relation for the rectangular waveguide may be derived by examining the wave velocities as a function of frequency (or wavelength) with the aid of Figure 13-9. Thus,

$$\left(\frac{\lambda_c}{2}\right)^2 = \left(\frac{\lambda}{2}\right)^2 + \left(\frac{\lambda_c}{2}\cos\alpha\right)^2 \tag{13-4}$$

or

$$1 = \left(\frac{\lambda}{\lambda_c}\right)^2 + \cos^2 \alpha \tag{13-5}$$

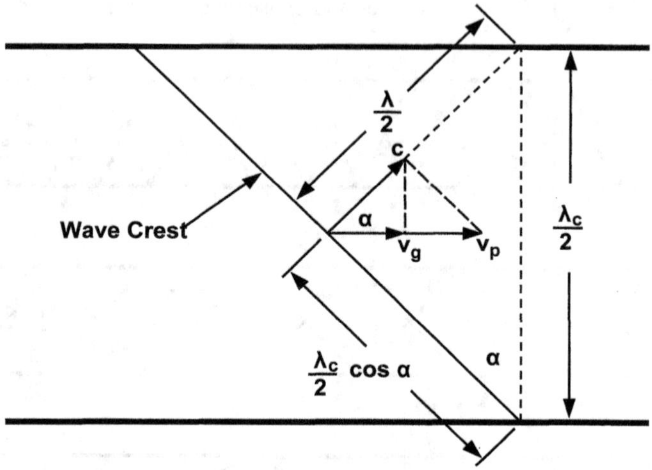

Figure 13-9 Quantities used in derivation of an ω-β relation.

Also,

$$\cos \alpha = \frac{c}{v_p} \tag{13-6}$$

but

$$\frac{\lambda}{\lambda_c} = \frac{c}{f}\frac{f_c}{c} = \frac{\omega_c}{\omega} \tag{13-7}$$

and

$$\frac{c}{v_p} = \frac{\omega}{v_p}\frac{c}{\omega} = \beta\frac{c}{\omega} \tag{13-8}$$

so

$$1 = \left(\frac{\omega_c}{\omega}\right)^2 + \beta\left(\frac{c}{\omega}\right)^2 \tag{13-9}$$

which may be rewritten to obtain the ω-β relation

$$\omega = \sqrt{\omega_c^2 + (\beta c)^2} \tag{13-10}$$

and which is plotted in Figure 13-10. Since $\beta = \omega/v_p$, v_p at any ω and β is simply the slope of a line drawn from the corresponding point on the ω-β curve through the origin. Note that the phase velocities for the various frequencies used in Figure 13-8 are indicated in Figure 13-10.

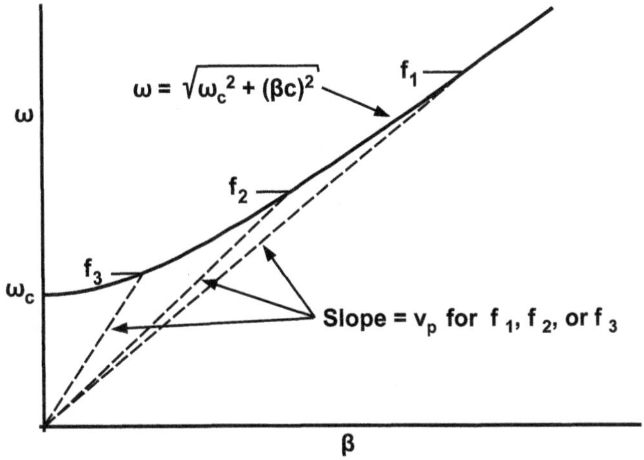

Figure 13-10 The ω-β diagram for rectangular waveguide.

The group velocity of the wave may also be displayed on an ω-β diagram. To determine how this is done, consider Figure 13-11, which shows the propagation constants k and β in vector form and the changes that occur (Δk and $\Delta\beta$) when the frequency is varied by $\Delta\omega$. First, from Figure 13-9,

$$\cos\alpha = \frac{c}{v_p} = \frac{c/\omega}{v_p/\omega} = \frac{1/k}{1/\beta} = \frac{\beta}{k} \tag{13-11}$$

which is why the relations of β, k, and α are as indicated in Figure 13-11. When ω is reduced by $\Delta\omega$, α increases by $\Delta\alpha$, k decreases by Δk, and β decreases by $\Delta\beta$. However, for a small $\Delta\omega$ and a small $\Delta\alpha$,

$$\frac{\Delta k}{\Delta \beta} \approx \cos\alpha \tag{13-12}$$

and, from Figure 13-10,

$$v_g = c\cos\alpha \tag{13-13}$$

so

$$v_g = \frac{\Delta\omega}{\Delta\beta} \tag{13-14}$$

Thus, group velocity can be shown on the ω-β diagram as indicated in Figure 13-12. The group velocity is the slope of the ω-β curve. Group velocities for the various wave configurations in Figure 13-8 along with the cutoff condition are also indicated in Figure 13-12.

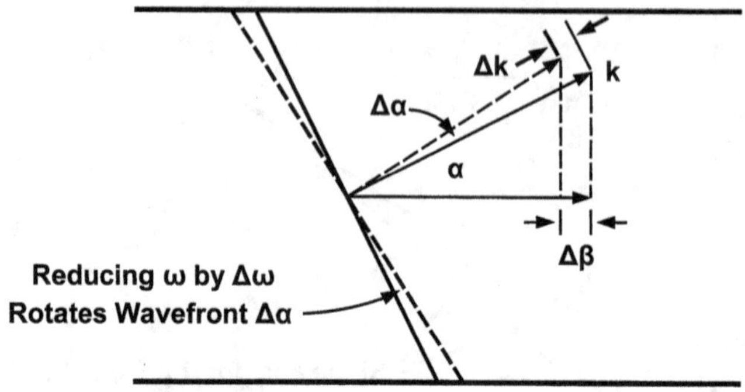

Figure 13-11 Changes in the propagation constant as frequency is varied.

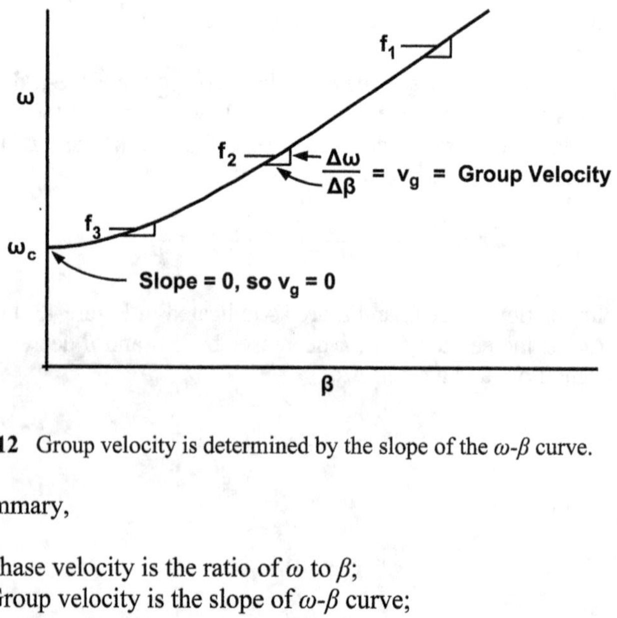

Figure 13-12 Group velocity is determined by the slope of the ω-β curve.

In summary,

1. Phase velocity is the ratio of ω to β;
2. Group velocity is the slope of ω-β curve;
3. For a waveguide, as $\omega \to \infty, \beta \to k$, $v_p \to$ c, and $v_g \to$ c.

In this section, we have treated the wave in a rectangular waveguide as though it were composed of two waves. As noted previously, there is only one wave, because each is a reflection of the other. The amplitude of the resulting electric field as functions of axial and transverse positions in the waveguide is shown in Figure 13-13. It should be noted that this is the fundamental TE_{01} mode of propagation. TE refers to the configuration of the electric field (E), which is transverse (T) to the axis of the guide. There are no variations of field across the narrow dimension of the guide (thus, the subscript 0) and one-half sinusoidal variation in the broad dimension (subscript 1). At higher frequencies, other modes not of interest at this point in the discussion can propagate.

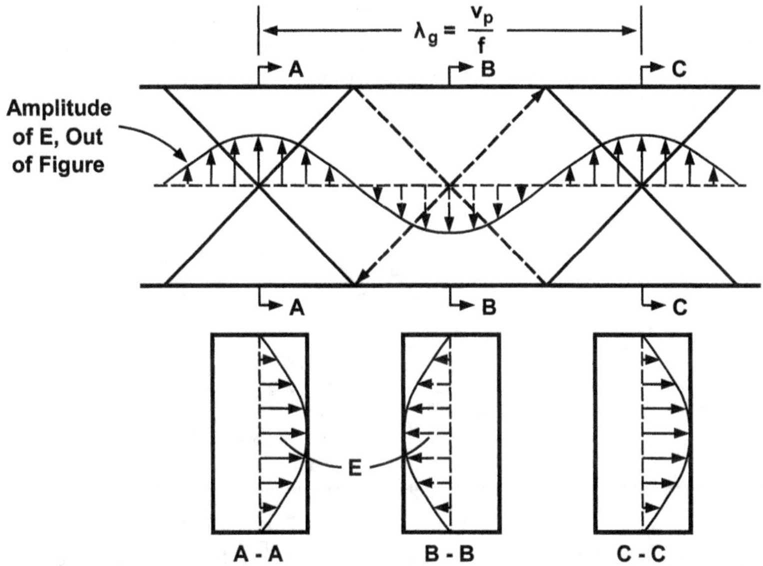

Figure 13-13 Electric field distributions in fundamental TE_{01} mode in rectangular waveguide.

The wavelength of the total wave is the guide wavelength, λ_g, which is

$$\lambda_g = \frac{v_p}{f} \tag{13-15}$$

The guide wavelength is longer than the free-space wavelength, λ, at all normal operating frequencies. In applications of waveguides, only the guide wavelength is of value. The free-space wavelength was an artifice of use in explaining waveguide behavior but plays no role in the use of waveguide. Thus, in the following discussions, in which wavelength in the waveguide is discussed, it is to be assumed that guide wavelength is the quantity is question even though the term *wavelength* is used.

Finally, in considering the ω-β diagram for waveguide, it is important to note that waves can propagate with identical characteristics in either direction and so the ω-β diagram is symmetrical about the ω axis as shown in Figure 13-14. At higher frequencies, in either direction, the group and phase velocities approach the velocity of light.

13.2.3 Periodically Loaded Waveguide

An obstacle, such as a metal vane in a waveguide, produces a change in impedance and is said to load the waveguide. This loading causes part of a propagating wave to be reflected. If equally spaced obstacles are used, the waveguide is said to be periodically loaded. In an infinitely long waveguide, if the reflections from the obstacles add in phase, complete reflection of a wave will eventually occur. Thus, the wave is "stopped" and, in a frequency range where complete reflection occurs, a "stopband" is produced.

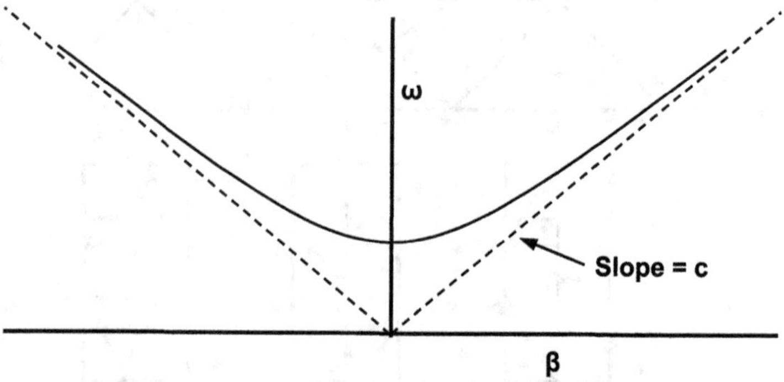

Figure 13-14 The ω-β diagram for waves propagating in either direction in a waveguide.

An example of a periodically loaded waveguide is shown in Figure 13-15. The obstacles are metal vanes attached to the top and bottom of the waveguide. The impedance produced by these vanes is capacitive. The phase shift that occurs in the electric field reflected from each vane is 180°. At frequencies near ω_c, the phase velocity is high and the wavelength is much longer than the spacing between vanes, L. Wave reflections from vanes do not add in phase, so there is little effect on wave propagation. As frequency is increased, the wavelength decreases and approaches $2L$ so that the phase shift between vanes is 180°. Then reflections from the vanes add in phase, as is illustrated in Figure 13-16.

In Figure 13-16(a), the sinusoid indicates the amplitude of the electric field in the waveguide. The forward and reflected waves at each of the vanes are indicated by vertical arrows. F_a, F_b, and F_c indicate the forward waves and R_a, R_b, and R_c indicate the reflected waves at vanes 1, 2, and 3.

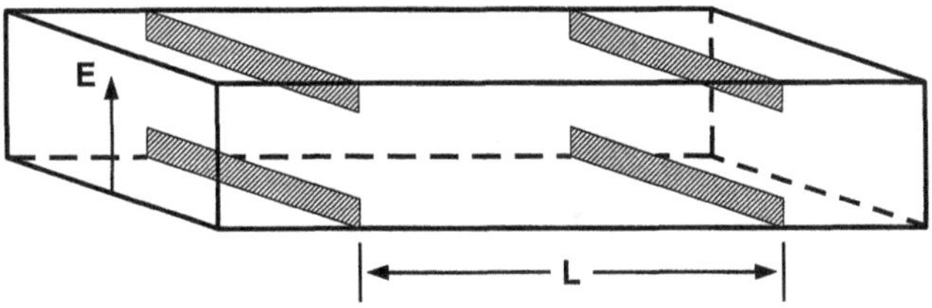

Figure 13-15 Use of vanes to periodically load waveguide.

For Figure 13-16(b), it is assumed that the forward wave has propagated a distance $\lambda/2 = L$ (π rad) to the right. At the same time reflections from the vanes in Figure 13-16(a) have propagated π rad to the left. Thus, for example, R_a arrives at vane 2 at the same time that F_c does and adds in phase with the reflection, R_c, from vane 2. Eventually, in an infinitely long waveguide, complete reflection of the forward wave occurs. There is no propagation so the group velocity goes to zero. The resulting effect on the ω-β diagram is shown in Figure 13-17. Here, instead of plotting β in radians/meter, the horizontal axis is βL, which is simply the phase shift in radians from vane to vane.

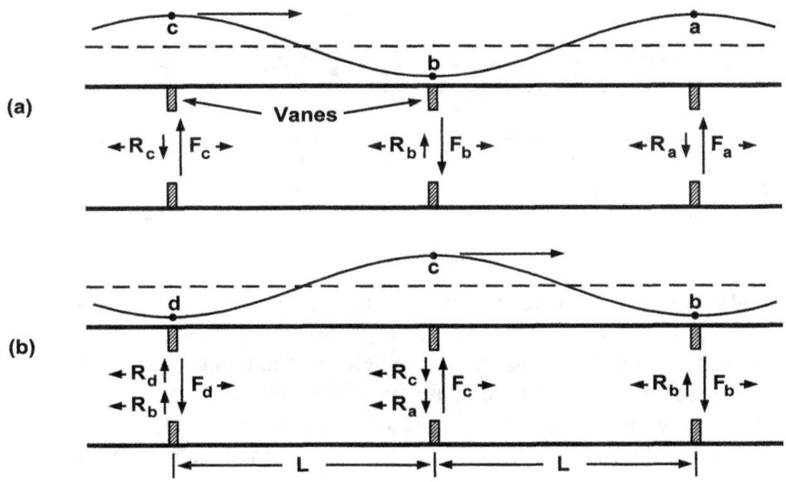

Figure 13-16 (a, b) Forward (F) and reflected (R) waves for $\beta L = \pi$.

The next problem to be considered is the shape of the ω-β characteristic for values of $\beta L > \pi$. Certainly, the curve does not stop at $\beta L = \pi$. Consider, first, a value of βL somewhat less than π. For this case, the wave pattern is as indicated in Figure 13-18(a). The phase velocity is higher than at $\beta L = \pi$ and the wavelength is longer.

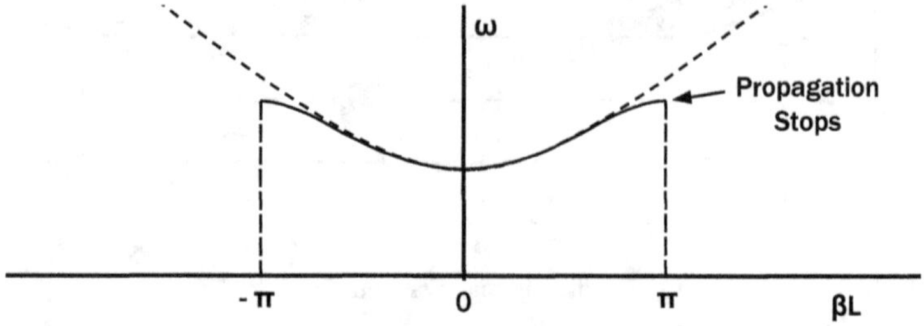

Figure 13-17 Effect of periodically loaded waveguide on an ω-β diagram.

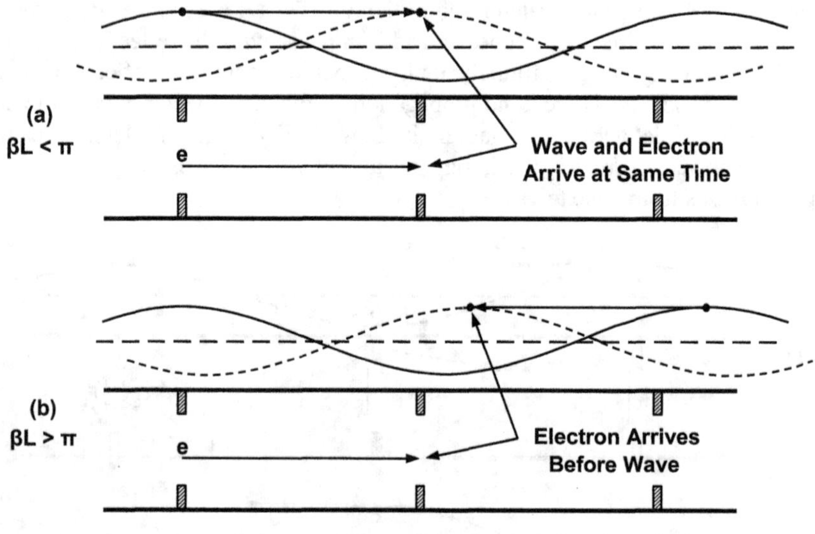

Figure 13-18 (a) Forward and (b) backward waves for βL near π.

It is useful, at this point, to consider an electron moving along the axis of the guide. If the electron moves at the phase velocity of the wave, then, of course, the electron will move from vane to vane in the same time that it takes the wave crest to move from vane to vane. The electron is said to maintain synchronism with the wave.

To consider a value of $\beta L > \pi$, assume, as indicated in Figure 13-18(b), that a wave with the same frequency as that for Figure 13-18(a) is traveling in the backward direction. The wavelength and phase velocity (in the backward direction) are the same as those for the forward wave in Figure 13-18(a). An electron moving to the right with the same velocity as in Figure 13-18(a) will arrive at the next vane before the next backward-moving wave crest does. The velocity of the electron must be reduced for the electron to arrive at the next vane when the next backward-moving wave crest does (for the electron to be synchronous with the backward wave). Thus, the phase velocity of the wave as seen by the electron in Figure 13-18(b) is lower than the phase velocity of the forward wave in Figure 13-18(a).

As a result of these considerations, it is seen that as βL increases above π, a backward wave exists and the ω-β curve must turn downward as shown in Figure 13-19. In fact, for values of βL in the range $\pi < \beta L < 2\pi$, the wave is a backward wave and the ω-β characteristic is the same as for $-\pi < \beta L < 0$.

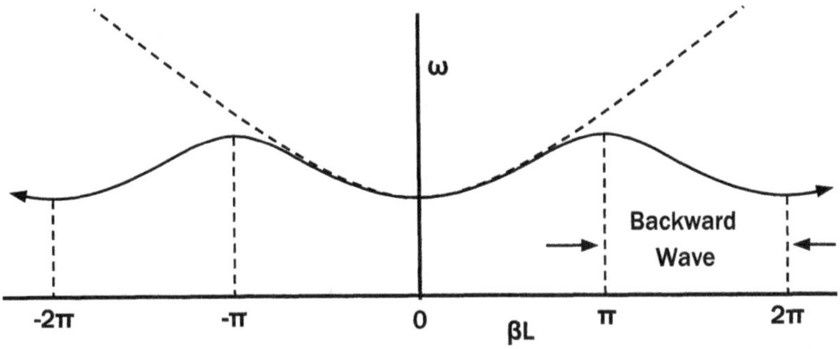

Figure 13-19 Extension of ω-β curve to $\pi < \beta L < 2\pi$.

As βL approaches 2π, the backward wave approaches the cutoff condition with wavelength and phase velocity (in the reverse direction) approaching infinity. For an electron velocity (and phase velocity in the forward direction) such that $\beta L = 2\pi$, the electron moves from vane to vane in synchronism with wave crests of the backward wave that are an infinite distance apart but moving with infinite velocity.

Finally, as βL increases to values greater than 2π, a forward wave again exists and the ω-β diagram turns upward again. The shape of the curve from $\beta L = 2\pi$ to $\beta L = 3\pi$ is the same as for $0 \leq \beta L \leq \pi$. The part of the ω-β diagram shown in Figure 13-19 is referred to as a *passband* because waves can propagate. This band will be of primary interest when coupled-cavity TWTs are considered.

Before coupled-cavity structures can be considered (in Chapter 15), it should be noted that there are other passbands for periodically loaded waveguides. There are bands of frequencies in which reflections from the vanes do not add in phase and wave propagation takes place. However, whenever the frequency approaches a condition where the distance between vanes is equal to several half wavelengths, a cutoff condition occurs and this places a limit on the upper or lower frequency of the band. As a result, the ω-β diagram contains an infinite set of passbands, four of which are shown in Figure 13-20.

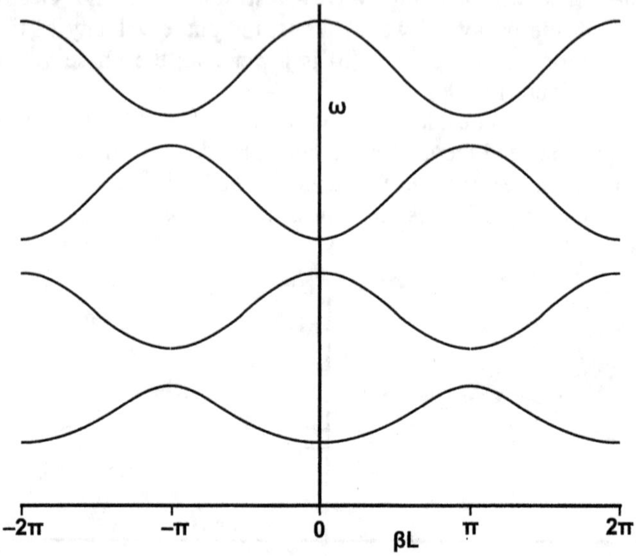

Figure 13-20 The ω-β diagram for a vane-loaded waveguide.

Chapter 14

Helix TWTs

The TWT invented by Lindenblad used a helix as the slow wave circuit. During the formative years of TWT development, hundreds of other propagation structures were investigated, but the helix remains as the circuit of choice when a broad bandwidth is required. Major applications for helix TWTs include radar, communications, and electronic countermeasures (ECM). The purpose of this chapter is to describe the characteristics of helix TWTs that have made them the mainstay of the TWT industry.

14.1 BANDWIDTH

The extremely broad bandwidth of the helix is the single most important property of this circuit. No other circuit has ever rivaled the helix in bandwidth and most do not even come close. Helix TWTs can now be made with a bandwidth of over 2 octaves, as shown in Figure 14-1.

The broad bandwidth of the helix results from its low dispersion. The helical circuit can be thought of as being derived from a single-wire transmission line for which the dispersion is zero. Shown in Figure 14-2(a) is a helix and in Figure 14-2(b) a view of the helix when it has been cut at the points marked by x and unrolled. A wave propagating along a wire with the speed of light, *c*, advances in the axial direction with phase velocity

$$v_p \approx c \sin \psi \tag{14-1}$$

The angle ψ is the helix pitch angle given by

$$\cot \psi = \frac{2\pi a}{p} \tag{14-2}$$

where a is the radius and p is the pitch of the helix.

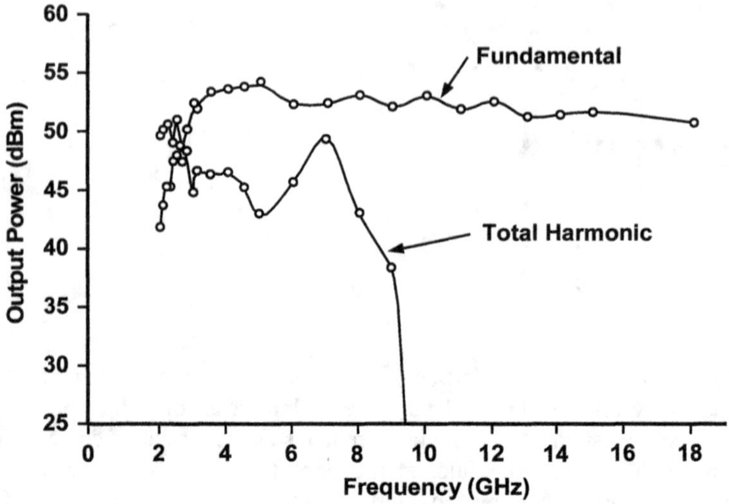

Figure 14-1 Saturated output power versus frequency for the Varian VTU-6292A1 helix TWT. (Courtesy of Varian Associates.)

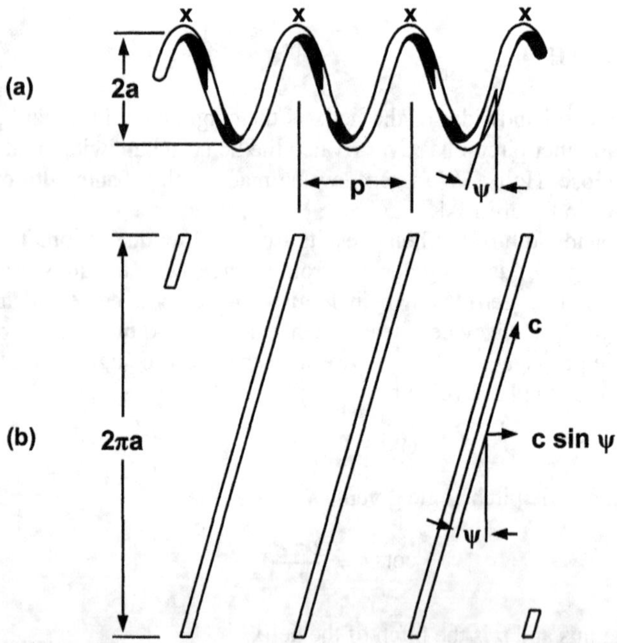

Figure 14-2 (a) Helix and (b) view of helix cut at each x and unrolled.

The ideal ω-β diagram for the helix is shown in Figure 14-3. The phase velocity, v_p, is represented by a straight line through the origin. Thus, in this case, the circuit has no dispersion. The dashed line represents the beam velocity, u_o, and has a slightly greater slope than the phase velocity line. This is so that the slow space charge wave (the electron bunches), which has a velocity slower than the electron beam, is synchronous with the phase velocity.

In theory, a circuit with the ω-β characteristic in Figure 14-3 could be used over an almost unlimited bandwidth. In practice, bandwidth is limited by several factors. First of all, for a given circuit, as the frequency of operation is decreased, the length of the circuit, expressed in wavelengths, decreases. Since gain is proportional to circuit length in wavelengths, gain decreases.

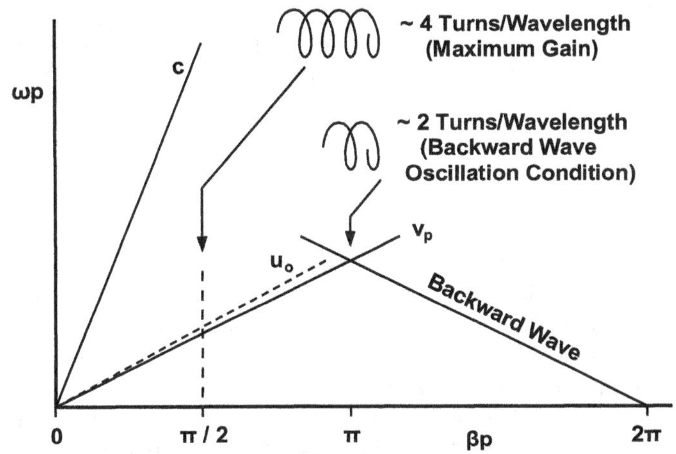

Figure 14-3 ω-β diagram for a helix with no dispersion.

As the frequency of operation is increased, the electric field pattern on the helix shrinks as is shown in Figure 14-4. As a result, the interaction of the field with the electron beam decreases and gain decreases. Maximum gain occurs at about four turns of the helix per wavelength (or $\pi/2$ radians per turn).

Another factor that limits bandwidth is dispersion. In reality, a helix with its supporting structure does have dispersion; this is discussed in Section 14.1.1.

A very serious factor that limits bandwidth and also the power capability of the helix is the backward wave oscillations (BWO) that occur at $\beta p = \pi$ radians. The reasons for these oscillations and techniques for controlling them are discussed later in this chapter.

If no techniques are used to control dispersion and oscillations, the bandwidth of the helix TWT is limited to about 1 octave at most. With dispersion control and BWO suppression, bandwidth can be increased to 2 or more octaves.

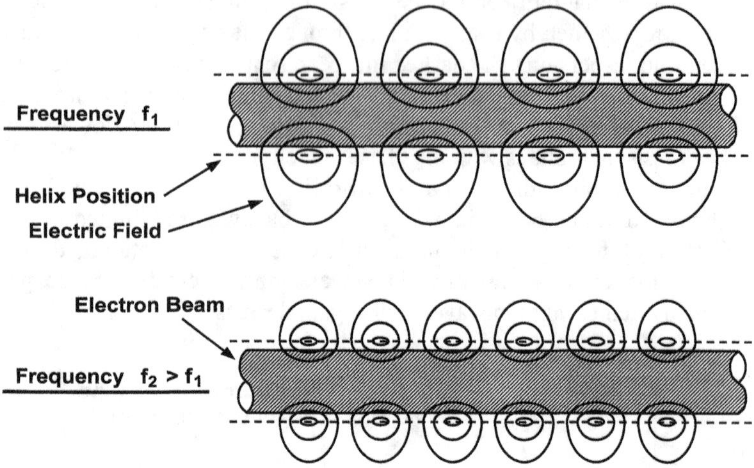

Figure 14-4 Change in the electric field that occurs as frequency is increased.

14.1.1 Dispersion

Unless corrective actions are taken, a helix has dispersion because phase velocity varies as shown in Figures 14-5 and 14-6. When compared with a coaxial line (with no dispersion), from which the helix can be considered to be derived, the helix has electric and magnetic coupling from turn to turn that is absent in the coax. This is one reason for the dispersion.

Another factor to consider is that, at high frequencies (and a relatively small number of turns per wavelength), the electric and magnetic fields are concentrated between turns of the helix. As the frequency is reduced and the number of turns per wavelength is increased, the fields extend farther from the helix, as is indicated in Figure 14-7. Therefore, a metal shell (barrel) surrounding the helix will intercept a greater portion of the fields at low frequencies than at high frequencies. (Note in Figure 14-7 that only the position of the barrel is shown. The distortion of the field patterns that the barrel produces is not shown.) As the barrel is brought closer to the helix, the effect increases, and the wave velocity on the helix decreases. The effect on velocity is greater at low frequencies than at high frequencies.

The helix is supported by an insulating structure that often takes the form of three equally spaced dielectric rods as indicated in Figure 14-5. These rods and the ways in which they are incorporated in a TWT are discussed in a following section. The effect of the dielectric material of the rods is also evident in Figure 14-5 [1]. Wave velocity is inversely proportional to the square root of the dielectric constant. Thus, when dielectric material is added in the region surrounding the helix, wave velocity decreases.

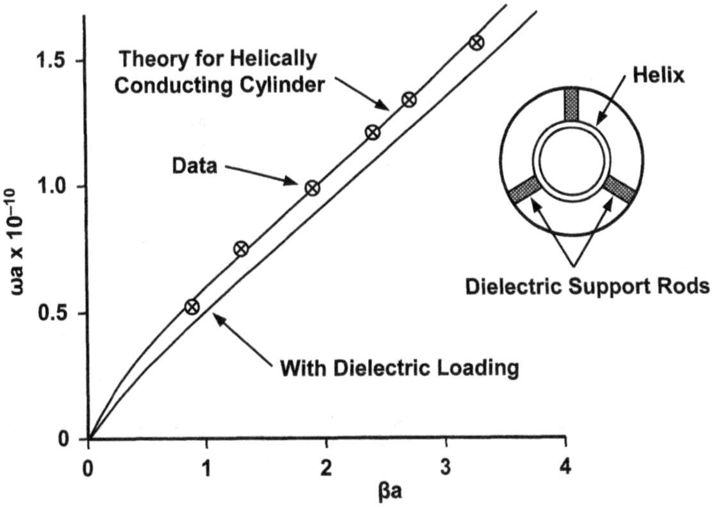

Figure 14-5 The ω-β diagram for a wire helix with a 10° pitch angle. (Adapted from: *Power Travelling Wave Tubes* by J. F. Gittins, published 1965 by American Elsevier, Inc.)

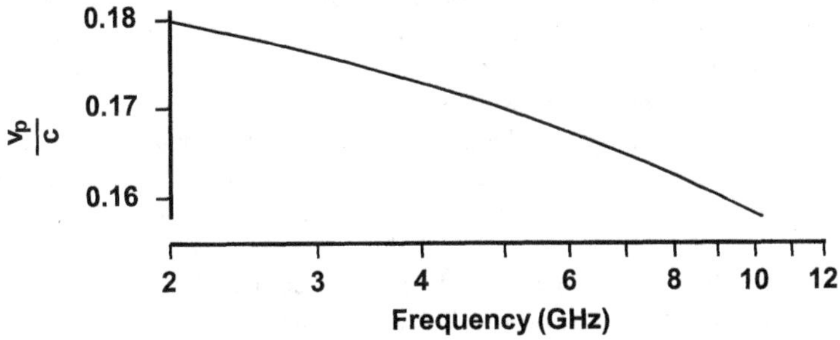

Figure 14-6 Variation of phase velocity with frequency for a circuit with no dispersion control.

14.1.2 Dispersion Control

Because the increase in phase velocity at low frequencies can be partially offset by bringing the barrel closer to the helix, dispersion is reduced. Unfortunately, there is an accompanying decrease in circuit interaction impedance that is undesirable because gain and efficiency are reduced. The decrease in impedance results from the increase in capacitive loading and inductive coupling of the helix currents to the barrel and the flow of current in the circumferential direction in the barrel.

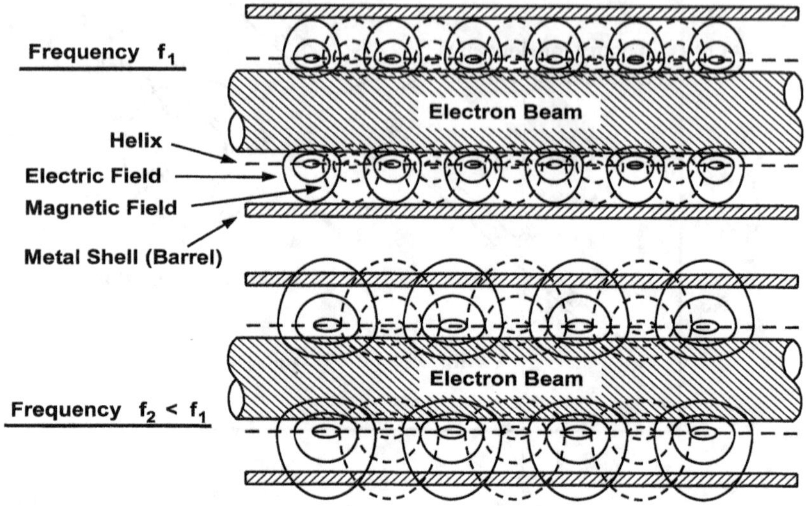

Figure 14-7 Variation of the electric and magnetic field patterns with frequency.

Ideally, if a metallic shell that conducted only in the axial direction could be placed around the helix, there would be no inductive coupling and no circumferential currents could flow in the shell. Still, the shell would have a greater effect on the electric field at low frequencies than at high frequencies so control of dispersion would be possible.

In practice, an axially conducting shell is approximated by one of the techniques illustrated in Figure 14-8 [2]. One of the four configurations shown (labeled unloaded) is used for comparison purposes and has only the outer metal shell, which supports the helix through dielectric support rods. In the other three configurations, either solid, vane-shaped, or T-shaped metallic elements, which are frequently referred to as anisotropic loading elements, are used to provide the desired loading.

Experiments have been performed to provide a basis for the choice of the best loading technique to use for controlling dispersion. Shown in Figure 14-9 are phase velocity and Pierce's velocity parameter as functions of frequency for three of the configurations in Figure 14-8. Tests were performed with two T-shaped elements having different spacings from the helix. The results were obtained in tubes containing these configurations, which were nearly identical in all other respects.

With tube number 1, phase velocity increases and b decreases as frequency is reduced. In tube number 2, the vanes produced a virtually flat phase-velocity characteristic over nearly the entire frequency range. In tubes number 3 and 4, the T-shaped segments produced negative dispersion characteristics. That is, the phase velocity decreased at low frequencies.

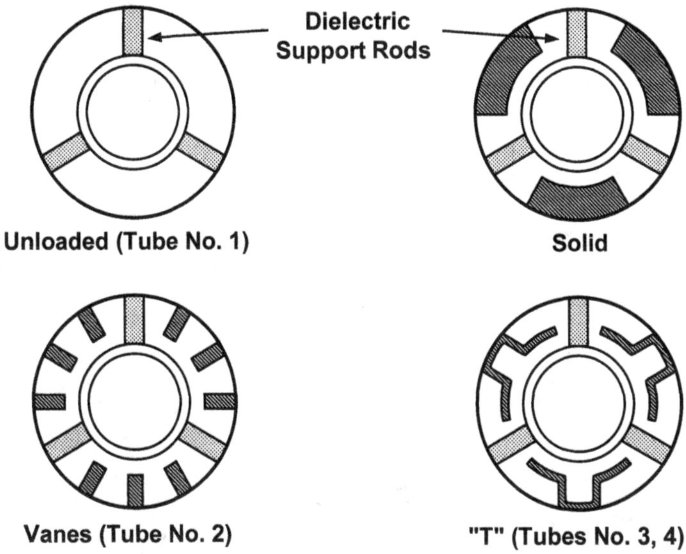

Figure 14-8 Techniques for dispersion control by approximating an axially conducting shell. (From: J. L. Putz and M. J. Cascone, 1979 IEDM. © 1979 IEEE.)

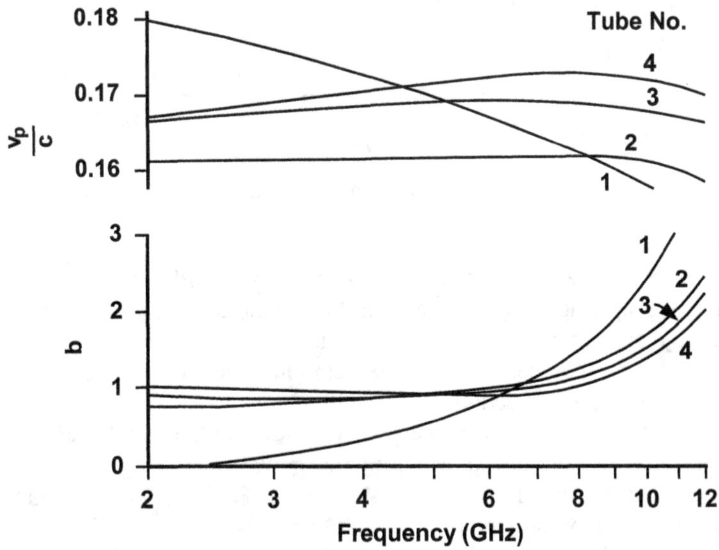

Figure 14-9 Normalized phase velocity and Pierce's velocity parameter as functions of frequency. (From: J. L. Putz and M. J. Cascone, 1979 IEDM. © 1979 IEEE.)

The small-signal gain characteristics of the four tubes tested are shown in Figure 14-10. The maximum gain for tube number 1 occurred in the 6- to 7-GHz range. This frequency corresponds to a wavelength such that there are about four helix turns per wavelength. This is in agreement with the prediction of the small-signal theory of helix TWTs. At higher frequencies, gain decreases because the field lines are concentrated closer to the helix (this was shown in Figures 14-4 and 14-7) and the component available for interaction with the beam decreases. At lower frequencies, gain decreases because the length of the helix, in terms of the number of wavelengths, decreases. Of the four tubes tested, tube number 4, which had a closer helix-to-loading-element spacing than tube number 3, had the best small-signal bandwidth (in excess of 2 octaves).

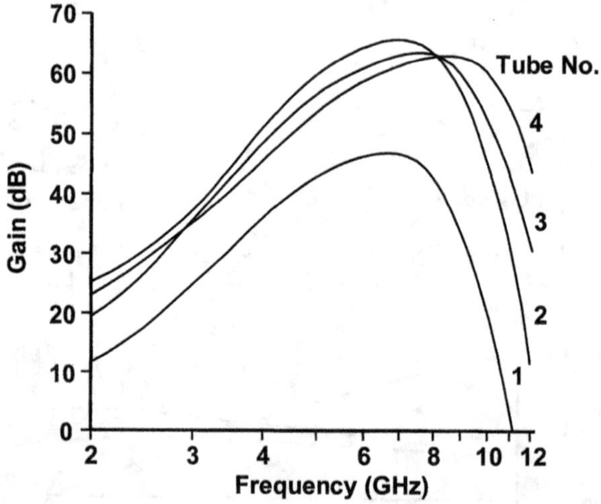

Figure 14-10 Small-signal gain for various loading configurations. (From: J. L. Putz and M. J. Cascone, 1979 IEDM. © 1979 IEEE.)

Shown in Figure 14-11 is a cutaway view of a PPM-focused helix TWT containing T-shaped loading elements. In this example, the tungsten tape helix is supported by anisotropic pyrolytic boron nitride (APBN) rods. The reason for using APBN and the techniques for assembling the helix and supported-rod structure are discussed in Section 14.3.2.

One of the problems with the very wide bandwidth that can be achieved with anisotropic loading is the high harmonic power that is generated at low frequencies. This is shown in Figure 14-1 and in Figure 14-12. The harmonics result from the velocity modulation and bunching processes that are highly nonlinear. At the low end of the operating band of a broadband helix TWT, the second- and even the third-harmonic signals generated in the beam are within the bandwidth of the tube and are amplified. At the lower end of the band, harmonic power can rival fundamental power.

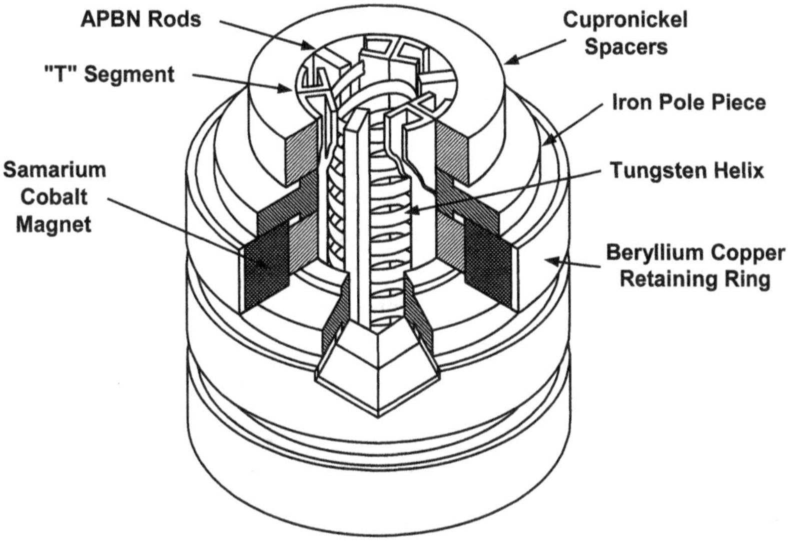

Figure 14-11 PPM-focused helix TWT with T-shaped loading elements. (Courtesy of Varian Associates.)

One way to reduce harmonic output power is with harmonic injection. An indication of the effectiveness of this technique is illustrated in Figure 14-12. When a second harmonic signal of the proper phase and amplitude is injected, along with the fundamental signal, the harmonic output power drops and the fundamental power increases. This occurs when the injected second-harmonic signal is 180° out of phase with the second harmonic signal generated by the beam-circuit interaction process. With proper design of the input circuit, the cancellation process can be made reasonably noncritical.

14.2 GAIN

At small signal levels, the gain produced by the interaction of the beam and the helix is proportional to CN where C is Pierce's gain parameter and N is the length of the helix in wavelengths (see Chapter 12). Clearly, for very long helices, gain can be enormous.

Here it is important to understand a distinction between a TWT and a klystron. In the klystron, there is no way for a wave to travel from the output to the input on the circuit so feedback by this mechanism cannot occur. However, in a TWT, one of the four waves on the circuit travels in the backward direction. This wave can carry power reflected from the load or output region backward to the input.

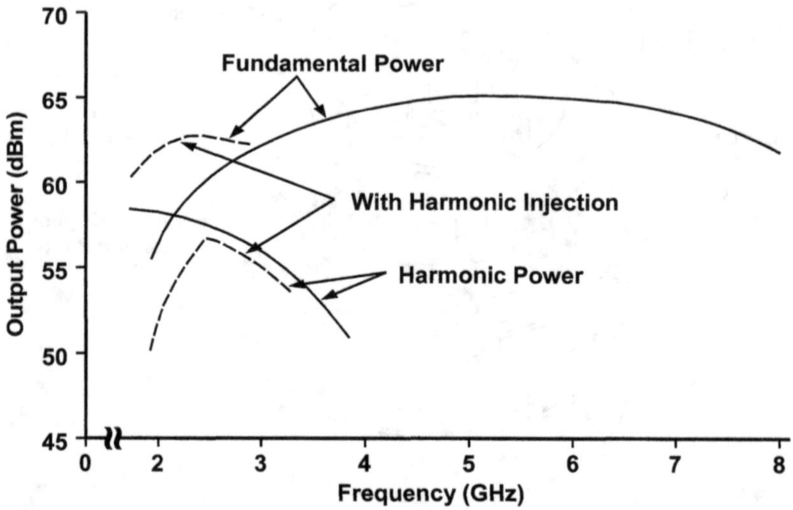

Figure 14-12 Effect of harmonic injection on output power for a wideband helix TWT. (From: Hughes Aircraft Co., *TWT and TWTA Handbook.*)

If there is a mismatch at the input, a portion of the signal will be reflected to provide a feedback signal. The phase of this signal depends on the frequency of operation and the path length through the tube. Depending on the phase, the signal can add to or subtract from the input signal to produce variations in gain ("gain ripple"). It is also possible for the feedback signal to cause oscillations.

To minimize feedback, the transitions from the transmission lines into and out of a TWT must be designed to provide very good impedance matches. Even if the output impedance match is perfect, the load impedance may not be matched to the output transmission line impedance and so reflections can occur. As a result, in addition to making sure that impedance matches are adequate, it is necessary to provide attenuation and/or severs within the TWT to prevent oscillations and to minimize gain ripple.

14.2.1 Transitions

The impedance of a helix varies with frequency, as is shown in Figure 14-13 [3]. In this example, the helix structure was designed for operation from 4.8 to 9.6 GHz (1 octave). At the center of the band, the impedance was about 50Ω. At the low end of the band, the impedance increased to over 80Ω, and at the high end it was about 35Ω. TWTs capable of operating over more than an octave have even greater impedance variations. As a result, the impedance transitions at the input and output of these TWTs must be able to match the wide variation in helix impedance.

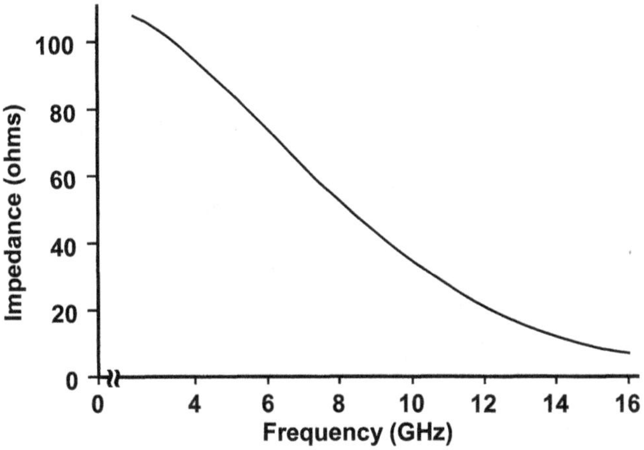

Figure 14-13 Helix impedance for an octave-bandwidth TWT. (From: A. S. Gilmour, Jr. et al., *IEEE Trans. Electron Devices*, October. 1979. © 1979 IEEE.)

Designs for broadband transitions are based on the idea illustrated in Figure 14-14. Here the tapered section of a coaxial line is used to match the low impedance of the coax on the left to the high impedance of the coax on the right. An excellent impedance match is realized if the tapered transition is long compared with the wavelength of the signal.

In a TWT, a very broadband impedance match to the helix could be made with the transition shown in Figure 14-15 [1]. A gradual change in geometry is made so that the helix slowly changes into a coaxial line. As long as the overall transition is long compared to the wavelength, there would be an excellent impedance match.

Unfortunately, transitions like that shown in Figure 14-15 have the very serious disadvantage of being long. If used, they would increase the length and weight of a TWT. In practice, transitions are normally made after the signal is removed from the helix. For example, the center conductor of the coaxial line can be tapered as shown in Figure 14-16.

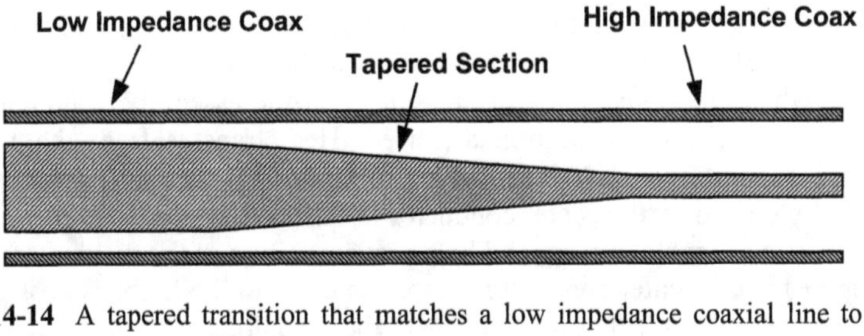

Figure 14-14 A tapered transition that matches a low impedance coaxial line to a high impedance line.

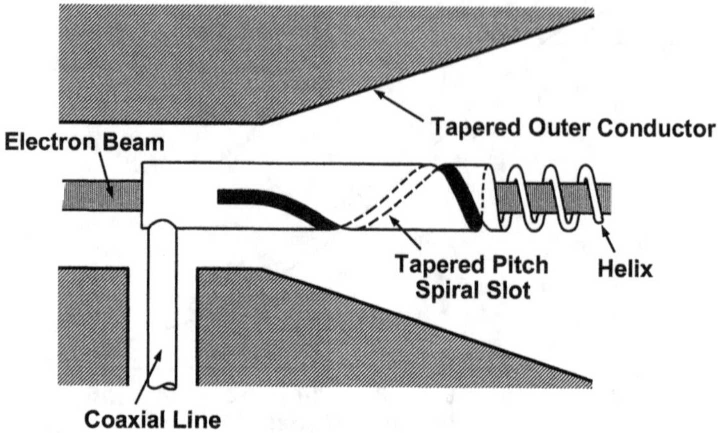

Figure 14-15 A very broadband transition from a helix to a coaxial line. (Adapted from: *Power Travelling Wave Tubes* by J. F. Gittins, published 1965 by American Elsevier, Inc.)

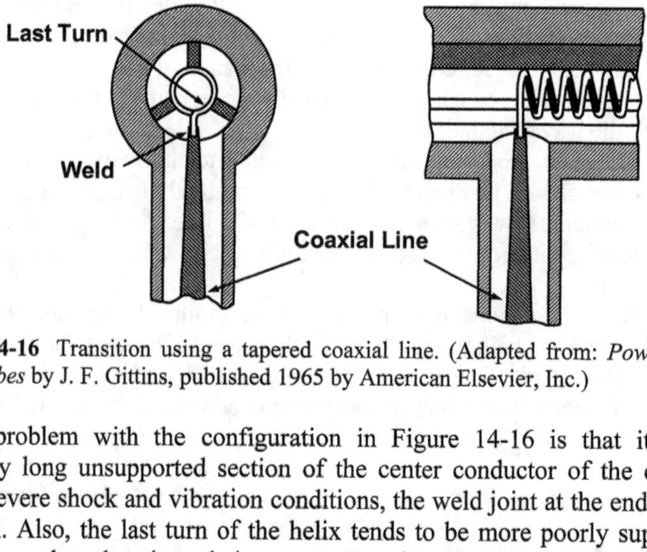

Figure 14-16 Transition using a tapered coaxial line. (Adapted from: *Power Travelling Wave Tubes* by J. F. Gittins, published 1965 by American Elsevier, Inc.)

A problem with the configuration in Figure 14-16 is that it requires a relatively long unsupported section of the center conductor of the coaxial line. Under severe shock and vibration conditions, the weld joint at the end of the helix may fail. Also, the last turn of the helix tends to be more poorly supported than other turns when the triangulation or pressure-insertion assembly techniques to be described in the following section are used. Heat transfer from the last turn through the support rods may be relatively poor so that the last turn must be at least partially cooled by the center conductor of the coaxial line.

To provide adequate cooling and help prevent failure of the weld, there should be a support for the center conductor located as close to the helix as possible. For example, in the output window of a TWT, the insulating support for the center conductor is usually only a fraction of a wavelength from the connection to the helix. The material of this support is usually beryllium oxide, which is a very good thermal conductor. The beryllium oxide is brazed to the center conductor and to the outer conductor and serves as part of the vacuum envelope.

Also, it should be pointed out that Teledyne uses a flexible strap to connect the end of the helix to the center conductor. This prevents movements of the center conductor (because of thermal expansion and contraction) from being transmitted to the helix.

14.2.2 Attenuators and Severs

The sketch in Figure 14-17 is meant to illustrate a TWT having no attenuators or severs. The limiting gain of this device may be readily estimated. If the reflection coefficients at the output and input terminals of the TWT are, respectively, ρ_o and ρ_i, then oscillations may occur when

$$G - L - \rho_o - \rho_i > 0 \tag{14-3}$$

where G (gain), L (loss of the circuit), and the reflection coefficients are expressed in decibels.

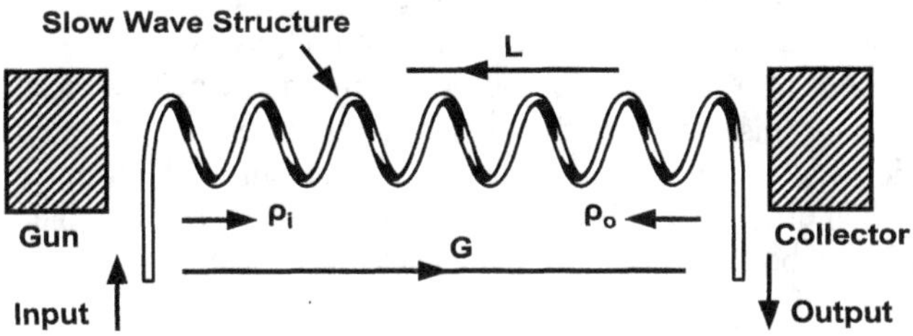

Figure 14-17 Quantities used to determine the oscillation condition. (Adapted from: *Power Travelling Wave Tubes* by J. F. Gittins, published 1965 by American Elsevier, Inc.)

Although the tube may be very well matched to the signal source and to the load, there will normally be reflections at input and output terminals because of the difficulty in making impedance transitions from the RF structure. As a result, it may be difficult to make ρ_o and ρ_i much greater than 10 dB (10% reflections). If the cold loss of the circuit is on the order of 6 dB, then the limiting gain is on the order of 26 dB.

To prevent large variations in gain with frequency, the actual gain cannot be as high as 26 dB and, in fact, is sometimes limited to about 20 dB. To permit increased gain, attenuators and severs are used in TWTs.

A common form of attenuation is shown in Figure 14-18. Portions of the support rods for a helix are coated with a lossy material. The thickness of the film is tapered to zero at the ends of the attenuation region to obtain a good impedance match over a wide range of frequencies.

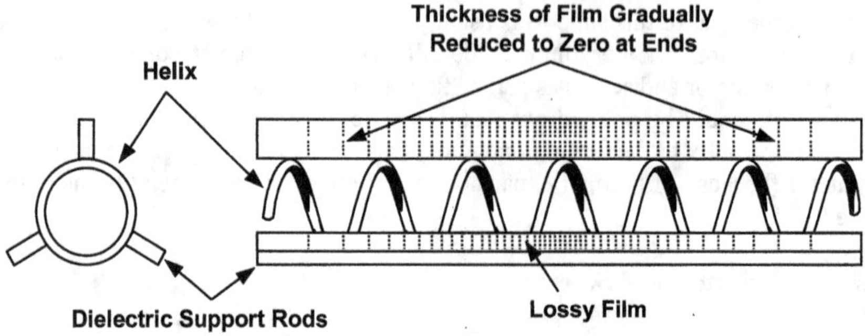

Figure 14-18 Lossy film attenuator for use with a helix.

Forward waves as well as backward waves are attenuated by the lossy film. However, we must remember that Pierce showed that the reduction in gain is only about one-third of the attenuation in the backward wave. Also, the loss in gain in the forward direction may be compensated for by an increase in length of the interaction region.

Unfortunately, if a good impedance match is to be realized, the attenuator must extend for a considerable length along the RF structure and, where there is significant attenuation, the amplitude of the circuit wave is small. As a result, the circuit fields control of the bunching process and the velocity spread in the beam increases. The increase in velocity spread results in less effective bunching after the attenuation and a reduction in efficiency. As a result, at high-power levels and, especially, when efficiency is important, one or more severs like that illustrated in Figure 14-19 may be used to suppress the backward wave.

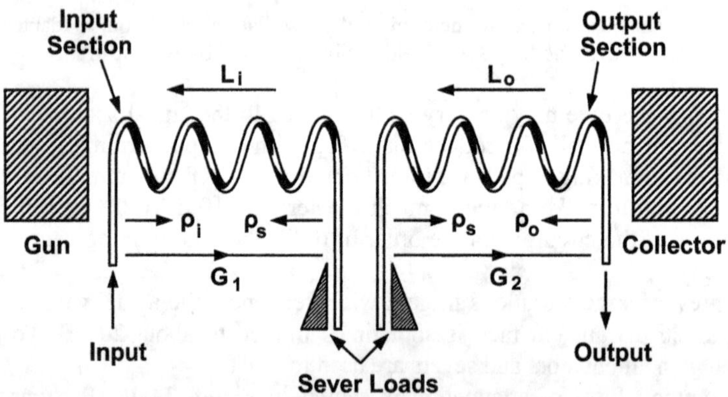

Figure 14-19 TWT containing a two-section severed helix. (Adapted from: *Power Travelling Wave Tubes* by J. F. Gittins, published 1965 by American Elsevier, Inc.)

The sever prevents reflections from the load and the output terminal from reaching the input terminal. Although the forward growing wave on the input section of the helix is lost at the sever, current and velocity modulation remain on the electron beam to carry the signal across the sever region for further amplification in the output section.

As the beam drifts across the sever region, bunching degrades and affects the final efficiency of the tube. As a result, the sever region should be as short as possible. Also, the amount of degradation depends on the intensity of bunching in the input section of the slow wave structure. Thus, for maximum efficiency, the gain of the input section should be relatively low and the gain of the output section should be as high as possible.

Just as there is a limit of about 20 dB to the practical gain of a tube without a sever, there is a similar limit to the gain of each section of a tube with severs. Referring again to Figure 14-19, the reflection coefficients at the output, ρ_o, at the sever, ρ_s, and at the input, ρ_i, along with the circuit attenuations L_o and L_i, must be taken into consideration in determining gain levels. Because it is usually possible to keep the reflection coefficients at the sever terminations less than those for the input and output, the gain per section of a severed tube can be higher than that for a single-section tube. On the other hand, there is some signal loss in the sever region so that an overall gain from a two-section tube of about 40 dB is practical. With additional severs, higher gains are possible

A cross-sectional view of a broadband TWT containing an impedance transition, attenuators, and a sever is shown in Figure 14-20. The impedance transition at the output is made by varying the height of the waveguide in steps. The attenuators are made by depositing lossy material on the dielectric support rods for the helix. The sever occurs where the circuit is physically divided into two sections. Lossy terminations are used to absorb signals reflected by the sever.

14.3 POWER

Although a peak power of 10 kW has been demonstrated for a helix TWT [4], peak power is usually limited to a few kilowatts. At the end of this chapter, TWTs with ring-bar and ring-loop slow wave structures are discussed. These structures are said to be derived from helices. Ring-bar and ring-loop TWTs are capable of peak powers of tens of kilowatts.

The average power of the helix TWT depends on the frequency of operation because size scales inversely with frequency. At low frequencies, average powers in the kilowatt range are possible. In the millimeter range, average power is limited to tens of watts.

In this section, the reasons for the peak and average power limitations are discussed.

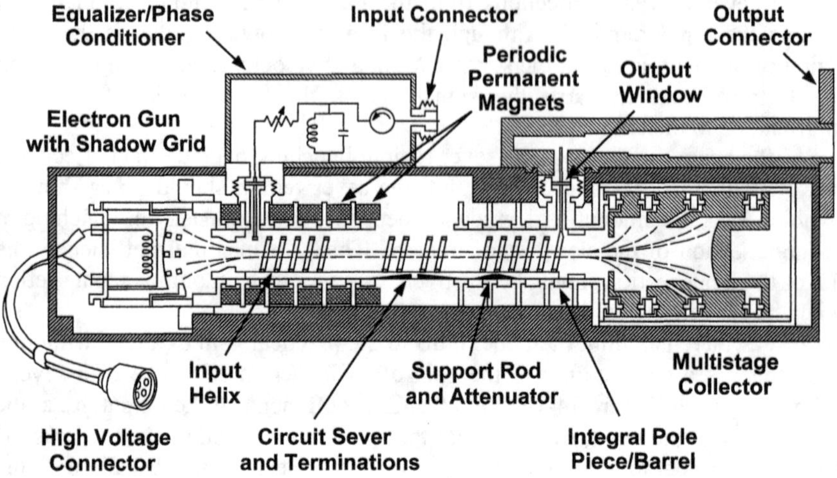

Figure 14-20 Helix TWT for electronic countermeasures use. (Courtesy of Varian Associates.)

14.3.1 Peak Power

A major problem that limits the peak power of helix TWTs is backward wave oscillations (BWO). When these oscillations occur, there is a reduction in power at the desired frequency of operation and so the oscillations produce in-band power holes. In addition, the BWO signal may induce oscillations or may intermodulate with the main signal to produce spurious signals inside the operating band. Before discussing the reasons for the limits on peak power, it is important to have an understanding of backward wave oscillations.

The BWO problem can be explained with the aid of Figure 14-21. At a frequency where the electron transit angle from turn to turn is 180°, there are two turns per wavelength or 180° per turn and a backward wave interaction can occur. Consider electrons near the helix moving to the right (designated by e) and a backward wave moving to the left. Electrons are initially accelerated as indicated by the arrow representing force (F) in Figure 14-21(a). These electrons experience essentially no force 180° later (Figure 14-21(b)) because the wave has moved to the left 180° and the nulls in the electric field are on the same side of the helix as the electrons. When the electrons have moved another 180° to the right while the wave has moved another 180° to the left (Figure 14-21(c)), acceleration again occurs.

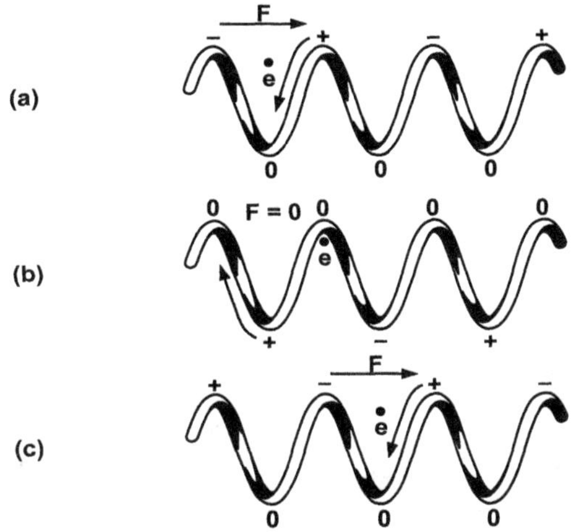

Figure 14-21 (a–c) Backward wave interaction on a helix at two turns per wavelength.

Similarly, and not shown in Figure 14-21, electrons that are initially decelerated experience no forces 180° later and then are decelerated when they have moved another 180°. Thus, there is bunching even though the wave on the helix moves in the opposite direction from the electrons, so backward wave interaction can take place. If the backward wave interaction is sufficiently intense, oscillations occur.

In Figure 14-21, to illustrate the BWO phenomenon, an electron was chosen that is moving close to the helix. However, notice that an electron traveling along the axis doesn't experience a net accelerating force (or decelerating force) as it moves from turn to turn because it experiences alternating fields.

From the foregoing discussion, we see that whether or not a tube oscillates in the BWO mode depends primarily on beam current and how close the beam is to the helix. An illustration of this dependency is shown in Figure 14-22 [5].

Now that the BWO problem has been discussed, the reasons why this problem limits peak power can be explained. For maximum gain, operation of the helix should be around four turns per wavelength, so the helix diameter is restricted. The beam diameter is further restricted to avoid oscillations and to limit beam interception. Beam current is limited to avoid oscillations. Current is also limited by the need to avoid an excessive current density and excessive space charge forces that limit the effectiveness of the bunching process and the efficiency of operation (see Chapter 12).

Since current is limited, the only way to achieve high-power operation is to operate at high voltages. However, as beam voltage is increased, the phase velocity of the helix must be increased, which means that the helix pitch must be increased. As pitch is increased, two things happen.

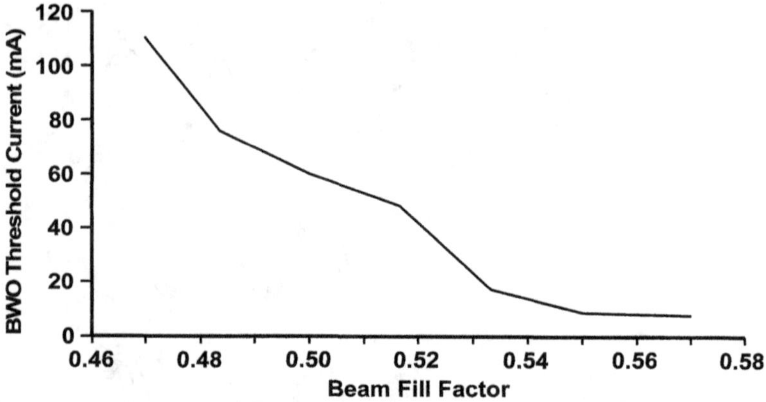

Figure 14-22 Dependence of start oscillation current on beam size. (From: P. Birtel et al., 2006 IEEE IVEC. © 2006 IEEE.)

First, to maintain about four turns per wavelength, the helix diameter must be decreased, so, for a given beam size, the possibility of oscillations is increased. This forces a reduction in beam size and current and a reduction in power capability. Second, as pitch is increased, the electric fields interacting with the beam are decreased, making the tube more prone to oscillations.

Before discussing techniques for suppressing BWO, it is interesting to observe that, because the BWO frequency is such that the beam has a 180° transit angle from turn to turn, the BWO frequency is tunable with beam voltage. In fact, tubes designed for operation in the BWO mode can be voltage tuned over a reasonably broad range of frequencies. Shown in Figure 14-23 are the essential elements of a helix backward wave oscillator, which is simply called a BWO.

Most significant, perhaps, is the electron-gun assembly with a ring-shaped cathode for generating a hollow electron beam. The use of a hollow electron beam traveling close to the helix structure enhances the backward wave interaction.

Helix BWOs are low-power devices typically generating tens of milliwatts. Operating frequencies extend well into the millimeter frequency range. These devices have, for the most part, been replaced by solid-state signal sources.

The most common technique for suppressing backward wave oscillations is to introduce a change in the slow wave circuit that changes the phase velocity. This can be done by changing the diameter of the helix, the number of helix support rods, or, as shown in Figure 14-24, the pitch of the helix [6]. In this case, the relatively small pitch of the right-hand side of the helix is such that energy is transferred from the beam to the backward wave and oscillations occur. As the backward wave travels to the left, it encounters a helix section with a larger pitch. Thus, the wave travels faster and, if the pitch is correct, the wave travels at the velocity of the fast space charge wave. As a result, energy is transferred from the backward wave back into the beam and the backward wave is attenuated.

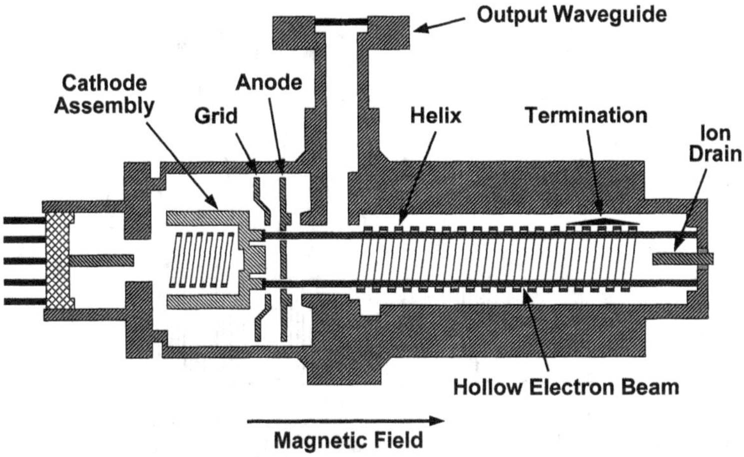

Figure 14-23 Helix-type backward wave oscillator. (Courtesy of Varian Associates.)

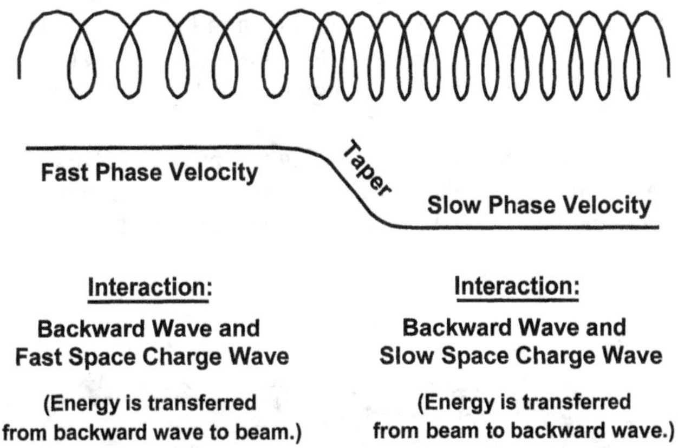

Figure 14-24 Illustration of the use of a pitch change to suppress backward wave oscillations. (From: A. Scott and M. J. Cascone, *Technical Digest*, 1978 IEDM. © 1978 IEEE.)

Another method for suppressing backward wave oscillations is the introduction of frequency-sensitive attenuation into the helix circuit. A technique developed by Varian Associates (now CPI) for producing frequency-sensitive loss is illustrated in Figure 14-25 [6]. Varian refers to this loss as resonant loss. Meander-line circuits are placed on the support rods in the regions between turns of the helix. The circuits are fabricated using standard microwave integrated circuit technology. The total length of each line is equal to half the wavelength at the BWO frequency so that the lowest resonant frequency is the BWO frequency.

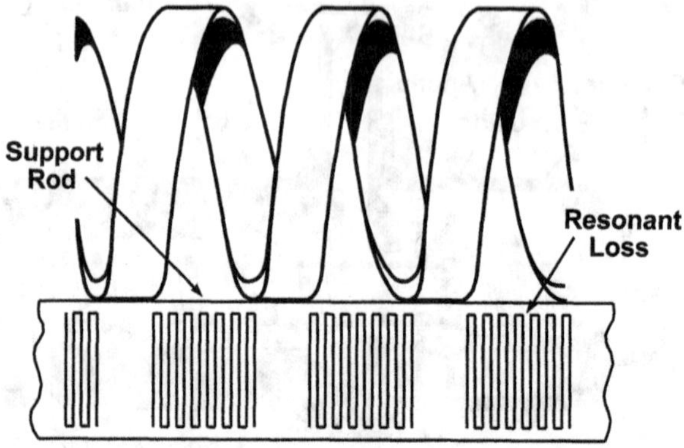

Figure 14-25 Use of resonant loss to produce frequency sensitive attenuation. (From: A. Scott and M. J. Cascone, *Technical Digest*, 1978 IEDM. © 1978 IEEE.)

The amount of loss that can be introduced with the resonant loss technique can be several tens of decibels. The power output for a 10-kW helix tube designed for operation in the 3.0- to 6.0-GHz range is shown in Figure 14-26 [4]. The BWO frequency of this tube was just above the top end of the band at about 8.0 GHz. Note the very sharp drop in power at the high end of the band where resonant loss becomes effective.

A problem that has been identified in using resonant loss is that a harmonic of the signal being amplified by the TWT may be at the frequency for which the resonant loss circuits have been designed. As a result, the resonant loss structures can be destroyed.

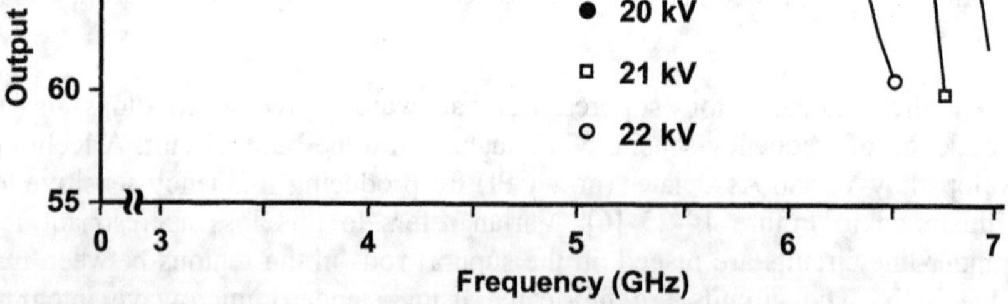

Figure 14-26 Saturated power output of 10-kW helix TWT with resonant loss. (From: Alan R. Jung, *Technical Digest*, 1978 IEDM. © 1978 IEEE.)

In Figure 14-27, estimates are shown of the peak output power that may be attainable using BWO suppression techniques. Note that peak power levels in excess of 100 kW are predicted to be possible in PPM-focused tubes using resonant loss.

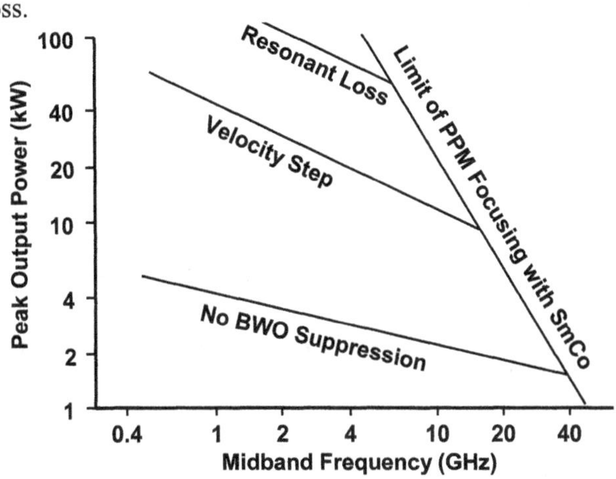

Figure 14-27 Obtainable peak power for helix TWTs with various oscillation-suppression techniques. (From: Alan R. Jung, *Technical Digest*, 1978 IEDM. © 1978 IEEE.)

14.3.2 Average Power

The helix of a TWT is a relatively delicate structure. As an example, a cross-section is shown in Figure 14-28 for the helix of the 10-kW TWT described above, which was designed to operate from 3.0 to 6.0 GHz. The helix is made of 0.010-inch by 0.075-inch tungsten tape. If the TWT is to be rugged and capable of operating at high average-power levels, this helix must be very well supported.

The use of metallic supports is out of the question because of their effect on the RF characteristics of the circuit. A dielectric cylinder completely surrounding the helix could conceivably provide excellent support and cooling. Unfortunately, the dielectric constants of suitable ceramic materials are high. As a result, their use in a cylinder would severely reduce the circuit interaction impedance of the helix. In fact, to minimize the loading effect of a ceramic support structure, it is necessary to minimize the amount of ceramic material used. As a result, thin, rectangular support rods like those shown in Figure 14-28 are often used. Even these load the helix to some extent, as is indicated by the data plotted in Figure 14-29.

Two dielectric materials are widely used for supporting the helices of TWTs: beryllium oxide (BeO) and anisotropic pyrolytic boron nitride (APBN). As shown in Figure 14-30, these two materials have relatively high thermal conductivity. It should be noted that diamond has an outstanding thermal conductivity and this is why there has been strong interest in using diamond for many years.

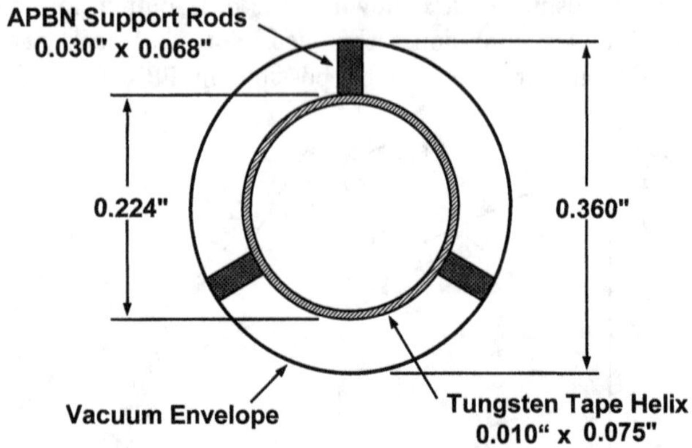

Figure 14-28 Cross-section of helix of a 10-kW, 3.0- to 6.0-GHz TWT. (From: Alan R. Jung, *Technical Digest*, 1978 IEDM. © 1978 IEEE.)

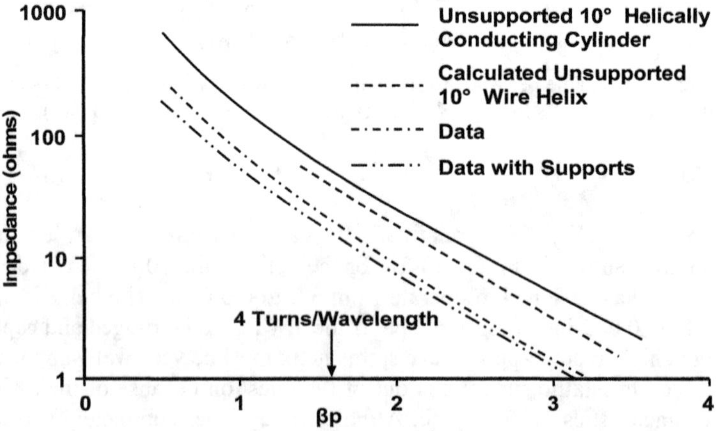

Figure 14-29 Coupled-interaction impedance of a helix with and without supports. (Adapted from: *Power Travelling Wave Tubes* by J. F. Gittins, published 1965 by American Elsevier, Inc.)

With the use of high thermal conductivity ceramics, the temperature drop across the support rods is not a severe problem in high-power helix TWTs. Instead, the major problems are the thermal resistances of the interfaces between the helix and the support rods and between the support rods and the barrel. In the example shown in Figure 14-31, the thermal drops across these interfaces account for 70% of the total temperature difference between the hottest point on the helix and the barrel. The temperature drops across the interfaces are extremely dependent on the pressure applied to the interface as is shown in Figure 14-32. Also, the surface finishes of the materials in contact are important.

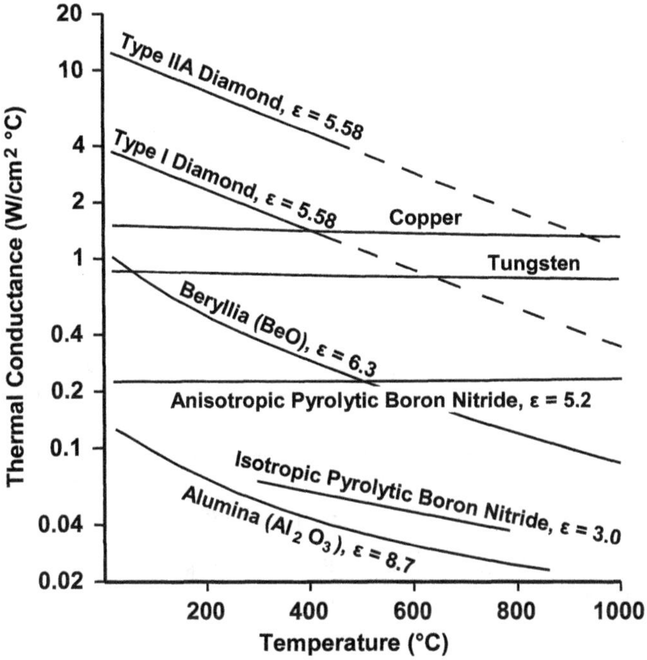

Figure 14-30 Thermal conductivities of some metals and ceramics. (From: A. E. Jacquez and A. Scott, *Final Report,* AFWAL-TR-81-1176, October 1981.)

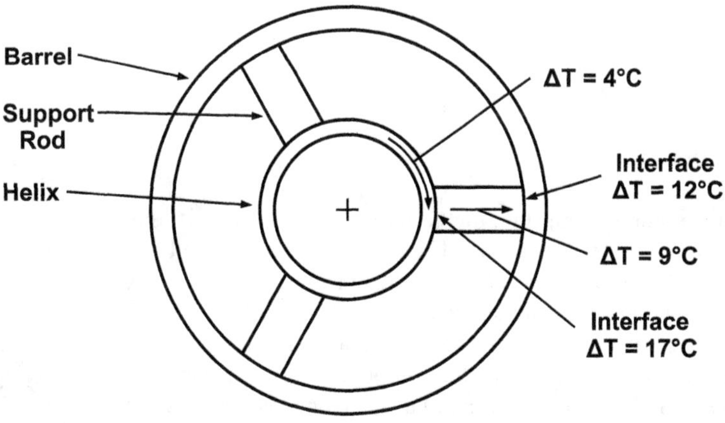

Figure 14-31 Temperature drops calculated for a tungsten helix and APBN rods dissipating 1 watt. (From: A. Scott and M. J. Cascone, *Technical Digest*, 1978 IEDM. © 1978 IEEE.)

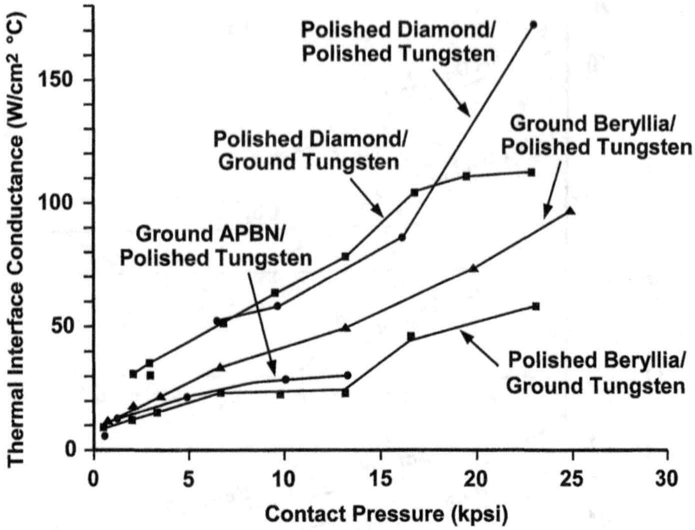

Figure 14-32 Thermal interface conductivities for several dielectrics in contact with tungsten. (From: A. E. Jacquez and A. Scott, *Final Report,* AFWAL-TR-81-1176, October 1981.)

Because of the large thermal drops at the support rod interfaces, the helix and support rod structures in high-power helix TWTs must be assembled using techniques that minimize the thermal interface resistances. Four of the techniques that are used are:

- Triangulation;
- Pressure or hot insertion;
- Wire wrapping;
- Brazing.

With triangulation, the metal barrel that surrounds the helix is slightly distorted into a triangular shape. The helix with three equally spaced support rods is then inserted into the barrel and the distorting forces on the barrel are removed. The higher the resulting forces are at the interfaces between the support rods and the helix or barrel, the lower the thermal resistances are. A limit is reached, however, when the forces are high enough to distort the helix excessively. For helix strength, the material normally used is tungsten or molybdenum.

It should be pointed out that with the triangulation technique the barrel is not a solid brazed structure (like that shown in Figure 14-11). Instead, it is a relatively thin metallic tube over which the magnetic pole pieces are slid after the helix has been inserted. With this technique, there is a relatively poor thermal interface between the barrel and the pole pieces. This problem is alleviated by making the pole pieces fit as tightly as possible on the barrel. Then, as the helix and barrel expand because of heating, the contact between the barrel and pole pieces improves so heat transfer improves.

With the insertion technique, the helix and rod assembly is inserted into a barrel structure like that shown in Figure 14-11. In some cases, the barrel is slightly expanded by heating before the helix and rod assembly is inserted. This is referred to as hot insertion. In other cases, the helix and rod assembly are machined with tolerances for an interference fit and then inserted under high pressure into the cold barrel. The temperatures given in Figure 14-31 are for a pressure-inserted structure. As with the triangulation technique, the helix materials used are tungsten or molybdenum.

With wire wrapping, a tungsten helix that may be copper plated is held in compression by support rods (usually four) that are wrapped with wire under high tension. This ensures that uniform pressure is applied to all portions of the helix. This technique simplifies fabrication and assembly of high-frequency tubes [7].

The most effective way of reducing the thermal interface resistances is by brazing. While brazing sounds attractive, it can be difficult to implement. If rods that have been metallized on the surfaces to be brazed are used, there are difficulties with thermal expansion problems. Also, the support rods must be brazed to every turn of the helix. There may be several tens of contact points of one support rod with a helix and several support rods (usually three). Every contact point of every rod must be brazed for proper protection of the helix. After brazing, the metallization of the support rods must be removed in the regions between turns of the helix.

With the brazing technique, a wider choice of helix materials is possible than with the triangulation or pressure-insertion techniques. In addition to tungsten or molybdenum, copper can be used. While tungsten and molybdenum can be copper-plated to reduce RF losses, only the use of pure copper throughout the helix leads to the maximum possible transfer of heat and the ultimate loss reduction. Finally, it should be pointed out that, in brazed structures, the helix is mechanically bonded to the dielectric support material. There is reduced probability, therefore, that helix failure will result from overheating of the helix and the resulting possible loss of thermal contact between the helix and the dielectric.

An alternative to the use of brazed rods is the block-supported structure in Figure 14-33 [8]. Individual ceramic blocks are brazed to the helix and, usually, to the barrel. The use of ceramic blocks eliminates the thermal expansion problems associated with rods.

Additional advantages of the block-supported structure are reduced dielectric loading (and higher interaction impedance) and a reduced effect of the dielectric material on phase velocity. As an example of the capability made possible by block-supporting, Thomson Tubes Electroniques (now Thales) produces a 1.5-kW average power, 6-GHz TWT for communications service.

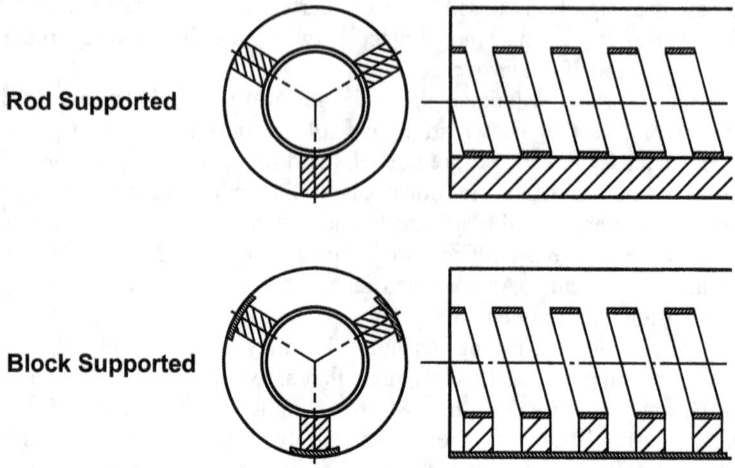

Figure 14-33 Comparison of rod-supported and block-supported helix structures. (From: Yukio Hiramatsu, *Technical Digest*, 1979 IEDM. © 1979 IEEE.)

Comparisons of the advantages of various techniques are often misleading because test conditions differ. Still, because of the continuing emphasis on higher powers at higher frequencies, it is important to compare the techniques. A comparison of the triangulation, pressure-insertion, and brazing techniques is given in Figure 14-34 [9]. The base-plate temperature for the test fixture in these tests was 100°C. Thus, the temperature rise of the brazed helix (which was copper) was about two-thirds of that for the insertion (shrunk) technique and about one half of that for the triangulation technique.

It is important to note that the results in Figure 14-34 were obtained by passing a dc current through the helices. RF losses and at least a small amount of additional heating are therefore not included. The additional heating may be larger in the metallized, brazed helix than in the other two.

In general, TWT dimensions scale with wavelength. At millimeter-wave frequencies, dimensions become extremely small and miniaturized helices and support rods are used. For example, rectangular support rods having cross-sectional dimensions of 0.010 inch by 0.020 inch, or smaller, are common. As helix assemblies become smaller, heat dissipation problems become more severe, and the thermal management techniques described in the previous sections become even more important than they are at lower frequencies.

The brazing technique, which offers excellent thermal capabilities, is very difficult to implement in the millimeter-wave range. As a result, at present, pressure insertion is widely used and average power levels are limited to about 100 watts.

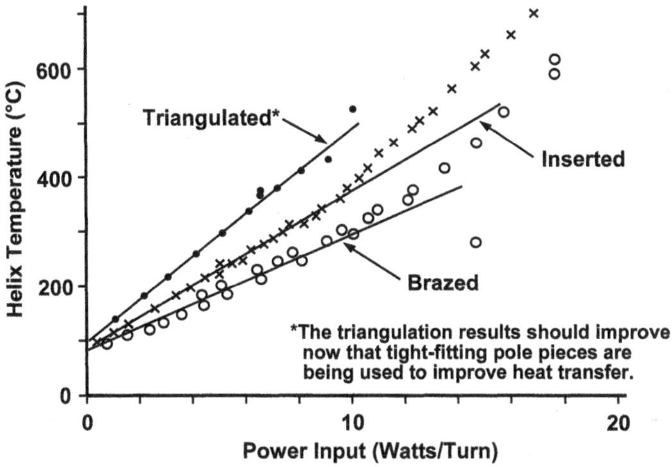

Figure 14-34 Comparison of thermal characteristics of triangulated, pressure-inserted (shrunk) and brazed helices. (From: T. Mosser et al., 1980 Microwave Power Tube Conference.)

The use of diamond may make possible significant improvements in average power capability. Type II-A diamond has a thermal conductivity several times that of copper. With diamond supports, it is predicted that average power levels as high as 200W may be achievable at frequencies as high as 50 GHz [10].

14.4 EFFICIENCY

In applications for TWTs where prime power is at a premium, efficient operation is essential. The efficiency of TWTs for space operation has increased steadily since their first use on satellites and, as is shown in Figure 14-35 for a Ku-band device, has passed the 70% mark [11]. Figure 14-36 shows 65% efficiency for an L-band device [12]. These are remarkable achievements because the basic electronic efficiency (the efficiency of conversion of beam power to RF power on the helix) for a typical helix TWT is on the order of 20% at best.

The impact of the high-efficiency devices in Figure 14-35 on the spacecraft power system is compared with solid-state power amplifiers (SSPAs) in Figure 14-37. For the same RF power output, the dc input required by the SSPA is nearly three times that of the TWTA (the TWT plus its power conditioning system). When radiation cooled TWTs are used, the waste heat that must be dissipated by the spacecraft is over an order of magnitude below that for SSPAs. When these power considerations are combined with the demonstrated superior reliability of TWTs compared with SSPAs [13], it is clear why TWTs are the amplifiers of choice for spacecraft applications.

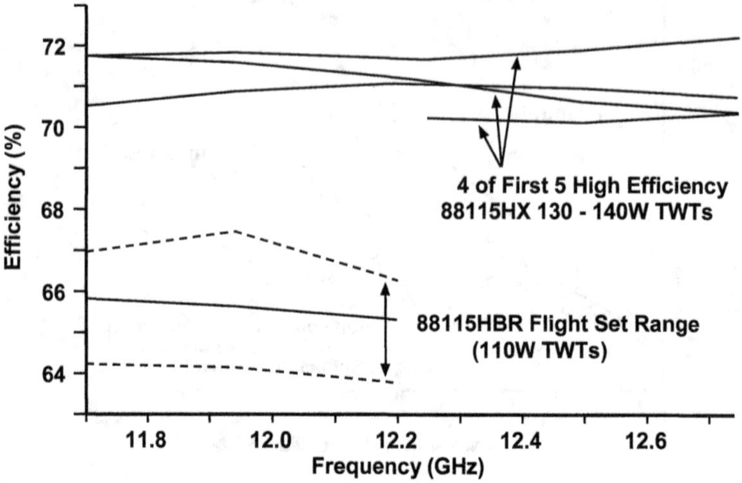

Figure 14-35 Overall efficiencies of the initial set of production Ku-band TWTs compared to the previous model. (From: W. Menninger et al., *IEEE Trans. Electron Devices*, May 2005. © 2005 IEEE.)

The achievement of 70% efficiency in a helix TWT requires great attention to detail in the design and fabrication of all aspects of the electron optics system and the RF structure. Most important, perhaps, is the use a multistage depressed collector for recovering the power remaining in the electron beam after it has interacted with the RF structure. These collectors are discussed in Chapter 16.

The design of the RF circuit must be optimized for several factors at the same time. These include electronic efficiency, circuit efficiency, collector efficiency (the energy distribution of the electrons entering the collector must be controlled), and linearity.

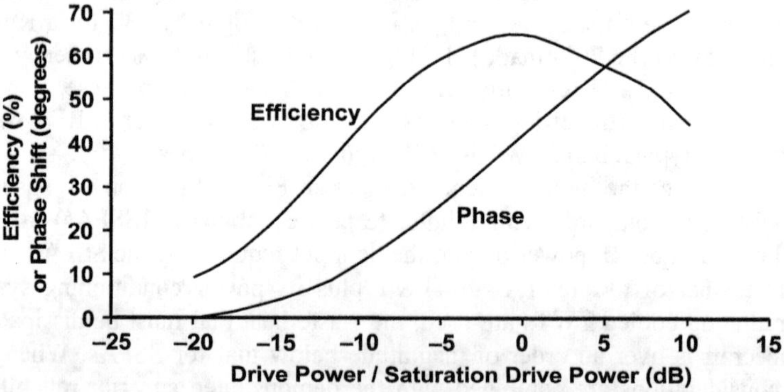

Figure 14-36 Efficiency and phase characteristics of the Thales Worldstar TWT at 1.65 GHz. (From: P. Ehret et al., *IEEE Trans. Electron Devices*, May 2005. © 2005 IEEE.)

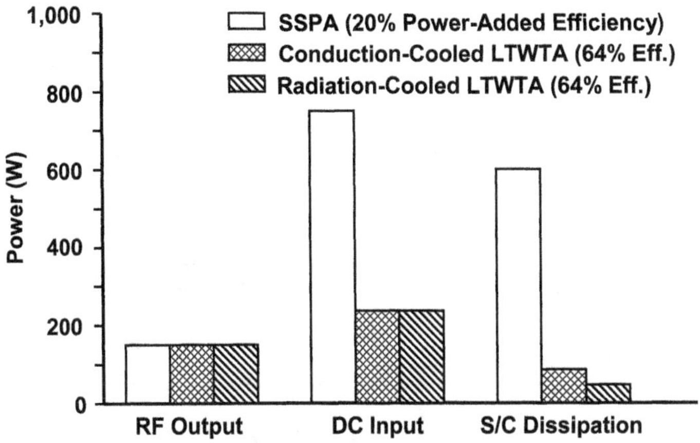

Figure 14-37 Amplifier output powers, dc input powers, and waste heat that a spacecraft must dissipate. (From: W. Menninger et al., *IEEE Trans. Electron Devices*, May 2005. © 2005 IEEE.)

Circuit efficiency is the efficiency with which the helix delivers the RF power generated in the TWT to the output port. Circuit efficiency is important because power lost to heating the circuit is no longer available in the electron beam for recovery by the collector. Collector efficiency is the efficiency with which the collector recovers power from the electron beam leaving the interaction region of the TWT. Figure 14-38 shows that an increase in circuit efficiency of 10% will result in an increase in overall efficiency of ~5% when an efficient collector is used.

Circuit efficiency is dependent primarily on the effective resistance of the circuit material. At high frequencies, dielectric losses are also important. Comparisons of measured and theoretical loss for a copper-plated helix are given in Figure 14-39.

The dashed curves of Figure 14-39 are calculated loss curves based on the bulk resistivity of copper (1.724×10^{-6} Ω-cm) and based on the best fit to experimental data at the frequencies where reflections in the measurement circuitry were small and a resistivity of 4×10^{-6} Ω-cm was assumed. The circled data points are measured losses and the solid curve is the calculated loss for a resistivity of 4×10^{-6} Ω-cm, taking into consideration reflections in the measurement circuitry. The excellent fit of the theoretical curves to data when a resistivity of 4×10^{-6} Ω-cm is assumed leads to the conclusion that 4×10^{-6} Ω-cm was the effective resistivity of the copper surface of the helix. This value is more than twice the bulk resistivity of copper.

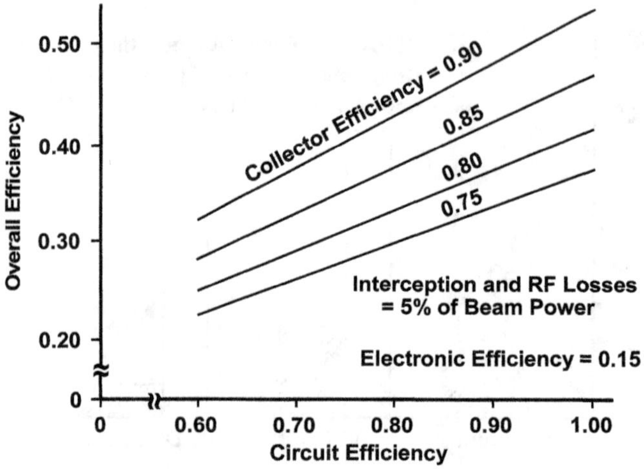

Figure 14-38 Effect of circuit efficiency on overall efficiency. (Adapted from: A. S. Gilmour, Jr. et al., *IEEE Trans. Electron Devices*, October 1979. © 1979 IEEE.)

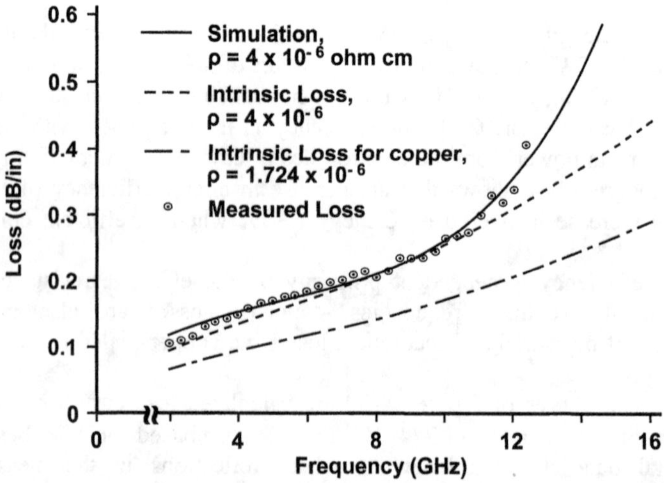

Figure 14-39 Measured and theoretical loss for a copper-plated helix. (From: A. S. Gilmour, Jr. et al., *IEEE Trans. Electron Devices*, October 1979. © 1979 IEEE.)

The reason for the high effective resistivity is indicated in Figure 14-40, which contains photomicrographs of the surface of the copper-plated helix [14]. The skin depth at 10 GHz was small compared with the sizes of the irregularities on the orange-peel textured copper surface. As a result, the path length for the RF current and the effective resistance were much larger than for a smooth surface. To minimize resistance and loss, it is essential to properly polish the helix surface.

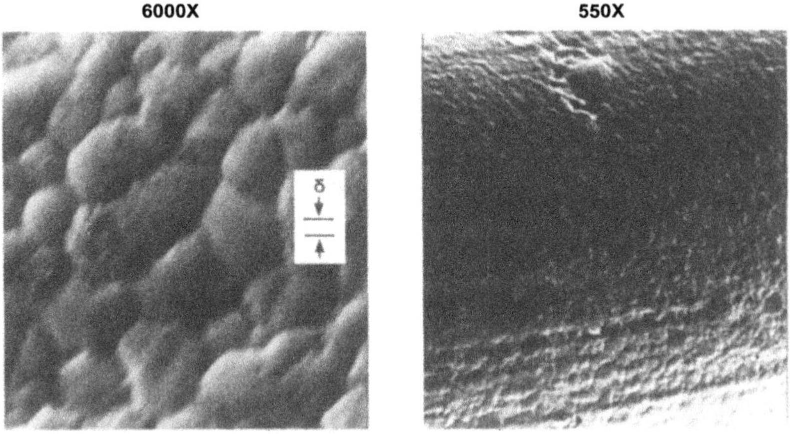

δ = Skin Depth at 10 GHz

Figure 14-40 Photomicrographs of a copper-plated helix. (From: A. S. Gilmour, Jr. et al., *IEEE Trans. Electron Devices*, October 1979. © 1979 IEEE.)

One of the very unfortunate aspects of the requirements placed on RF amplifiers for some communications applications, whether they are TWTs or solid state, is that there is very little tolerance for nonlinearities in the amplification process. To operate with acceptable linearity, operation of the amplifiers has to be "backed off." This means that the amplifiers must be operated at power levels well below their maximum capabilities. The amount of back-off depends on the RF modulation scheme as shown in Figure 14-41. If frequency modulation (FM) is used, operation at saturation is possible and the amplifiers can operate at maximum efficiency. For some of the phase shift keying (PSK) applications, the back-off required for acceptable operation is on the order of 4 dB. For various amplitude modulation schemes, which are coupled with phase modulation (QAM, for example), the back-off requirement can be on the order of 12 dB or more.

TWTs can be and are operated in the back-off mode, but the overall efficiency is below the values of 70% or more that are realized for operation at saturation. Examples of the degradation in efficiency that can occur at reduced power levels are shown in Figure 14-36 and Figure 14-42. (Here efficiency was not necessarily optimized for operation below saturation.)

At 10 or more decibels below saturation, power usage that would normally be considered to be nearly insignificant becomes important. An example is the power for heating the cathode. This could be eliminated with a cold cathode. Losses due to secondary emission effects in the collector would have to be reduced. Beam interception would have to be virtually eliminated. If these losses, as well as power conditioning losses could be all but eliminated, very efficient operation at 10 or more decibels below saturation could be achieved.

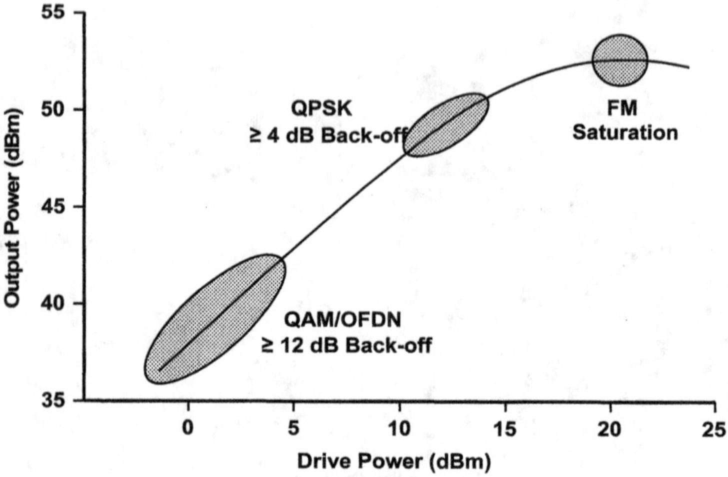

Figure 14-41 Reduction in TWT output power required for acceptable operation with various modulation techniques. (From: K. Tsutaki et al., *IEEE Trans. Electron Devices*, May 2005. © 2005 IEEE.)

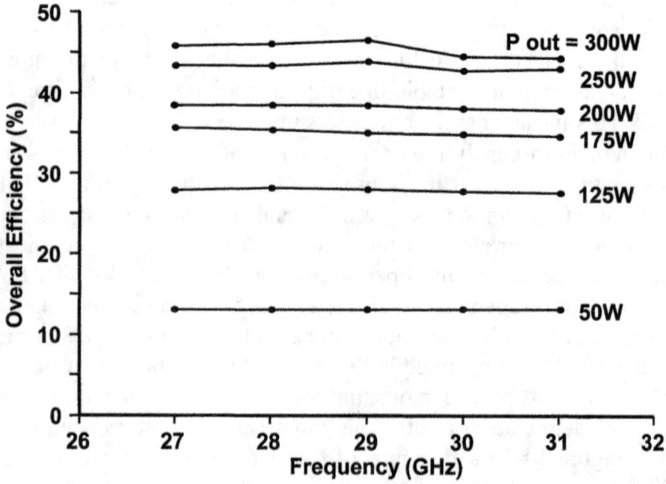

Figure 14-42 Reduction in TWT efficiency that results from operation below saturation. A. Gallien et al., 2006 IVEC. © 2006 IEEE.)

14.5 DUAL-MODE OPERATION

The purpose of a dual-mode TWT is to provide two different RF power levels (separated by as much as 10 dB) efficiently in a single device. This is important in airborne electronic countermeasures systems, for example, where volume and weight must usually be minimized. In these systems, the use of two tubes, each of which is designed for operation at one of the power levels, often results in excessive system weight and volume. There are basically three ways in which a single tube can be efficiently operated at two power levels:

1. Beam current and voltage are held constant and RF drive level is changed. Efficiency is obtained by using a multistage depressed collector to recover spent-beam power. This technique is discussed further in Chapter 16.
2. A complex tube, basically consisting of two tubes in one envelope, is used. A drawing of a device of this type is shown in Figure 14-43. The tube consists of a conventional TWT in series with a very short, low gain (10 dB), nearly lossless device. In the low-power mode, the beam in the 10-dB device (sometimes referred to as a transparent tube) is turned off and the signal from the conventional TWT passes through it with no amplification and little loss. In the high-power mode, the transparent tube is turned on to provide an additional 10 dB of gain. This device has the advantage of providing a full 10 dB of pulse-up capability and the disadvantage of being more complex and heavier than the single-tube device.
3. Beam current is varied with a gridded electron gun. (Beam voltage is not changed because of the weight and complexity of a modulator and because the beam velocity must remain near synchronism with the circuit phase velocity.)

The third technique (changing beam current) for achieving dual-mode operation is difficult to implement when the power separation between modes must be more than a few decibels. One difficulty encountered is that of focusing both the low-power and the high-power beams with the same magnet structure.

Assuming constant efficiency, beam power and beam current must vary in direct proportion to RF power. If beam diameter is held constant, then beam charge density varies directly with beam current. Finally, the radial electric field in the beam is proportional to the charge density and, therefore, to the power level. If the PPM structure is designed for the low-power mode, the magnetic field will be inadequate to offset the electric field in the high-power mode. Of course, the opposite situation exists if the PPM structure is designed for the high-power mode. This problem may be alleviated by changing the beam diameter when the beam current is changed.

The other difficulty encountered when the beam current is changed is that the beam impedance changes and this changes Pierce's gain parameter, C. As a result, the RF characteristics of the tube change with beam current, making it difficult to obtain the desired performance at two power levels that differ by a large amount.

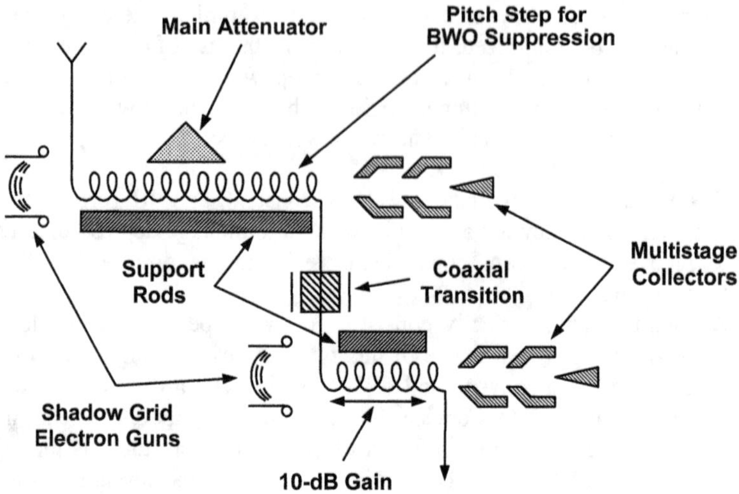

Figure 14-43 Dual-beam, dual-mode TWT. (Courtesy of Varian Associates.)

14.6 MICROWAVE POWER MODULES [15]

A solid-state amplifier (SSA) can have an extremely low noise figure but cannot efficiently produce high RF power. On the other hand, a TWT can efficiently deliver high power but does not have such a low noise capability. In a microwave power module (MPM), as shown in Figure 14-44, advantage is taken of both the SSA and the TWT. Gain is divided approximately equally between these two devices. The SSA and TWT are combined with a power conditioner in a single package that is highly miniaturized and can be designed to operate in the microwave or millimeter-wave frequency bands.

As shown in Figure 14-44, in addition to low noise, the SSA is small and has functionality, that is, it has strong signal processing capabilities. In addition to high power, the TWT is capable of operating efficiently over a wide band of frequencies. The power conditioner delivers low voltage to the SSA and high voltage to the TWT with high efficiency. Special transformers have been developed, which are very compact and have form factors that make it possible for them to fit into the very compact MPM packages.

The development of MPMs began in 1991 as a Tri-Service initiative. The original specifications that were set as a goal for the program are listed in Table 14.1. Notice, in particular, that the power levels were ~100W at noise figures of 10 dB or lower.

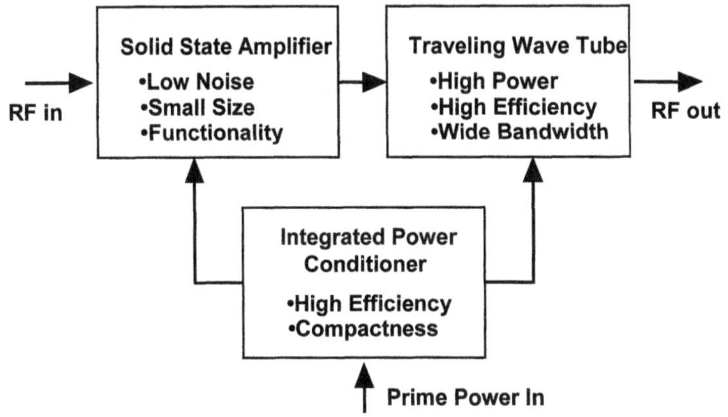

Figure 14-44 Basic components of a microwave power module (MPM).

As pointed out in [15], the Tri-Service investment has brought forth a new miniaturized RF amplifier capability for the RF electronics industry. Now, in addition to the military applications, MPMs are being developed for commercial use such as satellite uplinks, line-of-site terrestrial links, and local multipoint distribution service.

Table 14.1
Tri-Service MPM Goal Specifications

Parameter	ECM	Radar
Frequency	6–18 GHz	7–11 GHz
RF Power Output	50–100W	>100W
Duty Cycle	100%	50%
Saturated Gain	>50 dB	>50 dB
Efficiency	>33%	>40%
Noise Power	–105 dBm/Hz	–157 dBc/Hz at 1 kHz
Noise Figure	10 dB	<10 dB
Spurious	< –10 dBc	< –60 dBc
Size	4 × 6 × 0.32 inches	2 × 6 × 1 inches

An example of a wide band MPM, dubbed the "ultraband MPM" by Northrop Corp. is the 125W (minimum), 4.5–18-GHz device with dimensions of 8.3 inches by 4.8 inches by 1.0 inch and weighing 3.6 lbs [15]. The power and gain characteristics of the TWT used in this MPM are shown in Figure 14-45. The gain variation of this device is offset by the SSA to produce a nearly constant power output over the entire frequency range. As Figure 14-46 shows, the efficiency of the ultraband MPM is as high as 45%.

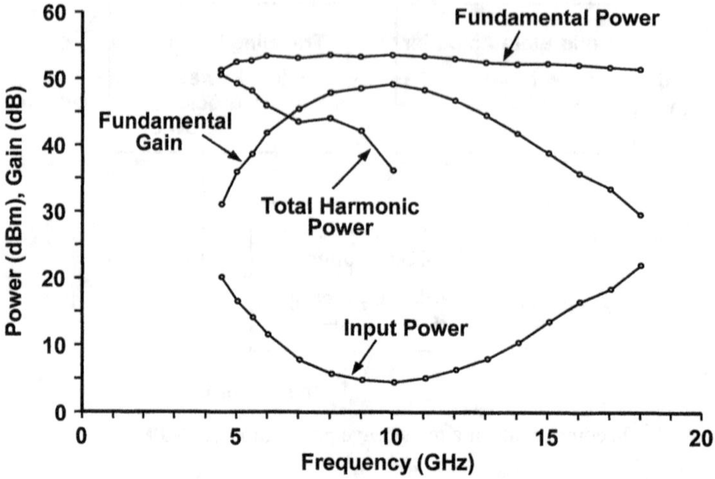

Figure 14-45 Power and gain characteristics of the TWT used in the Northrop ultraband MPM. (From: C. R. Smith et al., *Proc. IEEE*, May 1999. © 1999 IEEE.)

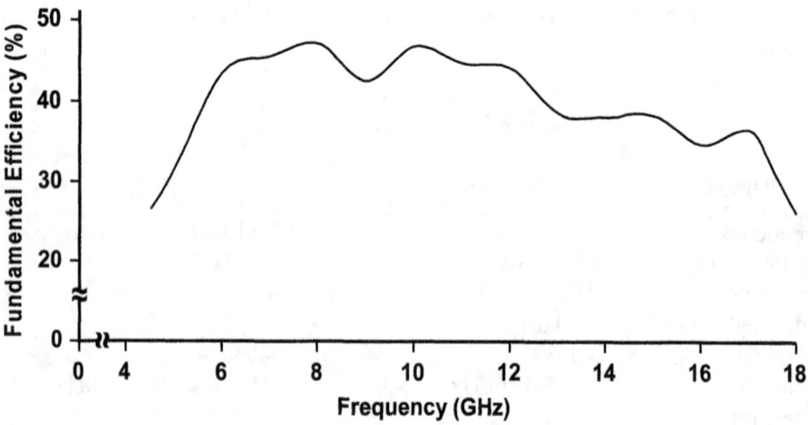

Figure 14-46 Efficiency of the Northrop ultraband MPM. (From: C. R. Smith et al., *Proc. IEEE*, May 1999. © 1999 IEEE.)

14.7 RING BAR and RING LOOP TWTs

The ring-bar circuit in Figure 14-47(a) was developed to make possible the high-power operation of a helix-type circuit. Basically, the ring-bar circuit is derived from the contrawound helix shown in Figure 14-47(b) and is far easier to fabricate. In another similar circuit, the ring-loop circuit, the rings are connected by loops rather than bars.

(a)

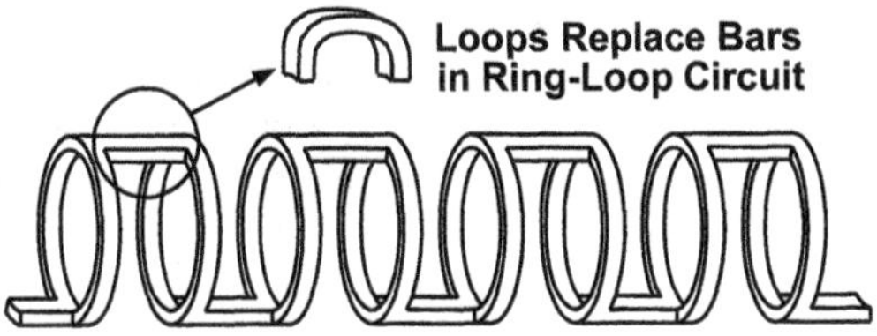

(b)

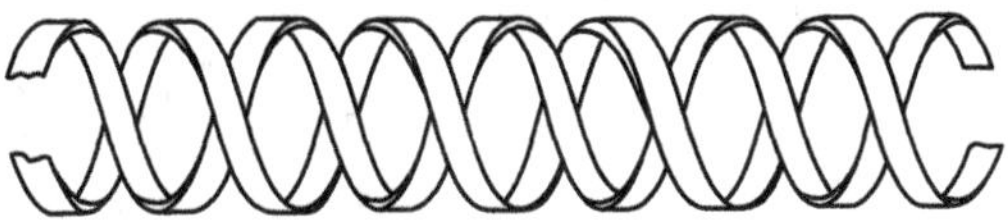

Figure 14-47 (a) Ring-bar and (b) two-tape contrawound helix circuits.

These circuits, originally described by Chodorow and Chu [16], suppress backward wave oscillations and can be operated at higher voltages than a single helix. Also, because BWO is suppressed, it is possible to fill more of the helix tunnel with the electron beam so a larger beam with more current can be used. The penalty for these advantages is that the bandwidth is less than that of a single helix.

To understand the operation of these circuits and how they suppress backward wave interactions, it is helpful to examine the bifilar helix shown in Figure 14-48 and to compare its characteristics with those of a single helix. The pitch of the helices making up the bifilar helix is the same as that of the single helix. For the moment, interactions between the two helices of the bifilar and single helices are assumed to be the same. As a result, for a given frequency, the voltage distributions indicated in Figure 14-48 for the single helix and the bifilar helix are the same. For the BWO discussion, these have been chosen so as to be those associated with two turns per wavelength.

The interaction of electrons with a backward wave on a single helix was discussed earlier in this chapter. To review, electrons that are initially accelerated (top left, Figure 14-48) experience essentially no forces 180° later (middle left) and then are accelerated again 360° later (bottom left). Similarly, and not shown in Figure 14-48, electrons that are initially decelerated experience no forces 180° later and then are decelerated 360° later. Thus, there is bunching and gain, even though the wave on the helix moves in the opposite direction from the electrons.

For the bifilar helix shown in Figure 14-48, it is assumed that the electrons and waves have the same velocities as for the single helix. In this case, however, electrons that are initially accelerated (top right) are decelerated 180° later (center right) and then accelerated again 360° later (bottom right). Because, on each cycle, the accelerating and decelerating forces tend to cancel each other, there is no net gain that can lead to backward wave oscillations. As a result, higher operating voltages, currents, and RF power levels are possible with the bifilar helix than with the single helix (with no suppression techniques being used). The same arguments apply to the contrawound helix and to the ring-bar and ring-loop circuits, so these circuits can operate without BWO at much higher voltages than can the single helix.

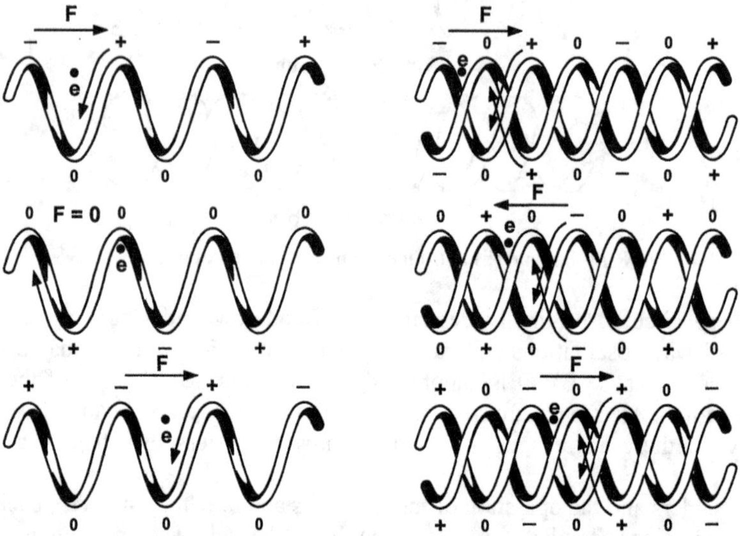

Figure 14-48 Backward wave interaction on a single and a bifilar helix for two turns per wavelength.

An ω-β diagram for a ring-bar circuit is shown in Figure 14-49 [17]. The forward wave characteristic is more dispersive that for a single helix. The point where the beam velocity, u_o, line crosses the forward wave line establishes the synchronous voltage and provides an indication of the bandwidth of the circuit which is typically 10% to 30%. This voltage is high enough so that the backward wave at the frequency f_π is not excited.

The design of the ring-bar and ring-loop circuits are more complex than a helix because the phase velocity is controlled by the width and length of the bar or the loop. Reference [18] describes progress in designing these circuits. Figure 14-50 shows the variation of phase velocity with frequency for a millimeter wave tube [17]. At the center of the operating band, the phase velocity is approximately 25% of the speed of light and the operating voltage is about 18 kV.

Figure 14-51 contains a plot of the power and bandwidth capabilities of an X-band ring-bar TWT. Note that the power is very constant over a bandwidth of nearly 20%. Ring-bar TWTs are capable of very high-powers at lower frequencies. Examples are the QKW1617(A) and the QKW1818, which were made by Raytheon and had power capabilities of 160-kW peak and 10-kW average [19].

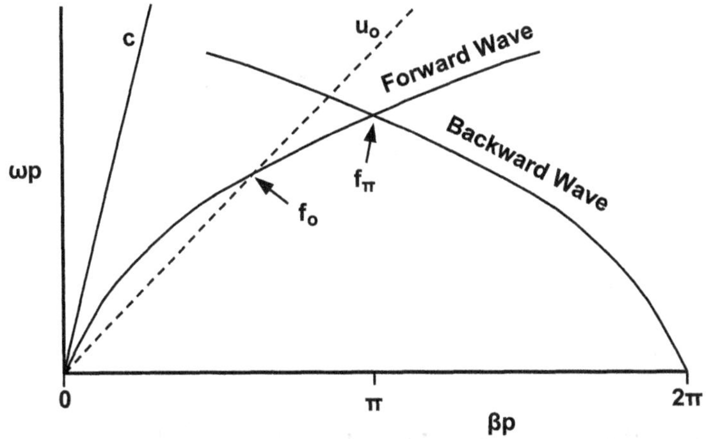

Figure 14-49 The ω-β diagram for a ring-bar structure. (Adapted from: R. LeBorgne et al., 1990 IEDM. © 1990 IEEE.)

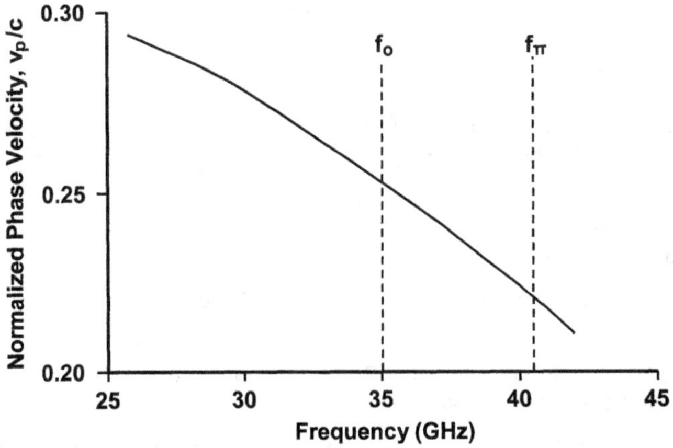

Figure 14-50 Phase velocity of the ring-bar circuit in a Ka-band TWT. (Adapted from: R. LeBorgne et al., 1990 IEDM. © 1990 IEEE.)

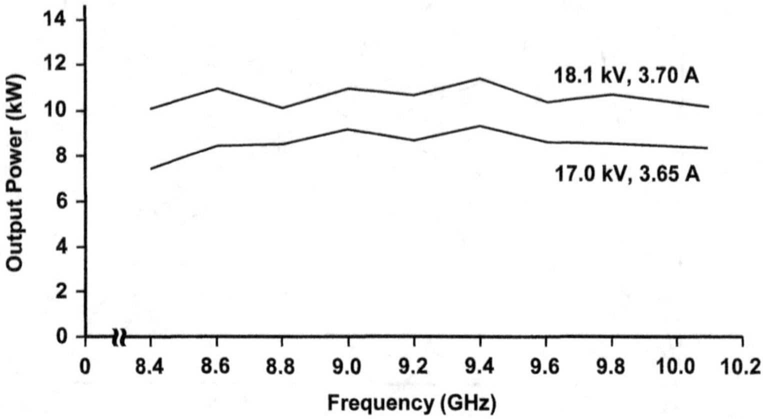

Figure 14-51 Output power characteristic for a ring-bar TWT. (From: Hughes Aircraft Co., *TWT and TWTA Handbook.*)

REFERENCES

[1] J. F. Gittins, *Power Travelling Wave Tubes*, New York: American Elsevier, 1965.

[2] J. L. Putz, and M. J. Cascone, "Effective use of dispersion shaping in broadband helix TWT circuits," *Technical Digest*, IEDM, 1979, pp. 422–424.

[3] A. S. Gilmour, Jr., M. R. Gillette, and C. Jenn-Tsung, "Theoretical and experimental TWT helix loss determination," *IEEE Trans. Electron Devices*, Vol. ED-26, No. 10, 1979, pp. 1581–1588.

[4] A. R. Jung, "10 kW and up, from a helix TWT?" *Technical Digest,* IEDM, 1978, pp. 530–553.

[5] P. Birtel et al., "Simulation of the BWO threshold current in a helix TWT," *Trans. IEEE IVEC*, 2006, pp. 403–404.

[6] A. Scott, and M. J. Cascone, "What's new in helix TWTs?" *Technical Digest,* IEDM, 1978, pp. 526–529.

[7] R. Harper, and M. P. Puri, "Heat transfer and power capabilities of EHF helix TWTs," *Technical Digest,* IEDM 1986, pp. 498–500.

[8] Y. Hiramatsu, "Study of block supported helix circuits for high-power TWTs," *Technical Digest,* IEDM, pp. 416–417, 1979.

[9] T. Mosser, W. Pinger, and D. Zavidil, "High-power brazed helix fabrication," *Microwave Power Tube Conference*, 1980.

[10] O. Sauseng, A. E. Manoly, and A. Hall, "Thermal properties and power capability of helix structures for millimeter waves," *Technical Digest, International Electron Devices Meeting*, 1978, pp. 534–537.

[11] W. L. Menninger et al., "70% efficient Ku-band and C-band TWTs for satellite downlinks," *IEEE Trans. Electron Devices*, Vol. 52, No. 5, pp. 673–678, 2005.

[12] P. Ehret et al., "L-band TWTAs for navigation satellites," *IEEE Trans. Electron Devices*, Vol. 52, No. 5, May 2005, pp. 679–684.

[13] J. M. Weekley, and B. J. Mangus, "TWTA versus SSPA: a comparison of on-orbit reliability data," *IEEE Trans. Electron Devices*, Vol. 52, No. 5, May 2005, pp. 650–652.

[14] A. S. Gilmour, Jr., "SEM analysis of TWT helix samples," *IEEE Trans. Electron Devices*, Vol. ED-24, No. 6, June 1977, pp. 774–775.

[15] C. R. Smith, C. M. Armstrong, and J. Duthie, "The microwave power module: a versatile RF building block for high-power transmitters," *Proc. IEEE*, Vol. 87, No. 5, May 1999, pp. 717–737.

[16] M. Armstrong, and E. L. Chu, "Cross-wound helices for traveling wave tubes," *Jour. App. Phy.*, Vol. 26, No. 1, January 1955, pp. 33–43.

[17] R. H. LeBorgne et al., "Development of an 800 W, Ka-band, ring-bar TWT," *Technical Digest,* IEDM, 1990.

[18] D. R. Dyson, M. J. Clark, and V. H. Smith, "Numerical analysis of ring-loop and ring-bar slow wave structures for traveling wave tubes," *Trans. IEEE International Vacuum Electronics Conf.*, 2003, pp. 46–47.

[19] M. Skolnik, G. Linde, and K. Meads, "Senrad: an advanced wideband air-surveillance radar," *IEEE Trans. Aerospace and Electronic Systems*, Vol. 37, No. 4, October 2001, pp. 1163–1175.

Chapter 15

Coupled-Cavity TWTs

In coupled-cavity TWTs [1, 2], several tens of klystronlike cavities are used for the slow wave structure. Except for sever regions, these cavities are electromagnetically coupled so that an RF signal can propagate along the structure. Coupling techniques take on widely differing forms and are discussed in this chapter. If an electron beam is passed through the structure and if it has a velocity near the phase velocity of the circuit, then amplification can occur.

With appropriate choices of cavity dimensions and cavity-coupling techniques, electron beam velocity can be high, and operation at high peak-power levels is possible. Also, coupled-cavity structures are mostly metallic (copper and iron) and have geometries such that thermal resistances are very low. As a result, high average powers are possible.

Shown in Figure 15-1 are the major components of a coupled-cavity TWT that is PPM focused. The operation of the electron gun was described in Chapter 6, and multistage collectors are discussed in Chapter 16. A qualitative explanation of the interaction of the electron beam with the coupled-cavity circuit is given in the following section.

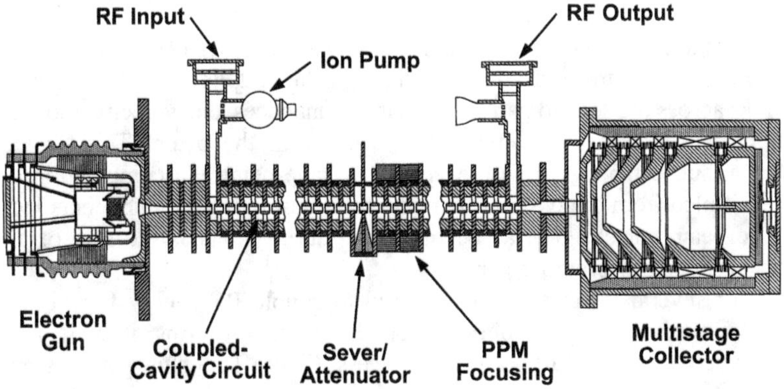

Figure 15-1 Coupled-cavity TWT.

15.1 BASIC OPERATING PRINCIPLES

Figure 15-2 is useful in describing the operation of a coupled-cavity TWT. The input circuit supplies a signal to the first cavity through a magnetic coupling loop or an impedance-matching section from a waveguide. That signal produces a voltage V_1 across the gap in the first cavity and also is coupled from the first cavity to the second (from A to B through one of several coupling structures to be described in the following paragraphs). The voltage across the first gap modulates the electron beam velocity so that an ac component of current develops as the beam moves toward the second cavity. The beam induces a signal into the second cavity, which splits into two equal wavelets, one traveling to the right and the other to the left. The reverse traveling waves from cavity 2 and from subsequent cavities are eliminated by loss or severs not shown in Figure 15-2.

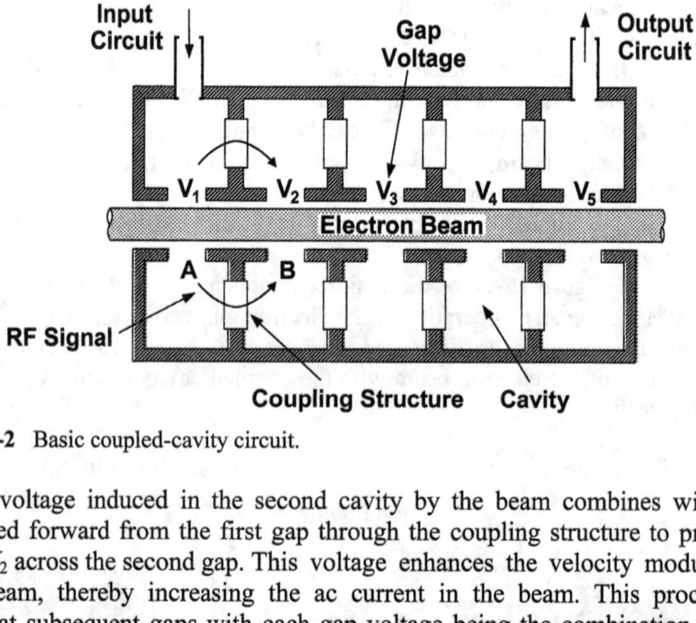

Figure 15-2 Basic coupled-cavity circuit.

The voltage induced in the second cavity by the beam combines with the voltage fed forward from the first gap through the coupling structure to produce voltage V_2 across the second gap. This voltage enhances the velocity modulation of the beam, thereby increasing the ac current in the beam. This process is repeated at subsequent gaps with each gap voltage being the combination of the voltage coupled from the previous gap and the sum of all the components induced by the ac beam current. The ac beam current at any point is the sum of all the currents produced by previous gaps.

One of several techniques can be used to couple RF energy from cavity to cavity. Before discussing coupling structures in general terms, it is of value to discuss the complexity of the coupling process. Figure 15-3 shows an example of cavity-to-cavity coupling. Here, the coupling is accomplished with slots in the plate between two cavities.

Assume, at first, that there is no signal in the cavities so all currents and electric fields are zero. When an RF signal is injected into cavity 1 (with a magnetic coupling loop for example), current I_1 flows and this produces electric field E_{s1} across the slot. The slot field in cavity 2 is the same as that in cavity 1. To produce this field, I_2 must flow in cavity 2. The charge separations that the current flows produce establish electric fields E_{g1} and E_{g2} across the cavity gaps. These are the fields that interact with the electron beam. It should be noticed that the field across gap 2 is opposite to that across gap 1. For a bunching action that is initiated in gap 1 to be reinforced in gap 2, the beam velocity must be such that the gap 2 voltage reverses phase before the bunch arrives.

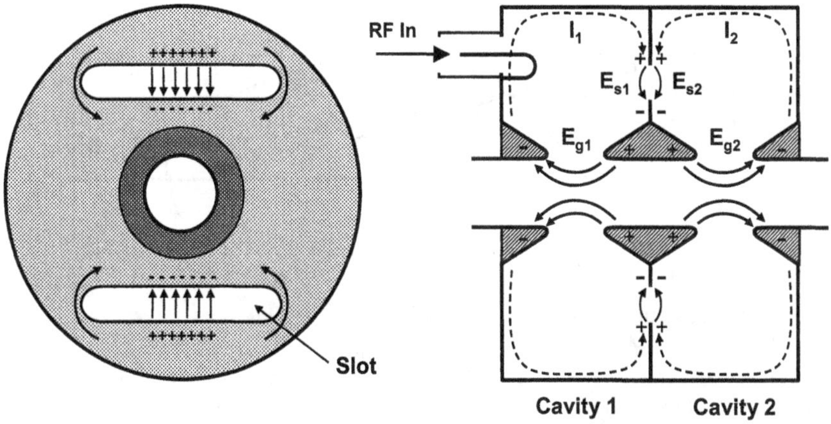

Figure 15-3 Example of coupling of signal from cavity to cavity.

In the TWT shown in Figure 15-1, each cavity is coupled to the following cavity and to the preceding cavity with a kidney-shaped slot. The two slots in each cavity are 180° apart in rotational position so this circuit can be viewed as a folded waveguide. In fact, many of the properties of this circuit are most easily understood by treating it as a folded waveguide. As a result, in the following discussions of ω-β characteristics, vane-loaded waveguides and folded waveguides are considered first.

While the waveguide approach is useful for a qualitative understanding of dispersion characteristics, a circuit model approach is most useful for a detailed examination of ω-β characteristics. The equivalent circuits described by Curnow [3, 4] and Gittins [5] are sometimes used for modeling coupled-cavity structures. The discussion of these circuits follows the waveguide discussion.

15.2 COUPLED-CAVITY STRUCTURES

15.2.1 Waveguide Approach

It is convenient to start the discussion of coupled-cavity structures with the dispersion characteristics of a waveguide. A vane-loaded waveguide discussed in Chapter 13 is shown in Figure 15-4. To examine the interaction of waves in the guide with an electron beam, consider a single electron following the serpentine path shown in Figure 15-4.

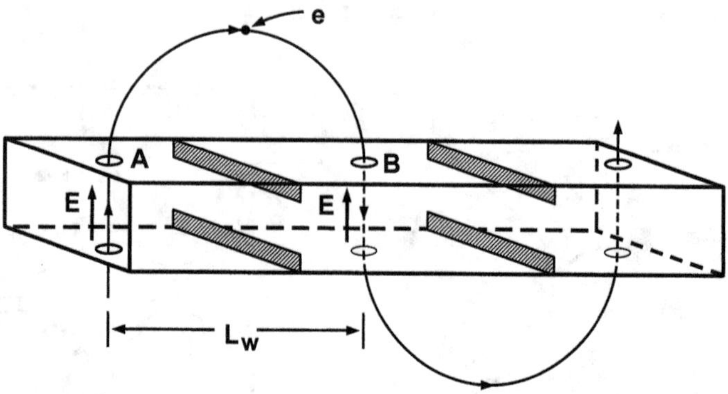

Figure 15-4 Serpentine electron path through vaned waveguide.

Assume first that the electron travels from A to B in the same time that the wave crest in the guide travels from A to B. With respect to the direction of motion of the electron, a 180° phase reversal takes place in the electric field. Thus, as "seen" by the electron, the ω-β diagram is shifted by π radians as shown in Figure 15-5. An obvious problem with the example given here is that the electron would have to travel faster than the speed of light to move from A to B in the time that the wave does. This problem can be alleviated by folding the waveguide as shown in Figure 15-6. Assuming that no additional impedance changes occur, then the ω-β diagram for the wave is not changed. From the perspective of the electron, the phase velocity of the wave and the electron velocity from A to B along path L is important, and so βL_w must be replaced by βL on the abscissa of Figure 15-5 to consider electron interaction.

An electron accelerated at A in Figure 15-6 can again be accelerated at B if:

1. The direction of wave propagation is reversed, and the electron velocity is adjusted so that the electron arrives at B when an inverted wave crest traveling from the right does (see Figure 15-7(a)). This is a backward wave interaction and is indicated by arrow 1 on the ω-β diagram in Figure 15-5(b). The transit angle of the electron for this interaction is about $\pi/2$ rad.

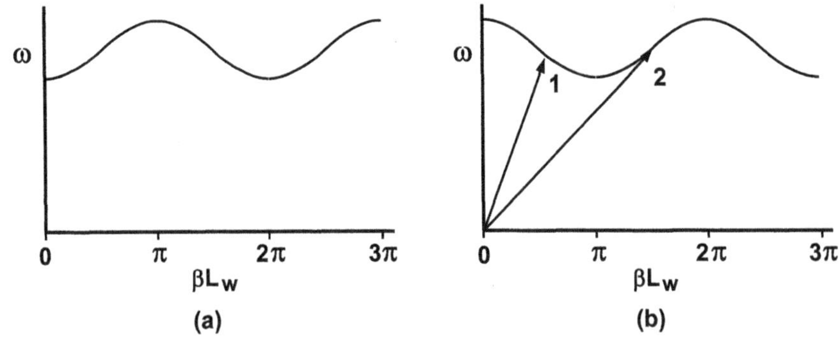

Figure 15-5 An ω-β diagram. (a) Vane loaded waveguide, and (b) as "seen" by an electron following the serpentine path in Figure 15-4.

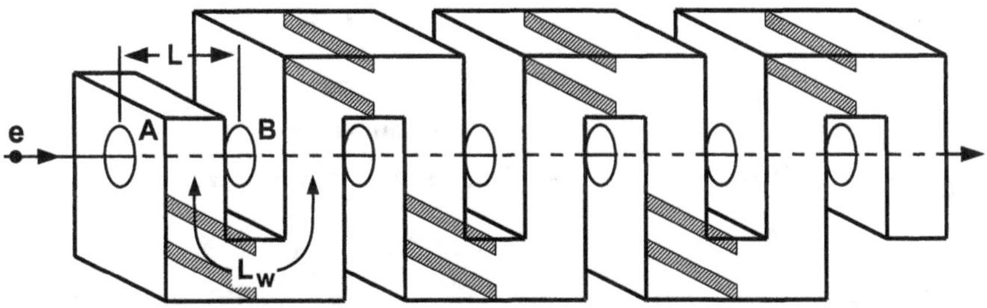

Figure 15-6 Folded waveguide that permits electron interaction below the velocity of light.

2. The electron velocity is reduced so that the next inverted wave crest traveling to the right arrives at B when the electron does. This is a forward wave interaction and is indicated by arrow 2 in Figure 15-5(b). For this interaction to occur, the transit angle of the electron is about $3\pi/2$ rad.

The operation of the folded waveguide discussed in the preceding paragraphs is similar to that of a large class of coupled-cavity TWTs, which includes the TWT shown in Figure 15-1. The circuits used in these TWTs are referred to as fundamental backward wave circuits.

Another class of coupled-cavity TWTs uses structures referred to as fundamental forward wave circuits. The operation of these circuits may be understood with the aid of Figures 15-8 and 15-9. In these circuits, there is a mechanism for shifting the phase of the RF signal by 180°. The 180° twist shown in Figure 15-8 is an obvious way to produce the desired phase shift but would be virtually impossible to implement in a folded waveguide structure.

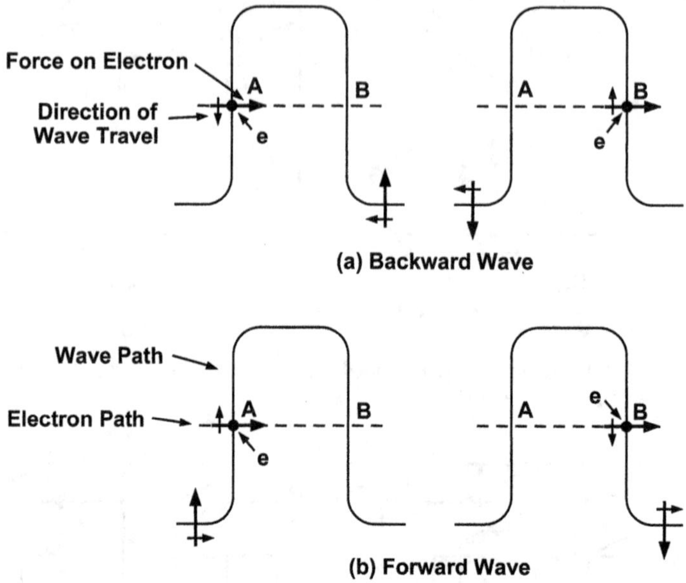

Figure 15-7 (a, b) Wave propagation that reinforces acceleration of electron.

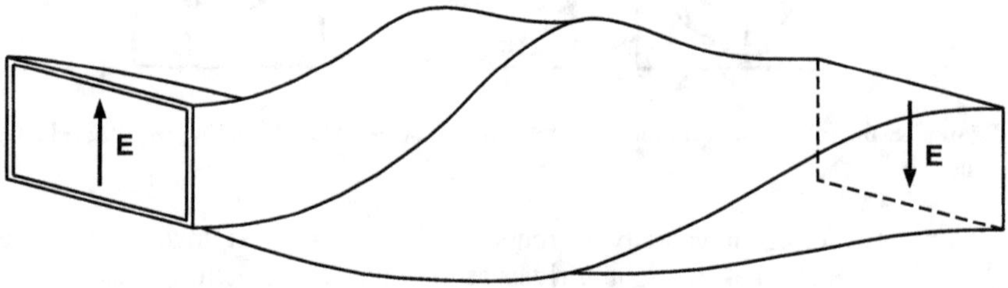

Figure 15-8 Waveguide with a 180° twist.

The dual-coupling loop shown in Figure 15-9(b) reverses the magnetic field at the coupling aperture and, as a result, produces a 180° phase shift in the electric field. This action may be understood by considering, first, the magnetic field at the aperture in the folded guide with vanes. As indicated in Figure 15-9(a), the direction of the *H* field is the same on the two sides of the coupling aperture. In Figure 15-9(b), the magnetic field on the left side of the aperture causes current to flow in the coupling loops. This current, in turn, produces a magnetic field on the right side of the aperture that is the reverse of that on the left. This reversal of the magnetic field also produces a reversal of the electric field. An electron that was accelerated at A and travels from A to B in the same time that the wave does will again be accelerated at B.

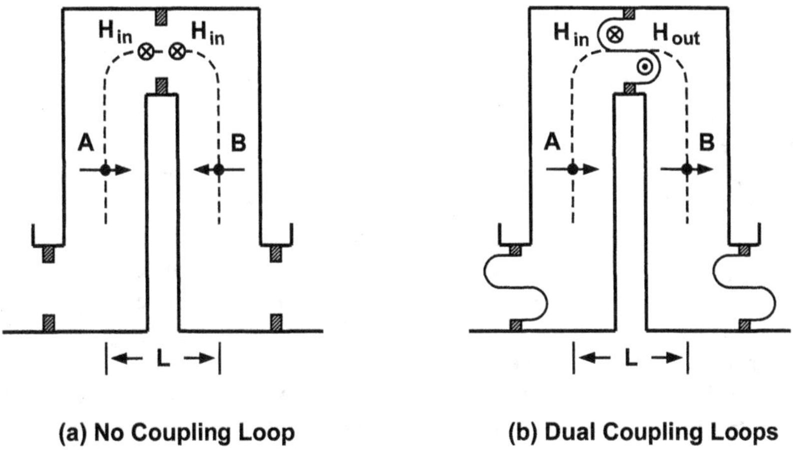

Figure 15-9 (a, b) Use of a dual coupling loop to reverse the direction of the *H* field.

The ω-β diagram for the loop coupled structure is shown in Figure 15-10. Interaction in this fundamental forward wave structure occurs with electrons at a transit angle near $\pi/2$ rad. It should be noted that this transit angle is much smaller than for the fundamental backward wave structure (~$3\pi/2$ rad). As a result, tubes with fundamental forward wave structures operate with faster and, therefore, higher voltage, higher-powered electron beams than fundamental backward wave tubes. Fundamental forward wave tubes can operate at power levels of several megawatts whereas fundamental backward wave tubes operate at kilowatts to tens of kilowatts.

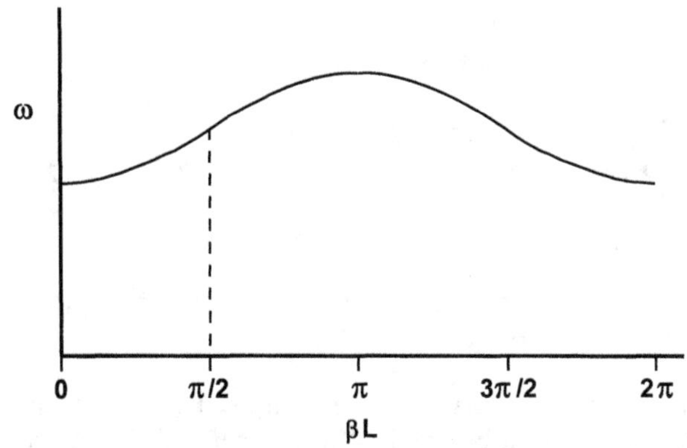

Figure 15-10 The ω-β diagram for a fundamental forward wave circuit.

15.2.2 Curnow-Gittins Equivalent Circuit Approach

The discussion of ω-β characteristics based on waveguide considerations is useful for a qualitative understanding of the dispersion characteristics of coupled-cavity circuits. However, detailed analyses, which necessarily use electromagnetic field theory, are very difficult because of the complex boundary conditions. Thus, the waveguide approach does not provide much insight or guidance concerning the dependence of dispersion characteristics on variations in the geometrical characteristics of the circuits. Even computer modeling does not help much in developing an intuitive understanding of the operation of coupled-cavity circuits. However, computer modeling, along with the use of equivalent circuits, can provide accurate predictions of the consequences of circuit changes along with an understanding of what is happening.

From the very early days of coupled-cavity TWTs, equivalent circuits have been used to help explain circuit operation. The most successful of these equivalent circuits have been those described by Curnow [3, 4] and Gittins [5]. The development of these circuits is based on an examination of the current paths in a cavity. In a cavity without coupling apertures, current is distributed uniformly as is indicated in Figure 15-11(a). Current flows back and forth from one side of the cavity gap to the other through the cavity walls. The gap is the capacitive portion of the cavity and the walls are the inductive portion, and so the equivalent circuit is in Figure 15-11(b). The resonant frequency is f_c.

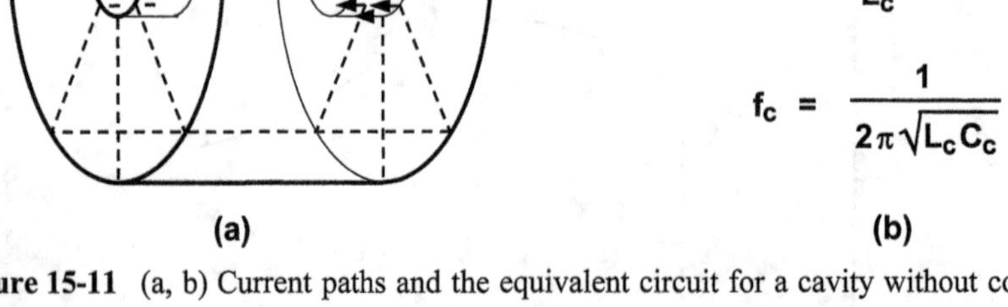

Figure 15-11 (a, b) Current paths and the equivalent circuit for a cavity without coupling slots.

When coupling slots are placed in the cavity walls, some of the current paths are intercepted, as shown in Figure 15-12. In general, there can be four types of current paths:

1. Unbroken by any coupling slot (I_1);
2. Broken by the slot coupling to the preceding cavity (I_2);
3. Broken by the slot coupling to the following cavity (I_3);
4. Broken by both coupling slots (I_4).

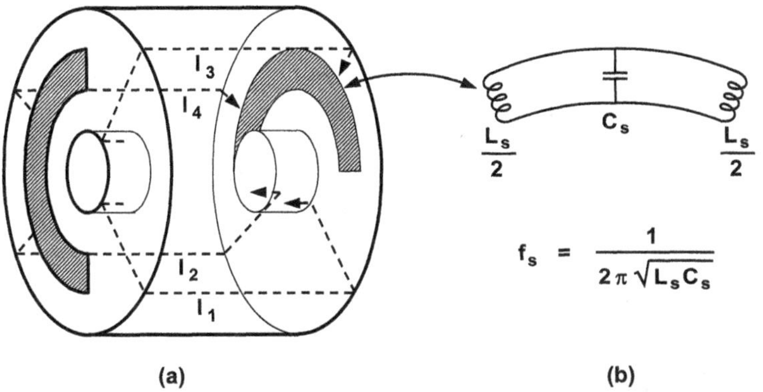

Figure 15-12 (a) Cavity currents used to define the inductance for the equivalent circuit, and (b) the equivalent circuit for a slot.

The equivalent circuit to represent this distribution of currents can be constructed by dividing the cavity inductance, L_c, into four parts:

1. L_c/p, where p is the fraction of the current not involved in coupling;
2. $2L_c/k$, where k is the fraction involved in coupling to the following cavity;
3. $2L_c/k$, where k is the fraction involved in coupling to the preceding cavity;
4. L_c/n, where n is the fraction involved in coupling both ways.

Here, it has been assumed, as is usually the case, that the two coupling slots are the same. The total of the four currents is the total cavity current and so

$$p + k + n = 1 \tag{15-1}$$

The coupling slots can be represented as parallel resonant circuits with resonant frequency f_s, as shown in Figure 15-12(b). Thus, the cavity with all of its equivalent circuit elements is as shown in Figure 15-13. The generalized equivalent circuit derived from this cavity is in Figure 15-14.

Using the circuit in Figure 15-14 an expression can be developed for the phase shift from cavity to cavity. Using measured values for the resonant frequencies of the cavity and the slots and estimates for the current coupling fractions, an ω-β diagram can be constructed. Also, the impedance of the circuit can be calculated. Then, based on comparisons with experimental ω-β results and impedance measurements, the current fraction can be revised so that the equivalent circuit is a reasonably accurate model of the cavity.

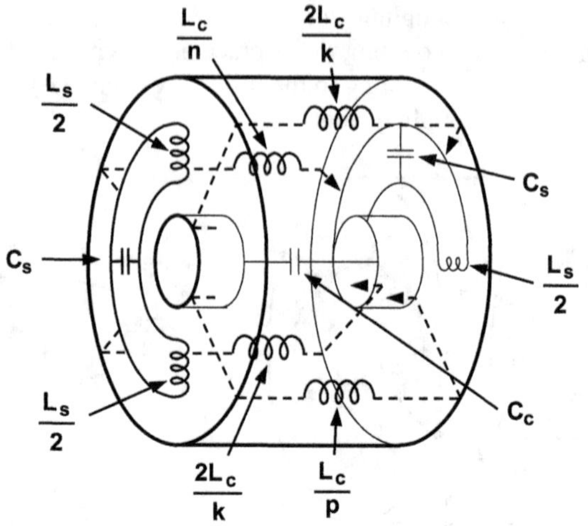

Figure 15-13 Equivalent circuit elements of a cavity.

One of the values of the Curnow-Gittins equivalent circuit is that it provides insight into the operation of the cavity. The other is that it provides a way to estimate the results in performance characteristics of changes to a coupled-cavity structure. For example, if the coupling slots are made longer, the fraction of the current that they intercept is increased and so the change in dispersion and impedance can be estimated.

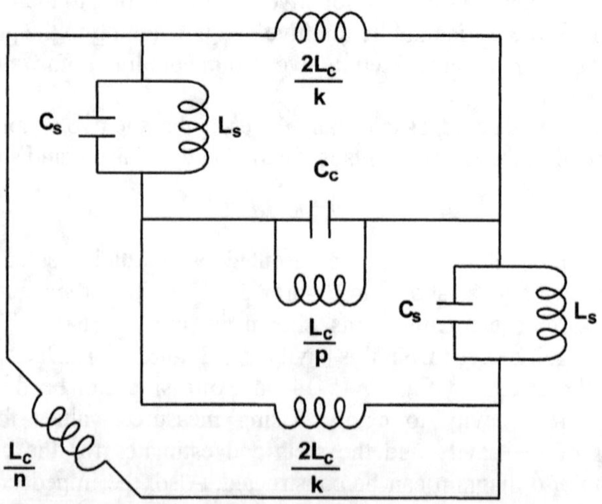

Figure 15-14 The equivalent circuit for a cavity.

15.2.3 Example of an Application of the Curnow-Gittins Circuit

As an example of the use of the Curnow-Gittins equivalent circuit, the staggered-slot circuit will be examined. This is the circuit used in the TWT shown in Figure 15-1, and shown more clearly in Figure 15-15. In this circuit, no current paths are intercepted by both slots. As a result, the fraction, n, used in the previous section is zero and so

$$p = 1 - k \tag{15-2}$$

where k is the total fraction of the current involved in coupling. The equivalent circuit is shown in Figure 15-16. Circuits for adjacent cavities are included to facilitate the analysis.

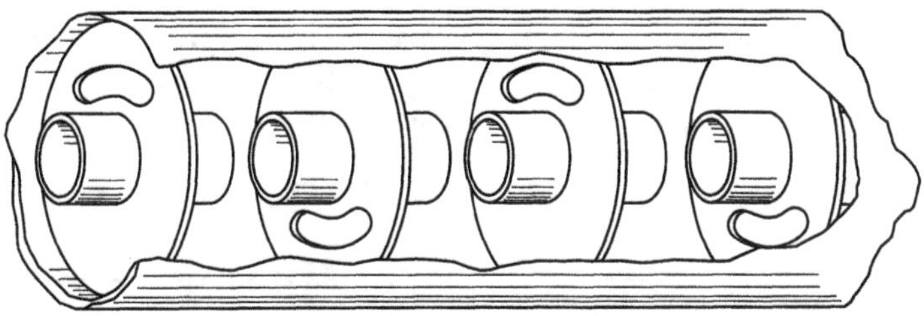

Figure 15-15 A staggered-slot fundamental backward wave circuit. (Adapted from: *Power Travelling Wave Tubes* by J. F. Gittins, published 1965 by American Elsevier, Inc.)

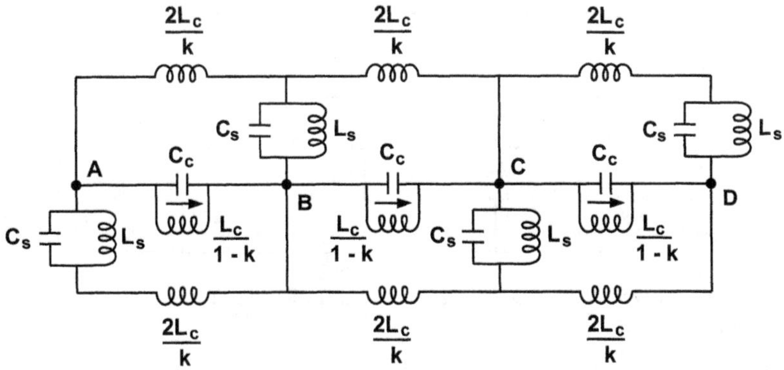

Figure 15-16 The equivalent circuit for the staggered-slot structure.

In manipulating the circuit so that it can be analyzed, it is important to keep track of the direction with which electrons pass through the circuit (across C_c). This direction is indicated by the arrow next to the circuit capacitance. By inspection, the circuit may be rearranged as is shown in Figure 15-17.

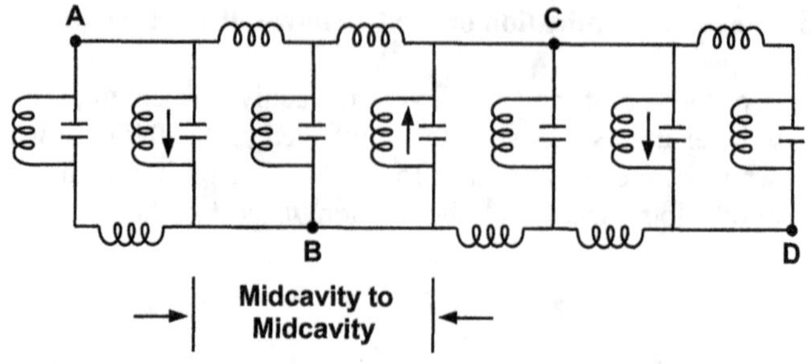

Figure 15-17 Rearranged circuit for staggered-slot structure.

One cell (midcavity to midcavity) of the equivalent circuit is shown in Figure 15-18(a) and with equivalent impedances, in Figure 15-18(b). This circuit is in a form so that the phase shift, φ, from cavity to cavity can readily be determined. Also, the impedance looking in at X-X is the impedance "seen" by the electron beam is the impedance needed for the calculation of Pierce's gain parameter.

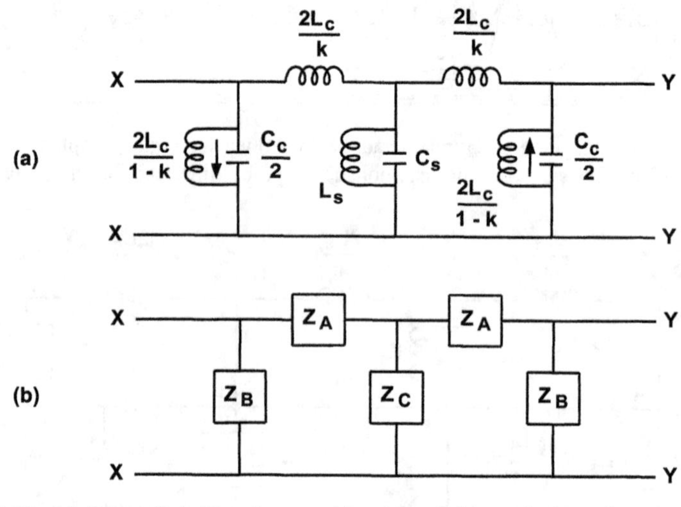

Figure 15-18 (a) One cell (midcavity to midcavity) of the equivalent circuit and (b) the equivalent impedances.

From the theory of filter circuits the equation for the phase shift, φ, which is needed for constructing the ω-β diagram, is

$$\cos\phi = \left(\frac{Z_{oc}}{Z_{oc} - Z_{sc}}\right)^{1/2} \tag{15-3}$$

where Z_{oc} is the open-circuit impedance looking in at X-X and Z_{sc} is the impedance at X-X with Y-Y short circuited. The phase shift "seen" by an electron, θ, is $\varphi + \pi$ (note the direction of the arrow indicating electron flow) and so

$$\cos\theta = -\left(\frac{Z_{oc}}{Z_{oc} - Z_{sc}}\right)^{1/2} \tag{15-4}$$

The impedance, K, of the circuit is

$$K = \left(Z_{oc}\ Z_{sc}\right)^{1/2} \tag{15-5}$$

After calculating Z_{oc} and Z_{sc} and making the following substitutions

$$\omega_c = \frac{1}{\sqrt{L_c C_c}} \tag{15-6}$$

$$\omega_s = \frac{1}{\sqrt{L_s C_s}} \tag{15-7}$$

and

$$k_c = \frac{k\, L_s}{L_c} \tag{15-8}$$

the ω-β characteristic can be plotted as shown in Figure 15-19. Here, it has been assumed that $k = 0.5$ and $k_c = 0.35$.

Notice that there are two frequency bands in which propagation can occur. The lower one is referred to as the cavity passband because its characteristics are controlled largely by the cavity-resonance condition. Similarly, the upper band is called the slot passband and its characteristics are controlled mostly by the slot resonance condition.

The solid curves in Figure 15-19 are the ω-β characteristics derived for one midcavity-to-midcavity cell of the equivalent circuit for the entire coupled-cavity structure. Because only one cell is used, results for other portions of the ω-β characteristic, such as those shown by the dashed lines in Figure 15-19, are not obtained. However, these can be inferred from the discussion of ω-β characteristics contained in Chapter 13. Thus, the curves from π to 2π rad are simply mirror images of those from 0 to π rad. For the staggered-slot circuit, the lower dashed curve, which indicates forward wave interaction, is the ω-β characteristic that is needed.

There are several variables in the Curnow-Gittins equations. When changes are proposed for a coupled-cavity structure, the effect of these changes on one or more of the variables can be estimated, and the change in the dispersion characteristic (and interaction impedance) can be predicted.

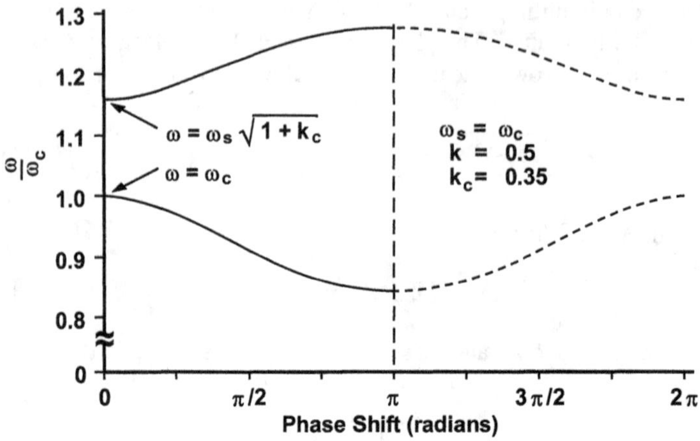

Figure 15-19 The ω-β diagram for the equivalent circuit for the staggered-slot structure.

As an example, changes in the angle, θ, of the coupling slot shown in Figure 15-20 change the percentage of the cavity current involved in coupling (k), change the slot inductance (in k_c), and change the slot resonant frequency (ω_s). The effect of these changes on the dispersion characteristics is shown in Figure 15-21. These changes, in turn, affect the cold bandwidth (bandwidth of circuit with no electron beam) as shown in Figure 15-22. Changes in the coupling slot angle also change the relative positions of the slot and cavity passbands as shown in Figure 15-23. Note that, for large coupling angles, the slot passband becomes the lower passband.

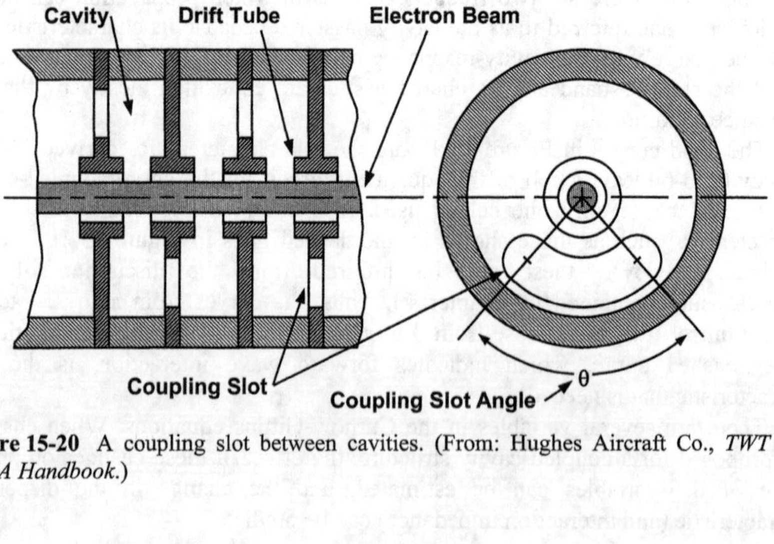

Figure 15-20 A coupling slot between cavities. (From: Hughes Aircraft Co., *TWT and TWTA Handbook.*)

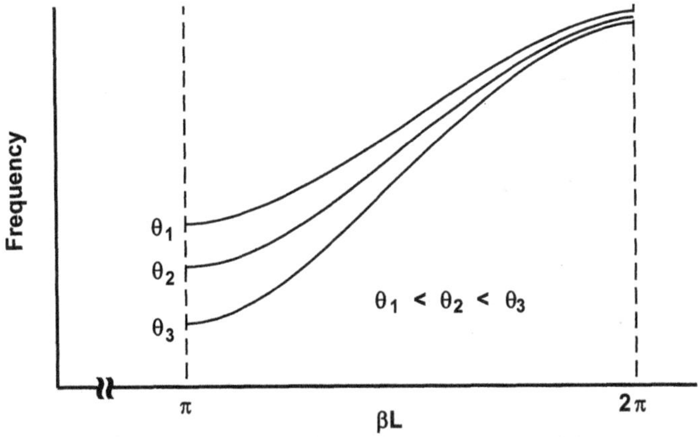

Figure 15-21 Effect of coupling angle on the ω-β diagram. (From: Hughes Electron Dynamics Division.)

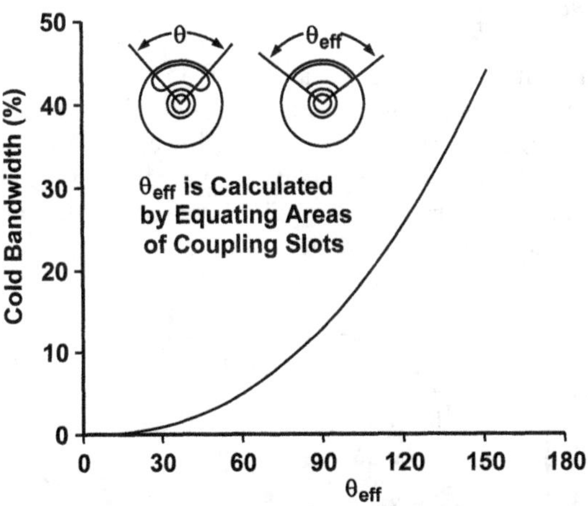

Figure 15-22 Cold bandwidth versus coupling slot angle. (From: Hughes Electron Dynamics Division.)

In arriving at a useful expression for impedance, K, the loaded Q of the circuit must be taken into consideration. This is done by making the substitution

$$\left(\frac{L_c}{C_c}\right)^{1/2} = \frac{R}{Q} \tag{15-9}$$

where R/Q is found from circuit measurements.

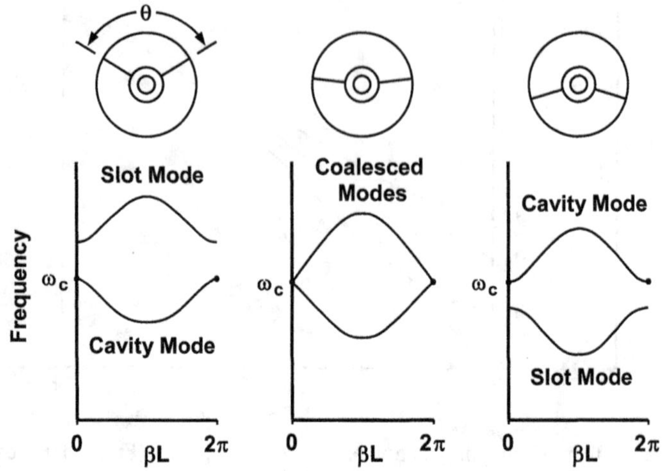

Figure 15-23 Change in passbands with increasing coupling slot angle. (From: J. R. Frey and I. Tammaru, 1981 IEDM. © 1981 IEEE.)

Again, using values of k = 0.5 and k_c = 0.35 and assuming that $R/Q = 37$, impedance can be plotted as shown in Figure 15-24. Note that impedance goes to infinity at all cutoff frequencies except for the lower slot-band cutoff frequency, where it is zero. For the values of constants used here, the impedance in the cavity band where normal operation occurs is a few hundred ohms, which is higher than what is normally encountered in staggered-slot TWTs.

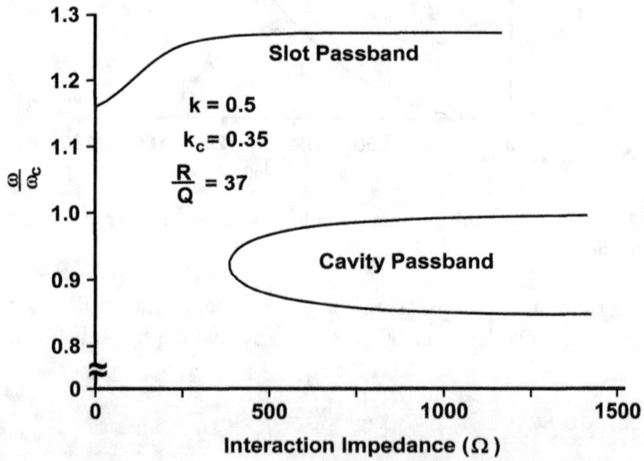

Figure 15-24 Interaction impedance derived from the circuit model for the staggered-slot structure.

Normally, the cavity passband is used for interaction with the electron beam, and this was the case for the results in the previous section. It is possible to use the slot passband, and tubes designed to operate in this mode have existed for many years [6]. An important property of the slot passband is that the interaction impedance at the upper cutoff frequency is zero (see Figure 15-24). This results from the vanishing axial electric field component on the axis, as is indicated in Figure 15-25. As a result, no special oscillation-suppression techniques are required at this frequency.

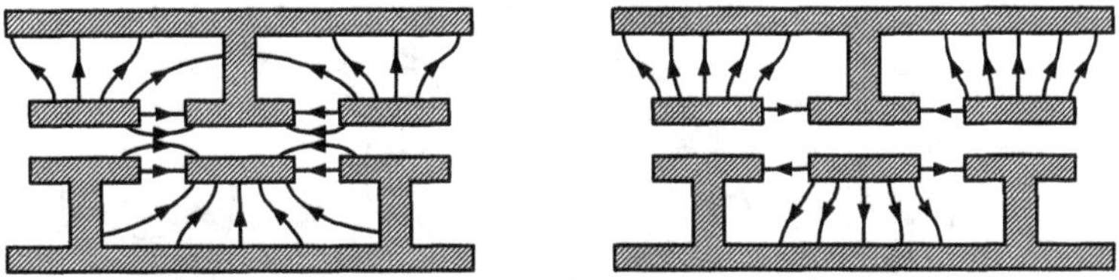

Figure 15-25 Electric field patterns for the slot-band cutoff frequencies. (From: J. R. Frey and I. Tammaru, 1981 IEDM. © 1981 IEEE.)

Attempts have been made in [7] to coalesce the cavity band and slot band as indicated in Figure 15-23 in an effort to achieve octave-bandwidth operation in staggered-slot TWTs. While the ω-β diagram indicates near synchronism of the circuit wave and the beam over the desired bandwidth, two difficulties are in-band backward wave instability and a large gain variation with frequency.

Once the equivalent circuit model with appropriate component values has been established for a coupled-cavity structure, the analysis of the interaction with the electron beam can be carried out. The steps of the analysis are conceptually the same as those used by Gittins in the vector diagram description of coupled-cavity TWT operation. That is, the electron beam induces current into the appropriate elements (cavity gaps) of the equivalent circuit, and the electric field of the circuit excites current and velocity modulation of the beam. The shapes of the electric fields of the circuit and the space charge fields in the beam must be accurately determined so that the analysis can properly predict tube performance.

15.3 FUNDAMENTAL BACKWARD WAVE OPERATION

Two commonly used fundamental backward wave circuits are illustrated in Figure 15-26. In each of these, apertures are used to couple the fields from one cavity to the next. The staggered-slot structure in Figure 15-26(a) was discussed in the previous section.

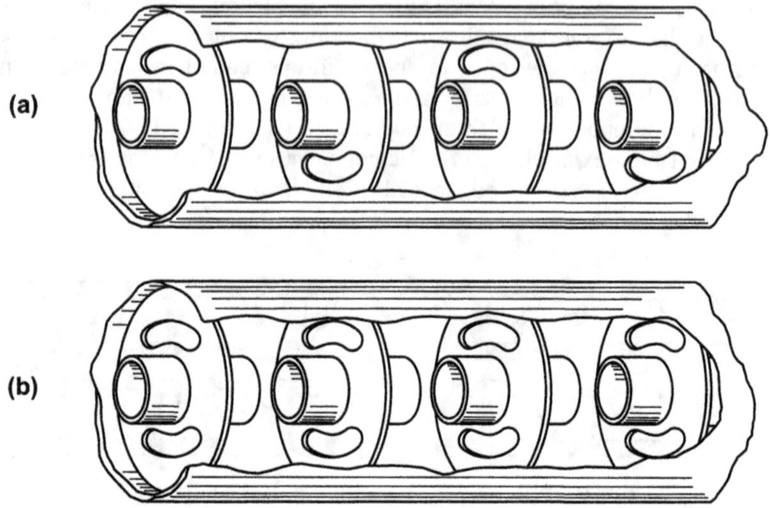

Figure 15-26 Staggered slot (a) and aligned slot (b) fundamental backward wave circuits. (Adapted from: *Power Travelling Wave Tubes* by J. F. Gittins, published 1965 by American Elsevier, Inc.)

When the slots are aligned, as shown in Figure 15-26(b), all of the circulating cavity current that couples to the previous cavity also couples to the following cavity. As a result, the aligned-slot structure has properties that are quite different than those for the staggered-slot structure. The primary difference is that the aligned-slot structure has a smaller bandwidth. However, the shape of the ω-β characteristic is such that, for a given frequency of operation, larger cavity dimensions are possible with the inline structure and so higher powers are possible than with the staggered-slot structure.

An example of the capability of a device with a staggered-slot circuit is shown in Figure 15-27. Confined flow solenoid focusing and a depressed collector are used. The solid curves show the power output at various drive levels (referenced to the 0 dB saturation drive level of 40 mW). The overall efficiency is in the 20% to 30% range.

Coupled-cavity structures are ideally suited for use in PPM focused TWTs. As shown in Figure 15-28, the magnetic pole pieces can be made integral with the cavity structure. The pole pieces are copper plated, and copper spacers are used to form cavities of the required height. Liquid cooling channels may also be included in the coupled-cavity structure. In some cases, the coolant is channeled around the drift tubes between cavities to carry away the heat resulting from interception of the electron beam.

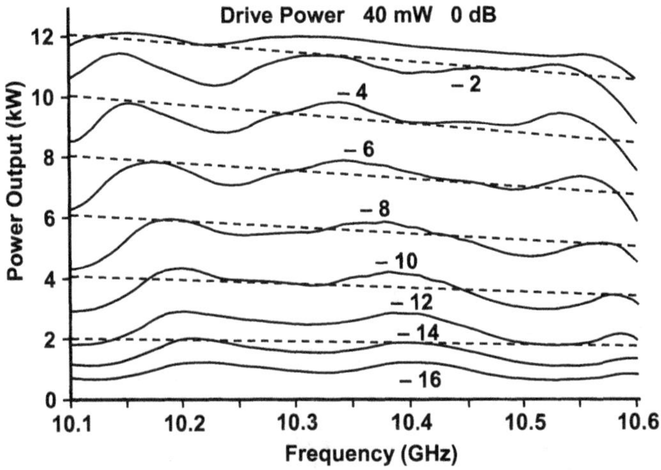

Figure 15-27 Example of the power capability of a CW coupled-cavity TWT. (From: Armand Staprans et al., *Proc. IEEE*, March 1973. © 1973 IEEE.)

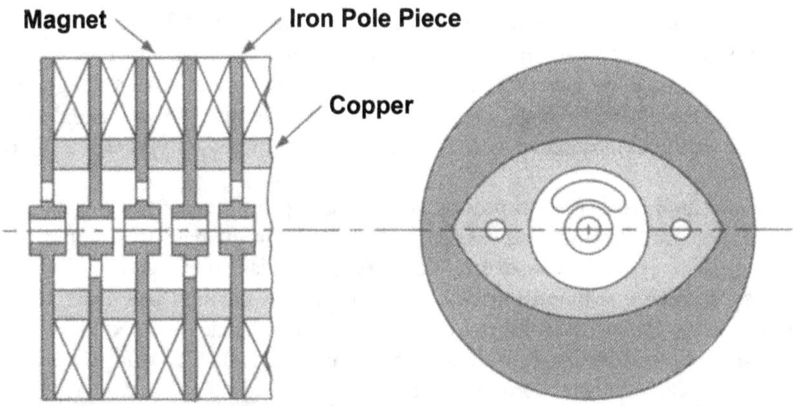

Figure 15-28 A coupled-cavity circuit with integral periodic permanent-magnet focusing. (From: Hughes Aircraft Co., *TWT and TWTA Handbook.*)

The TWT, for which the power output is given in Figure 15-29, is an example of what can be achieved in a staggered-slot circuit when PPM focusing was used. In this case, confined flow PPM focusing was utilized to minimize beam interception. With 35G at the cathode, the maximum interception current was less than 500 mA (out of a cathode current of 7.65A) so the RF beam transmission was 93.4% or better. A single-stage depressed collector was used, and the overall efficiency at the midband was 33.7%. It is estimated that, with a five-stage collector, the efficiency would be 50%.

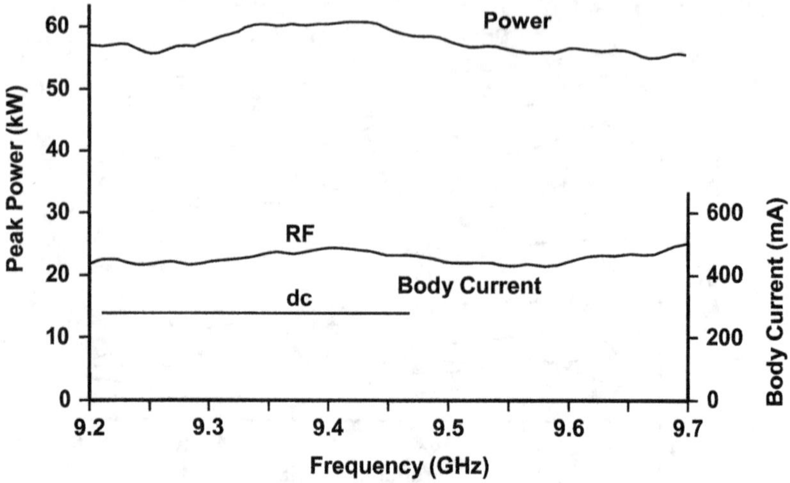

Figure 15-29 Performance of a pulsed PPM-focused coupled-cavity TWT at 10% duty cycle. (From: Hughes Electron Dynamics Division, *Microwave Jour.*, April 1989.)

When periodic permanent magnets are used, the electron beam is not as well focused as it is when a solenoid is used. As a result, the percentage of the beam current intercepted by the circuit is relatively high, and so the average power capability is relatively low. The reduction in average power capability of PPM-focused tubes compared with solenoid-focused tubes is typically a factor of four or more.

When coupled-cavity technology is extended to millimeter-wave frequencies, cavity dimensions become very small. The TWT circuit based on conventional cavity technology requires several dozen copper disks per section. Each disk contains a coupling slot and ferrules, which form the cavities. The disks are lathe turned, preferably by diamond-point machining. The kidney-shaped slots are cut by EDM (electric discharge milling). These disks are stacked and heated under pressure to braze or sinter them together.

At millimeter-wave frequencies, the problem with this conventional construction technique is that it requires the parts to be machined to extremely tight tolerances. One reason is that errors in cavity lengths may add cumulatively when the cavities are stacked. To maintain synchronism between the beam and circuit wave, extremely tight tolerances on the height of each cavity are necessary. A second reason is that the cavity resonant frequency is critically dependent on the cavity gap width. Machining tolerances on this dimension may be 70 millionths of an inch or smaller.

The problem of using a large number of parts machined to extremely tight tolerances can be circumvented by using the ladder-circuit designs described later in this chapter [8]. With these designs, tolerances are relaxed to 200 millionths of an inch. The cost, in turn, is reduced by a factor of about 2.

The ladder circuit is fabricated by machining notches in a slab of copper to form a "ladder" or "comb" structure as is indicated in Figure 15-30. The beam tunnel is formed by cutting half-round grooves in the tips of the comb teeth. Two identical structures are then brazed together. The circuit is completed by brazing cover plates to the faces of the ladder assembly. Because the ferrules, which form the drift-tube tips in the cavities, are omitted with the ladder-assembly technique, gain per cavity is reduced and so the interaction circuit length is increased. Still, the advantage of needing only a few parts per section rather than several dozen is considerable.

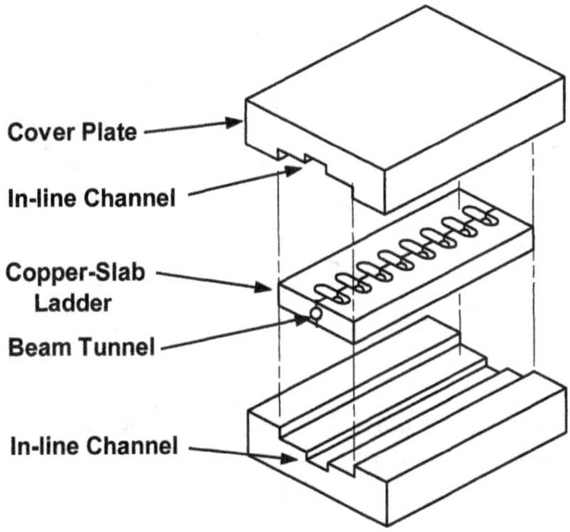

Figure 15-30 Ladder-core structure with dual in-line coupling channels. (Adapted from: Bill G. James, *MSN & CT,* September 1986.)

In [8] three versions of ladder circuits are described:

1. In the aligned-slot structure shown in Figure 15-30, the coupling slots from cavity to cavity are formed by machining channels in the cover plates.
2. In the staggered-slot structure shown in Figure 15-31, slots are machined into the edge of every other ladder rung, with the skipped rungs notched on the other side. The cover plates are flat slabs of copper.
3. In the double-staggered ladder, two coupling slots spaced 180° apart are used in each cavity wall. The pair of coupling slots is rotated 90° from cavity to cavity.

As shown in Figure 15-32, there are significant differences between the ω-β characteristics for these structures. These differences lead to advantages (or disadvantages) that govern the applications of the tubes.

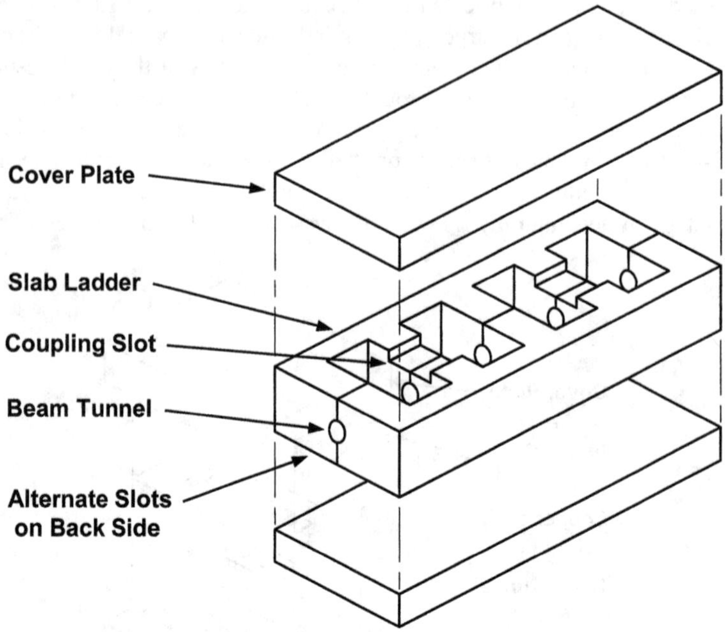

Figure 15-31 Ladder-core structure with staggered coupling slots. (Adapted from: Bill G. James, *MSN & CT*, September 1986.)

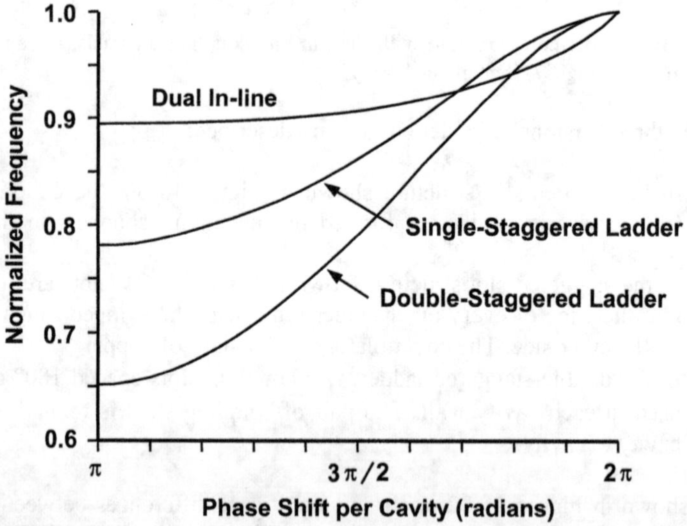

Figure 15-32 ω-β characteristics for ladder structures. (Adapted from: Bill G. James, *MSN & CT*, September 1986.)

With the aligned-slot design, the slot and cavity modes are coalesced (see the discussion of ω-β characteristics previously in this chapter), which allows the use of a reduced cavity-resonant frequency and larger cavity dimensions than for the staggered-slot circuit. In addition, because of the nature of the ω-β characteristic, a larger value of βL can be used and so the cavities can be longer than for the staggered-slot circuit. These two properties essentially double the power-handling capability of the circuit, making it one of the highest power linear-beam millimeter-wave circuits available. Figure 15-33 contains power and efficiency curves for a tube with a dual aligned-slot circuit. The bandwidth for this tube is approximately 1%. Theoretically, it is possible to achieve bandwidths of over 3%, depending on operating voltage and perveance.

The double-staggered ladder is the broadest bandwidth circuit of the three ladder structures discussed here. This circuit has a demonstrated 20% operating bandwidth over the 80- to 100-GHz range with a power output of 100-W CW at 90 GHz.

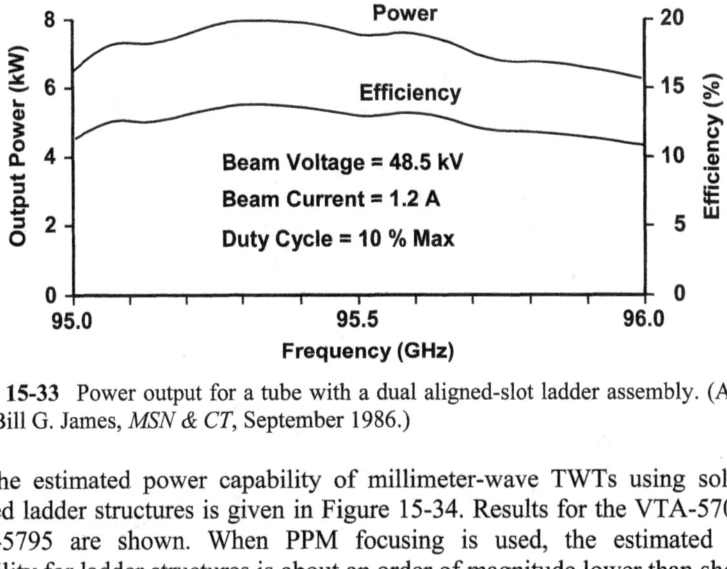

Figure 15-33 Power output for a tube with a dual aligned-slot ladder assembly. (Adapted from: Bill G. James, *MSN & CT*, September 1986.)

The estimated power capability of millimeter-wave TWTs using solenoid-focused ladder structures is given in Figure 15-34. Results for the VTA-5700 and VTW-5795 are shown. When PPM focusing is used, the estimated power capability for ladder structures is about an order of magnitude lower than shown in Figure 15-34.

The 1989 status of power capabilities of fundamental backward wave coupled-cavity TWTs with moderate bandwidth (few percent) are summarized for pulsed tubes in Figure 15-35 and for CW tubes in Figure 15-36 [9]. For larger bandwidths, the maximum power capability is lower than shown here. In Figure 15-35, E_k is the electron gun voltage and A_c is the area compression ratio of the electron gun. Note that, in these plots, the power capabilities of PPM tubes are about a factor of 3 below those for solenoid-focused tubes.

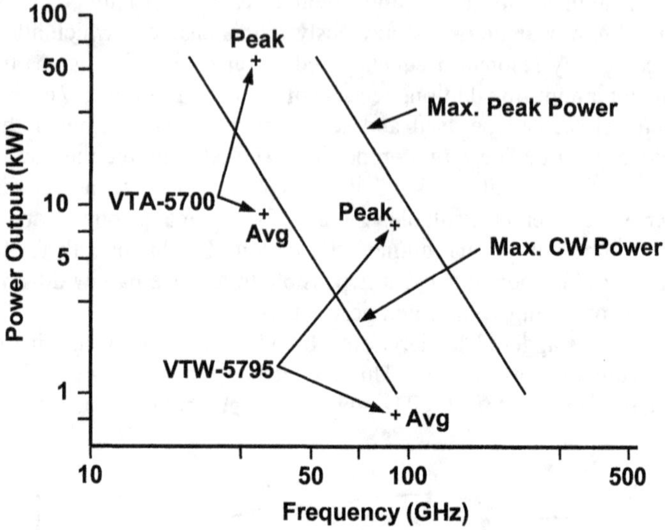

Figure 15-34 Practical power limits for millimeter-wave TWTs with solenoid-focused ladder structures. (Adapted from: Bill G. James, *MSN & CT*, September 1986.)

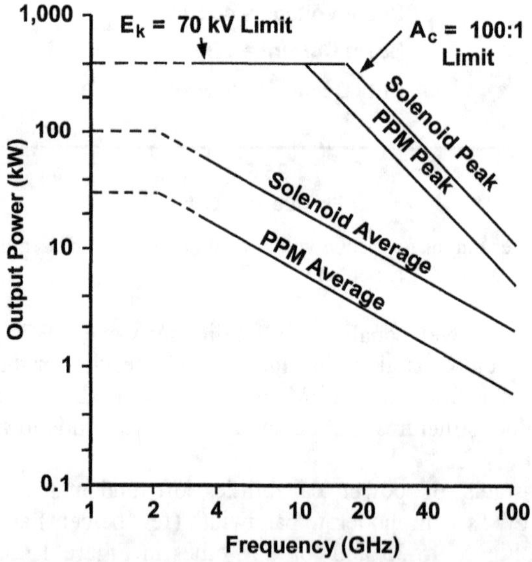

Figure 15-35 Output power for pulsed coupled-cavity TWTs with moderate bandwidths. (From: James W. Hansen, *Microwave Jour.*, 1989 State of the Art Reference.)

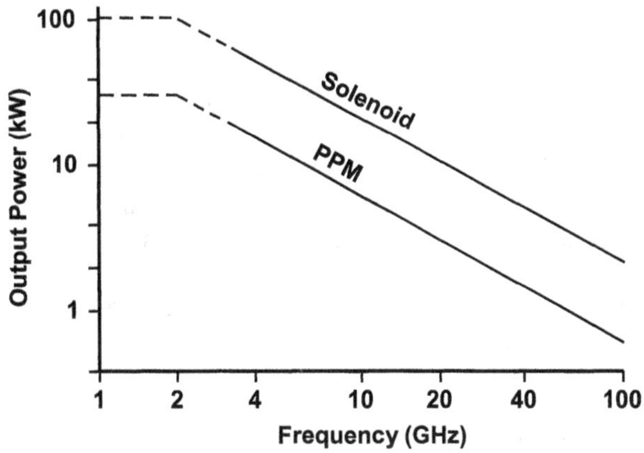

Figure 15-36 Output power for CW coupled-cavity TWTs with moderate bandwidths. (From: James W. Hansen, *Microwave Jour.*, 1989 State of the Art Reference.)

15.4 FUNDAMENTAL FORWARD WAVE OPERATION

The centipede (interlaced) structure shown in Figure 15-37 is an implementation of the coupling loop scheme discussed at the beginning of the chapter for achieving fundamental forward wave operation [10]. Many coupling loops are used to maximize coupling and bandwidth. As shown in Figure 15-37(b), the height of each cavity must be adequate to accommodate the large coupling loops required to maximize coupling. Since the beam transit angle from cavity to cavity is only about 90°, the electron beam velocity must be high, so the centipede structure is suitable for high-power operation.

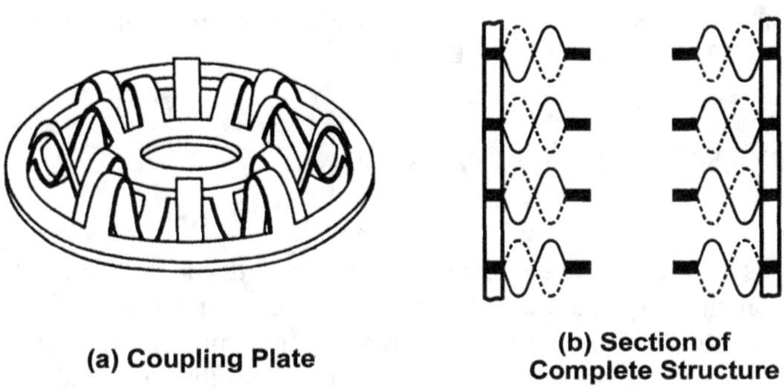

Figure 15-37 (a, b) Centipede (interlaced) structure. (Adapted from: *Power Travelling Wave Tubes* by J. F. Gittins, published 1965 by American Elsevier, Inc.)

The cloverleaf structure shown in Figure 15-38, is a fundamental forward-wave circuit [10]. This structure was also described Chapter 11 where output circuits for klystrons were discussed. The four-lobed cavity distorts the magnetic field lines into a cloverleaf pattern. Radial components of the field are coupled from cavity to cavity by radial slots. By rotating each cavity 45° with respect to adjacent cavities, the fields in adjacent cavities are in phase except for a delay of approximately 90° associated with coupling the signal from cavity to cavity.

The cloverleaf structure is used in high-power TWTs. An example, which was developed many years ago and which is shown in Figure 15-39, is the Varian VA-125 for which the power output is given in Figure 15-40. Note that output powers in excess of 4 MW are obtained.

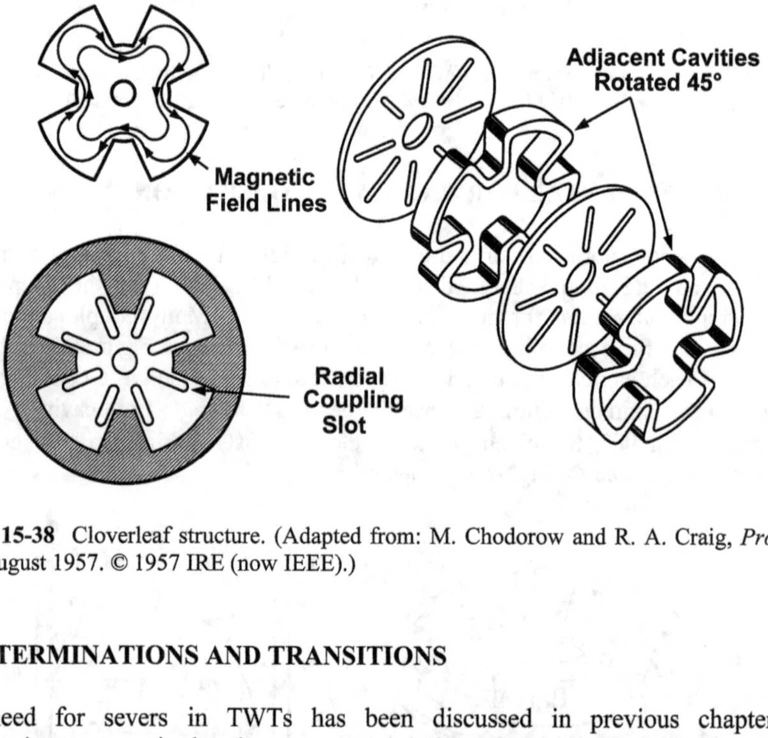

Figure 15-38 Cloverleaf structure. (Adapted from: M. Chodorow and R. A. Craig, *Proc. IRE*, August 1957. © 1957 IRE (now IEEE).)

15.5 TERMINATIONS AND TRANSITIONS

The need for severs in TWTs has been discussed in previous chapters. Terminations are required at the severs to minimize reflections. The RF power that a termination must dissipate can be appreciable in coupled-cavity TWTs. As a result, the termination is normally a lossy ceramic material capable of high-temperature operation and appropriately bonded to the surrounding metallic structures to achieve the required heat dissipation capability. An example of a termination for a fundamental forward wave cloverleaf structure is shown in Figure 15-41.

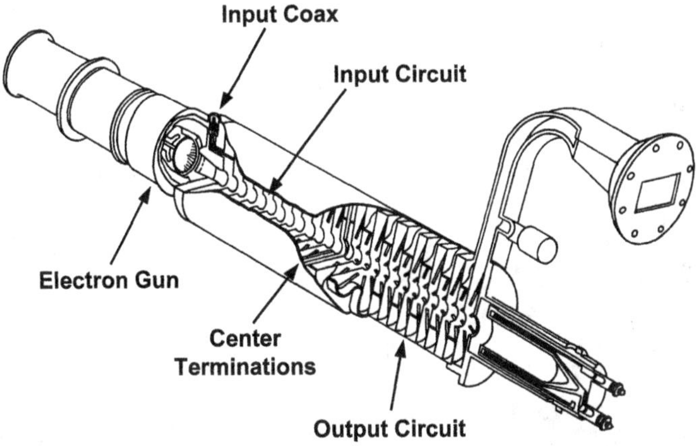

Figure 15-39 The use of a cloverleaf structure in the Varian VA-125 high-power TWT. (Adapted from: *Power Travelling Wave Tubes* by J. F. Gittins. © 1965 American Elsevier, Inc.)

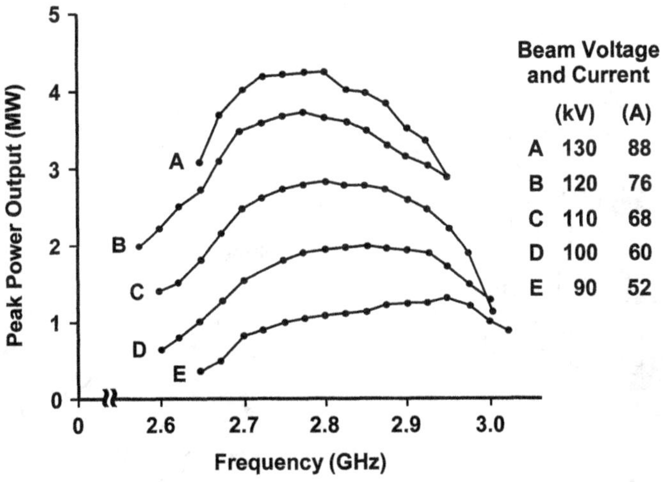

Figure 15-40 The power output for a VA-125A TWT. (From: Armand Staprans et al., *Proc. IEEE*, March 1973. © 1973 IEEE.)

As shown in Figure 15-42, reentrant loss pellets can be used in a fundamental backward wave structure to reduce small-signal gain ripple. According to Hughes, the buttons absorb about three quarters of the reflected power; however, they complicate matching but still they are widely used [11]. Another example of the use of loss pellets is shown for a double-slotted cavity in Figure 15-43.

Examples of termination assemblies at a sever are shown in Figure 15-44. According to Hughes, the double-wedge design has been shown to reduce small-signal gain ripple significantly [11]. Also, the double wedge has twice the thermal capability of a single wedge.

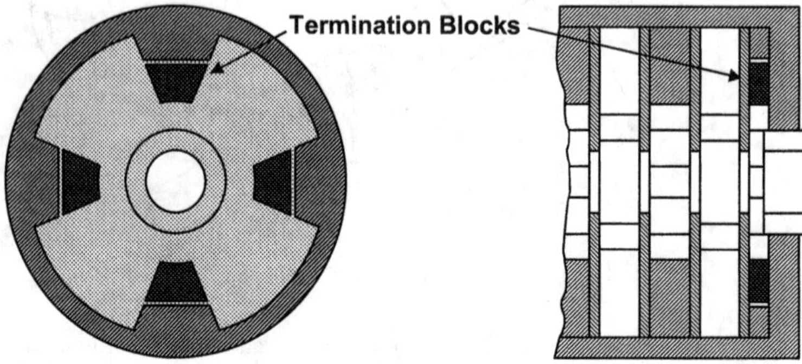

Figure 15-41 Termination for a cloverleaf structure. (Adapted from: *Power Travelling Wave Tubes* by J. F. Gittins. © 1965 American Elsevier, Inc.)

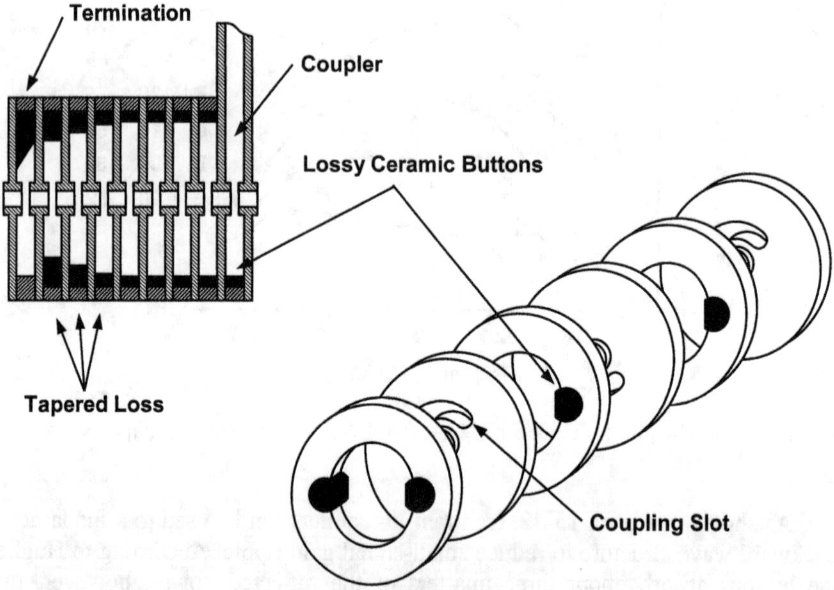

Figure 15-42 Termination and lossy ceramic buttons. (From: Hughes Aircraft Co., *TWT and TWTA Handbook.*)

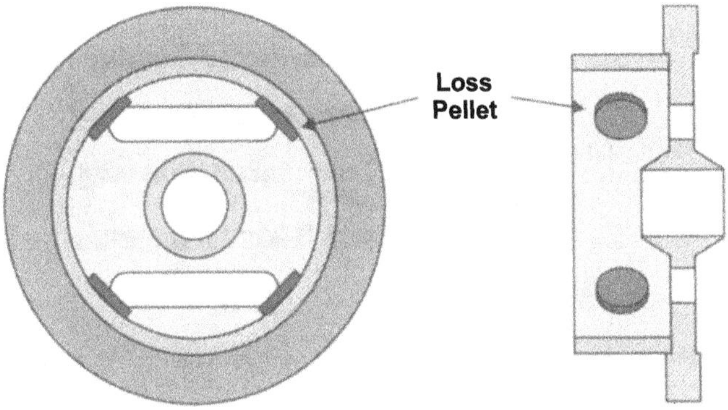

Figure 15-43 Loss pellets in a slotted cavity.

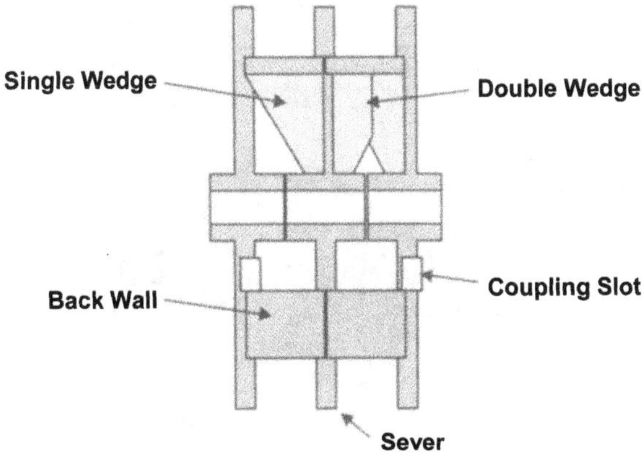

Figure 15-44 Wedge terminations. (Courtesy of Hughes Electron Dynamics Division.)

Two types of transitions are shown for the high-power coupled-cavity TWT in Figure 15-39. At the input, where the power level is low, a coaxial line is used to couple to a cavity, and at the output, a transition from the cavity to a waveguide is used. The use of a coaxial line and a coupling loop is common at low power levels. A coaxial line and coupling loop for a coupled-cavity structure are illustrated in Figure 15-45.

At X band the small size of the input cavity (for example, 15.4 mm inside diameter by 3.8 mm height [12]) makes it difficult to use a coupling loop because the loop is so small. Because the use of a coupling loop seemed to be very complicated and nonrepeatable, Wiejak used the cloverleaf antenna arrangement shown in Figure 15-46. The cloverleaf was attached directly to the ferrule of the cavity [12]. The impedance match to the input coax can be adjusted by varying the area and thickness of the clover leaf and the shape of the transformer.

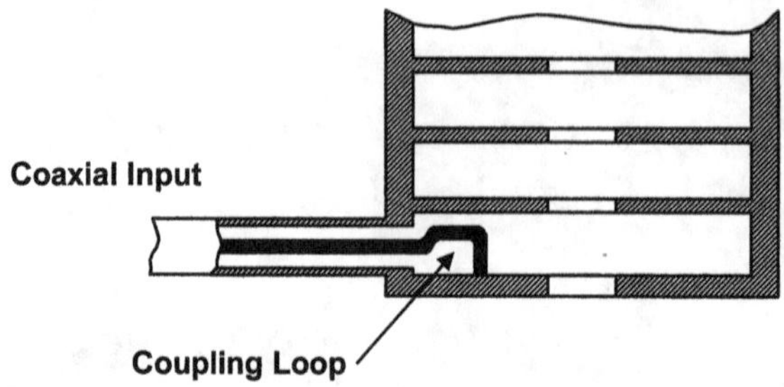

Figure 15-45 Coaxial-line transition to a coupled-cavity structure. (Adapted from: *Power Travelling Wave Tubes* by J. F. Gittins. © 1965 American Elsevier, Inc.)

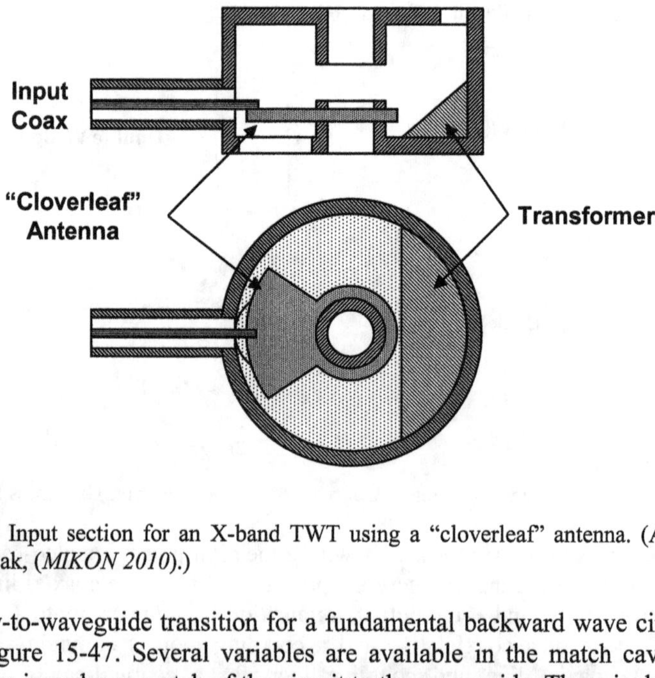

Figure 15-46 Input section for an X-band TWT using a "cloverleaf" antenna. (Adapted from: W. Wiejak, (*MIKON 2010*).)

A cavity-to-waveguide transition for a fundamental backward wave circuit is shown in Figure 15-47. Several variables are available in the match cavity for optimizing the impedance match of the circuit to the waveguide. These include the ferrule height, the distance of the backwall from the ferrule, and the shape of the backwall. Then there is a stepped transformer to match the cavity waveguide to the load waveguide.

A transition from waveguide to the cloverleaf circuit of a fundamental forward wave TWT is shown in Figure 15-48. For impedance matching, it is often necessary to use inductive vanes in the waveguide. These elements offset the capacitive reactance of the cavity gap.

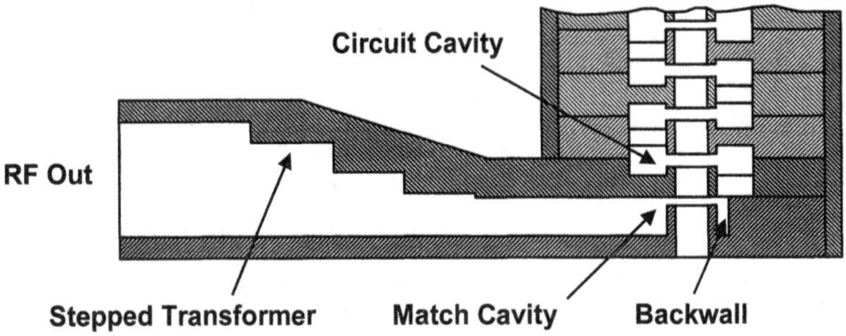

Figure 15-47 Output section for X-band TWT, Courtesy of Hughes Electron Dynamics Division.

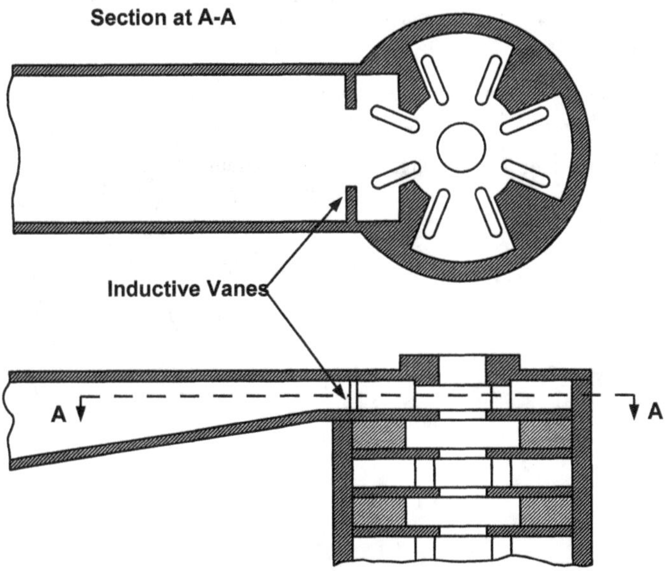

Figure 15-48 Transition from waveguide to clover-leaf structure. (From: *Power Travelling Wave Tubes* by J. F. Gittins. Published 1965 by American Elsevier, Inc.)

REFERENCES

[1] J. T. Mendel, "Helix and coupled-cavity traveling wave tubes," *Proc. IEEE*, Vol. 61, No. 3, March 1973, pp. 280–298.

[2] A. Staprans, E. W. McCune, and J. A. Ruetz, "High-power linear-beam tubes," *Proc. IEEE*, Vol. 61, No. 3, March 1973, pp. 299–330.

[3] H. J. Curnow, "A general equivalent circuit for coupled-cavity slow wave structures," *IEEE Trans. Microwave Theory Tech.*, Vol. MTT-13, No. 9, September 1965, pp. 671–675.

[4] H. J. Curnow, "A new equivalent circuit for coupled-cavity structures," *Proc. 5th International Congress on Microwave Tubes*, (conference held in Paris, France), New York and London: Academic Press, September 14–18, 1964.

[5] J. F. Gittins, Power *Travelling Wave Tubes*, New York: Elsevier, 1965.

[6] J. R. Frey, and I. Tammaru, "A coupled-cavity TWT operating in the inverted slot mode," *Technical Digest*, IEDM, 1981, pp. 504–506.

[7] A. Karp, and W. R. Ayers, "Design concepts for an octave bandwidth coupled-cavity TWT," *Technical Digest*, IEDM, 1978, pp. 546–549.

[8] B. G. James, "Coupled-cavity TWT designed for future mm-wave systems," *MSN Microwave Systems News & Communications Technology*, Vol. 16, No. 10, September 1986, pp. 105–116.

[9] J. W. Hansen, "State of the art reference, US TWTs from 1 to 100 GHz," *Microwave Jour.*, 1989, pp. 179–193.

[10] M. Chodorow, and R. A. Craig, "Some new circuits for high-power traveling-wave tubes," *Proc. IEEE*, Vol. 45, No. 8, August 1957, pp. 1106–1118.

Chapter 16

Collectors

There are, basically, two major considerations in the design of collectors for linear-beam tubes. These are:

1. Power dissipation;
2. Power recovery.

When power dissipation is the dominant consideration, the collector is a single electrode at the voltage of the body of the tube (usually ground) and is designed to dissipate as heat all of the power in the spent beam (that is, the beam leaving the RF interaction structure of the tube). These collectors are usually used on very large devices in applications where power consumption is not a critical consideration.

When power recovery is most important, the collector may have a single electrode or it may have two or three or more electrodes. To recover power, the voltages on the electrodes must be negative with respect to the body of the tube so that the beam is slowed down before it is collected. The energy recovered in slowing the beam is returned to the power supply and the heat that must be dissipated by the collector is greatly reduced. These collectors are used where overall efficiency is an important consideration.

16.1 POWER DISSIPATION [1]

In a linear-beam tube, in the absence of RF power, the entire beam energy must be dissipated in the collector. At first glance this does not appear to be a difficult problem, since, in principle, the collector may be made as large as desired. However, it is usually desirable to keep the collector as small as intercepted power densities will allow. This requires a detailed understanding of the electron trajectories inside the collector.

In most collectors, the magnetic field is very small. The magnetic field is terminated in a pole piece between the end of the interaction structure and the collector. Ideally, for all focusing schemes where the cathode is also in a field-free region, such as for Brillouin flow, the beam enters the collector with only an axial component of velocity. Thus, spreading of the beam in the collector results only from space-charge forces. If magnetic flux threads the cathode, then it also threads the collector. Then there is rotation of the beam in the collector, which is directly proportional to the amount of flux through the cathode. This rotation spins the electrons out radially and so it contributes to beam spreading in the collector. The effect of the magnetic field may be shown by universal beam-spread curves in the presence of the magnetic field threading the cathode, as shown in Figure 16-1. Beam spreading is increased substantially for confined-flow beams, which may use a focusing field of about three times the Brillouin value. For constant charge density in the beam, the spreading of the beam is laminar and, in principle, power density is known everywhere.

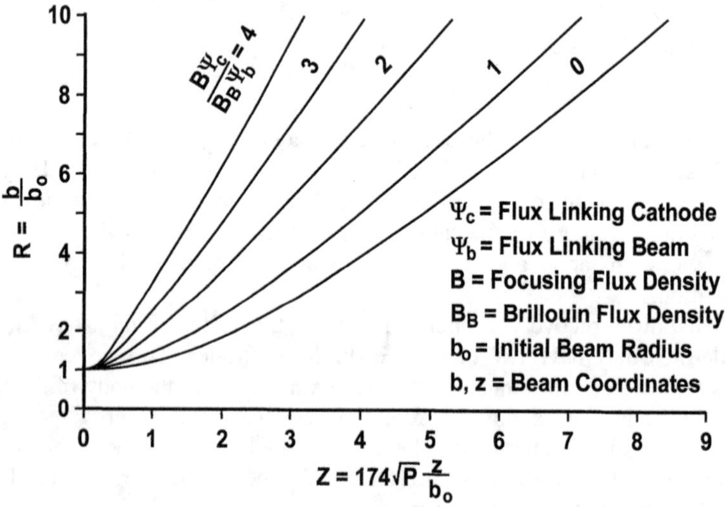

Figure 16-1 Beam spread curves for magnetically confined beams in a field-free collector. (Adapted from: *Linear-Beam Tubes* by A. Staprans, Internal Varian Report, September 1968.)

In the ideal case, to minimize the size of a collector, the objective is to design the inside shape so that the intercepted power density is constant at the maximum tolerable value. The shape of the resulting collector (Figure 16-2(a)) is impractical because it would be difficult to fabricate and because small changes in the beam pattern would vary power dissipation at the walls and cause overheating in some regions.

A collector is generally designed approximately as shown in Figure 16-2(b). As shown in Figure 16-2(c), two critical areas of high dissipation exist:

1. Where the beam first impinges on the collector wall (dissipation is maximum);
2. At the end of the collector.

The errors in the idealized analysis of trajectories inside a collector are mostly such that they make the beam less divergent than occurs in an actual tube. Causes may be:

1. Leakage magnetic field from the focusing magnet inside the collector;
2. A gradual instead of the idealized sharp cutoff of the magnetic focusing field at the entrance to the collector;
3. The presence of positive ions which neutralize the electron space-charge forces.

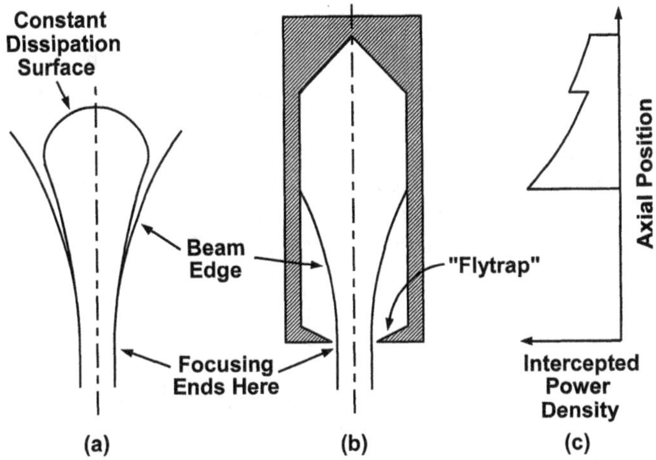

Figure 16-2 (a) Collector for constant dissipation, (b) practical collector, and (c) intercepted power density. (Adapted from: *Linear-Beam Tubes* by A. Staprans, Internal Varian Report, September 1968.)

In practice the cylindrical wall of the collector is designed according to the best estimate of actual beam trajectories. However, the collector is made longer than the calculation would indicate, and a conical end is often used to increase the intercepting surface area. Typically, the ratio of average to maximum (where the beam is first intercepted) wall dissipation will vary between 0.25 and 0.5.

The maximum allowable intercepted power density is affected by:

1. Collector wall material and thickness;
2. Method of heat transfer from the wall to the cooling medium;
3. Whether the intercepted power is continuous or pulsed.

Collectors for high-power tubes are usually made of copper because of its high thermal conductivity. If the collector wall is relatively thin, less than 1 cm, the temperature rise in the wall from the average beam power is less important than the mechanism of heat transfer to the cooling medium. For forced-flow water cooling with water channels of ordinary design, the average dissipation is limited to a working value of about 1 kW/cm^2. Dissipations up to 18 kW/cm^2 have been realized over small areas in some crossed-field devices by the use of laminated wall materials, very small water channels, and extremely high water pressures [2]. It has not been necessary to resort to these measures in linear-beam tubes.

The limits of pulse heating are less well understood than average heating. Since pulse lengths do not usually exceed a few milliseconds, the problem is essentially a surface phenomenon. During a pulse, the temperature rises to a high value in a very thin surface layer. Two problems may arise:

1. The surface may melt during the pulses.
2. Stresses caused by the expansion and contraction of the metal during the repeated heating and cooling may gradually destroy the surface.

Since a very thin boundary layer is involved, the analysis of pulse heating can be dealt with as a one-dimensional problem. If all intercepted beam power is assumed to be stored by diffusion into the collector walls, the pulse temperature rise as a function of peak intercepted power density, P, and pulse duration, t, is [3]:

$$\Delta T = \frac{0.27\, P \sqrt{t}}{\sqrt{K\, c\, \rho}} \quad ^\circ\text{C} \tag{16-1}$$

Here K, c, and ρ are the thermal conductivity, specific heat, and specific gravity. This gives a somewhat larger surface temperature rise than what actually occurs, because electrons penetrate a small distance into the material. The penetration distance increases rapidly with beam voltage. An analysis of pulse heating, including electron penetration into the material, has been provided by Crapuchettes [4]. He has also derived the stresses induced in the surface of the material by the repeated expansion and contraction due to pulse heating. He concludes that the allowable temperature rise due to thermal stresses (to avoid surface cracking) is below 100°C for copper. Practice has shown that this is a pessimistic value and that a somewhat higher pulse temperature rise is permissible.

Since cyclic thermal stresses present a more severe limitation than does melting of the surface, the materials other than copper for the boundary layer are sometimes used. For example, molybdenum has a lower coefficient of thermal expansion and much higher strength than copper, which more than compensates for its poorer thermal conductivity. As a result, molybdenum is commonly used on the vane tips of crossed-field devices.

In high-power tubes it is important to design the tubes to reduce the possibility of "slow" electrons (which occur at RF saturation) and of secondary electrons (from the collector) striking the tube body. Therefore, the entrance to the collector is made as small as possible without actual interception of primary beam electrons. The entrance section of the collector, often of a conical shape, is sometimes referred to as the "fly trap."

16.2 POWER RECOVERY

A primary consideration in selecting microwave tubes for many applications is the overall efficiency in converting dc input power to RF output power at the desired frequency. One very important part of overall efficiency is electronic efficiency, which is the efficiency of conversion of power in the electron beam to RF power. Electronic efficiency has received considerable attention since the invention of microwave tubes. In this section, it will be assumed that the electronic efficiency has been established in the design of the RF characteristics of a tube. Then, with electronic efficiency fixed, the enhancement of the overall efficiency with the collector will be discussed.

16.2.1 Power Flow

The various elements of power flow in a linear-beam tube are illustrated in Figure 16-3. Power is supplied for the heater, the electron beam, and the solenoid. The RF input power is normally at least 30 dB below the RF output power and is assumed to be negligible.

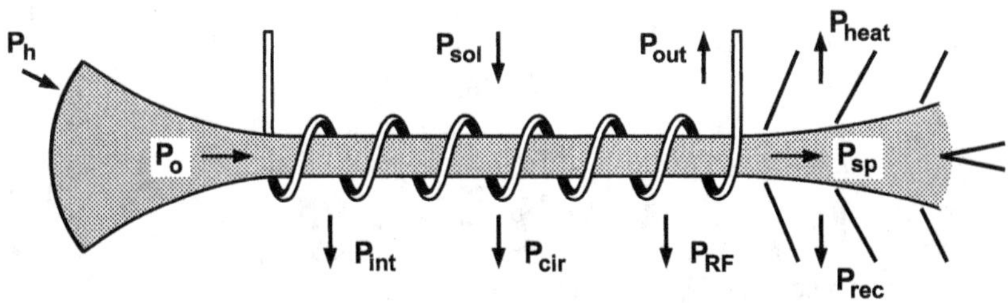

P_h = Heater Power

P_o = Beam Power from Gun

P_{int} = Interception Loss

P_{cir} = Circuit Loss

P_{RF} = Harmonics, Intermods, etc.

P_{sol} = Solenoid Power

P_{out} = RF Output

P_{sp} = Power in Spent Beam

P_{heat} = Heat in Collector

P_{rec} = Power Recovered by Collector

Figure 16-3 Power flow in a linear-beam tube.

Power is lost when electrons are intercepted by the circuit and by RF losses within the tube. In addition, some of the RF power that is generated is lost because it is not at the desired frequency (harmonics, intermodulation products, and so forth).

Finally, after the RF output is removed from the tube, the beam leaving the RF circuit (the spent beam) contains power that is delivered to the collector. Some of this power is converted to heat by electrons striking the collector. If the collector is designed properly, the electrons can be slowed down before they strike the collector surfaces, and the collector can thereby recover power from the spent beam [5].

Using the various elements of power identified in Figure 16-3, it is straightforward to derive power flow and efficiency relationships. The overall tube efficiency, η_{ov}, is simply the ratio of the RF output power to the total input power, P_{in}. That is,

$$\eta_{ov} = \frac{P_{out}}{P_{in}} \tag{16-2}$$

where

$$P_{in} = P_o + P_h + P_{sol} - P_{rec} \tag{16-3}$$

or

$$P_{in} = P_{ot} - P_{rec} \tag{16-4}$$

where

$$P_{ot} = P_o + P_h + P_{sol} \tag{16-5}$$

is the total power associated with generating and focusing the electron beam. As a result,

$$\eta_{ov} = \frac{P_{out}}{P_{ot} - P_{rec}} \tag{16-6}$$

The power recovered by the collector is

$$P_{rec} = \eta_{coll} P_{sp} \tag{16-7}$$

where η_{coll} is the efficiency of the collector. The spent beam power is beam power from the gun less the power removed from the beam as it passes through the tube. That is,

$$P_{sp} = P_o - P_{int} - P_{cir} - P_{RF} - P_{out} \tag{16-8}$$

so

$$P_{coll} = \eta_{coll}\ (P_o - P_{int} - P_{cir} - P_{RF} - P_{out}) \tag{16-9}$$

Two additional efficiency terms are useful in completing the analysis. One is the electronic efficiency, η_e, which, as pointed out earlier, is the efficiency of conversion of beam power to RF power at the desired frequency, that is,

$$P_{out} + P_{cir} = \eta_e\, P_o \tag{16-10}$$

Values for η_e range from a few percent to well over 50%.

The other efficiency is the circuit efficiency, η_{cir}, which is the efficiency of the RF circuit in delivering the generated RF power at the desired frequency to the output connector. Thus,

$$P_{out} = \eta_{cir}\,(P_{out} + P_{cir}) = \eta_{cir}\,\eta_e\, P_o \tag{16-11}$$

Values for η_{cir} are usually in the 75 to 90% range.

Now, by combining (16-11) and (16-9) with (16-6), the overall efficiency is

$$\eta_{ov} = \frac{\eta_{cir}\ \eta_e}{\dfrac{P_{ot}}{P_o} - \eta_{coll}\left(1 - \eta_e - \dfrac{P_{int} + P_{RF}}{P_o}\right)} \tag{16-12}$$

The following plot of (16-12) illustrates the significances of various efficiencies and powers.

In Figure 16-4, the effect of P_{ot} is shown on the 90% collector efficiency plot. This same range of variations would apply to other η_{coll} plots, but is shown for only the 90% plot. Similarly, the effects of variations in $P_{int} + P_{RF}$ are shown on the 70% η_{coll} plot, but would apply to other collector efficiency plots. The effects of variations in circuit efficiency are shown on the 50% η_{coll} plot but would apply to other η_{coll} plots.

Perhaps the most striking aspect of Figure 16-4 is the strong dependence of overall efficiency on collector efficiency, especially for high values of collector efficiency. This leads directly to the next section, which deals with the design of efficient collectors.

The importance of circuit efficiency is also evident in Figure 16-4. Note that a 10% increase in η_{cir} increases η_{ov} about five percentage points. This leads to considerations of circuit efficiency, which were discussed in Chapter 14.

Finally, in considering (16-12) and Figure 16-4, heater power and RF losses (harmonics and intermodulation products) are normally small. Thus, P_{ot} contains primarily the power penalty resulting from the use of a solenoid for focusing. Increases in η_{ov} that result from decreases in solenoid power (and P_{ot}) can readily be traded off against the decrease in η_{ov} that results from an increase in P_{int} (interception power), which, in turn, normally results from a decrease in solenoid power.

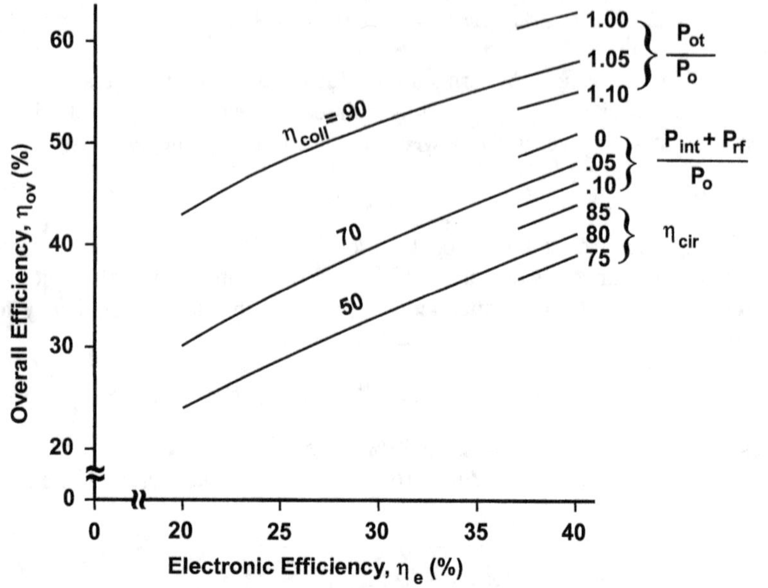

Figure 16-4 Variations of overall efficiency with the various elements of efficiency and power.

16.2.2 Power Recovery with a Depressed Collector

A collector for a linear-beam tube is shown in Figure 16-5. As the beam passes through the pole piece, the magnetic focusing field is removed, and so space charge forces cause the beam to expand as it enters the collector.

With the collector at the same potential as the body of the tube, electrons strike the collector at a relatively high velocity. Electron energy is converted to heat at the collector surface. With a collector like that shown in Figure 16-5, a relatively large percentage (normally 50% or more depending on electronic efficiency) of the power input to the tube is dissipated in heat in the collector.

By reducing (depressing) the voltage on the collector below the body potential, the velocities of electrons striking the collector and the heat generated are reduced. As a result, the collector recovers some of the power in the spent electron beam.

The concept of power recovery by a collector is sometimes confusing. The following example should help in understanding power recovery. Assume that a TWT with a 10-kV, 1-A electron gun produces an RF power of 2 kW as is indicated in Figure 16-6. Assume that, other than interception, there are no losses, heater power is negligible, and PPM focusing is used. With the collector at the same potential as the body and the helix (ground potential) as shown in Figure 16-6(a), the overall efficiency is

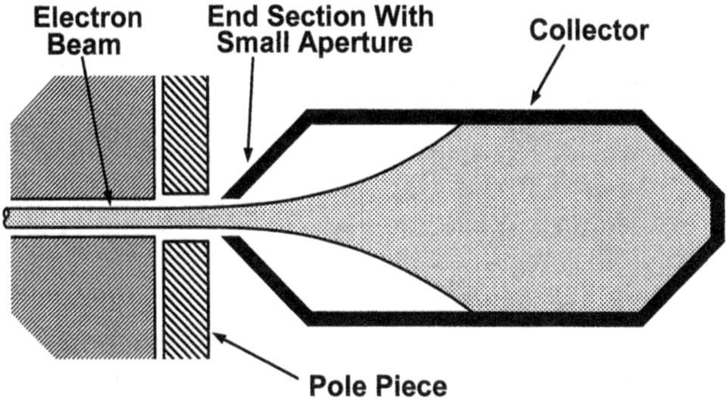

Figure 16-5 Collector for a linear-beam tube.

$$\eta_{ov} = \frac{2\,\text{kW}}{10\,\text{kV} \times 1\text{A}} = 20\% \qquad (16\text{-}13)$$

All of the current intercepted by the circuit and collected by the collector returns through the cathode power supply. It should also be noted that, to regulate the RF output, the full 10 kW of input power must be regulated.

(a)

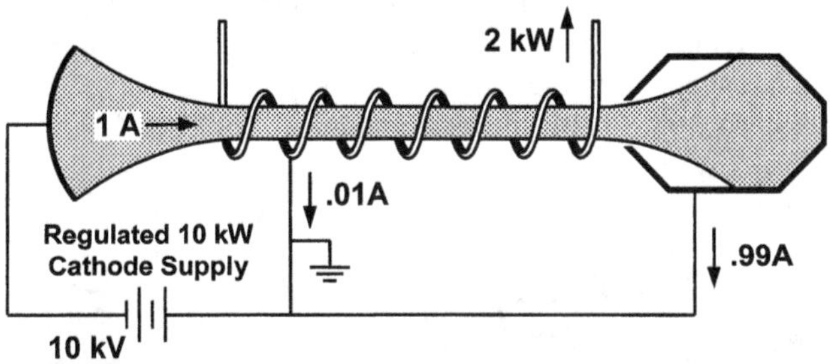

(b)

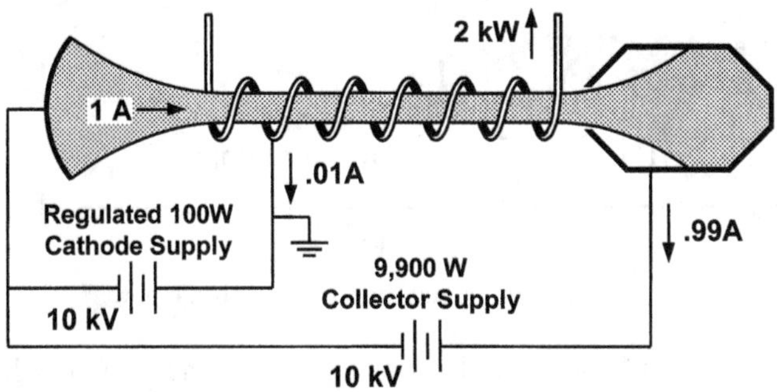

Figure 16-6 Power supply connections for collectors at ground potential: (a) common power supply and (b) separate power supplies.

In Figure 16-6(b), a second 10-kV power supply has been provided for the collector. The total power supplied to the tube is the same as in Figure 16-6(a) and so the efficiency is still 20%. However, to regulate the RF output power, only the cathode supply must be regulated and the power provided by that supply is now only 100W. The regulation requirement for the collector supply of 9,900W is normally far less severe than for the cathode supply. As a result, while there has been no improvement in efficiency in Figure 16-6(b), a substantial reduction in power supply complexity may have been achieved.

In Figure 16-7, the voltage of the collector supply has been reduced to 6,000V. The current provided by this supply is still the full collector current of 0.99A; however, the power provided is only 5,940W. The RF output power is still 2 kW and so the overall efficiency is now

$$\eta_{ov} = \frac{2{,}000\text{W}}{100\text{W} + 5{,}940\text{W}} = 33\% \tag{16-14}$$

which is a substantial improvement over the 20% obtained in Figure 16-6.

The improvement in efficiency in Figure 16-7 has been achieved by reducing (depressing) the voltage of the collector below the body and helix voltage and, thereby, slowing down the electrons before collecting them. As a result, we say that the collector has "recovered" power from the electron beam. It should be noted here that the power supply engineer considers collector depression to be a technique for reducing power input rather than for recovering power because the power was not supplied to the tube in the first place. In fact, of course, both points of view are correct.

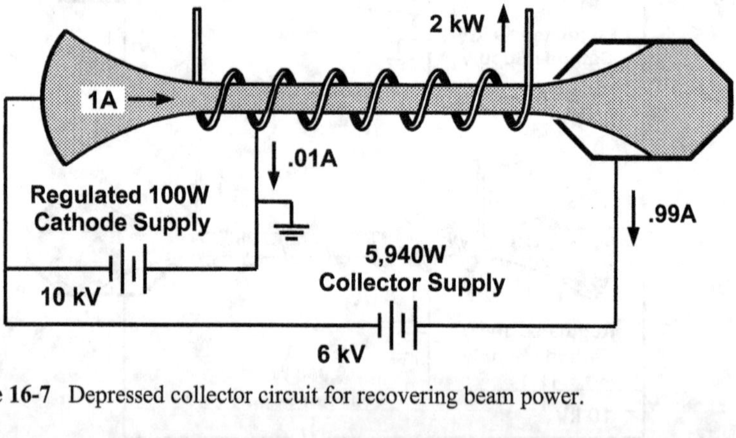

Figure 16-7 Depressed collector circuit for recovering beam power.

At the risk of belaboring this point, consider the example shown in Figure 16-8. Here, the collector is at cathode potential and there is no RF circuit to extract power from the electron beam. It is also assumed that no current is intercepted by the anode. With 10 kV applied to the anode, a 10-kW electron beam is generated.

However, as this beam enters the collector, it is slowed down and the electron velocity drops to zero as the electrons land on the collector. No heat is generated in the collector and there are no other losses and so no power is dissipated. Thus, a 10-kW beam has been generated by the electron gun and all of this power has been recovered by the collector. No power is provided by the 10-kV supply because there is no current flowing to it.

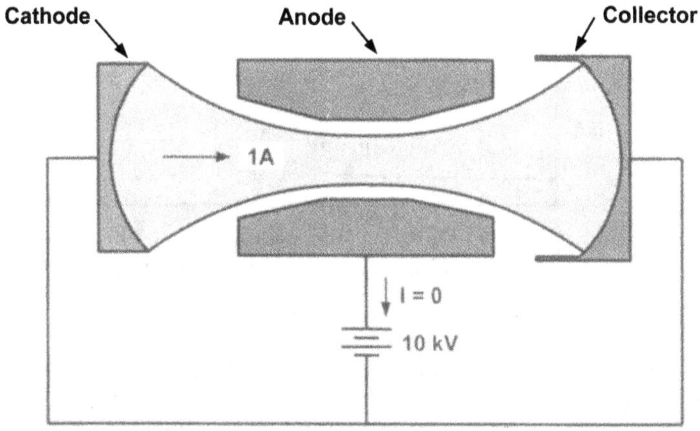

Figure 16-8 Electron beam device with a 10-kW beam and full beam power recovery.

16.2.3 Electron Energy Distribution

In Figure 16-7, the collector voltage was 6 kV, or 4 kV below the body potential. The collector was "depressed" 4 kV, and this raised the efficiency from 20% to 33%. A logical question to ask is: Why not depress the collector further and improve the efficiency more? The problem is that the voltage of a depressed collector supply like that shown in Figure 16-7 is highly dependent on the distribution of electron velocities in the beam. The slower electrons will be reflected by the collector if the potential is depressed too far. With increasing collector voltage, as shown in Figure 16-9, collector current will decrease and body current will increase. Excessive body current will damage the RF structure. Also, reflected electrons traveling back through the tube induce noise in the RF circuit and can provide a signal feedback path to the TWT input.

Single-element collectors, like that in Figure 16-7, recover roughly 30% to 40% of the power in a spent beam (dependent on tube type). To recover a large fraction (over 80%) and, at the same time, prevent electrons from being reflected, the electric fields in a collector must be shaped so that electrons are nearly stopped just as they strike the collector electrodes. In addition, the fields must be shaped to prevent secondary electrons from being accelerated (thus wasting energy) or leaving the collector.

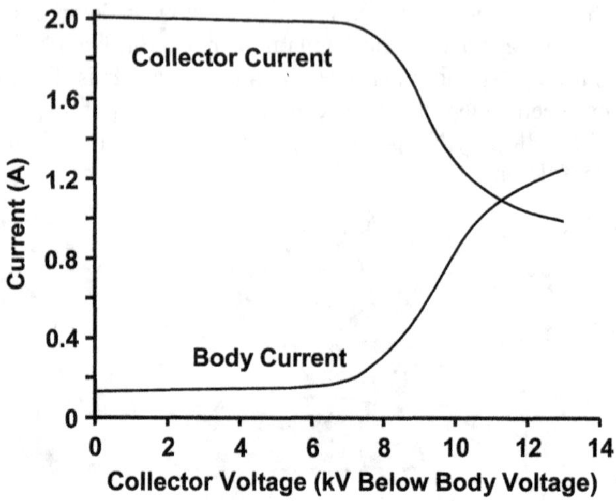

Figure 16-9 Effect of collector voltage on body current.

To better understand the problem of building a highly efficient collector, it is necessary to examine the energies of the electrons in the beam entering the collector. For purposes of illustration, it will be assumed that the collector power supply is connected as shown in Figure 16-10 and that the collector voltage can be varied. (Note that this is not a practical way to apply voltage to the collector because power is always flowing into the supply. Thus, the supply is acting like a battery that is continually being charged.)

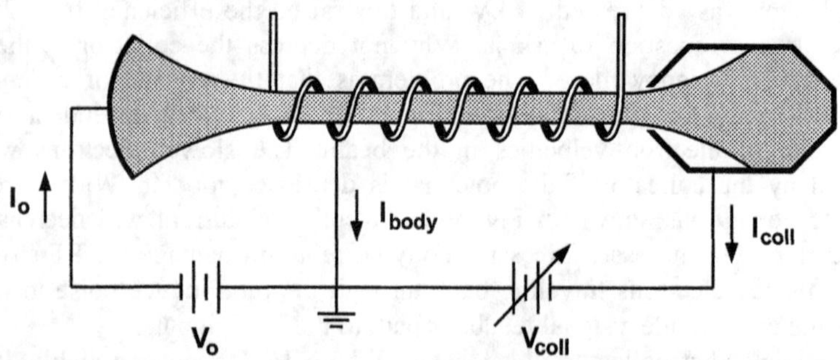

Figure 16-10 Power supply configuration for consideration of energy distribution of spent beam.

In Figure 16-11, the current to the collector is plotted as a function of the amount by which the collector voltage is depressed, V_{coll}. It is assumed that the body current is zero as long as the collector is not depressed too far. Also, the secondary electron current is ignored. When V_{coll} = 0, all of the cathode current reaches the collector and so $I_{coll} = I_o$. With no RF drive, no energy is extracted from the beam and so all of the electrons entering the collector have the same energy that they had when leaving the electron gun. As a result, as the collector voltage is depressed, all of the electrons can reach the collector until $V_{coll} = V_o$. As V_{coll} becomes larger than V_o, suddenly none of the electrons can reach the collector and so I_{coll} drops to zero.

When the RF drive is applied to the tube, energy in varying amounts is extracted from (or in some cases, added to) electrons during the interaction of the RF wave on the circuit with the beam. In a helix TWT, some electrons lose as much as 30% to 40% of their initial energy (when they left the gun) and some may gain as much as 30%.

With the RF drive, as the collector voltage is depressed from the body potential, all of the cathode current reaches the collector until the collector voltage reaches the point where the slowest electrons can no longer reach the collector surface. This voltage is indicated by V_{knee} in Figure 16-11. As the collector voltage is depressed below V_{knee}, more and more electrons are reflected and so I_{coll} decreases. At some voltage V_{max}, which is actually more negative that the cathode potential, V_o, the collector current drops to zero. The small "tail" of the curve at voltages between V_o and V_{max} is caused by those electrons that were accelerated rather than decelerated as the beam interacted with the RF wave.

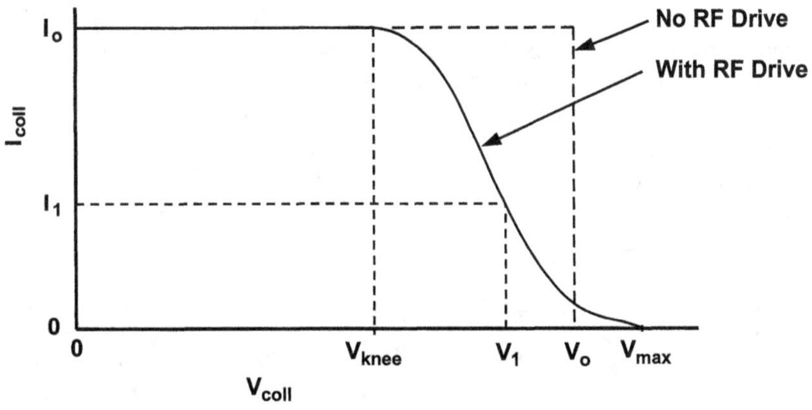

Figure 16-11 Collector current versus collector voltage.

In general, any current, such as I_1 in Figure 16-11, consists of electrons with energy eV_1 or greater. With the collector at potential V_1, the maximum current that could be collected would be I_1. The current, $I_o - I_1$, would be reflected and returned to the body of the tube.

The shape of the curve in Figure 16-11 depends on the nature of the interaction between the beam and the RF circuit. For example, Figure 16-12 contains curves for a helix TWT and for a high-efficiency klystron. (More will be said later about the reasons for normalizing the current and voltage to the cathode values.) For the klystron, the interaction is so intense that some electrons lose nearly all of their energy while others have their energy nearly doubled so a collector that was depressed even a small amount would reflect some current.

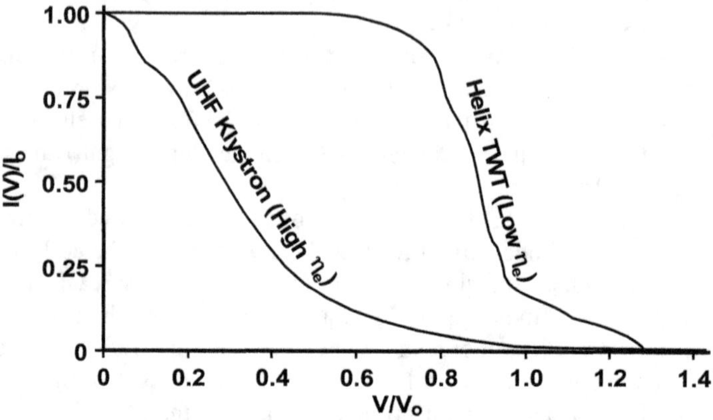

Figure 16-12 Beam energy distributions for tubes with high and low electronic efficiency.

16.2.4 Spent Beam Power

In Figure 16-11, current is plotted versus voltage and so the area under a curve represents power. For example, the area under the No RF Drive plot is simply $I_o \times V_o$, which is the dc beam power. When the RF drive is applied, as shown by the shaded area in Figure 16-13, the difference between the no-drive rectangle and the curve with drive is nearly the beam power that is converted to RF power. Some power is lost to the $V_{max}–V_o$ tail and there are RF losses, so the RF power out of the tube is somewhat smaller than is shown by the shaded area.

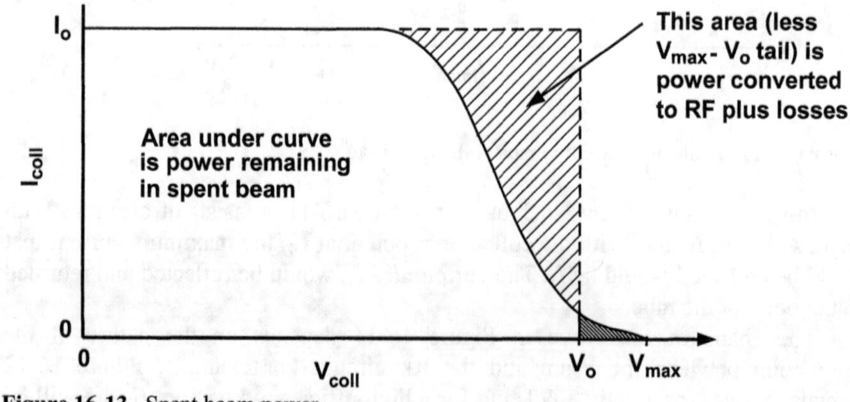

Figure 16-13 Spent beam power.

In Figure 16-13, the area under the solid curve is the power remaining in the spent beam. This is the maximum power that could possibly be recovered by an ideal collector. To recover part of that power, the power supply configuration must be as shown in Figure 16-14 (and as was shown in Figure 16-7). The collector supply voltage, V_c, is the cathode voltage, V_o, less the amount by which the collector voltage is depressed, V_{coll}. The maximum power that could be recovered by a single-electrode collector is shown by the shaded area in Figure 16-15. (The collector with a single electrode is often referred to as a single-stage collector. Collectors with several electrodes, to be discussed in following paragraphs, are called multistage collectors.)

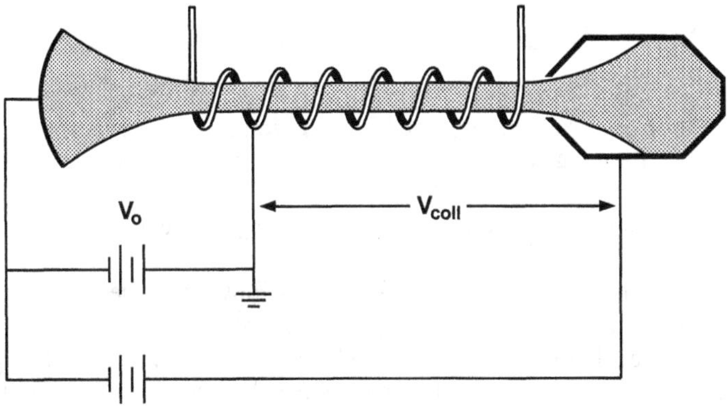

Figure 16-14 Power supply configuration for depressed collector operation.

When V_c is smaller than what is shown in Figure 16-15, electrons are reflected by the collector because the amount by which the collector is depressed, V_{coll}, is greater than the knee voltage. When V_c is greater than what is shown in Figure 16-15, all electrons are collected. In practice, it is not possible to operate a single-stage collector as close to the knee of the curve as is indicated in Figure 16-15. One reason is that V_c is not normally highly regulated and so V_c must be high enough to avoid electron reflections under all operating conditions. Another reason is that the knee of the curve is normally not as sharply defined as is shown in Figures 16-11, 16-13, and 16-15 and so excessive electron reflection (and body current) occurs with collector voltages well above that indicated in Figure 16-15.

16.2.5 Effect of Body Current [6]

The effect of body current on the energy distribution curves is shown in Figure 16-16. Here, currents and voltages have been normalized with respect to the cathode values. This normalization is often used because body current and collector currents are often expressed as percentages of the cathode values.

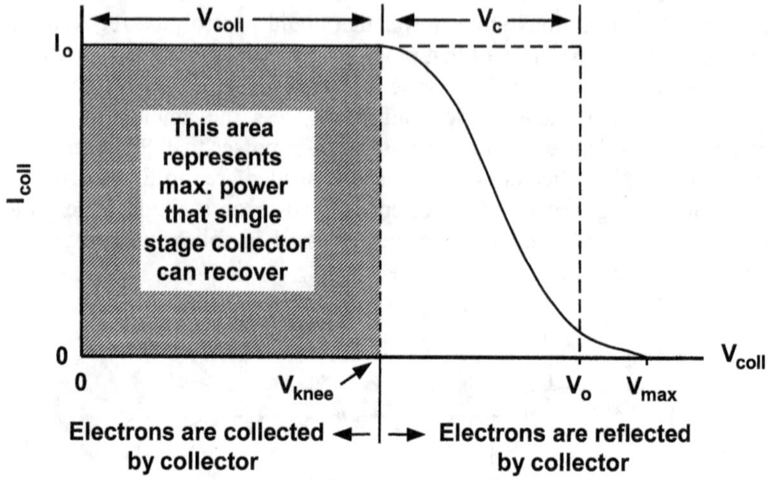

Figure 16-15 Maximum power recovered by a single-stage collector.

The three plots given in Figure 16-16 show how increases in body current affect the energy distribution curves as the RF drive is increased from zero (dc beam) to a low level (low basic efficiency) and then to a high level (high basic efficiency). Body current reduces the areas under the curves, and this reduces the power available for recovery by a collector.

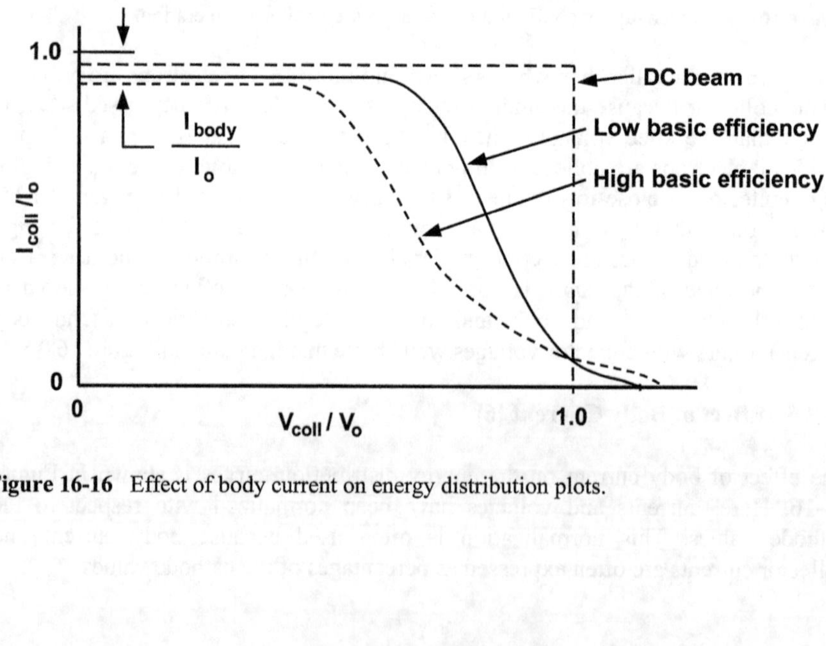

Figure 16-16 Effect of body current on energy distribution plots.

16.2.6 Multistage Depressed Collectors

The preceding discussion provides the basis for the design of efficient collectors. Highly efficient collectors use several electrodes at different potentials to selectively collect electrons at low energies. These collectors are multistage depressed collectors (MDCs).

In the previous section, energy distribution curves were shown for the electron beam entering the collector. It was pointed out that the area under a curve represents the power available in the spent beam for recovery by the collector. To recover a large fraction of this power, the collector must be designed to sort the electrons in the spent beam into various energy classes. Then electrons in each energy class must be collected on an electrode at a voltage that recovers as much of that energy as possible. Figure 16-17 illustrates the recovery of a substantial amount of spent beam power by a four-stage collector. Here, as in previous discussions, the effects of secondary electrons are not yet being considered.

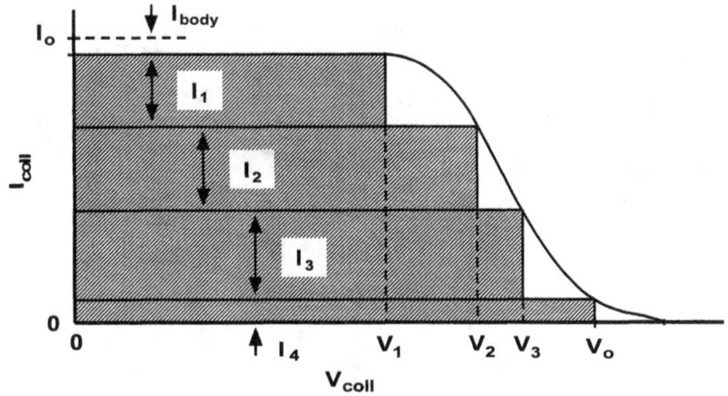

Figure 16-17 Power recovered by a four-stage depressed collector.

A collector electrode depressed to the cathode potential, V_o, can collect that current, I_4, consisting of electrons with energies in the range eV_{max}–eV_o. An electrode depressed to potential, V_3, can collect current, I_3, but not I_4 because that current goes on to the electrode at V_o. The total power recovered by the four-stage collector is represented by the shaded area in Figure 16-17 and is

$$P_{rec} = I_1 V_1 + I_2 V_2 + I_3 V_3 + I_4 V_o \tag{16-15}$$

The research and development for the collectors that operate as indicated in Figure 16-17 was led by NASA. The original impetus for the work was the high premium on efficiency that resulted from the high cost of power generation and heat reflection on satellites. Now the NASA collector technology is used in a broad range of applications where high microwave power is required and prime power is limited. An example is airborne electronic countermeasures systems [7].

The collector-electrode configurations for the NASA MDCs are illustrated in Figures 16-18 and 16-19. The first NASA collectors were derived from mathematical solutions to the problem of both slowing down and dispersing the electron beam entering the collector. As shown in Figure 16-18, the equipotential surfaces were designed to be concave toward the aperture where the beam entered the collector. This forced electrons to move radially outward. Collector electrodes were designed to coincide with the equipotential surfaces that were derived mathematically. The aperture sizes in the electrodes were chosen so that electrons would land on electrodes at the appropriate voltage. Since most beam electrons landed on the back sides of the electrodes, secondary electrons were emitted into a negative potential region and, therefore, were immediately returned to the electrode surfaces.

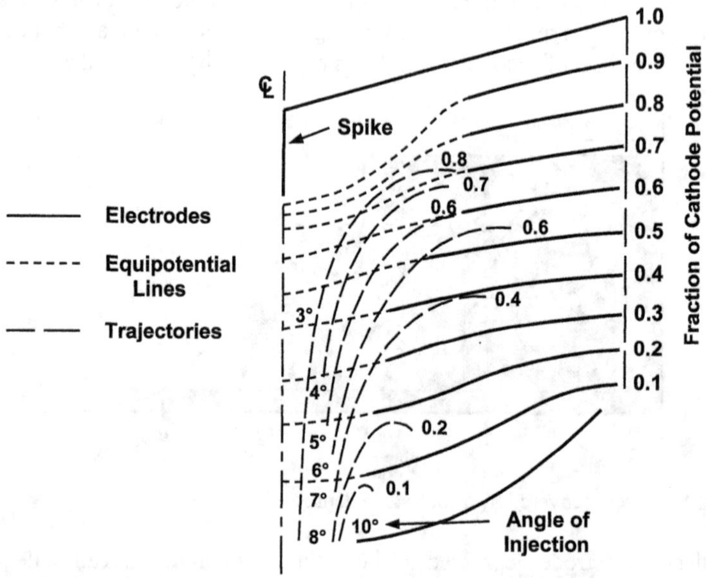

Figure 16-18 Axisymmetric MDC with equipotential surfaces determined mathematically. (From: H. G. Kosmahl, *Proc. IEEE*, November 1982. © 1982 IEEE.)

In all microwave tubes, some electrons are out of phase with the RF voltage and are accelerated, rather than decelerated, by the interaction process. These electrons leave the interaction region with energies greater than the original beam energy. In the NASA collector in Figure 16-18, the sharp spike at the back of the collector helps shape the fields to disperse high energy electrons. In addition, the surface area presented to high energy electrons is minimized. Thus, the generation of reflected electrons is minimized. The efficiencies of collectors patterned after the design in Figure 16-18 are very high (> 85%) and the current reflected from the collector is very small. Complexity and size are disadvantages.

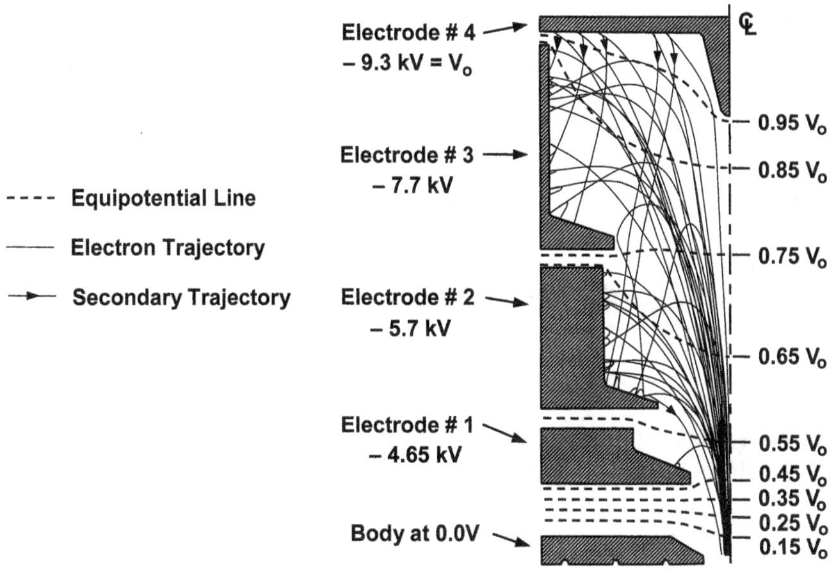

Figure 16-19 Equipotential surfaces and electron trajectories in "individual lens" collector. (From: J. A. Dayton, Jr., AIAA 12th International Communication Satellite Systems Conference.)

Collectors of the type shown in Figure 16-19 were originally developed during a joint NASA/USAF program. Collector size was reduced significantly with a small compromise in efficiency so that MDCs could be used in airborne applications, where size and weight are important considerations. In these collectors, called individual lens collectors, each electrode and its associated fields were optimized to produce maximum collector efficiency while minimizing the current resulting from reflected electrons.

For an MDC to operate efficiently, the electron beam entering it must not diverge rapidly. Rapid divergence leads to large radial components of electron velocities and to energy components that cannot be recovered by the collector.

To reduce the divergence of a beam, it is sometimes necessary to expand its diameter so that space charge forces are reduced. A magnetic refocusing system, one version of which is shown in Figure 16-20, is used for this purpose. An additional benefit of the refocusing system is an improvement in the laminarity of the beam entering the collector.

Improvements in overall efficiency that have been demonstrated by using an MDC on a dual-mode helix TWT for electronic countermeasures applications are shown in Figure 16-21. An overall efficiency improvement by a factor of three or more was demonstrated across a bandwidth of an octave. The collector efficiency was nearly constant at about 85%, as shown in Figure 16-22.

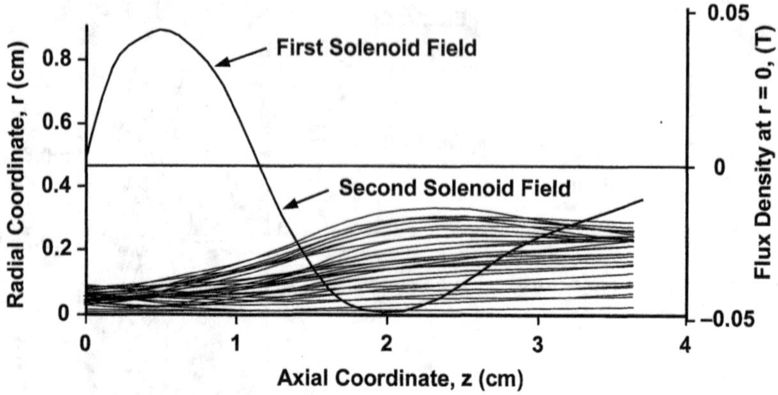

Figure 16-20 Magnetic refocusing system for reducing beam divergence and improving laminarity. (Adapted from: H. G. Kosmahl, *Proc. IEEE*, November 1982. © 1982 IEEE.)

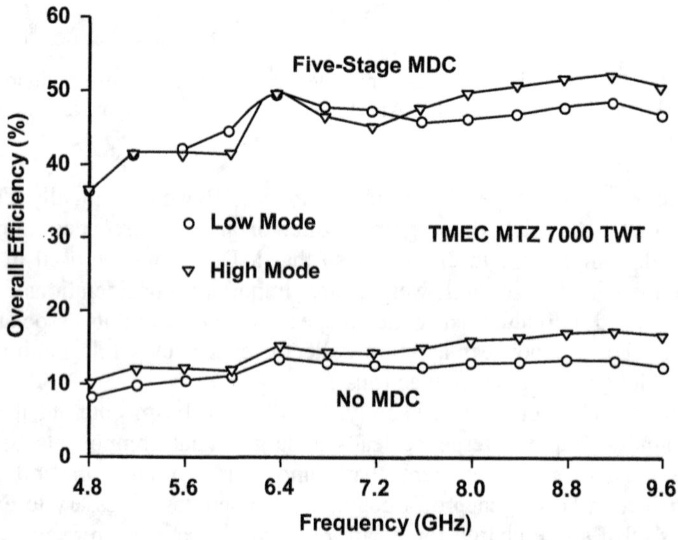

Figure 16-21 Overall efficiency of a dual-mode ECM helix TWT with and without an MDC. (Adapted from: H. G. Kosmahl, *Proc. IEEE*, November 1982. © 1982 IEEE.)

A very important application for MDCs is in tubes, such as satellite tubes, where efficient operation is necessary when operating well below saturation. As shown in Figure 16-23, the efficiency of the MDC increases significantly as the input signal level is reduced. For the example in Figure 16-23, the MDC efficiency exceeds 97% when the drive signal level (and basic efficiency) are reduced to zero.

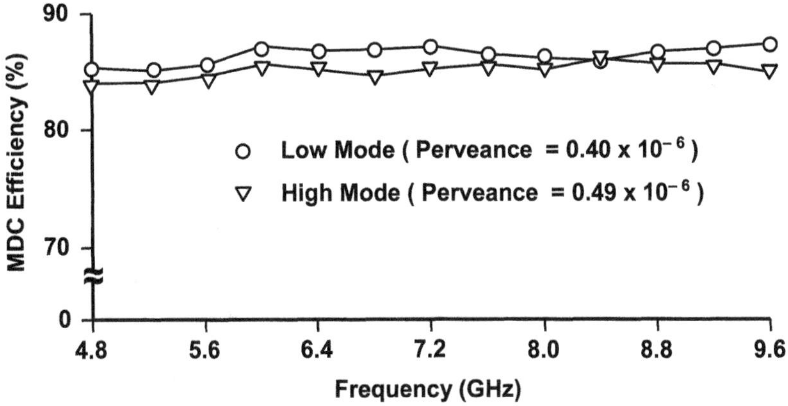

Figure 16-22 Collector efficiency at saturation. (Adapted from: H. G. Kosmahl, *Proc. IEEE*, November 1982. © 1982 IEEE.)

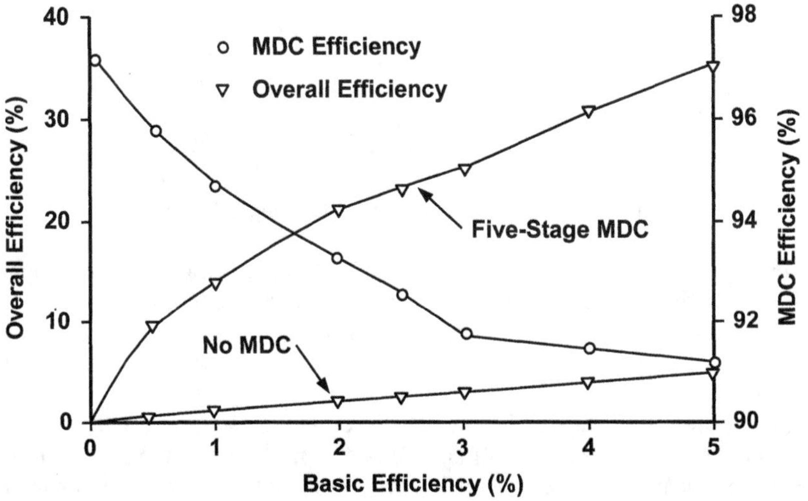

Figure 16-23 Overall and MDC efficiency versus basic (electronic) efficiency in the small-signal region. (Adapted from: H. G. Kosmahl, *Proc. IEEE*, November 1982. © 1982 IEEE.)

At a drive level low enough so that the basic (electronic) efficiency and the overall efficiency (without MDC) is 1%, use of an MDC increases the overall efficiency to 14%. At a basic efficiency of 4%, the MDC increases the efficiency to 30%. Thus, using a MDC, efficient operation over a wide range of drive levels is possible.

A power supply configuration for a MDC is shown in Figure 16-24. As was the case for the single-element depressed collector, the cathode-to-anode (and helix) supply must be regulated, but it must supply only the power resulting from the interception current. The tapped power supply for the collector is essentially an unregulated supply. This is because variations in voltages to collector elements may cause some electrons to move from one collector element to another, but the overall effect on efficiency is negligible.

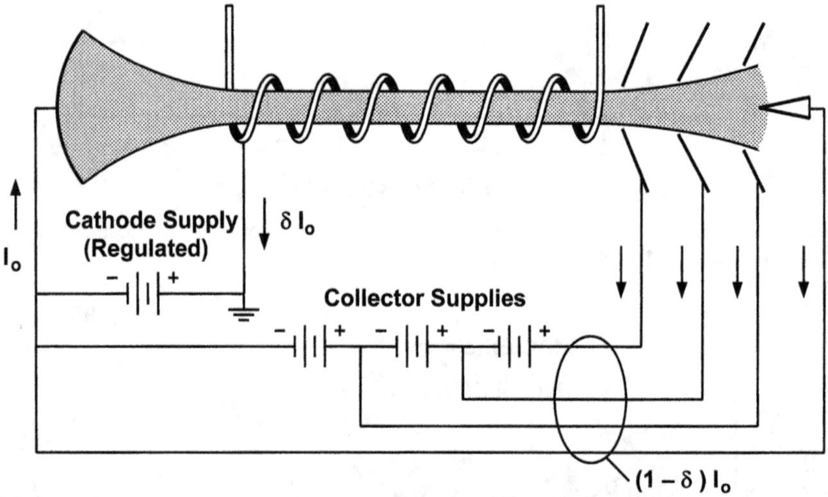

Figure 16-24 Power supply connection with a multistage depressed collector.

In Figure 16-18, the number of stages in the MDC was 10 and, in Figure 16-19, it was four. The optimum number is determined by considering the increase in tube efficiency that can be achieved with a large number of electrodes and trading this off against the increased complexity of the power supply and of tube construction.

As Figure 16-25 shows, the increase in efficiency achieved with more than about three or four stages is small and Thus, for most applications, no more than three or four stages are used. The efficiencies in Figure 16-25 were computed and are based on a knowledge of the spent beam characteristics of each type of tube.

16.2.7 Secondary Electrons in Depressed Collectors

Excessive secondary electron emission in depressed collectors can cause two problems:

1. Escape of secondaries from the collector, which may cause excessive noise, signal distortions, and heating of the circuit;

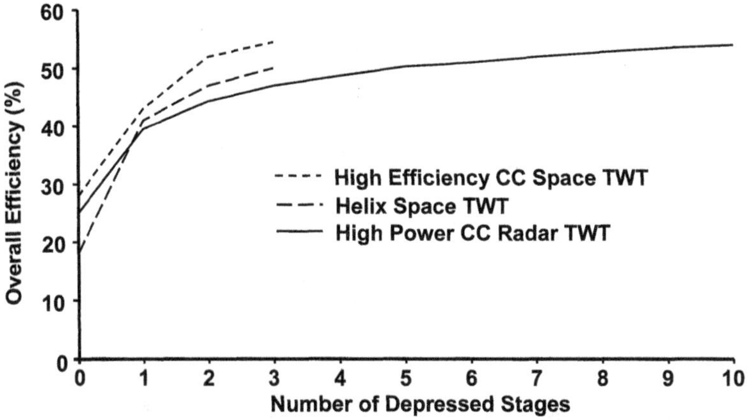

Figure 16-25 Efficiency versus the number of depressed collector stages. (Courtesy Hughes Electron Dynamics Division.)

2. Current flow between collector electrodes and excessive heating of electrodes, which, in turn, results in the reduction of collector efficiency.

The first of these problems is discussed in Chapters 23 and 24. The second problem arises when secondary electrons released from the front surface of a collector electrode are collected by a more positive electrode. An example was shown in Figure 16-19 where trajectories for five secondary electrons released from electrode #4 are plotted. Three of these electrons strike electrode #3 after being accelerated through 1.6 kV (9.3–7.7 kV). The other two strike electrode #2 after being accelerated through 3.6 kV.

The effect of secondary electrons on efficiency is shown in Figure 16-26 [8]. Tests were performed with a four-stage MDC having various electrode surface treatments. Using electrodes with textured isotropic graphite surfaces rather than untreated copper, collector efficiency increased about 10%. The overall efficiency increased over 5%.

The technique of texturing materials for collector surfaces has been developed by the NASA Lewis Research Center [9]. The texturing is accomplished by bombarding the surface with argon ions. For the textured graphite surfaces described in [9], the ion accelerating potential was 1,000–1,500V dc, the surface current density of the sample was 3–5 mA/cm^2, and the texturing period was 1–4 hours. A scanning electron microscope photomicrograph of a fully developed textured carbon surface on a copper substrate is shown in Figure 16-27. The surface appearance in this figure is also typical of fully developed ion-textured pyrolytic and high purity isotropic graphite. The larger spires have average heights of 6–12 μm and average spacings of 3–6 μm.

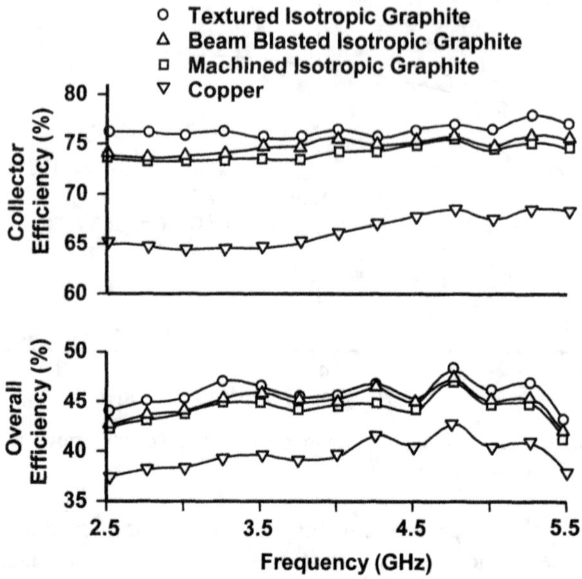

Figure 16-26 Efficiencies for various surfaces in a four-stage MDC on a TWT with $\eta e = 16\%$. (From: P. Ramins and B. T. Ebihara, *IEEE Trans. Electron Devices*, November 1986. © 1986 IEEE.)

Figure 16-27 SEM photomicrograph of textured carbon on a copper substrate. (From: A. N . Curren, *IEEE Trans. Electron Devices*, November 1986. © 1986 IEEE.)

The secondary emission characteristics of textured carbon on a copper substrate are shown in Figure 16-28. The true secondary emission coefficient for normal incidence of the primary beam is from 0.2 to about 0.25. This is significantly below the value for soot, which has, in the past, been the lowest yield surface available. Soot has not been used in microwave tubes because of its poor physical characteristics. The reflected primary yield index, π, for normal incidence ranges from about 0.5 to about 1.5. The basis for this index is the primary yield for soot with a primary electron energy of 1 keV and normal incidence. For these conditions, the value of π is unity.

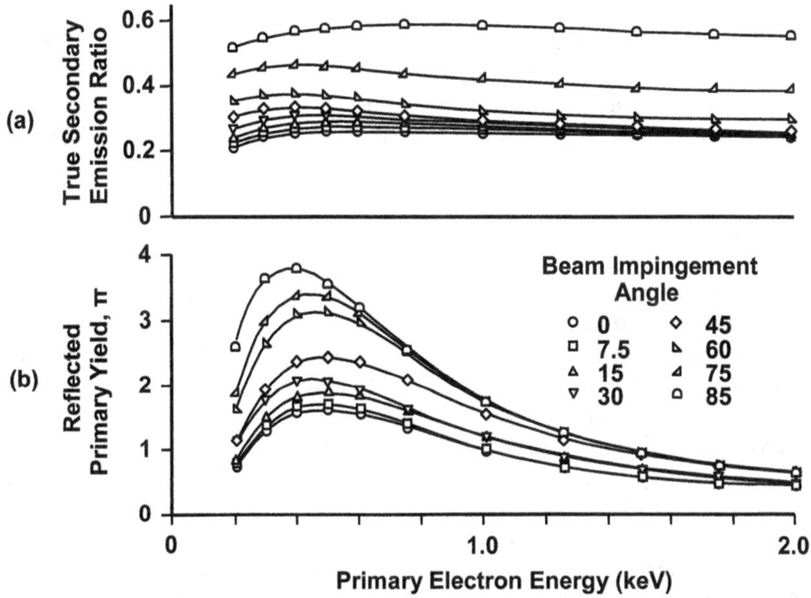

Figure 16-28 (a, b) Secondary emission characteristics of textured carbon on a copper substrate. (From: A. N. Curren, *IEEE Trans. Electron Devices*, November 1986. © 1986 IEEE.)

In addition to texturing various carbon surfaces, NASA has developed techniques for texturing copper [10]. With one technique, sputtered molybdenum is used as a texture-inducing masking film. The true secondary electron emission ratios for these surfaces range from 0.3 to 0.5 and are a factor of 3 below ratios for untreated copper surfaces. The reflected primary yield index is a factor of about 3 to 7 below that for an untreated copper surface. The results reported in [10] were for three samples of ion textured copper. One conclusion of the NASA work is "with further development, the molybdenum-masked, ion-textured OFHC copper surface will be a promising material for high efficiency MDC electrodes."

16.3 COLLECTOR COOLING [1]

16.3.1 Conduction Cooling

Heat transfer by conduction to a heat sink is the simplest form of cooling where the power level of the tube is low enough to permit it. Little maintenance is normally required. However, it is necessary to make sure that tubes are tightly bolted to the heat sink and that no gaps exist between surfaces.

16.3.2 Convection Cooling

Devices that generate up to a few watts of average power may require no cooling other than natural air convection. To take the best advantage of this type of cooling, the surface exposed to the air is often many times larger than the actual inside area of the collector. About 1 watt/cm^2 of surface exposed to open air can be handled by this approach.

16.3.3 Forced-Air Cooling

Although forced-air cooling is capable of dissipating power densities nearly as high as water cooling can, its use is limited because of practical considerations. A typical air-cooled collector, which has been used to dissipate a power density of 200 watts/cm^2, is shown in Figure 16-29. The cooling fins are brazed to the body of the collector to provide a large area multiplication factor (larger than 10) from the power-intercepting area to the air-exposed surface. One of the limitations of air cooling is blower power. To minimize the input to the air pump, it is necessary to have a large number of short air passages in parallel. An optimum cooling fin arrangement, shown in Figure 16-29, requires a blower power of from 5 to 10% of the average collector power. This is a much higher level than for water cooling. Another limitation of air cooling is that it is not easily applicable to other portions of the tube such as the interaction structure. Because of these disadvantages, forced-air cooling is generally used only for tubes up to the few kilowatts level. At higher-power levels the circulating water system is preferred.

16.3.4 Forced-Flow Liquid Cooling

This is by far the most often used method for cooling collectors for medium- and high-power tubes. An effective method is simply to place a water jacket around the collector and remove heat by conduction from the collector body to the water. The amount of heat that can be transferred to the water is directly proportional to the wetted surface area and the heat transfer coefficient. Both are optimized when many narrow channels with large perimeters, but small cross-sectional areas are used together with high velocity flow.

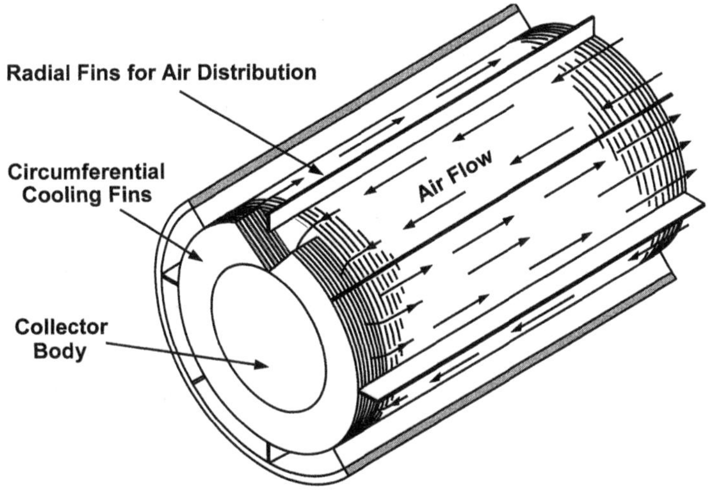

Figure 16-29 Flow patterns in an air-cooled collector. (Adapted from: *Linear-Beam Tubes* by A. Staprans, Internal Varian Report, September 1968.)

To avoid excessively high temperature rises in the collector walls, water channels must be located near the intercepting surface. As a result, the wetted surface area is generally limited to a factor of about three or four times the interior collector area. Higher factors could be achieved, but this requires machining of channels of a more complicated cross-section than is convenient with ordinary techniques. Some collector wall sections showing water channel shapes are shown in Figure 16-30.

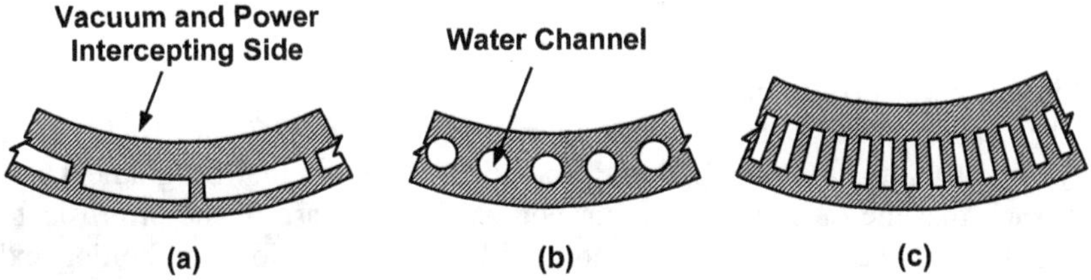

Figure 16-30 (a–c) Collector water cooling channels with increasing area multiplication factors from (a) to (c). (Adapted from: *Linear-Beam Tubes* by A. Staprans, Internal Varian Report, September 1968.)

The heat transfer coefficient used is usually computed on the basis of allowing no boiling of water at the surface. The heat transfer coefficient will characteristically be on the order of 1 cal/cm^2/sec/°C temperature rise in the surface film of water. Much higher heat transfer coefficients can be achieved if the water is allowed to boil, as is frequently the case with high-power collectors. Care must be taken to insure that operation occurs in the nucleate boiling region, which has a high heat transfer capability and does not enter into the film boiling region, for which the surface temperature must be very high and might cause melting of the collector material.

It is customary to design water cooling to be operable at pressure drops of less than 100 lb/in^2 and flow rates commensurate with average temperature rises from 30°C to 50°C. If a specific design has been made for water, cooling will generally suffer when other fluids are used because of differences in thermal conductivity, heat capacity, viscosity, and other characteristics. Ethylene glycol is used for low-temperature applications, principally to avoid the problem of freezing during storage. It is inferior to water as a coolant. Other fluids have been considered but are seldom used in practice, even though some have advantages with respect to water. Cost and availability are the most frequent deterrents to the use of most of the special fluids.

16.3.5 Vapor Phase Cooling

This method takes advantage of the latent heat of evaporation of water by allowing the water to boil at the collector surface. The steam is then condensed in a heat exchanger and returned to the collector, forming a closed system. To achieve reasonable dissipation densities, operation must be either in or near the transition between film boiling and nucleate boiling. Since this would ordinarily represent a runaway region, it can be used only when boiling equilibrium is maintained.

Beurtheret [11] pioneered this type of cooling for vacuum tubes in the "Vapotron"-type collectors. To understand this, the curve shown in Figure 16-31 is helpful. Heat transfer increases very rapidly with the surface temperature until the discontinuous region is reached where heat transfer actually drops as nucleate boiling changes into film boiling. At this point the surface temperature will undergo a temperature increase of about 1,000°C for the same heat flux, which is prohibitive because of burnout.

The vapor-phase-cooled collector (a section is shown in Figure 16-32) has an array of protruding bars, which, during operation, develop a thermal gradient along them such that the outer end does not boil at all. The base of the protrusions is always low in the film boiling region, and all other regions of boiling exist continuously between the two extremes. Maximum heat transfer occurs near the middle of the bar. Any small change of heat transfer in the system will shift the point of maximum heat transfer along the bar, thus readily absorbing the fluctuation without a discontinuous jump in the temperature. The detailed shape of the collector protrusion is crucial in the functioning of the system.

Vapor-phase-cooled collectors are used in high-power tubes where dissipation densities are on the order of 1 to 200 $watts/cm^2$ and where it is preferable to avoid a forced-flow system with water pumps. Examples are CW klystrons designed for UHF television transmitters in the 30- to 50-kW range.

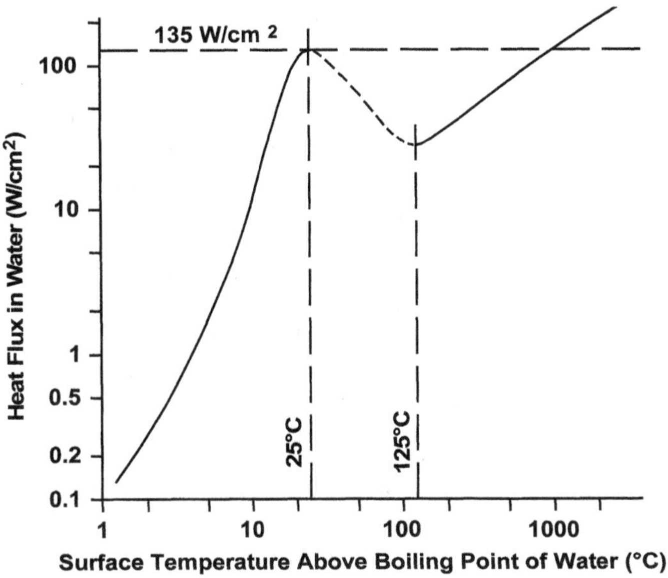

Figure 16-31 Heat transfer by boiling. (Adapted from: *Linear-Beam Tubes* by A. Staprans, Internal Varian Report, September 1968.)

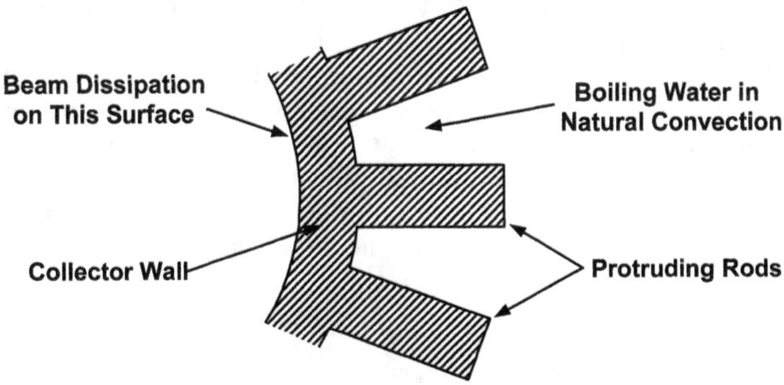

Figure 16-32 Partial cross section of a vapor-phase cooled collector. (Adapted from: *Linear-Beam Tubes* by A. Staprans, Internal Varian Report, September 1968.)

16.3.6 Radiation Cooling

One approach to radiation cooling is to allow the collector to run at a high enough temperature so that heat is lost by radiation directly into space. This approach, though attractive because of its simplicity, poses certain problems. The collector temperature must be in the neighborhood of 1,000°C to provide a reasonable dissipation capability. In order to minimize temperature rise on the inside as well as to suppress lateral flow of heat to the tube body, a thin membrane-type collector is necessary. This raises problems with mechanical strength as well as gas diffusion through the collector. The latter, at least in principle, may be avoided if the tube is intended for use in a space environment. However, some tube testing at atmospheric pressure is always necessary. Thus, it is preferable to enclose the collector in a vacuum envelope essentially transparent to radiated heat. Some of the possible collector materials are the refractory metals, such as tungsten, molybdenum, and tantalum.

Black-body radiation at 1,000°C is approximately 12 watts/cm^2. A reasonable value of emissivity with proper carbonization might be 80%, reducing the above value to about 10 watts/cm^2. So it appears that, because radiated power varies as the fourth power of temperature, operation between 1,000°C and 2,000°C might yield cooling capabilities on the order of 100 watts/cm^2. One of the largest problems is back radiation to the tube, which dictates the use of a fairly large collector with a small beam entrance aperture. Also, problems may arise from collector evaporation and deposition inside the tube as well as on the radiation window.

A radiation-cooled collector used by Lien et al. [12] in a lightweight space communications tube is shown schematically in Figure 16-33. A reflector is used inside as well as outside the vacuum envelope to protect the tube from back radiation. A sapphire radiation window was chosen because of its high transparency to infrared. This collector was operated at a power level of about 300 watts at a temperature of 1,500°C. The collector material was tungsten. The measured radiation-cooling efficiency was approximately 75%, indicating that 25% of the beam power striking the collector was returned to the tube body. Although this was far from being an optimized device, the latter percentage is indicative of some of the problems with this type of cooling. It appears that radiation cooling will only find application in a narrow range of special-purpose tubes where considerations of system weight are paramount.

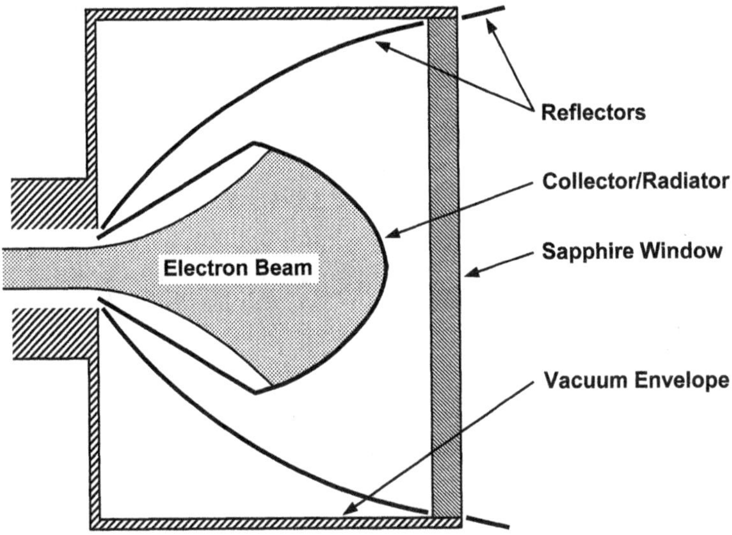

Figure 16-33 Schematic layout of radiation-cooled collector. (Adapted from: *Linear-Beam Tubes*, by A. Staprans, Internal Varian Report, September 1968.)

REFERENCES

[1] A. Staprans, *Linear-Beam Tubes*, Internal Varian Report, September 1968.

[2] J. F. Skowron, "Ultra high-power amplitron," *Final Report*, RADC-TDR 64-389, Vol. 1, May 1965.

[3] H. S. Carslaw, and J. C. Jaeger, *Conduction of Heat in Solids*, 2nd ed., Oxford: Clarendon Press, 1959.

[4] P. W. Crapuchettes, "Cooling of anodes subjected to long impulses of high peak power," Conference Paper No. CP 61-172, *AIEE Winter General Meeting*, New York, January 29–February 3, 1961.

[5] H. G. Kosmahl, "Modern multistage depressed collectors—a review," *Proc. IEEE*, Vol. 70, No. 11, Nov 1982, pp. 1325–1334.

[6] I. Tammaru, "Collector lecture notes," Hughes Electron Dynamics Division.

[7] H. G. Kosmahl, and P. Ramins, "Small size 81 to 83.5 percent efficient 2 and 4 stage depressed collectors for octave bandwidth high performance TWTs," *IEEE Trans. Electron Devices*, Vol. ED-24, No. 1, January 1977, pp. 36–44.

[8] P. Ramins, and B. T. Ebihara, "Improvements in MDC and TWT overall efficiency through the application of carbon electrode surfaces," *IEEE Trans. Electron Devices*, Vol. ED-33, No. 11, November 1986, pp. 1915–1924.

[9] A. N. Curren, "Carbon and carbon-coated electrodes for multistage depressed collectors for electron-beam devices—A technology review," *IEEE Trans. Electron Devices*, Vol. ED-33, No. 11, November 1986, pp. 1902–1914.

[10] A. N. Curren, K. A. Jensen, and R. F. Roman, "Secondary electron emission characteristics of molybdenum-masked, ion-textured OFHC copper," *NASA Technical Paper 2967*, January 1990.

[11] C. Beurtheret, "The Vapotron technique," *Revue Technique* C.F.T.H, No. 24, December 1956, pp. 53–82.

[12] E. L. Lien, A. Mizuhara, and D. I. Boilard, "Electrostatically-focused extended-interaction S-band klystron amplifier," *Proc. 6th International Conference on Microwave and Optical Generation and Amplification,* Cambridge, England, September 1966.

Chapter 17

Crossed-Field Tubes

In crossed-field tubes, the dc electric field is perpendicular to the magnetic field as is indicated in Figure 17-1(a). The general motion of the electron beam is perpendicular to both fields. This is in contrast to linear-beam tubes (such as klystrons and TWTs, Figure 17-1(b)) where the general direction of beam motion is aligned with the fields. Crossed-field tubes are referred to as "M-type" tubes after the French TPOM, which in English means tubes for propagation of waves in a magnetic field.

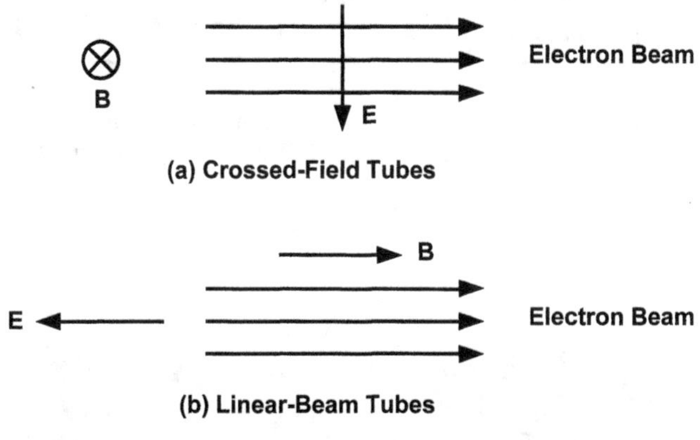

Figure 17-1 Comparison of field configurations in (a) crossed-field and (b) linear-beam tubes.

Many types of crossed-field tubes have been developed and can be categorized as indicated in Figure 17-2. With the exception of injected-beam crossed-field amplifiers, these devices are in use in deployed systems. Injected-beam crossed-field amplifiers, while highly developed, have not found application in the United States and will not be discussed.

17.1 BASIC CONFIGURATION OF CROSSED-FIELD TUBES

The basic configuration of a magnetron and of a crossed-field amplifier (CFA) is shown in Figure 17-3. Magnetrons are oscillators and CFAs are amplifiers. The CFA is derived from the magnetron and, in fact, the original CFA invented by Brown [1] was a magnetron that had been modified so that it could amplify a signal.

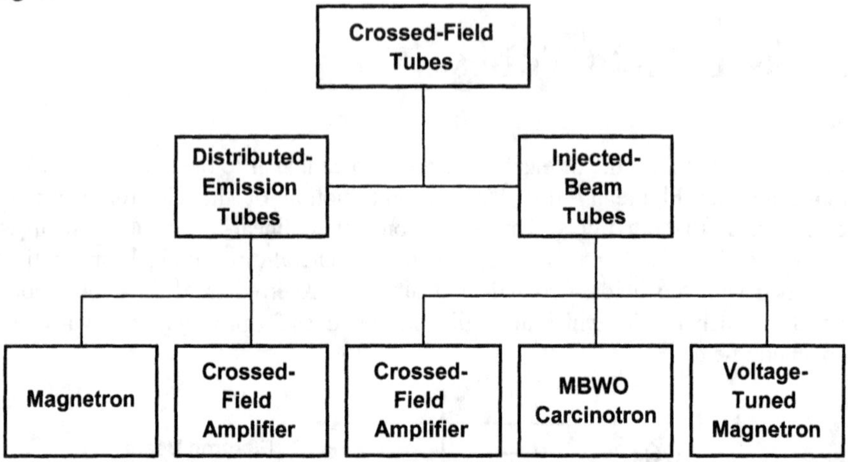

Figure 17-2 Family tree of crossed-field tubes. (From: *Microwave Tube Manual* by Varian Associates, Air Force Publication Number T.O.00-25-251, October 1979.)

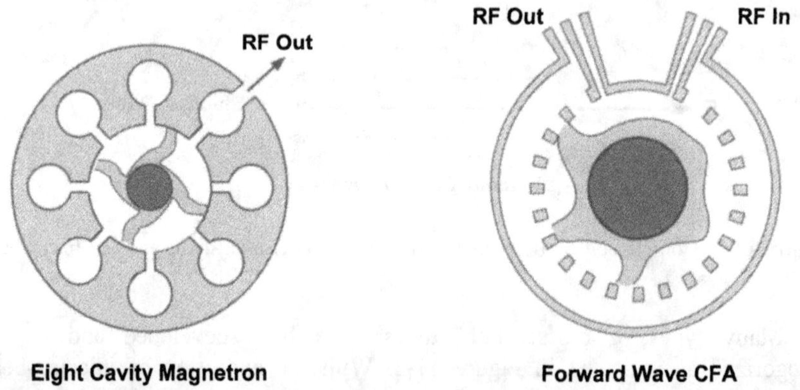

Figure 17-3 Comparison of magnetron and crossed-field amplifier configurations.

The magnetron in Figure 17-3 has eight cavities. As shown in Figure 17-4, the equivalent circuit for one of the cavities is a parallel resonant LC circuit. The hole of the cavity is the inductive portion and the slot is the capacitive portion. In an eight-cavity magnetron, the equivalent circuit has eight of these parallel resonant circuits in series.

When a magnetron is first turned on, the only signal that is present is the noise (random fluctuations in electron velocity and density) in the electron cloud circling about the cathode. However, the noise consists of signals of all frequencies including the frequency at which the cavities resonate. The oscillations in a magnetron build up very rapidly (in nanoseconds) by amplification of the noise signal of the correct frequency.

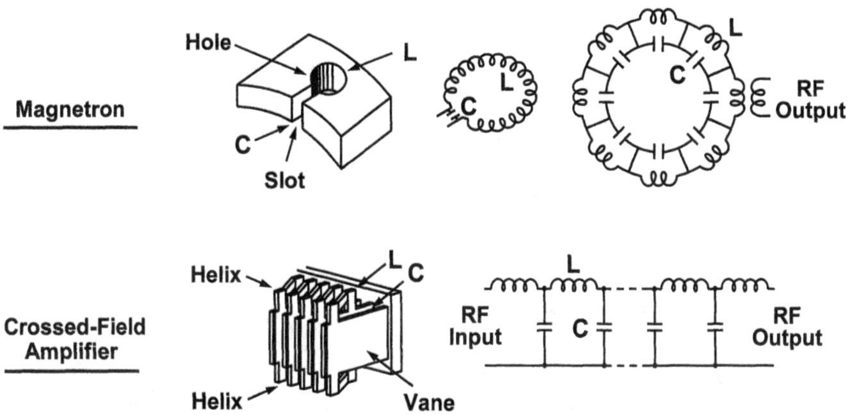

Figure 17-4 Equivalent circuits for magnetrons and crossed-field amplifiers.

The RF (slow wave) circuit in a CFA can have the configuration of any one of many circuits. These can be either forward wave or backward wave circuits. These circuits are sometimes modeled as LC transmission lines, in contrast to the LC parallel circuits of magnetrons. The CFA circuit shown in Figure 17-4 is a forward wave circuit similar to the helix in a TWT and consists of two helices supported by copper vanes. In addition to providing support for the helices, the vanes carry heat away from the helices.

17.2 ELECTRON FLOW WITH NO RF FIELDS

The crossed-field devices considered in this book are all cylindrical. In nearly all cases, the cathode is a cylinder, which is concentric with the anode. If there is no RF circuit on the anode, the configuration of the device is that of the smooth bore magnetron in Figure 17-5.

In the absence of RF fields, and ignoring space charge forces, electron trajectories can readily be found. Since the electrodes are concentric, the electric field is radial with no θ variations. The magnetic field, which is perpendicular to the electric field, is axial. The force components are:

$$F_r = m(\ddot{r} - r\dot{\theta}^2) \tag{17-1}$$

and

$$F_\theta = m(2\dot{r}\dot{\theta} + r\ddot{\theta}) \tag{17-2}$$

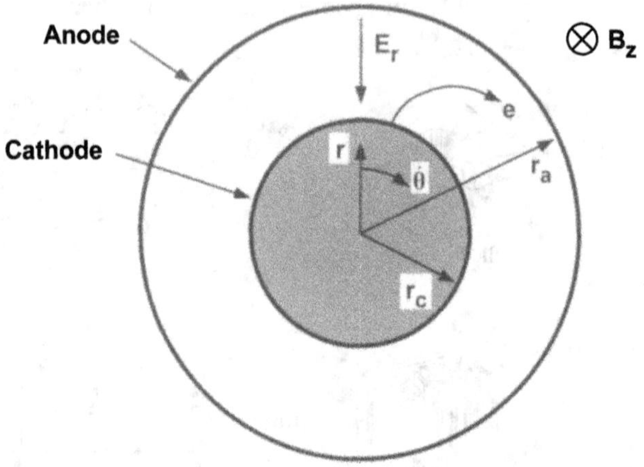

Figure 17-5 Configuration of a smooth-bore magnetron.

As electrons leave the cathode and move toward the anode, they are deflected in the θ direction. In cylindrical coordinates, if r is radially outward from the axis and θ is clockwise about the axis, then z is into the plane of r and θ and the θ directed force is

$$F_\theta = e\dot{r}B_z \tag{17-3}$$

Equations (17-1) and (17-2) can be combined and integrated to obtain

$$r^2\dot{\theta} = \frac{\omega_c}{2}r^2 + \text{constant} \tag{17-4}$$

Next, assume that the initial condition for an electron at the cathode is $\dot{\theta} = 0$ at $r = r_c$. Then, at the anode, where $r = r_a$, the angular velocity of the electron is

$$\dot{\theta}_a = \frac{\omega_c}{2}\left(1 - \frac{r_c^2}{r_a^2}\right) \tag{17-5}$$

If the anode is at a potential V_a with respect to the cathode such that an electron just barely grazes the anode (that is, $\dot{r} = 0$) then the velocity of the electron is

$$u_a = \sqrt{2\eta V_a} = r_a \dot{\theta}_a \tag{17-6}$$

This can be combined with (17-5) to obtain

$$V_a = \frac{\omega_c^2 r_a^2}{8\eta}\left(1 - \frac{r_c^2}{r_a^2}\right)^2 \tag{17-7}$$

Thus, the voltage that causes the electron to just graze the anode depends on the dimensions r_a and r_c and on the flux density (in ω_c) but not on the way in which the voltage varies between the cathode and anode. The electron trajectory is shown in Figure 17-6(a). The anode voltage at which this occurs is called the Hull cutoff voltage and is denoted here by V_H. When $V_a > V_H$, the electron strikes the anode as indicated in Figure 17-6(b). When $V_a < V_H$, the electron returns to the cathode (Figure 17-6(c)).

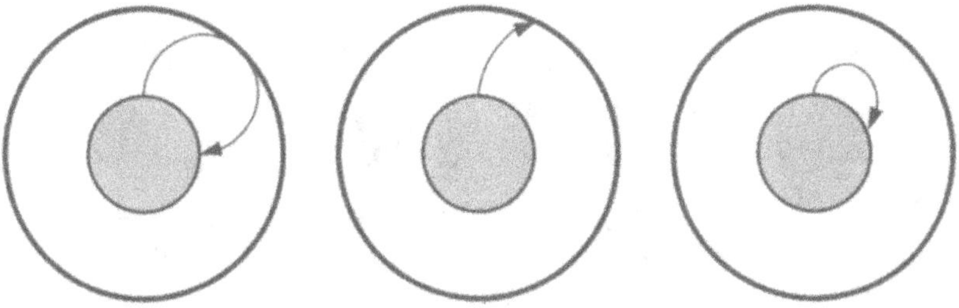

$V_a = V_H$ (a) $V_a > V_H$ (b) $V_a < V_H$ (c)

Figure 17-6 (a–c) Electron trajectories for various anode potentials.

If the anode voltage is held constant at V_H, then, as the magnetic field is decreased, current flows to the anode and as the magnetic field is increased, no current flows. Thus, it is the critical combination of electric and magnetic fields known as the Hull cutoff condition and shown in Figure 17-7 at which electrons just touch the anode.

Before leaving the subject of the Hull cutoff condition, it is interesting to examine the current-voltage characteristic of the ideal smooth bore magnetron. As shown in Figure 17-8, when the anode voltage is small ($< V_H$), no current flows to the anode. As the anode voltage is increased and reaches the Hull cutoff value, the current increases from zero to the full amount that is available. As the anode voltage increases beyond the Hull cutoff value, all of the available current continues to flow. Thus, when the smooth bore magnetron is conducting, it is a constant current device. In fact, in real magnetrons and crossed-field amplifiers, while current is not constant, it does vary relatively slowly as voltage is varied.

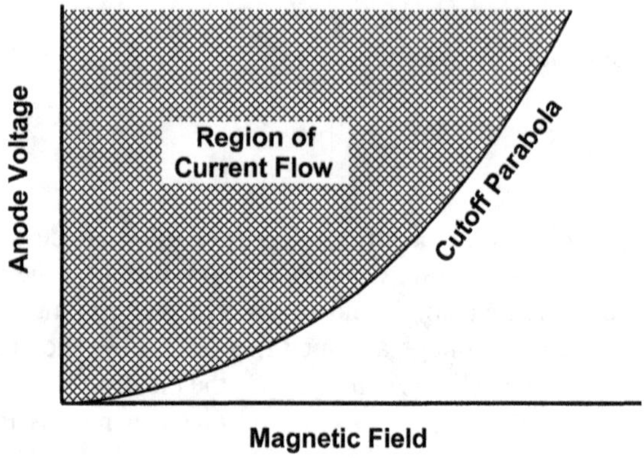

Figure 17-7 Hull cutoff condition.

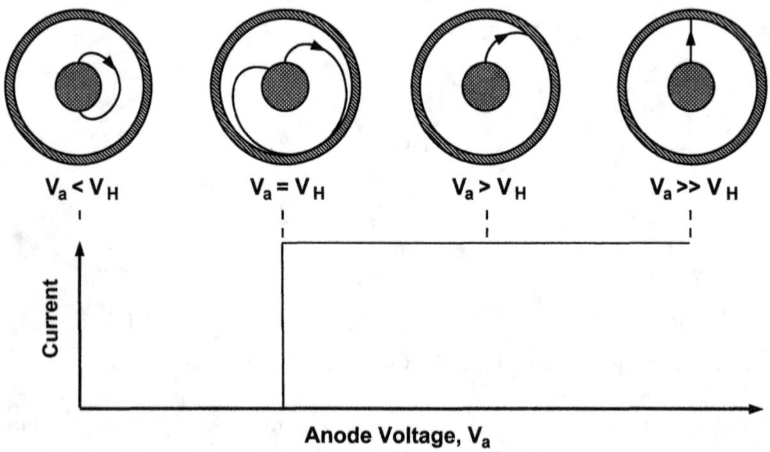

Figure 17-8 Current flow as a function of anode voltage in an ideal smooth-bore magnetron.

The first magnetrons that were described in the literature were the cyclotron resonance magnetron and the negative resistance magnetron. These operated very close to the Hull cutoff condition but were very inefficient. Their operation is described in Chapter 19. On the other hand, traveling wave magnetrons and CFAs, which are described in Chapters 19 and 20, are very efficient and operate at voltages and magnetic fields that are well below the Hull cutoff parabola. If there was no RF interaction, no current would flow in the devices. However, with RF, current does flow and the high efficiency of the devices is related to the energy that the electrons acquire and then give up as they move from the cathode to the anode.

REFERENCE

[1] W. C. Brown, "Description and operating characteristics of the platinotron—a new microwave tube device," *Proc. IRE,* vol. 45, 1957, pp. 1209–1222.

Chapter 18

Cathodes for Crossed-Field Tubes

In crossed-field tubes, electrons are emitted from the cathode by secondary electron emission, which may be combined with thermionic emission. The thermionic emission process was discussed in detail in Chapter 5. In this chapter, the secondary emission process and its role in the operation of crossed-field devices are discussed.

18.1 INTRODUCTION

As shown in Figure 18-1, when electrons bombard a surface, they may cause other electrons to be emitted from that surface. The bombarding electrons are called primary electrons and the emitted electrons are called secondary electrons. Secondary emission is of extreme importance in the operation of cathodes of crossed-field tubes where the electric and magnetic field configurations cause primary electrons to bombard the cathode. The way in which this occurs is indicated in Figure 18-2, which shows the RF field distribution in a magnetron. The path of an electron emitted at point A on the cathode is shown. After emission from the cathode, the electron is accelerated toward the anode, which, in crossed-field devices, contains an RF circuit (eight resonant cavities in Figure 18-2). Because a magnetic field is present (the direction is into paper), the electron moves in a curved path in the clockwise direction. The direction of the RF field is such that it accelerates this particular electron and drives it back into the cathode surface. If the energy with which the electron strikes the cathode surface is high enough, then it can cause as many as several secondary electrons to be emitted.

Often, all or at least a major portion of the electrons from the cathode of a crossed-field device are emitted by secondary emission and the remainder are thermionically emitted. An example is the Philips YJ 1183 spin-tuned magnetron in which it is estimated that thermionic emission produces 3 A/cm^2 and secondary emission produces 23 A/cm^2 for a total cathode emission of 26 A/cm^2.

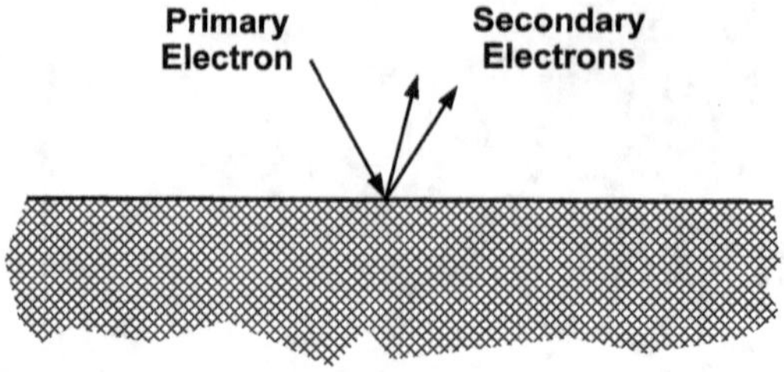

Figure 18-1 Secondary emission.

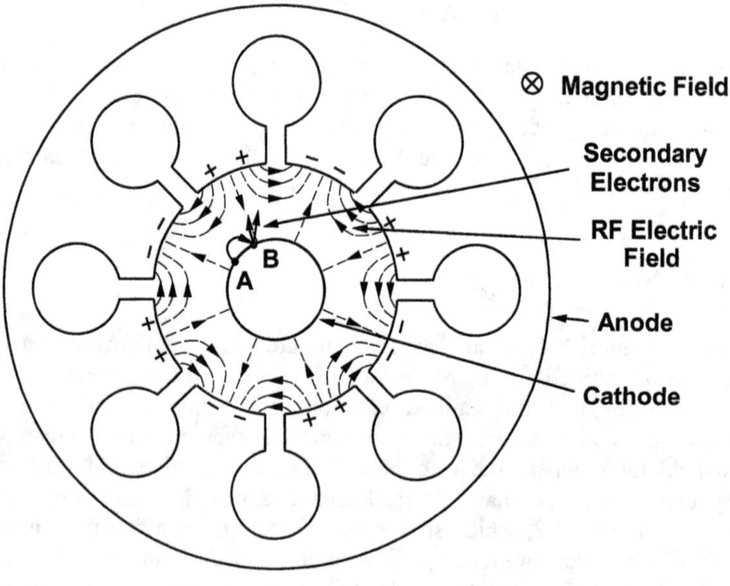

Figure 18-2 Secondary emission from the cathode in a crossed-field device.

18.2 CHARACTERISTICS OF SECONDARY EMISSION

In the secondary emission process, when a primary electron strikes a solid, it gives up energy by repelling electrons in the solid. The electrons in the solid gain energy and may or may not be emitted, depending on how much energy they receive, how close they are to the surface, and the direction of their motion. The number of secondary electrons emitted for each primary electron is called the secondary emission coefficient and is usually denoted by δ or σ. We will use δ. The value of δ depends on several quantities:

1. Energy of the impacting primary electrons;
2. Angle with which the primary electrons strike the surface;
3. Secondary emitting properties of the surface.

18.2.1 Energy of Impacting Primary Electrons

The typical variation of δ with primary electron energy is shown in Figure 18-3. At very low energy levels, δ approaches zero. As primary electron energy is increased, δ increases, crosses unity, and reaches a maximum. Then, as primary electron energy is increased beyond this point, δ decreases and becomes less than unity again. The reason for this decrease in δ is that the primary electrons penetrate so deeply into the surface that their energy is dissipated in electrons that cannot reach the surface.

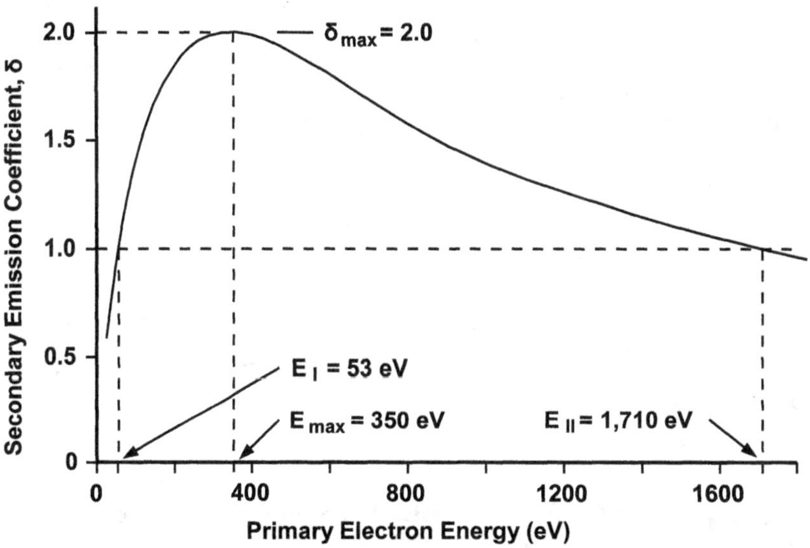

Figure 18-3 Secondary emission ratio as a function of primary electron energy.

There are four quantities associated with the secondary emission property of a surface that are usually of significance. These are:

1. First crossover energy, E_I (53V in Figure 18-3);
2. Maximum secondary emission coefficient, δ_{max} (2.0 in Figure 18-3);
3. Primary electron energy, E_{max}, at which δ_{max} occurs (350V in Figure 18-3);
4. Second crossover energy, E_{II} (1,710V in Figure 18-3).

The first crossover energy, E_I, is the lowest primary electron energy that produces a δ of unity. The maximum secondary emission coefficient, δ_{max}, is a convenient way of characterizing the secondary emissivity of a surface. The second crossover energy, E_{II}, is the highest electron energy that produces a δ of unity.

18.2.2 Angle of Incidence of Primary Electrons

The angle of incidence, θ, of the primary electrons has a strong effect on the secondary emission coefficient of a surface. This dependency is shown in Figures 18-4 and 18-5 for molybdenum [1]. Note that δ_{max} increases from <1.4 to about 1.9 or about 40% as θ increases from 0° to 60°. The reason for the increase in δ with θ is that, as θ increases, primary electrons remain closer to the surface and so they give up their energy to electrons closer to the surface of the secondary emitter. As a result, there is a greater probability that the excited electrons will be emitted from the surface.

It is interesting that the first crossover energy of about 100 eV is nearly independent of θ. Also, it is clear, although it is not shown in Figure 18-4, that the second crossover energy increases significantly as θ increases.

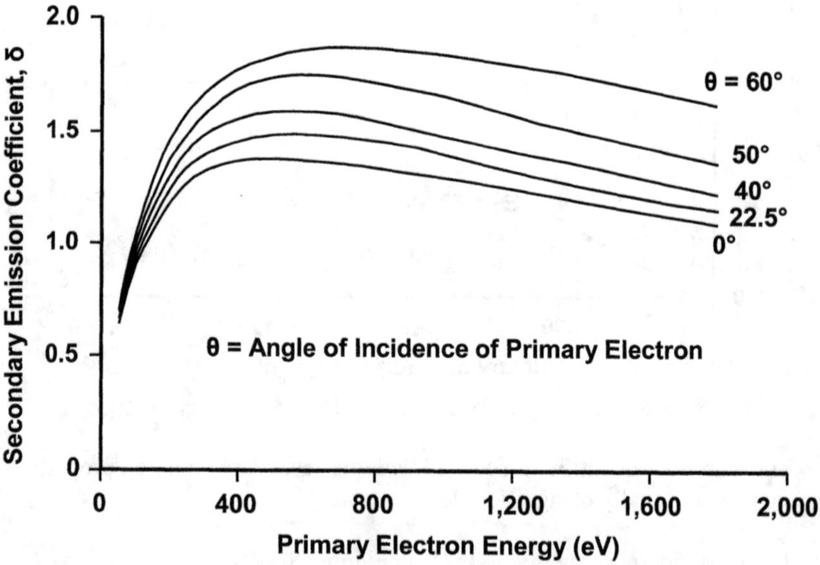

Figure 18-4 Secondary emission coefficient of molybdenum as functions of primary electron energy and angle of incidence. (From: Arnold Shih and Charles Hor, *IEEE Trans. Electron Devices*, April 1993. © 1993 IEEE.)

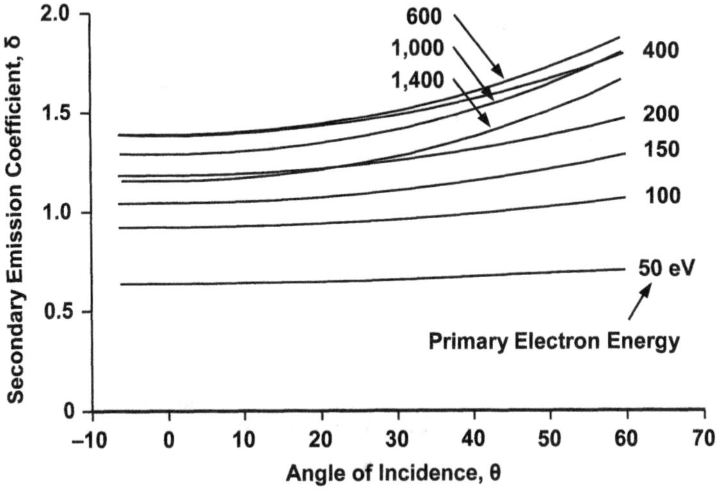

Figure 18-5 Secondary emission coefficient of molybdenum as functions of angle of incidence and primary electron energy. (From: Arnold Shih and Charles Hor, *IEEE Trans. Electron Devices*, April 1993. © 1993 IEEE.)

18.2.3 Secondary Emitting Properties of Surfaces

The secondary emitting properties of surfaces vary widely from material to material and are critically dependent on the conditions of the surface. Examples are shown in Figures 18-6 through 18-10. In each case, δ is shown for an incidence angle of 0°. A material with an extremely low δ ($\delta_{max} \approx 0.25$) is shown in Figure 18-6. This material is textured carbon on a copper surface and was developed by NASA for possible use in multistage depressed collectors where it is important to inhibit secondary electron emission.

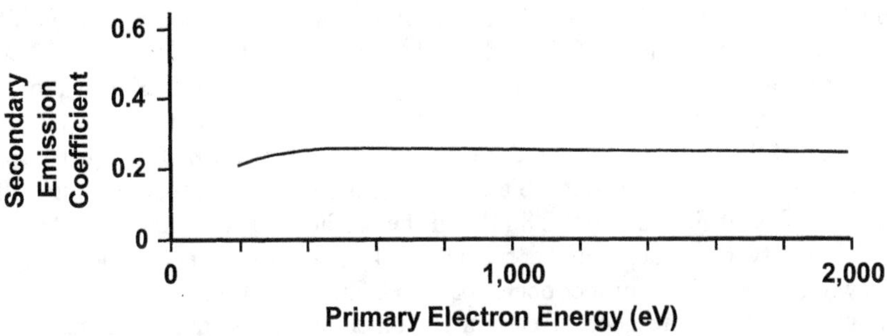

Figure 18-6 Secondary emission coefficient of textured carbon on a copper substrate. (From: A. N. Curren, *IEEE Trans. Electron Devices*, November 1986. © 1986 IEEE.)

In Figure 18-7, the extremely high δ ($\delta_{max} > 20$) of a diamond film on a molybdenum substrate is shown. When this film was subjected to ion bombardment so that the surface was converted to carbon, the value of δ_{max} dropped to about unity. Diamond films have been considered for use in the cathodes of crossed-field amplifiers, but the degrading effects of ion and electron bombardment make their use questionable.

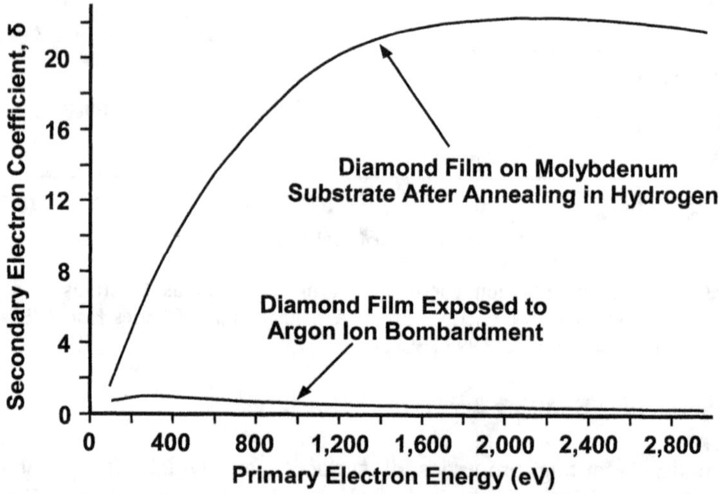

Figure 18-7 Secondary emission characteristics of a diamond film on a molybdenum substrate. (Adapted from: T. L. Bekker et al., *Final Report*, NSWC/Crane Division Diamond Film Study, August 1993.)

In Figure 18-8, the secondary emission characteristics of tungsten are shown [private communication from A. Dallos et al., Raytheon]. The value of δ_{max} is about 1.4. The presence of oxygen on the surface of the tungsten increases δ_{max} to about 2.3. This clearly illustrates the necessity for careful surface preparation if the true secondary emission coefficient of the surface is to be determined.

Beryllium oxide (BeO) is frequently used as a cathode material in crossed-field amplifiers. The secondary emission coefficient for this material depends on the nature of the BeO surface and the way in which it was formed. For grit-blasted beryllium oxidized by exposure to air for several days, δ is shown in Figure 18-9 [2]. The value of δ_{max} is about 2.8 (1% of the surface was covered by carbon). A small quantity of carbon on the surface (1/4 of a monolayer) reduces δ_{max} to somewhat over 2. Since proper operation of a CFA depends on the value of δ_{max}, it is obvious that the presence of carbon (or other materials) that contaminate the cathode surface can degrade tube operation.

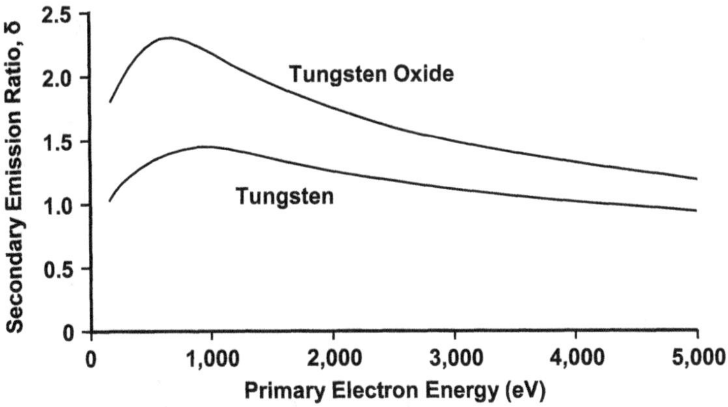

Figure 18-8 Secondary emission characteristics of tungsten and tungsten oxide.

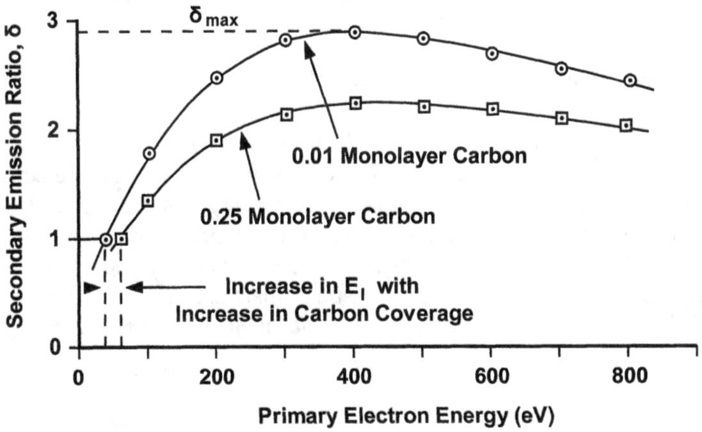

Figure 18-9 Secondary emission coefficient of oxidized beryllium with and without carbon contamination. (Adapted from: V. H. Ritz, et al., *Surface and Interface Analysis*, November 1988.)

Figure 18-9 also shows that the first crossover energy, E_I, increases rapidly with carbon contamination. This is important in the operation of CFAs that use cold cathodes because there is no thermionic emission and tube operation depends entirely on secondary emission. Electron emission is initiated by the interaction of the RF input signal with stray electrons. E_I is important because, the lower E_I is, the lower the RF input signal level has to be to start the tube properly. Figure 18-10 shows the dependence of δ_{max} and E_I on carbon. It was noted by Ritz et al. that a 0.03 difference in carbon atomic fraction will change δ_{max} by 0.1 and shift E_I by 7 volts. Again, this illustrates the strong dependency on carbon of the secondary emitting properties of a cathode most important to CFA operation.

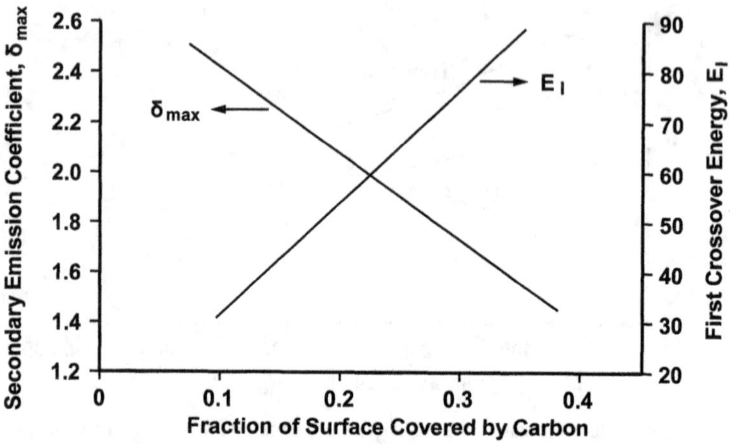

Figure 18-10 Effect of carbon coverage on beryllium oxide. (Adapted from: V. H. Ritz, et al., *Surface and Interface Analysis*, November 1988.)

18.2.4 Energy Distribution of Secondary Electrons

The energy distribution of secondary electrons is shown in Figure 18-11. The majority of the electrons are emitted with energies of a few tens of electron volts. These are the electrons in the solid that acquire energy as the result of collisions with the primary electrons. A second peak in the secondary emission plot occurs at energy levels that are near the primary electron energy (near 100%). The electrons in this peak are those that experience elastic collisions at the solid surface and are reflected from that surface with an energy that is at or near the primary electron energy.

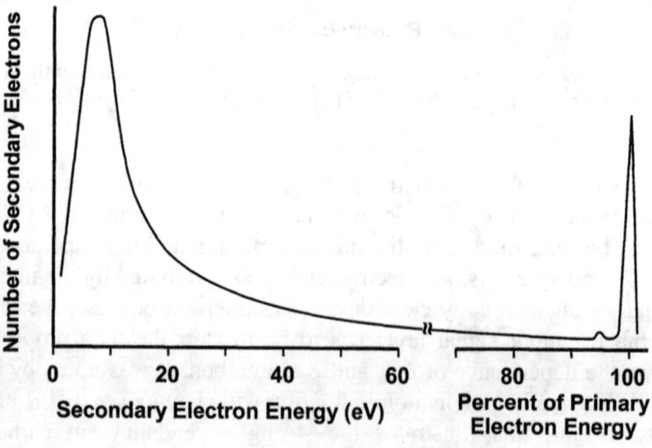

Figure 18-11 Distributions of energies of secondary electrons.

Perhaps the most significant aspect of the secondary electron energy distribution is the high values of electron energies (compared with thermionically emitted electrons). Thermionic electrons have energies of a small fraction of an electron volt. Secondary electrons have energies two orders of magnitude higher. This is part of the cause of the high noise output of crossed-field devices.

18.2.5 Modeling of Secondary Emission Characteristics

Vaughan developed simple empirical formulas that appear to describe quite accurately the variations of secondary emission coefficient with primary electron energy and with angle of incidence [3, 4]. These formulas are of value in computer codes used to model the operation of crossed-field tubes. The first step in using the formulas is to calculate $E_{max}(\theta)$ and $\delta_{max}(\theta)$ as functions of the impact angle, θ, from those quantities when $\theta = 0$. The equations for this calculation are:

$$E_{max}(\theta) = E_{max}(0)\left(1 + k_{sE}\,\theta^2/2\pi\right) \tag{18-1}$$

and

$$\delta_{max}(\theta) = \delta_{max}(0)\left(1 + k_{s\delta}\,\theta^2/2\pi\right) \tag{18-2}$$

where k_{sE} and $k_{s\delta}$ are factors, the values of which depend on the smoothness of the surface being considered. They should each be assigned a default value of 1.0, which, Vaughan said, is appropriate for typical dull surfaces in a working tube environment. Lower values would be appropriate for deliberately roughened surfaces, down to zero for textured carbon. Higher values up to about 2.0 will be appropriate for exceptionally smooth, clean, oxide-free surfaces.

The values of E_{max} and δ_{max} are then used to find δ as a function of E using

$$v = \frac{E - E_0}{E_{max} - E_0} \tag{18-3}$$

and

$$\frac{\delta}{\delta_{max}} = \left(v\,e^{1-v}\right)^k \tag{18-4}$$

where

$$k = 0.56 \quad \text{for } v < 1 \tag{18-5}$$

$$k = 0.25 \quad \text{for } 1 < v \leq 3.6 \tag{18-6}$$

and

$$\frac{\delta}{\delta_{max}} = 1.125/v^{0.35} \quad \text{for } v > 3.6 \tag{18-7}$$

The quantity E_o is the minimum impact energy at which any secondary electrons at all are generated and is taken as 12.5 eV.

Shih and Hor [1] showed that Vaughan's formulas agree extremely well with their experimental results for molybdenum. Careful measurements similar to those of Shih and Hor are needed for other materials to provide a more stringent test of Vaughan's formulas.

18.3 OPERATION OF CATHODES IN CROSSED-FIELD DEVICES

Some crossed-field devices use hot cathodes and others use cold cathodes. The cold cathode devices are crossed-field amplifiers. Cathode emission is initiated by the interaction of the RF input signal with a few stray electrons (at least 10 "good" electrons according to one CFA engineer). The source of these stray electrons is not clear. Field emission and multipactor are both possible sources. The RF input signal drives these electrons into the cathode surface, causing secondary emission. If the secondary emission coefficient of the cathode is high enough, then emission current quickly (in nanoseconds) builds up to the operating current level.

In either hot-cathode or cold-cathode devices, the environment for the cathode is extremely harsh and many degradation mechanisms exist. First of all, as shown in Figure 18-12(a), there is intense electron bombardment of the surface. This causes heating and, if the temperature is not properly controlled, excessive evaporation of cathode constituents. In addition, electron bombardment leads to dissociation of emissive materials, such as barium oxide, on the surface of the cathode. The only cathodes free from the deleterious effects of electron bombardment are those fabricated from pure metal or from carburized thoriated tungsten. Electron bombardment also decomposes carbon monoxide and leaves a deposit of carbon on the cathode surface.

When arcs occur in crossed-field tubes, arc products are deposited on the cathode surface as is indicated in Figure 18-12(b). The arc products contaminate and reduce the secondary emission capability of the cathode. As a result, the noise figure and jitter of the tube increase. There must be some built-in mechanism in the tube to restore emission. In some tubes, it is the residual gas that provides ions during tube operation that clean the cathode surface. In tubes with BeO cathodes, an atmosphere of oxygen at a pressure of about 10^{-6} torr is used to maintain the oxidized surface of the cathode. This oxygen also becomes ionized and the positive ions bombard and clean the cathode surface. Also, the oxygen serves to oxidize carbon on the cathode surface and this goes off as CO or CO_2. In some tubes with platinum cathodes, the anode vane tips are also coated with platinum. This material is eroded from the anode during tube operation and is deposited on the cathode so that the platinum surface is maintained.

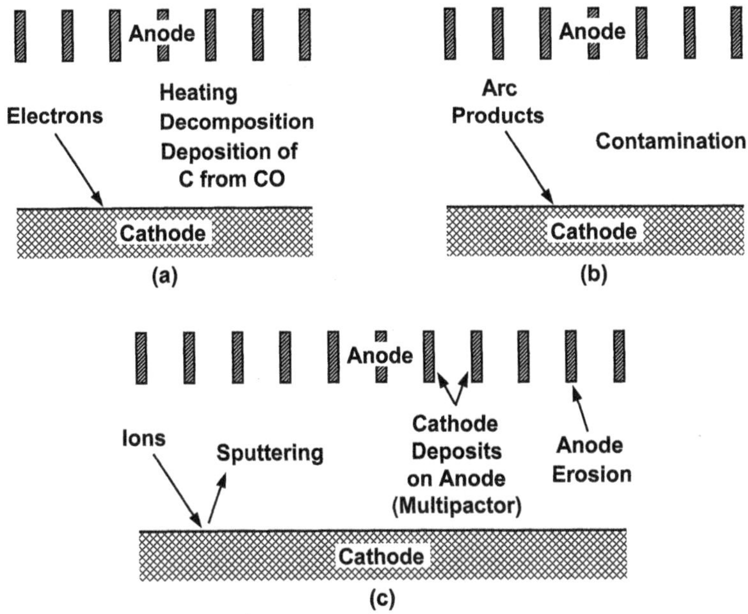

Figure 18-12 (a–c) Cathode degradation mechanisms.

Ions serve to clean the cathode surface, but they also sputter emissive material from the surface as indicated in Figure 18-12(c). Not only does this degrade emission, but also it results in the deposition of material on the anode. The secondary emission coefficient of that material is high so multipactor discharges may occur between elements of the RF circuit on the anode.

Another degradation mechanism indicated in Figure 18-12(c) is the erosion of the anode and this can be produced by decomposition products from the cathode. For example, Vaughan reported magnetron anode erosion in the presence of cathodes containing oxides [5]. He cited an example in which a thoria-molybdenum cathode was replaced by a tungsten-thorium cathode and this completely eliminated a severe anode erosion problem. It was not clear whether it was O or CO that was causing the erosion.

REFERENCES

[1] A. Shih, and C. Hor, "Secondary emission properties as a function of the electron incidence angle," *IEEE Trans. Electron Devices*, Vol. 40, No. 4, April 1993, pp. 824–829.

[2] V. H. Ritz et al., "Secondary electron emission characteristics of oxidized beryllium cathodes," *Surface and Interface Analysis*, Vol. 11, No. 6-7, April 1988, pp. 389–397.

[3] J. R. M. Vaughan, "A new formula for secondary emission yield," *IEEE Trans. Electron Devices,* Vol. ED-36, No. 9 pt. 2, September 1989, pp. 1963–1967.

[4] J. R. M. Vaughan, "Secondary emission formulas," *IEEE Trans. Electron Devices*, Vol. 40, No. 4, April 1993, p. 830.

[5] J. R. M. Vaughan, "Magnetron anode erosion in the presence of cathodes containing oxides," *IEEE Trans. Electron Devices*, Vol. ED-17, No. 4, April 1970, p. 377.

Chapter 19

Magnetrons

In magnetrons [1–7], the anode and cathode are cylindrical and concentric. They operate in a magnetic field that is perpendicular to the electric field between the cathode and anode. The magnetic field causes electrons from the cathode to move in curved paths. Electron motion in crossed dc fields was discussed in Chapters 4 and 17.

In addition to the dc field effects, RF electric fields and space-charge forces combine to alter the trajectories indicated in Chapters 4 and 17. The RF circuit and field patterns along with operating characteristics for magnetrons are described in the following sections.

19.1 TYPES OF MAGNETRONS

There are three basic types of magnetrons:

1. Cyclotron-frequency magnetrons;
2. Negative-resistance magnetrons;
3. Traveling wave magnetrons.

Cyclotron-frequency and negative-resistance magnetrons are no longer used. However, their basic principles of operation will be described to help form a foundation for the understanding of traveling wave magnetrons, which are now in use.

19.1.1 Cyclotron-Frequency Magnetrons

The first crossed-field tube, shown in Figure 19-1, was the cyclotron-frequency magnetron invented by A. W. Hull and reported in 1921 [8]. This and other early magnetrons used cylindrical anodes with no RF structure. Efficiencies and power capabilities were very low.

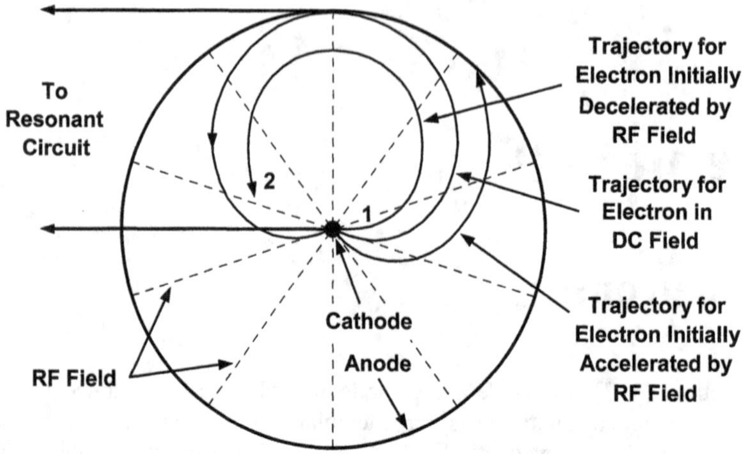

Figure 19-1 Electron trajectories in a cyclotron frequency magnetron.

Cyclotron-frequency magnetrons made use of interaction between orbiting electrons in a magnetic field and a resonant circuit tuned to the cyclotron frequency. In the simplest form of these magnetrons, the cathode and anode were coaxial cylinders and were operated in a magnetic field parallel to the electrode axis. The resonant circuit was a quarter-wavelength parallel-line resonator connected between the cathode and anode as indicated in Figure 19-1.

To produce oscillations, the voltage and magnetic field were adjusted close to the cutoff condition so that electron trajectories were like that for the dc-field case. Electrons emitted when the RF electric field aided the dc electric field landed on the anode, and so these, which extracted energy from the RF field, were removed from interaction. Electrons emitted when the RF field opposed the dc field gave up energy to the RF field and missed the anode. If the frequency of rotation of the electrons (cyclotron frequency) was close to the RF frequency, then electrons that initially gave up energy (in Region 1) arrived in Region 2 as the field reversed and again gave up energy to the field. So these electrons contributed energy to the RF oscillation. The process continued as long as the phase relationship with the field persisted or until the electrons were removed from interaction. Cyclotron frequency magnetrons operated at frequencies up to the millimeter-wave range, but had very low power output and efficiency.

19.1.2 Negative-Resistance Magnetrons

Negative-resistance magnetrons were split-anode devices as shown in Figure 19-2. A magnetic field was applied parallel to the axis and a dc voltage was applied between the cathode and the anode segments. The magnetic field was sufficient to keep electrons from reaching the anode when both segments were at the same potential as shown in Figure 19-2(a). In the absence of an RF signal, electrons would leave the cathode, move toward the anode, and then be turned back to the cathode by magnetic deflecting forces.

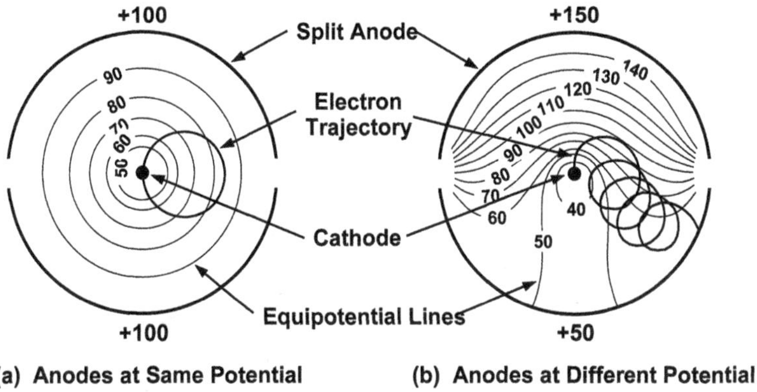

Figure 19-2 (a, b) Electron trajectories in a negative-resistance magnetron with a split anode.

When an RF voltage was present and the anode segments were at different potentials, much more complicated trajectories were formed as are illustrated in Figure 19-2(b). Electrons moving in a crossed-field region tend to move along equipotential lines. As indicated in Figure 19-2(b), as one electrode segment is made negative with respect to the other, more electrons are collected on that electrode. When the voltages on the anode segments are reversed, more electrons are collected on the other electrode so a negative resistance effect occurs. When a resonant circuit was connected between the two anode segments, it was possible to obtain sustained oscillations. Negative resistance magnetrons generally operated below 500 MHz, were low power devices, and had low efficiency.

19.1.3 Traveling Wave Magnetrons

Early traveling wave magnetrons were split-anode devices similar to negative resistance magnetrons. An example is shown in Figure 19-3. The primary difference between the two was the magnitude of the magnetic field. In the negative resistance magnetron, the magnetic field was near the Hull cutoff and the frequency of operation was low enough so that an electron moved in a single cycloid-like path all the way from the cathode to the anode, while the voltage on the anode segments remained nearly constant (as indicated in Figure 19-2(b)). In traveling wave magnetrons, the magnetic field is well above the Hull cutoff field. The frequency with which the alternating voltage on the anode segments changes can be made near the rate of rotation of the electrons about the cathode.

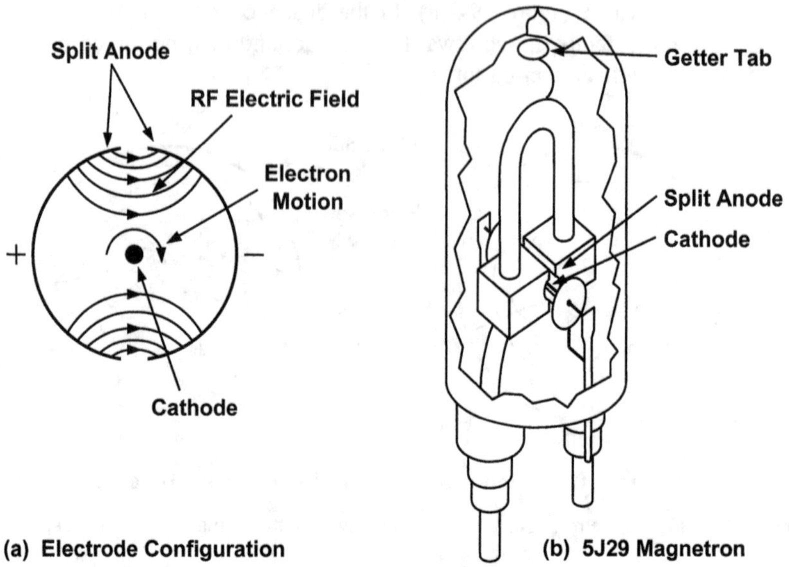

Figure 19-3 (a, b) Configuration of a traveling wave magnetron with a split anode.

In 1935, Posthumous deduced theoretically that, in magnetrons with multisegment anodes and large magnetic fields, RF power should be generated with high efficiency. He then demonstrated an efficiency of about 50% in a magnetron operating at a wavelength of 50 cm (600 MHz).

In 1936, Samuel proposed the hole-and-slot resonator arrangement for use in a magnetron anode at high frequencies, but apparently did not reduce its use to practice.

By about 1938, a magnetron with a copper anode and hole-and-slot resonators had been built by Russian scientists Alekseroff and Malearoff. A tungsten helix cathode was used and the power generated was about 100-W CW at a wavelength of 10 cm.

In late 1939, Boot and Randall at the University of Birmingham (United Kingdom), who were apparently unaware of the work of Posthumous, Samuel, and the Russians, set about to combine the cylindrical-anode-magnetron work with the klystron's advantage of using resonators for confining the RF fields. By early 1940, a magnetron with an anode containing six resonant cavities had been built and operated. By May 1940, an experimental radar containing a pulsed 10-cm cavity magnetron was in operation and by September 1940 a submarine periscope could be detected at a range of 7 miles.

At some point during the early development of the magnetron, an oxide-coated cathode was supplied by Warnecke and Ponte of CSF (France) to replace the carburized thoriated-tungsten cathode being used at the time. The peak power capability increased by about an order of magnitude to 100 kW.

In an article by J. J. Coupling (a pseudonym of J. R. Pierce) concerning the magnetron, the following opening statement appeared [10]:

> "The two most important weapons of the recent war were the atomic bomb, and radar . . . and it was Maggie, the magnetron, that made Allied radar incomparably superior to the Nazi's boat."

Modern magnetrons are all traveling wave magnetrons. The primary differences between these devices and the 5J29 split-anode magnetron shown in Figure 19-3(b) are in the RF circuit. In the 5J29, the part of the resonant circuit internal to the tube consisted of an inductive loop (to which the getter is attached in Figure 19-3(b)) and the capacitance between the anode segments. The RF circuit in all modern magnetrons consists of several resonant cavities similar to those shown in Figure 19-4, which is an early version of a pulsed cavity magnetron. In this tube, the anode contains eight resonators and is concentric with an oxide-coated cathode. A permanent magnet (not shown) was used for applying a magnetic field parallel to the cathode surface.

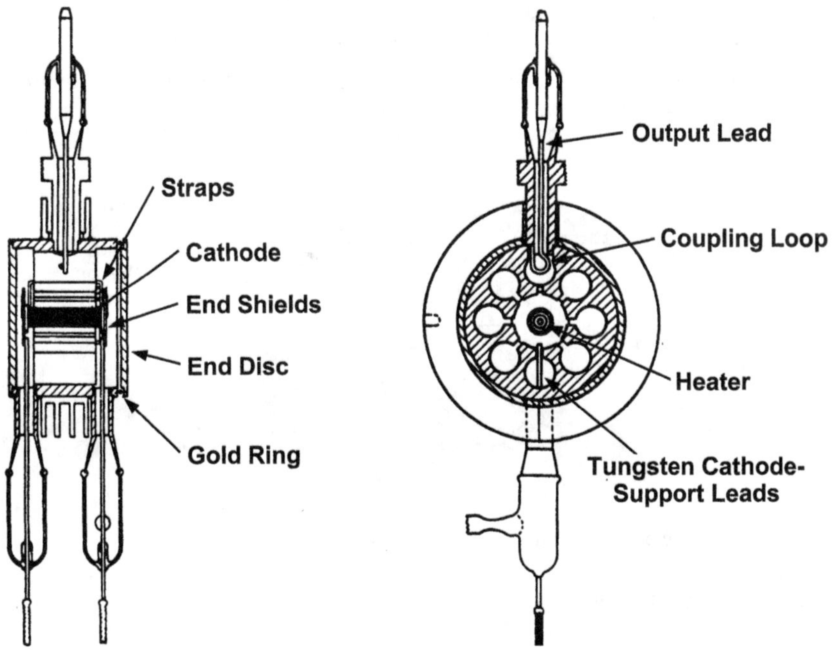

Figure 19-4 Early sealed-off pulsed cavity magnetron. (From: H. A. H. Boot and J. T. Randall, *IEEE Trans. Electron Devices*, July 1976. © 1976 IEEE.)

19.2 OPERATION OF THE TRAVELING WAVE MAGNETRON

When voltage is first applied to a traveling wave magnetron, there are no RF fields and the electrons form a space charge cloud as is shown in Figure 19-5 that rotates about the cathode. This cloud is often referred to as a space charge hub.

Because of fluctuations in the emission process and noise-amplifying effects in the space-charge hub, electron velocities and the charge density in the rotating space-charge cloud fluctuate randomly with time and produce a noise component to the rotating current in the space charge hub. The frequency of the fluctuations varies over such a wide range that the noise is referred to as "white noise," which implies fluctuations over an infinite frequency range. While the frequency range of fluctuations is not infinite, it is rich in components in the microwave range, and so there are components present to initiate any of the frequencies that may be supported by the cavity structure of a magnetron.

This noise current in the hub induces noise current in the cavities. When the component of this noise current that is at the resonant frequency of the cavities passes through a high impedance of the cavities at one of the resonant frequencies, RF voltages are produced across the cavity gaps and these amplify the RF current in the hub. RF current and power build up rapidly and saturate when the power supplied to the magnetron equals the power to the load plus losses.

In the following sections, the formation of the hub, the buildup of RF current, and the characteristics of the cavity resonator are discussed.

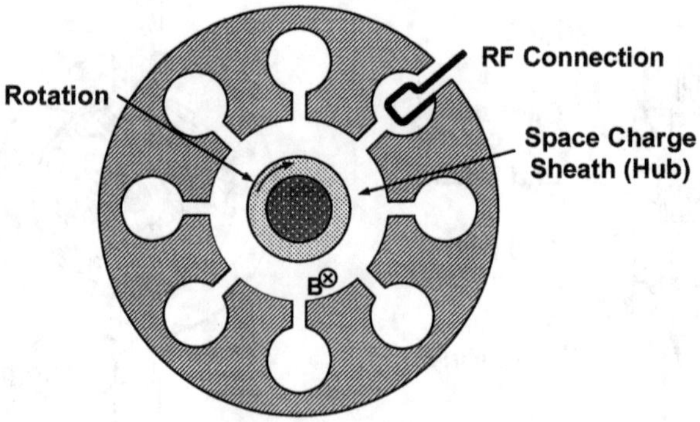

Figure 19-5 Space charge cloud that is formed when RF is absent.

19.2.1 Hub Formation

Part of the behavior of the electron hub can be explained by considering the motion of a single electron when no RF fields are present.

Referring to Figure 19-6, it is seen that there are two electron motions to consider. With a radial electric field, E, and an axial magnetic field, B, two forces act on an electron. The force from the electric field, $-eE$, is toward the anode. The force exerted by the magnetic field, $-e(u \times B)$, is perpendicular to the motion of the electron. The resulting path of the electron is approximately represented by combining a slow rotation around the cathode at radius R_o and a faster circular motion with a smaller radius r_o. The result of these two motions corresponds roughly to the motion of a point on the circumference of a wheel as it rolls around a circle somewhat smaller than the cathode in such a way that its center moves in a circle of radius R_o.

The speed of the slow R_o rotation is given approximately by the ratio E/B. The fast r_o rotation corresponds to the cyclotron frequency so it is determined only by B. Although the angular velocity of this r_o rotation is constant, the magnitude of r_o depends on the initial kinetic energy of the electron so it may vary for different electrons. With no RF, the maximum distance that any electron can move toward the anode ($R_o + r_o$) is fixed by the ratio E/B and, in practice, is about half of the distance from the cathode to the anode.

This description of the path of a single electron is only accurate for small r_o when RF fields and space charge are negligible.

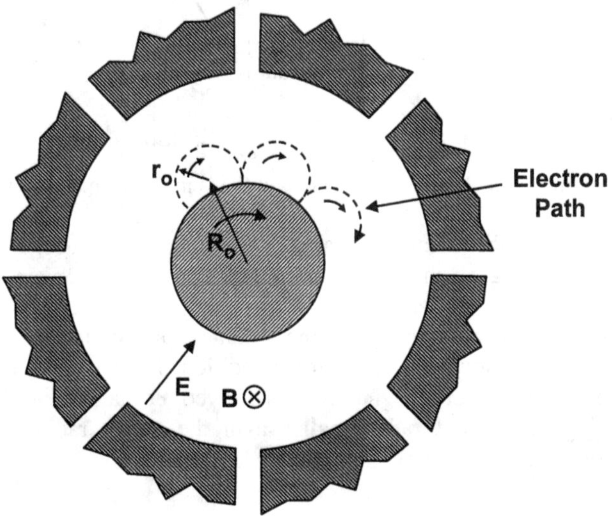

Figure 19-6 Path of a single electron when space charge effects and RF fields are absent.

One approach to understanding more about the electron hub is to examine a theoretical electron flow in a crossed-field region that is not attainable in a practical magnetron, but is useful for consideration and visualization of electron interaction with anode circuit waves. This is called laminar flow in a linear format and a Brillouin hub in a circular format. This flow is illustrated in Figure 19-7. In the linear format, layers of electrons are all moving parallel to the negative electrode. In the circular format, layers of electrons are moving in concentric circles about the cathode.

Because the electron flow is in crossed fields, the velocities with which the electron streams move are given by *E/B* (E_y for the linear format and E_r for the circular format). Because of space charge, the electric field near the cathode is lower than in the absence of space charge and so the velocities of the electron streams are reduced. This results in a condition called slipping flow.

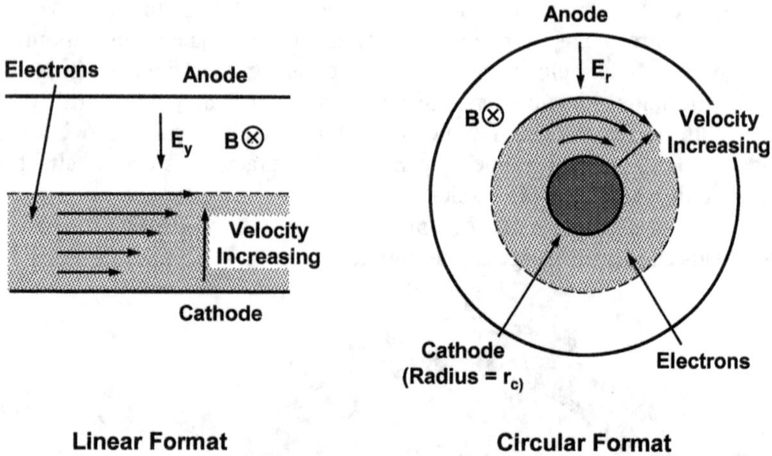

Figure 19-7 Theoretical electron flow in crossed fields in linear and in circular formats.

The velocities of the electron streams are of interest because, for interaction with the RF circuit in a magnetron, there must be near synchronism between the electron velocity and the phase velocity of the circuit. The electron velocities are given by *E/B* and *E* can be found from the voltage that produces tangential flow. In Chapter 4, the condition (the Hull cutoff condition) for tangential electron flow at the anode of a smooth bore magnetron was derived. For tangential flow like that indicated in Figure 19-7 at radii smaller than the anode radius, the relation between rate of rotation, magnetic flux density, and radius is

$$\dot{\theta}(r) = \frac{\eta B}{2}\left(1 - \frac{r_c^2}{r^2}\right) \tag{19-1}$$

and the corresponding relation for voltage is

$$V(r) = \frac{\eta B^2 r^2}{8}\left(1 - \frac{r_c^2}{r^2}\right)^2 \tag{19-2}$$

The electric field is

$$E(r) = \frac{\eta B^2 r}{4}\left(1 - \frac{r_c^2}{r^2}\right) \tag{19-3}$$

so the electron velocities as a function of radial position are

$$v(r) = \frac{\eta B r}{4}\left(1 - \frac{r_c^2}{r^2}\right) \tag{19-4}$$

Although neither laminar flow nor a Brillouin hub can be obtained in practice, they are often considered to be present to simplify the problem of analysis. In these procedures, the top layer of space charge is assumed to be moving in near synchronism with an RF traveling wave on the anode circuit.

19.2.2 The Hartree Voltage

The Hull cutoff condition is derived when no RF fields are present. This condition defines the anode voltage as a function of the magnetic field that prevents the flow of electrons to the anode. In the previous sections, we have said that a magnetron operates at a voltage well below the Hull cutoff voltage so the electrons are constrained to a hub rotating about the cathode. However, under the right circumstances, electrons interact with the RF wave that is rotating about the cathode, leave the hub, and flow to the anode. This happens when the anode voltage is such that the electrons are rotating faster than the RF wave. Then electrons are slowed to the RF wave rotation rate by interaction with the wave. In the process, the electrons give up energy to the wave and amplify the wave.

If the anode voltage is reduced, the rate at which the electrons tend to rotate is reduced so they give up less energy to the rotating wave. There is a critical voltage at which the rate of electron rotation equals the rate of rotation of the wave. At that voltage, the electrons can no longer give up energy to the wave and so the magnetron stops operating. Below that voltage, the electrons rotate more slowly than the RF wave and so the magnetron cannot operate. The critical voltage at which energy transfer to the wave ceases and the magnetron stops operating is called the *Hartree voltage*. We will show that the Hartree voltage varies linearly with the magnetic field and so the Hartree voltage and the Hull cutoff voltage can be plotted as shown in Figure 19-8. The significant voltages for a magnetron are between the Hartree and Hull cutoff voltages.

The Hartree voltage can be found as follows. Consider an RF wave in a magnetron rotating at an angular rate of $\dot{\varphi}$ and an electron rotating at $\dot{\theta}$ as indicated in Figure 19-9. As long as $\dot{\theta} > \dot{\varphi}$, the electron can give up energy to the wave. The energy, W, which can be transferred from the electron to the wave is

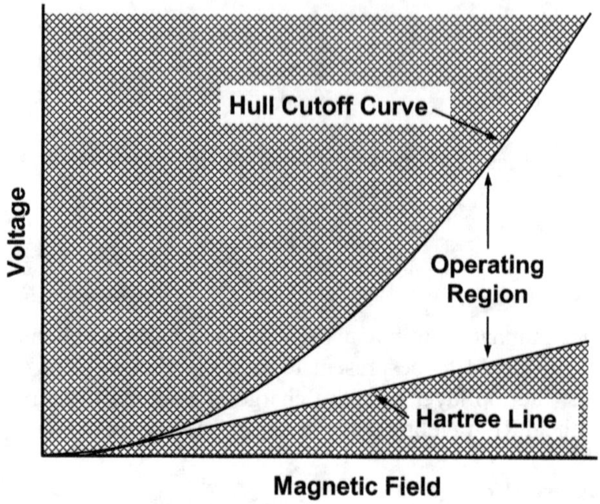

Figure 19-8 Operating region of a magnetron.

$$W = \frac{1}{2} m r^2 \left(\dot{\theta} - \dot{\varphi}\right)^2 \qquad (19\text{-}5)$$

or

$$W = \frac{1}{2} m r^2 \left(\dot{\theta}^2 - 2\dot{\theta}\dot{\varphi} + \dot{\varphi}^2\right) \qquad (19\text{-}6)$$

where r is the radius of the electron path.

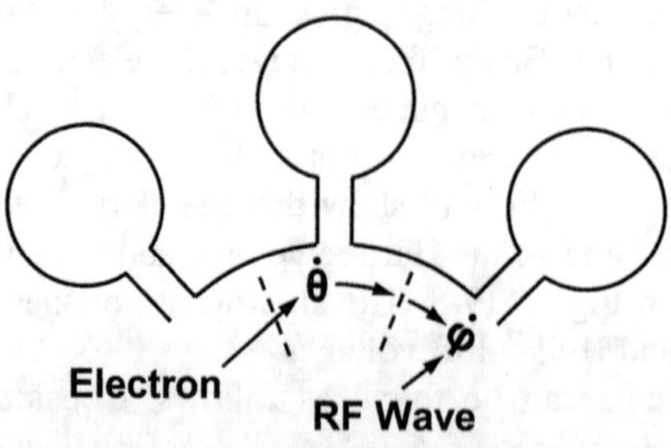

Figure 19-9 Electron and wave rotation in a magnetron.

Now, the condition for which the energy that can be transferred decreases to zero is

$$\frac{1}{2}mr^2\dot{\theta}^2 = \frac{1}{2}mr^2\left(2\dot{\theta}\dot{\varphi} - \dot{\varphi}^2\right) \tag{19-7}$$

If we assume that the amplitude of the RF wave is small, then

$$\frac{1}{2}mr^2\dot{\theta}^2 = eV_r \tag{19-8}$$

where V_r is the voltage (19-2) that causes the electron to rotate at $\dot{\theta}$. This means that the voltage for which the energy transfer goes to zero (the Hartree voltage, V_H) is

$$V_H = \frac{1}{2}\frac{m}{e}r^2\left(2\dot{\theta}\dot{\varphi} - \dot{\varphi}^2\right) \tag{19-9}$$

where $\dot{\theta}$ was given by (19-1). For the normal π mode of magnetron operation, there are $N/2$ cycles of the RF wave around the anode where N is the number of cavities. This means that the rate of rotation of the wave must be the operating frequency of the magnetron divided by the number of cycles or

$$\dot{\varphi} = \frac{2\omega}{N} \tag{19-10}$$

Substituting for $\dot{\theta}$ and $\dot{\varphi}$, the Hartree voltage can be found,

$$V_H = \frac{1}{2}B\left(r_a^2 - r_c^2\right)\frac{2\omega}{N} - \frac{1}{2\eta}r_a^2\left(\frac{2\omega}{N}\right)^2 \tag{19-11}$$

By making the following substitutions, this can be rewritten in a form that leads to a straightforward physical interpretation [9].

Length of spoke, ℓ, $= r_a - r_c$

Mean radius of spoke, r_m, $= (r_a - r_c)/2$

Mean spoke velocity, v_m, $= r_m\dot{\varphi}$

Velocity of spoke at anode, v_a $= r_a\dot{\varphi}$

so

$$V_H = B\ell v_m - \frac{1}{2\eta} v_a^2 \tag{19-12}$$

The first term on the right is the voltage, sometimes referred to as the "back EMF," produced across a conductor of length ℓ moving with velocity v_m through magnetic flux density B. The second term is the voltage equivalent of the kinetic energy of an electron striking the anode with velocity v_a.

The Hartree relation provides insight into tube designs. The second term should be small compared to the first term because it represents wasted energy. This means that the overall behavior of a crossed-field tube depends mostly on the first term. From this, one can deduce the following four general properties of a crossed-field tube.

1. The anode voltage will be proportional to the magnetic field.
2. It will be proportional to the anode-cathode spacing (the spoke length).
3. It will be proportional to the RF frequency (contained in v_m).
4. It will be inversely proportional to the number of spokes, or the mode number.

Items 2 and 4 are dimensional features built into the tube and cannot be changed externally. Items 1 and 3 are externally controllable.

19.2.3 Spoke Formation

In considering the interactions of electrons with RF electric fields, it is helpful to review one aspect of the electron trajectories in a dc electric field, E. That is, as is shown in Figure 19-10, electrons with velocity E/B move in a straight line. Those electrons with $u > E/B$ are deflected downward in the direction of the dc electric field toward the negative electrode. Those electrons with $u < E/B$ are deflected upward in the direction opposite to the dc electric field toward the positive electrode.

When the interactions of single electrons with RF fields are considered, the paths of the electrons can be traced in a relatively straightforward manner. For example, consider Figure 19-11. An electron leaving the cathode at point **A** enters an accelerating field and is driven back into the cathode with appreciable energy.

An electron leaving the cathode at **B** enters a decelerating field and moves toward the anode. With the correct frequency of operation, this electron will remain in synchronism with the decelerating field as it passes successive anode segments and will eventually strike the anode. In moving from the cathode to the anode, this electron acquires dc energy, which it gives up to the RF field.

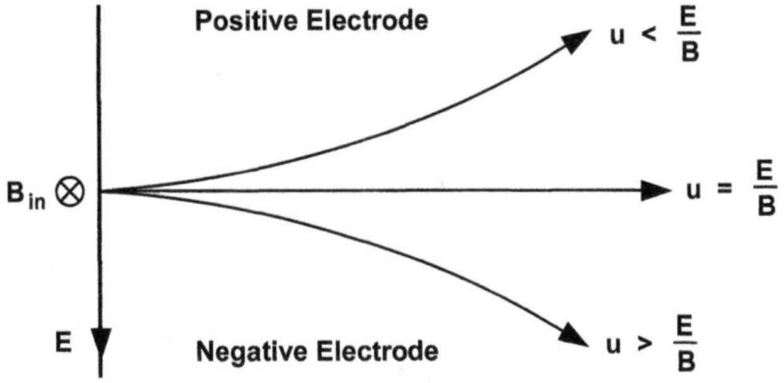

Figure 19-10 Effect of electron velocities on curvature of electron trajectories.

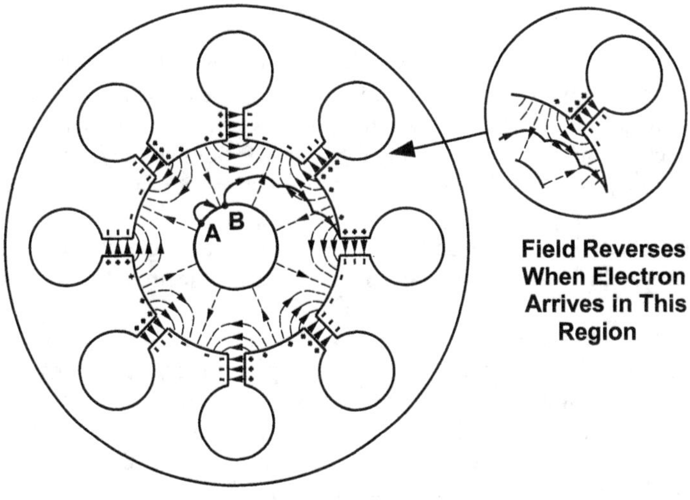

Figure 19-11 Paths followed by electrons in a magnetron. (Adapted from: *Microwave Magnetrons* by George B. Collins, copyright 1948 by McGraw-Hill, Inc.)

Next, consider the interaction of a single stream of electrons with an RF field. It will be shown in the following section that the electric field pattern on the anode of a magnetron is a standing wave that can be resolved into two waves of equal magnitude traveling in opposite directions. Figure 19-12 shows one wavelength of the traveling electric field pattern and an electron stream with E/B adjusted so that it is moving in synchronism with the traveling wave. The directions of the forces on the electrons (opposite to the field directions) are indicated at the bottom of the figure and the resulting deflections of the electrons are indicated by small arrows along the electron stream.

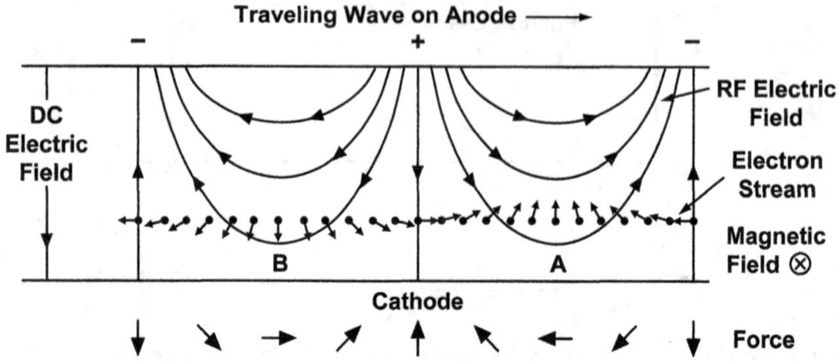

Figure 19-12 Electron deflections produced by an electric field traveling in synchronism with an electron stream.

Electrons near **A** experience decelerating forces so they move toward the anode. As they move toward the anode, the RF fields become stronger and the electrons are pushed progressively into spoke-like projections that extend to the anode. Electrons near **B** are accelerated by the RF field so they move toward the cathode. This produces depressions in the electron stream that extend to the cathode.

As with individual electrons, the interaction of a single stream of electrons with an RF field is straightforward and individual electron trajectories can be determined.

The space-charge cloud surrounding the cathode greatly complicates the analysis of electron trajectories. However, qualitatively, the behavior of the electrons is very similar to that which has just been described. The rotating space-charge cloud is distorted into the spoke-like pattern shown in Figure 19-13.

Electrons in decelerating field regions move through the spokes from the cathode to the anode. During synchronous interaction, electrons at the surface of the hub and those transported to the anode move with a velocity approximately equal to that of the traveling wave on the anode.

The spoke-like cloud rotates with an angular velocity such that it remains in synchronism with the alternating RF voltage on the anode segments. As the spokes pass the slots between anode segments, the RF fields provide a focusing action that tends to further bunch the electrons. As indicated in Figure 19-14, an electron at position **A** on the leading edge of a spoke is in a region where the radial component of the RF field is in opposition to the dc field. As a result, from $u = E/B$, its angular velocity will be reduced so that it will fall back into the spoke. An electron at point **B** will experience an increased field and will be accelerated into the spoke. This same focusing action helps maintain synchronism between the rotating spokes and the alternating field so that the operating conditions of the magnetron (magnetic field, voltage, and so forth) are not as critical as they would appear to be at first.

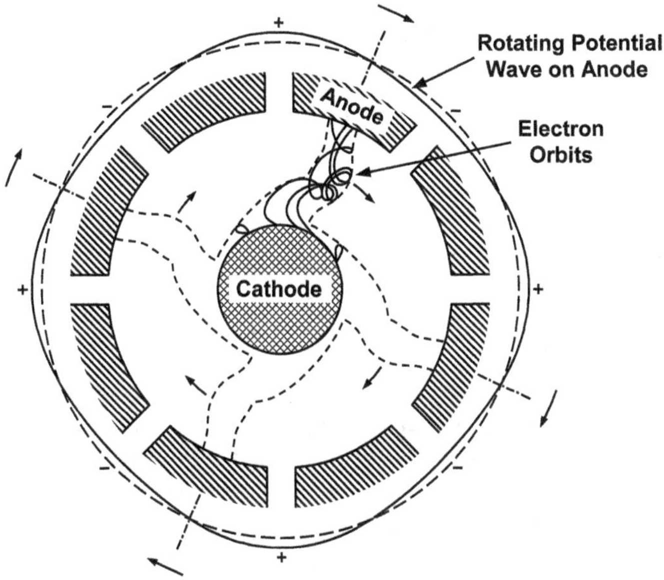

Figure 19-13 Spoke-like electron cloud in eight-cavity magnetron. (From: H. A. H. Boot and J. T. Randall, *IEEE Trans. Electron Devices*, July 1976. © 1976 IEEE.)

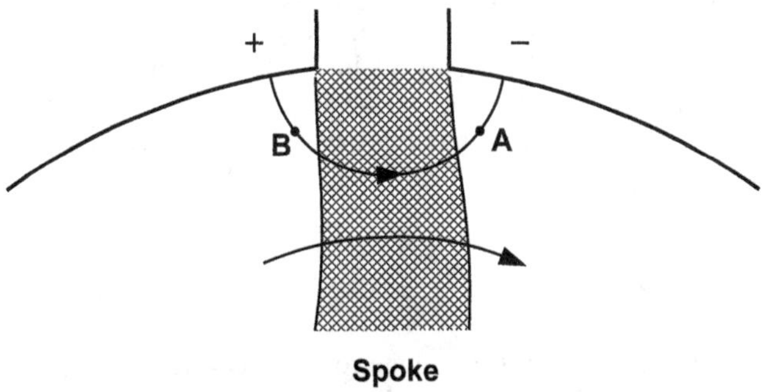

Figure 19-14 Focusing action of RF fields on space charge.

During transport from the hub to the anode, the kinetic energy of the electrons remains approximately constant. The kinetic energy is

$$eV_o = \frac{1}{2}mv_p^2 \tag{19-13}$$

where v_p is the phase velocity of the RF circuit and V_o is called the *synchronous voltage*.

The energy extracted from the power supply by an electron is the sum of its potential energy and its kinetic energy. Since the kinetic energy remains nearly constant, it is the potential energy that is converted to microwave power. This is different from a linear-beam tube where the kinetic energy of the beam is converted to microwave power and the beam slows down as a result of the energy exchange.

The operating voltage of a magnetron is typically several times the synchronous voltage. In the ideal situation, the electrons strike the anode at the synchronous velocity and so the anode dissipation is eV_o for each electron. Then the maximum theoretical electronic efficiency is

$$\eta_e = \frac{eV_a - eV_o}{eV_a} = 1 - \frac{V_o}{V_a} \tag{19-14}$$

In practice, electrons follow cycloidal motions as they move through a spoke. An electron moving at the top of a cycloid is moving at twice the synchronous velocity. An electron striking the anode with that velocity dissipates four times eV_o so the electronic efficiency on average is

$$\eta_e = 1 - \frac{kV_o}{V_a} \tag{19-15}$$

where $1 \leq k \leq 4$.

Electrons in accelerating field regions are forced back to the cathode surface. When they strike the cathode, they have sufficient energy to produce secondary emission. In a typical magnetron, the secondary electron emission is far greater than the primary emission. Secondary yield is great enough to replenish the hub and make up for the electrons that go to the anode. The back bombardment power delivered to the cathode results in heating the cathode surface. In a typical crossed-field tube, approximately 5–10% of the dc input power is dissipated in heating the cathode.

19.2.4 RF Circuit Operation

Each cavity of the resonator in a magnetron can be visualized as a parallel LC circuit, as indicated in Figure 19-15. The resonant frequency of a resonator is determined by its dimensions as well as the reactive effect of any perturbation. For example, the presence of electrons near the vane tips affects the capacitance of the resonator. The equivalent circuit for an eight-cavity resonator is eight LC circuits in parallel (Figure 19-16).

The magnetic and electric field patterns for each cavity are indicated in Figure 19-17. In the desired mode of operation, some of the magnetic flux lines of each cavity link adjacent cavities (two lines are shown). The fields in adjacent cavities are 180° out of phase. This is referred to as the π mode of operation. The resulting field configurations in an eight-cavity structure are shown in Figure 19-18.

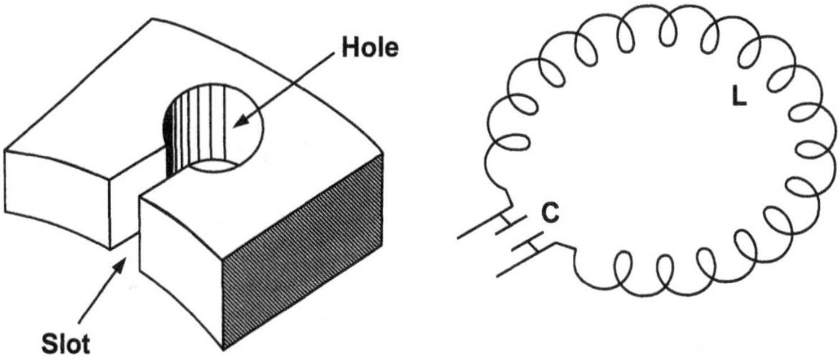

Figure 19-15 Equivalent circuit of hole-and-slot cavity.

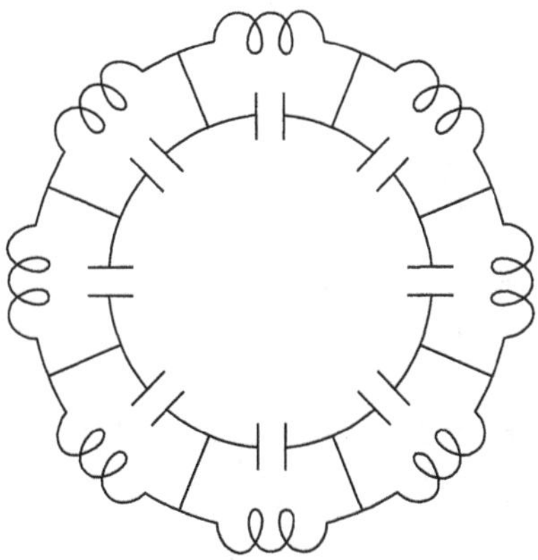

Figure 19-16 Equivalent circuit for an eight-cavity resonator.

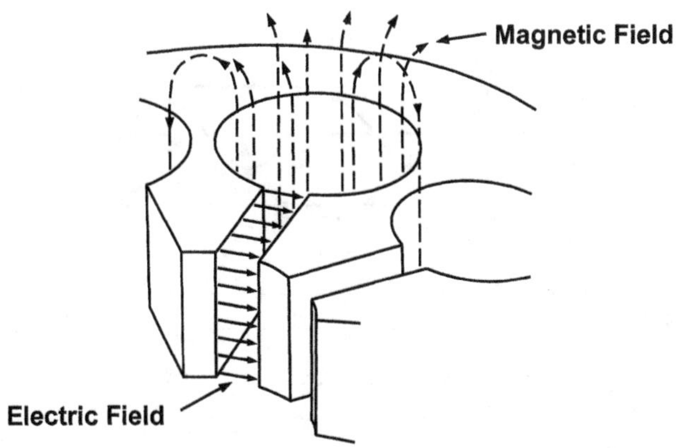

Figure 19-17 Magnetic and electric field patterns in a hole-and-slot cavity. (Adapted from: *Microwave Magnetrons* by George B. Collins. © 1948 McGraw-Hill, Inc.)

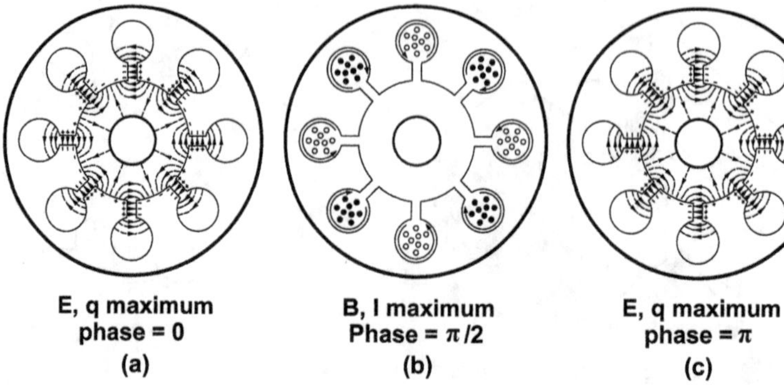

Figure 19-18 (a–c) Charges, fields, and currents for the π-mode in an eight-cavity resonator. (Adapted from: *Microwave Magnetrons* by George B. Collins. © 1948 McGraw-Hill, Inc.)

Now, the reason why these magnetrons are referred to as traveling wave magnetrons can be explained. In Figure 19-19 the cavities of an eight-cavity structure are shown in a linear configuration. The electric field pattern is a standing wave with the direction of the electric field lines across the cavity gaps oscillating back and forth as the RF voltage on the anode segments oscillates.

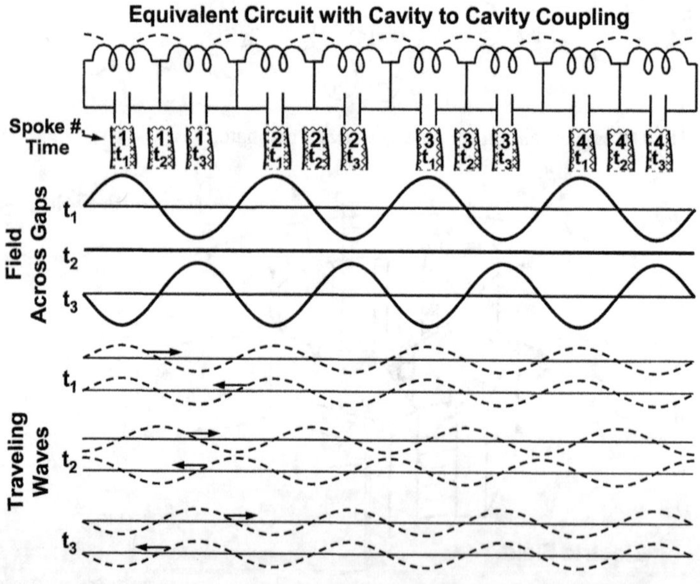

Figure 19-19 Illustration of how traveling waves combine to produce the field across the gaps. Wave positions and gap fields at times t_1, t_2, and t_3, are shown along with the positions of four spokes.

The standing wave, in turn, can be thought of as consisting of two waves, one moving to the right (rotating clockwise in the magnetron) and the other moving to the left (rotating counterclockwise). At time, t_1, the left-moving and right-moving waves combine to maximize the electric field across the gaps of the first and every other cavity. At this time, forces that compress the spokes are maximum.

At time t_2, the phases of the wave moving to the right and the wave moving to the left are such that the fields of the waves are opposite and so they cancel. As the electron spokes move by the gaps between cavities, the electric field is zero.

At time t_3, the fields from the two waves again add, but this time the field is maximum at the second, fourth, sixth, and eighth gaps. Again, the forces that compress the spokes are maximum. Thus, it is clear that, to optimize the interaction of the wave on the anode with the movement of the spokes, the velocity of the wave on the anode must be near the spoke velocity. Then energy exchange from the electrons to the RF field is maximum.

19.3 MODING

Modes and field patterns other than those associated with the π mode can exist and result in magnetron operation at frequencies other than the desired frequency. These other modes and the steps taken in attempts to prevent their occurrence are discussed here.

In his science-fiction article on magnetrons [10], Pierce leads into a discussion of moding by commenting on the various problems faced by early magnetron engineers and the humorous names that they applied to the peculiarities of performance encountered.

> ...The Civil War Effect, or North-South Effect had to do with differences in performance when the direction of the magnetic field was reversed. The differences proved to be due to asymmetries in the tubes, and disappeared when the tubes were made more accurately. The Blowing Out of Brains Effect had simply to do with the melting of a copper end cap which covered the end of the anode block of a small magnetron to form part of the vacuum envelope. The Gnawing Away of the Anode Effect was similar. The Gardarens Swine Effect has to do with the particularly precipitous descent of certain contours on a performance chart, and the Valley of the Shadow was a region of low efficiency lying close to a high-efficiency region known as the Heights of Abraham.
>
> Undoubtedly the worst effect was one perhaps too serious for a fancy name. It was called, simply, moding. An engineer in using an early magnetron would be adjusting the load, that is, the impedance into which the magnetron fed power, in order to get more power. Suddenly, the power,

which had been rising, would drop, and he would find the tube to be oscillating feebly at some new and unwanted frequency a little removed from the old. Or, he would be gradually raising the voltage to get more power, and the same thing would happen. The tube was said to be operating in another mode of oscillation....

Moding occurs because the RF circuit of a magnetron consists of many resonators coupled together. An analogous situation exists when several identical pendulums are suspended from a common support that is not rigidly held in place. For example, in Figure 19-20, three pendulums are shown along with the three forms of motion (modes of oscillation) that they can have. The frequencies at which the pendulums swing back and forth are different for each mode. Similarly, a magnetron may have several modes of oscillation that depend upon the coupling between cavities.

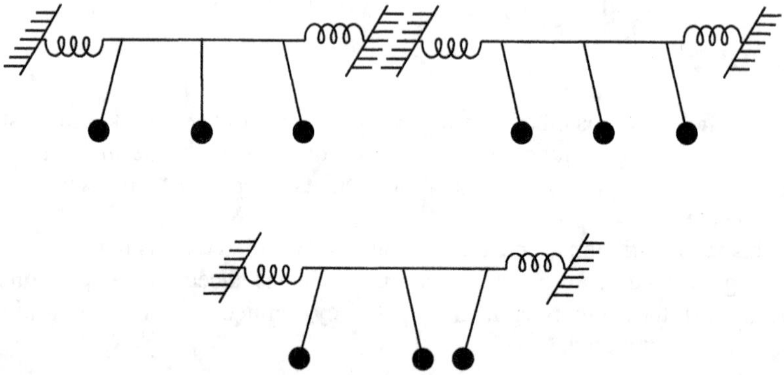

Figure 19-20 Modes of oscillation of three pendulums supported by a movable bar.

Field distributions for the modes in an eight-cavity magnetron are illustrated in Figure 19-21. The modes are designated by a mode number *n*, which is the number of times the RF field pattern is repeated in going around the anode once. The largest value of *n* is *N*/2, where *N* is the number of cavities. This mode is called the π mode and is always the desired mode of operation.

In addition to there being several modes of oscillation for a magnetron, there is, quite literally, an infinite number of electron velocities that can excite any given mode. As an example, the situation existing in the π mode is illustrated in Figure 19-22. The equivalent circuits of an eight-cavity magnetron are across the top of the figure. The electric field distribution along the anode is shown for various time intervals separated by one-half of a period. An electron, such as one at point **A** and time *t* is in a decelerating region and gives up energy to the RF field. An electron in the region between **A** and **B** is considered to be essentially shielded from the RF field by the anode segment.

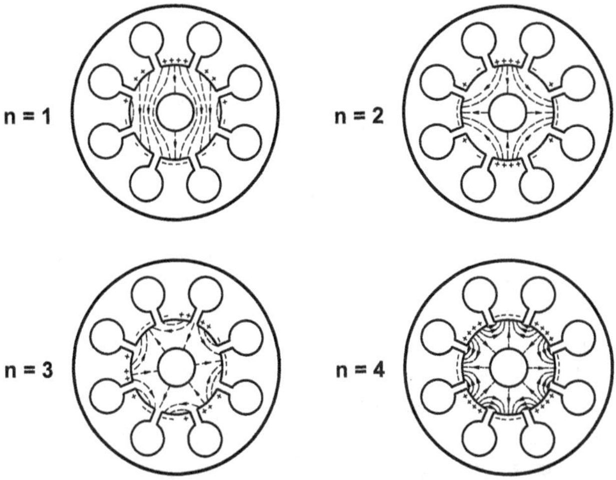

Figure 19-21 Field distributions for the modes in an eight-cavity magnetron. (From: H. A. H. Boot and J. T. Randall, *IEEE Trans. Electron Devices*, July 1976. © 1976 IEEE.)

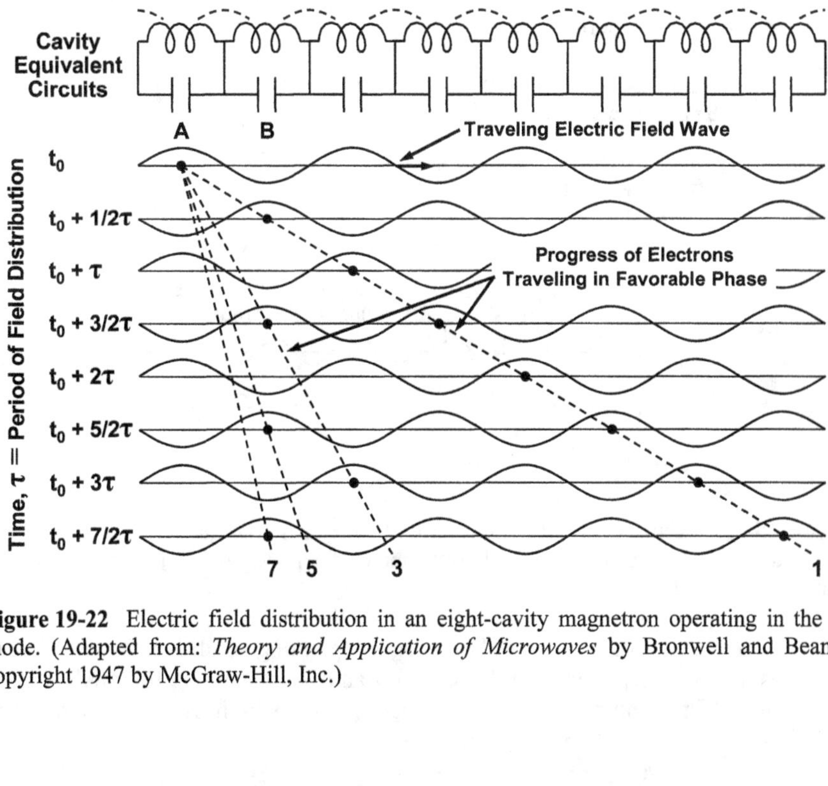

Figure 19-22 Electric field distribution in an eight-cavity magnetron operating in the π mode. (Adapted from: *Theory and Application of Microwaves* by Bronwell and Beam, copyright 1947 by McGraw-Hill, Inc.)

The dashed lines represent the progress of electrons traveling in favorable phases (continually giving up energy) with the RF field. For example, electrons represented by line 1 have a transit angle from cavity to cavity of π radians and travel in synchronism with the traveling field. Electrons represented by line 3 travel at one-third of the velocity of those represented by line 1 and have a transit angle of 3π radians. A complete cycle of oscillation of the RF circuit occurs while these electrons pass by each anode segment in traveling from one cavity to the next. Still, these electrons are synchronous with the RF field and give up energy on every third cycle of oscillation of the RF field. This situation is analogous to keeping the pendulums in Figure 19-20 moving by giving them a push on every third swing.

Close to the resonator gaps, the electric field is a square wave as shown in Figure 19-23. If this field is broken into its harmonic components, odd-numbered harmonics (1, 3, 5, 7, and so forth) result, and it is found that the electrons represented by line 3 are in synchronism with the third harmonic component of the field. Similarly, electrons represented by lines 5, 7, in Figure 19-22, and an infinite set not shown are synchronous with still higher order harmonic components of the field. In principle, any of these harmonic interactions can produce oscillation in the π mode. Each of the other magnetron modes can be broken into an infinite set of harmonic components and can, therefore, in principle, be driven by electrons having an infinite range of velocities.

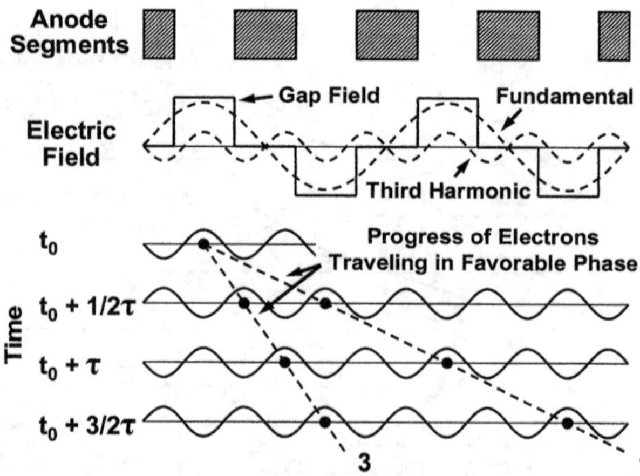

Figure 19-23 Electric field distribution across four gaps in a magnetron. The progress of electrons traveling in phase with the third harmonic component of the field as well as the fundamental component is shown.

The resulting possible array of magnetron operating conditions is bewildering and this is, perhaps, what led Boot and Randall in their 1976 article [4] to comment: "If this point had been appreciated in 1939 it is doubtful that the magnetron would ever have been made." Fortunately, there are several selection processes, both natural and by design, which limit the range of operating conditions of a magnetron. For example, the amplitudes of the harmonic fields decrease rapidly as the harmonic number increases so that the probability of interaction at harmonics decreases. For a square wave like that in Figure 19-23, the amplitude of the third harmonic is 1/3 of the fundamental amplitude.

As an increasing anode voltage is applied to a magnetron, increases occur in both the radius and angular velocity of the space charge cloud surrounding the cathode. When the angular velocity is such that the cloud is rotating in near synchronism with one of the RF modes, or one of the harmonic components of an RF mode, noise currents on the space charge cloud may cause oscillations to start in that mode. The voltages at which the oscillations are initiated in various modes and various harmonics are the Hartree threshold voltages. Some of these voltages are plotted as a function of magnetic flux density for an eight-cavity magnetron in Figure 19-24. The dashed line represents a harmonic of the $n = 3$ mode, which has a threshold voltage nearly identical to the $n = 4$ mode, which is the π mode. As a result, unless precautions are taken, operation may occur in the $n = 3$ mode, rather than the π mode.

In considering precautions that can be taken to force a magnetron to operate in the desired mode, it is important to realize that the frequencies of the various modes are very close together. This is indicated in Figure 19-25 by the unstrapped line. The separation between the $n = 3$ and $n = 4$ modes is only 1% to 2%. For structures having more than eight cavities (in practice, the number of cavities usually ranges from 8 to 18), the separation in frequency between the π mode and its nearest neighbor becomes even smaller than 1% to 2%. Because of the frequency instabilities when the mode separation is small, structural changes are made to the resonator to increase this separation.

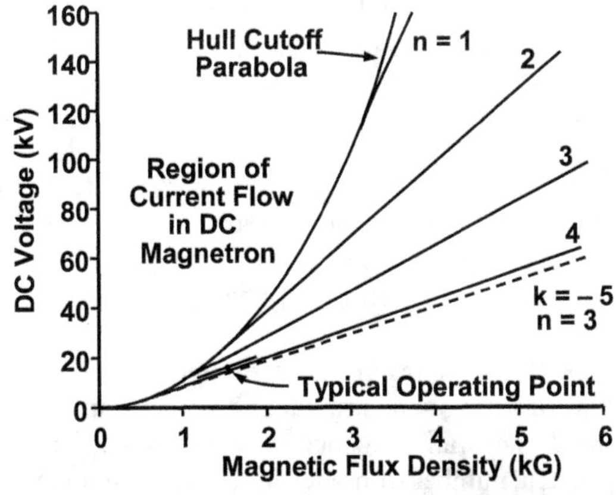

Figure 19-24 Hartree threshold voltage diagram for an eight-cavity magnetron. (From: H. A. H. Boot and J. T. Randall, *IEEE Trans. Electron Devices*, July 1976. © 1976 IEEE.)

Two techniques, the use of straps and the rising-sun structure, for increasing the mode separation are discussed at this point. The third, the use of a coaxial cavity, is discussed in Section 19.4 on coaxial magnetrons.

In the desired π mode of operation, alternate anode segments are at the same RF potential so, if alternate segments are connected together by wires (straps), no additional inductance will be introduced because the ends of the straps are at the same potentials. The straps add capacitance to the circuit, so the π-mode frequency will be altered.

In modes other than the π mode, the voltage differences between alternate anode segments is not zero, so the straps introduce inductance as well as capacitance resulting in different frequency shifts than occur for the π mode. The effect of strapping on mode separation in 18-cavity magnetrons is shown in Figure 19-25.

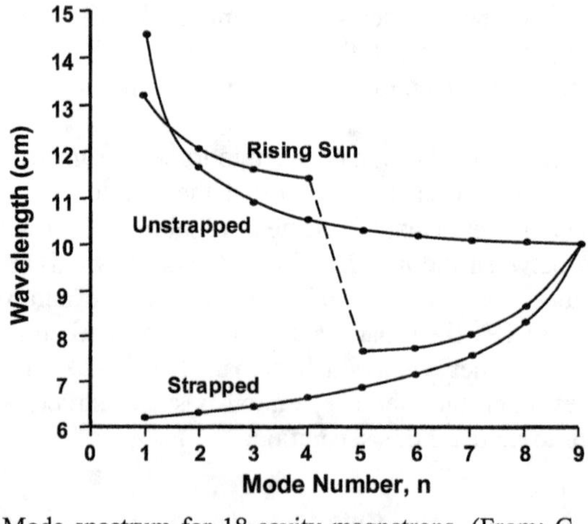

Figure 19-25 Mode spectrum for 18-cavity magnetrons. (From: G. K. Farney, *Crossed-Field Microwave Devices,* Technology Service Corporation/Crane Division Naval Surface Warfare Center, 1995.)

In low-power tubes, mode separations of 25% to 35% are achieved with strapping. In high-power tubes, the mode separation is normally about 10% to 14%. In either case, the undesired modes are at frequencies far enough removed from the π mode so that the tube can usually be prevented from operating in these modes.

Shown in Figure 19-26 are examples of strapped structures and an unstrapped resonator known as a rising-sun structure. The way in which the rising-sun structure causes mode separation to occur is not as easy to understand as the strapping technique. The number of modes in the rising sun magnetron is the same as in the strapped magnetron. However, in the rising sun magnetron, half the modes are such that the large cavities control the frequency, and so the frequencies associated with these modes are relatively low. For the higher-numbered modes (except the π mode), the field patterns are such that the small cavities control the frequency, and so the frequencies associated with these modes are relatively high. In the π mode, the frequency is midway between the resonant frequencies of the two cavity configurations. Thus, the rising sun structure increases the frequency separation between the π mode and the two adjacent modes (one above π mode and one below π mode). By comparison, in a strapped system, all the modes are resonant at frequencies below the π mode.

The mode separation characteristic of rising sun structures has facilitated the design and fabrication of millimeter-wavelength magnetrons where dimensions are small and it is difficult, if not impossible, to fabricate strapped resonators. However, the number of resonators is limited by the mode separation requirements and the size of the cathode is limited by the number of resonators. Cathode size also decreases as the frequency of operation is increased. This limits the current and the output power.

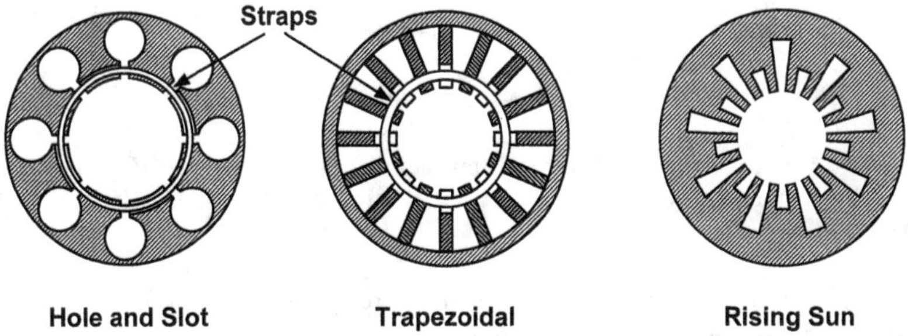

Figure 19-26 Techniques for achieving mode separation.

19.4 COAXIAL MAGNETRONS [11]

An indication of the significance of the development of the coaxial magnetron is that it is listed by Skolnik [12] along with the cavity magnetron and the high-power klystron as a significant event in the development of radar technology. The coaxial magnetron was developed as an alternative to the strapped cavity magnetron.

In the conventional magnetron, the strap-vane structure results in anode segments that are sometimes small and operate at an excessively high-power density. Close interelectrode spacing leads to occasional voltage breakdown and the cathode loading is often excessive. As a result, in early radars, the conventional magnetron was often the least reliable component of the system.

As Figure 19-27 shows, the coaxial magnetron contains an anode resonator structure very much like that in a conventional magnetron. This structure is surrounded by a single high-Q cavity operating in the TE_{011} mode. Slots in the back walls of alternate cavities of the anode resonator structure tightly couple the fields in these resonators to the surrounding cavity. In the π mode, the fields in every other cavity are in phase, and they couple in the same direction into the surrounding cavity. Thus, the surrounding coaxial cavity helps stabilize the magnetron in the desired π mode of operation.

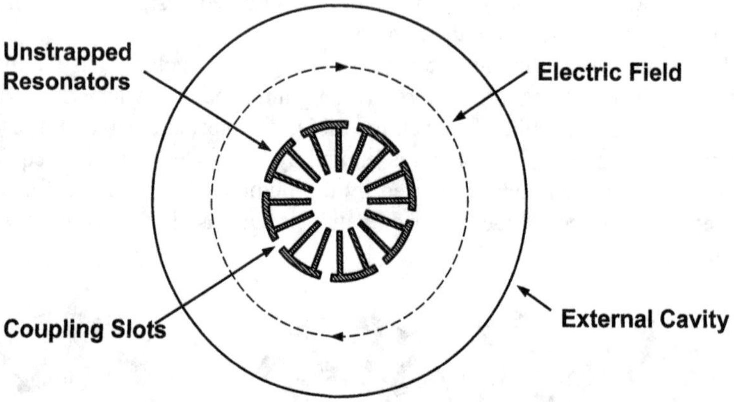

Figure 19-27 Cross section of coaxial magnetron. (Adapted from: *Microwave Tube Manual* by Varian Associates, Air Force Publication Number T.O.00-25-251, October 1979.)

Because straps are not required, the anode resonator for the coaxial magnetron can be larger and less complex than for the conventional strapped structure. A comparison of anode sizes for the same operating frequency and power is shown in Figure 19-28. Because the anode is larger, the cathode and interelectrode spaces can also be larger. As a result, cathode loading is lower and voltage gradients are reduced. These factors all combine in the coaxial magnetron to produce relatively conservative magnetron designs having enhanced reliability compared with conventional magnetrons.

In the coaxial magnetron, the multiple-cavity anode structure has a series of modes like those of the conventional magnetron. In addition, there are various modes of oscillation of the coaxial cavity. Fortunately, the use of the TE_{011} mode makes possible the use of attenuators to damp unwanted modes.

In the desired TE_{011} mode, the RF electric field exists in circular form within the cavity as indicated in Figures 19-27 and 19-29, and decreases to zero at the walls of the cavity. Current flow in the TE_{011} mode is in the walls of the cavity in circular paths about the axis of the tube. An attenuator placed as shown in Figure 19-29 at the junction of the inner cylinder to the fixed end of the cavity couples very little to TE_{011} currents, but couples heavily to other modes, thereby suppressing these unwanted modes of oscillation.

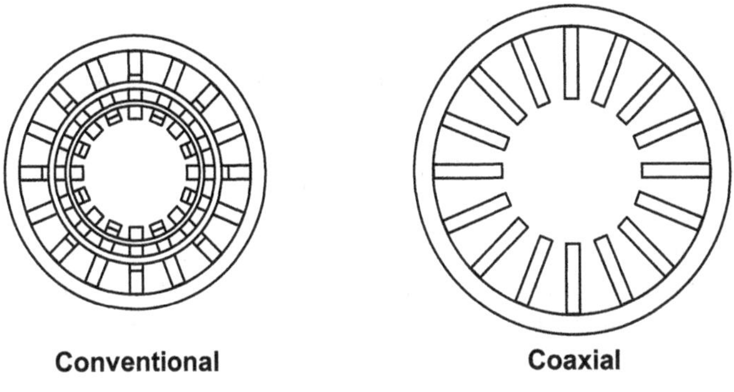

Figure 19-28 Comparison of anodes of conventional strapped magnetron and coaxial magnetrons. (From: *Microwave Tube Manual* by Varian Associates, Air Force Publication Number T.O.00-25-251, October 1979.)

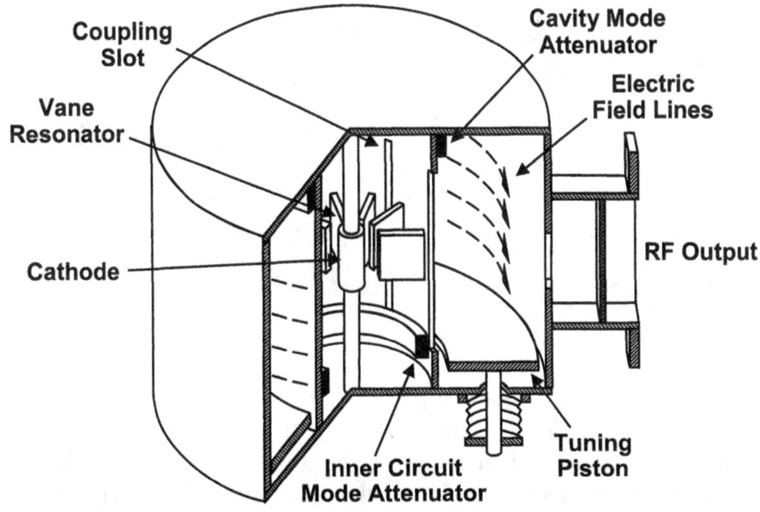

Figure 19-29 Cutaway view of coaxial magnetron. (From: *Microwave Tube Manual* by Varian Associates, Air Force Publication Number T.O.00-25-251, October 1979.)

Similarly, when the fields near the coupling slots from the resonator to the coaxial cavity are examined, it is found that the ends of the slots are free of TE_{011} current. Thus, an attenuator within the inner cylinder near the ends of the coupling slots does not couple to the TE_{011} mode currents, but it does dampen undesired modes of the vane structure.

Another advantage of the TE_{011} mode is that one of the cavity end plates can be made movable (tuning piston in Figure 19-29) without interrupting the flow of current. As a result, tuner reliability is high for coaxial magnetrons.

19.5 INVERTED MAGNETRONS

A further advantage of the coaxial-magnetron configuration is that it can be turned inside out, or inverted, as shown in Figure 19-30. In this figure, an inverted structure is compared with a conventional rising-sun magnetron designed to operate at the same frequency. In comparing the two structures, it should be noted that the cathode size in the inverted configuration is, perhaps, an order of magnitude larger than the cathode in the conventional magnetron. Also, the number of anode vanes and cavities can be large in the inverted structure because moding is controlled by the center coaxial cavity.

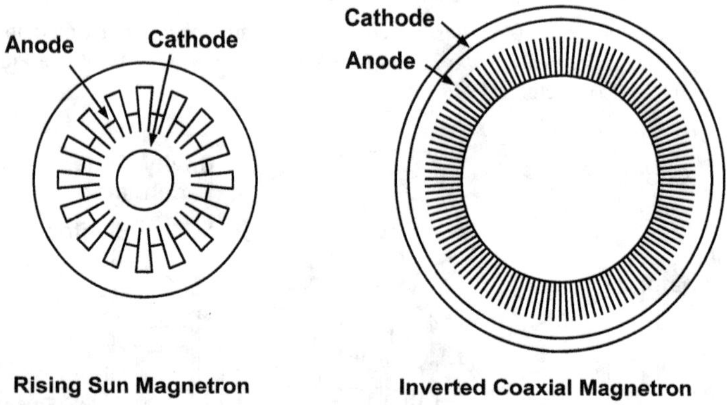

Figure 19-30 Comparison of rising-sun and inverted coaxial magnetrons designed for the same frequency. (From: *Microwave Tube Manual* by Varian Associates, Air Force Publication Number T.O.00-25-251, October 1979.)

The inverted coaxial configuration has application in the millimeter-wave region where dimensions are extremely small. In particular, the power of the rising-sun magnetron is limited by the high cathode loading. This problem is alleviated in the inverted structure. In addition, the center TE_{011} cavity lends itself to coupling to an output waveguide operating in the circular electric mode, which has extremely low loss. It should also be noted that experiments have been performed with inverted magnetrons at S-band to determine if they can generate extremely high power [13].

19.6 MAGNETRON TUNING

Many techniques are used for changing the frequency (tuning) of magnetrons. In one way or another, these tuning techniques employ changes in the capacitance or the inductance of the resonant cavities. For example, as shown in Figure 19-31, the capacitance of the cavity may be varied by the movable metallic element (or dielectric element) in the vicinity of the slot portion of the cavity. The inductance of the cavity may be decreased by inserting a metallic plunger into the hole portion of the cavity. Configurations of capacitive and inductive tuning elements are shown in Figure 19-32.

Coaxial magnetrons can be tuned by changing the volume of the cavity with a movable end plate as indicated in Figure 19-29. A fine-tuning technique using a compressible ring in the wall of the coaxial cavity is also shown in Figure 19-33.

In another technique for tuning a coaxial magnetron, rotatable dielectric paddles are inserted into the high-electric-field region of the coaxial cavity. The lowest frequency occurs when the paddles are aligned with the electric field. The highest frequency occurs when the paddles are perpendicular to the electric field lines.

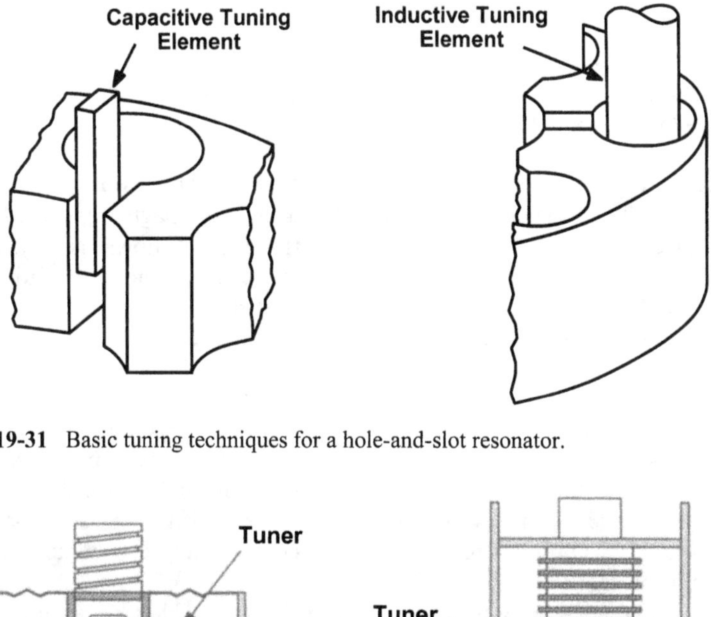

Figure 19-31 Basic tuning techniques for a hole-and-slot resonator.

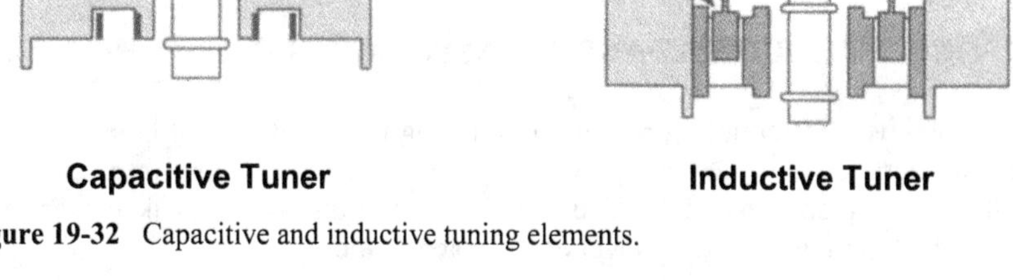

Figure 19-32 Capacitive and inductive tuning elements.

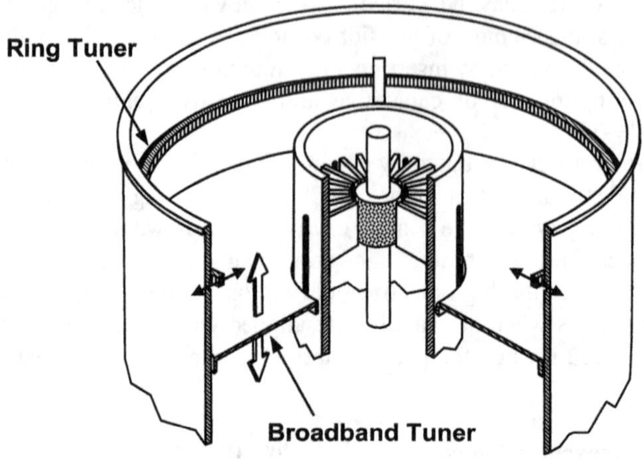

Figure 19-33 End-plate (broadband) tuning and ring tuning of a coaxial magnetron. (From: *Microwave Tube Manual* by Varian Associates, Air Force Publication Number T.O.00-25-251, October 1979.)

There are many ways for positioning tuning elements in a magnetron. Examples are the mechanical gear arrangements on early magnetrons in which the position of the tuning element was adjusted through a bellows using a gearing arrangement. Tuning speed, backlash, and drive torque are limiting factors. Early frequency-agile magnetrons were tuned by a hydraulic system that replaced the mechanical gear system. Tuning was slow and maintenance difficult, so these systems are becoming obsolete.

There are various electromagnetic techniques for positioning tuning elements. In one, the tuner operates much like the voice coil of a speaker. Current through a coil produces a magnetic field that acts with the field of a permanent magnet for axial positioning of a shaft to which the tuning element is attached. Another example is the use of an electric motor that is coupled through a gear-and-cam arrangement to position a tuning plunger. Accurate tuning can be accomplished through a resolver.

Some additional tuning techniques that make it possible to change frequency very rapidly are described in Section 19.10.2.

19.7 OUTPUT COUPLERS AND TRANSFORMERS

The method used for coupling power out of a magnetron differs with the frequency range of operation and the peak power level from the tube. A low-frequency tube usually uses a loop and a coaxial transmission line unless the peak power level exceeds the tolerable operating level of the coaxial line.

Figure 19-34 shows an example of loop coupling to a coaxial line. The coaxial line extends through the anode wall, and the center conductor is bent back to the outer conductor forming a loop. The loop couples to the RF magnetic fields in the inductive portion of a resonator. The size of the coupling loop and its orientation affect the magnitude of the coupling. Sometimes the coupling loop is located above the inductive region of the anode in the end space region of the magnetron. This kind of coupling is called "halo" loop coupling.

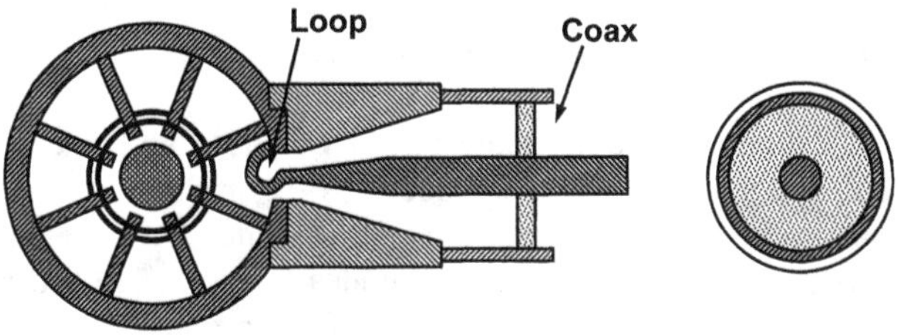

Figure 19-34 Loop coupling to a coaxial line.

At higher frequencies, iris coupling from the back of a resonator to a wave guide is used. In the example in Figure 19-35, the slot at the back of the resonator is coupled into the ridge, or capacitive region, of a ridge waveguide. The end of waveguide is attached to a flat surface on the anode with the center at the slot opening. The cross section of the ridge waveguide is shown relative to the full-size waveguide output for the tube. The waveguide portion is called the "H" section for obvious reasons. The ridge waveguide section is approximately a quarter wave in length and has the appropriate impedance to optimize the coupling.

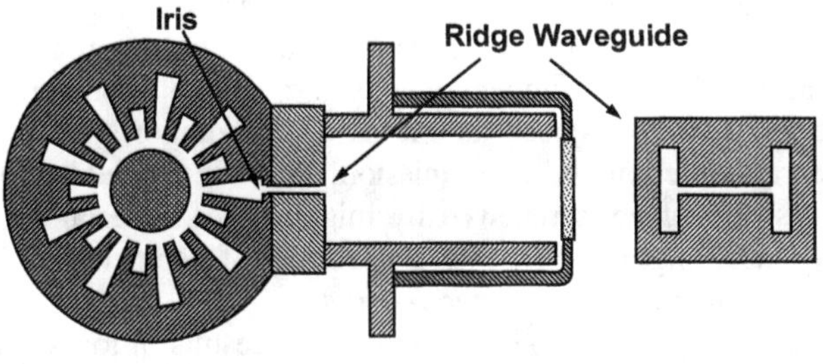

Figure 19-35 Iris coupling to waveguide.

Although these transitions have been called matching sections, they do not provide a match in the sense of having no reflected power. Contrarily, the proper impedance provides the necessary coupling to load the magnetron resonator with the correctly transformed load impedance that will allow oscillation with high circuit efficiency. The typical resonator system is overcoupled to do this.

The matching impedance looking into the cavity resonator of a magnetron is very low. The exact magnitude of the impedance is difficult to calculate or to measure. It is always much smaller than that of the coaxial line or waveguide output from the magnetron. Hence, some form of impedance-matching transformer must be used. A convenient method with coaxial outputs is to use a linearly tapered center conductor to obtain both impedance and a geometry transitions. An example is shown in Figure 19-34.

19.8 CATHODE AND HEATER OPERATION

In crossed-field devices, electrons that are out of phase with the RF fields are driven back into the surface of the cathode (see Figure 19-11). These electrons produce secondary emission and heating of the cathode. The secondary emission process was discussed in the chapter on cathodes (Chapter 18). In particular, its significance was illustrated by the Philips YJ 1183 spin-tuned magnetron in which it is estimated that thermionic emission is 3 A/cm^2 and secondary emission is 23 A/cm^2 for a total cathode emission of 26 A/cm^2.

In the microwave oven magnetron, the cathode is carburized thoriated tungsten for which the secondary emission coefficient is low. As a result, thermionic emission is the main emission mechanism. Carburized thoriated tungsten is used to achieve a long tube life.

The importance of cathode heating by out-of-phase electrons cannot be overstated. Typically, 5% to 10% of the anode power input may go into cathode heating. In some magnetrons, this effect is such that the filament voltage can be removed once the tube is operating and the cathode temperature is maintained by bombarding electrons. Figure 19-36, which is a more typical case, shows how the magnetron filament voltage should be reduced as average anode current is increased if essentially constant cathode temperature is to be maintained.

The primary effect of cathode overheating, which occurs if the filament voltage is not adequately reduced, is excessive evaporation of cathode material, resulting in early depletion and low emission. The effect of underheating is that, since the emissive surface is renewed by migration of material from below the surface and, since migration rate is a function of temperature, depletion of emissive material on the cathode surface occurs.

Either overheating or underheating eventually results in low emission and, in turn, low output power. Low emission may also lead to moding, which is evidenced by higher voltages (resulting from improper loading of the modulator), lower currents, much lower output power, and a shift in frequency. Finally, low emission may also lead to arcing because of the higher voltages or because of the increased electric field at the cathode surface in the absence of adequate space charge. Arcing results in gas evolution, the destruction of the cathode surface and, along with overheating, the deposition of emissive material on critical areas.

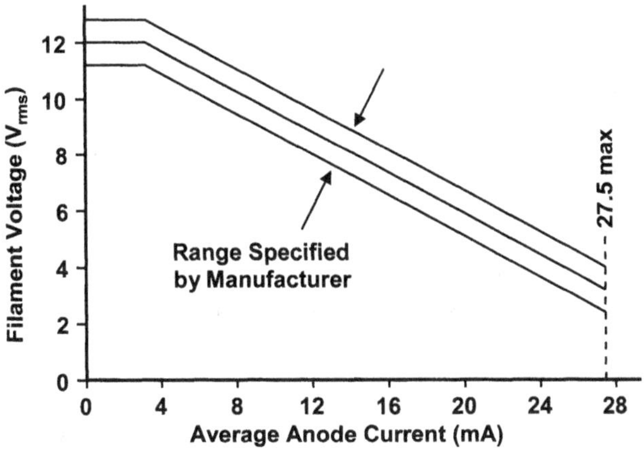

Figure 19-36 Magnetron filament voltage as a function of average anode current.

An example shown in Figure 19-37 of a particularly critical area is the end hat on the cathode in a spin-tuned magnetron. If precautions are not taken, the deposition of emissive material on the end hat leads to electron emission from it. The resulting electron current impinges on the magnetic pole piece and may heat it to the Curie temperature at which point it becomes nonmagnetic. As a result, the magnetic field collapses and the anode current may increase to the point where the anode melts.

Finally, the disastrous effect of overheating a magnetron cathode might possibly be best illustrated by the actual example in which the filament voltage schedule given in Figure 19-36 was not followed. In this case, the manufacturer of the equipment utilizing the magnetron did not properly understand the effect of cathode overheating and failed to reduce filament voltage adequately at high average anode currents. The result was a mean time between failures (MTBF) of approximately 4 hours for a magnetron costing over $30,000 (in 1980).

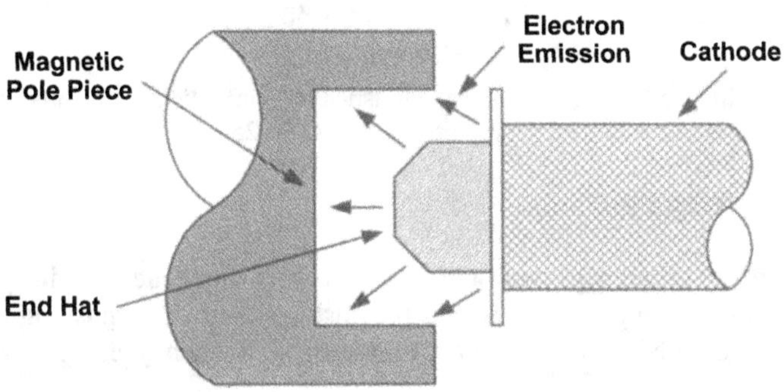

Figure 19-37 End-hat emission resulting from migration of cathode material.

19.9 PERFORMANCE

19.9.1 Voltage-Current Characteristic

The voltage-current characteristics of a magnetron are sometimes shown as in Figure 19-38. Magnetic flux density is fixed and so operation is along one of the constant Gauss lines. Normally, the optimum operating condition is indicated (26 kV and 38A). As anode voltage is varied, current changes, and so power output and, to some extent, efficiency changes. As current changes, the electronic impedance varies, which affects the resonant frequency of the resonator, and so the frequency of oscillation changes.

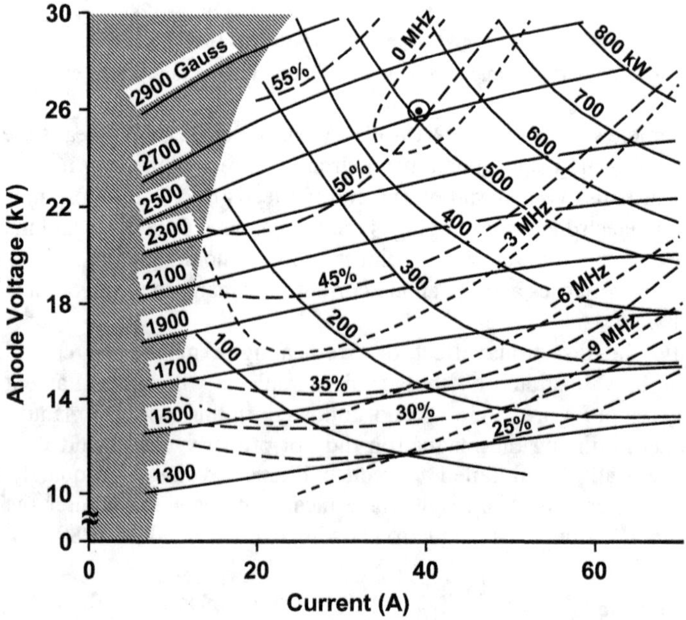

Figure 19-38 Magnetron operating characteristics. (From: *Microwave Tube Manual* by Varian Associates, Air Force Publication Number T.O.00-25-251, October 1979.)

19.9.2 Frequency Pushing

The change in frequency with current is called frequency pushing and the frequency shift per ampere is known as the pushing figure. The pushing figure of a magnetron is defined as the change in magnetron frequency due to a change in current.

We noted earlier in this chapter that the resonant frequency of a vane resonator is determined by its dimensions as well as the reactive effect of electrons. The presence of electrons near the vane tips affects the capacitance of the resonator by an amount proportional to the density of the electrons. Since electron density is related to current, changes in current will produce changes in output frequency.

The pushing figure is usually expressed in MHz/Amp and is represented by the slope of a frequency versus current curve plotted for a particular magnetron type as shown in Figure 19-39. Notice that the slope is not a constant so it is necessary to specify the range of current over which it applies. Also, since power output is proportional to current, the pushing figure at current levels well below the normal operating point are usually unimportant because the power output at these current levels is low. The primary importance of a low pushing figure is that the pushing figure will determine intrapulse FM, and thereby will affect the spectral quality of the transmitting pulse.

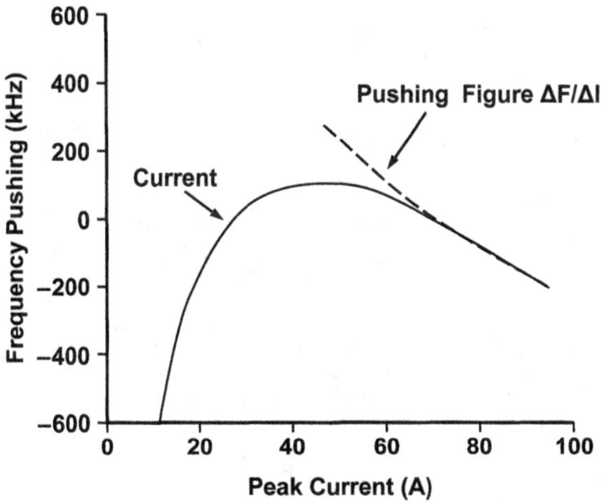

Figure 19-39 Example of magnetron pushing. (Adapted from: CPI magnetron brochure.)

19.9.3 Frequency Pulling

Varying load impedances reflected into a magnetron affect the resonance condition and with it the operating frequency and power output. A Rieke diagram, is used to show the effects of load impedance on performance. As shown in Figure 19-40, a Rieke diagram is a Smith chart on which contours of constant power and frequency are plotted. It should be noted that, for a given load mismatch and resulting voltage standing wave ratio (VSWR), the power and frequency are highly dependent on the electrical distance between the tube and load. This is because a load may appear to be inductive, resistive, or capacitive depending on its distance from the tube. Thus, in Figure 19-40, for example, for a load VSWR of 1.5, the frequency may vary from –0.7 to +0.75 MHz relative to the value at the VSWR of 1.0, and the power may vary from 1.07 to 1.44 MW.

If a magnetron operates into a load, the impedance of which varies with time, then frequency and power will vary with time. As an example, a rotating antenna will usually present a varying impedance to a magnetron, so frequency and power will vary with antenna position. The use of ferrite isolators significantly reduces this variation. The resulting variation in frequency with changes in load conditions is called frequency pulling. The pulling figure is the change in output frequency that results when a fixed amplitude mismatch, located in the output waveguide of the magnetron, is moved through a distance of one half wavelength.

During the design of a magnetron, the coupling of the output waveguide to the resonator structure is selected to optimize certain performance parameters. Strong coupling increases output power and efficiency but also increases jitter and sensitivity to changes to load mismatch. Usually, the coupling is chosen to obtain the best compromise between efficiency and stability.

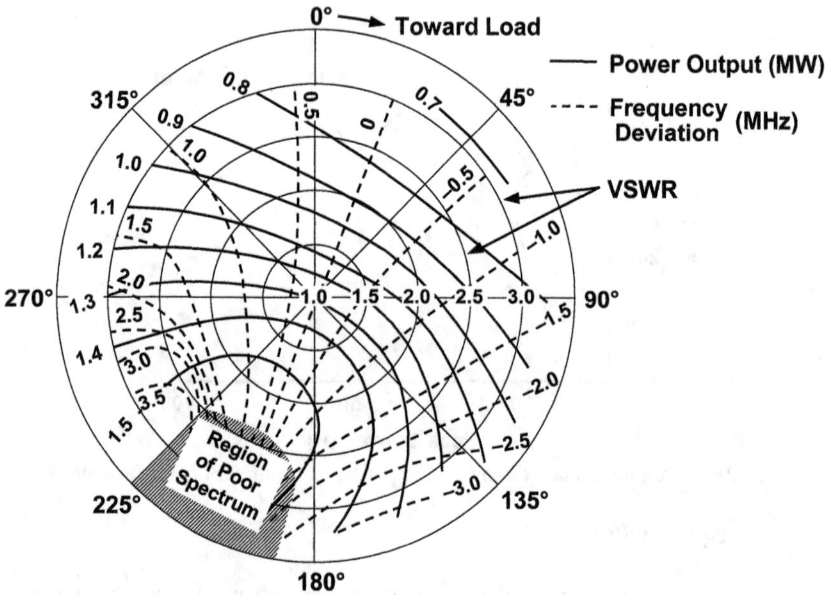

Figure 19-40 Rieke diagram for displaying the effect of load on tube operation. (From: *Microwave Tube Manual* by Varian Associates, Air Force Publication Number T.O.00-25-251, October 1979.)

The pulling figure is measured by operating the tube into a load with a specified VSWR (usually 1.5:1) and determining the frequency shift as the mismatch is moved through a distance of one half wavelength. In very high-power magnetrons a 1.3:1 VSWR is often used. When referring to the pulling figure of a magnetron the VSWR value used in the measurement should always be specified.

19.9.4 Thermal Drift

In addition to frequency pulling and frequency pushing, it should be pointed out that the operating frequency of a magnetron is subject to variations in tube temperature. When high voltage is first applied to a magnetron, the tips of the anode vanes begin to heat due to electron bombardment. Then the entire anode/cathode structure undergoes a transient change in thermal profile. During the time required for each part of the magnetron to stabilize at its normal operating temperature, the output frequency of the magnetron will "drift." This is because cavity size determines frequency and cavity size is affected by temperature. The curve of output frequency versus time during the period following initial turn on is called the thermal drift curve.

Generally speaking, the maximum drift occurs during the first few minutes after turn on, and slowly approaches equilibrium. In a low-frequency magnetron, the RF circuit elements are relatively massive and the time required to reach thermal equilibrium may be as much as an hour, as indicated by Figure 19-41. At X-band, cavity dimensions are small and the time to reach equilibrium is a few minutes.

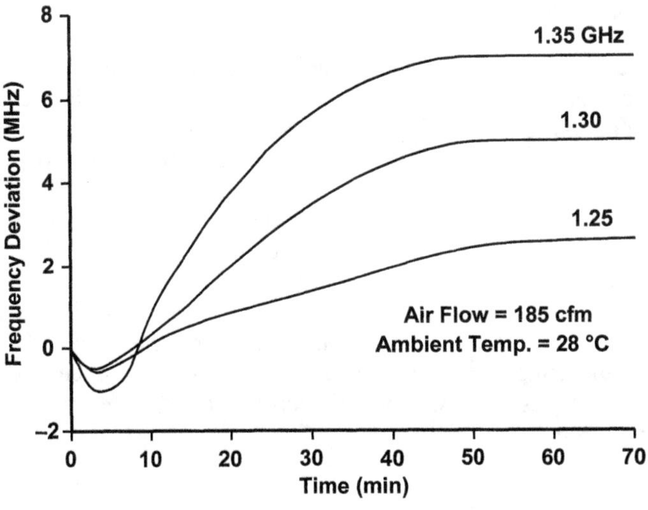

Figure 19-41 Effect of temperature on a 1.3-GHz magnetron. (From: *Microwave Tube Manual* by Varian Associates, Air Force Publication Number T.O.00-25-251, October 1979.)

Thermal drift curves for a variety of magnetron types operating at the same frequency and power can differ radically. Each type is usually designed for a particular application and small differences in internal configurations can produce large differences in the thermal drift curves. It should be noted that a thermal drift effect will also occur whenever the input power to the magnetron is changed, for example, when a change of pulse duration, or PRF occurs. Figure 19-42 shows thermal drift curves for a magnetron plotted as a function of duty cycle. The dashed line indicates the effect of a change in duty from 0.001 to 0.0005 after thermal equilibrium has been initially achieved.

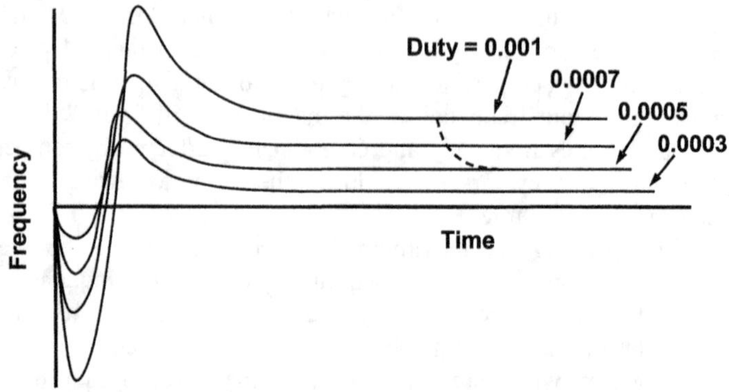

Figure 19-42 Example of thermal drift characteristic as a function of duty cycle. (Adapted from: CPI Magnetron brochure.)

In some devices, compensation techniques are used to offset the temperature effects. For example, bimetallic disk elements can be used in the cavity of a coaxial magnetron to correct for cavity expansion. Similarly, bimetallic elements can be used in the drive linkage to the tuner to compensate for temperature-induced frequency drift.

After a tube has reached thermal equilibrium and a stable operating frequency has been achieved, a change to ambient conditions, which causes a change in the magnetron temperature, will produce a change in the output frequency. The change in frequency for each degree change in temperature is the temperature coefficient for the magnetron and is usually expressed in MHz/°C. For most magnetrons the temperature coefficient is a negative (frequency decreases as temperature increases) and is essentially constant over the operating range of the magnetron.

19.10 APPLICATIONS OF MAGNETRONS

19.10.1 Conventional Magnetrons

Conventional magnetrons are considered here to be fixed frequency and mechanically tuned tubes intended for use in radar transmitters that operate primarily at one frequency. The purpose of the tuning capability is to allow operation at any selected frequency within the tuning range of the tube. Because of this, mechanical tuning by hand or by relatively slowly driven automatic means is sufficient. Most of the discussions up to this point have been related to conventional magnetrons.

19.10.2 Frequency Agile Magnetrons [1, 14]

The frequency agile magnetron is in another class of tubes intended to be tuned at very rapid rates. They are more complex than the conventional magnetrons and are used in applications for more sophisticated missions than, for example, search-and-find radars.

The requirement for frequency agile magnetrons stems from the requirement to change the output frequency of a radar with sufficient speed to produce a pulse-to-pulse frequency change greater than the amount required to effectively obtain decorrelation of adjacent radar echoes. There are many applications and requirements for frequency agile systems and many of these are explained in the CPI booklet entitled "The Frequency Agile Magnetron Story."

A frequency agile magnetron differs from the basic tunable magnetron in terms of tuning speed. The tuning speed must be rapid enough to produce the required pulse-to-pulse frequency difference and yet not be so rapid as to produce a significant frequency change during a single pulse.

In the CPI literature, frequency agile magnetrons are divided into four classes:

1. *Dither magnetrons*: Output RF frequency varies periodically with a constant excursion, a constant rate, and a fixed-center frequency.
2. *Tunable/dither magnetrons*: Output RF frequency varies periodically with a constant excursion and a constant rate. The center frequency may be slowly tuned by hand or by external servomotor drive to any point within the tunable band.
3. *Accutune magnetrons*: Output RF frequency variations are determined by the wave shape of an externally generated, low-level voltage signal. With appropriate selection of a tuning wave shape, the Accutune magnetron combines the features of dither and tunable/dither magnetrons.
4. *Accusweep magnetrons*: The output RF tuning rate and wave shape are infinitely variable within the design limits of each device. User inputs are typically any waveform from random to square wave and a +5-volt command.

The spin-tuned magnetron is a frequency-agile magnetron that falls under the dither magnetron category. The frequency can be varied 5% about a center value at over 1,000 tuning cycles per second. A rotary tuner mounted on a ball-bearing system in vacuum is used for very fast inductive tuning (Figure 19-43). The tuner is driven by an electric motor through the vacuum envelope by a magnetic coupling system.

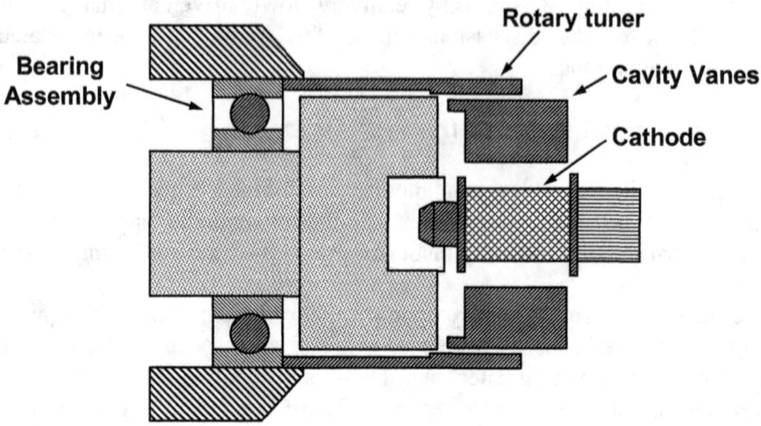

Figure 19-43 Ball bearing support assembly for the rotary tuner in a spin-tuned magnetron.

The tuner is a copper cylinder that forms part of the inductive portions of the cavity resonator. As the tuner is rotated, holes in the cylinder move into and out of alignment with the cavities, and thereby inductively tune the cavities. In X-band versions of the spin-tuned magnetron, there are 16 cavities and a rotor with 16 holes (Figure 19-44). With a nominal rotor speed of 4,000 rpm, the frequency of the magnetron is changed over 1,000 complete tuning cycles per second. The frequency excursion of the magnetron is approximately 5% of the carrier frequency.

Because the ball bearings supporting the rotary tuner must operate in vacuum, bearing life would appear to be a matter of considerable concern. In the Philips versions of the spin-tuned magnetrons, bearing life is over 4,000 hours and, in most cases, exceeds cathode life.

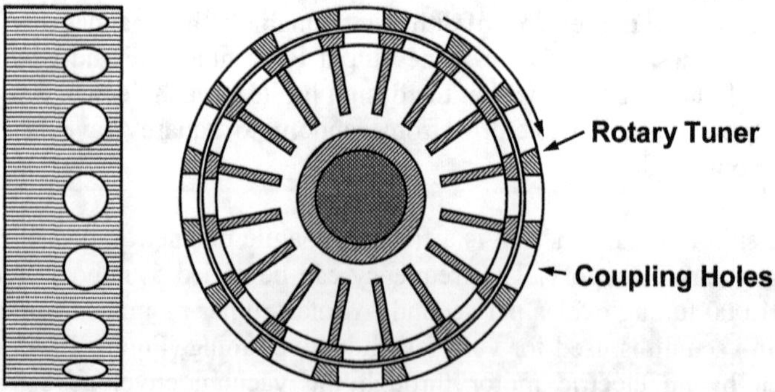

Figure 19-44 Resonator and rotary tuner for a 16-cavity spin-tuned magnetron.

The coaxial magnetron in Figure 19-45 is a tunable dither magnetron. The dither mechanism is mounted on the magnetron tuner and uses a motor with a small cam, or eccentric, ground into its shaft. As the motor shaft revolves, the eccentric imparts an up and down sinusoidal motion through a ball bearing to a shaft connected to the tuning plunger. Thus, the standard tuning plunger is used to dither the frequency.

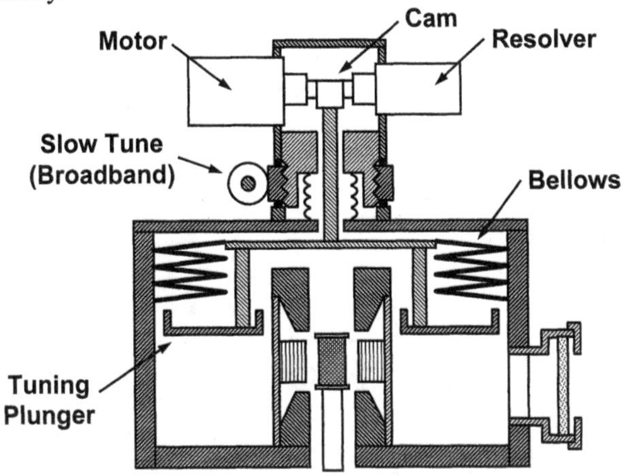

Figure 19-45 Coaxial magnetron that is dither tunable and broadband tuned.

Broadband tuning is accomplished in the normal manner using a geared drive. In dither-tuned tubes, the broadband tuning mechanism may move the dither mechanism as well as the tuning plunger as shown in Figure 19-45 or, in some types, the broadband tuning mechanism may move only the plunger, through a spline connection. In either case, the broadband tuner can operate independently of the narrowband tuner. The broadband tuner may be adjusted either by an external drive motor or manually.

19.10.3 Signal Injected Magnetrons

19.10.3.1 Injection Locking

Several types of radar systems (for example, MTI, and pulse Doppler) use phase coherence between pulse signals to enhance target detection and to improve clutter attenuation. Phase coherence is inherent in radar transmitters that use amplified signals obtained initially from a reference source such as a stabilized, low-level oscillator. It is more complicated to obtain pulse-to-pulse phase coherence using a magnetron oscillator because oscillations build up during the rise of the applied pulse voltage from noise signals inherent in the circulating electron stream. The start time of oscillation is random, so there is no pulse-to-pulse phase coherence.

The only RF connection to a magnetron is the output port from which the full power output of the tube is obtained. However, a signal can be injected into the magnetron through the output port to help control operation of the magnetron. The signal injection can be achieved by means of a circulator as shown in Figure 19-46. The injection signal is sent into one port of the circulator, which is then redirected through a second port of the circulator into the output window of the magnetron.

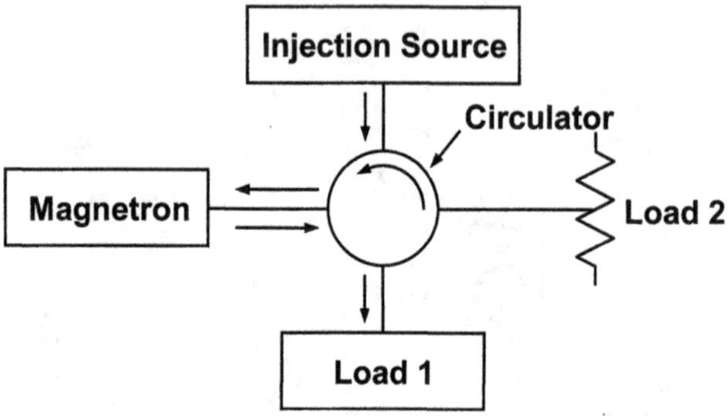

Figure 19-46 Circuit for injecting a signal into a magnetron.

The typical magnetron oscillator is heavily overcoupled for optimum efficiency so the impedance looking into the magnetron through the output port is a mismatch to the waveguide impedance. Only a portion of the injection signal is coupled into the tube. The remainder is reflected back from the magnetron and diverted to a load (load 1) through a third port of the circulator. The portion of the input signal that enters the magnetron can be used for controlling the magnetron frequency.

The magnetron power output is also directed to a load (load 2) by the circulators. Hence, the circulator provides the necessary isolation and protection of the input signal source from the magnetron power output. The maximum control occurs when the injection signal is at the same frequency as the free-running oscillation frequency of the magnetron. Injection locking can be achieved at other frequencies, but the required signal power level for proper operation increases with frequency departure from the free-running frequency.

Typical locking bandwidths are in the range of 1%. The locking bandwidth can be made somewhat larger by increasing the coupling to the magnetron circuit. However, this will further increase the mismatch looking into the output port and requires a larger injection signal power for proper frequency locking.

The ratio of the magnetron power output to the required input power is called the locking gain. Typical locking gains for narrow bandwidths are about 10–15 dB. This is comparable to the levels of gain available from crossed-field amplifiers but only over a significantly smaller bandwidth than the amplifier. Hence, injection locked magnetrons are used only in those cases where satisfactory operational requirements can be achieved with the restricted bandwidths. In these cases, the lower costs of an injection-locked oscillator are preferred to the larger cost of an amplifier.

19.10.3.2 Injection Priming

Injection locking is the terminology applied to the circuit of Figure 19-46 when the magnetron output signal is fully slaved to the input signal. The magnetron, for all practical purposes, duplicates the input signal in frequency and phase throughout the entire pulse. The limited gain and bandwidth attainable with this technique has resulted in limited use of this technology. However, there is an alternate injection technique that offers many of the advantages of locking, while avoiding many of the problems that have limited the use of full injection locking. This technique is called injection priming. Injection priming differs, primarily, from injection locking in that much smaller signals are used. Also, the priming signal is slightly lower in frequency than that of the free-running oscillator. The lower frequency is at the magnetron's "start oscillation frequency." Since this frequency is lower than that of the free running magnetron, its synchronous voltage will be lower than the normal operating voltage for the tube. Therefore, the magnetron passes through the critical voltage for the start oscillation frequency in a very short time during the rise of the applied voltage pulse.

Priming has a significant effect on the magnetron during the start-oscillation period because this is when space charge bunching is most susceptible to external influences. The priming signal produces a nonresonant, π mode, electric field pattern on the anode. This pattern begins to phase sort the random motions of the circulating space charge into a spoke formation that is spatially separated for π mode operation.

As the applied voltage continues to increase, the phase-sorted spoke pattern rotates faster and serves as an enhanced electronic input signal that is more effective than a noise signal in exciting the resonant frequency of the magnetron. As the fields on the RF circuit build up, the resonant π mode is able to capture and phase lock the circulating space charge to that field configuration. Then sustained oscillations are produced at the free-running magnetron frequency. After this, the field pattern produced by the priming source will have no further influence on tube performance and so the priming signal can be turned off.

Useful performance has been achieved with injection priming with input signals 30 to 40 dB below the magnetron output power. In some cases, effective control has been reported with priming signals 60 dB below the magnetron output power. Some early experiments were performed using low-power magnetrons, but more recent applications have been with high-power tubes.

19.10.4 Beacon Magnetrons

Beacon magnetrons are small conventional magnetrons with peak power output less than 4 kW and average power output of less than 5W. Typically, they weigh 8 ounces. The technical requirements for these magnetrons demand precise frequency control. The temperature stability factor is of great importance since it allows frequency control without additional electronics in the radar transponder. The magnetron must be tunable but must have the properties of a fixed-frequency magnetron after adjustment and locking. As a result, the techniques used for temperature compensation must work over a band of frequencies. Also, frequency stability is generally that of missile and aircraft electronic systems. This means that the tube must be stable over a temperature range that is typically –65°C to +100°C, a shock of 100G, and a vibration of 15G.

19.10.5 Microwave Oven Magnetrons

Microwave oven magnetrons supply 500–1000W at 2,450 MHz and an efficiency of ~ 70%. They are produced in quantities of tens of millions annually at a price of about $7 each. In 1984 [6], Brown noted that there were an estimated 50 million microwave ovens in the world. If turned on at the same time, they would produce 25 GW of microwave power. No recent estimate of these numbers has been found.

Microwave oven magnetrons are extremely reliable devices. There is a good chance that a microwave oven will operate for 10 years or more without requiring repairs, and at a performance level that is indistinguishable from when it was new.

The essential elements of one of these magnetrons are shown in Figure 19-47. Pole pieces carry the magnetic flux from ring-shaped magnets to the interaction region between the cathode and resonator. The resonator usually consists of 10 to 12 cavities. A 12-cavity resonator is shown in Figure 19-48. The RF output lead is connected directly to one of the resonator vanes. The cathode is directly heated carburized thoriated tungsten and comes up to operating temperature within seconds of the time that it is turned on. Even though the magnetron is very efficient, several hundred watts must be carried away by the aluminum cooling fins.

The power circuit for the magnetrons in most microwave ovens, which is shown in Figure 19-49, is a model of simplicity. The anode and chassis of the magnetron are at ground. When the transformer voltage tends to drive the cathode of the magnetron positive, the conducting diode charges the capacitor to the peak voltage of the transformer. At the peak of the negative swing of the transformer voltage, the voltage applied to the magnetron cathode is the peak voltage of the transformer plus the voltage across the capacitor. As a result, the voltage on the magnetron is nearly twice the peak output voltage of the transformer or ~5,000V. Thus, this circuit is a half-wave voltage doubler.

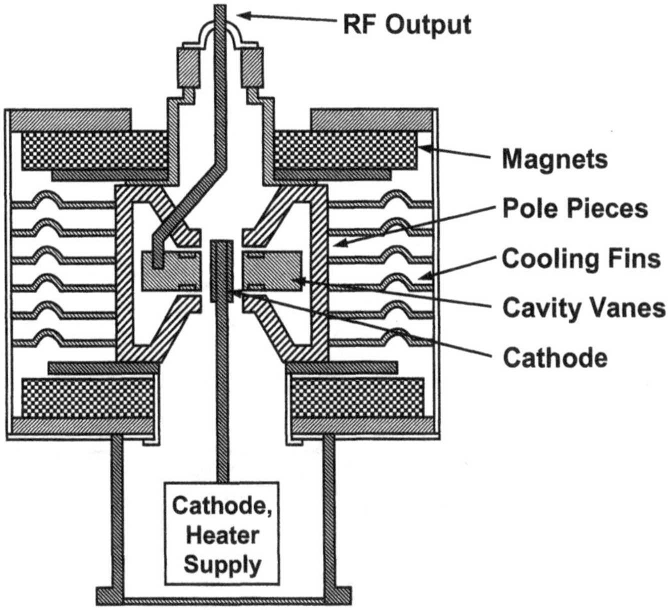

Figure 19-47 Essential elements of a microwave oven magnetron.

Figure 19-48 Resonator assembly for a 12-vane oven magnetron.

The voltage waveform applied to the magnetron looks like a sinusoid with a p-p voltage equal to the p-p voltage of the transformer with its positive peaks at chassis ground. The negative peaks are negative with respect to the chassis and get distorted when loaded by the magnetron. The waveform applied to the operating magnetron is like that shown in Figure 19-50.

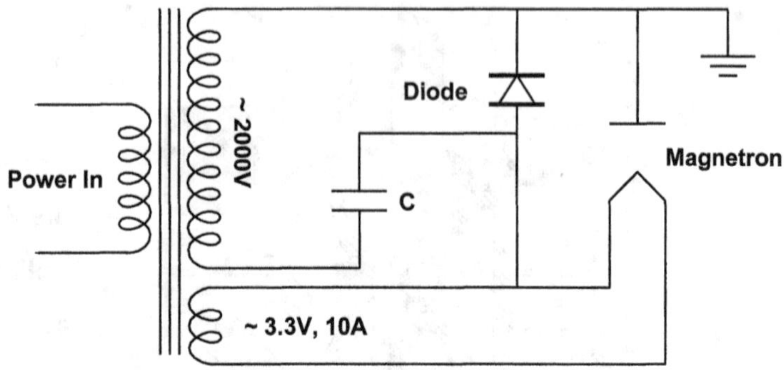

Figure 19-49 Power supply for an oven magnetron.

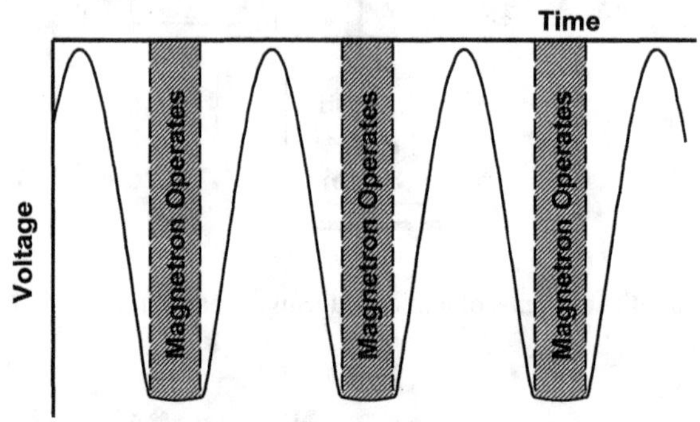

Figure 19-50 Voltage waveform applied to an oven magnetron.

19.10.6 Industrial Heating Magnetrons

Microwave heating using magnetrons has gained wide acceptance in industry because of its cleanliness, its high processing speed, high efficiency, accurate control, and flexibility. In addition to food processing, applications include product sterilization, wood and rubber curing, tempering, and controlled evaporation. Two frequencies, L-band and 2,450 MHz, are available for use. L-band systems can provide up to 100 kW from a single magnetron. The largest 2,450-MHz units use 30-kW magnetrons. A sketch of a commercially available 100-kW L-band magnetron, without a magnet, is shown in Figure 19-51. It should be noticed that the center portion of this device, which contains the resonator and is where the conversion from dc power to RF takes place, is less than 5 inches in diameter and less than 5 inches long.

The operating parameters for the 100-kW magnetron are listed in Table 19.1. Perhaps the most significant characteristic is efficiency, which is typically 88%.

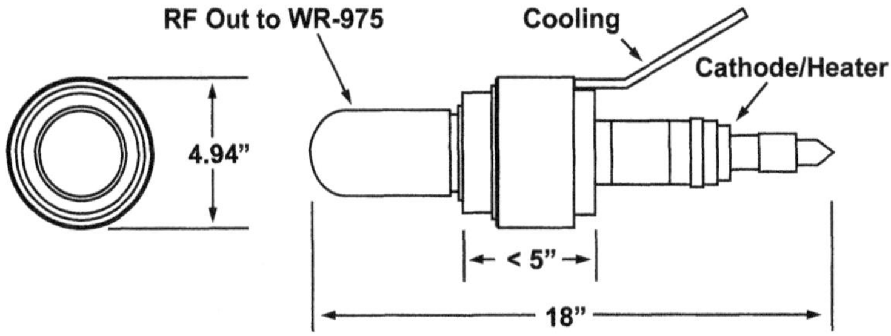

Figure 19-51 100-kW L-band industrial heating magnetron without a magnet. (Adapted from: L-3 Communications California Tube Laboratory brochure.)

Table 19.1

100-kW, CW, L-Band Magnetron

Characteristic	Specification	Typical
Frequency (fixed)	896, 915, 922, 929 MHz	Custom Available
Frequency Variation	+/– 10 MHz	+/– 5 MHz
CW Power Output, minimum	100 kW	100 kW
Anode Voltage	20 kV	19.5 kV
Anode Current	6.0A	5.8A
Efficiency	83%	88%
Filament Voltage (standby)	12.6V	12.0V
Filament Current	115A	112A
Load VSWR	1.1:1 Circulator Required	
RF Launch	Coaxial into WR-975	

Although the operating life of a 915-MHz commercial magnetron can be greater, Industrial Microwave Systems of Morrisville, North Carolina, recommends that the tube be replaced after 8,000 hours of operation. This translates to a maintenance cost of about $1 per operating hour.

19.10.7 Low-Noise Magnetrons

19.10.7.1 Emission Limiting

The carburized thoriated tungsten cathode in the oven magnetron is fundamentally a primary emitter as distinguished from other magnetron cathodes that are either pure secondary emitters or partially so. It is not generally known that, when the ordinary microwave oven magnetron is operated from a source of filament power that is removed after it is used to start tube operation and from a well-filtered dc power supply, it has amazingly low noise properties. By comparison, when the tube is operated normally in a microwave oven, the noise levels may be 60 to 100 dB higher [15]. In the low-noise mode, the cathode temperature is maintained and regulated by an internal feedback mechanism involving electron bombardment alone.

Experimentally, it has been found that the tube automatically operates with the cathode in an emission-limited condition. This is thought to be part of the reason for the low-noise performance because thermionically emitted electrons leave the cathode with very low velocities compared to the velocities of electrons emitted by secondary emission during normal magnetron operation. The experimental evidence that the cathode is operating temperature limited is shown in Figure 19-52 [16]. In this figure, the observed temperature of the cathode is plotted against the anode current of the tube. This experimental relationship is seen to nearly follow the slope that is predicted by the Richardson-Dushman equation for temperature limited emission. The two curves were matched to each other at a Brightness temperature of 1,500°C. It was assumed that the magnetron current and emission current density are related by a multiplying constant.

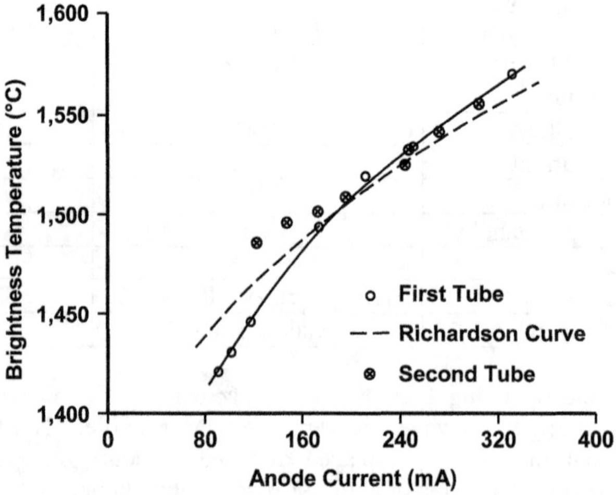

Figure 19-52 Comparison of emission limited-magnetron data with Richardson-Dushman thermionic-emission curve. (From: W. C. Brown, *IEEE Trans. Electron Devices*, November 1984. © 1984 IEEE.)

When a magnetron is operating temperature limited, it is observed that the signal-to-noise ratio is very high. For example, it was reported by Brown that, with external microwave circuitry on the support for the cathode, spectral noise densities as low as 196 dB below the carrier were found at frequencies (other than harmonics) removed 50 MHz or more from the carrier [16].

A second benefit, in addition to low noise, of operating the cathode at the lowest possible temperature, is that it is possible to design tubes for lives of decades. This is because the life of a carburized thoriated tungsten filament is highly dependent upon the temperature, being much greater as the temperature is reduced [15, 16]. The emission current density requirements to supply the anode current of the magnetron are fortunately low and consistent with the low temperatures needed for a very long life. Also, the internal feedback mechanism that controls cathode temperature makes it unnecessary to externally regulate cathode temperature.

One of the early applications proposed for low-noise oven magnetrons was in the 1980s for the Solar Power Satellite Program. These amplifiers are ideal for combining with slotted-waveguide radiators to form radiating modules in a low-cost, electronically steerable phased array for beamed power transmission.

Another application pointed out by Brown was that one of these tubes can be combined with external circuitry to convert it into a phase-locked amplifier with 30-dB gain, without compromising its low-noise properties.

19.10.7.2 Magnetic Priming

The normal oven magnetron generates a high level of noise and spurious oscillations. Noise generated in the magnetron can leak out through filament or cathode leads and then is radiated outside from an ac power line. Solving the noise problem in magnetrons could potentially have important practical implications such as reducing unwanted interference with 2.4-GHz wireless systems and possibly providing a technique to reducing CFA noise.

Neculaes et al. [17–19] demonstrated that azimuthally varying axial magnetic fields can be used on standard oven magnetrons to produce rapid startup, low noise, and mode control with an efficiency loss on the order of 10%. The new magnetic field geometry was implemented by placing a number of small perturbing magnets on the perimeter of one of the existing annular magnets of the magnetron, providing an azimuthally varying axial magnetic field. The two annular permanent magnets create the baseline magnetic field of about 1.7 kG on the axis of the magnetron. The perturbing magnets were SmCo, with ~3.5–4 kG on the face. Experiments were performed with perturbing magnets on one existing magnet and, as shown in Figure 19-53, perturbing magnets on both existing magnets.

Neculaes et al. called this technique "magnetic priming." For an N-cavity magnetron operating in the pi-mode, it consists of the imposition of N/2 magnetic field perturbations in the azimuthal direction. The perturbing magnets are about three times stronger than the annular magnets of the magnetron. This is a clear indication of the strong azimuthal magnetic perturbation introduced with magnetic priming.

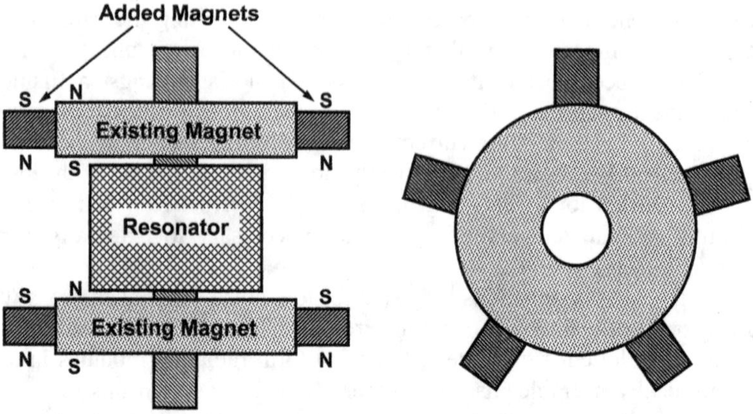

Figure 19-53 Placement of perturbing magnets on oven magnetron. (From: Neculaes et al., *IEEE Trans. Electron Devices*, May 2005. © 2005 IEEE.)

Using azimuthally varying axial magnetic field techniques, two magnetic priming configurations have been employed; the four perturbing magnet configuration (nonoptimized) and the five perturbing magnet configuration (optimized) to stimulate the pi-mode in the 10-cavity oven magnetron, which is characterized by the existence of five electron spokes. It is clear that these magnetic priming configurations distort the axial symmetry and introduce radial and azimuthal components in the magnetic field.

The effect of magnetic priming on noise is shown in Figure 19-54. The microwave spectrum for an aged magnetron without magnetic priming is in Figure 19-54(a). The microwave signal exhibits a broadband (up to 300 MHz from the carrier) noise spectrum. The microwave noise level is some 60 dB below the carrier (20 dB above the noise floor of the spectrum analyzer). When magnetic priming is imposed, as seen in Figure 19-54(b), the microwave noise is reduced at least 20 dB to the limit of the spectrum analyzer sensitivity both close to the carrier and far from the carrier.

This magnetic priming concept has been extended for application to relativistic magnetrons. Two-dimensional and three-dimensional simulations indicate that "magnetic priming" could significantly decrease the startup time for these magnetrons.

19.10.8 Relativistic Magnetrons

Relativistic magnetrons are capable of producing gigawatts of RF power for tens of nanoseconds. The operating voltages are hundreds of kilovolts and the currents are kiloamperes. Applications include testing of equipment for electromagnetic susceptibility and, possibly, the destruction of anthrax and other biological substances. A description of these magnetrons is beyond the scope of this book.

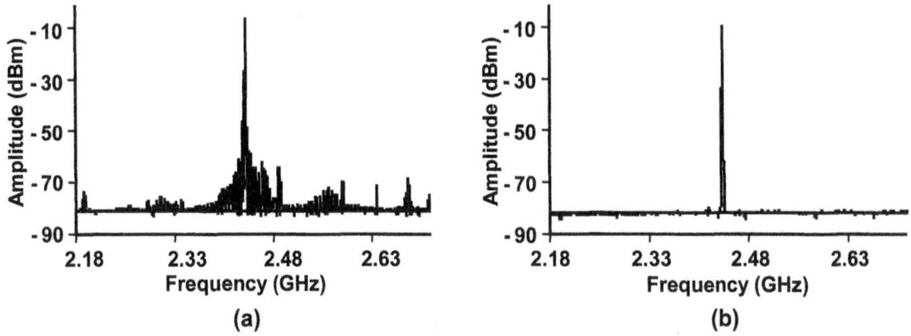

Figure 19-54 Microwave spectra (a) without and (b) with magnetic priming. (From: Neculaes et al., *IEEE Trans. Electron Devices*, May 2005. © 2005 IEEE.)

19.11 SUMMARY OF POWER CAPABILITIES

The overall publishable power characteristics of magnetrons are shown in Figure 19-55. Peak power levels well in excess of 1 MW have been available since the early days of magnetron development. In the laboratory, peak powers in the gigawatt range have been generated.

Average and CW power levels are limited because the beam power must be dissipated by the RF structure. This matter is discussed further in Chapter 20 concerning crossed-field amplifiers. Still, average powers approaching the megawatt level are available. Magnetron conversion efficiency from dc to RF is normally in excess of 50% and may be as high as 90%.

Fixed-frequency and tunable magnetrons are available that weigh as little as a few ounces or as much as about 200 lbs. These weights may be reduced in weight-critical applications through the use of samarium cobalt magnets. Possible magnet configurations are shown in Chapter 20 in discussions concerning crossed-field amplifiers.

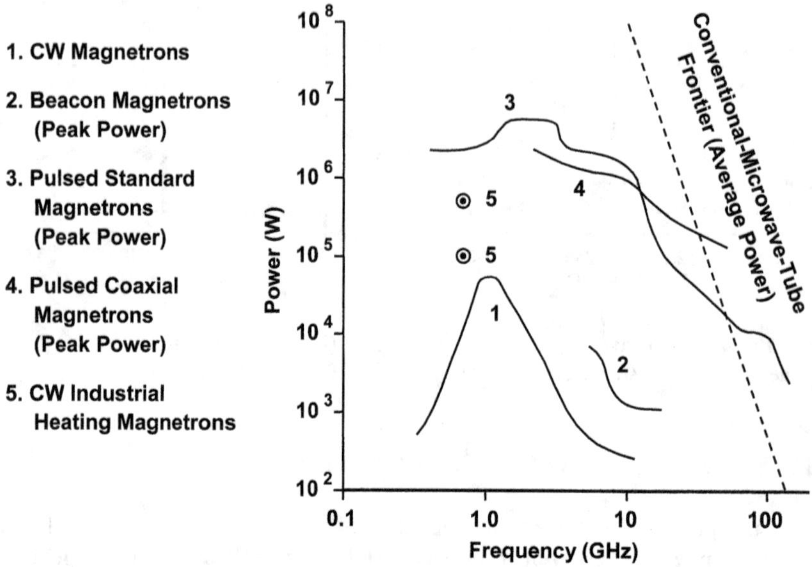

Figure 19-55 Magnetron power capabilities. (From: J. Skowron, (MPTD), Raytheon Co., Workshop on High-Power, Space-Based Microwave Systems, Los Alamos, NM, March 1985.)

REFERENCES

[1] G. K. Farney, *Crossed-Field Microwave Devices,* Technology Service Corporation/Crane Division Naval Surface Warfare Center, 1995.

[2] G. B. Collins, *Microwave Magnetrons*, New York: McGraw-Hill, 1948.

[3] E. Okress, *Crossed-Field Microwave Devices*, New York and London: Academic Press, 1961.

[4] H. A. H. Boot and J. T. Randall, "Historical notes on the cavity magnetron," *IEEE Trans. Electron Devices*, Vol. ED 23, No. 7, July 1976, pp. 724–729.

[5] G. R. Kilgore, "Recollections of pre World War II magnetrons and their applications," *IEEE Trans. Electron Devices*, Vol. ED-31, No. 11, November 1984, pp. 1593–1595.

[6] W. C. Brown, "The microwave magnetron and its derivatives," *IEEE Trans. Electron Devices,* Vol. ED 31, No. 11, November 1984, pp. 1595-1605.

[7] J. F. Hull, "A young engineer enters the microwave tube field during World War II," *IEEE Trans. Electron Devices*, Vol. ED 31, No. 11, November 1984, pp. 1611–1613.

[8] A. W. Hull, "The effect of uniform magnetic fields on the motion of electrons between coaxial cylinders," *Phys. Rev.*, Vol. 18, No. 31, 1921.

[9] J. F. Skowron, "The continuous cathode (emitting sole) crossed-field amplifier," *Proc. IEEE*, Vol. 61, No. 3, March 1973, pp. 330-336.

[10] J. J. Coupling, a pseudonym of J. R. Pierce, *Astounding Science Fiction*, Street and Smith, 1946.

[11] Varian brochure, "Introduction to Coaxial Magnetrons."

[12] M. I. Skolnik, "The future of microwave engineers: from the technical perspective," *Microwave Jour.*, Vol. 18, 1985, pp. 24–36.

[13] W. M. Black et al., "A high-power magnetron for air breakdown studies," *Technical Digest,* IEDM, 1980.

[14] CPI brochures, "Magnetrons."

[15] W. C. Brown, "The magnetron—a low noise, long life amplifier," *Applied Microwave*, Summer 1990.

[16] W. C. Brown, *Satellite Power System (SPS) Magnetron Tube Assessment Study*, NASA Contractor Report 3383, 1981.

[17] V. B. Neculaes et al., "Magnetron microwave noise reduction and magnetic priming by azimuthally varying axial magnetic fields," Fifth IEEE IVEC, 2004, pp. 158–159.

[18] V. B. Neculaes et al., "Magnetic priming effects on noise, startup, and mode competition in magnetrons," *IEEE Trans. Plasma Science*, Vol. 33, No. 1, February 2005, pp. 94–102.

[19] V. B. Neculaes et al., "Magnetic perturbation effects on noise and startup in DC-operating oven magnetrons," *IEEE Trans. Electron Devices*, Vol. 52, No. 5, May 2005, pp. 864–871.

Chapter 20

Crossed-Field Amplifiers

20.1 INTRODUCTION

The crossed-field amplifier (CFA) is a broadband amplifier of RF signals at microwave frequencies. It is a capable of providing relatively large amounts of output power with high efficiency. The electronic bandwidth of the CFA is nominally 10.0% to 15.0%. Power levels of many megawatts peak and tens of kilowatts average are currently being obtained from production CFAs. Efficiency figures in excess of 70% are not uncommon in some crossed-field amplifiers that utilize backward wave circuits and operate at low anode voltages.

Crossed-field amplifiers are classified as either injected-beam or distributed-emission devices. In an injected-beam CFA, an electron gun generates a ribbon-shaped electron beam that passes through the device near an RF circuit. Operation of the injected-beam CFA is similar to a TWT. Like a magnetron, the distributed-emission CFA contains a cathode adjacent to the full length of the RF structure.

20.1.1 Injected-Beam CFAs

In the late 1940s and early 1950s, there was much interest in extending the crossed-field interaction principle of the magnetron oscillator to the development of a crossed-field amplifier [1, 2]. The initial work was done largely in France on devices that were similar to traveling wave tubes. These were injected-beam, crossed-field amplifiers with configurations like that in Figure 20-1. These devices were similar to the TWT in that they used a separate electron gun to form an electron beam. The electron beam was ribbon shaped and was launched into the interaction region between the anode delay line (slow wave circuit) and the negative electrode, which was called the sole by the French tube engineers.

Interaction between the RF input signal on the delay line and the electron beam produced phase sorting of the beam into spokes much like those in a magnetron. The electrons in the spokes moved toward the anode, and most were collected on it just as they are in a magnetron. Those that were not collected on the anode would exit the interaction space and be intercepted by the collector. There was no intended back bombardment of or emission from the sole electrode.

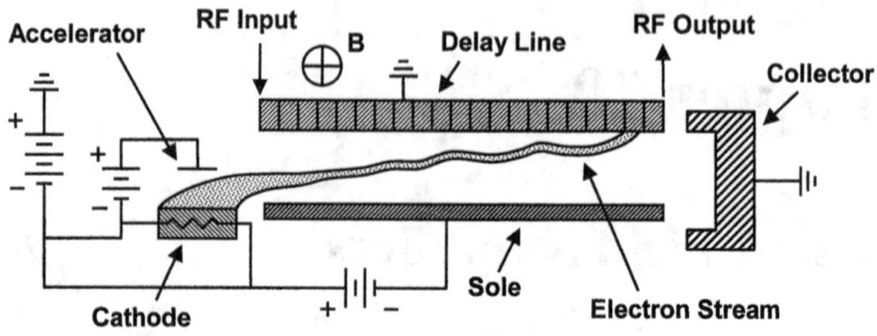

Figure 20-1 French injected-beam crossed-field amplifier.

The basic elements of a backward wave injected-beam CFA with a circular format are shown in Figure 20-2. Very-high-power pulsed devices were developed in France. In the United States, much development work was devoted to CW and low peak power injected-beam CFAs for very wideband use in airborne electronic countermeasures systems. In spite of significant development efforts, injected-beam CFAs are not currently being produced because the applications for which they might be suited are better filled by other devices. For this reason, only distributed-emission CFAs are discussed in this book and henceforth they are referred to simply as CFAs.

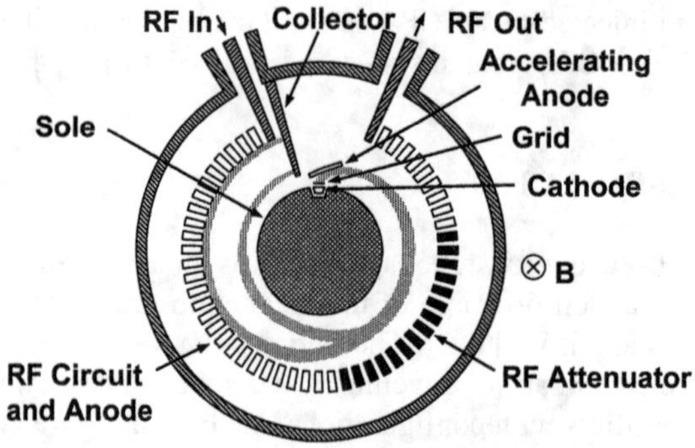

Figure 20-2 Basic injected-beam crossed-field amplifier. (From: J. Skowron, *Proc. IEEE*, March 1973. © 1973 IEEE.)

20.1.2 Distributed Emission CFAs [3]

The first practically successful distributed emission CFA was introduced by Brown in 1953 [4]. The device was similar to a magnetron, except that, as shown in Figure 20-3, the RF circuit was interrupted to provide input and output connections. The RF structure was a strapped-vane network similar to that in a magnetron. The tube had a gain of 13 dB over a frequency range of 3,780 to 3,850 MHz. The peak power output was 24 kW and the efficiency was 40%.

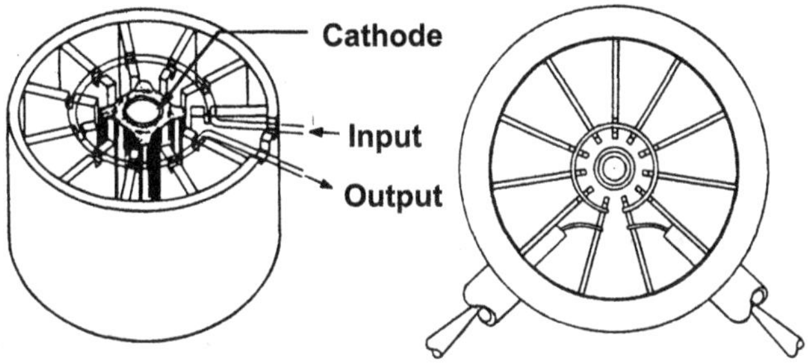

Figure 20-3 Schematic of the first device to demonstrate the principles of converting a magnetron into an amplifier. (From: W. C. Brown, *IEEE Trans. Electron Devices*, November 1984. © 1984 IEEE.)

Since the first demonstration of the distributed-emission CFA, nearly every conceivable configuration has been examined, as indicated in Figure 20-4 [5]. The circular format is the one shown in Figures 20-3 and 20-6. In the linear format device, as shown in Figure 20-5, the cathode and RF circuit are parallel [6].

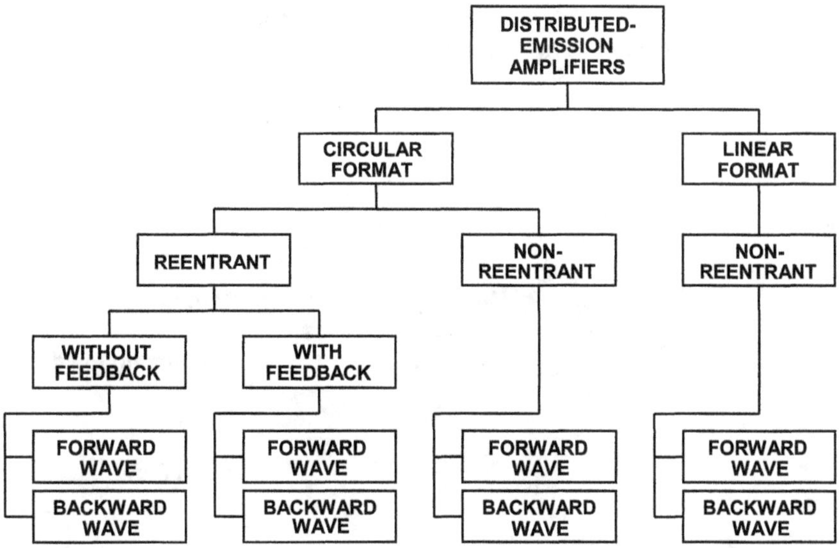

Figure 20-4 Family tree of distributed-emission CFAs. (From: *Microwave Tube Manual* by Varian Associates, Air Force Publication Number T.O.00-25-251, October 1979.)

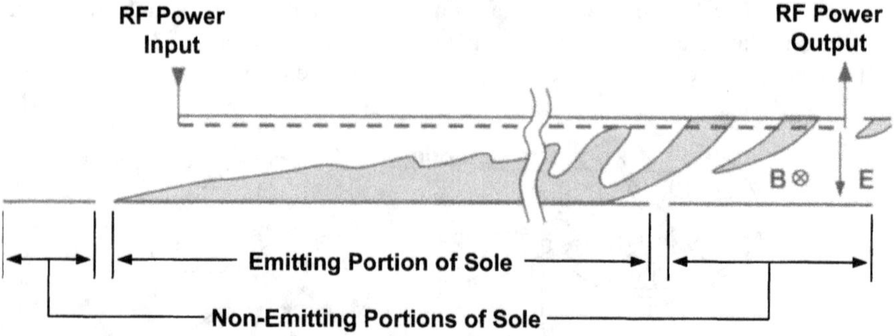

Figure 20-5 The DEMATRON, a linear format distributed-emission crossed-field amplifier. (From: G. E. Pokorny, et al., *IRE Trans. Electron Devices*, July 1962. © 1962 IRE, now IEEE.)

In the reentrant circular format, the electron cloud circles around the cathode as it does in a magnetron until electrons give up sufficient energy to reach the anode. This contributes to high efficiency and, it is believed, to improved turn on and starting. In some devices, a drift region is used between the output and input ports (see Figure 20-6) so that space charge forces cause electron bunches to disperse. This removes most of the modulation from the reentering cloud of electrons, which minimizes feedback from the output to the input.

As Figure 20-4 shows, either forward wave or backward wave operation is possible with any of the CFA configurations. The characteristics of both forward wave and backward wave devices will be discussed in this chapter.

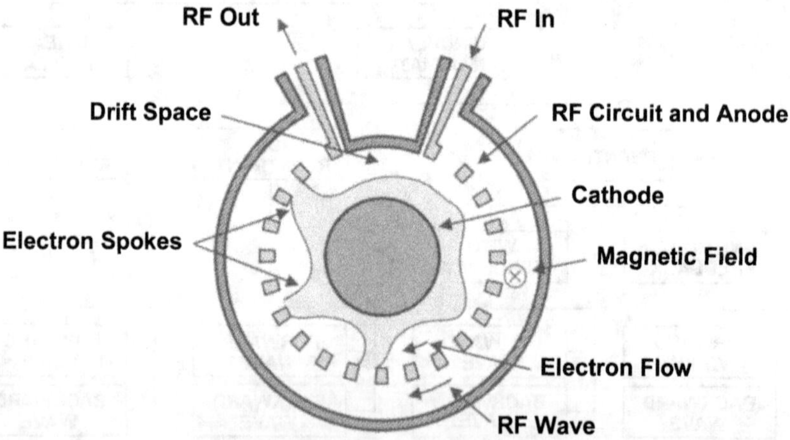

Figure 20-6 Basic distributed emission crossed-field amplifier. (From: J. Skowron, *Proc. IEEE*, March 1973. © 1973 IEEE.)

20.2 CFA OPERATION

A cross-sectional view of a CFA is given in Figure 20-7 [5]. The cathode assembly and slow wave structure are in the center of the device. Notice that, even though these elements are the heart of the device and are responsible for providing the operating characteristics of the device, they make up only a very small part of the volume of the device. The magnets and magnetic circuit, the RF output structure, and the cooling apparatus make up well over 90% of the volume of the device. This section contains a qualitative description of the interaction between the electrons from the cathode and the wave on the slow wave structure.

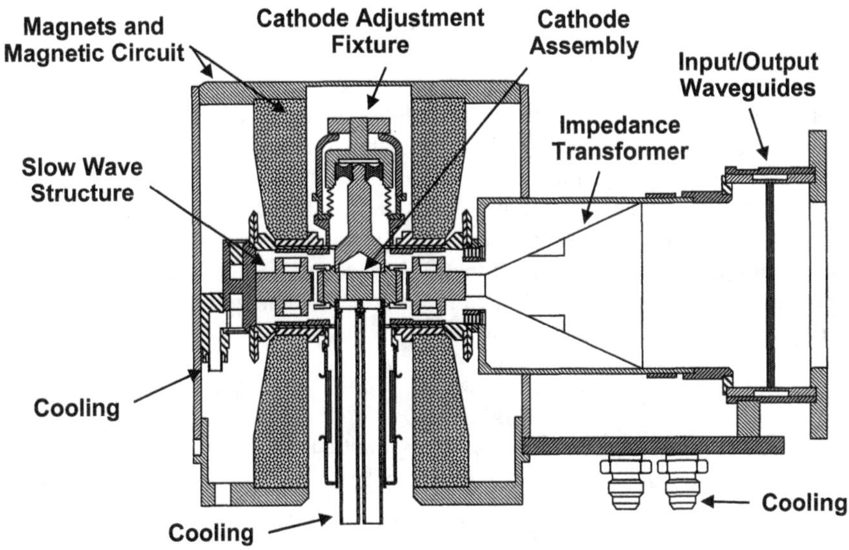

Figure 20-7 Cross section of a forward wave distributed-emission crossed-field amplifier.

20.2.1 Electron Emission and Hub Formation

Indirectly heated thermionic cathodes and cold secondary emission cathodes are used in CFAs. The operating conditions for each type of CFA such as pulse width, power output, duty cycle, and so forth, affect the choice of cathode. The cathode selected may be an oxide-coated type, an impregnated tungsten matrix type using barium or thorium compound impregnants, pure metal such as platinum, or variants of these types. Each has its advantages and disadvantages.

When cathodes are used that require heater power to raise their temperature for starting, a word of caution is necessary. These cathodes may become overheated when the back-heating power is added to the starting heater power.

Extreme care must be taken to follow specified instructions on reducing heater power after the tube has been brought to operating power level. If this is not done, the cathode can quickly become overheated and destroyed.

The cold cathode technique is used primarily in high-power amplifiers with ≥ 10-kW output, although lower power levels are possible. At lower power levels, the RF electric fields are less likely to impart sufficient energy to electrons to produce the required secondary emission.

Many high-power backward wave CFAs use liquid-cooled platinum emitters. Liquid cooling is used to dissipate the back-bombardment power. Platinum is generally used where the secondary emission ratio requirement is less than 1.80. Platinum cathodes are inert, easily mounted, and inexpensive, and can withstand high temperatures. Because the thin layer of platinum (0.010 inch) is so easy to mount on its base metal support, the platinum cathode is less expensive to build than a thermionic cathode.

Beryllium is another cold emitter used in CFAs. This cathode requires an accompanying oxygen dispenser to maintain the oxidized surface of the beryllium. Tubes using this type of emitter have operated for tens of thousands of hours without experiencing any emission limitations.

With the cold-cathode technique, emission is initiated by the interaction of the RF input signal with a few stray electrons. As was observed in Chapter 18, there has been no proven explanation yet that fully accounts for the source of the initial electrons. In some way, as is indicated in Figure 20-8, energy has to be supplied to the cathode surface to produce emission

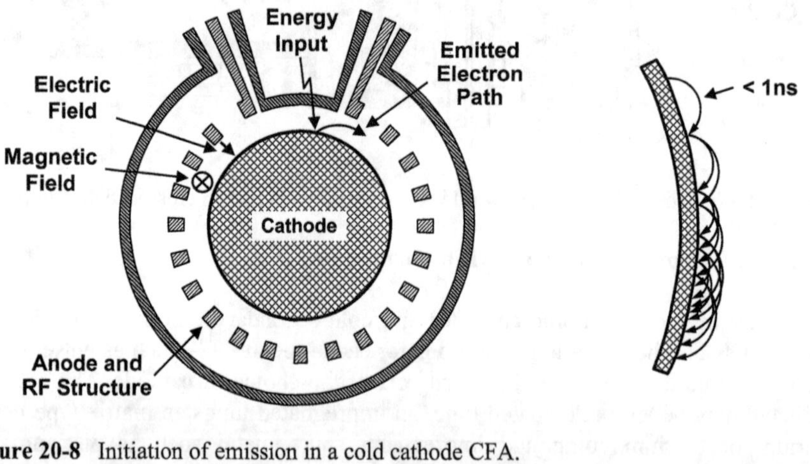

Figure 20-8 Initiation of emission in a cold cathode CFA.

. Field emission and the very low level of thermionic emission that occurs at or near room temperature are possibilities. Cosmic ray ionization of internal gas molecules providing some free electrons and single surface multipactor discharge from the cathode surface have been suggested. None have been proven conclusively. However, many hundreds of thousands of hours of reliable starting have been demonstrated on many different CFAs, and the phenomenon is now accepted as reproducible and reliable even if not fully understood.

The cathode in the CFA shown in Figure 20-7 is cold beryllium that is oxidized so that the secondary emission coefficient is ~ 2 to 3. As is indicated in Figure 20-9, this cathode is liquid cooled. Because of intense electron and ion bombardment, oxygen is driven from the beryllium oxide surface from BeO and so, if corrective action is not taken, the secondary emission coefficient decreases and emission becomes inadequate. To maintain the correct state of oxidation of the cathode surface, oxygen is supplied by a heated cupric oxide source. The internal pressure of the CFA is controlled at about 10^{-6} torr with a 0.2 liter/sec ion pump. Surprisingly, the life of the cathode and of the CFA is many tens of thousands of hours.

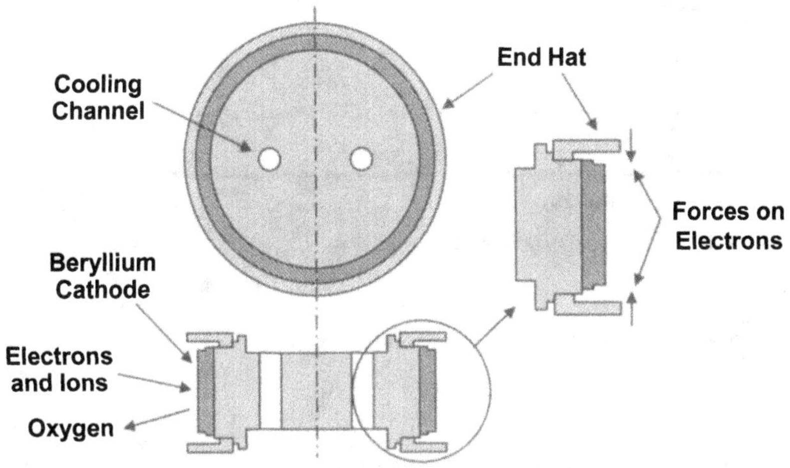

Figure 20-9 Cathode assembly for the CFA shown in Figure 20-7.

End hats are a necessary part of a CFA or magnetron cathode. These operate at cathode potential and are designed to distort the equipotential lines near the ends of the interaction space so that inwardly directed forces are applied to the electrons. As a result, electrons are urged to move toward the center of the cathode and this prevents electron leakage from the interaction space.

20.2.2 Spoke Formation and Growth

The basic interaction that occurs between the electron cloud near the cathode surface and the wave traveling on the slow wave circuit is like that in a magnetron.

Referring to Figure 20-10, we assume that the wave on the slow wave circuit is traveling to the right and that the magnitudes of the dc electric field and the magnetic field have been adjusted so that the electron velocity ($u = E/B$) is the same as the wave velocity. Near A, the electric field is from left to right and so the force on the electron stream is from right to left and is tending to reduce the velocity of the electrons. As is indicated in Figure 20-11, when $u < E/B$, electrons are deflected upward and the electron cloud expands toward the anode. Near B, The force is such that electrons are being accelerated and so the electrons are deflected toward the cathode.

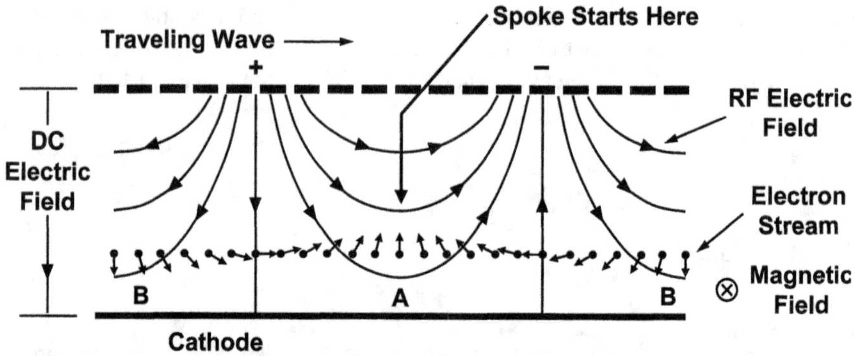

Figure 20-10 Electron deflection by RF fields.

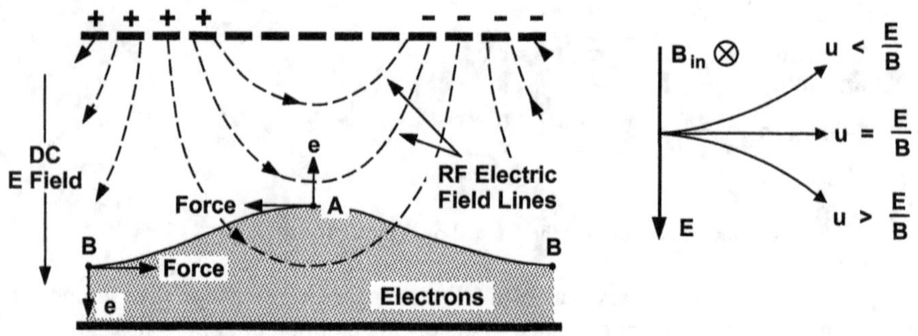

Figure 20-11 Start of spoke formation in a CFA.

As this process continues, the electrons in region A eventually reach the slow wave structure (and anode) as shown in Figure 20-12. As long as the spoke and wave travel at nearly the same velocity (the phase velocity of the circuit), the wave and spoke interact to produce amplification.

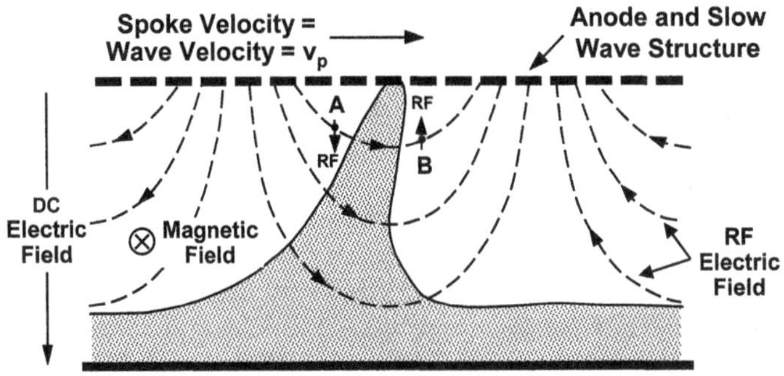

Figure 20-12 Forces that hold the spoke together.

It is reasonable to ask why the spoke and wave should travel in synchronism. This can be understood by examining the forces on electrons on each side of a spoke as indicated in Figure 20-12. At A, RF and dc electric fields add to accelerate electrons toward the anode. The magnetic field directs electrons toward spoke with $v > v_p$ so electrons catch up with spoke. At B, the RF field subtracts from the dc field and this decelerates electrons moving toward anode. The magnetic field directs electrons in the direction of spoke travel, but with $v < v_p$ so the spoke traveling at $v = v_p$ catches up with electrons. As a result, the spoke is forced to travel in near synchronism with the traveling wave.

As a spoke travels along, there are two mechanisms by which it causes RF current to flow in the slow wave circuit. First, as shown in Figure 20-13, electrons in a spoke repel electrons on the circuit just as a bunch in the beam in a TWT repels electrons on the circuit. Second, and unlike a TWT, electrons move through a spoke from the cathode to the anode just as they do in a magnetron. In effect, a spoke acts as a conduit carrying RF current from the cathode to the anode.

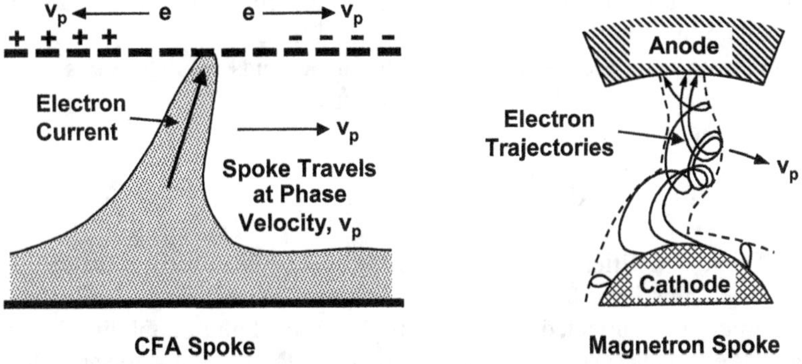

Figure 20-13 RF current flow in a CFA.

Electrons on the circuit travel right and left at phase velocity, v_p. Relative to the spoke, electrons on the left move left at $2v_p$. The deficiency of electrons makes the circuit positive. Relative to the spoke, electrons on the right move at a velocity near 0. As a result, electrons accumulate and make the circuit negative. As the charges and current on the circuit increase, the electric field increases and the RF power on the circuit increases.

At the RF output port as shown in Figure 20-14, the circuit is terminated and RF power is removed from the tube. However, electrons remaining in the drift space continue to travel around the cathode to the RF input where they again participate in the RF interaction process. The electron stream is partially debunched in the drift space, enough so that the RF feedback from the output to input is held to an acceptable level.

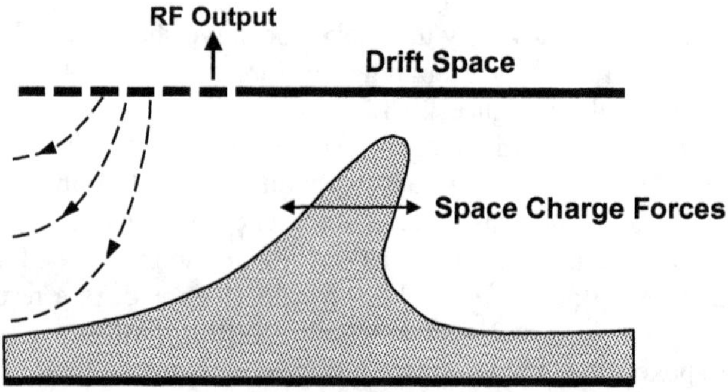

Figure 20-14 Space charge dissipation of RF bunching in the drift space region.

The complete picture of the spoke behavior in a forward wave reentrant CFA is shown in Figure 20-15. The drift space between the output and input port provides a feedback path that helps to limit gain to about 15 dB. There are reports of CFA configurations, in which part of the RF circuit is on the cathode. This helps isolate the output from the input and, reportedly, makes a gain of 30 dB or more possible. While considerable development funds went into this effort, it is not clear that any practical device resulted from it.

20.3 CFA SLOW WAVE CIRCUITS

Crossed-field amplifiers can be forward wave or backward wave devices and many RF circuits have been evaluated for these devices. The selection of a slow wave circuit is dominated by the form factor and the suitability for cooling, because the circuit serves the dual function of electron collector and slow-wave network.

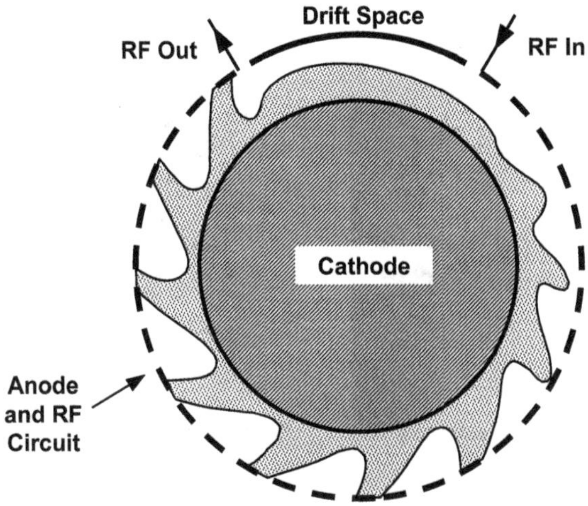

Figure 20-15 Spoke configuration in a forward wave reentrant-beam CFA.

The requirements for good bandwidth and flat dispersion limit the selection of circuits for forward wave operation to some version of a helix or a meander line. Adapting these for use in a CFA will usually compromise either the bandwidth or its power-handling capability because of the necessity to deform the circuit to accommodate either conductive or liquid cooling. The helix-coupled vane and its counterpart, the stub-mounted helix, provide satisfactory compromises by retaining the good electrical features of the helix and providing cooling through the vane or stub. Ceramic-mounted circuits have provided a method for conduction cooling through the ceramic and are particularly suited to high frequencies. Figure 20-16 shows a liquid-cooled meander line [6] and a ceramic-mounted circuit.

A network used by many backward wave CFAs and which satisfies the requirements of slowing down the RF wave and providing the proper phase relationships for electron interaction is shown in a general form in Figure 20-17.

An implementation of this network is the strapped bar line, which is similar to the strapped magnetron. The original CFA built by Brown was of this form [4]. This network is used in many backward wave tubes that are in radar systems in the field.

The circuit as seen by the RF wave entering from the coupler is a parallel plane transmission line consisting of two straps periodically loaded by a "bar" or "ladder" line, the alternate bars or vanes being connected to the same strap. As "seen" by the electrons, the network consists of a series of vanes or "bars" exposing RF fields, which are properly phased to extract energy from the moving electrons in the tubes interaction region.

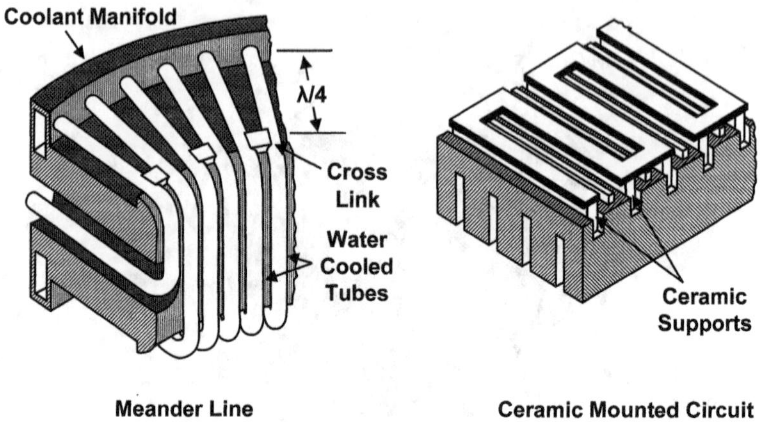

Figure 20-16 Forward wave slow wave circuits.

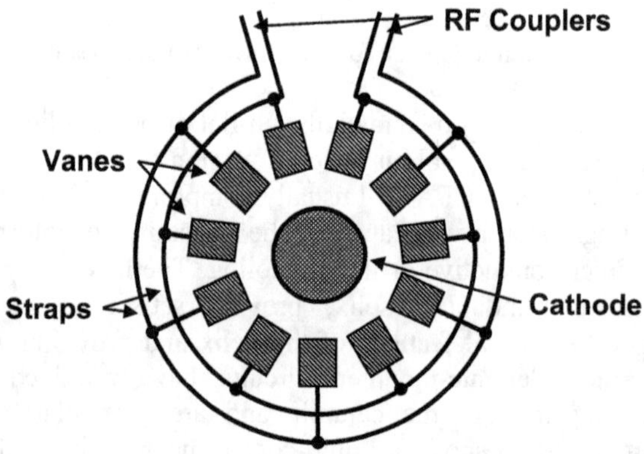

Figure 20-17 Generalized vane-strap slow wave circuit used in many backward wave CFAs.

A circuit which has been used extensively in backward wave tubes is the parallel-plane transmission line consisting of two straps periodically loaded with bars. It is a broadband, nonresonant structure directly connected at both ends to RF input and output couplers. Because of this direct connection and its circuit characteristics, the CFA acts as a passive low loss element in both directions when no high voltage is applied.

A helix-coupled vane forward wave circuit and parallel-plate backward wave circuit are shown in Figure 20-18. The forward wave structure has a helix on each side of the vanes. The vanes are ¼ λ long where λ is the wavelength of the frequency at the center of the operating band. By making the vanes ¼ λ long, the impedance of the supporting structure is infinite so the helices are electrically isolated from the back plate, which is at ground potential. The helices function just as the helix in a TWT does. The RF wave can be envisioned as following a helical path. The velocity of the propagation along the circuit (the phase velocity) is the same as the velocity of the spokes and so the interaction of the wave and spokes takes place as described previously.

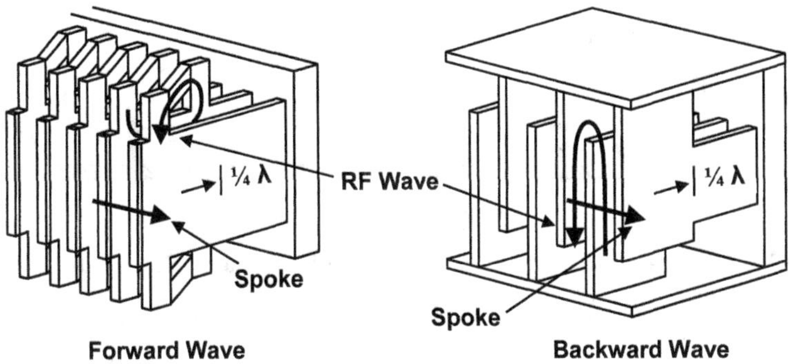

Figure 20-18 Helix-coupled vane forward wave and parallel-plate backward wave slow wave circuits.

The backward wave structure shown in Figure 20-18 is a folded parallel-plate transmission line. This line is also supported by vanes ¼ λ long. The operation of the backward wave circuit is not as easy to understand as the forward wave circuit is. Assume that the RF wave follows the serpentine path indicated in Figures 20-18 and 20-19. If the spoke and the wave travel in the same direction (Figure 20-19(a)), then, as the spoke and wave go from A to B, the direction of the electric field and the phase reverse (indicated by the dashed sine wave) so any modulation placed on the spoke at A is canceled at B. For backward wave operation (Figure 20-19(b)), as a spoke goes from A to B, the wave on the circuit goes from C to B and the resulting forces are in phase with those from A so the modulation from A is reinforced.

There are significant differences between the performance characteristics of forward wave CFAs and backward wave CFAs. The dispersion curves for these devices in Figure 20-20 show the primary differences. Recall that the slope of a straight line from the origin to any point on the dispersion curve is equal to the phase velocity of the circuit wave at that frequency, that is, $v_p = \omega/\beta$. The slope of the curve at any point is the group velocity, $v_g = \Delta\omega/\Delta\beta$, and is equal to the velocity of energy flow on the slow wave circuit at that frequency. When both v_p and v_g have the same algebraic sign, the direction of the energy flow and the direction of the phase velocity of the circuit wave are the same. When v_p and v_g have opposite signs, then the circuit phase velocity and the velocity of the energy flow on the circuit are in opposite directions.

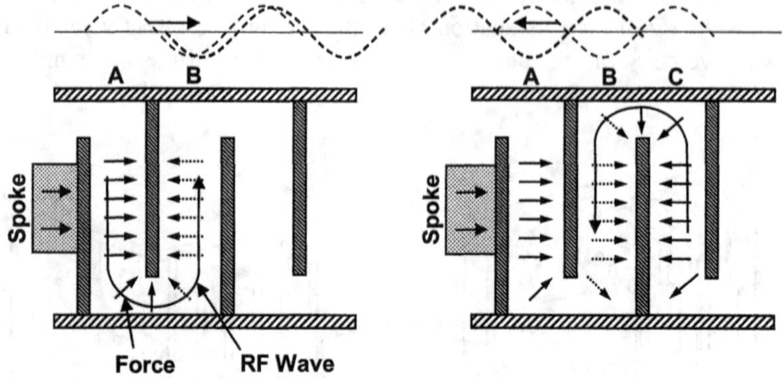

Figure 20-19 (a) Forward and (b) backward wave interaction with the wave on a folded transmission line.

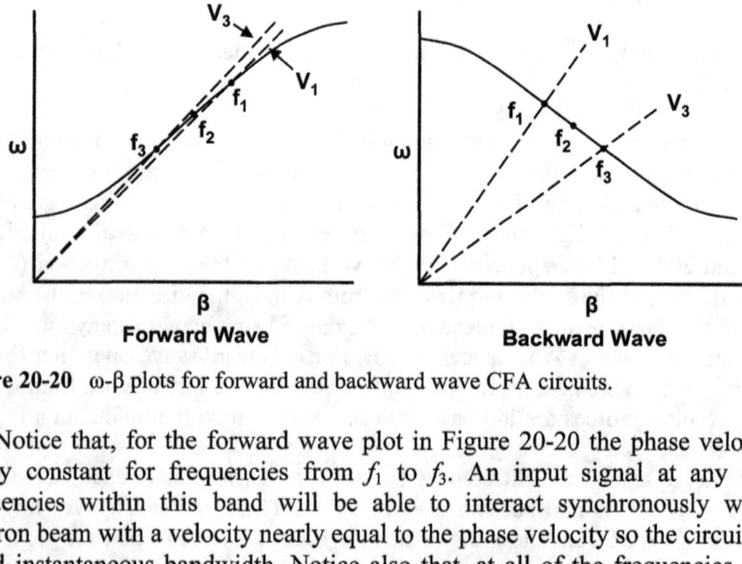

Figure 20-20 ω-β plots for forward and backward wave CFA circuits.

Notice that, for the forward wave plot in Figure 20-20 the phase velocity is nearly constant for frequencies from f_1 to f_3. An input signal at any of the frequencies within this band will be able to interact synchronously with an electron beam with a velocity nearly equal to the phase velocity so the circuit has a broad instantaneous bandwidth. Notice also that, at all of the frequencies within this band, the circuit has nearly the same phase and group velocity.

The backward wave dispersion curve shown in Figure 20-20 has a different phase velocity for each frequency as shown by f_1 and f_3 at V_1 and V_3, respectively. Synchronous interaction between the circuit wave and the electron beam can only occur for a narrow range of frequencies at any particular beam velocity, so the amplifier has a very narrow instantaneous bandwidth. Amplification at different frequencies requires a voltage change to the amplifier to change the electron beam velocity. Hence the backward wave tube is a voltage-tunable amplifier.

20.4 CFA PERFORMANCE

Like a magnetron, a CFA is a diode with a magnetic field applied. Voltage-current (V-I) characteristics are shown in Figure 20-21. Because of the magnetic field, the tube is essentially cut off until a relatively high voltage level is reached. As the magnetic field is increased, voltage must be increased to maintain the same current. Factors other than the magnetic field that affect the V-I characteristic are frequency (and the dispersion characteristic of the slow-wave circuits) and RF drive power.

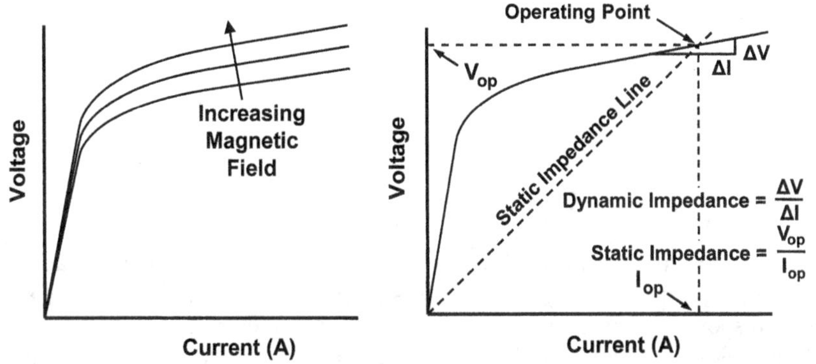

Figure 20-21 Voltage-current and impedance characteristics for a CFA.

The impedance characteristics of CFAs are of importance when considering their operation with a power supply. As Figure 20-21 shows, there is a significant difference between static impedance, which is simply the operating voltage divided by the operating current, and the dynamic impedance, which is the rate of change of voltage with current. The dynamic impedance may be an order of magnitude smaller than the static impedance. With the exception of dc-operated CFAs (to be discussed in Section 20.4.3), current control in CFAs, forward wave or backward wave, is accomplished by controlling (pulsing) the cathode voltage.

As indicated in Figure 20-22, the RF drive pulse is applied before high voltage is applied to the cathode. There are two primary reasons for this. First, if the amplifier uses a cold cathode and if RF drive is absent, the emission process may fail to build up. As a result, the modulator providing the cathode voltage will be essentially unloaded, so excessive voltage may be applied to the tube. This can result in internal arcing and damage to the tube.

The second reason for applying RF drive before cathode voltage is important in CFAs using cathodes heated by a heater or by back bombardment so that thermionic emission occurs. In these devices, RF drive is required to help control the space charge and to help prevent oscillations at the band edge or the generation of broadband noise.

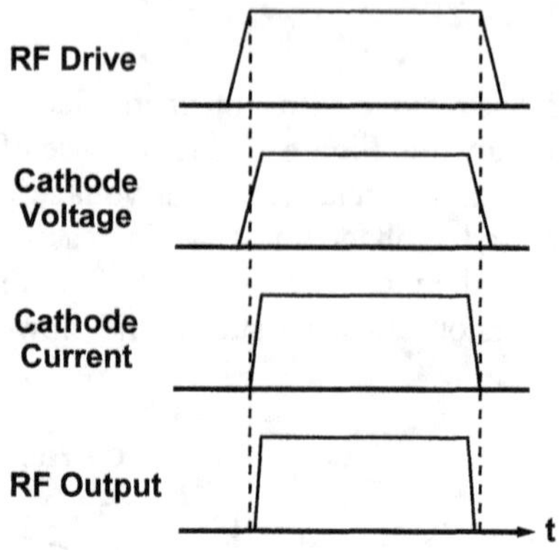

Figure 20-22 Pulse timing in a CFA. RF drive is applied first and then cathode voltage is applied.

20.4.1 Forward Wave CFAs

The ω-β plot for the forward wave CFA in Figure 20-20 shows that, when frequency is changed from f_1 to f_3, there is a small change in phase velocity and in the corresponding optimum operating voltage. As frequency is reduced, voltage increases slowly. As a result, the V-I characteristic for a forward wave amplifier is as shown in Figure 20-23.

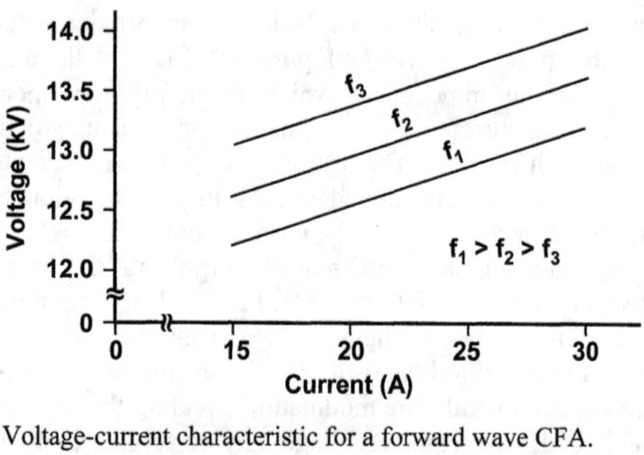

Figure 20-23 Voltage-current characteristic for a forward wave CFA.

Because voltage is nearly constant as frequency is changed, the current in a forward wave CFA should be held nearly constant if power output is to be held constant. Figure 20-24 shows current and voltage as functions of frequency for a forward wave CFA. For this tube type, voltage is adjusted at f_3 to obtain a specified current (22A). Then, as frequency changes, voltage changes slowly to maintain a nearly constant current.

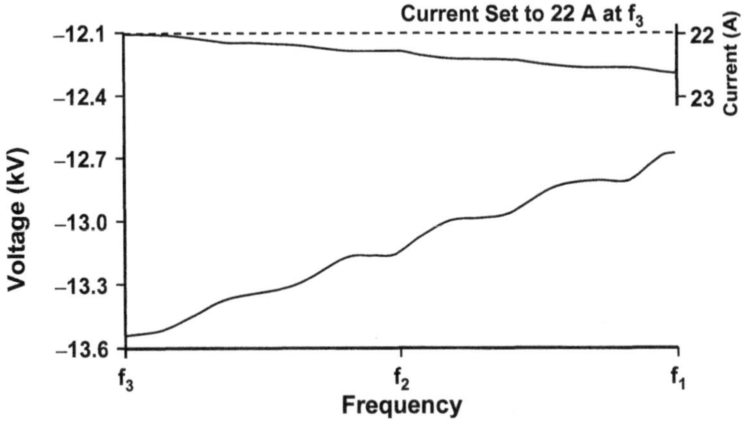

Figure 20-24 Current and voltage as functions of frequency for a forward wave CFA.

An example of the power circuit used to control voltage as well as to turn the forward wave CFA on and off is shown in Figure 20-25. When the CFA is turned off, the full voltage of the power supply is applied to the tetrode. As shown in Figure 20-26, which is the constant current plot for the tetrode, the voltage that across the tetrode when the CFA being considered in this example is turned off is 17.4 kV. To apply cathode voltage to the CFA, the control grid voltage for the tetrode is switched from –608V to ~+150V (with respect to the tetrode cathode). As the CFA frequency is varied from f_1 to f_3, the voltage across the switch tube varies automatically to apply the required voltage to the CFA while maintaining a nearly constant current.

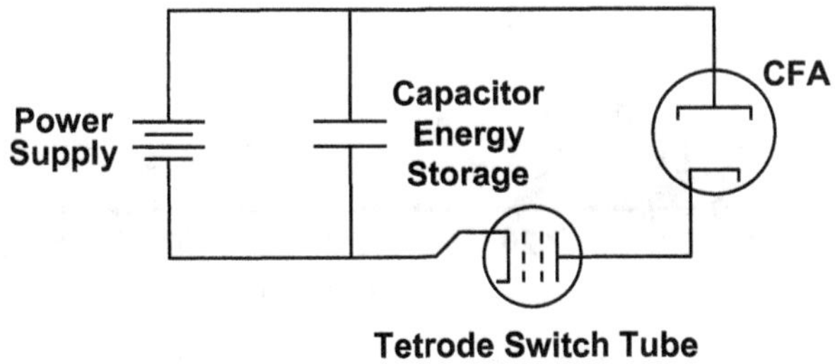

Figure 20-25 Example of the power circuit that can be used for a forward wave CFA.

20.4.2 Backward Wave CFAs

The voltage-current characteristic for a backward wave CFA is shown in Figure 20-27. From the ω-β plot in Figure 20-20, it would be expected that the CFA voltage required for proper operation would decrease as frequency is decreased from f_1 to f_3 and that is what is observed in practice.

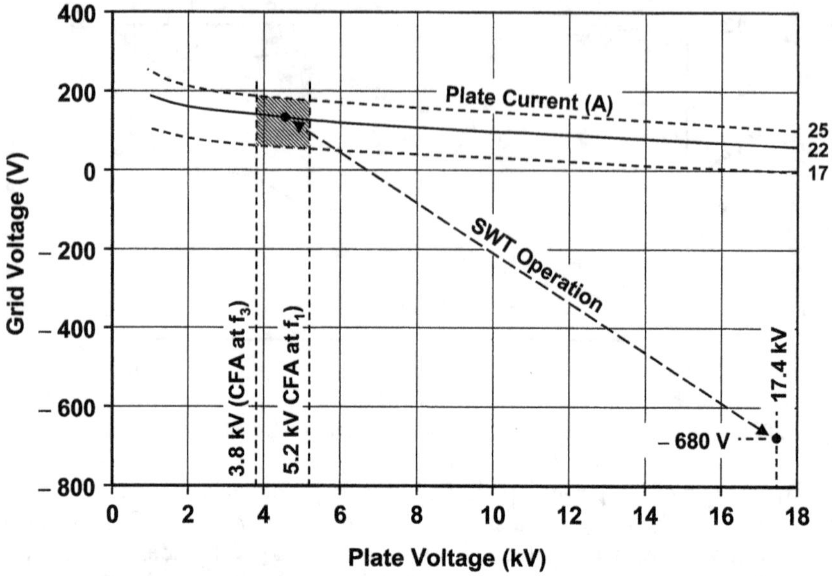

Figure 20-26 Constant current characteristic for the switch tube being considered in this example.

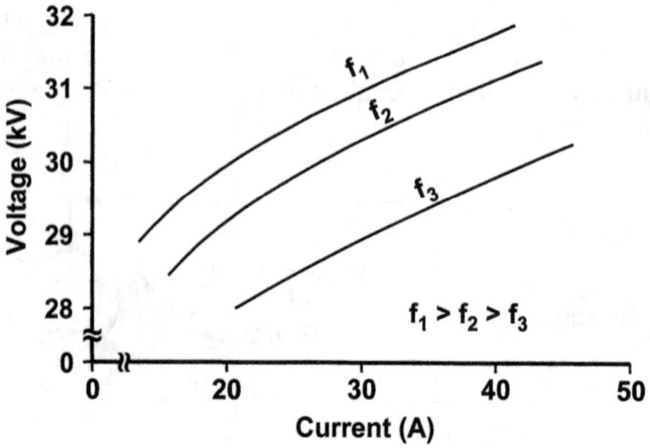

Figure 20-27 Voltage-current characteristic for a backward wave CFA.

Another chart showing the voltage-current characteristics for a backward wave CFA is shown in Figure 20-28. This chart shows the power output and efficiency of the device and is similar to a portion of a magnetron performance chart. The magnetic flux density is not normally a variable to the user of a CFA so the user is likely to receive a "one-line" version of this chart (as shown in Figure 20-29).

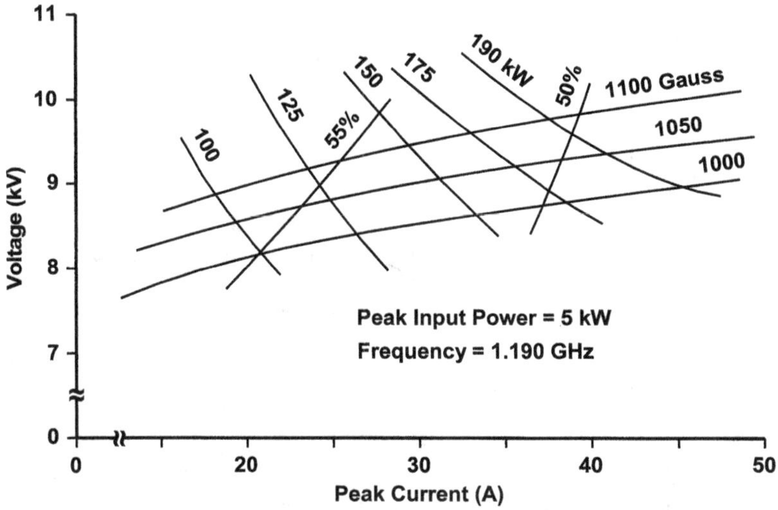

Figure 20-28 CFA performance chart. (From: *Microwave Tube Manual* by Varian Associates, Air Force Publication Number T.O.00-25-251, October 1979.)

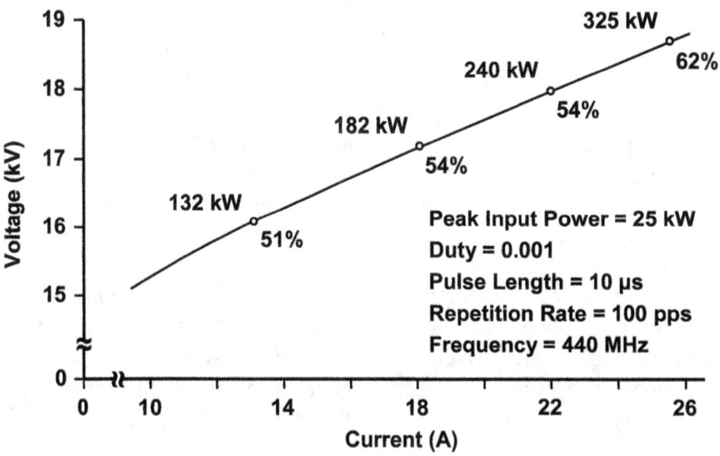

Figure 20-29 CFA "one-line" performance chart. (From: *Microwave Tube Manual* by Varian Associates, Air Force Publication Number T.O.00-25-251, October 1979.)

When voltage and current are dependent on frequency of operation, as in a backward wave CFA, the performance chart will show the frequency dependence as illustrated in Figure 20-30. With this chart, the user can determine the variations in power, current, and voltage that will result from the modulation technique used in tuning the amplifier across the required frequency range.

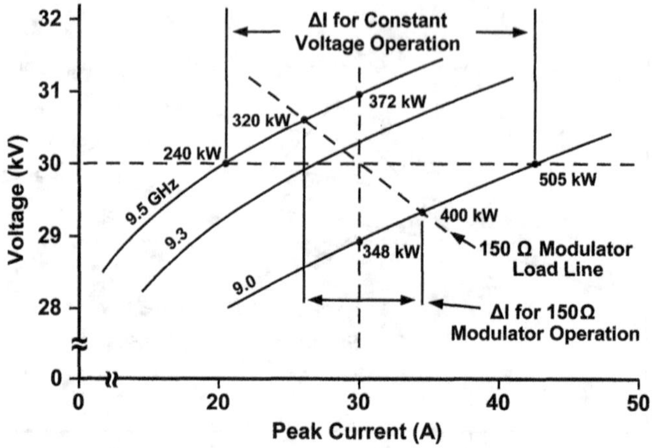

Figure 20-30 Frequency dependence of voltage and current. (From: *Microwave Tube Manual* by Varian Associates, Air Force Publication Number T.O.00-25-251, October 1979.)

20.4.3 DC Operation

CFAs are diodes and as such do not have control elements like the grid in a TWT to turn them on and off. There have been attempts to use an interdigital structure on the cathode for current control but, to the best of the author's knowledge, this has not found application. It is also possible to turn a CFA on or off by pulsing the magnetic field, but this is a relatively slow process because of the inductances involved. While CFAs do not have control elements, it is possible to use the RF pulse along with a cutoff electrode to turn the tube on and off. These tubes are referred to as dc-operated CFAs.

In a CFA with a cold secondary emitting cathode, it is possible to apply anode voltage without drawing current. Upon application of the RF input pulse, out-of-phase electrons are driven into the cathode surface and the secondary emission process is initiated. As shown in Figure 20-31, the emission process builds up rapidly as is indicated by the fact that the RF output power is at its maximum value within about 10 ns.

At the end of a pulse, space charge in the cathode-anode region must be cleared; otherwise, the device can oscillate or generate noise. In nonreentrant CFAs, like the dematron, this occurs automatically, but efficiency is low, so these devices have not found application. In reentrant CFAs, a control or cutoff electrode is pulsed positive with respect to the cathode to collect the space charge. The configuration of a CFA containing a cutoff electrode is shown in Figure 20-32.

The sequence of pulses applied to a CFA during dc operation is shown in Figure 20-32. The applied RF drive pulse produces anode current and an RF output pulse. At the end of the RF pulse, the voltage pulse to the cutoff electrode produces a short pulse of current corresponding to the charge stored in the interelectrode space.

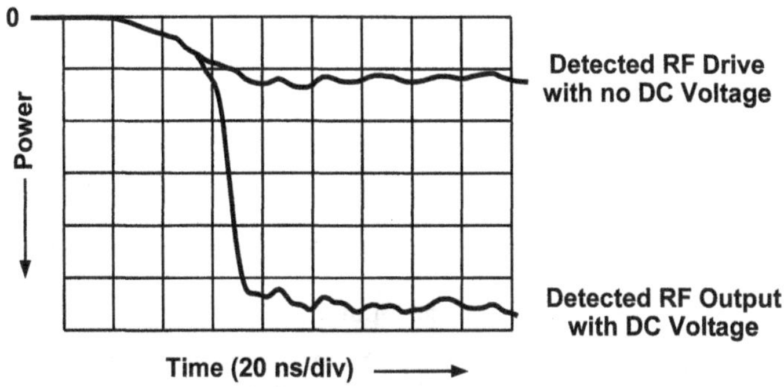

Figure 20-31 Leading edge of RF drive and output pulse on a dc-operated CFA. (From: J. Skowron, *Proc. IEEE*, March 1973. © 1973 IEEE.)

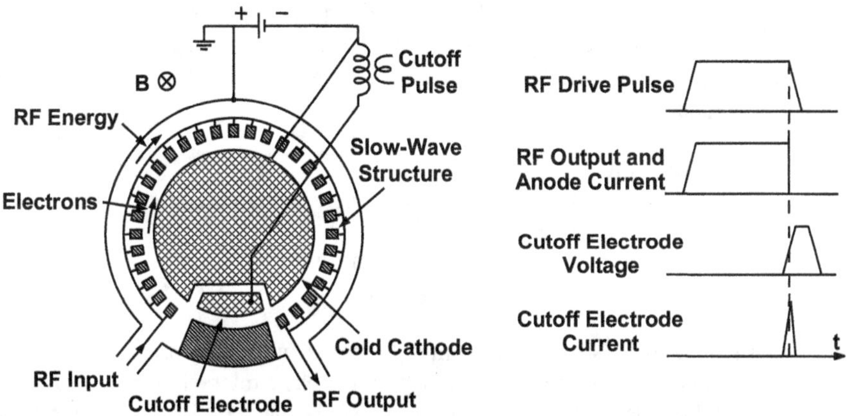

Figure 20-32 DC-operated CFA with cutoff electrode. (From: J. Skowron, *Proc. IEEE*, March 1973. © 1973 IEEE.)

20.4.4 Gain and Operating Limits

The gain of a CFA is often shown on a plot of output power versus input power (referred to as a compression curve) like that shown in Figure 20-33. Values of gain shown in Figure 20-33 are typical, and are low compared with those for linear-beam tubes. It should also be noted that the CFA does not exhibit any region in which operation is linear (constant gain).

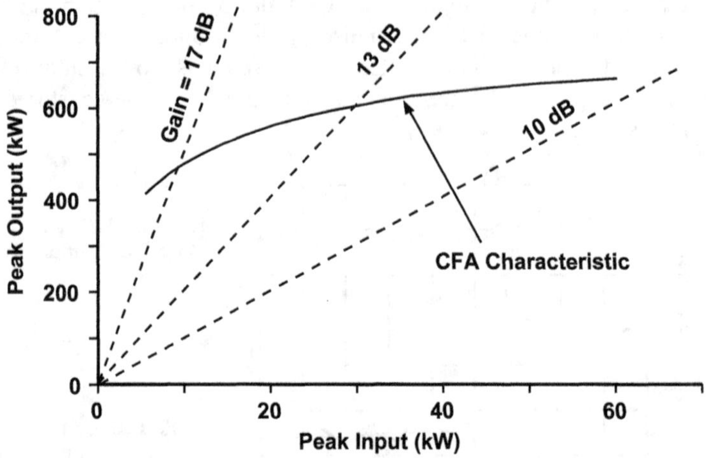

Figure 20-33 Compression curve for CFA showing lines of constant gain. (From: *Microwave Tube Manual* by Varian Associates, Air Force Publication Number T.O.00-25-251, October 1979.)

The gain-limiting factors in CFAs are varied and complex and lead to the chart in Figure 20-34, showing upper and lower boundaries of output power for given values of input power. The upper boundary can be caused by any one of the following factors:

1. A limitation on cathode current;
2. The onset of oscillation because the synchronous voltage of a competing mode has been reached;
3. Sufficiently large output powers so that the drive level is insufficient to retain lock in the main amplifying mode.

The first two of these limitations may be eased with appropriate design, but the third is an intrinsic limit resulting only from the basic interaction process.

The lower boundary may also result from oscillation because the synchronous voltage of another mode has been reached. As with the upper voltage competing mode, this lower voltage limitation may be eased with proper design procedures.

At low drive levels, the gain may be 20 to 30 dB or more and operation is limited by feedback (for example, by imperfect matching or through the electron stream). At low currents, if the RF circuit is designed to avoid resonances, the tube will produce a broad noisy output. In a backward wave device, the anode voltage determines the center of the frequency band and so a voltage-tunable noisy output results. In the forward wave case, the output is broadband noise. Crossed-field noise generators are built on this concept.

As the frequency of operation is varied, changes occur in the various factors leading to the upper and lower boundaries of operation of a CFA. As a result, the boundaries change with frequency. In the example shown in Figure 20-35, the relationship of the actual operating current to the currents that would lead to upper- or lower-mode instabilities is shown. In this case, a limiting factor in bandwidth is the intersection of the operating current line with the upper-mode characteristic.

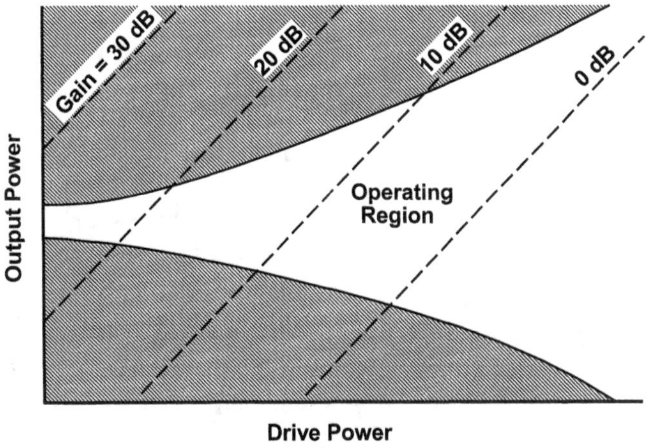

Figure 20-34 Upper and lower boundaries of output power of a CFA. (From: J. Skowron, *Proc. IEEE*, March 1973. © 1973 IEEE.)

In some CFAs it is possible to improve operating characteristics by introducing an asymmetry into the interaction region. As an example, for the forward wave CFA discussed earlier in this chapter, with the cathode centered, performance at the high end of the band is marginal. The problem is that the spokes are out of phase with the RF field. The spoke phase can be controlled with voltage and current and current density can be controlled with cathode position. Thus, by pushing the cathode off-center, the spoke phase can be corrected. The result is that the high-frequency performance is improved and, it turns out, intrapulse noise is reduced. Figure 20-36 shows the cathode positioning device in the CFA in Figure 20-7 [5].

In practice, mode interference is often the limiting factor in determining the useful range of CFA operation. Mode interference is an oscillation that takes time to build up. In pulsed operation, the rate of cathode voltage rise has a significant effect on mode interference. Care must be taken in modulator design, however, because a fast rate of rise, which prevents a lower mode from starting, may initially result in insufficient tube current to load the modulator. As a result, voltage can swing through the values appropriate for operation in the amplifying mode and cause oscillation in an upper mode.

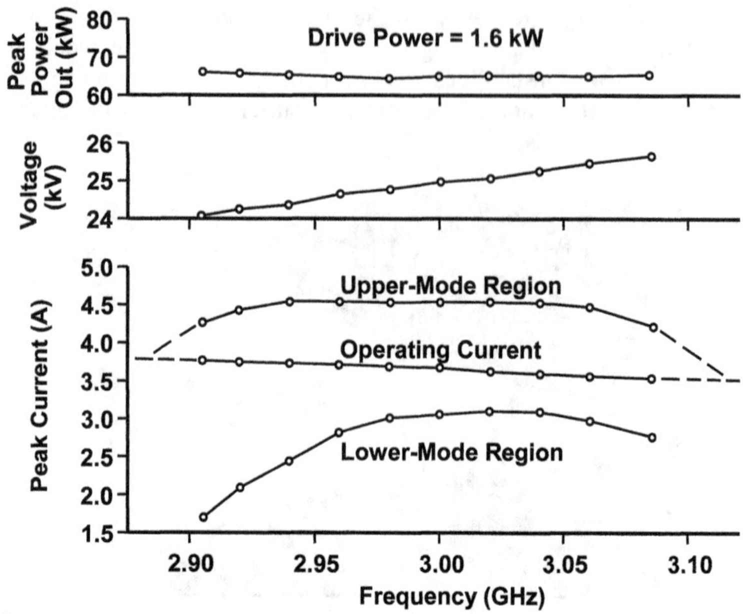

Figure 20-35 Relation of operating current to currents that lead to upper- or lower-mode instabilities in a backward wave CFA. (From: J. Skowron, *Proc. IEEE*, March 1973. © 1973 IEEE.)

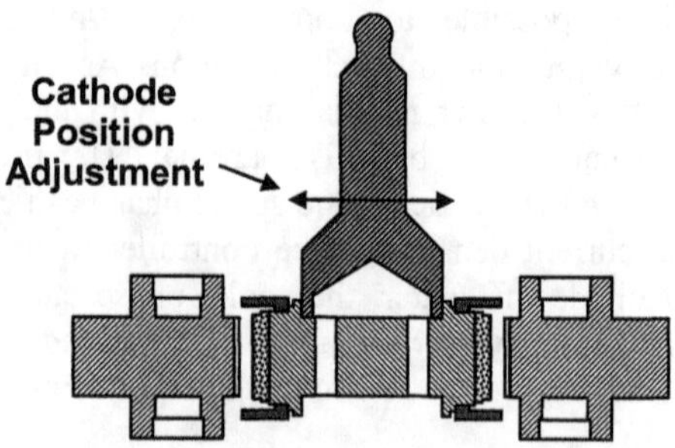

Figure 20-36 Cathode positioning device in a forward wave CFA.

In connection with a discussion of mode interference, it is instructive to examine the impedance characteristics of a CFA. The input VSWR as a function of frequency is typically as shown in Figure 20-37. Within the operating band, the VSWR is low, typically between 1.2:1 and 1.5:1. The corresponding return loss (magnitude of reflected power) is between 20 and 14 dB below the input power.

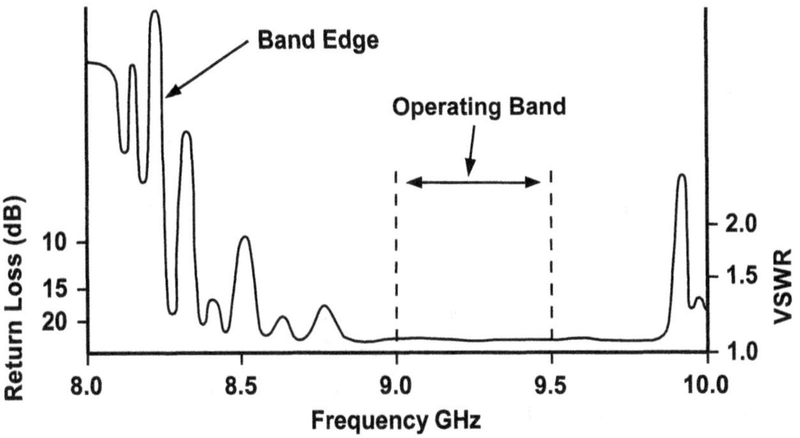

Figure 20-37 Input VSWR for CFA circuit with output terminated. (From: *Microwave Tube Manual* by Varian Associates, Air Force Publication Number T.O.00-25-251, October 1979.)

Outside of the operating band of the amplifier, the frequency regions where the VSWR increases to very high values are referred to as the band edges of the circuit. It is at one of these band edges (the low-frequency edge in a backward wave amplifier) that oscillations may occur during the rise and fall of the voltage pulse. The resulting spurious outputs of the tube are frequently referred to as "rabbit ears," because of the way they appear on a time display of the RF output.

In addition to the input VSWR being very low in the operating band, it should be pointed out that the transmission loss of the signal through a CFA in the absence of beam current may be very low. This is indicated in Figure 20-38 for a low-frequency device. As a result, the amplifier can be considered essentially transparent, and this is of use in some system applications. Transmission losses are referred to as insertion losses and are related to electrical characteristics like resistivity as well as mechanical characteristics such as roughness and size of the circuit.

20.4.5 CFA Phase Characteristics

Three types of phase sensitivities are normally of interest in applications of microwave tubes. These are:

- Phase sensitivity to voltage or current;
- Phase sensitivity to RF drive;
- Phase linearity with frequency.

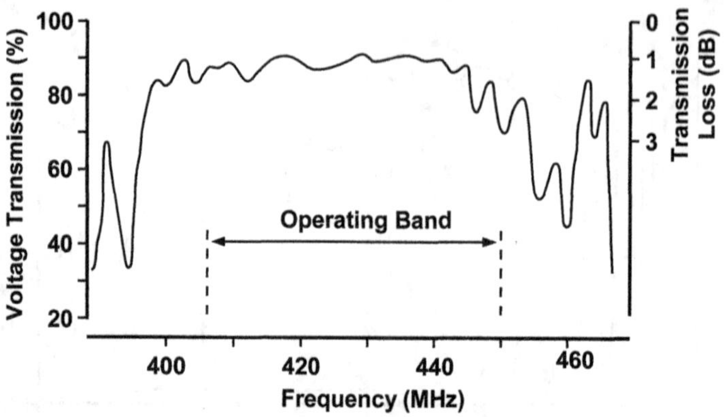

Figure 20-38 Transmission characteristic of a CFA. (From: *Microwave Tube Manual* by Varian Associates, Air Force Publication Number T.O.00-25-251, October 1979.)

In a linear-beam tube, phase shift through a tube is proportional to the transit time of the beam which is inversely proportional to beam velocity. Beam velocity, in turn, varies with the square root of beam voltage. For a TWT, phase changes about 30° for a 1% change in beam voltage. Current varies with voltage to the three-halves power, so the phase changes about 45° for a 1% change in current.

In a CFA, the rotational rate of the spokes in the space-charge cloud is locked to the velocity of the signal on the RF circuit. The speed of the spokes tends to be controlled by the propagation characteristics of the circuit and the signal level in the circuit. Also, the electrical length of the RF circuit is short. As a result, phase sensitivity to changes in the anode voltage would be expected to be small.

In a linear-beam tube, because of the three-halves power relationship between voltage and current, the phase shift for a 1% change in voltage or current is of the same order of magnitude. In a crossed-field device, as can be seen in the voltage-current plots, a small change in voltage is accompanied by a large change in current. Thus, the phase shift for a 1% change in voltage is much larger than the phase shift for a 1% change in current.

The phase variations with voltage for three RF drive levels in an X-band CFA are shown in Figure 20-39. The phase changes indicated are for a 1% change in anode voltage. Thus, at low drive levels and high gain, the phase shift (15°/1% ΔV) is much larger than at high drive levels and low gain (3.8°/1% ΔV).

In Figure 20-40, the phase shift per 1% change in current is shown for a low-power CFA with a thermionic cathode and for a medium-power, cold-cathode device. In both cases, the phase shift at all drive levels is small. While the comparison is not strictly valid because of the large difference in devices, the phase sensitivity to voltage in Figure 20-39 is much larger than the phase sensitivity to current in Figure 20-40, as would be expected. In all cases, the phase sensitivity is much smaller than for linear-beam devices.

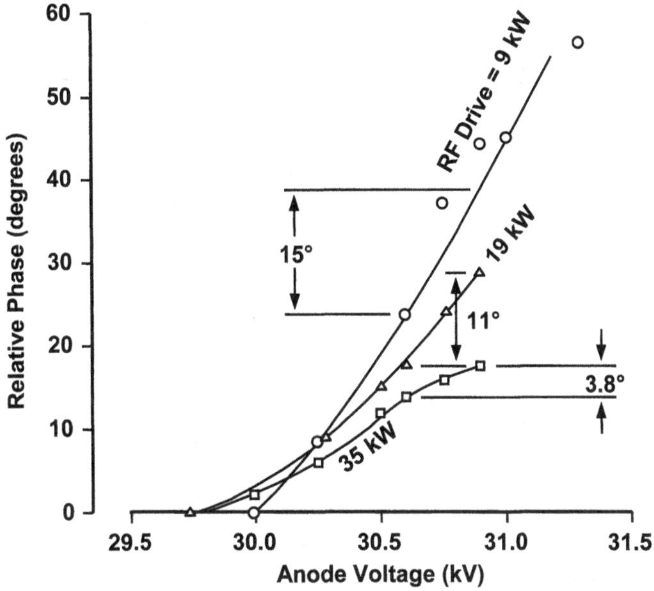

Figure 20-39 Phase variation with voltage for three RF drive levels for an X-band CFA. (From: *Microwave Tube Manual* by Varian Associates, Air Force Publication Number T.O.00-25-251, October 1979.)

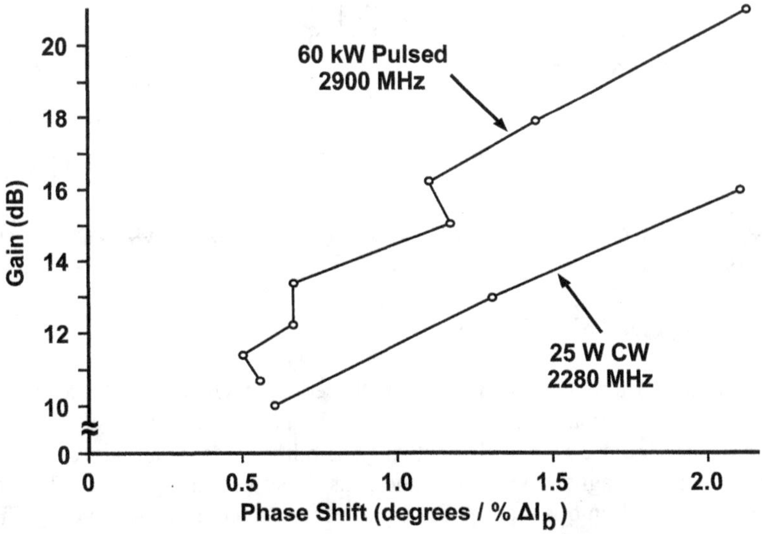

Figure 20-40 Effect of gain on phase sensitivity for low- and medium-power backward wave CFAs. (From: J. Skowron, *Proc. IEEE*, March 1973. © 1973 IEEE.)

The phase linearity of a CFA with frequency is largely determined by the circuit reflections and the gain of the device. The effect of circuit reflection depends on electrical length and gain. Because the CFA is electrically short (compared with linear-beam devices) and gain is low, phase deviations from linearity of less than ±5° can be achieved.

The favorable phase characteristics of CFAs lead to their use in systems where arrays of devices are used and phasing is important. This is indicated in Figure 20-41. Also indicated are other applications where the transparent characteristic is used (for interpulse feedthrough) or where tubes are readily connected in parallel to upgrade power capability.

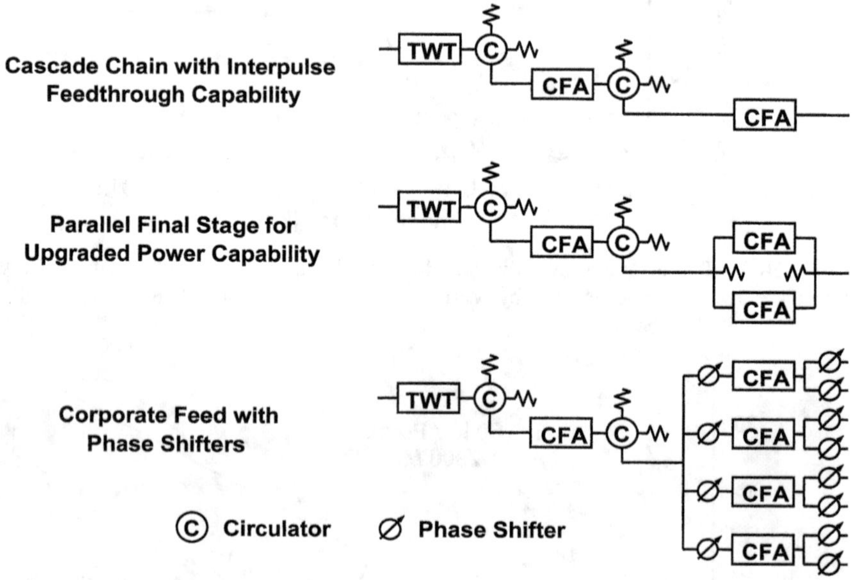

Figure 20-41 Configurations for applications of CFAs. (From: J. Skowron, *Proc. IEEE*, March 1973. © 1973 IEEE.)

20.4.6 Weight and Size Considerations

The use of samarium cobalt can significantly reduce magnet and CFA weight. For example, a size comparison between Alnico V and samarium cobalt magnets is illustrated in Figure 20-42. Still another possibility in cold-cathode CFAs is the placement of the magnet inside the cathode as illustrated in Figure 20-43. The use of samarium cobalt in this manner has been demonstrated in low-frequency CFAs.

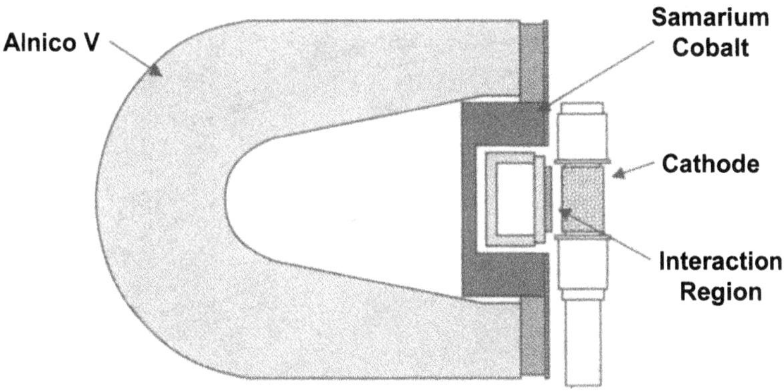

Figure 20-42 Size comparison between Alnico V and samarium cobalt magnets for a CFA. (From: J. Skowron, (MPTD), Raytheon Co., Workshop on High-Power, Space-Based Microwave Systems, Los Alamos, NM, March 1985.)

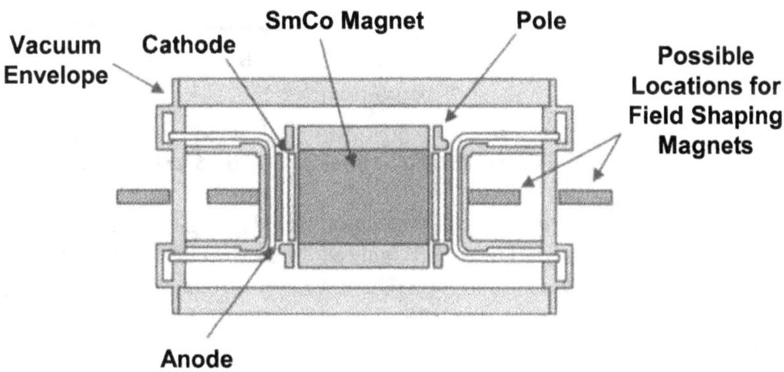

Figure 20-43 Placement of samarium cobalt magnet inside the cathode in cold-cathode CFA. (From: J. Skowron, *Proc. IEEE*, March 1973. © 1973 IEEE.)

20.5 POWER CAPABILITIES

In practice, CFAs are capable of impressive performance. For example, Figure 20-44 shows an efficiency of approximately 50% over a 10% bandwidth for a pulsed C-band amplifier having a power output of about 1 MW.

An important feature of CFAs is the relatively low voltage at which they operate. The voltage of less than 40 kV for the megawatt device in Figure 20-44 is a factor of 2 to 3 below the voltage required by a single-beam linear-beam device at the same power level. (New multiple-beam linear-beam devices operate at voltages comparable to CFAs.)

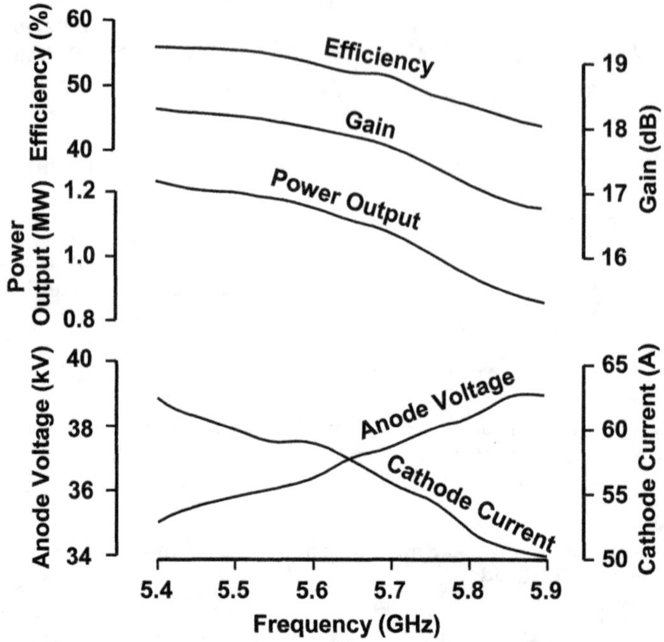

Figure 20-44 CFA performance over a 500-MHz band. (From: *Microwave Tube Manual* by Varian Associates, Air Force Publication Number T.O.00-25-251, October 1979.)

The performance characteristics of an extremely high-power ("super power") CW CFA operating at 3 GHz with a bandwidth of over 5% are shown in Figure 20-45. Output power was as high as 425 kW and efficiency exceeded 70%. The performance characteristics of a pulsed super power CFA are shown in Figure 20-46. These high-power tubes were developed for use in very high-power radar systems, which were replaced by the development of active phased array technology using large numbers of lower powered tubes.

The overall state of the art of publishable power capabilities of CFAs at the present time is comparable to that in 1985 and is shown in Figure 20-47. Note that megawatt power levels are available in pulsed devices at frequencies in excess of 10 GHz.

20.6 THERMAL CONSIDERATIONS

The electron beam in an efficient crossed-field device is collected on the RF circuit so the circuit must be capable of dissipating the spent-beam power. In addition, there may be a significant amount of heating of the cathode by back bombardment so that the cathode may have to be cooled.

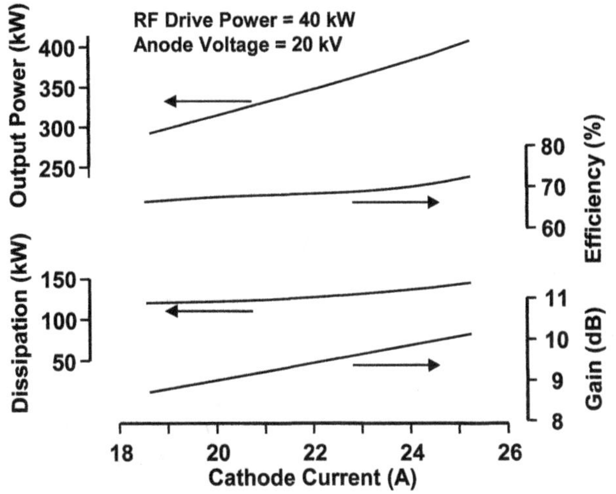

Figure 20-45 Performance of a super power CW CFA. (From: J. Skowron, Raytheon Co., Workshop on High-Power, Space-Based Microwave Systems, Los Alamos, NM, March 1985.)

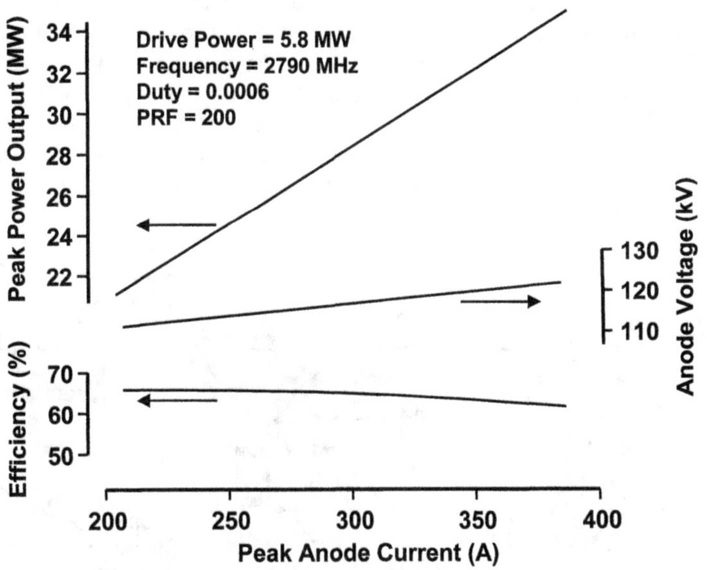

Figure 20-46 Performance of a super power pulsed CFA. (From: J. Skowron, Raytheon Co., Workshop on High-Power, Space-Based Microwave Systems, Los Alamos, NM, March 1985.)

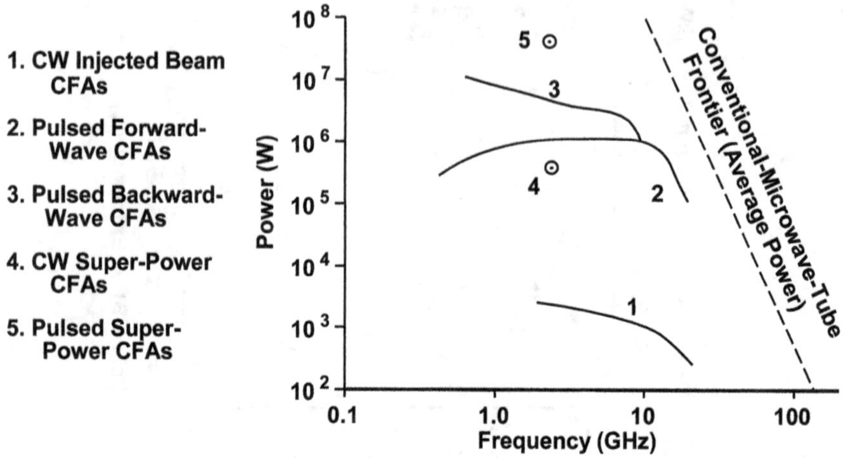

Figure 20-47 State of the art of CFAs. (From: J. Skowron, Raytheon Co., Workshop on High-Power, Space-Based Microwave Systems, Los Alamos, NM, March 1985.)

The cooling arrangement used in a helix-coupled vane circuit is shown in Figure 20-48. In Figure 20-48(a), the thermal path for heat from the electron bombardment of the vane tips is through the entire length of the vanes. For operation at high duty cycles, increased cooling of the vane tips is required and this is provided, as is shown in Figure 20-48(b), by injecting coolant into hollow chambers in the vanes. The interception power that is capable of being removed is several kilowatts. The reason for using molybdenum vane tips is that, for pulsed operation, the instantaneous heating of the vane tips is so great that copper vane tips are destroyed.

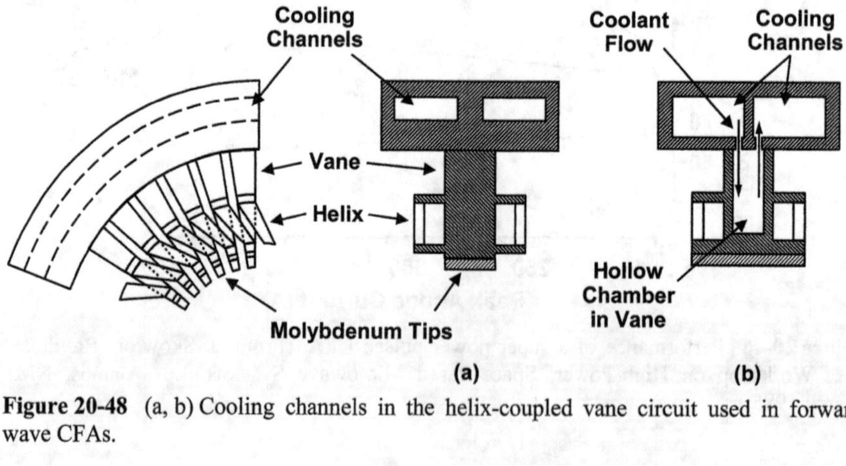

Figure 20-48 (a, b) Cooling channels in the helix-coupled vane circuit used in forward wave CFAs.

A cooling arrangement for many CFA circuits, forward wave and backward wave, was shown in Figure 20-16. The circuit is fabricated from tubes that are attached to a coolant manifold so coolant is forced through the tubes in parallel. The circuit shown in Figure 20-16 is a forward wave circuit. By changing the support for the circuit as well as the technique for interconnecting the tubes, the circuit can operate in the backward wave mode.

The highest dissipation ever achieved in a microwave tube was in the Super Power CFA program carried out by the Raytheon Company [7]. Dissipations of nearly 20 kW/cm^2 were demonstrated. The technique used was a convection process involving a change in phase from a liquid to vapor of the coolant being used. In this process, as the heat is being transferred from the vane to the coolant, it raises the temperature of the coolant to its boiling point, forming a vapor bubble. The bubble then displaces the liquid above, which rushes back to fill the void left by the bubble when it left the hot surface. Although the bubble carries away only a small amount of the total heat being dissipated, it acts somewhat like a pump, carrying hot liquid away from the hot surface and creating intense turbulence within the tubing. For this technique to be used, velocities of approximately 100 feet per second or vortex flow are needed.

The vanes of the CFA were constructed as indicated on Figure 20-49. Copper-clad stainless steel tubes were inserted between the inner and outer splines, short sections of which are shown in Figure 20-49. The resulting spline and tube assembly was then drawn to the final dimension and brazed together. The vane was given a full 360° twist so that each of the six tubes would receive an equal portion of the electron bombardment. (The twist is shown in Figure 20-49 without the splines for clarification.) The twist also gave the water a spiral or vortex rotation, which provided better scrubbing of the bubbles from the wall than could be achieved with straight flow.

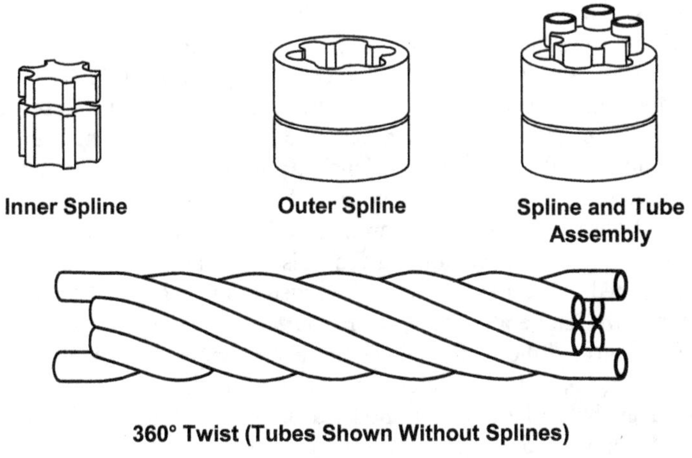

Figure 20-49 Elements of vane construction for the 400 kW CW super power CFA.

As an example of the way in which power may be divided in a CFA, the power distribution is given in Table 20.1 for the CW CFA, the performance characteristics for which were shown in Figure 20-45. Note that the anode dissipation is over 100 kW and the cathode dissipation is nearly 25 kW.

Table 20.1

Distribution of Power in "Super Power" CW CFA

RF power generated	71.7%	350.0 kW*
Anode dissipation	21.4%	104.5 kW
Cathode dissipation	5.0%	24.6 kW
Transmission line	1.9%	9.3 kW
Totals	100.0%	488.4 kW

*Note: Drive power is added to power generated for a total output power of 405 kW.

Source: J. Skowron, (MPTD), Raytheon Co. Workshop on High-Power, Space-Based Microwave Systems, Los Alamos, NM, March 1985.

The severity of the cooling problem becomes apparent when it is realized that the total cross-sectional area of the anode segments ranges from a fraction of a square centimeter in high-frequency devices to a few tens of square centimeters in low-frequency devices. These relatively small areas may result in power densities in excess of a kW/cm^2 that must be dissipated by the anodes.

Table 20.2 contains a listing of the cooling capabilities possible with various techniques. Over 10 kW/cm^2 can be removed by high-velocity water when nucleate boiling occurs on the surfaces being cooled. As a result, this is the cooling technique used in extremely high-power applications.

Shown in Figure 20-50 are predicted power capabilities of CW CFAs based on anode dissipation capability. The Raytheon QKS 849, which used nucleate boiling to achieve an anode dissipation of 3,600 W/cm^2 is used as a design reference point. Based on this power density, power outputs in excess of 1 MW should be possible at frequencies below about 1,500 MHz.

In pulsed CFAs, electron bombardment of the anode results in a transient temperature rise of the surface during the pulse. The amount of energy that can safely be imparted to the anode is directly related to the heat capacity of the anode material and to the temperature to which the surface may rise.

The calculation of the surface temperature and temperature distribution within an anode during a pulse is relatively straightforward. Because the heat is concentrated in the surface of the anode, a one-dimensional model like that in Figure 20-51 can be used.

Table 20.2
Cooling Capability of Various Media ($\Delta T = 200°C$)

Black body radiation (T ~ 200°C)	0.25 W/cm^2
Air, natural convection to finned forced convection	0.5–110 W/cm^2
Organic oils, natural convection to forced convection, boiling	5–90 W/cm^2
Water, pool boiling	125 W/cm^2
Water, pressurized, nonboiling	155 W/cm^2
Liquid metals, forced convection nonboiling	450 W/cm^2
Water, high velocity, nucleate boiling	
Operational	10,000 W/cm^2
Maximum achieved	20,000 W/cm^2

Source: J. Skowron, (MPTD), Raytheon Co. Workshop on High-Power, Space-Based Microwave Systems, Los Alamos, NM, March 1985.

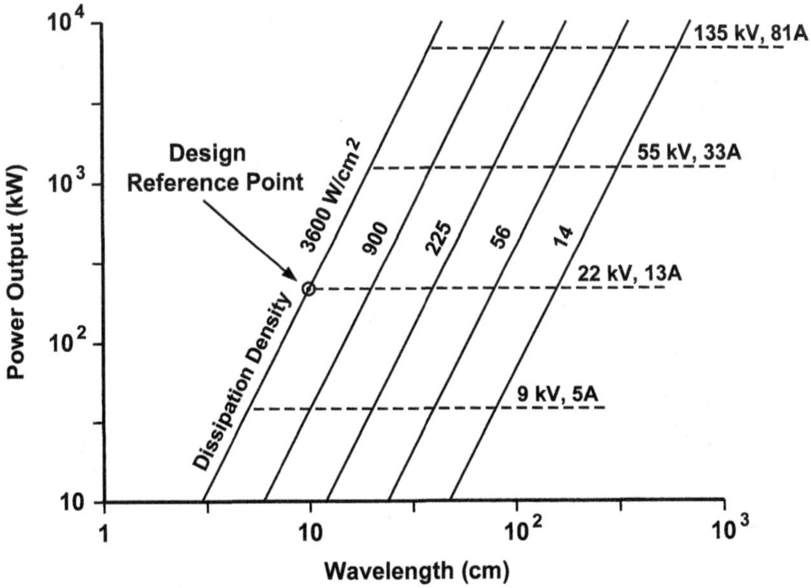

Figure 20-50 Predicted power capabilities of CW CFAs based on anode dissipation power densities. (From: J. Skowron, (MPTD), Raytheon Co., Workshop on High-Power, Space-Based Microwave Systems, Los Alamos, NM, March 1985.)

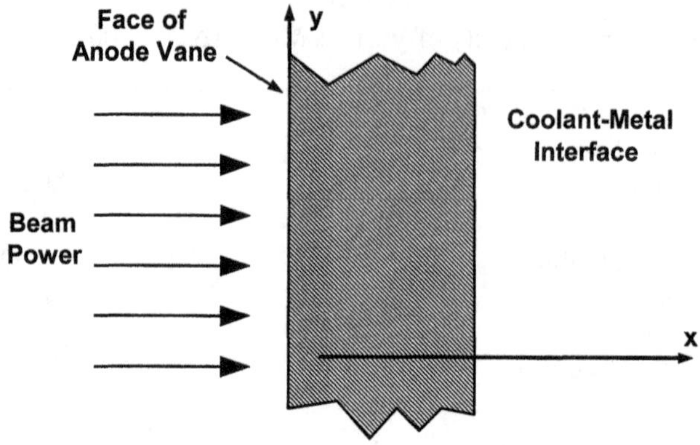

Figure 20-51 Model used for solution of one-dimensional heat flow equation. (From: J. Skowron, *Proc. IEEE*, March 1973. © 1973 IEEE.)

Shown in Figures 20-52 and 20-53 are temperatures calculated for the case of the 200-μs pulse length, 100 pulses per second, and a peak power density of 67,850 W/cm^2. Notice that the temperature of the face of the anode vane reaches nearly 500°C during the pulse and that the temperature only 0.010 inch into the vane is just 200°C. During the interpulse period, the temperature of the entire vane nearly drops to that of the coolant-metal interface. It is because of the high vane surface temperature reached during a pulse that the vane tips of CFAs are sometimes fabricated from a high temperature metal such as molybdenum.

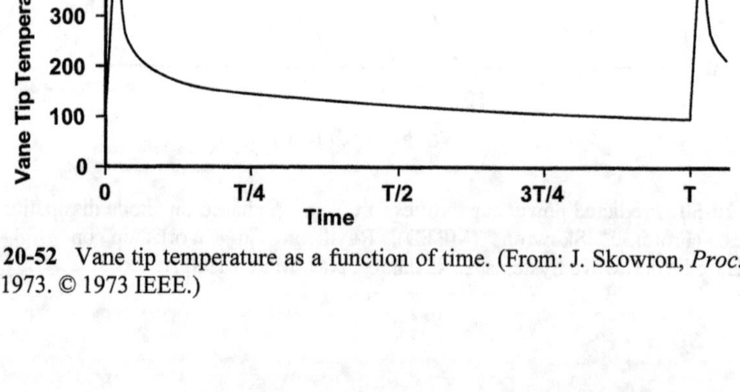

Figure 20-52 Vane tip temperature as a function of time. (From: J. Skowron, *Proc. IEEE*, March 1973. © 1973 IEEE.)

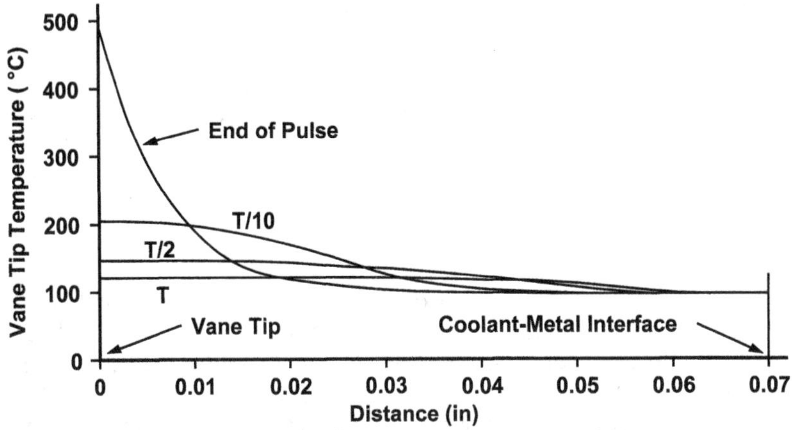

Figure 20-53 Temperature as a function of distance into vane tip. (From: J. Skowron, *Proc. IEEE*, March 1973. © 1973 IEEE.)

The predicted power capabilities of pulsed CFAs based on the temperature of the vane-tip surface are given in Figure 20-54. For a given surface temperature, as pulse duration is reduced, the peak-power density can be increased before the limiting energy input is reached. At 3,000 MHz, 20 MW has been achieved and is used as the design reference point.

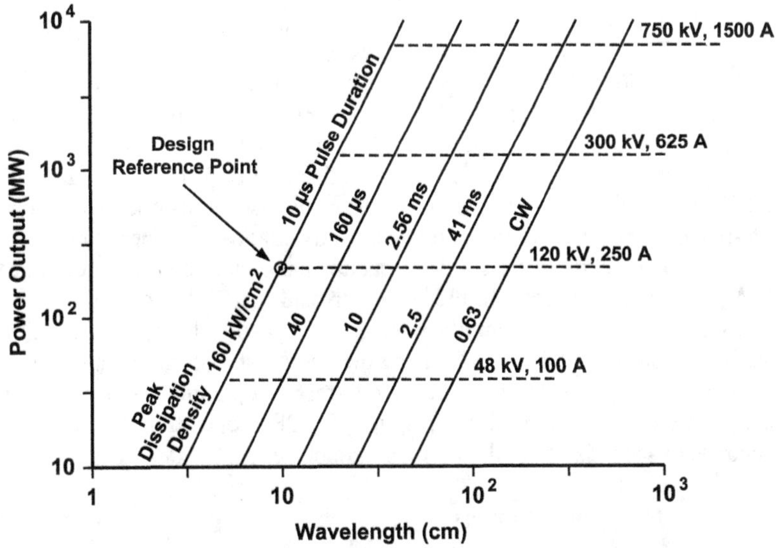

Figure 20-54 Predicted power capabilities of pulsed CFAs based on anode surface temperature limitations. (From: J. Skowron, Raytheon Co., Workshop on High-Power, Space-Based Microwave Systems, Los Alamos, NM, March 1985.)

20.7 CFA POWER SUPPLY CONSIDERATIONS [8]

The decision concerning whether to use a CFA for a particular application or whether it's a forward wave device, backward wave device, pulsed, or dc operated is based on many factors. Some considerations are the complexity of the power supply, the RF power capability, bandwidth, impedance, cost, weight, and size.

20.7.1 DC-Operated Supplies

Concerning the power supply complexity, the dc-operated CFA would appear to have a significant advantage over cathode pulsed CFAs. As was described previously, the dc-operated CFA can be operated directly from a dc high-voltage power supply (HVPS) without a high-power modulator. Application of RF drive causes the tube to start drawing current. The tube is then turned off by removal of the RF drive concurrently with application of a short "quench" pulse to the control electrode to collect the electrons and interrupt the secondary-emission process. In spite of its apparent simplicity, the dc-operated CFA has not found wide usage. Some reasons for this are the following.

First of all, dc-operated backward wave CFAs are not usable with frequency agility, because the steep voltage versus frequency characteristic results in extremely large power-output variations as frequency is changed when these tubes are tied directly to a dc supply. For typical backward wave CFA characteristics, the power output would vary 100% for a 10% change in frequency. A suitable regulator tube that would be capable of overcoming this variation on a pulse-to-pulse basis could also serve as a cathode pulser.

Second, the low dynamic impedance of a CFA sets a limit of only ~1% voltage droop on the HVPS if the RF power-output droop is to be limited to about 10%. A regulator used between the HVPS and the CFA to absorb HVPS droop (and to interrupt arcs) would have to be rated for the full peak current and voltage of the CFA. Thus, the regulator would be just as big as a hard-tube modulator for a cathode-pulsed CFA and would eliminate any advantage of dc operation.

A regulator preceding the HVPS capacitor bank can be considered, but then the CFA is connected directly to the capacitor bank. A CFA arc will draw a very large current that must be promptly interrupted to avoid damage to the CFA. A crowbar can protect the CFA by diverting the energy stored in the capacitor bank, but the power supply must then be recycled to restore operation. As a result, an arc in dc operation may result in interruption of CFA operation for a period of milliseconds to seconds rather than for a single pulse.

20.7.2 Cathode Pulsing Supplies

If cathode-pulsing is used, it doesn't matter very much to the modulator designer whether the voltage of the CFA varies over the band or not. As will be discussed shortly, this conclusion is reached because the types of modulators that would normally be used with cathode-pulsed CFAs can readily absorb the voltage variations of the backward wave CFA, and they still hold the operating power of the CFA adequately constant. The basis for choice between forward wave and backward wave CFAs for cathode pulsing would thus be other characteristics such as availability, efficiency, cost, and life.

For cathode-pulsed CFAs, several types of high-power modulators may be used. For one or a few fixed pulse widths, low duty cycle, and moderate PRF, a line-type modulator is suitable and has been widely used. Regulation of the CFA operating power, as well as attenuation of HVPS ripple, can be accomplished in several ways. Line-type modulators readily tolerate CFA arcing.

Among the hard-tube modulator alternatives, constant-current operation is attractive because the modulator acts as a regulator and because it limits the effects of CFA arcs, as in the line-type modulator. Besides attenuating power-supply ripple, the regulating effect of a constant-current, hard-tube modulator tends to hold CFA current constant in spite of HVPS voltage droop during the pulse. This makes it possible for a smaller capacitor bank to be used. Therefore, either a line-type modulator or a constant-current, hard-tube modulator is generally preferred for cathode-pulsed CFAs. As noted earlier, both of these types of modulators are able to tolerate pulse-to-pulse CFA voltage variations. The constant-current, hard-tube modulator maintains constant CFA current, while the line-type modulator (being a matched source) maintains constant power to the CFA as its voltage varies.

REFERENCES

[1] G. K. Farney, "Crossed-field microwave devices," Technology Service Corporation/ Crane Division Naval Surface Warfare Center, 1995.

[2] R. R. Warnecke et al., "Magnetron-type traveling-wave amplifier tube," *Proc. IRE*, Vol. 38, No. 5, May 1950, pp. 486-495

[3] J. F. Skowron, "The continuous cathode (emitting sole) crossed-field amplifier," *Proc. IEEE*, Vol. 61, No. 3, March 1973, pp. 330–336.

[4] W. C. Brown, "Description and operating characteristics of the platinotron-a new microwave tube device," *Proc. IRE*, Vol. 45, 1957, pp. 1209–1222.

[5] NAVSEA Crane, "AEGIS microwave tubes," NAVSURFWARCENDIV.

[6] Litton, "Operation of the crossed-field amplifier (CFA)," Crossed-Field Amplifier Dept. 589, Litton Electron Tube Division, Williamsport, PA.

[7] Raytheon Company, "Ultra high-power amplitron," Vol. 1, CW Amplitron Development, Rome Air Development Center, Griffiss Air Force Base, NY, Technical Report No. RADC-TDR 64-389, May 1965.

[8] T. A. Weil, "Comparison of CFAs for pulsed-radar transmitters," *Microwave Jour.*, No. 3, June 1973.

Chapter 21

Gyrotrons

21.1 INTRODUCTION

Russian scientists at IAP Nizhny Novgorod, R.A.S., operated the first gyrotron in September 1964. The name *gyrotron* was originally used by the Russians for a single-cavity oscillator, now often referred to as a gyro-monotron. The name now refers to a class of devices including both oscillators and amplifiers. As shown in Figure 21-1, there is a one-to-one correspondence between linear-beam (O-type) tubes (klystrons, TWTs, and so forth) and gyro-devices [1].

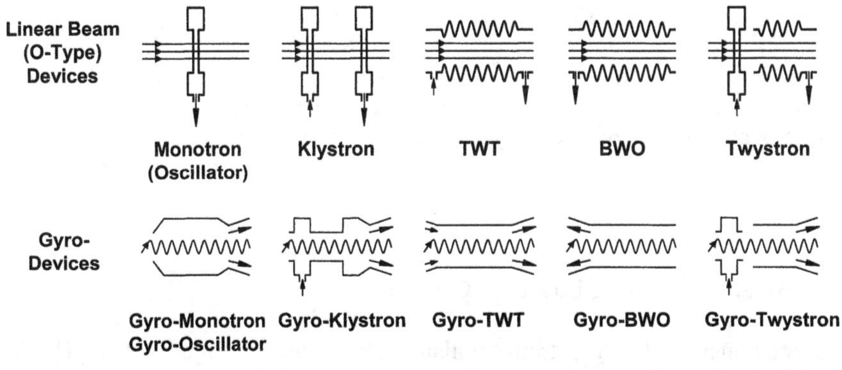

Figure 21-1 Linear-beam devices and corresponding gyro-devices. (From: V. A. Flyagin, et al., *IEEE Trans. Microwave Theory and Techniques,* Vol. 25, No. 6, 1977. © 1977 IEEE.)

Gyrotrons fill an important gap in the spectrum of electromagnetic sources. As is indicated in Figure 21-2, the possible operating capabilities of gyrotrons far exceed those of conventional microwave tubes in both frequency and power [2]. The upper limit, indicated by the dashed line, is estimated with the relation that is valid for conventional microwave tubes, that is that the product of power (MW) times frequency squared (GHz^2) is a constant. The value of 5×10^4 shown in the figure is based on the development of a 170-GHz, 2-MW CW gyrotron now in process for the International Thermonuclear Experimental Reactor (ITER) [3].

Either microwave tubes or gyrotrons could operate in the shaded region between these devices. As gyrotrons develop and the needs arise, it is becoming more likely that gyrotrons will be the devices of choice for operation in this region.

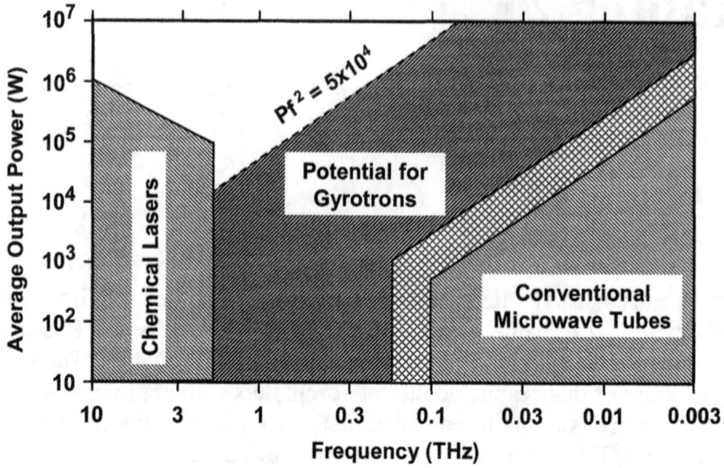

Figure 21-2 The operating region for gyrotrons. (Adapted from: H. Jory et al., Gyrotron introduction for ECRIS 2008.)

Gyrotrons and conventional microwave tubes are vacuum electron devices in which and electron beam interacts with an electromagnetic field. To explain why gyrotrons are capable of operating at power levels and frequencies far above the capabilities of microwave tubes, it is necessary to first explain how a gyrotron works.

21.2 BASIC INTERACTION MECHANISM

The components of a gyrotron oscillator are shown in Figure 21-3 [4]. The electron gun generates a hollow electron beam, which is injected into a very strong magnetic field. Flux densities of several Tesla are normally required and this usually necessitates the use of superconducting magnets. The beam is then passed through a cavity where an interaction with an electromagnetic field occurs and RF power is generated. The beam and the RF power leave the cavity and travel through a cone-shaped expanding waveguide. The beam is collected on the walls of the waveguide and the RF propagates out of the gyrotron through the output waveguide and window.

The electron gun and beam-focusing fields are designed so that the rotational velocities of the electrons are normally 1.5 to 2 times the axial velocity. As a result, most of the electron energy is rotational. Because the magnetic field is very large, the rate of rotation of the electrons (the cyclotron frequency) is very high and the orbit diameter for the electrons is small. Because the orbit diameter is small, as is indicated in Figure 21-4, the thickness of the hollow beam may be several times the diameter of the electron orbit. In effect, the hollow beam contains a large number of small beams, often referred to as beamlets [5].

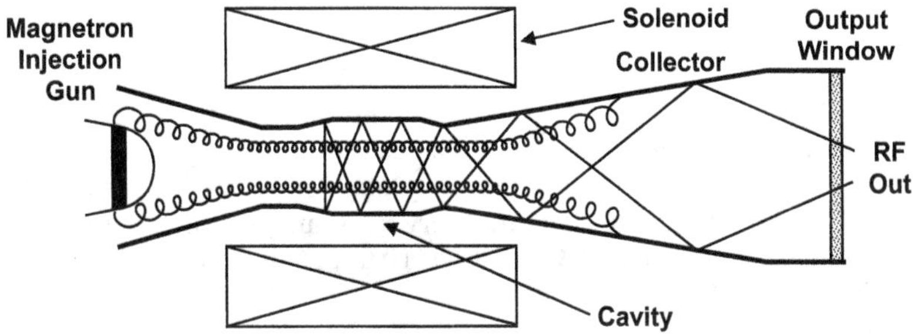

Figure 21-3 Gyrotron oscillator. (Adapted from: A. Grudiev, Hamburg University of Technology, 2010.)

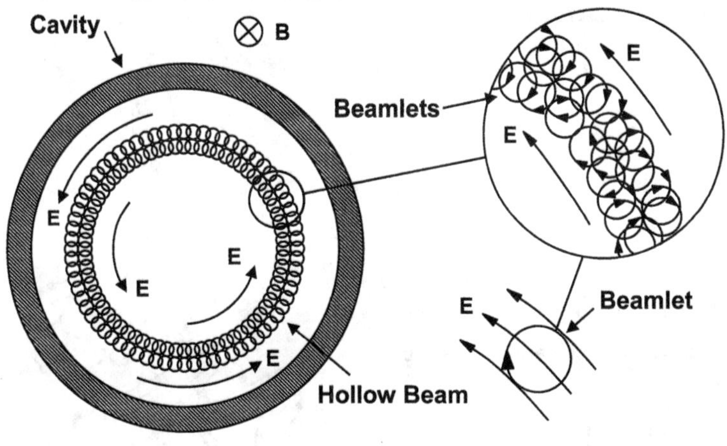

Figure 21-4 Hollow electron beam in a gyrotron cavity. (Adapted from: G. Dohler, *Int. J. Electronics,* 1983.)

As is the case in any oscillator, at the initiation of oscillations, an electromagnetic wave originates from the broadband noise that the electron beam generates. In that noise spectrum, there is a frequency component at the resonant frequency of the cavity. In the cavity, the rotating electrons interact with the circular electric field of the electromagnetic wave. The way in which electrons interact with the wave is at the heart of the operation of all gyro-devices.

The basic operating mechanism may be explained by examining the interaction of the electrons in a single beamlet like that in Figure 21-5 with the electric field. Assume that the electrons in the beamlet are uniformly distributed along a single helical path prior to interaction. Also, assume for this example that the electrons rotate in the counterclockwise direction. The angular rotational frequency of the electrons is the cyclotron frequency, ω_c, where B is the applied dc magnetic field, e is the electron charge, and m is the electron mass.

$$\omega_c = \frac{e}{m} B \tag{21-1}$$

Some values of the frequency of rotation, $f_c = \omega_c / 2\pi$, are listed in Table 21.1 along with the corresponding magnetic flux density. Since the frequency with which a gyrotron operates is the cyclotron frequency (or, sometimes, a multiple of the cyclotron frequency) it is evident that very large magnetic fields are required for their operation. Note that a flux density of 10 tesla (100,000 gauss) is required for operation at 280 GHz.

Table 21.1
Magnetic Flux Density and Corresponding Cyclotron Frequency

Flux Density (T)	f_c (GHz)
0.1	2.8
0.3	8.4
1	28
3	84
10	280

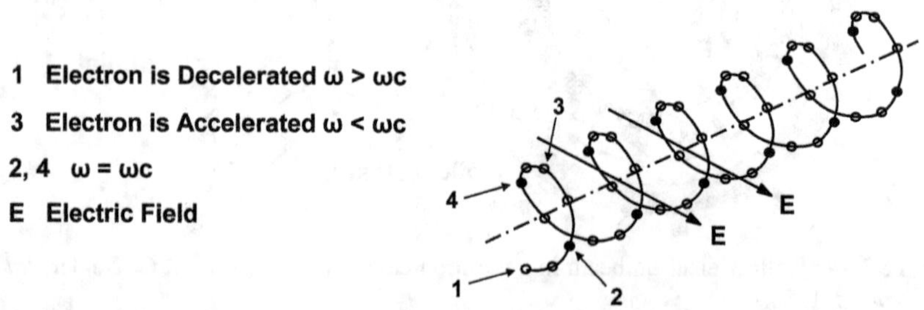

Figure 21-5 Affect of electric field on electron motion.

In a gyrotron, the speed of the electrons is a large enough fraction of the speed of light so that their mass, m, is greater than their mass at rest, m_o. The dependence of m/m_o on electron energy is shown in Figure 21-6. Now, electrons in position number 1 along the helical beam are moving in the direction of the electric field, so they are being decelerated and their mass is being decreased. As a result, their frequency of rotation is being increased. Electrons in positions 2 and 4 are moving perpendicular to the field so their mass is not being affected. Electrons in position 3 are being accelerated and so their frequency of rotation is decreased. Because the rate of rotation is decreased for some electrons and is increased for others, orbital bunching occurs as is indicated in Figure 21-7 if the electrons are permitted to drift [6].

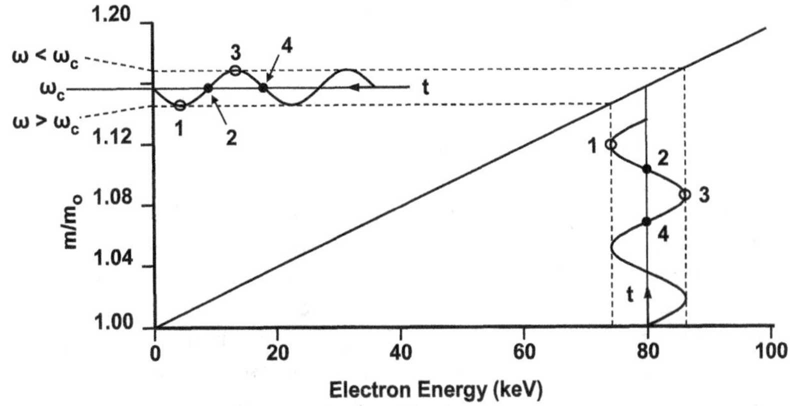

Figure 21-6 Variation of electron mass and frequency of rotation with electron energy.

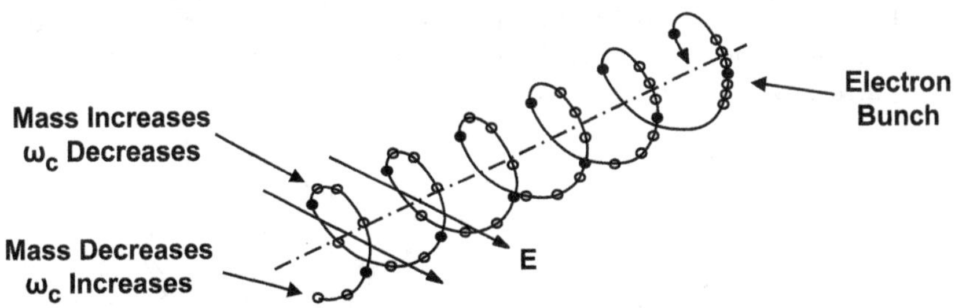

Figure 21-7 Orbital bunching of electrons. (Adapted from: *Advances in Electronics and Electron Physics,* Vol. 55, by R. S. Symons and H. R. Jory, copyright 1981 by Academic Press.)

If the cyclotron frequency is somewhat lower than the frequency of the RF field, then the position of the bunches along the helical orbit is delayed with respect to the phase of the applied field as is indicated in Figure 21-8. At this point, it should be noted that the bunched electrons are all moving in the direction of the electric field, and so they all give up energy to the field. As the electron bunches rotate in near synchronism with the alternating RF field, they continue to give up energy on each half-cycle of rotation.

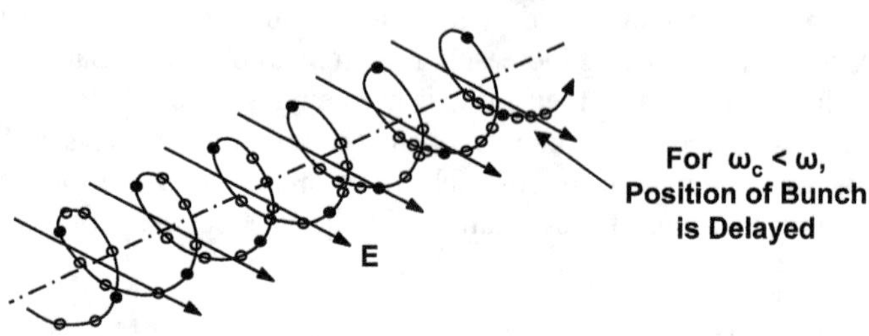

Figure 21-8 Orbital bunching if the cyclotron frequency is lower than the frequency of the RF field. (Adapted from: *Advances in Electronics and Electron Physics,* Vol. 55, by R. S. Symons and H. R. Jory, copyright 1981 by Academic Press.)

The interaction that has just been described for the electrons in a single beamlet in a gyrotron also takes place in the other beamlets. Thus, beamlets become synchronized and the electron distribution becomes as indicated in Figure 21-9. As the direction of the electric field alternates, the direction of motion of the electrons also alternates, so electrons throughout the hollow beam are being decelerated by the electric field. As a result, the electrons give up energy on each half-cycle of operation, and thereby produce amplification of the field.

We can now understand why a gyrotron can operate at power levels and frequencies far above the capabilities of conventional microwave tubes. Figure 21-10 shows a comparison of the sizes of a gyrotron cavity and a klystron cavity for the same frequency of operation. For operation with higher-order modes than the TE_{01} mode, gyrotron cavities are even larger than the TE_{01} cavity.

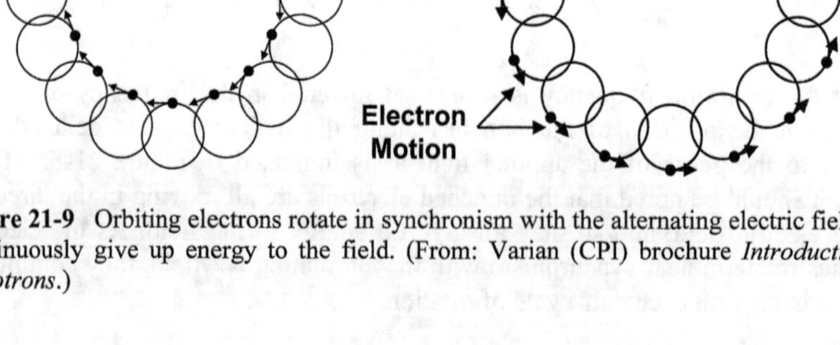

Figure 21-9 Orbiting electrons rotate in synchronism with the alternating electric field and continuously give up energy to the field. (From: Varian (CPI) brochure *Introduction to Gyrotrons*.)

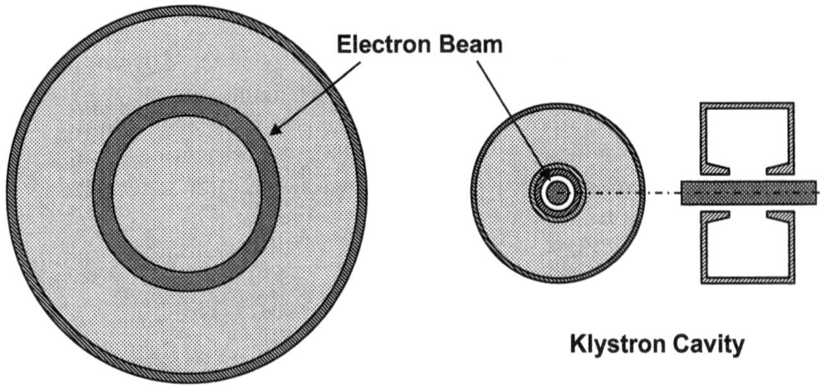

Figure 21-10 Comparison of a gyrotron cavity and a klystron cavity for the same frequency of operation.

The following is a comparison of some of the characteristics of gyrotrons and conventional microwave tubes.

1. In conventional tubes, the RF fields with which the electrons must interact are most intense near the circuit surrounding the beam, so the beam must travel as close to the circuit as possible. This leads to beam interception and heating of the circuit.
2. In conventional tubes, circuit heating limits power capability.
3. In conventional tubes, circuit dimensions are on the order of the wavelength of the operating frequency and scale with wavelength so power is severely limited at high frequencies.
4. In conventional tubes, the RF circuit must slow the phase velocity of the signal approximately to the beam velocity. These tubes are referred to as slow wave tubes.
5. In gyrotrons, the field with which the beam must interact is most intense some distance from the cavity surface. Beam interception problems are far less severe than in linear-beam tubes.
6. Because of the mode structure of the electric field in gyrotrons, the cavities in which interaction takes place can be very large (many wavelengths in diameter).
7. The large dimensions and absence of beam interception in gyrotrons make operation possible at extremely high powers and frequencies.
8. The beam-wave interaction is transverse rather than axial as in conventional tubes so there is no requirement for the axial beam velocity to be near the phase velocity of the circuit. In fact, the phase velocity is

greater than the speed of light and so gyrotrons are referred to as fast wave devices.

9. Finally, and perhaps most importantly, the operation of gyrotrons is even more critically dependent than in linear-beam tubes on the characteristics of the cathode. Gyrotron cathodes operate temperature limited and so they don't have the benefit of space charge limiting of emission, which virtually eliminates the effects of variations in cathode temperature and work function in linear-beam tubes.

21.3 MIG CONFIGURATIONS AND REQUIREMENTS

In Section 21.2 the basic gyrotron interaction process was described. To make this process work properly, an annular electron beam with very specific characteristics is required. To generate this beam, most gyro-devices use an electron gun that is referred to as a magnetron injection gun (MIG). The reference to magnetron arises because some versions of these guns have a cathode, which is similar to the cathode in a magnetron. As shown in Figure 21-11, the cathode operates in a magnetic field that is nearly axial and an electric field that is nearly radial so that electrons follow cycloidal orbits near the cathode as they do in a magnetron. However, unlike a magnetron, the electric field in a MIG has an axial component so that electrons move axially away from the cathode and form an annular beam.

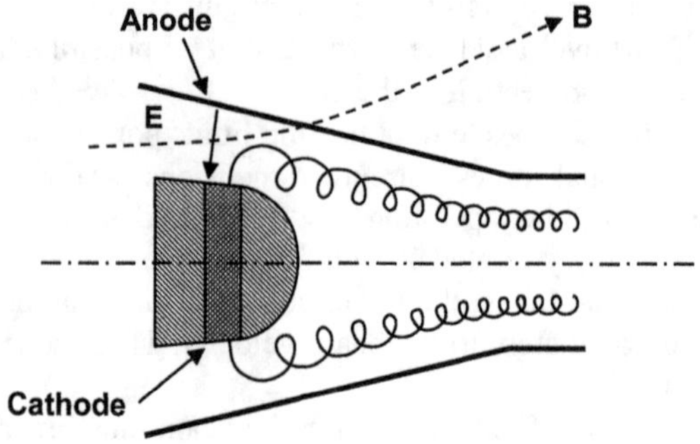

Figure 21-11 Electric and magnetic fields near the cathode in a MIG.

21.3.1 MIG Configurations

The type of interaction cavity and the beam required for the desired mode of interaction in that cavity determine, to a large extent, the configuration of the MIG. A MIG can have the conventional design used in most gyrotrons in which the interaction cavity has no center conductor (referred to as a hollow cavity) or it can be a coaxial MIG for use in a gyrotron with a coaxial interaction cavity.

Examples of cathodes for conventional MIG designs are shown in Figure 21-12 [7]. Depending on the slant angle of the cathode surface, the electron flow can be nonlaminar as shown in Figure 21-12(a, b) where electron trajectories cross) "laminar" as in Figure 21-12(c). (Electron flow is never truly laminar because of the emitter characteristics such as surface roughness, work function variations, and thermal emission velocities.)

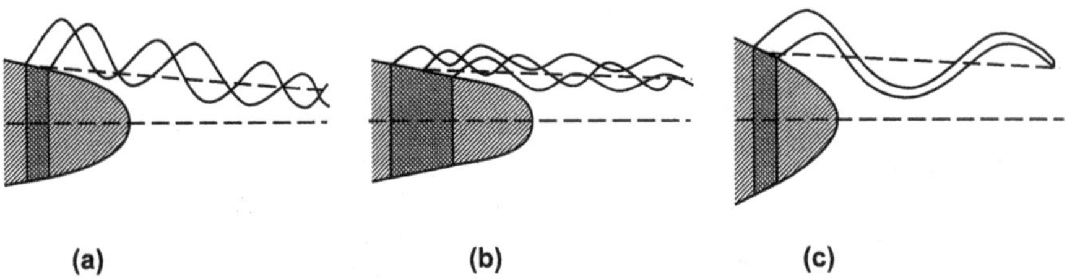

Figure 21-12 (a–c) Cathodes for conventional MIGs. (Adapted from: *Introduction to the Physics of Gyrotrons*, by G. Nusinovich, © 2004 The Johns Hopkins University Press.)

The gun in Figure 21-13 has a cathode with a very steep slant angle (50°) [8]. This angle was critical in minimizing the transverse velocity spread of the electrons. Also, the beam is relatively insensitive to electrode misalignments and field errors. A transverse relative velocity spread of 1.6% and a transverse-to-axial velocity ratio of 1.5 was obtained in simulations for a 6-A, 65-kV beam.

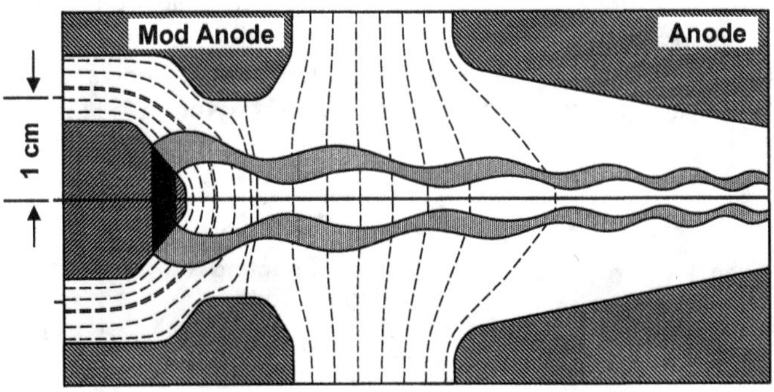

Figure 21-13 MIG design for a 94-GHz gyro-klystron. (Adapted from: Nguyen et al., *IEEE Trans. Plasma Science*, 1998. © 1998 IEEE.)

The MIG in Figure 21-13 was designed for a 94-GHz, 80-kW peak power, 10-kW average power gyro-klystron, the cavities of which operated in the TE_{01} mode. This MIG design employs an optimized double-anode geometry. The modulating anode is used to ease modulator design.

Two coaxial MIGS are shown in Figure 21-14 [9, 10]. For the gun on the left, emission current flows radially outward from the cathode. The gun on the right is referred to as an inverted coaxial MIG (ICMIG) and emission current flows radially inward from the cathode.

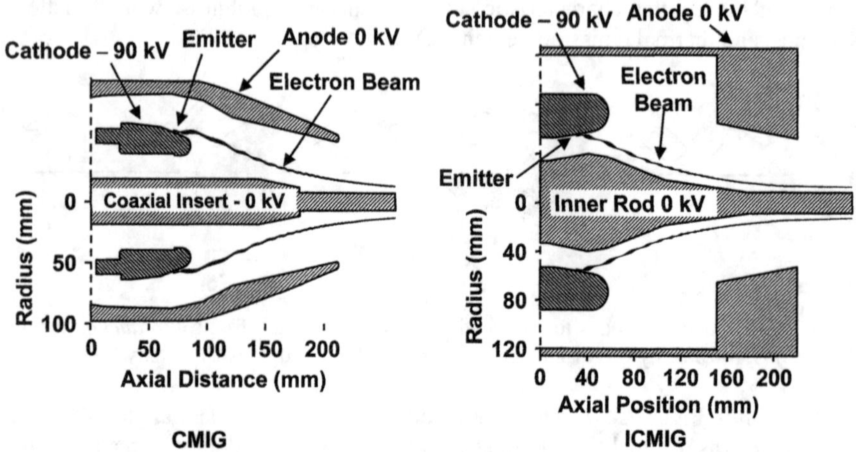

Figure 21-14 MIGs for use with coaxial gyrotrons. (Adapted from: B. Piosczyk, *IEEE Trans. Electron Devices*, 2001, and Dumbrajs, *IEEE Trans. Plasma Science*, 2004. © 2001 and 2004 IEEE.)

The magnetic field in a gyrotron focuses the electron beam and also establishes the frequency of operation. The variation in the field as the beam travels from the MIG to and through the interaction region is enormous. This variation is indicated in Figure 21-15 for a 140-GHz gyrotron [11].

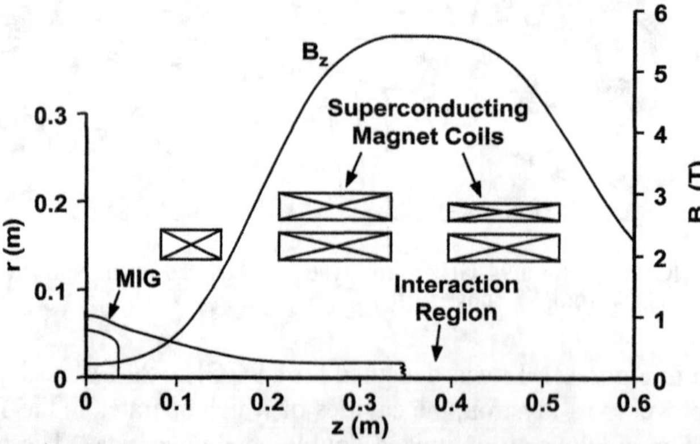

Figure 21-15 Magnetic flux density as a function of axial position for a 140-GHz gyrotron. (Adapted from: *Gyrotrons*, by M. V. Kartikeyan et al., © Springer-Verlag 2004.)

The rotation frequency of the electrons varies over an extremely wide range as the beam is compressed and travels from the MIG to the interaction region. Because of this, the beam can couple to and excite resonances in the MIG and in the tunnel (compression zone) leading to the interaction region. To suppress these resonant interactions, absorbing material is placed on the walls of the tunnel to the interaction region as is indicated in Figure 21-16.

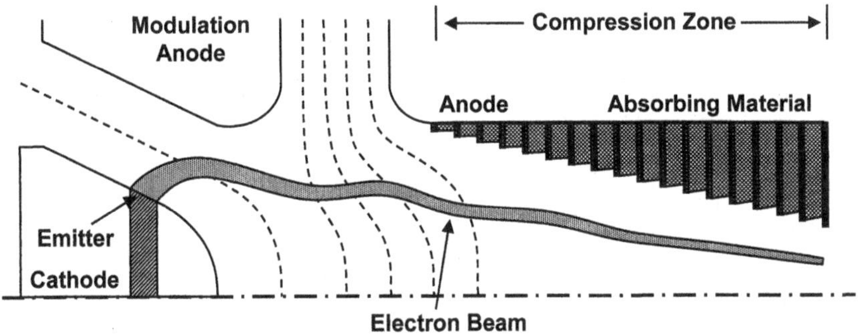

Figure 21-16 Placement of absorbing material for suppressing resonances. (Adapted from: *Gyrotrons*, by M. V. Kartikeyan et al., © Springer-Verlag 2004.)

21.3.2 First-Order Design Procedure

As is the case with linear-beam tubes, the design of the electron gun depends on numerical simulation to determine beam performance and ensure proper interface with the RF circuit. Also, it is very important to have a good starting design before proceeding with the simulation. This is the value of the Vaughan synthesis procedure for Pierce guns (Chapter 6).

Baird [12, 13] and Lawson [13] provided seven equations for obtaining a first-order design for the MIG configuration shown in Figure 21-17. The following material is from their analysis.

There are five MIG design parameters, which are required by the RF interaction region[1]:

---

[1] By comparison, there are three Pierce-gun design parameters, which are set by the RF interaction region; beam voltage, beam current or perveance, and beam radius. These are comparable to items 1, 2, and 4. The magnetic field is established by these three quantities and is used only for focusing the beam. In the gyrotron, the magnetic field guides the beam and determines the frequency of operation so it must be specified in the gun design. The only place where the transverse velocities are of considerable importance in linear-beam tube design is in confined flow PPM-focused beams. There, a considerable amount of energy can be expended in transverse velocities, which do not contribute to RF performance and so tube efficiency suffers. In the gyrotron, the efficiency of the interaction depends on the fraction of beam power that is in the transverse motion of the electrons.

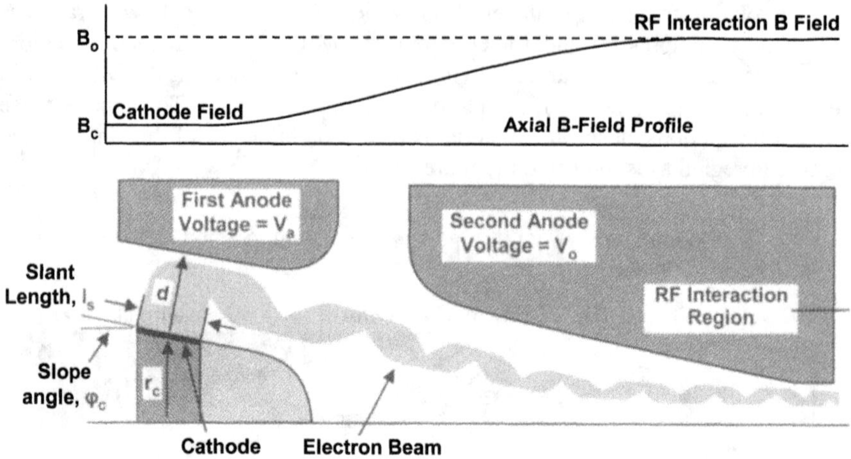

Figure 21-17 MIG configuration used by Baird and Lawson in obtaining a first-order design [13].

1. Beam power;
2. Electron energy;
3. Required cyclotron frequency;
4. Average radius of the electron guiding centers;
5. Transverse-to-longitudinal velocity ratio.

These parameters are usually well known and relatively fixed in value. They determine the RF interaction region values of the voltage and current (V_o and I_o), the transverse and longitudinal velocity components $v_{\perp o}$ and v_{zo}, the magnetic field B_o, and the position and size of the orbits (r_{go} = mean guiding center radius and r_{Lo} = Larmor radius). The transverse velocity and the radii are shown in Figure 21-18, which is a cross section of the electron beam downstream from the MIG.

In addition to the design parameters required by the RF interaction region, there are additional variables in the MIG design that can be used to form a satisfactory beam. Baird and Lawson selected the following four variables: [2]

1. r_c, the average cathode radius;
2. J_c, the cathode current density;
3. φ_c, the slope angle of the cathode;
4. D_F, the spacing factor between the cathode and first anode of the MIG.

---

[2] In the Pierce gun design, only two additional design parameters, r_c and J_c, are required.

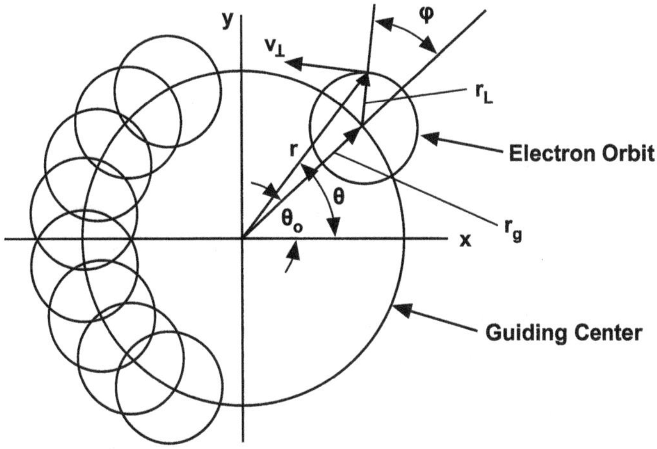

Figure 21-18 Model used for the cross section of the beam in the interaction region [13].

The most important trade-off variable is the cathode radius, r_c. Baird and Lawson noted that it is usually very helpful in arriving at a suitable cathode radius to plot all the calculated variables together on a single plot versus r_c. Identification of the acceptable ranges of the calculated variables then makes it possible to quickly identify the acceptable design range for r_c. Also, in some cases it may be found that there is no acceptable range for r_c due to a high level of space charge in the gun.

21.3.2.1 Magnetic Field Compression Ratio

The starting point in the analysis is the conservation of angular momentum which gives the magnetic field compression ratio.[3] This is the reduction in the beam cross-sectional area that results from the increase in magnetic field as the beam moves from the gun to the interaction region. Conservation of angular-momentum for electron trajectories in cylindrically symmetric dc fields E(r, z) and B(r, z), can be written (assuming the near axis approximation $B_z = B_z(z)$ only) as

$$\left(\gamma m_o r^2 \dot{\theta} - eB_z r^2 / 2\right) = \text{constant} \tag{21-2}$$

where γ is the relativistic mass factor, $\dot{\theta}$ the angular velocity, and r is the instantaneous radius of cycling electron orbits.

[3] In linear-beam tubes, Busch's theorem, which is an adaptation of the conservation of momentum, is used, primarily in the electron gun, to find the rate of beam rotation as electrons cross magnetic field lines and enter the main focusing field.

The constant is readily found by evaluating the equation at the cathode where $\dot{\theta} = 0$. Thus,

$$\text{constant} = eB_{zc}r_c^2 / 2 \tag{21-3}$$

where the subscript c indicates that the variable is evaluated at the cathode.

When the angular momentum is interpreted in the beam drift region (see Figure 21-16) where the flux density is B_o, the frequency of rotation, $\dot{\theta}$, is the Larmor frequency, $(e/2\gamma m_o)B_o$, and the radius of the electron orbit is the Larmor radius r_{Lo}. In the beam drift region, the second term in the conservation of momentum relation accounts for the rotation of the electron orbit about the gun axis so the value of that quantity is $(e/2)B_o r_{go}^2$.

Now, the momentum equation reduces to

$$B_{zc}r_c^2 = B_o\left(r_{go}^2 - r_{Lo}^2\right) \tag{21-4}$$

By normalizing the distance to r_{Lo} and the magnetic field to B_{zc}, this equation yields Baird and Lawson's equation for the magnetic field compression ratio, F_m,

$$F_m = \frac{R_c^2}{R_g^2 - 1} \tag{21-5}$$

where R_c is the normalized cathode radius.

21.3.2.2 Slant Length of Cathode

The emitting area of the cathode in the MIG must be large enough to supply the specified beam current, I_o, at the selected cathode current density, J_o. This leads to the second Baird and Lawson equation for determining the ratio L_s/R_c where $L_s = l_s/r_{Lo}$ is the normalized slant length of the cathode. This comes from area calculations of a conical segment of the cathode.

21.3.2.3 Spread in Guiding Centers

By combining the slant length with the magnetic field compression ratio, the third Baird and Lawson design equation for the spread δR_g in the guiding centers in the interaction region of the beam can be determined. This is an important design quantity, which must be controlled in a MIG because a large spread in the guiding centers is usually associated with a decrease in the RF interaction efficiency.

21.3.2.4 Distance from Cathode to Anode

The fourth Baird and Lawson equation relates the slant distance, d, from the cathode to the anode (see Figure 21-17) to the Larmor radius at the cathode. This equation says that the increase in distance from the cathode to the anode along the path d is proportional to r_{Lc}. When $d < 2r_{Lc}$, beam interception on the first anode occurs. As d is made larger than $2r_{LC}$, the clearance between the beam and the first anode increases and a higher voltage on the first anode is required.

21.3.2.5 Anode Voltage

The voltage, V_a, which is required on the first anode to produce the specified transverse beam velocity, is estimated from the fifth Baird and Lawson relation. This equation is derived using conservation of angular momentum from the cathode surface to the first peak in the Larmor orbit above the cathode. A relativistic Hull cutoff equation is found, and this establishes the potential at the first beam peak (peak of the first Larmor orbit). Then, using modified coaxial line equations, the voltage on the anode is determined.

21.3.2.6 Electric Field at Cathode and Lower Bound on Cathode Radius

The final two design equations provide the most important information required to assess the viability of any given design choices. The sixth relation gives the electric field required at the cathode to achieve the desired beam performance. If this field is above the acceptable design limit of about 100 kV/cm, the MIG is in danger of arcing during operation. (The actual field levels resulting from field enhancements should be limited to 100 kV/cm. The field on the nose of the cathode electrode tends to be the highest stress point.) If the electric field for a given design is too high, the primary remedy is to increase r_c. Thus the sixth relation sets a *lower bound* on acceptable values of the cathode radius.

21.3.2.7 Upper Bound on Cathode Radius

Baird and Lawson's seventh equation alternatively sets an *upper bound* on r_c. This equation gives the ratio of the beam current density J_c to the space charge limiting current density, J_L. As the cathode radius is increased, the space charge limiting current decreases very rapidly and the ratio J_c/J_L increases quickly to its maximum value of unity. MIG design, however, is based on negligible space-charge effects. It is, therefore, mandatory that J_c/J_L be kept below about 15–20% to rely on these design rules. The seventh equation, therefore, imposes an upper bound on r_c, which cannot be exceeded without finding some method for including the effects of space-charge in the design process.

The MIG in Figure 21-19 is an example of the application of Baird and Lawson's analysis to the design of a 600-kW, 70-kV MIG for use at 35 GHz. Even though the gun design has two anodes, full beam voltage was applied to the first anode to create a smoothly varying E-field in the cathode-anode region to better approximate adiabatic flow conditions.

21.3.3 MIG Performance

Gyrotrons share an extremely important characteristic with other electron devices that operate in vacuum. That is, the electron beam must be, and must remain for the duration of the life of the device, nearly perfect. For optimum gyrotron performance, there are very stringent requirements on the MIG and on the electron beam that it provides, including the following.

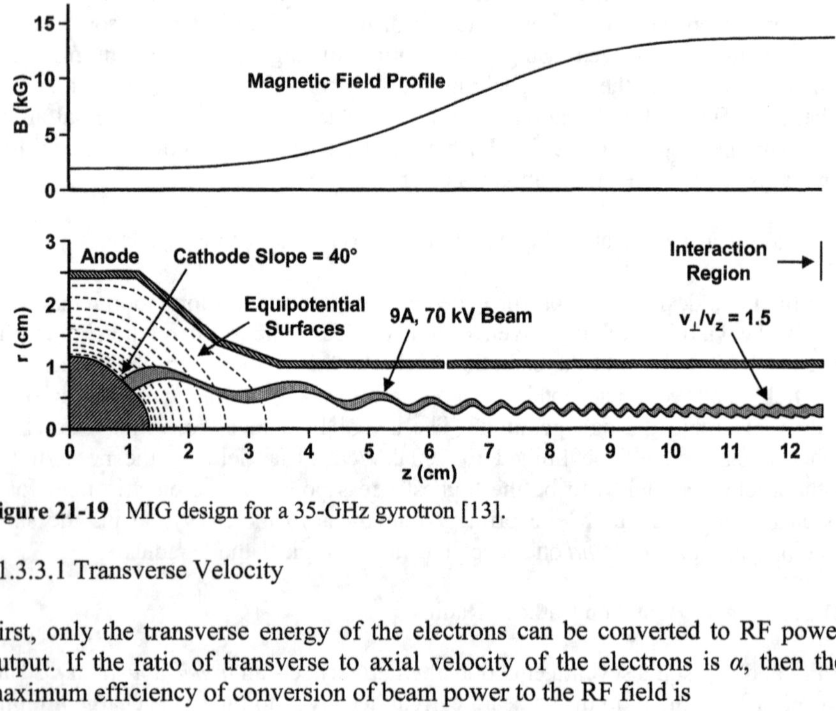

Figure 21-19 MIG design for a 35-GHz gyrotron [13].

21.3.3.1 Transverse Velocity

First, only the transverse energy of the electrons can be converted to RF power output. If the ratio of transverse to axial velocity of the electrons is α, then the maximum efficiency of conversion of beam power to the RF field is

$$\eta = \frac{\alpha^2}{1+\alpha^2} \tag{21-6}$$

This means that the transverse to axial velocity ratio should be as high as possible. Ratios between 1.5 and 2 are typical.

The transverse velocity spread must be as low as possible to help ensure phase-bunching stability. Velocity spread is a function of many factors:

1. MIG cathode slant angle. Figure 21-20 shows an example calculated for a 94-GHz gyrotron [8].
2. The presence of space charge. The cathodes in MIG guns are operated temperature limited to avoid instabilities that occur in the space charge cloud that surrounds the cathode in magnetron-type cathodes. Brown [14]

reported a reduction in noise by 60 dB or more from a magnetron when cathode temperature was reduced so that operation was temperature limited rather than space-charge limited (see Chapter 19).

3. Rack velocity noise, shot current, flicker current, and surface roughness.

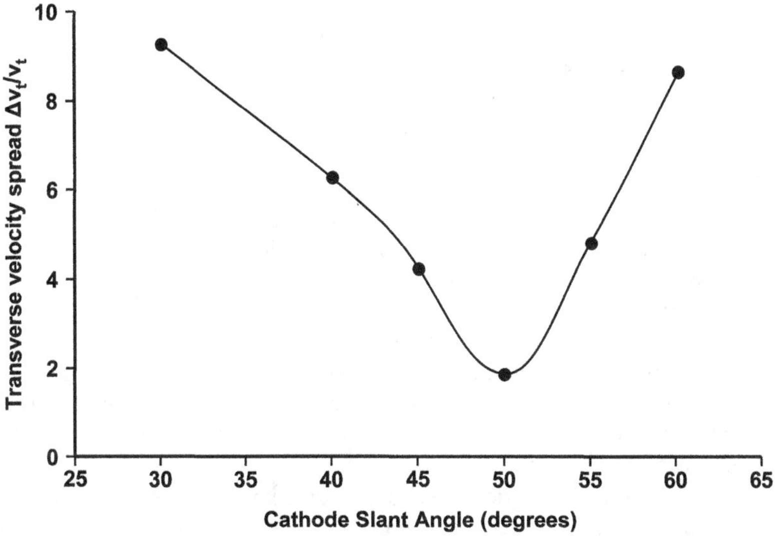

Figure 21-20 Dependence of transverse velocity spread on cathode slant angle for MIG designed for 94-GHz TE_{01} gyrotron. (From: K. T. Nguyen et al., *IEEE Trans. Plasma Science*, 1998. © 1998 IEEE.)

21.3.3.2 Emission

The cathodes in MIGs operate temperature limited (TL) with a current that is 15–20% of the space charge limited current. Thermionic emission is strongly dependent on temperature and work function. From the Richardson-Dushman equation, emission current changes by a factor of 2 when cathode temperature changes 40K or work function changes 0.06 eV.

In most cases, the first indication of an emission problem occurs after the cathode is installed and testing has begun. Then it becomes difficult to determine if the problem was caused by the cathode or something else. Problems that are thought to have resulted from nonuniform cathode temperature or work function are [15]:

1. Communications and Power Industries, Inc. reported 10:1 azimuthal variations in power density of the electron beam in the collector for a 140-GHz gyrotron [16].

2. MIT found that asymmetries in emission for a coaxial gyrotron produced a 6:1 variation in current with angular position [17]. When the electron gun was replaced with one having more uniform emission (1.2:1), the output power doubled.
3. Researchers in Germany reported problems with a 1-MW, long-pulse gyrotron at 140 GHz built by Thales Electron Devices. The prototype device failed to achieve the power specification even though a preprototype gyrotron exceeded the power requirement. The only relevant difference between the two gyrotrons was a change in the cathode. It is suspected that nonuniform emission in the prototype gyrotron caused the reduced performance [15].
4. At Saint Petersburg State Polytechnic University [18], Gennadi et al. developed a gyrotron with internal probes in the collector and anode for measuring the azimuthal current distribution in the beam. Their research demonstrated that significant degradations in efficiency occur when there are variations in the azimuthal current distribution [15].

Nonuniform heating, thermal shorts, poor thermomechanical design, or flaws or voids in heater potting can cause temperature variations. Fortunately, these problems can often be detected prior to installation by energizing the cathode heater in a vacuum bell jar.

Variations in work function are a larger problem that have been battled by the tube industry for decades. Often, it is not clear what is causing these problems. Possibilities include pore smearing, nonuniform impregnation, nonuniform porosity, braze contamination, and incomplete infiltrant removal.

Before leaving the subject of work function, we should return to the work of Miram et al. on emission spread, which was discussed in Chapter 5 [19]. Emission spread is defined at a specific TL operating point (Miram used 40% of FSCL current) and is the difference between the calculated (Richardson-Dushman) work function that produces the emitted current and the lowest measurable work function for the cathode emission characteristic. For a typical dispenser cathode, emission spread is ~0.04 eV. By comparison, as is indicated in Figure 21-21, the emission spread determined in tests on a scandate cathode was 0.02 eV. (The lowest measurable work function in a test for a scandate cathode was 1.52 eV.)

A small emission spread like that shown here indicates that, at a temperature of 700°C, the work function distribution (WFD) is very narrow. It has been reported that WFD variation is one of the main reasons for emission nonuniformity, which leads to reduced gyrotron efficiency and parasitic oscillations [20].

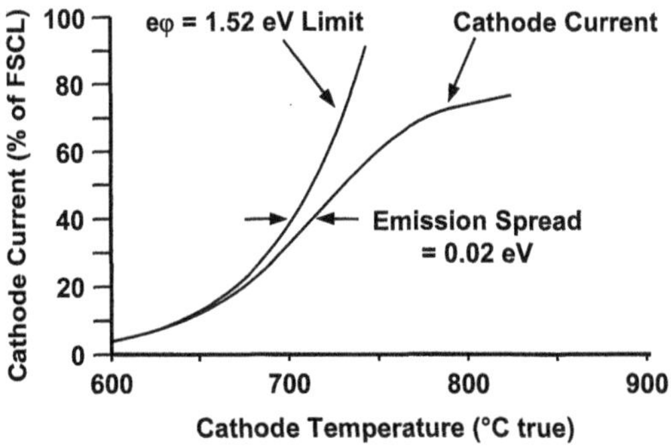

Figure 21-21 Emission spread for a scandate cathode. (From: G. V. Miram et al., *Technical Digest,* Fifth IEEE IVEC, 2004. © 2004 IEEE.)

21.4 BEAM-WAVE INTERACTION

The operating mode that we have discussed so far has been the TE_{01} mode (Figures 21-9, 21-10, and 21-19) and the cavities have been hollow cavities, that is, there is no center conductor in the cavities. In this section we discuss coaxial cavities as well as hollow cavities and modes of operation that are far more complex than the TE_{01} mode. MIGs for supplying beams to hollow and to coaxial cavities were discussed in the previous section.

21.4.1 Hollow Cavities

Most high-power gyrotrons use highly overmoded cavities. The TE_{01} cavity mode used in the discussion in the previous section has constant electric field in the azimuthal direction and one maxima in the radial direction. Overmoded cavities can have several cycles of variation in the field in the azimuthal direction and several maxima in the radial direction. Cavities may be hollow or coaxial.

Examples of radial field variations in hollow cavities are shown in Figure 21-22 [21]. Five different waveguide modes are shown. The placement of hollow electron beams for maximum interaction with these modes at the fundamental cyclotron frequency is also shown. In general, TE_{0n} modes ($n > 1$) are used to increase the power handling capacity of the device while retaining the favorable field geometry and mode control properties of these modes. The TE_{52} and TE_{91} modes are used because the fields are localized near the outside wall of the cavity. These are called "whispering-gallery" modes. With the fields localized near the wall of the cavity, it is possible to use a large radius electron beam and, at the same time, preferentially excite the desired operating mode. These modes, therefore, permit a measure of mode control in a large overmoded cavity with an enlarged (high-power) electron beam.

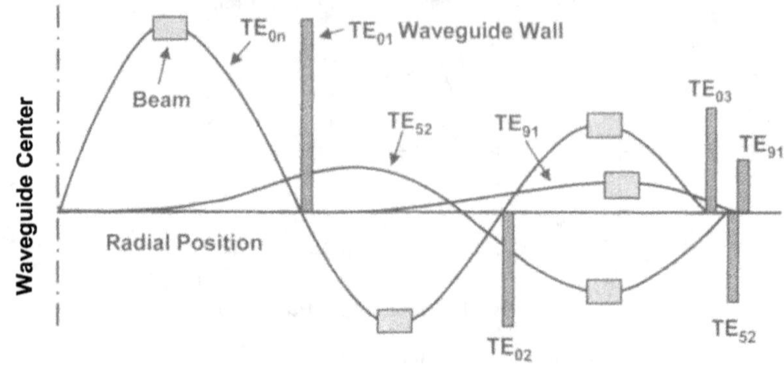

Figure 21-22 Radial field variations for several hollow cavity modes. (Adapted from: J. M. Baird and W. Lawson, *Int. J. Electron,* 1986.)

An example of the field pattern in an overmoded hollow cavity is shown in Figure 21-23 [2]. This is a plot of the magnitude of the square of the electric field as functions of the radial and azimuthal directions for the TE_{62} mode. Electron orbits and electric field lines for maximum interaction with the electrons are shown.

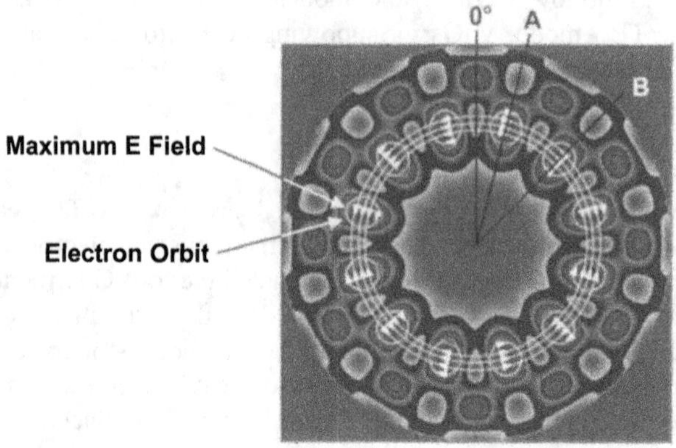

Figure 21-23 The magnitude of the square of the electric field for the TE_{62} mode is shown here with optimum placement of orbits of electrons. (Adapted from: H. Jory, et al.[2].)

It is helpful in explaining the operation of an overmoded cavity like the cavity with the TE_{62} mode in Figure 21-23, to "unroll" the field pattern and show it in a linear manner as in Figures 21-24 and 21-25. The field has six cycles of variation in the azimuthal direction and one cycle (two maxima) in the radial direction.

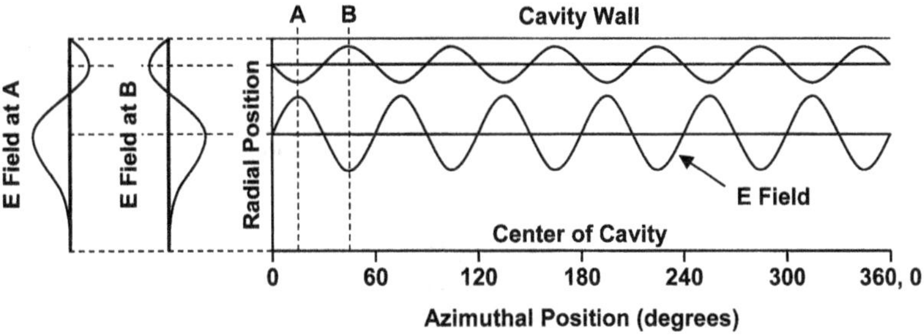

Figure 21-24 E field in Figure 21-23 as functions of radial and azimuthal positions.

As a function of time, the fields vary as shown in Figure 21-25. Electron orbits are shown to rotate in synchronism with the alternating field. If one particular phase of the wave is chosen, such as the phase of the peak field, the wave appears to move to the right (rotate in the clockwise direction in the cavity). Notice that the phase of the field peaks could have been chosen so that the wave would appear to rotate counterclockwise. The TE_{62} mode is referred to as a rotating mode and, in fact, all TE_{mn} modes with $m \neq 0$ are rotating modes.

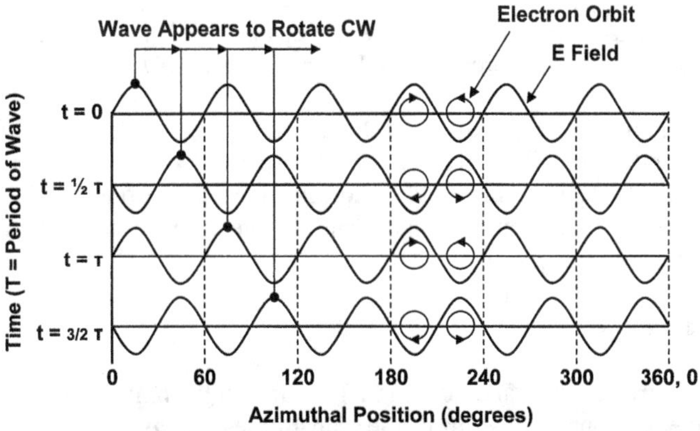

Figure 21-25 E field as a function of time. When one particular phase is selected, the field appears to rotate clockwise (or counterclockwise).

The apparent rotation of the mode in the gyrotron cavity can be compared to the "traveling wave" in a magnetron. Figure 21-26 depicts an eight cavity magnetron that has been unrolled [22]. Only the anode segments that make up the capacitive portion of the cavities are shown. As the cavity fields oscillate, the potential across the gaps alternates. For an electron spoke that is rotating about the cathode to continuously interact with the anode field, the spoke must travel at the speed of the traveling anode wave. Thus, this magnetron, like all modern magnetrons, is referred to as a traveling wave magnetron.

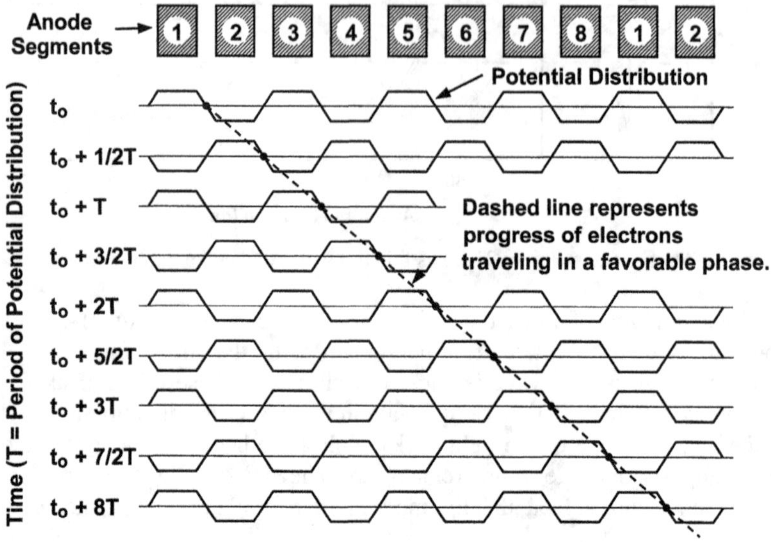

Figure 21-26 Eight-cavity traveling wave magnetron. (Adapted from: *Theory and Application of Microwaves* by Bronwell and Beam, copyright 1947 by McGraw-Hill, Inc.)

So we see that the magnetron and the TE_{mn} mode gyrotron ($m \neq 0$) are both traveling wave devices. The difference between them is that in the magnetron, the electrons must travel with the field, whereas in the gyrotron the electrons stay relatively stationary and spin in small orbits in synchronism with the alternating field as the field travels past them.

21.4.2 Coaxial Cavities

At very high frequencies and ultrahigh powers, high ohmic wall losses and/or mode competition problems limit the capabilities of gyrotrons. There is an argument to be made for gyrotrons with coaxial cavities because mode competition is reduced, so it possible to use even higher-order modes than in hollow resonators [11]. Also, the inner rod helps in limiting the potential depression caused by space charge effects and ultrafast frequency step tuning is possible by applying an appropriate voltage to the coaxial insert.

In a coaxial cavity, ohmic losses can be controlled by corrugating the center conductor as is indicated in Figures 21-27 and 21-28. If axial rectangular groves are placed in the center conductor, which are $1/4\lambda$ deep, the impedance of the surface of the center conductor at the frequency of interest appears to be infinite, so currents and losses are very low [23].

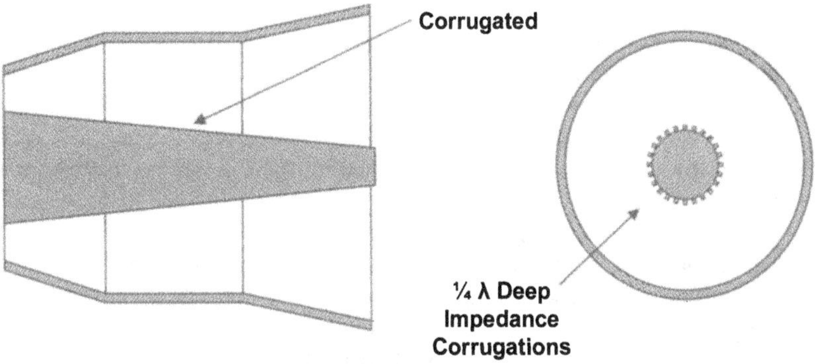

Figure 21-27 Coaxial gyrotron cavity. (From: B. Piosczyk et al., IAEA Technical Meeting on ECRH Physics and Technology for ITER, 2003.)

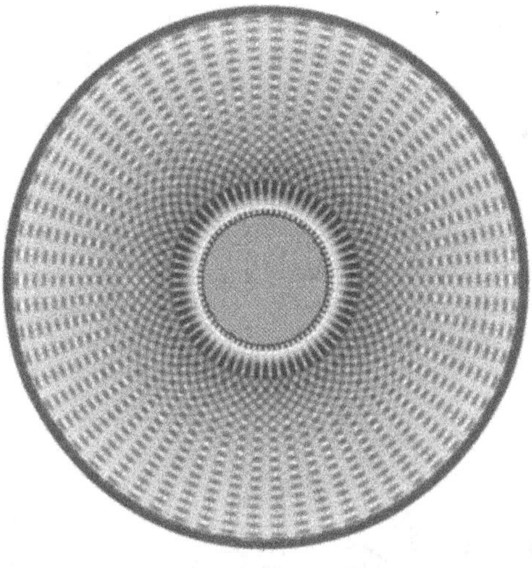

Figure 21-28 High-order mode in a coaxial cavity. (From: A. Grudiev, Hamburg University of Technology, 2010.)

In either a hollow cavity or a coaxial cavity, the tunnel expands from the cavity to the output window so RF propagates towards the window. The tunnel contracts toward the gun and this prevents propagation of the RF wave toward the gun.

Figure 21-28 shows the transverse field of the $TE_{31,17}$ mode in a high-order mode coaxial cavity [4]. The corrugations in the center conductor are 0.45 mm deep and 0.35 mm wide and the number of grooves is 72.

21.4.3 Mode Converters [24, 25]

The output mode of a high-power gyrotron is a TE_{0n} or TE_{mn} mode in a circular waveguide. For most applications, these modes are not suitable for use because of serious diffraction and polarization losses. To be of use, they must be converted into the Gaussian mode or lower-order linearly polarized mode (HE_{11}). The conventional waveguide mode converters that were used in early gyrotrons are in the form of rippled-wall or corrugated circular waveguides and serpentine structures. These devices have high efficiency but are large and have narrow bandwidth, so they are not appropriate for use in high-power gyrotrons.

The short-wavelength, high-output power and the rotating asymmetric high-order cavity modes have led to the need for quasi-optical mode conversion concepts, which were first developed in Russia [26] and are known as Vlasov converters. The Denison quasi-optical converter [27], which is a modification of the Vlasov converter is compact and can be placed between the gyrotron cavity and the output window. This is a highly efficient way to transform the complex circular cavity mode into a linearly polarized Gaussian beam. Efficient conversion is crucial for high-power gyrotrons to avoid severe cooling problems and, possibly, to avoid device destruction. An additional benefit of these quasi-optical converters is that they make it possible to separate the RF beam from the electron beam. This will be discussed in Section 21.5.1.

The basic principles of quasi-optical mode converters for circular symmetric (TE_{0n}) modes and rotating asymmetric (TE_{mn}) modes are shown in Figures 21-29 and 21-30. As in Figures 21-24 and 21-25, the field patterns are unrolled to simplify the visualization of the operation of the mode launchers that are part of the quasi-optical converters.

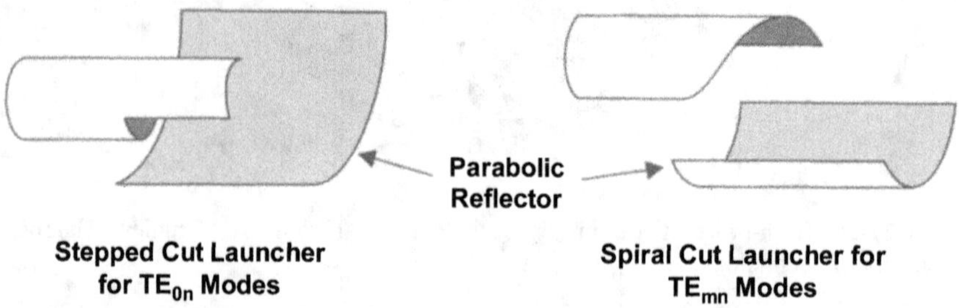

Figure 21-29 Launchers for TE_{0n} and TE_{mn} gyrotron cavity modes. (Adapted from: *Gyrotrons*, by M. V. Kartikeyan et al., © Springer-Verlag 2004.)

Because the TE_{0n} modes have no variations in the field in the azimuthal direction, all parts of a wave crest arrive at the launcher simultaneously and so the launcher has a square-cut end for launching the wave. In contrast to this, TE_{mn} modes have cyclical variations in the field in the azimuthal direction. By cutting the launcher at the appropriate angle as shown in Figure 21-30, wave fronts, all parts of which are in phase, can be launched. After being reflected by an appropriately shaped mirror (quasi-parabolic) a beam of parallel rays is formed (Figure 21-31).

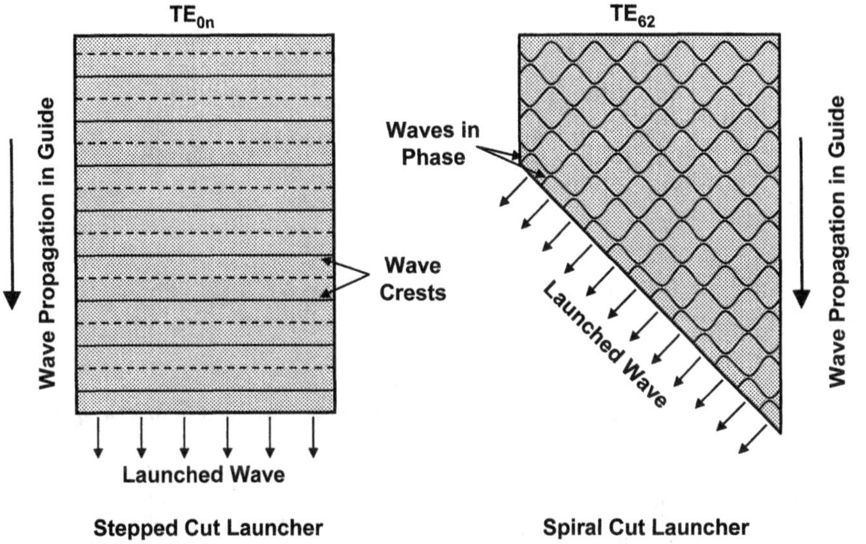

Figure 21-30 Unrolled wave patterns at the launcher for TE_{0n} and TE_{mn} gyrotron cavity modes.

A quasi-optical mode converter has been theoretically and experimentally investigated at Karlsruhe Institute of Technology (KIT). The converter was developed for the 2-MW, 170-GHz, $TE_{34,19}$ mode coaxial-cavity ITER gyrotron. The converter, shown in Figure 21-32, consists of a launcher, a quasi-elliptical mirror, and two toroidal mirrors [28].

Theoretical and experimental results show that the TEM_{00} fundamental gaussian mode content (FGMC) of the RF beam from the ITER gyrotron is about 96.7% at the window plane. To improve the FGMC further, and then to decrease the stray radiation in the tube, two adapted phase correcting mirrors have been designed. The simulation results show that with these mirrors, the FGMC can be increased to 99.1% at the output window plane and the stray radiation can be decreased by 0.6%. A CVD diamond output window that is used at high-power levels with a quasi-optical converter is described in Chapter 22.

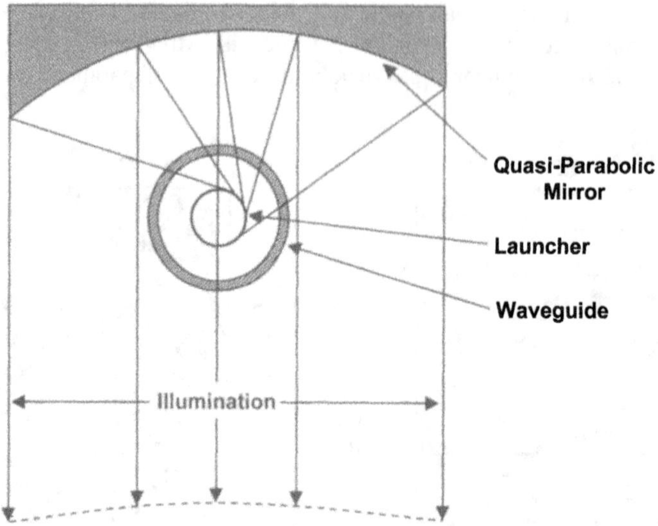

Figure 21-31 Wave propagation from a spiral cut launcher. (Adapted from: *Gyrotrons*, by M. V. Kartikeyan et al., © Springer-Verlag 2004.)

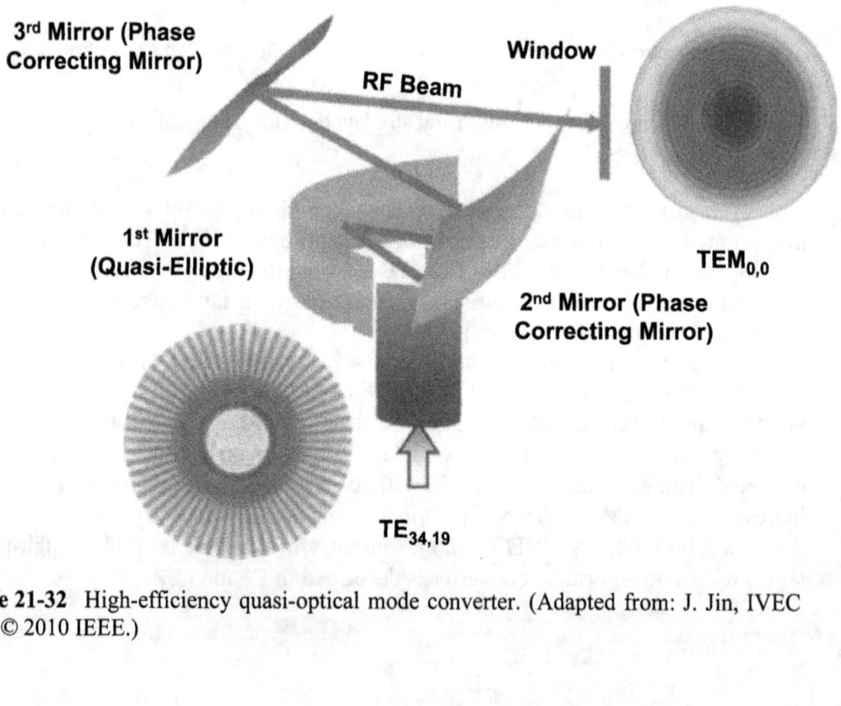

Figure 21-32 High-efficiency quasi-optical mode converter. (Adapted from: J. Jin, IVEC 2010. © 2010 IEEE.)

21.4.4 Harmonic Operation

With the proper shape for the RF field, it is possible for harmonic interactions with the electrons to occur. In the example shown in Figure 21-33, the field varies at twice the cyclotron frequency. The direction of the field reverses in the center of the electron orbit.

Thus, an electron that is initially decelerated by the field Figure 21-33(a), is moving transverse to the field when the field reverses Figure 21-33(b) and so its orbital energy is not changed. By the time the field reverses again, the electron has moved another 90° around its orbit and is again in a decelerating field. Thus, during each orbit of the electron, the RF field goes through two complete cycles. For a given operating frequency, the cyclotron frequency is half the value used in the fundamental mode of operation. As a result, the magnetic field is reduced by a factor of 2.

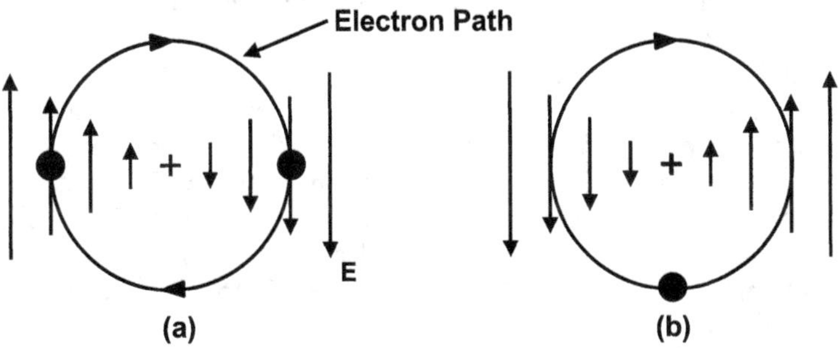

Figure 21-33 (a, b) Second harmonic interaction.

21.4.5 Collectors

Collectors, including depressed collectors, for linear-beam tubes were discussed in Chapter 16. There is usually very little magnetic field in those collectors and the shape of the beam entering the collector is controlled by space charge forces and electric field configurations so the design of the undepressed collectors is relatively straightforward. In TWTs, the interaction of the beam with the circuit is relatively weak and the energy spread of the electrons in the spent beam is about 30–40% of the initial beam energy. This makes it possible to design depressed collectors having as many as 10 stages of depression and to achieve overall efficiencies over 70%.

The author of this book is not aware of undepressed collectors for linear-beam tubes that operate at power levels above 1 MW or depressed collectors for linear-beams that operate at power levels above 100 kW.

For high-power gyrotrons, on the other hand, collectors and even depressed collectors are being designed to operate at power levels over 2-MW CW. The design of these collectors is particularly challenging because the magnetic field in the collector may be several tens of gauss and the energy range of the electrons in the spent beam can be 35–140% of the incident beam energy [11]. Still, single-stage depressed collectors have been used on high-power gyrotrons and have increased the overall efficiency to above 50% [29–31]. Although these experiments were limited to single stage collectors they demonstrated the value of depressed collectors in reducing the power loading, X-ray generation, and, possibly, of using less costly power supplies.

In a high-frequency gyrotron the magnetic flux density in the interaction region is several tens of tesla. In going from the interaction region to the collector, the flux density decreases to several tens of gauss. Electrons follow the expanding flux lines in the collector and land on a ring-shaped band on the collector surface, the width of which can be estimated. Thus, knowing the beam current and voltage, the power density imparted to the collector surface (the wall loading) can be estimated. In high-power gyrotrons, the diameter of the collector has to be excessively large to keep the wall loading within the technological limit of ~500 W/cm^2 [11]. It should be noted here that in linear-beam tube design, the wall loading limit is ~1 kW/cm^2 (Chapter 16).

If the radius of the collector is to be limited, the width of the band on which the electrons land must be made larger. Auxiliary collector coils can be used to widen the interception band. In the collector in Figure 21-34, these coils are located in the insert in the center of the collector [32]. The second way to reduce loading is to use sweeping coils to move the electron beam axially in the collector. In one example [33] the sweep frequency was 10–15 Hz and the magnetic induction was 100–500 gauss.

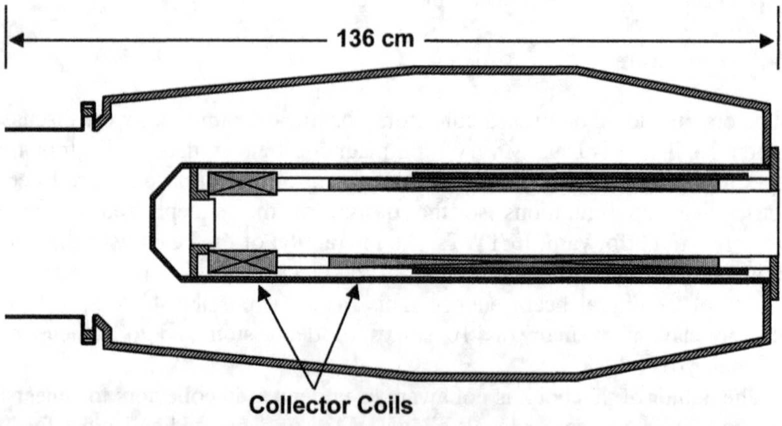

Figure 21-34 Single-stage depressed collector for a 2.2-MW CW gyrotron. (From: B. Piosczyk, Conference on Infrared and Millimeter Waves, 2001.)

21.5 GYRO-MONOTRONS (OSCILLATORS)

Gyro-monotrons are usually referred to simply as gyrotrons. Cross-sectional drawings are shown in Figures 21-35 through 21-37. The device depicted in Figure 21-35 is typical of gyrotrons that are capable of modest RF output powers [34]. At very high-power levels, the drawing in Figure 21-36 is a cross-sectional view of the device [35]. The main differences between these devices are illustrated in Figure 21-37 [36].

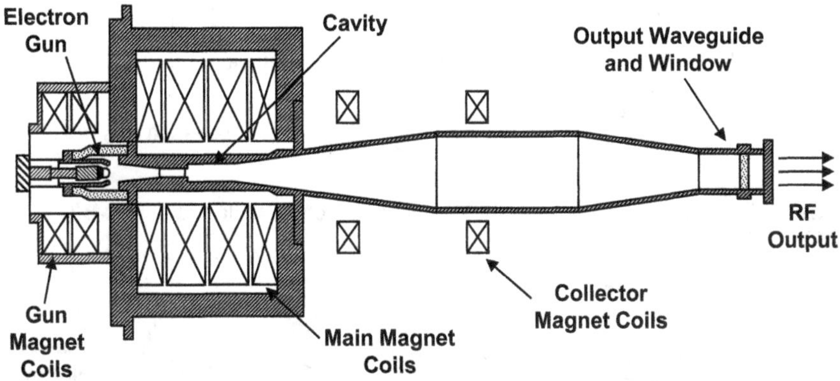

Figure 21-35 Cross section of a gyrotron capable of generating modest RF power levels. (From: Varian, Introduction to Gyro-Devices.)

21.5.1 RF Output Coupling

Early 200-kW TE_{0n}-mode gyrotrons (e.g., TE_{02} mode) having the classical cylindrical symmetric configuration employ standard output coupling. As shown in Figure 21-37(a), the electron beam collector also serves as the output waveguide. For CW gyrotrons operating at the megawatt level, the quasi-optical mode converter makes it possible to separate the spent electron beam from the outgoing RF power as in Figure 21-37(b). This makes it possible to increase the collector beam interception area [36]. Also, it makes it possible to depress the collector.

Gyro-monotrons can produce extremely high powers at very high frequencies. A good example is the device being developed for ITER. To fulfill the needs of ITER, gyrotrons with a hollow waveguide cavity with 1-MW output power are currently under development in Japan and Russia. Also for ITER, the EU gyrotron consortium is investigating a coaxial-cavity gyrotron with the potential to produce 2-MW CW RF output power. An industrial prototype of a CW coaxial gyrotron built by Thales has already been tested at Centre de Recherches en Physique des Plasmas (CRPP) Lausanne. Problems with the high-voltage stability did not allow operation of the gyrotron at nominal parameters. As a result, an RF output power of only 1.4 MW has been achieved, instead of the design goal of 2 MW [37].

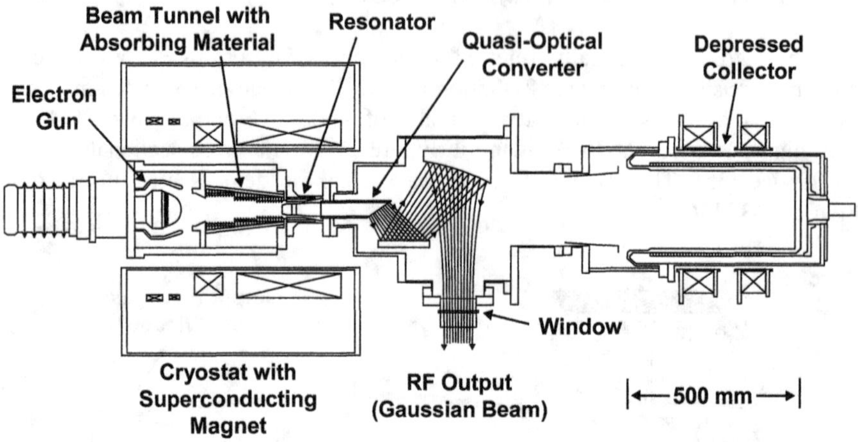

Figure 21-36 Primary components of a high-power gyrotron. (Adapted from: M. Thumm, 2007.)

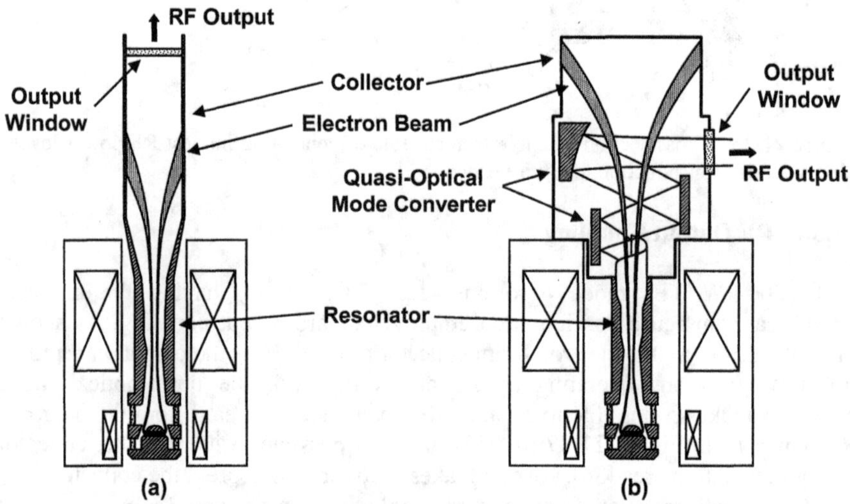

Figure 21-37 (a, b) RF output coupling schemes. (From: M. Thumm, IVEC 2009. © 2009 IEEE.)

The reason for the limitations remains unclear and is currently under investigation. To support the activities of this industrial prototype, studies of a short pulse (≤ few milliseconds) experimental 170-GHz coaxial cavity gyrotron ("preprototype") are being performed at Karlsruhe Institute of Technology (KIT), formerly the Forschungszentrum Karlsruhe (FZK). The European Gyrotron Consortium has demonstrated short-pulse operation at 2.2 MW of this preprototype 170-GHz gyrotron. A critical component of this tube is the quasi-optical RF output system, which converts the $TE_{34,19}$ cavity mode into a free space TE_{00} beam [37].

21.5.2 Second-Harmonic Gyrotrons

Figure 21-38 shows the cross section of a second-harmonic gyrotron. The operating frequency was 34 GHz, the RF power output was 10-kW CW, and the efficiency was 40%. Because the tube operated at the second harmonic, the cyclotron frequency was about 17 GHz and the magnetic field was low enough so that a room-temperature solenoid could be used [38].

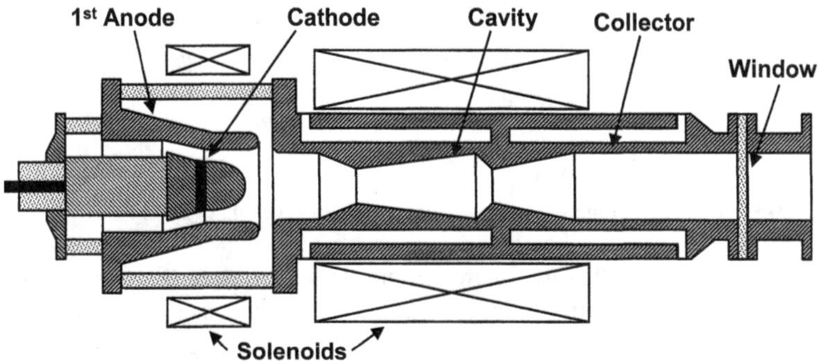

Figure 21-38 Second-harmonic gyrotron with room-temperature solenoid. (From: D. V. Kisel et al., *Radio Engineering and Electronic Physics*, 1974.)

21.5.3 Permanent Magnet Gyrotrons

Calabazas Creek Research, Inc. (CCR) is developing a permanent magnet gyrotron for a W-band active denial, nonlethal weapon system (Figure 21-39) [39]. The current RF source for these systems is a 95-GHz gyrotron producing approximately 100 kW and using a super-conducting magnet. The cost and logistics issues associated with the source and magnet impose considerable operational and maintenance requirements. A particular disadvantage to the current system is the 12-hour start-up time for the magnet. This could be eliminated by use of a permanent magnet.

Rare earth permanent magnet technology allows construction of compact magnets with peak fields up to 1.9 Tesla. This is sufficiently high to support a second harmonic interaction at 95 GHz. However, the mass of such a magnet would be prohibitively high so the third harmonic interaction was chosen, thereby reducing the field requirement to 1.2 Tesla. The magnet mass for this approach is 890 kG.

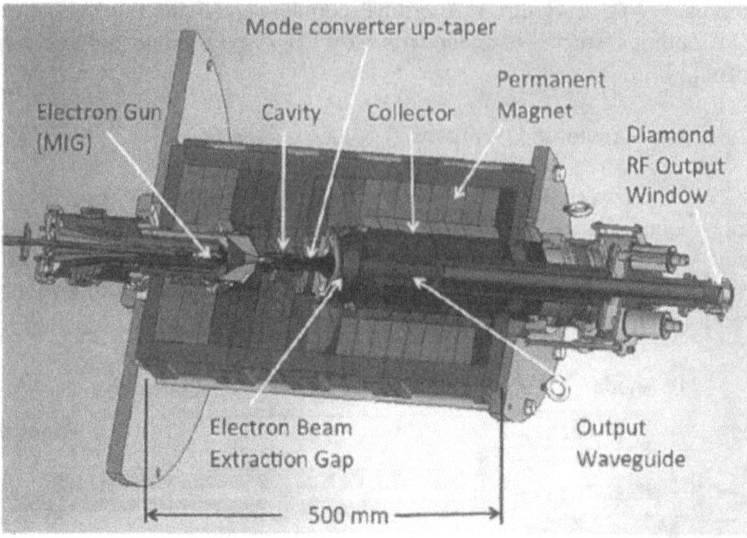

Figure 21-39 Permanent magnet gyrotron. (From: J. Neilson et al., IVEC 2009. © 2009 IEEE.)

The 2009 state-of-the-art for megawatt-class gyrotrons was complied by M. Thumm and is presented in Tables 21.2 and 21.3. Gyrotrons capable of long-pulse operation are listed in Table 21.2. Short-pulse megawatt gyrotrons are listed in Table 21.3.

Table 21.2
State of the Art in 2009
Industrial Megawatt-Class Long-Pulse Gyrotrons with TEM_{00} Output*

Company	Frequency (GHz)	Cavity Mode	Power (MW)	Efficiency (%)	Pulse Length (minutes)
CPI	140	$TE_{28,7}$	0.90	35 (SDC)	30
TED/EU	140	$TE_{28,8}$	0.92	44 (SDC)	30
GYCOM/IAP	170	$TE_{25,10}$	0.95	53 (SDC)	1.7
			0.64	44 (SDC)	5
Toshiba/JAER	170	$TE_{31,8}$	1	55 (SDC)	13.3
			0.6	46 (SDC)	60

*M. Thumm, IVEC 2009

(SDC) Single Stage Depressed Collector

TED/EU Thales/European Consortium

GYCOM Consortium with contributors from the Kurchatov Institute in Moscow, the Institute of Applied Physics in Nizhny Novgorod and industrial partners in the Russian microwave tube industry.

JAER Japan Atomic Energy Research Institute

Table 21.3
State of the Art in 2009
Advanced Short-Pulse 1.5–2 MW Gyrotrons with TEM_{00} Output*

Company	Frequency (GHz)	Cavity Mode	Power (MW)	Efficiency (%)
CPI/MIT/GA	110	$TE_{22,6}$ cylindrical	1.5	48 (SDC)
FZK/EFDA	165	$TE_{31,12}$ coaxial	2.2	48 (SDC)
GYCOM/IAP	170	$TE_{28,12}$ cylindrical	1.44	41 (SDC)
TED/EFDA	170	$TE_{34,19}$ coaxial	1.8	28
Toshiba/JAEA	170	$TE_{31,12}$ cylindrical	1.56	27

*M. Thumm, IVEC 2009

CPI/MIT/GA Communications and Power Industries/Massachusetts Institute of Technology/General Atomic

FZK/EFDA Forschungszentrum Karlsruhe/European Fusion Development Agreement

21.6 GYRO-AMPLIFIERS

The gyro-amplifiers to be discussed are indicated in Figure 21-40. These are the gyro-klystron, gyro-twystron, and gyro-TWT. Gyro-amplifiers are being developed for applications requiring phase coherence and instantaneous bandwidth. A primary application is radar where they are of considerable interest for future high-performance applications [40].

Gyro-amplifiers have larger weight and volume than state-of-the-art coupled-cavity TWTs, but they can provide significantly higher powers than the TWTs. The wide variety of radar applications for which gyro-amplifiers are now being considered are both those existing millimeter-wave radar applications and new radar applications previously not considered due to the absence of powerful amplifiers at the required frequencies.

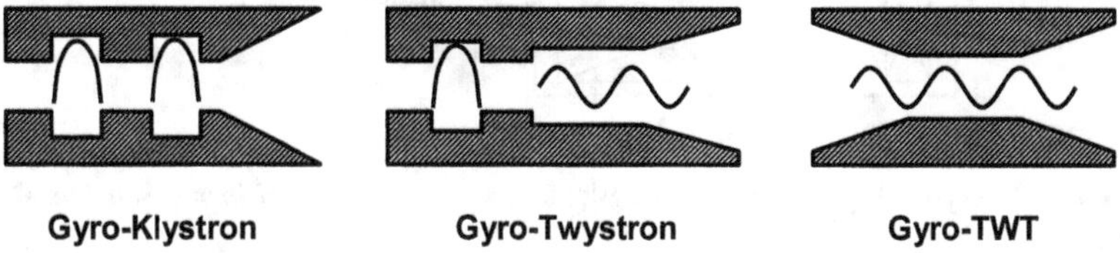

Figure 21-40 Gyro-amplifiers.

The frequency bands in which gyro-amplifiers are expected to find important radar applications are primarily Ka-band (26.5–40 GHz) and W-band (75–110 GHz). Within these bands, radars are generally designed to take advantage of the atmospheric windows at frequencies of approximately 35 GHz and 94 GHz.

The bandwidth requirement of the system determines the tube type:

- Gyro-klystron for < 1% bandwidth;
- Gyro-twystron for 1% to 2% bandwidth;
- Gyro-TWT for bandwidths > 2%.

These devices are described in Sections 21.6.1 through 21.6.3.

21.6.1 Gyro-Klystrons

Conceptually, the only difference between the gyro-klystron and the gyrotron oscillator is the addition of the input cavity. The operation of a gyro-klystron is similar to a conventional klystron except that electron bunching occurs in the azimuthal direction rather than in the axial direction. In a gyro-klystron, a signal to be amplified is fed into an input cavity where the cyclotron bunching process is initiated. Then, the beam is permitted to drift while bunches form. As the beam passes through a second cavity, the amplified signal may be removed or the bunching process may be enhanced and the signal removed in a subsequent cavity. A sketch of a four-cavity gyro-klystron is shown in Figure 21-41 [41].

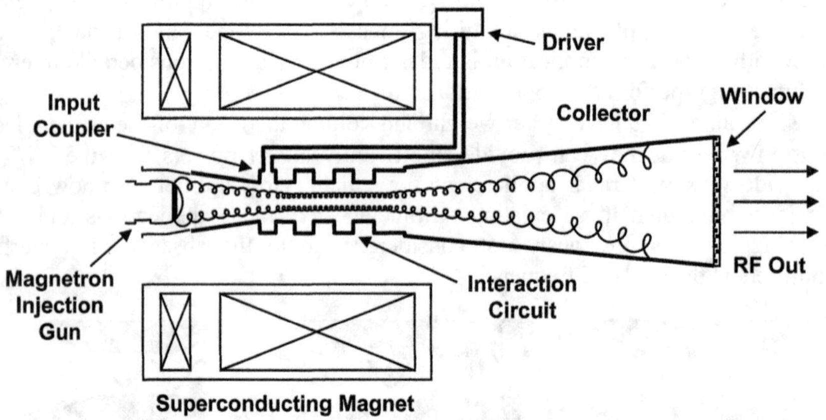

Figure 21-41 Gyro-klystron. (From: M. Blank et al., *IEEE Trans. Plasma Science.* © 2002 IEEE.)

An example of a gyro-klystron is the CPI VGB-8194. This five-cavity amplifier operates at 93.5 to 94.5 GHz with a 3-dB bandwidth of 700 MHz. The peak RF output power is 100 kW and the average output power is 10 kW. The amplifier is available with a cryogen-free 4-tesla superconducting magnet [42].

Among other applications, the VGB-8194 gyro-klystron is the key component of the NRL WARLOC radar and is what made this development unique [43]. WARLOC is a high average power, coherent W-band (94-GHz) radar operating at the NRL Chesapeake Bay Detachment (CBD) site in Maryland. Developed by the Navy as a research radar for imaging and tracking applications, WARLOC also brings unprecedented sensitivity to the measurement of cloud parameters (including phase, morphology, and particle size and motion) at spatial resolutions unparalleled by contemporary radar systems.

21.6.2 Gyro-Twystrons

The gyro-twystron, like the conventional twystron, is a hybrid device with a modest bandwidth capability. It is derived from the gyro-klystron by using the gyro-klystron interaction in the cavities near the RF input and by replacing the output cavity with a slightly tapered waveguide section as in the gyro-TWT to be described in the next section. The output section is excited by the electron beam, which is bunched by the interaction in the klystron section.

The configuration of the gyro-twystron can prevent the problem of breakdown at high-power levels because the RF power density in the output waveguide can be much smaller than in the gyro-klystron output cavity.

An application for gyro-twystrons is the Haystack radar upgrade. The Haystack Ultra-wideband Satellite Imaging Radar (HUSIR) development is being conducted at the Lincoln Space Surveillance Complex by MIT Lincoln Laboratory. For the Haystack upgrade, it is planned that 16 CPI VGB 8193 W-band gyro-twystrons with 1.5-GHz bandwidth at 55-kW peak RF output and 10% duty will be used to cover four frequencies in the 92–100-GHz band [2].

21.6.3 Gyro-TWTs

There has been substantial interest and development effort in gyro-TWTs. The gyro-TWT is an attractive candidate for use as the transmitter power amplifier in millimeter-wave radars. Gyro-TWTs are capable of a much broader bandwidth than gyro-klystrons.

In these devices, a nonresonant RF structure is used to produce a traveling wave interaction. As in other gyro-devices, a spiraling electron beam immersed in an axial magnetic field is used. Traveling waves are launched into the interaction space by an input coupler. Axial phase synchronism is required between the traveling wave and the rotating electrons as indicated in Figure 21-42.

As the electromagnetic wave and the spiraling electron beam move through the interaction space, the kinetic energy of the electron beam is transferred into the electromagnetic fields, creating RF amplification. For typical operation near the cutoff frequency of the interaction waveguide, it is primarily the transverse component of electron motion that interacts with the electromagnetic wave.

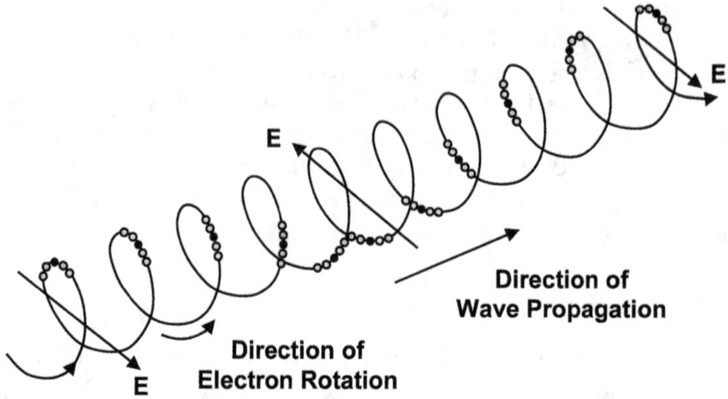

Figure 21-42 Interaction of beamlet with a traveling wave. (Adapted from: *Advances in Electronics and Electron Physics,* Vol. 55, by R. S. Symons and H. R. Jory, copyright 1981 by Academic Press.)

An important problem that must be dealt with in gyro-TWT design is maintaining stability and preventing backward wave oscillations (BWO) in the interaction space. At the same time, of course, acceptable performance, including high overall gain, stability to local reflective oscillations, and high average power capability must be achieved [44].

To ensure stability, the majority of this interaction space should exhibit a moderate amount of electromagnetic loss per unit length. This distributed loading approach has been shown to have superior stability characteristics compared to using localized lossy severs. Finally, the downstream part of the interaction space is completed with a short, unloaded cylindrical tunnel in which the final, highest powered portion of the amplification takes place.

The necessity for the distributed loss was clearly illustrated by Varian in a program to develop a C-band gyro-TWT [45, 46]. Three tubes were built, the first two of which had no distributed loss. These tubes had low gain and exhibited very large powers emanating from the *input* coupler. In order to increase the stable gain and reduce problems with reflections, the third tube was built with distributed loss. Kanthal was applied to the first two-thirds of the circuit. The C-band tube is illustrated in Figure 21-43. This tube was basically a large scale model of a 94-GHz TWT. The development effort for this tube was undertaken in preparation for the development of the 94-GHz gyro-TWT.

The interaction waveguide used in the C-band tube was a circular waveguide operating in the TE_{11} mode. Two rectangular to circular waveguide input transitions at 90° to each other were used to match into the two orthogonal TE_{11} modes that can exist in the circular waveguide. The output window was at the end of the collector. Transverse field permanent magnets deflected electrons to the collector wall before they could reach the window.

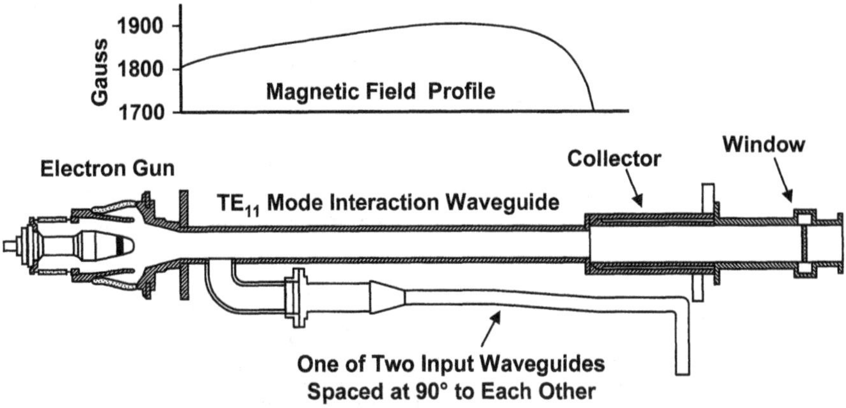

Figure 21-43 C-band gyro-TWT. (Adapted from: P. Ferguson and R. Symons, *Technical Digest, 1980* IEDM. © 1980 IEEE.)

The tube had two gun solenoids and four main field solenoids. As indicated by the magnetic flux density plot, optimum results were obtained by using a slightly tapered field. The power output characteristics for this device are shown in Figure 21-44. Modulating anode pulsing was used and, at a beam voltage of 65 kV, the power output was 120 kW at an efficiency of 26%. The 3-dB bandwidth was 7.25%.

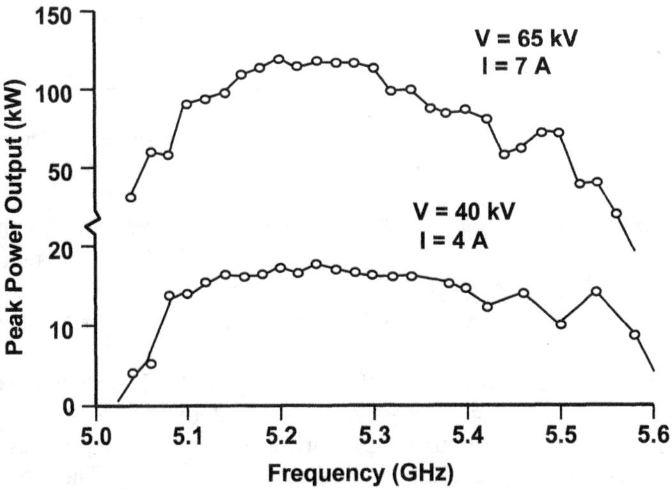

Figure 21-44 Peak power output for a gyro-TWT for two operating conditions. (Adapted from: P. Ferguson and R. Symons, *Technical Digest, 1980* IEDM. © 1980 IEEE.)

In another early effort to develop a 95-GHz gyro-TWT, Varian used a Pierce gun and a contra-wound helical winding along the beam axis to produce the required hollow electron beam [34, 47]. This winding, along with a straight magnetic field along the axis provides a force, which is perpendicular to the beam axis and which rotates in azimuth with distance along the beam axis. This force gives the electrons a transverse velocity as they move through this section of the tube. A schematic drawing of the device is shown in Figure 21-45 and the power output is plotted in Figure 21-46. The tube demonstrated output power in excess of 20-kW peak.

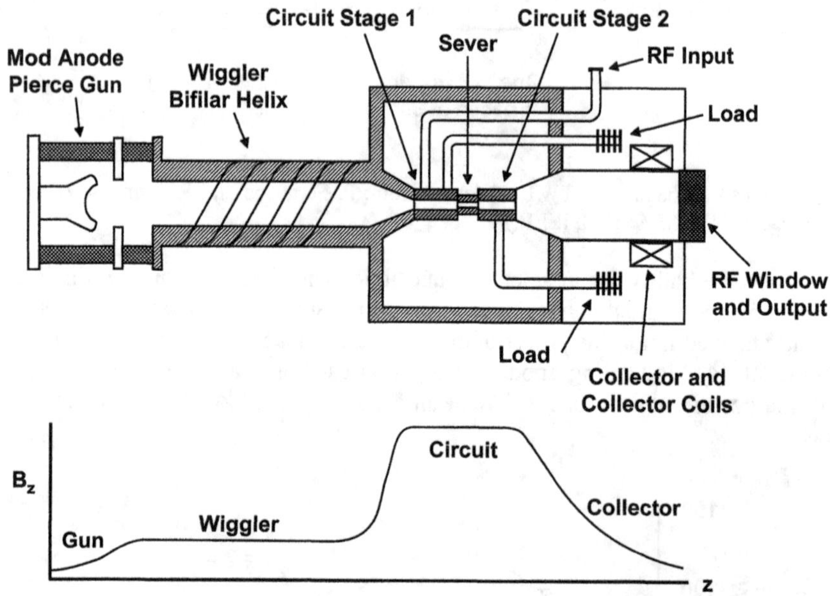

Figure 21-45 The VGT-8195A gyro TWT using a Pierce gun and a wiggler magnet for operation at 95 GHz and powers up to a 20-kW peak.

Now, CPI's VGB-8190 gyro-TWT has demonstrated the ability to produce 2-kW peak power (0.8 kW average) with a bandwidth of 6.5 GHz around a center frequency of 95.5 GHz. A gyro-TWT similar to this (8-GHz bandwidth, 500-W peak, >250-W average from 92–100 GHz) is used as the driver for twystrons in the Haystack radar [2].

A novel wideband, 140-GHz, 1-kW confocal gyro-TWT is being developed at MIT [48].The waveguide geometry and mode pattern used in this device are shown in Figure 21-47. The curved cylindrical mirror geometry consists of two mirrors to make up the cylindrical confocal waveguide system. The total width of each mirror can be adjusted in order to induce distributed diffractive loss.

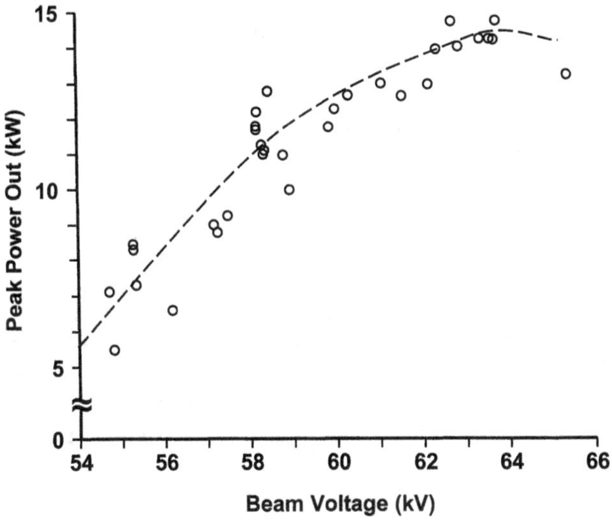

Figure 21-46 Power output of the VGT-8195A gyro-TWT.

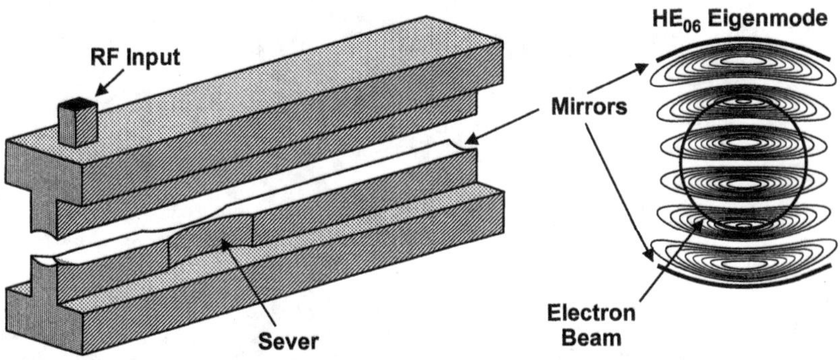

Figure 21-47 Confocal gyro-TWT. (From: C. D. Joye et al., *IEEE Trans. Electron Devices,* Vol. 56, No. 5, 2009. © 2009 IEEE.)

Power contours for the HE_{06} mode and the geometry of the hollow annular electron beam are also shown. The electron beam interacts primarily with the second and fifth maxima to produce a Gaussian-like higher-order mode. Also, the mode spectrum is sparser than in a cylindrical waveguide and the device is suitable for harmonic operation.

The device has demonstrated a linear gain of 34 dB at 34.7 kV and 2.7A and produced an output power of over 820W at 37.7 kV and 2.7A. With a slight change in operating point, the amplifier achieved a saturated bandwidth of over 1.5 GHz with 570-W output power at 38.5 kV and 2.5A. The device has been shown to be applicable to short-pulse spectroscopy. Although the experiments were carried out at a 2-μs pulse length, pulses as short as 4 ns, which is the limit of the present input source, were amplified with no noticeable pulse broadening. These nanosecond-scale pulses were used to diagnose the system by a novel TDR technique. This method provided valuable insights to the nature of echoes, resonances, and reflections in the system, which could be pinpointed inside of the vacuum tube without the need to ever open the vacuum envelope.

A concept for an ultrawideband distributed gyrotron traveling-wave amplifier for millimeter and submillimeter waves is shown in Figure 21-48 [49]. The radius of the waveguide in the interaction region is increased along the axis. Amplification occurs over ΔL, which moves along the circuit as frequency changes. The strength of the dc magnetic field is decreased in such a way that the wave cutoff frequency is kept nearly equal to the electron cyclotron frequency. The basic principle of operation, peak gain, and saturated efficiency have been analyzed. It is shown that instantaneous bandwidth over at least 2 octaves is theoretically possible.

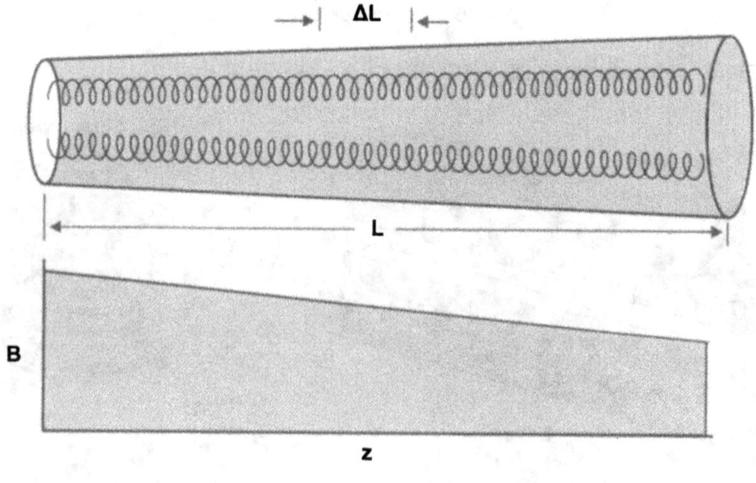

Figure 21-48 Technique proposed to obtain a broad bandwidth. (Adapted from: K. R. Chu, et al., *IEEE Trans. Electron Devices,* July 1981. © 1981 IEEE.)

21.7 TERAHERTZ GYROTRONS

The magnetic field required for fundamental-frequency operation at 1 THz is nearly 40T (400,000G). A present, this is beyond the capability of superconducting magnets, so pulsed magnets that are wound directly on the gyrotron are being used. Glyavin and Luchinin reported the generation of 50T with the basic circuit shown in Figure 21-49 [50]. Output powers of 5 kW at 1 THz and 0.5 kW at 1.3 THz have been obtained with fundamental operation at a pulse duration of 40 microseconds with a 30 kV/5A electron beam. The problems of selecting suitable materials for the gyrotron envelope and for the solenoid are discussed in [50].

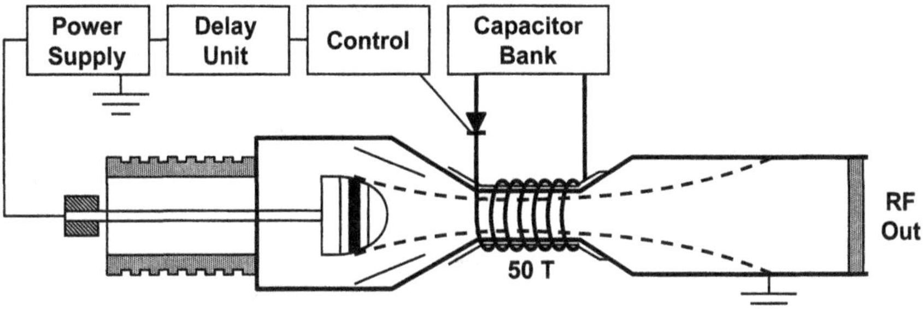

Figure 21-49 Basic elements of a pulsed terahertz gyrotron circuit. (Adapted from: M. Glyavin et al., *Terahertz Science and Technology*, 2009.)

Other results reported in [50] are second-harmonic operation at 100-W CW at 395 GHz, 2.2-kW CW at 300 GHz, and third-harmonic operation at 371–414 GHz in 10-μs pulses with powers of 10–20 kW.

Nusinovich et al. described a program to develop a 30–35% efficient 670-GHz, 300-kW gyrotron pulsed with a 28-T magnetic field. Much of the work on this device is expected to be done by GYCOM in Russia [51]. The application for this gyrotron is for creating localized breakdown in air when nearby radioactive ionizing sources produce an increase in the density of free electrons.

REFERENCES

[1] V. A. Flyagin et al., "The gyrotron," *IEEE Trans. Microwave Theory and Techniques,* Vol. 25, No. 6, 1977, pp. 514–521.

[2] H. Jory et al., "Gyrotron introduction for ECRIS 2008," 2008.

[3] B. Piosczyk, S. Alberti, and D. Bariou, "170 GHz, 2 MW, CW coaxial cavity gyrotron for ITER—status and experimental results," *MTDC Meetings,* Vol. IT-4Ra, 2006.

[4] A. Grudiev, *Numerical Study of Nonstationary Phenomena in Gyro-Oscillators*, Hamburg University of Technology, 2010.

[5] G. Dohler, and W. Friz, "Physics and classification of fast-wave devices," *Int. J. Electronics,* Vol. 55, No. 4, 1983, pp. 505–522.

[6] R. Symons, and H. Jory, "Cyclotron resonance devices," *Advances in Electronics and Electron Physics,* Vol. 55, 1981.

[7] G. S. Nusinovich, *Introduction to the Physics of Gyrotrons*, Baltimore and London: The Johns Hopkins University Press, 2004.

[8] K. T. Nguyen et al., "Electron gun and collector design for 94-GHz gyro-amplifiers," *IEEE Trans. Plasma Science,* Vol. 26, No. 3, 1998, pp. 799–813.

[9] B. Piosczyk, "A novel 4.5-MW electron gun for a coaxial cavity gyrotron," *IEEE Trans. Electron Devices,* Vol. 48, No. 12, 2001, pp. 2938–2944.

[10] O. Dumbrajs, and G. S. Nusinovich, "Coaxial gyrotrons: past, present, and future (review)," *IEEE Trans. Plasma Science,* Vol. 32, No. 3, 2004, pp. 934–946.

[11] M. V. Kartikeyan, E. Borie, and M. Thumm, *Gyrotrons*, Berlin: Springer, 2003.

[12] J. M. Baird, and A. C. Attard, *Gyrotron Gun Study Report TR-3-476*, Naval Research Laboratory, 1981.

[13] J. M. Baird, and W. Lawson, "Magnetron injection gun (MIG) design for gyrotron applications," *Int. J. Electron,* Vol. 61, No. 6, 1986, pp. 953–967.

[14] W. C. Brown, "The magnetron—a low noise, long life amplifier," *Applied Microwaves*, 1990.

[15] R. L. Ives et al., "Improved magnetron injection guns for gyrotrons," *IEEE Trans. Plasma Science,* Vol. 36, No. 3, 2008, pp. 620–630.

[16] K. Felch et al., "Results of 10-sec pulse tests on a 110 GHz gyrotron," IVEC 2000.

[17] R. Advanti et al., "Multimegawatt gyrotrons for ECRH " *Conf. Digest, 24th Int. Conf. on Infrared and Millimeter Waves,* 1999.

[18] O. Louksha et al., "Experimental study of gyrotron efficiency enhancement by improvement in electron beam quality," *Proc. Joint 32nd Int. Conf. on Infrared and Millimeter Waves and 15th International Conference on Terahertz Electronics*, 2007, pp. 880–881.

[19] G. V. Miram et al., "Emission spread in thermionic cathodes," *Technical Digest*, Fifth IEEE IVEC, 2004, pp. 303–304.

[20] N. P. Venedictov, et al., "Experimental study of emission spread at gyrotron cathodes by volt-ampere characteristics," *Radio Physics*, 1997.

[21] J. M. Baird, "Survey of fast wave tube developments." IEDM 1979, pp. 156–163.

[22] A. B. Bronwell, and R. E. Beam, *Theory and Application of Microwaves*, New York: McGraw-Hill, Inc. 1947.

[23] B. Piosczyk et al., "A 2 MW, CW, 170 GHz coaxial cavity gyrotron for ITER," *IAEA Technical Meeting on ECRH Physics and Technology for ITER*, Kloster Seeon, 2003.

[24] S. Sabchevski et al., "Quasi-optical converters for high-power gyrotrons: a brief review of physical models, numerical methods and computer codes," *Jour. of Physics,* Vol. 44, 2006, pp. 102–109.

[25] M. K. Thumm, and W. Kasparek, "Passive high-power microwave components," *IEEE Trans. Plasma Science,* Vol. 30, No. 3, 2002, pp. 755–786.

[26] S. N. Vlasov, L. I. Zagryadskaya, and M. I. Petelin, "Transformation of a whispering gallery mode propagating in a circular waveguide into a beam of waves," *Radio Eng. Electron. Phys.*, Vol. 12, 1975.

[27] G. G. Denisov et al., "110 GHz gyrotron with built-in high-efficiency converter," *Int. J. Electronics,* Vol. 72, 1992, pp. 1079–1091.

[28] J. Jin et al., "2.2: Design of phase correcting mirror system for coaxial-cavity ITER gyrotron," *IEEE IVEC*, 2010, pp. 29–30.

[29] B. Piosczyk et al., "Single-stage depressed collectors for gyrotrons," *IEEE Trans. Plasma Science,* Vol. 24, 1996, pp. 579–585.

[30] T. A. Hargreaves et al., "Depressed collector performance on the NRL quasi-optical gyrotron," *Conf. Digest, 15th Int. Conf. on Infrared and Millimeter Waves*, 1990, p. 330.

[31] K. Sakamoto et al., "Major improvements in gyrotron efficiency with beam energy recovery," *Phys. Rev. Lett.,* Vol. 73, 1994, p. 3532.

[32] B. Piosczyk, "A 2.2 MW CW collector for a coaxial cavity gyrotron," *Conf. Digest, 26th Int. Conf. on Infrared and Millimeter Waves*, 2001.

[33] G. Dammertz, "Beam tunnel, resonator, collector and RF window," *Gyrotron Oscillators, Their Principles and Practice*, C. Edgecombe, (ed.), London: Taylor and Francis, 1993.

[34] Varian brochure, "Introduction to gyro-devices," 2000.

[35] M. Thumm, "State-of-the-art of high-power gyro-devices and free electron masers update 2006," 2007.

[36] M. Thumm, "History, presence and future of gyrotrons," *IEEE IVEC,* 2009, pp. 37–40.

[37] T. Rzesnicki et al., "2.2-MW record power of the 170-GHz European preprototype coaxial-cavity gyrotron for ITER," *IEEE Trans. Plasma Science,* Vol. 38, No. 6, 2010, pp. 1141–1149.

[38] D. V. Kisel et al., "An experimental study of a gyrotron, operating at the second harmonic of the cyclotron frequency, with optimized distribution of the high frequency field," *Radio Engineering and Electronic Physics* Vol. 19, No. 4, 1974, pp. 95–100.

[39] J. Neilson, M. Read, and L. Ives, "Design of a permanent magnet gyrotron for active denial systems," *IEEE IVEC,* 2009, pp. 92–93.

[40] K. L. Felch et al., "Characteristics and applications of fast-wave gyrodevices," *Proc. IEEE,* Vol. 87, No. 5, 1999, pp. 752–781.

[41] M. Blank et al., "Development and demonstration of high-average power W-band gyro-amplifiers for radar applications," *IEEE Trans. Plasma Science,* Vol. 30, No. 3, 2002, pp. 865–875.

[42] CPI, "VGB-8194, gyrotron amplifier."

[43] G. J. Linde et al., "WARLOC: A high-power coherent 94 GHz radar," *IEEE Trans. Aerospace and Electronic Systems,* Vol. 44, No. 3, 2008, pp. 1102–1117.

[44] J. P. Calame et al., "Gyrotron-traveling wave-tube circuits based on lossy ceramics," *IEEE Trans. Electron Devices,* Vol. 49, No. 8, 2002, pp. 1469–1477.

[45] R. S. Symons et al., "An Experimental Gyro-TWT," *IEEE Trans. Microwave Theory and Techniques,* Vol. 29, No. 3, 1981, pp. 181–184.

[46] P. Ferguson, and R. Symons, "A C-band gyro-TWT," *Technical Digest, IEDM,* 1980, pp. 310–313.

[47] P. E. Ferguson, and R. S. Symons, "A gyro-TWT with a space-charge limited gun," *Technical Digest, IEDM,* 1981, pp. 198–201.

[48] C. D. Joye et al., "Demonstration of a 140-GHz 1-kW confocal gyro-traveling-wave amplifier," *IEEE Trans. Electron Devices,* Vol. 56, No. 5, 2009, pp. 818–827.

[49] L. R. Barnett et al., "An experimental wide-band gyrotron traveling-wave amplifier," *IEEE Trans. Electron Devices,* Vol. ED-28, 1981, pp. 872–875.

[50] M. Y. Glyavin, and A. G. Luchinin, "Powerful terahertz gyrotrons based on pulsed magnets," *Terahertz Science and Technology,* Vol. 2, No. 4, December 2009, pp. 150-155.

[51] G. S. Nusinovich et al., "Development of THz gyrotrons with pulse solenoids for detecting concealed radioactive materials," *IEEE IVEC,* 2009, pp.197-198.

Chapter 22

Windows

22.1 BACKGROUND

The problem of transmitting RF energy from the vacuum environment inside a tube to an external load is common to all types of microwave tubes. The number of possible window configurations seems to be limited only by the ingenuity of the designer [1, 2]. Many fairly distinct designs in coaxial lines and waveguides have been used successfully. All of these techniques have been highly dependent upon materials and vacuum seal technology. Over the last 70 years, windows have followed a gradual evolution from simple mica membranes in waveguides to high purity alumina, beryllia, sapphire, and now, diamond windows. The works of Churchill [3–6] provide information on high-power windows.

22.2 COAXIAL WINDOWS

In the case of coaxial windows, the mechanical design and fabrication can be difficult. Two vacuum seals are required (inner conductor to ceramic and outer conductor to ceramic). Also, at high frequencies, the center conductor in the ceramic becomes very small and metallization and brazing become difficult. The RF design of coaxial windows is relatively easy because the transmission line uses the TEM mode and the ceramic is a disk as shown in Figure 22-1.

The impedance of a coaxial line operating in the TEM mode is given by

$$Z_o = \frac{60}{\sqrt{\varepsilon}} \ln \frac{b}{a} \tag{22-1}$$

where ε is the relative dielectric constant of the media between the center and outer conductors. For vacuum or air where $\varepsilon = 1$, the ratio $b/a = 2.3$ for an impedance of 50Ω. The ceramic that is used for the window is nearly always Al_2O_3 or, when high thermal conductivity is required, BeO. The relative dielectric constant of alumina is 9.3 and so, for an impedance of 50Ω, $b/a = 12.7$. This means that the radius of the center conductor in alumina (a') must be about 20% of that in vacuum or air (*a*). For beryllia, the relative dielectric constant is 6.4 so the radius of the center conductor must be about 30% of that in vacuum or air. Figure 22-2 shows the relative sizes of conductors for a coaxial window used in practice. Notice that the radius of the center conductor in the ceramic is about one-third of the radius in air or vacuum. It should also be noticed that the reduced diameter is extended a short distance from the ceramic. This is done to improve the impedance match.

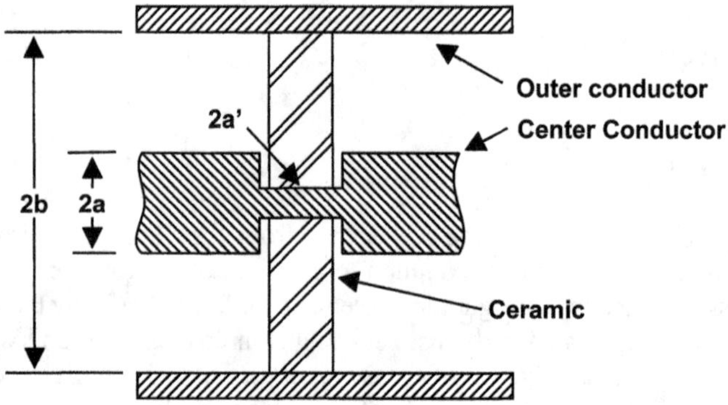

Figure 22-1 Basic configuration of a coaxial window.

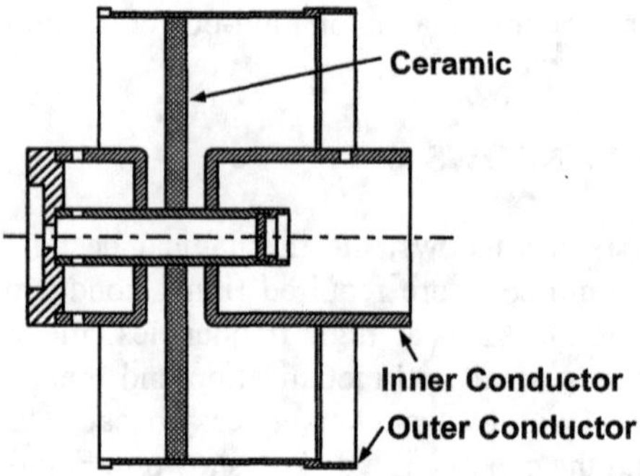

Figure 22-2 Coaxial window for operation at 1.25 GHz. (Adapted from: *Power Klystrons Today* by M. J. Smith and G. Phillips, copyright 1995 by Research Studies Press, Ltd.)

Hughes Electron Dynamics Division has described the integral coax-to-waveguide adapter window shown in Figure 22-3. This was used on a coupled-cavity TWT with a waveguide input. The interface with the customers system was coaxial. In this case, the design was empirical. Impedance match adjustments could be made by varying the waveguide backwall position, the inner-conductor diameter, and the depth of penetration of the inner conductor in the waveguide. The vacuum seal was made to the outer conductor only.

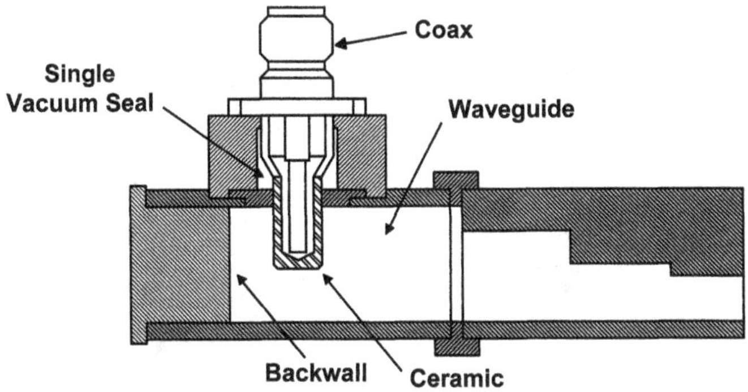

Figure 22-3 Coax window with waveguide adapter. (Courtesy of Hughes Electron Dynamics Division.)

22.3 WAVEGUIDE WINDOWS

The simplest window consists of an inductive iris into which is sealed a capacitive dielectric plate. An example is shown in Figure 22-4. The impedance is adjusted so that it is resonant and the shunt admittance is made zero at the band center. The frequency at which the window appears to be matched can be adjusted by changing the dielectric thickness or diameter. The window can be matched over a broader band by making the inductive hole (and the dielectric) elliptical, but still the design has restricted bandwidth. This design can be refined as shown in Figure 22-5 by extending the support for the window and using chokes to isolate the support for the window. Iris windows are limited on peak power handling capability (for example, about 1 MW at 3 GHz and less at higher frequencies).

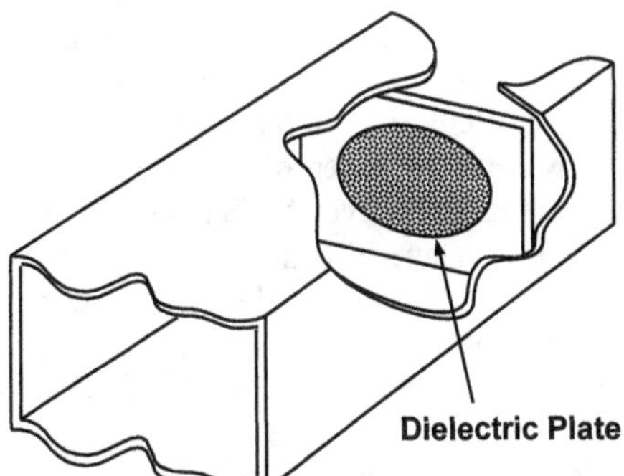

Figure 22-4 Iris window. (Adapted from: *Power Klystrons Today* by M. J. Smith and G. Phillips, copyright 1995 by Research Studies Press, Ltd.)

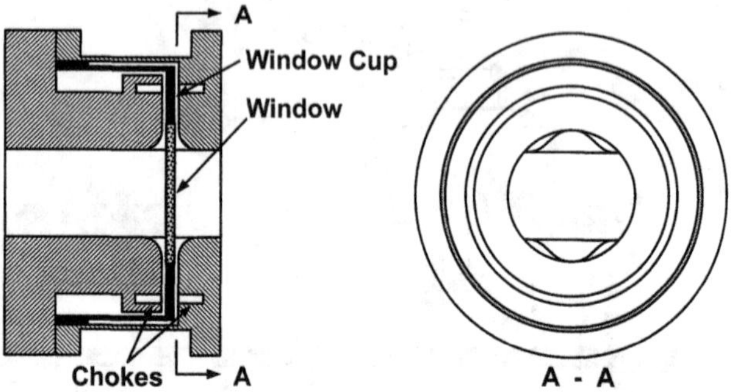

Figure 22-5 Iris window with choke. (Adapted from: *Power Klystrons Today* by M. J. Smith and G. Phillips, copyright 1995 by Research Studies Press, Ltd.)

The two most generally used window designs suitable for medium- and high-power tubes consist of either a rectangular ceramic block in the output waveguide, or a circular ceramic disk (sometimes referred to as a poker chip) in a section of round waveguide, which is then adapted to rectangular waveguides on both ends. The window thickness is almost always one-half wavelength or less (although the thickness is somewhat arbitrary, since almost any configuration can be designed for good transmission). The most important factors are the methods of design used, the average and peak power handling capabilities, the bandwidth available, the elimination of unwanted resonances, and the basic technology of manufacture.

The block window is not used as often as the disk window, but it is easier to design and is useful in illustrating some of the characteristics and problems in design and use. The most common block window is half a wavelength thick and fills the entire cross section of the waveguide. These windows are well matched at the resonant frequency and are suitable for narrow bandwidths without modification (see Figure 22-6). If irises or other matching elements are added, the window can be made into a filter with a bandwidth of 10% to 20%. It is mainly a high-power window, particularly suitable for high average power. The transmission loss is low, on the order of 0.2%, and, because it is a filter, the reflection from it is essentially reactive. A VSWR plot for one of these windows is shown in Figure 22-7. The electrical design is straightforward and can easily be adapted to computer approaches where the matching elements are determined automatically for a given bandwidth [7].

Probably the most severe problem for block windows is "ghost modes." Forrer and Jaynes described a method for determining these resonances, which arise from stored energy in higher-order waveguide modes inside the dielectric [8]. These modes are beyond cutoff in the waveguide and behave as high-Q cavities that are loosely coupled to the dominant propagating mode. These resonances, although barely visible on a VSWR plot, are hazardous for high-power operation because of the possibility of high-power absorption and subsequent failure of the window at ghost-mode frequencies.

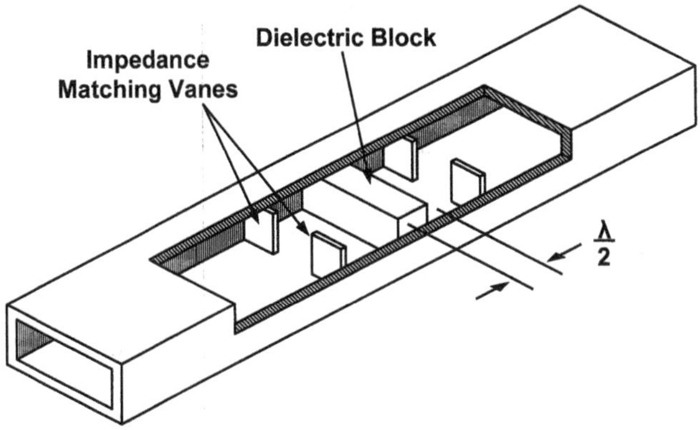

Figure 22-6 Half-wave resonant block window in rectangular waveguide. (Adapted from: *Linear-Beam Tubes* by A. Staprans, Internal Varian Report, September 1968.)

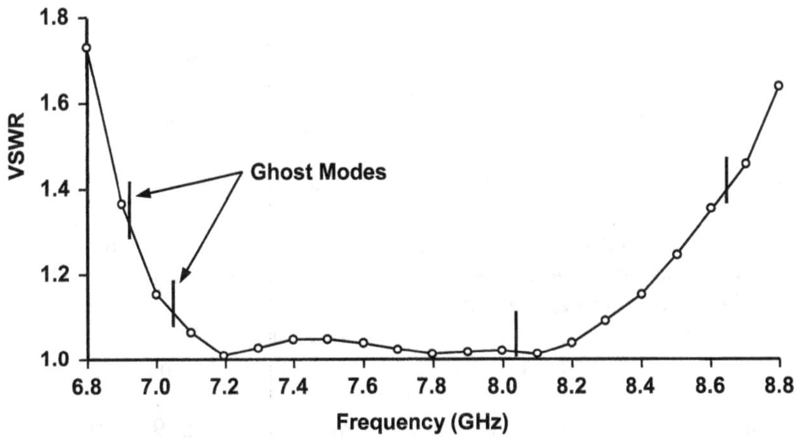

Figure 22-7 Impedance match of a half-wave resonant block window in a rectangular waveguide. (Adapted from: *Linear-Beam Tubes* by A. Staprans, Internal Varian Report, September 1968.)

For a standard window configuration, using alumina ceramic, more than 10 such resonances, irregularly spaced, will be present in the ordinary passband of the waveguide. The choice of the center frequency of the window allows some adjustment of these resonant frequencies. A more powerful method is the shaping of the window cross section. This usually entails the reduction of window height, and sometimes the width, somewhat below those in the waveguide. This may be done by either a small abrupt step or a gradual taper. A small adjustment of the window aspect ratio will allow up to 15% mode-free bandwidth.

These windows have been used in a great variety of high-power tubes. The highest-power examples are X-band, CW power of 263 kW in a resonant-ring experiment, and 125 kW in the VA-949AM. The dielectric window material for both was beryllia because of its high thermal conductivity. For lower-power tubes, alumina ceramics are adequate.

The most widely used window configuration is the disk. It consists of a flat disk in a section of circular waveguide which is adapted on both sides to rectangular waveguide. The transition from round to rectangular waveguides, which may be either abrupt or gradual, often incorporates matching elements. Although all thicknesses of the ceramic are possible, almost all of these windows belong to one of two categories.

1. The ceramic is made as thin as mechanical stresses and the fear of electrical breakdown will allow (about one-tenth of a wavelength).
2. The window is made self-resonant (a half-wave long at the center frequency).

Disk windows are substantially easier to build than rectangular windows. However, trapped-mode resonances are a problem. These resonances are due to modes that propagate in the round part of a waveguide, but are cut off at the transitions to the rectangular waveguides. In addition, ghost modes may also be present in the ceramic. Their density increases with the thickness of the ceramic. Since a thicker dielectric is usually used in round than in rectangular windows, the ghost-mode population in round windows is usually higher than in rectangular windows.

The simplest and most commonly used form of the disk window is a thin ceramic in a short circular waveguide section joined abruptly to the rectangular waveguides. This window, which in its basic form was patented by Symons [9], is shown in Figure 22-8. It is an easy window to make and has been used for tubes at all power levels. It can be made exceptionally broadband if trapped modes can be tolerated as is the case in many low- and medium-power tubes. The VSWR of a broadbanded version of this window is shown in Figure 22-9. Notice that the trapped-mode population is relatively high, although only one of the modes couples very strongly to the propagating wave. By proper adjustment of all dimensions, these windows can be made completely mode-free in a 15% to 20% bandwidth.

Since most output windows rely on cooling at the edge only, the disk window is inferior to the rectangular window for average power because of the longer heat path to the periphery. Nevertheless, by taking advantage of extremely low-loss materials, such as synthetic sapphire, very high average powers have been realized in these windows. The thin-disk window is also suitable for high peak power. One of its many applications was in the 25-MW peak power S-band klystron for the Stanford linear accelerator [10].

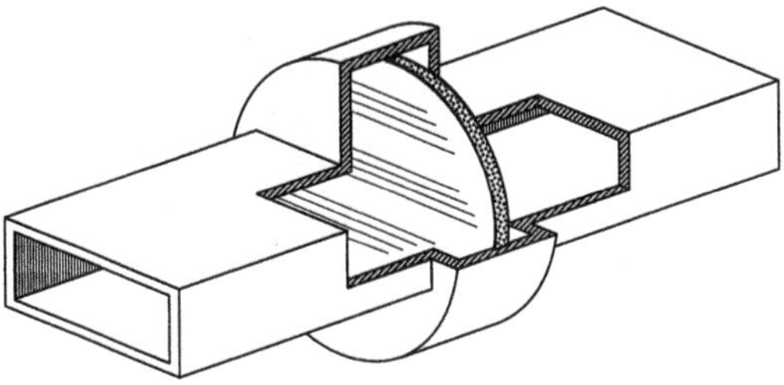

Figure 22-8 Thin-disk, pillbox-type window. (Adapted from: *Linear-Beam Tubes* by A. Staprans, Internal Varian Report, September 1968.)

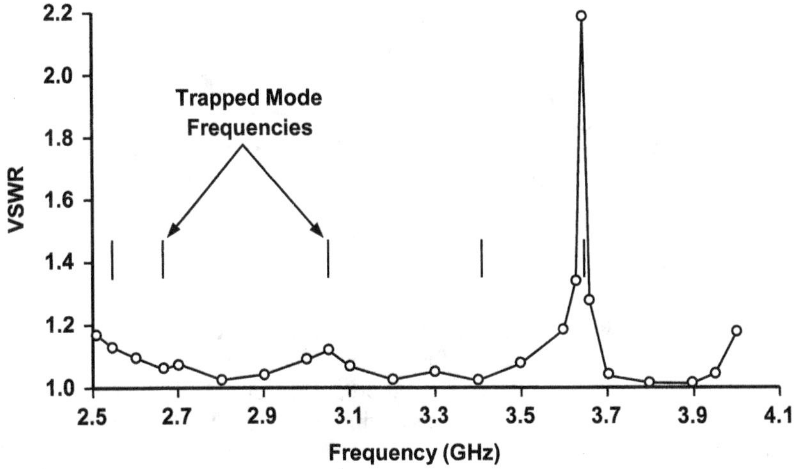

Figure 22-9 Impedance match of a thin-disk, pillbox-type window. (Adapted from: *Linear-Beam Tubes* by A. Staprans, Internal Varian Report, September 1968.)

Where a large bandwidth is not necessary, an excellent high-power window is the half-wave-thick disk type shown in Figure 22-10. Its bandwidth is generally limited by the ghost-mode resonances to about 5%. An example for such a window is its application for a 500-kW CW low S-band tube [11]. This window uses a beryllia ceramic.

Although many windows are made of alumina, the very highest powers (average or continuous) obtained have been with beryllia, which has a very high thermal conductivity, sapphire, which is extremely low-loss and has high strength, or CVD diamond. The capabilities of disk windows are, perhaps, best described by Goldfinger [12]. His results indicated that powers as high as 1-MW average at X-band through a thin-disk sapphire window are possible. For gyrotrons, face-cooled double disk sapphire windows as shown in Figure 22-11 can be used at power levels of 0.3–0.4 MW [13].

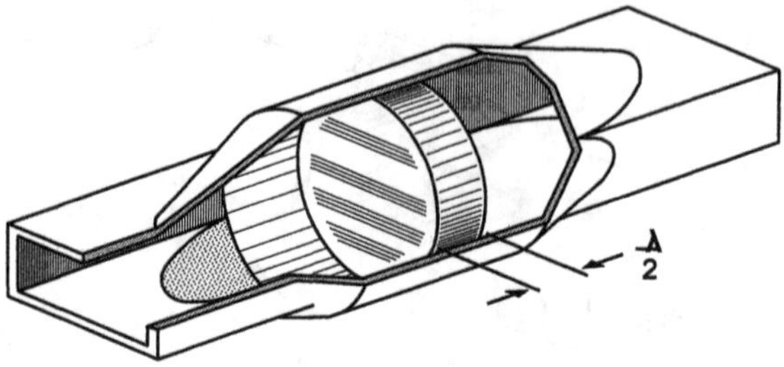

Figure 22-10 Half-wave, resonant, thick-disk window. (Adapted from: *Linear-Beam Tubes* by A. Staprans, Internal Varian Report, September 1968.)

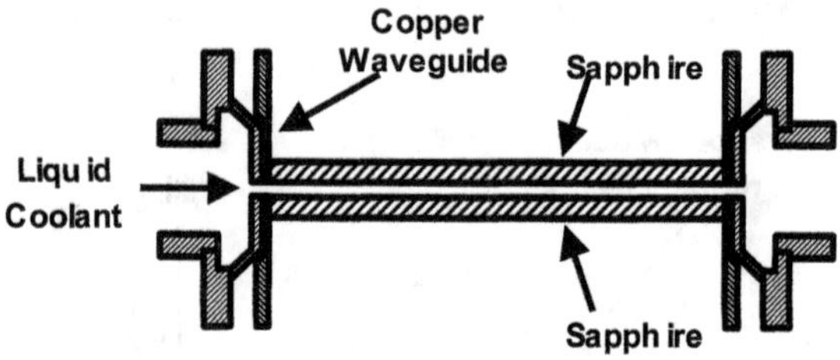

Figure 22-11 Face-cooled double-disk sapphire window.

During the past two decades, there has been strong interest in the development of diamond windows [14]. Reasons for this interest are evident from some of the information given in Table 22.1 where the properties of chemical vapor deposited (CVD) diamond are compared with other window materials. Properties of natural diamond are also given. Notice that its properties are far superior to the other window materials. While the properties of CVD diamond are not equal to those of natural diamond, CVD diamond can be made in the form of a window, whereas it would virtually impossible to use natural diamond for that purpose.

The properties of CVD diamond are far more favorable for use in high-power tubes than either BeO or sapphire. The relatively low dielectric constant of CVD diamond allows wider bandwidth designs than either BeO of sapphire. The low-loss tangent coupled with the extremely high thermal conductivity make it an excellent material for high-power applications.

As an example of the improved thermal performance of CVD diamond compared to BeO, [14] gives a comparison of calculations of the temperature rise in the center of Ku-band windows made of the two materials. The power transmitted through the windows was 10 kW. The calculation for the BeO window showed 19W of loss and a 3.7°C gradient between the center and the edge of the window. For the CVD diamond, the loss was only 4.71W and the temperature gradient was 0.546°C. The temperature gradient in the CVD diamond was a factor of 6.7 below that in BeO.

Table 22.1
Properties of Diamond Compared to Selected Materials [14]

Material	RF Loss Tangent $\times 10^{-4}$	Dielectric Constant	Thermal Conductivity W/cm K	Tensile Strength kpsi
Natural Diamond	1.4	5.64	20	109–1,090
CVD Diamond	2.0–6.8	5.7	12–18	33–59
BeO	4.0	6.8	2.6	22–29
Alumina	2-8	9.8	0.2	—
Sapphire	2	9.4–9.8	0.2–0.5	29–54
Quartz	0.4–3	3.8–4.4	0.01–0.06	10–17

The output window for a 140-GHz, 900-kW gyrotron is shown in Figure 22-12. This is an edge-cooled CVD diamond disk. The clear aperture diameter of is 88 mm. The thickness is 1.8 mm. This is two wavelengths in the diamond at 140 GHz. A high-temperature braze technique was used in the fabrication of the window assembly [15].

Figure 22-12 CVD diamond disk window for a gyrotron. (From: Felch et al., *Jour. of Physics*, 2005.)

22.4 SCALING OF WINDOWS

The most common failure mode of a window is fracture due to thermal stresses. As a result, the limitation of thermal stresses is one of the most important factors in window design. The average power capability of a window can be scaled to another frequency with reasonable accuracy. If the loss tangent and dielectric constant are held constant and the size of the window is scaled directly with wavelength, the average power handling capability is inversely proportional to frequency.

At a fixed frequency, the power handling capability is virtually insensitive to the window cross-section dimensions. This is because, even though dissipation density is decreased for a larger window, a longer path for the heat flux is created. This results in essentially the same maximum temperature at the center of the window regardless of the window size. As a result, there is no advantage from an average-power standpoint in going to a very large window. However, problems associated with operation at high peak-power levels, such as breakdown, will be alleviated in a larger window.

REFERENCES

[1] A. Staprans, *Linear-Beam Tubes*, Internal Varian Report, September 1968.

[2] M. J. Smith, and G. Phillips, *Power Klystrons Today*, Taunton, Somerset, England: Research Studies Ltd. 1995.

[3] D. Churchill, "Investigation of microwave window failure mechanisms and their elimination," Contract No. DA-36-039-SC-78314, Sperry Report No. NA-8240-8182-5, November 1960.

[4] D. Churchill, "High average-power RF window study," Final Report, Contract No. AF 30(602)-2428, Sperry Report No. NA-8240-8299, February 1962.

[5] D. Churchill, "Investigation of microwave window failure mechanisms and their elimination," Contract No. DA-36-039-SC-87389, Sperry Report No. NA-8240-8338, February 1963.

[6] D. Churchill, "Superpower microwave windows," Final Report, Contract No. DA 36-039 AMC-02161(E), Sperry Report No. NA-8240-8372, January 1964.

[7] F. O. Johnson, "Computer synthesis of filter networks as applied to waveguide windows," *Proc. International Conference on Microwave Circuit Theory and Information Theory*, Tokyo, Japan, September 7–11, 1964.

[8] M. P. Forrer, and E. T. Jaynes, "Resonant modes in waveguide windows," *IRE Trans. on Microwave Theory and Techniques*, Vol. MTT-8, No. 1, March 1960, pp. 147–150.

[9] R. S. Symons, and H. E. Schoennauer, "Sealed waveguide window," U.S. Patent 2,958,834, November 1, 1960.

[10] J. H. Jasberg, and J. V. Lebacqz, "High-power microwave windows," *Proc. 5th International Congress on Microwave Tubes*, Paris, September 1964, pp. 246–250.

[11] R. H. Giebler, "A 500 kW CW klystron at S-band," IEEE IEDM, Washington, D.C., October 1967.

[12] A. Goldfinger, "High-power RF window study," Varian Associates, Eimac Division, Final Report, RADC-TR-66-657, January 1967.

[13] M. V. Kartikeyan, E. Borie, and M. Thumm, *Gyrotrons*, Berlin: Springer, 2003.

[14] B. G. James, "Diamond windows for high-power broadband microwave tubes," Communications and Power Industries, Final Report, Contract No. N00164-94-C-004, December 1997.

[15] K. Felch et al., "Recent ITER-relevant gyrotron tests," *Jour. of Physics: Conference Series,* Vol. 25, No. 1, 2005, pp. 13–23.

Chapter 23

Noise

There are many sources of noise in a microwave tube. Some are unavoidable, but most can be minimized or eliminated through proper design and construction. In this chapter, an attempt has been made to provide discussions of important noise sources and techniques that can be used to minimize their effects. The discussions center on noise in linear-beam tubes where noise phenomena are well understood. Noise phenomena in crossed-field devices are still not understood. The chapter starts with a description of thermal agitation noise, which is the noise basis against which all other noise is compared, and a review of the definitions of noise figure. The chapter ends with a discussion of the minimum noise figure that is obtainable in a beam-type device.

23.1 THERMAL AGITATION NOISE

Thermal agitation noise results from the minute voltage generated by the random motion of free electrons in conductors. The voltage produced by thermal agitation has energy distributed throughout the entire frequency spectrum. In the frequency band $f_2 - f_1$, the effective voltage produced, e, is given by

$$e^2 = 4kT\int_{f_1}^{f_2} R\,df \tag{23-1}$$

where k = Boltzmann's constant, T = absolute temperature, and R = resistive component of impedance. For R independent of frequency and, letting B be bandwidth,

$$e^2 = 4kTR(f_2 - f_1) = 4kTRB \tag{23-2}$$

Thermal agitation noise sets the ultimate limit to the smallest signal that can be amplified without being lost in a background of noise. For example, for R = 1/2 MΩ, T = 300K, and B = 5 kHz, the effective noise voltage, e, is 6.4 μV. Signal amplitudes on the order of this voltage level (6.4 μV) or smaller would be distorted or lost in the noise.

The thermal power, N_g, generated in a resistance is

$$N_g = \frac{e^2}{R} = 4kTB \tag{23-3}$$

If R is connected to a matched load resistance, R_L, maximum noise transfer occurs. Then the noise voltage, e, divides equally between R and R_L and so the noise power, N_d, delivered to R_L is

$$N_d = \frac{\left(\frac{e}{2}\right)^2}{R} = kTB \tag{23-4}$$

This is the theoretical minimum noise level. At room temperature, its value is

$$N_d \text{ (room temp)} = -144 \text{ dBW/ MHz} \tag{23-5}$$

That is, for a 1-MHz bandwidth, N_d is 144 dB below 1W (i.e., 4×10^{-15} W).

23.2 DEFINITIONS OF NOISE FIGURE

For an amplifier or receiver that

1. Is noise free;
2. Has gain G;
3. Is perfectly matched to its input termination;

the output noise power is $kTBG$. The noise figure, F, for a real amplifier or receiver that generates noise is

$$F = \frac{\text{actual output noise power at 290K}}{\text{theoretical minimum noise power at 290K}} \tag{23-6}$$

or

$$F = \frac{N_1}{kT_oBG} \tag{23-7}$$

where N_1 is the measured output noise power at temperature T_o = 290K. The amount of noise added, N_a, by an amplifier is

$$N_a = N_1 - kT_oBG = (F-1)kT_oBG \tag{23-8}$$

For a noise-free amplifier, $N_a = 0$ and $F = 1$.

Noise figure can be expressed in decibels as follows

$$F_{dB} = 10\log_{10} F \tag{23-9}$$

For a noise-free amplifier, $F_{dB} = 0$.

The equivalent noise temperature, T_e, of an amplifier is defined using

$$N_a = k\,T_e\,B\,G \tag{23-10}$$

and is

$$T_e = \frac{N_a}{k\,B\,G} = (F-1)\,T_o \tag{23-11}$$

23.3 OVERVIEW OF NOISE PHENOMENA

There are three unavoidable sources of noise in a microwave tube:

- Shot noise;
- Velocity noise;
- Thermal agitation noise.

Shot noise results from the discrete nature of the electron and the fact that electron emission is a random process. Velocity noise occurs because of the wide distribution (Maxwellian) in velocities of electrons emitted by the hot cathode. This noise is often referred to as Rack velocity noise, named after A. J. Rack, who first described a technique for analyzing the effects of this noise. Thermal agitation noise is present because the RF circuit, which has loss, is normally at room temperature or above.

In addition to these three unavoidable noise sources, there are many other sources of noise that depend largely on some aspect of design and construction. Most of these can be made insignificant through the use of careful design and construction procedures. Some of these noise sources result from:

- Partition effect;
- Flicker effect;
- Collision ionization;
- Secondary and reflected electrons;
- Lens effects;
- Noise growth;
- Poor insulation;
- Varying insulator charges;
- Microphonics resulting from vibration;
- Multipactor;
- Power-supply induced noise, such as hum.

The most significant noises in a linear-beam tube are usually those produced by the cathode (shot and velocity). These noises are modified in significant ways as the electron beam travels through the electron gun and into the RF region of the tube.

An extremely important characteristic of noise impressed on an electron beam is that noise propagates like any other RF signal on the beam. That is, noise travels as fast and slow space charge waves. The two sources that produce these waves are the shot effect current and the velocity noise. In the general case, the two sources are assumed to be uncorrelated, and so each source produces fast and slow waves. As these waves propagate along the beam, noise current and velocity standing waves are independently produced.

In low-power, low-noise TWTs, some degree of correlation between the current and velocity noise sources is produced with a special low velocity drift region in the electron gun. Then care is taken in accelerating the electron beam in the gun so that the amplitudes of the standing waves that develop are minimized. The design of the acceleration region in the gun is analogous to the design of an impedance matching section of a transmission line. Finally, as the beam leaves the gun, the axial position of the entrance of the beam into the helix is adjusted (by properly positioning the helix) to coincide with a noise minimum.

With no correlation between current and velocity noise, the minimum noise figure for a TWT is predicted to be 6.5 dB. With correlation (produced by the low velocity drift region in the gun) noise figures below 3 dB have been achieved. It has also been found that the use of a large axial magnetic field in the acceleration region of the gun and in the input section of the RF structure further inhibits noise. With large magnetic fields in these regions, noise figures as low as 1 dB have been achieved and it has been predicted that, with additional refinements, a noise figure of 0 dB could be approached.

Some of the discussions in this chapter relate primarily to low power, low noise TWTs that are no longer used in modern systems. However, the noise phenomena in these TWTs exist, to some extent, in all TWTs so it is of value to discuss them.

23.4 NOISE IN ELECTRON GUNS

For a discussion of noise properties, it is convenient to divide the electron gun into four regions as shown in Figure 23-1. The unavoidable sources of noise at the cathode surface (Region I) are velocity noise as described by Rack and shot noise. There may also be flicker noise but this is a low frequency noise that is negligible at microwave frequencies. Flicker noise will be discussed at the end of this section.

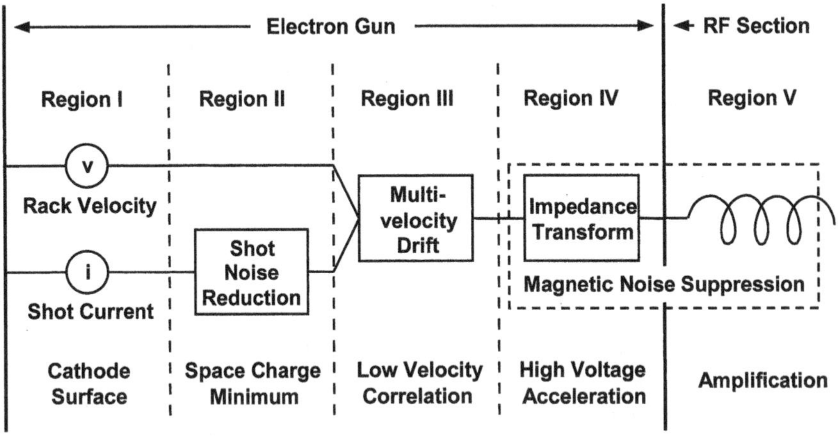

Figure 23-1 Noise regions in an electron gun.

Because of the initial emission velocities of the electrons, the potential of the space charge cloud (Region II) near the cathode surface is slightly below cathode potential. Low velocity electrons are returned to the cathode by this potential minimum. The amount by which the potential is depressed fluctuates as the cathode current fluctuates (shot effect). Thus, the number of electrons returned to the cathode fluctuates with time and this, in turn, changes (reduces at low microwave frequencies) the shot noise current passing the potential minimum. Additional noise may be introduced at the potential minimum by ions and reflected electrons from the collector. It turns out that the velocity distribution of the electrons is unaffected by the potential minimum.

Region III exists in low-noise TWTs through careful design of the cathode configuration and the electrodes near the cathode. In the gun of a high-power TWT this region, if it exists at all, is so short that it has negligible effect on noise performance.

In low-noise TWTs, electrons drift at low velocities (energy <1 eV) for an extended distance in Region III. Noise current and velocity become correlated to some degree, and this reduces noise figure by about 3 dB. The reason for this correlation is not clear.

When the electron beam emerges from Region III and is accelerated in Region IV, space charge waves are launched and a standing wave pattern is produced. As is the case in an RF transmission line, a sudden change in impedance produces an increase in the standing wave ratio. In a low-noise TWT, an increase in the standing wave ratio results in an increase in noise figure. To avoid this problem, several electrodes are used in Region IV to control the rate of voltage rise (and the rate of impedance change). In high-power microwave tubes, the function of the electron gun is to produce a well-focused high-perveance beam and no attempt is made to control the rate of impedance change.

As the electron beam emerges from the electron gun into Region V, care is taken in low-noise tubes to position the entrance of the RF structure at the noise minimum. A strong magnetic field in Regions IV and V has been found to reduce noise. The phenomena occurring in Regions I–IV are described in Sections 23.5 to 23.7.

23.5 NOISE GENERATION AT THE CATHODE

23.5.1 Shot Noise

Shot noise [1] results from the random emission of electrons from the cathode. The randomly emitted electrons are a component of the noise current in the electron beam. This noise current is coupled to nearby RF circuits as discussed in Chapter 8 and illustrated in Figure 23-2. The circuit current can be found from a Fourier analysis of the induced current pulses. The mean square of the current fluctuations, i^2, depends only on emitted current, I_o, and bandwidth, B, as follows:

$$i^2 = 2\,e\,I_o B \qquad (23\text{-}12)$$

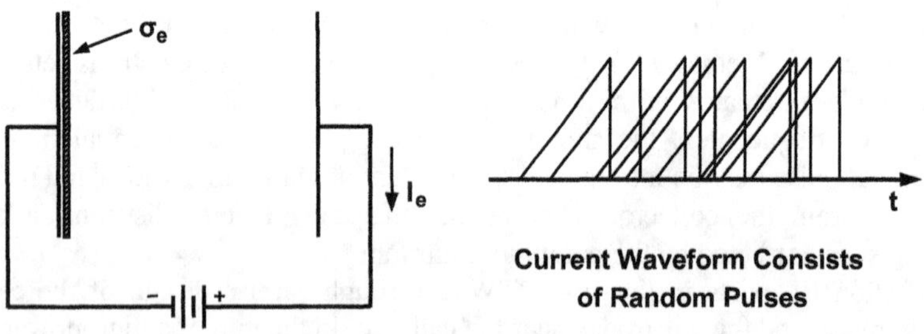

Figure 23-2 Random emission of electrons from cathode produces shot noise.

If this current is passed through a resistance, R, the noise power generated is

$$i^2 R = 2\,e\,I_o B\,R \qquad (23\text{-}13)$$

For example, assuming R = 100Ω, I_o = 1A, and B = 1 MHz, the noise power generated is

$$i^2 R = 3.2\times10^{-11}\ \text{W/ MHz} = -105\ \text{dB/ MHz} \qquad (23\text{-}14)$$

The resulting noise figure is

$$F = 39\ \text{dB} \qquad (23\text{-}15)$$

Thus, shot noise can be a major source of noise in a TWT.

23.5.2 Velocity Noise

Velocity noise arises from the distribution in velocities of the electrons emitted from a hot cathode. The distribution is Maxwellian and, at 888°C, is as shown in Figure 23-3.

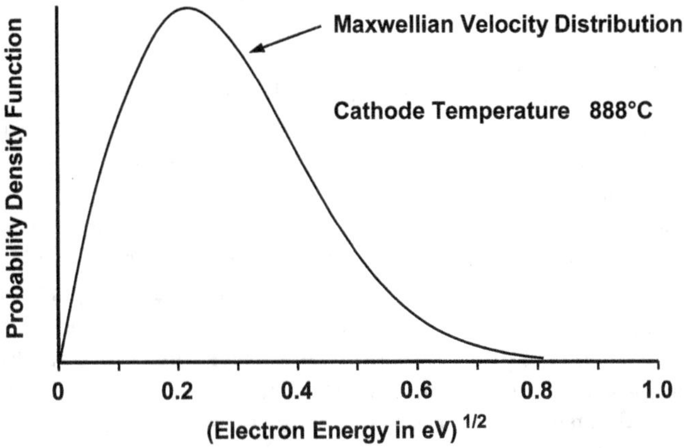

Figure 23-3 Velocity distribution of electrons from a cathode at a temperature of 888°C.

The multivelocity flow of electrons from the cathode that results from the Maxwellian velocity distribution is difficult, if not impossible, to treat analytically. Rack [2] showed that, at low frequencies, this multivelocity flow can be replaced by a single-velocity flow with the same mean square fluctuation in velocity, which is

$$v^2 = (4 - \pi)\eta \frac{k T_c}{I_o} B \tag{23-16}$$

where T_c is the temperature of the cathode in Kelvins.

We will discuss in Section 23.8.1 how Peterson and Pierce used the Rack velocity expression to predict noise behavior at microwave frequencies. Cutler and Quate found surprisingly good experimental agreement with the Pierce prediction.

23.5.3 Other Noise Generation Mechanisms

At low frequencies, two other (other than shot and velocity) noise generation mechanisms occur. These are hum produced by the filament and flicker noise. The reason for the hum was discussed in Chapter 5 on thermionic cathodes. Magnetic flux from the filament linking through the cathode surface disturbs the trajectories of electrons and results in modulation of the beam shape.

The flicker effect produces a large amount of noise over and above shot noise. Figure 23-4 contains representative plots of flicker noise for directly and indirectly heated cathodes [3]. Note that flicker noise can be orders of magnitude above shot noise. Note also that flicker noise drops rapidly as frequency is increased. There appear to be many possible causes for flicker noise [4], depending on the type of cathode being used. For dispenser cathodes of the types normally used in TWTs, the original theory of flicker noise proposed by Schottky may apply.

Schottky attributed the flicker effect to foreign atoms arriving and departing from the cathode surface at random. For temperature-limited cathodes, he predicted that the mean square noise current should be proportional to the square of emission current density and, at high frequencies, inversely proportional to the square of frequency. Flicker noise was predicted to be even more strongly suppressed by space charge than shot noise.

A dispenser cathode consists of a monoatomic layer of barium on oxygen on the substrate metal (tungsten in a B cathode, osmium in an M cathode, and so forth). During operation of the cathode, there is continual evaporation of atoms from the surface and migration of additional atoms from the tungsten substrate to the surface. The evaporation of atoms from the surface is a random thermal process that produces fluctuations in work function and, thereby, in emission current. These fluctuations are thought to be the primary source of flicker noise in a dispenser cathode.

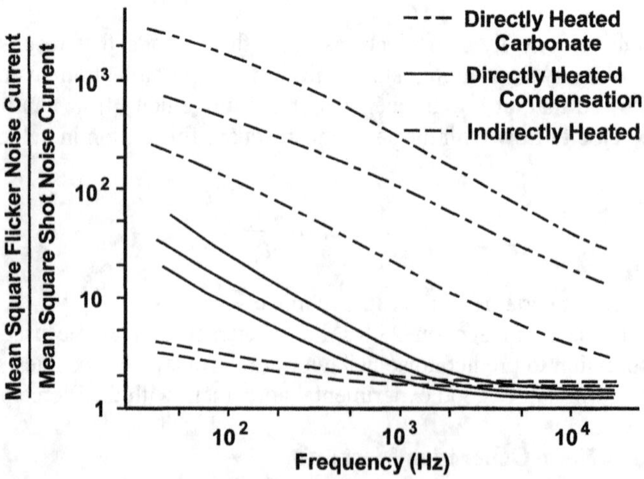

Figure 23-4 Flicker noise for some directly and indirectly heated cathodes. (Adapted from: *The Oxide-Coated Cathode* by G. Herrman and P. S. Wagener, Chapman and Hall, 1951.)

23.6 THE SPACE CHARGE MINIMUM REGION

The electron emission velocities were ignored when we examined space charge limited flow. Emission velocities are included in the analysis of noise behavior. Because of these velocities, there is a small depression in the potential near the cathode surface. This depression, which is indicated in Figure 23-5, reflects low energy (< ~0.2 eV) electrons and returns them to the cathode. At the potential minimum, the energies of all other electrons are reduced by ~0.2 eV.

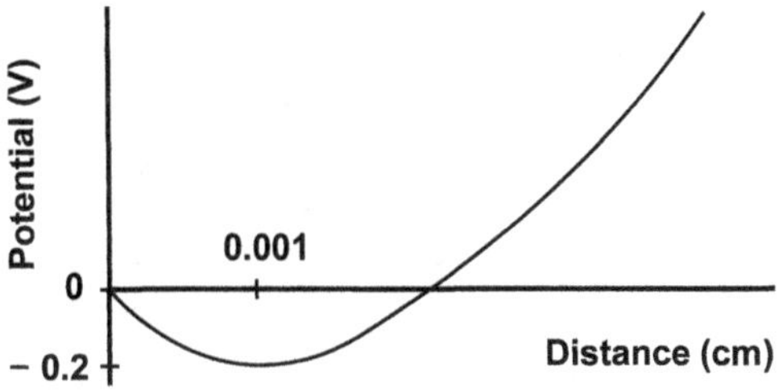

Figure 23-5 Initial emission velocities of electrons result in depression of potential near the cathode.

Figure 23-6 shows plots of electron positions as functions of time [5]. Each plot also shows the distance from the cathode of an electron at time t and the time at which the electron must have left the cathode surface (depending on its initial velocity) to arrive at that position. Note that the time required for electrons to travel to the potential minimum (located at 0.0075 mm) is only ~0.02 ns.

23.6.1 Rack Noise Invariance

Because the initial velocity distribution is Maxwellian and because the velocities of all electrons that reach the potential minimum are reduced by the same amount, the velocity distribution at the potential minimum is Maxwellian. Figure 23-7 shows velocity distributions at the cathode surface and at the potential minimum.

23.6.2 Shot Noise Reduction

Because emission is random (shot effect), the potential minimum produced by the electrons fluctuates with time. The increase in depression produced by a burst of electrons is shown in Figure 23-8. The additional depression increases the number of reflected electrons, thereby reducing the number of electrons in the burst that pass the minimum. As a result, the potential minimum reduces the shot noise current. The amount of reduction depends on frequency and approaches 100% at low frequencies. Fluctuations in the potential minimum do not change the average velocity or the velocity distribution of the electrons.

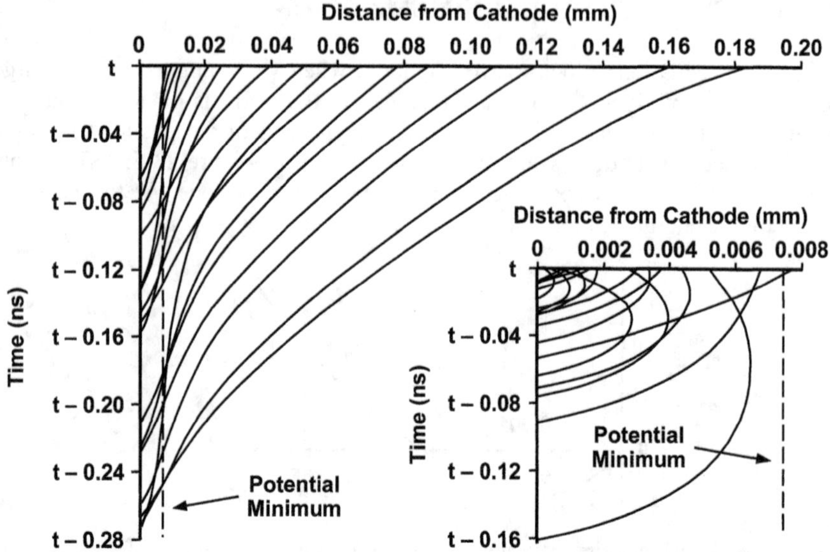

Figure 23-6 Electron velocity plots resulting from the Maxwellian velocity distribution in Figure 23-3. (From: P. K. Tien, and J. Moshman, *Jour. App. Phys.*, September 1956.)

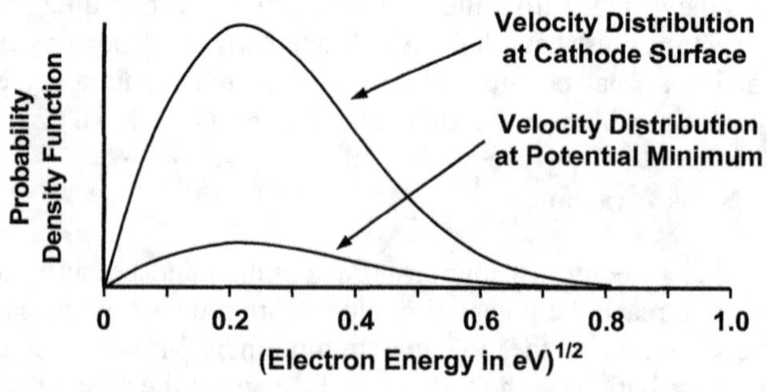

Figure 23-7 Electron velocity distributions at the cathode surface and at the potential minimum.

At frequencies approaching the electron plasma frequency, a resonance condition exists that enhances shot noise. One explanation for this resonance may be understood with the aid of Figure 23-9 [6]. The equivalent circuit for the cathode-to-potential-minimum region may be thought of as the capacitance of this region in parallel with an inductance that represents the inertial effect of the space charge cloud at the minimum. The quantity, G, is the conductance of the retarding field diode (cathode to potential minimum). The plasma frequency at the potential minimum for the current densities used in some low noise TWTs is in the low microwave frequency range (3–4 GHz).

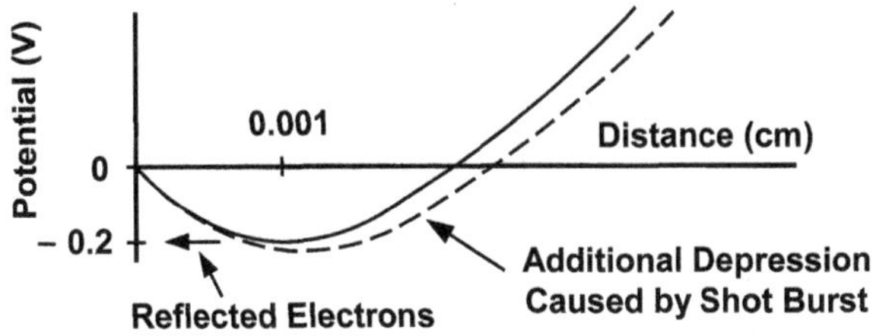

Figure 23-8 Perturbation of potential minimum by shot burst.

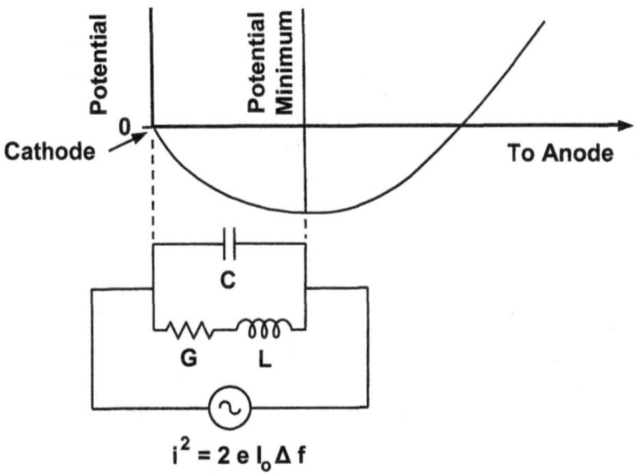

Figure 23-9 Circuit analog for potential minimum region. (From: R. W. Peter, *Noise in Electron Devices*, published by MIT and John Wiley & Sons, 1959.)

Predictions of the shot noise reduction factor Γ^2 (the amount by which the shot current is reduced) are shown in Figure 23-10. For these predictions, the ratio of emitted current, I_s, to the current passing the potential minimum, I_o, was 5. (The remainder of the current consisted of low-velocity electrons that were reflected by the potential minimum and returned to the cathode.) For frequencies near the plasma frequency, shot noise current is increased by a factor of about 2.

23.6.3 Other Noise Effects

Charged particles other than electrons from the cathode will also affect the potential minimum. For example, positive ions generated in or near the electron gun will travel toward the cathode. As they pass through the potential minimum they cause the voltage to rise, so additional current from the cathode can pass the minimum region. Since the ion production process is random, the fluctuations of the potential minimum that they produce are random and the resulting current fluctuations are random.

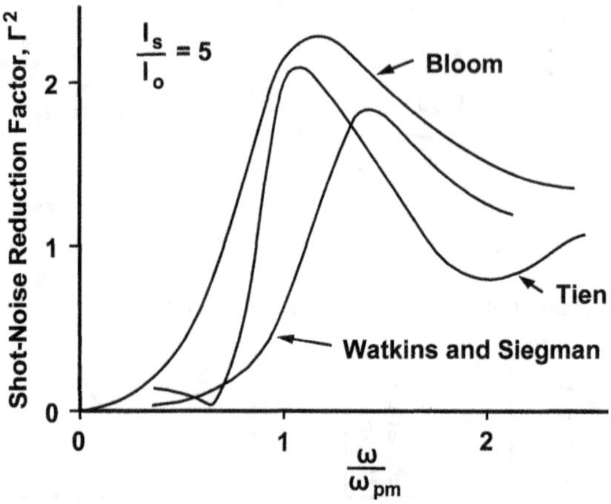

Figure 23-10 Predictions of shot noise reduction factor. (Adapted from: R.W. Peter, *Noise in Electron Devices*. MIT and John Wiley & Sons. 1959)

Another source of charge that can affect the potential minimum is high-velocity secondary electrons, primarily from the collector. These electrons are guided through the tube in the reverse direction by the magnetic focusing field. If their interaction with the RF fields is such that they gain (or at least do not lose) energy, they will be able to reach the potential minimum region. The secondary emission process is a noisy process and so secondary electrons in the potential minimum region produce random fluctuations of the potential minimum, which, in turn, produce an additional component of noise modulation on the emission current passing the potential minimum.

Finally, it is noted that there is evidence that noiselike fluctuations of the space charge cloud in the potential minimum region may occur. These fluctuations occur on a large scale in the vircator (virtual cathode oscillator). In this device a dense beam of electrons is used to produce a fluctuating potential minimum. Reasonably efficient conversion of beam energy to generally noisy RF power takes place.

23.7 LOW-VELOCITY CORRELATION REGION

The biggest breakthrough on low-noise tubes was made by Currie [7] when the possibility of producing correlation between shot current and velocity noise was made apparent. Currie's low noise electron gun is shown in Figure 23-11. The cathode was an annulus surrounded by an electrode, called the profile-shaping electrode. In Currie's early experiments, there was also a portion of the profile shaping electrode in the center of the annular cathode. This was later shown to not be needed. Anode 1, which played an important role in shaping the potential profiles near the cathode, contained an annular opening through which the hollow beam from the cathode could pass. Anodes 2, 3, and 4 were used to provide the voltages for the desired impedance transformation to minimize the standing wave ratio of the of the space charge waves (to be described next).

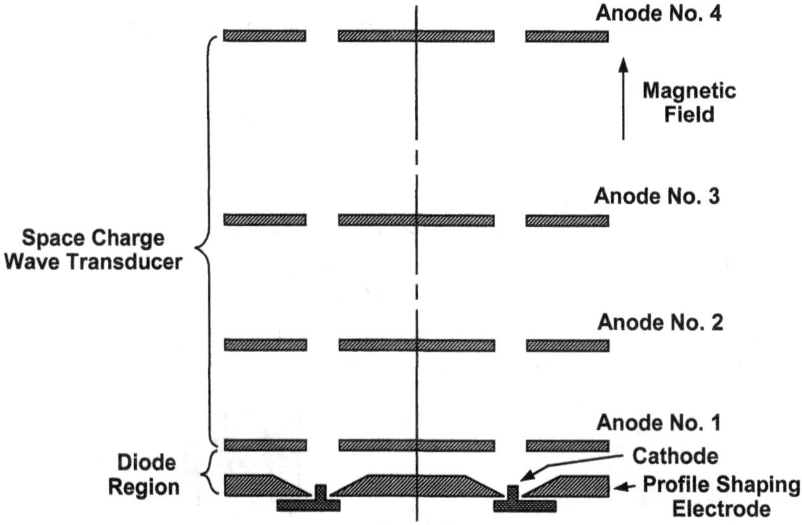

Figure 23-11 Currie's low noise electron gun. (From: M. R. Currie and D. C. Forster, *Jour. App. Phys.*, January 1959.)

The excess noise (noise produced by the beam) with beam conditions adjusted for minimum beam noisiness is shown in Figure 23-12. By comparison, the minimum excess noise predicted by theory with no correlation is shown by the dashed line. The noise figure for the TWT at 2.54 GHz was 3.6 dB. Noise figures as low as 2.9 dB were obtained.

The potential profiles in the cathode region when voltages were adjusted for near-optimum conditions are shown in Figure 23-13. The electric fields near the cathode were such that most of the electron emission was from the edges of the cathode. A strong axial magnetic field was used to form the electrons into a very thin hollow beam.

From Figure 23-13, it is seen that after the electrons left the region near the edge of the cathode and moved toward the anode, they drifted through a region of very slowly varying potentials. Thus, the situation created by Currie in his electron gun forced the electrons to drift for a considerable distance at a low average velocity that was comparable to the noise velocities of the electrons. As a result of this multivelocity flow, the basic noise quantities in the beam were transformed and some degree of correlation between noise current and velocity was achieved. The physical mechanism(s) for this transformation have not been described.

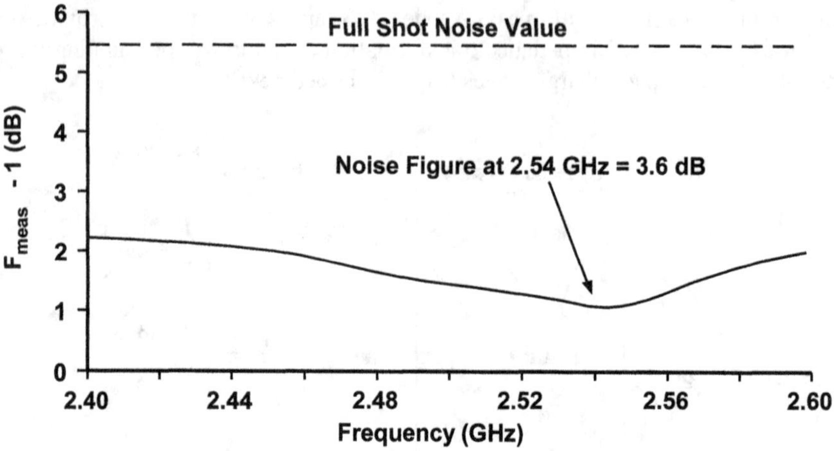

Figure 23-12 Excess noise with beam conditions adjusted for minimum beam noisiness. (From: M. R. Currie and D. C. Forster, *Jour. App. Phys.*, January 1959.)

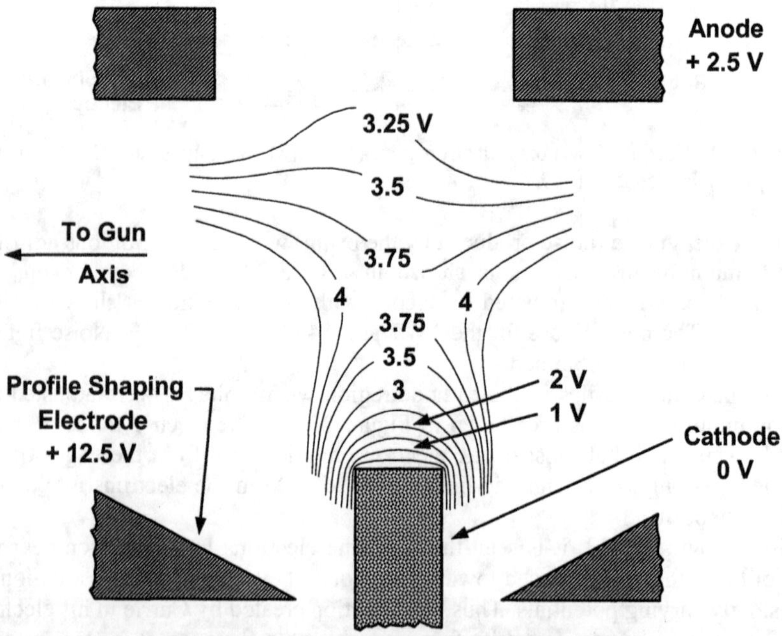

Figure 23-13 Near optimum potential distribution for Currie's low-noise gun. (From: M. R. Currie and D. C. Forster, *Jour. App. Phys.*, January 1959.)

23.8 HIGH-VOLTAGE ACCELERATION REGION

To understand the rational for the design of the high voltage acceleration region in a low-noise TWT, it is necessary to understand the nature of the noise space charge waves that propagate on the beam. In this section, we will first discuss the propagation of noise space charge waves and then will discuss the impedance transformation in an electron gun necessary to avoid large noise standing wave ratios. At the end of this section, a discussion of lens effects on noise is included. Lens effects cause radial components of noise to be converted to axial noise and can thereby significantly increase the noise figure of a TWT.

23.8.1 Noise Space Charge Waves

The propagation of noise as fast and slow space charge waves was predicted by Pierce [8] and by Peterson [9]. They carried out analyses of the noise resulting in an electron beam after acceleration of the electrons in the gun. They used the equations of Llewellyn and Peterson [10] for a space charge limited diode. We will not reproduce the Llewellyn-Peterson equations here, but will point out that two of the initial conditions for these equations are the noise current and velocity at the potential minimum. Because the dc velocity of the electrons is zero at the potential minimum, the equations predict that the effect of shot noise current vanishes. This is equivalent to using a shot-noise reduction factor of 0, which, in turn, is equivalent to assuming that fluctuations of the potential minimum produced by the shot current completely eliminate shot noise current beyond the potential minimum. Thus, only Rack velocity noise was used to predict noise in an electron beam.

During acceleration of the electrons in the gun, some of the Rack velocity fluctuations result in current noise, and so both current and velocity noise exist in the electron beam. Pierce used the noise current and velocity at the exit from the gun as input conditions to a TWT and used his TWT theory to predict fast and slow noise space charge waves on an electron beam.

The theories of Peterson and Pierce (and of Rack) met with considerable skepticism. However, the December 1950 issue of *Physical Review* [11] published C. C. Cutler and C. F. Quate's paper "Experimental Verification of Space Charge and Transit Time Reduction of Noise in Electron Beams." The noise-measuring apparatus used by Cutler and Quate is shown in Figure 23-14. A cavity was used to detect a portion of the noise current in the electron beam. A shutter could be moved across the beam to investigate partition current (to be discussed later in this chapter).

The key result of the Cutler-Quate experiment is shown in Figure 23-15. Measured noise power was compared with the predictions of Pierce. The existence of noise space charge waves was clearly verified. In addition, the variation of the position of the minimum with voltage and with the transit angle of the electrons in the gun agreed with the theoretical prediction. The reason that the position of the minimum differs from the predicted minimum by 0.4 in was thought to be that calculations were based on an infinite plane diode configuration rather than a cylindrical beam configuration in the gun.

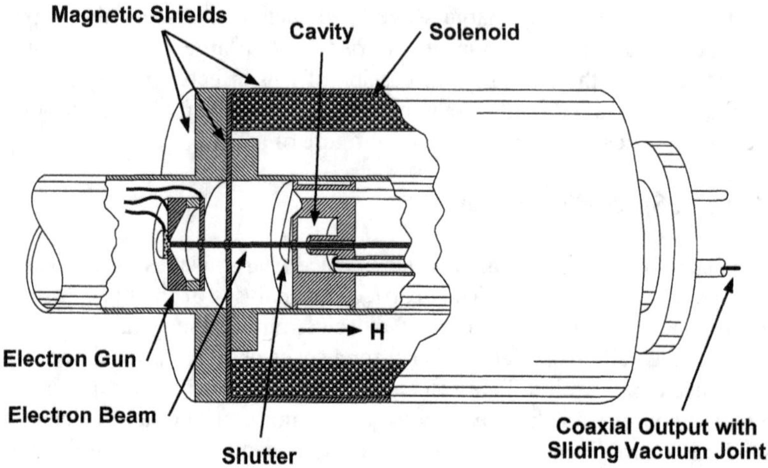

Figure 23-14 Cutler and Quate's noise measuring apparatus. (From: C. C. Cutler and C. F. Quate, *Phys. Rev.*, December 1950.)

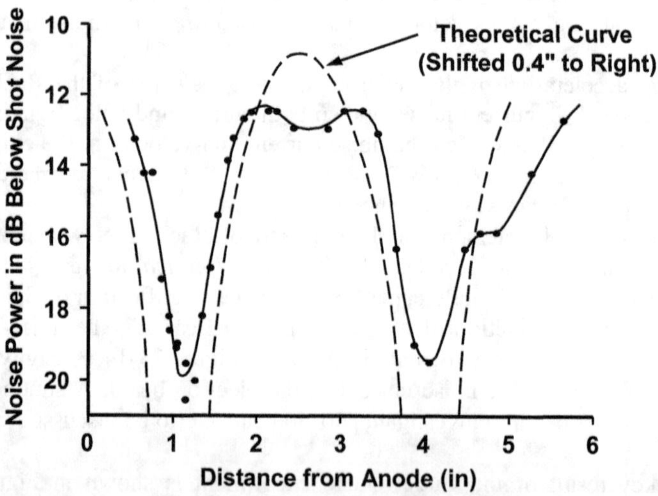

Figure 23-15 Noise power versus distance from the anode. (From: C. C. Cutler and C. F. Quate, *Phys. Rev.*, December 1950.)

The absolute values of the noise levels in Figure 23-16 were determined by comparing them with the temperature-limited shot noise. This was done by reducing the cathode temperature and measuring the noise power as a function of beam current as shown in Figure 23-16. The slope of the curve at low beam currents shows the shot noise power to be proportional to current as expected (23-12). At larger beam currents, the noise power was reduced by space charge. The greatest total reduction of noise power occurred at the first minimum and was 20 dB below the shot noise level.

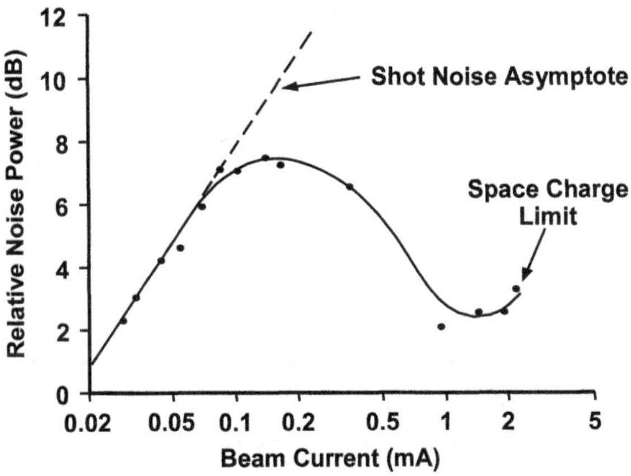

Figure 23-16 Noise power versus beam current for temperature limited emission. (From: C. C. Cutler and C. F. Quate, *Phys. Rev.* December 1950.)

23.8.2 Impedance Transformation for Low-Noise Tubes

The electron beam acts as a transmission line to carry space charge waves launched by shot current and Rack velocity noise generators in the potential minimum region. From the shot current source, fast and slow current waves emerge and produce a current standing wave. Associated with this current wave is a velocity standing wave that is displaced one quarter of a plasma wavelength from the current wave.

The Rack velocity noise generator produces fast and slow velocity waves that form a velocity standing wave. Associated with this velocity wave is a current standing wave that is displaced one quarter of a plasma wavelength from the velocity wave. These standing waves produced by the velocity generator are shown for a space charge limited planar diode in Figure 23-17 [12]. Velocity and current waves for two space charge conditions are shown. For a high space charge a velocity minimum and current maximum occur near the potential minimum (where the relative distance through the diode is zero). Subsequently, as the electrons are accelerated through the diode, the noise velocity wave grows and the noise current wave decreases. For the low space charge condition, the velocity minimum and noise current maximum occur near the anode.

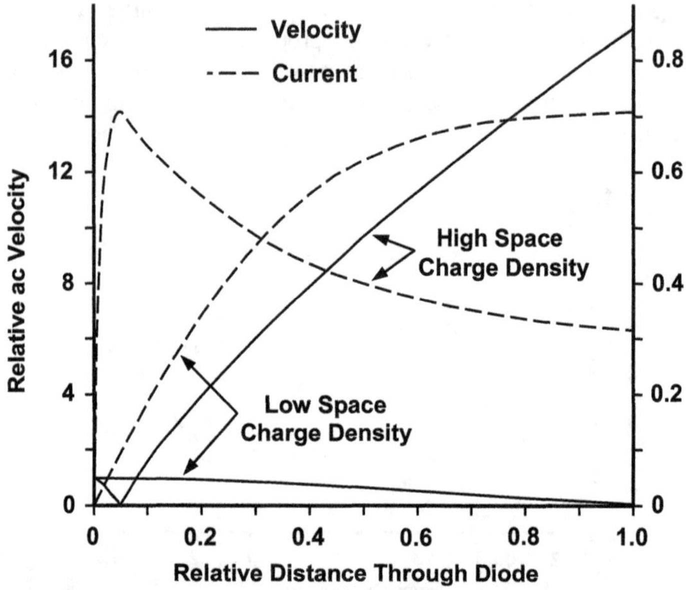

Figure 23-17 Noise velocity and current in a space charge limited diode. (Adapted from: S. Bloom and R. W. Peter, *RCA Review*, March 1954, copyright 1954 by RCA.)

The two current standing waves (one from the shot noise current source and the other from the Rack velocity noise source) are separated by a one-quarter plasma wavelength (assuming no correlation). As a result, the current distributions along these two waves at the exit from the gun are as shown in Figure 23-18. Each of these waves contributes to the noise figure of a TWT. Thus, in the design of a low noise TWT, it is necessary to:

1. Control the amplitudes of the standing waves;
2. Locate the entrance to the helix at a position that minimizes the noise induced by the two waves.

The amplitudes of the standing waves in the electron gun are governed by the impedance of the electron beam. This impedance is very low at the potential minimum where the electron velocity is low. As the beam is accelerated and the velocity increases, the impedance increases (see Figure 23-18). The impedance at the input to the helix may be over two orders of magnitude greater than the impedance at the potential minimum.

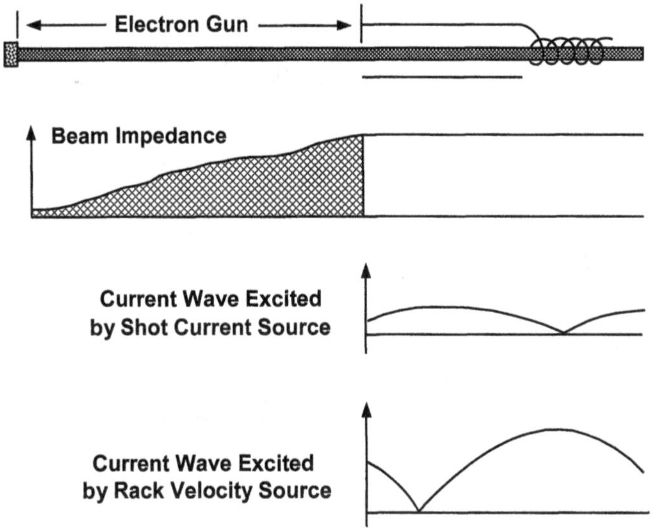

Figure 23-18 Beam impedance in gun and current waves that emerge from gun.

To control the current standing waves so that minimum noise is introduced in the RF section, it is necessary to avoid rapid changes in beam impedance. This may be accomplished with an impedance transformer like that shown in Figure 23-19 [6]. Several electrodes are used in the electron gun to control the rate of acceleration of the beam and, thereby, to control its impedance. In the example shown in Figure 23-19, the impedance variation as a function of distance from the potential minimum to the helix is nearly exponential. This is analogous to the situation in acoustics or RF transmission lines, where an exponential variation in impedance is used to match a low impedance source to a high impedance load.

23.8.3 Lens Effects

The transverse and axial velocities of electrons emitted from a thermionic cathode are assumed to be the same. In a low-noise gun, as electrons are accelerated, the axial velocity fluctuations decrease (see the low space charge curve in Figure 23-17, for example). However, the transverse components of velocity are not appreciably affected. Thus, after acceleration, the transverse velocity components are much larger than the axial components.

As electrons pass through a lens, their transverse and axial velocities are altered. If, as the electrons enter the lens, the fluctuations in their transverse velocities are larger that those in the axial velocities, then the axial velocity fluctuations are increased by the lens. This was proven theoretically and experimentally by Knechtli [13].

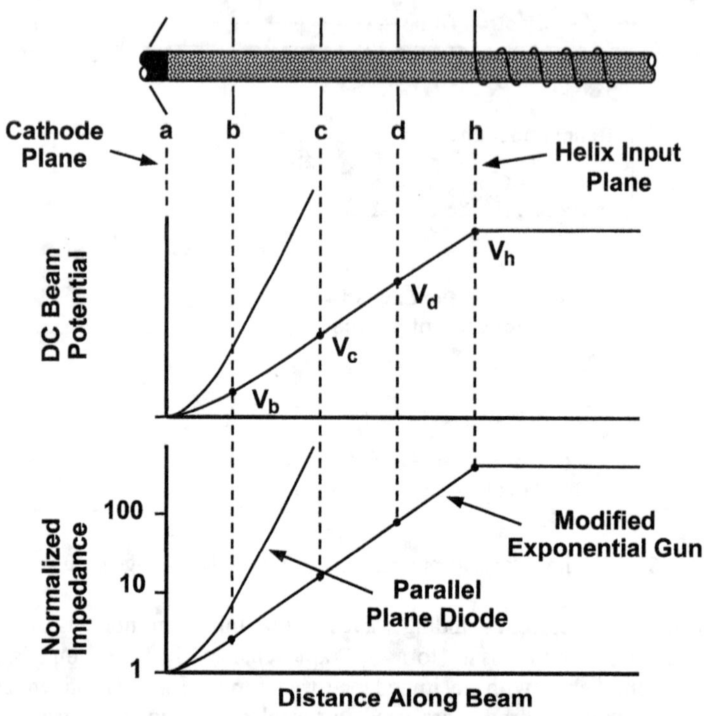

Figure 23-19 Potential and impedance for a low-noise gun. (Adapted from: R.W. Peter, *Noise in Electron Devices*, published by MIT and John Wiley & Sons, 1954.)

The situation illustrated in Figure 23-20 will help in providing a qualitative understanding of the effect of a lens on transverse velocities. The trajectories of three electrons with different transverse velocities are shown. To simplify the illustration, it is assumed that the electrons have no axial velocity fluctuations when entering the lens and that they all pass some plane, z_o, at the same time.

The electrons all pass through the lens at the same time and the trajectory of each is deflected through an angle, φ. At some later time, the electrons arrive at positions 1, 2 and 3, which are separated in the axial direction. For this to happen, the electrons must have different axial velocities. Thus, in passing through the lens, transverse velocity components have been converted to axial velocity components. Thus, it is clear that discontinuities in accelerating or decelerating voltage increase beam noisiness (as long as the transverse velocity fluctuations are larger than the axial ones), and so these discontinuities, which act as electrostatic lenses, should be avoided in low-noise TWTs.

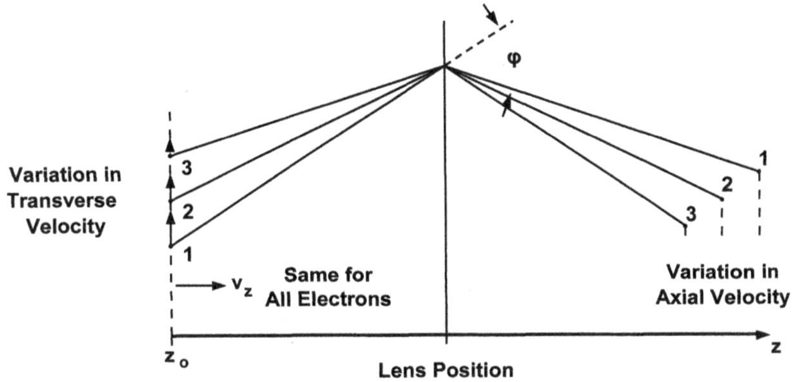

Figure 23-20 Conversion of transverse electron velocities to axial velocities by a lens.

23.9 RF SECTION NOISE PHENOMENA

23.9.1 Circuit Loss

There are two effects of circuit loss on the noise figure. The smaller of the two is the thermal agitation noise that is generated by the resistance of the helix. The larger affect occurs because the noise figure depends on the gain of a TWT and gain, in turn, is affected by circuit loss. The affect of circuit loss on the noise figure amounts to, at most, a few decibels as is shown in [14] and is important in low-noise TWTs. As an example of the effect of helix material on the loss and noise figure, assume that the Pierce loss parameter, d, is 0.35 and so, from Figure 23-21, the contribution to the noise figure is about 2.3 dB. Compared with tungsten, if copper with an excellent surface finish was used for the helix, the value of d would drop to about 0.2 and the improvement in the noise figure would be 0.9 dB.

23.9.2 Partition Noise

Partition noise arises because transverse components of thermal velocity cause fluctuations in the current intercepted by electrodes other than the collector. Transverse components of velocity may induce electron motion in a helix and produce a noise analogous to partition current even though no electrons actually strike the helix. A similar effect can occur in the grid cathode region, where space charge fluctuations induce grid signals. The noise added to the beam by partition current equals the shot noise contained in an equal amount of temperature-limited current [8, 15]. Also, the partition current adds a fully random velocity component to the beam [15].

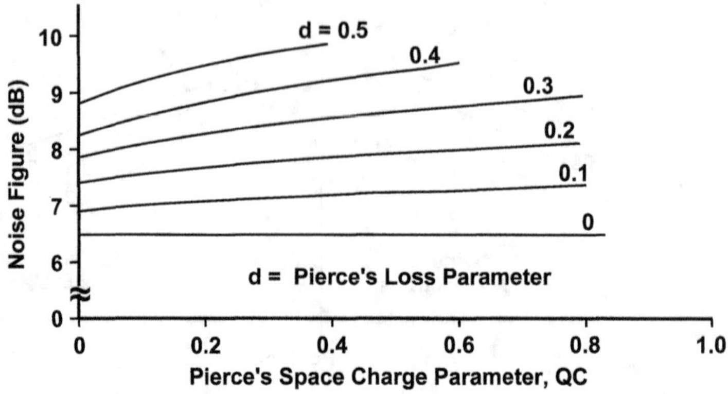

Figure 23-21 Predicted noise figure for low-noise TWTs including the effect of circuit loss.

In the experiments of Cutler and Quate [11], a shutter was used (see Figure 23-14) to intercept a small fraction of the beam current and, thus, control the partition current. The resulting noise data given in Figure 23-22 show a direct correlation between noise power and partition current.

23.9.3 Secondary Electron Interactions

The electron beam striking the collector generates a large number of slow secondary electrons and a small percentage of fast secondaries (often referred to as reflected electrons) that have energies nearly equal to the incident electron energy. The secondary electron current contains noise of the primary beam plus shot noise from the secondary emission process. If a portion of the secondary current escapes from the collector, it can introduce noise on the RF circuit through interception or by interacting with reverse traveling waves on the RF circuit. This noise is amplified and, at the input of the RF circuit, may be partially reflected and thus contribute to the total noise output.

In multistage depressed collectors, great care is taken to minimize the generation of secondary electrons and to prevent their escape from the collector. Still, some secondaries are generated and those that do leave the collector have a wide range of energies. This is because those electrons were generated from electrodes at depressed potentials. Thus, those electrons are accelerated as they travel toward the exit from the collector (and into the interaction region). The problem of noise generation by secondaries from collectors is greatly reduced with properly designed multistage depressed collectors.

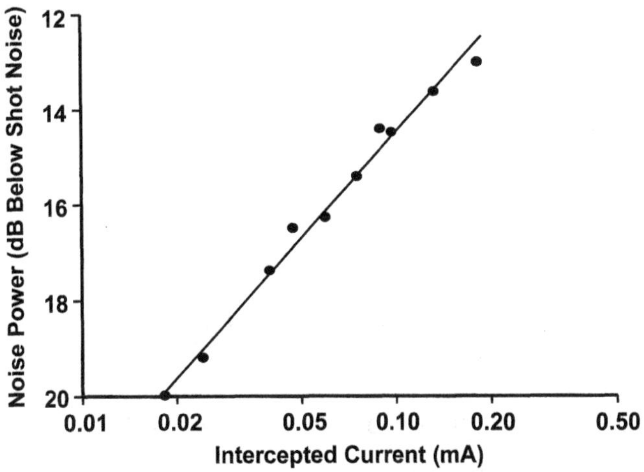

Figure 23-22 Noise power resulting from partition effect in Cutler and Quate's experiment. (From: C. C. Cutler and C. F. Quate, *Phys. Rev.*, December 1950.)

23.9.4 Noise Growth

Measurements of noise on a beam usually show a standing wave pattern like that found by Cutler and Quate [11]. Sometimes, however, noise increases very rapidly up to very large saturation values, as shown in Figure 23-23 [6]. The results of measurements by Gilmour [16] show (see Figure 23-24) that a large increase in peak velocity of the electrons may also occur. Voltage equivalents of electron velocities ~100V above the dc beam voltage were observed. In both Peter's and Gilmour's experiments, the electron gun was shielded and the magnetic field at the cathode was near zero. In measurements made at MIT [17] an auxiliary magnetic field was applied to the cathode region. The growing noise could be eliminated by making the auxiliary field have the same direction as the axial component of the main focusing field at the cathode. The opposite polarity of the auxiliary field increased the noise. As of yet, no theory adequately explains the noise growth phenomena. Scalloped beam amplification has been proposed but the variation of noise growth with the magnetic field that is predicted is far different from the measured results.

23.9.5 Magnetic Noise Suppression

A high magnetic field applied to the helix input region of a low-noise TWT can reduce the noise figure to a very low value. For example, in [18], Hammer and Thomas reported TWT noise figures of 1.0 dB at S-band. The equivalent beam noise temperature was only 57K, and the ratio of this noise temperature to room temperature was 0.19.

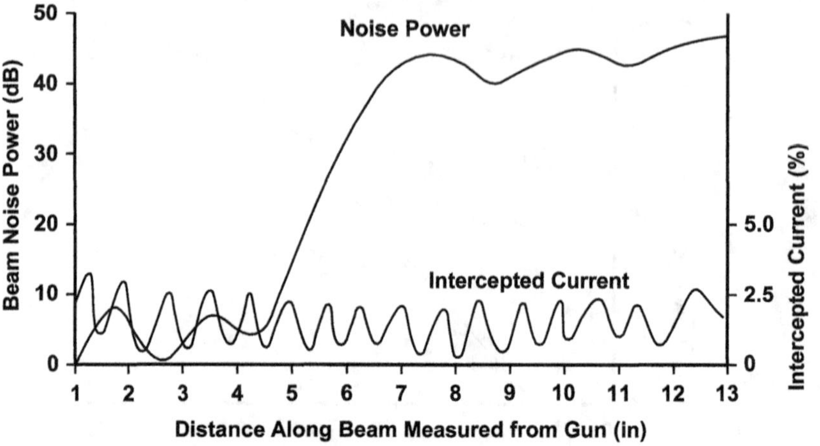

Figure 23-23 Noise growth on an electron beam. (From: R. W. Peter, *Noise in Electron Devices*, published by MIT and John Wiley & Sons, 1954.)

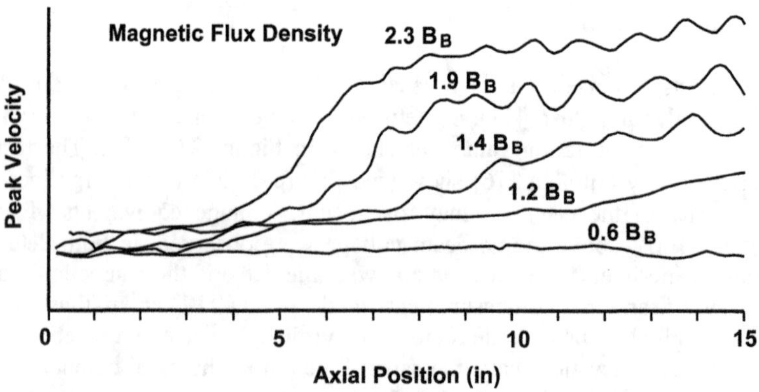

Figure 23-24 Peak electron velocity for various beam focusing conditions. (From: Technical Report No. RADC-TR-64-563, April 1965.)

In a subsequent paper [19], Hammer and Wen showed that the beam noise temperature varied inversely with the magnetic field as shown in Figure 23-25. They stated that these results were obtained after some degradation in noise performance. Before degradation, the straight line through the data passed through the origin, indicating a beam noise temperature of 0K and a noise figure of 0 dB. The Hammer and Wen study indicated that the magnetic field suppresses transverse noise fluctuations that couple to the helix. The transverse fluctuations originate in the current emitted from a thermionic cathode and are statistically independent of the axial fluctuations.

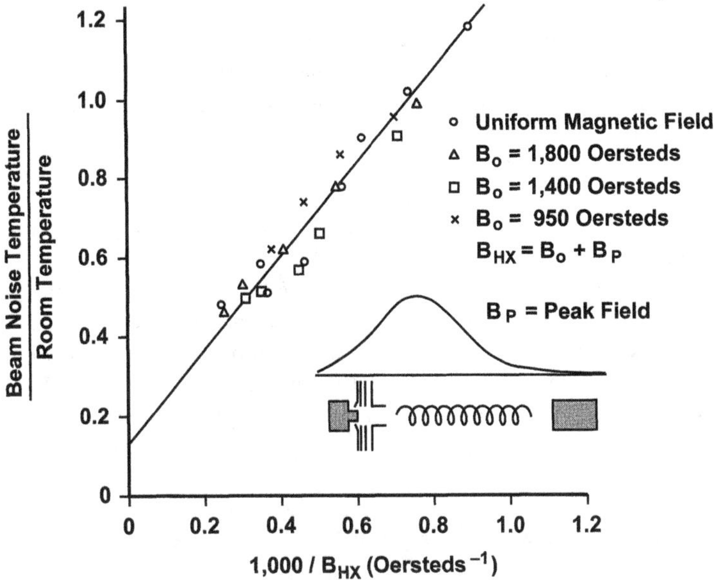

Figure 23-25 Reduction of noise with a high magnetic field. (Adapted from: J. M. Hammer and C. P. Wen, *RCA Review*, December 1964, copyright 1964 by RCA.)

23.10 OTHER NOISE SOURCES

There are several other sources of noise in a microwave tube. Some of these are the following:

1. One or more grids are often used in an electron gun to control current. Some interception of emitted current occurs, and this introduces partition noise. In addition, field distortions occur near the grid, and these introduce transverse velocity components to electron trajectories. The overall affect of the grid is to raise the apparent temperature of the cathode. In one report in which a computer code was used to simulate a shadow gridded gun with a fictitious cathode temperature, the best correlation between computed and measured beam interception was obtained with an apparent cathode temperature of 10,750K.
2. Hum results from inadequate filtering of power supply voltages and can result in RF modulation at power frequencies and their harmonics. This will be discussed further in Chapter 24. The hum produced by an ac filament magnetic field that links the cathode surface was discussed previously in this chapter.

3. Faulty insulation can result in breakdown, corona, or partial discharges that produce fluctuations in voltages and modulation of the RF signal.
4. Multipactor is a form of RF vacuum breakdown that occurs across vacuum gaps or on insulator surfaces. Noiselike impedance fluctuations occur that modulate the RF signal.
5. Time-varying charge accumulations on insulator surfaces, such as the support rods for the helix, can modulate the RF signal.
6. Movement of the cathode, grid, helix, and so forth as the result of shock and vibration produces modulation of the RF signal.
7. "Gas noise" is generated by ions that produce fluctuations of the potential minimum at the cathode surface and, thereby, produce noise. In addition, ions can produce oscillations, beam compression and reduced power output, RF phase shift and phase noise and reduced cathode activity, which, in turn, results in patchy emission and increased flicker and shot noise.

23.11 MINIMUM NOISE FIGURE OF A TWT

An expression for the minimum noise figure of a TWT has been derived by several researchers [20–22] with the assumption that the noise at the potential minimum is uncorrelated shot noise current and Rack velocity noise. With the additional assumptions that the gain is high and that circuit losses are negligible, the minimum noise figure is

$$F_{min} = 1 + \sqrt{4 - \pi}\,\frac{T_c}{T_o} \tag{23-17}$$

where T_c is the cathode temperature and T_o is room temperature. For typical cathode temperatures, F_{min} is in the range of 6.4 to 7.3 dB. This is the minimum noise figure if noise reduction (but not correlation) is provided to minimize noise figure. When helix losses are considered, the minimum noise figure is somewhat higher, as was shown in Figure 23-21.

The derivation of (23-17) is reasonably straightforward, but does not provide a means for taking into account correlation between the shot current and Rack velocity noise. However, the elegant noise theory for linear-beam devices developed by Haus [23, 24] and Robinson [20] takes correlation into consideration.

REFERENCES

[1] W. Schottky, "Spontaneous current fluctuations in electron streams," *Ann. Physik*, Vol. 57, December 1918, pp. 541–567.

[2] A. J. Rack, "Effect of space charge and transit time on the shot noise in diodes," *Bell Sys. Tech. Jour.*, Vol. 17, October 1938, pp. 592–619.

[3] G. Herrman and P. S. Wagener, *The Oxide-Coated Cathode*, London: Chapman and Hall, 1951.

[4] A. van der Ziel, "Low-frequency noise in vacuum tubes (flicker effect)," *Noise in Electron Devices*, New York: The Technology Press of MIT, and John Wiley & Sons, 1959.

[5] P. K. Tien and J. Moshman, "Monte Carlo calculation of noise near the potential minimum of a high-frequency diode," *Jour. App. Phys.*, Vol. 27, No. 9, September 1956, pp. 1067–1078.

[6] R. W. Peter, "Low-noise traveling-wave tubes," *Noise in Electron Devices*, New York: published jointly by The Technology Press of MIT and John Wiley & Sons, 1959.

[7] M. R. Currie and D. C. Forster, "New mechanism of noise reduction in electron beams," *Jour. App. Phys.*, Vol. 30, No. 1, January 1959, pp. 94–103.

[8] J. R. Pierce, *Traveling Wave Tubes*, Princeton, NJ: Van Nostrand, 1950.

[9] L. C. Peterson, "Space-charge and transit-time effects on signal and noise in microwave tetrodes," *Proc. IRE*, November 1947, pp. 1264–1272.

[10] F. B. Llewellyn and L. C. Peterson, "Vacuum-tube networks " *Proc. IRE*, March 1944, pp. 144–166.

[11] C. C. Cutler and C. F. Quate, "Experimental verification of space charge and transit time reduction of noise in electron beams," *Phys. Rev.*, Vol. 80, No. 5, December 1950, pp. 875–878.

[12] S. Bloom and R. W. Peter, "Transmission-line analog of a modulated electron beam," *RCA Review*, March 1954, pp. 95–112.

[13] R. C. Knechtli, "Effect of electron lenses on beam noise," *IRE Trans. Electron Devices*, April 1958, pp. 84–88.

[14] S. Bloom, "The effect of initial noise current and velocity correlation on the noise figure of traveling-wave tubes," *RCA Review*, Vol. 16, No. 2, June 1955, pp. 179–196.

[15] W. R. Beam, "Interception noise in electron beams at microwave frequencies," *RCA Review*, Vol. 16, No. 4, December 1955, pp. 551–579.

[16] T. G. Mihran et al., "The growth of peak velocity, noise, and signal along O-type beams from shielded guns," *Conference on Electron Device Research*, Cornell University, 1964.

[17] L. D. Smullin and C. Fried, "Microwave noise measurements on electron beams," *IRE Trans. Electron Devices*, Vol. ED-1, No. 4, 1954, pp. 168–183.

[18] J. M. Hammer and E. E. Thomas, "Traveling-wave tube noise figures of 1.0 dB at S-band," *Proc. IEEE*, Vol. 52, No. 2, February 1964, p. 207.

[19] J. M. Hammer and C. P. Wen, "Effect of high magnetic field on electron-beam noise," *RCA Review*, December 1964, pp. 785–789.

[20] F. N. H. Robinson, "Space-charge smoothing of microwave shot noise in electron beams," *Phil. Mag*., Vol. 43, 1952, p. 51.

[21] S. Bloom and R. W. Peter, "A minimum noise factor for the traveling-wave tube," *RCA Review*, Vol. 15, 1954, p. 252.

[22] J. R. Pierce and W. E. Danielson, "Minimum noise figure of traveling-wave tubes with uniform helices," *Jour. App. Phys*., Vol. 25, 1954, p. 1163.

[23] H. A. Haus, "Noise in one-dimensional electron beams," *Jour. App. Phys*., Vol. 26, No. 5, May 1955, pp. 560–571.

[24] H. A. Haus and F. N. H. Robinson, "The minimum noise figure of microwave beam amplifiers," *Proc. IRE*, Vol. 43, No. 8, August 1955, pp. 981–991.

Chapter 24

Nonlinearities and Distortion

This chapter addresses the problems of nonlinearities and distortion in amplifiers, which have reasonably constant gain over a wide dynamic range. Crossed-field amplifiers have no linear range of operation and are not considered. At small signal levels, the power output versus power input curve for a linear-beam amplifier is reasonably linear and signal distortion is small. As a device is driven toward saturation, the transfer characteristic becomes more and more nonlinear. This nonlinearity results in distortions of amplitude modulated (AM) or phase modulated (PM) signals for several reasons:

- AM/AM conversion;
- AM/PM conversion;
- Harmonic generation;
- Intermodulation distortion;
- Signal capturing.

In addition, other signal distortions occur, which are not related to saturation effects. Noise, discussed in Chapter 23, is one example. Also, the following are often of concern:

- Variations of gain and power output with frequency;
- Phase nonlinearities or time delay distortion;
- Power supply and load effects (referred to as pushing and pulling).

24.1 DISTORTION RESULTING FROM SATURATION EFFECTS

24.1.1 AM/AM Conversion

The curve in Figure 24-1 shows the RF power output versus input for a TWT. The distortion of a signal is also indicated. A sinusoidal input signal to the TWT is assumed and the resulting output signal is shown. The shape of the output signal was determined graphically, as is indicated by the dashed lines, by plotting output power levels for various input signal levels as a function of time. The distortion resulting from gain compression and saturation is apparent in the flat-topped shape of the output curve.

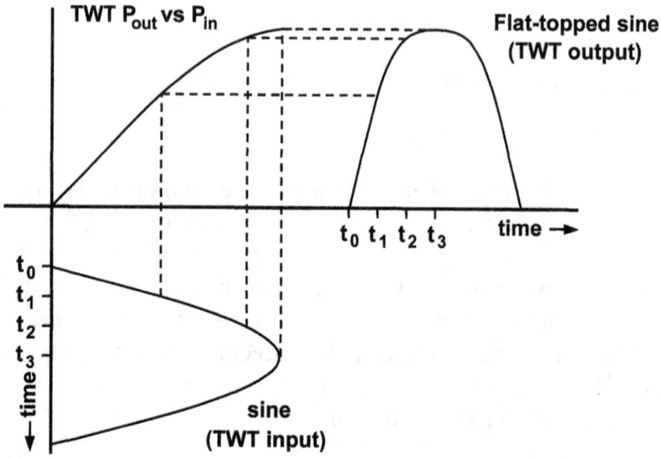

Figure 24-1 AM/AM distortion.

AM/AM conversion is a measure of the change in RF power output that results from a change in RF power input and is usually defined by the slope of the curve of RF power output versus RF power input expressed in dB/dB. Thus,

$$\text{AM/ AM conversion} = \frac{\Delta\ \text{RF power output (in dB)}}{1 - \text{dB change in input power}} \tag{24-1}$$

AM/AM conversion and gain are shown as functions of RF power input in Figure 24-2. Gain is defined as the ratio of RF power output to RF power input. Two values of gain are usually of importance. These are the small signal gain and the saturated gain. Saturated gain is usually 5 to 6 dB less than small signal gain.

In the linear (small signal) region, the AM/AM conversion is very nearly 1 dB/dB. At saturation, where no change in output occurs as the input is varied, the AM/AM conversion is 0 dB/dB.

AM/AM conversion is important in communication applications of TWTs. The specification of AM/AM conversion depends on the type of modulation used. For example, if amplitude modulation is employed and if amplitude distortion must be avoided, then it may be necessary to operate a TWT at a maximum power level well below saturation. For this operating condition, we say that the power level must be "backed off" from saturation.

The amount of "back off" depends on the amplitude distortion that can be tolerated. Most amplitude modulation communication systems can tolerate no less than 0.9 dB/dB of AM/AM conversion [1]. If no provisions are made for linearizing a TWT, it may be necessary to back off as much as 6–8 dB to meet this AM/AM conversion requirement. Through the use of a pitch taper in a TWT [2, 3] and an external linearizing circuit, distortion is significantly reduced and it is possible to operate much closer to the saturation power level. Backed off operation is discussed in more detail later in this chapter.

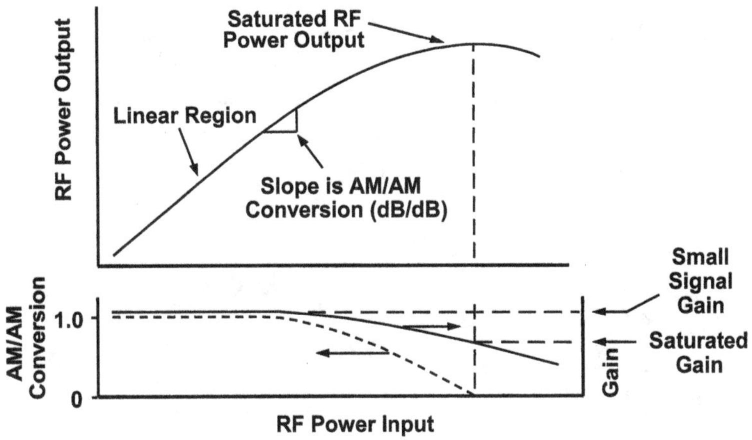

Figure 24-2 Gain and AM/AM conversion versus RF power input. (Adapted from: *L3 Communications TWT and TWTA Handbook.* 2007.)

24.1.2 AM/PM Conversion

AM/PM conversion is a measure of the change in tube phase length that results from a change in RF drive level. This occurs because, as drive level is increased, more power is extracted from the electron beam and the velocity of the beam is reduced. As beam velocity decreases, the velocity of the RF wave on the circuit is reduced and this increases the phase length of the tube. It should be noted here that the cathode voltage supply may aggravate the phase delay. As drive increases, body current usually increases, which in turn increases the load on the cathode supply causing a reduction in voltage and an increase in phase length.

A curve showing the variation of phase length with RF input power is given in Figure 24-3. In the linear region of the RF power output curve, phase shift is relatively insensitive to drive level. As drive is increased toward saturation, the phase length increases rapidly and then becomes constant again as the TWT saturates. The slope of the phase length curve in degrees/dB and shown in Figure 24-4 is the AM/PM conversion.

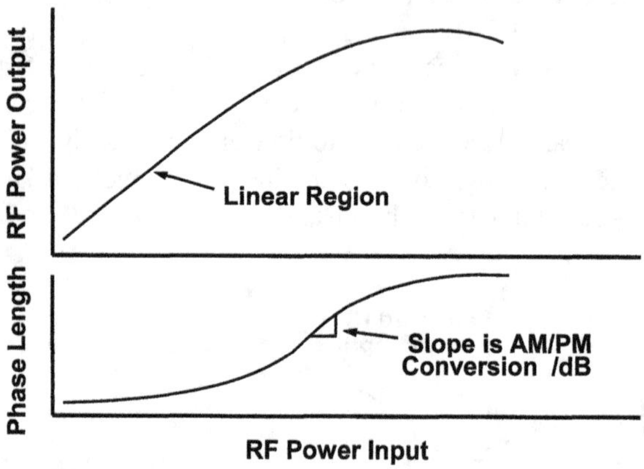

Figure 24-3 Change of phase length with drive power.

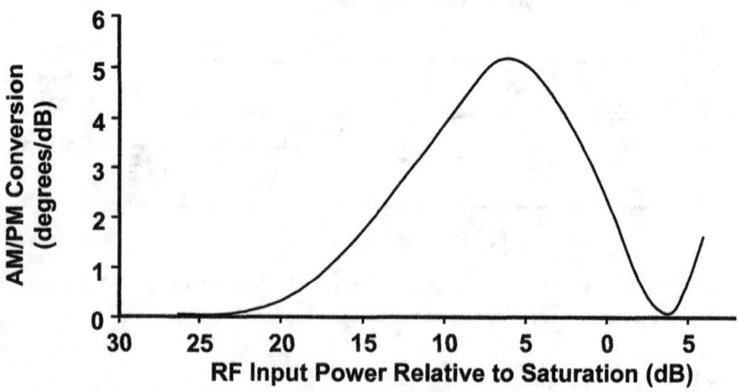

Figure 24-4 Example of variation of AM/PM conversion with RF drive. (Adapted from: *L3 Communications TWT and TWTA Handbook*. 2007.)

The maximum value of AM/PM conversion is usually the value that is specified. This occurs at 3 to 10 dB below the input power that produces saturation. The type of modulation employed determines the need for specifying AM/PM conversion. For example, in amplitude-modulated systems, AM/PM conversion can result in amplitude distortions. In multiple carrier systems in which both AM and PM signals are present, AM/PM conversion can cause distortion or interference with the phase- or frequency-modulated signals. Finally, amplitude variations in phase- or frequency-modulated systems can cause distortions or interference.

24.1.3 Harmonic Generation

Near saturation, the electron bunches in a TWT become sharply defined, and so the RF current is rich in harmonics. In addition, as saturation is approached and the output signal becomes distorted, harmonics are generated. This is apparent from Figure 24-5 where a sinusoidal input signal has become flat-topped because of saturation. The fundamental and third and fifth harmonic components of the distorted output signal are shown in Figure 24-5. It will be shown in the next section that these harmonics are one of the sources of the intermodulation distortion produced by an amplifier.

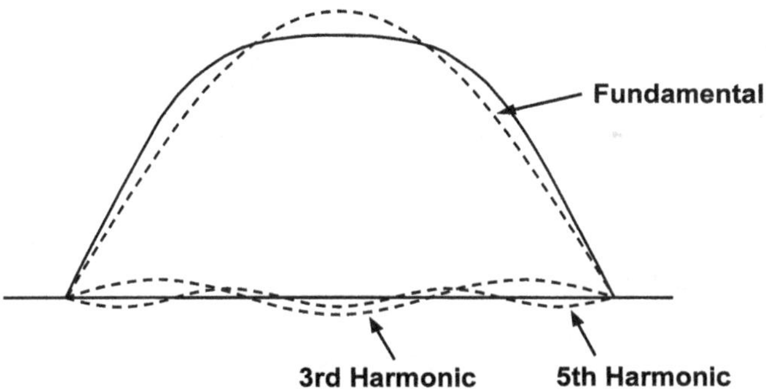

Figure 24-5 Harmonics in flat-topped sine wave.

Harmonics are also produced as the result of AM/PM conversion. This is illustrated in Figures 24-6 and 24-7. As a TWT is driven into saturation and the beam slows down, the phase of the output signal is delayed compared with the phase at small drive levels. As a result, the output curve for a sinusoidal input becomes a tilted sine wave. This tilted wave contains harmonics as shown in Figure 24-7. The third and fifth harmonics are phase shifted relative to those produced by amplitude distortion during saturation. These harmonics also contribute to intermodulation distortion.

At saturation, helix TWTs used for communication applications usually produce second harmonic power that is 8 to 10 dB below the fundamental power. Third harmonic power at saturation is usually 23 to 18 dB below the fundamental power [2].

In some extremely broadband ECM applications, TWTs are used that operate with bandwidths greater than an octave. There have been cases where the second harmonic power output was equal to or greater than the fundamental power when one of these TWTs was driven to saturation at the low end of the frequency band. One of the reasons for this high harmonic output is that, at the low end of the band, the harmonic frequency is within the amplifying band of the TWT.

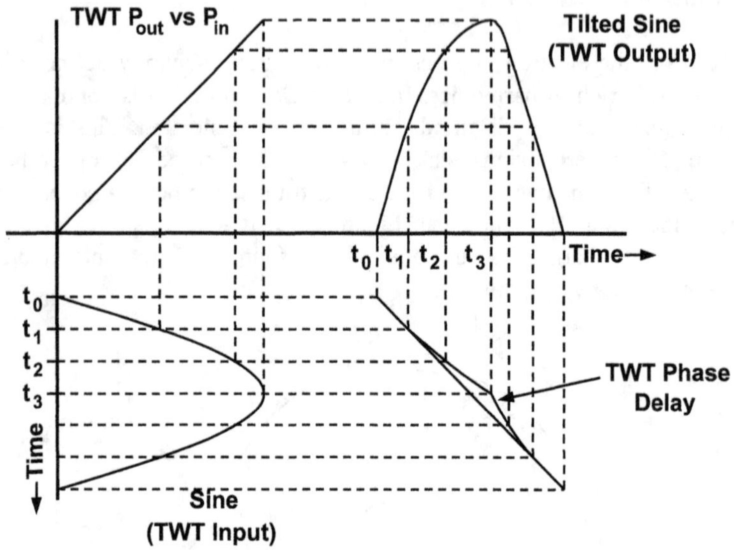

Figure 24-6 Distortion resulting from AM/PM conversion.

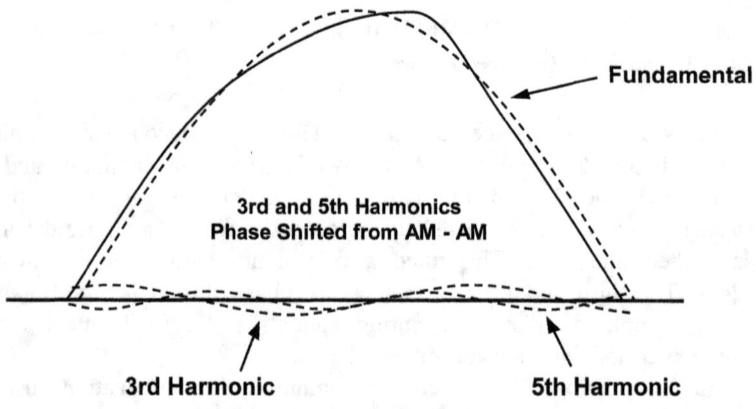

Figure 24-7 Harmonics generated by AM/PM conversion.

In coupled-cavity TWTs, the slow-wave structure is a filter that normally does not propagate harmonics. Even if the structure does contain a higher-order mode at a harmonic frequency, the field configuration is such that the interaction with the beam is weak. As a result, the harmonic output from a coupled-cavity TWT may be 30 dB or more below the fundamental power.

24.1.4 Intermodulation Products

When two or more carrier signals are being amplified by a TWT, intermodulation occurs and signals (intermodulation products or intermods) are produced that are displaced from the carriers by multiples of the frequency separation of the carriers. This intermodulation occurs because of the nonlinearities in the amplification process. The third-and fifth-order intermods are the most important because they are closest to the signal frequencies and largest in amplitude.

The origin of the intermods can be understood by examining operation with two carrier signals and the distortion of those signals during the amplification process. Consider, first, two carrier signals of equal amplitude as follows:

$$V_1 = V \cos \omega_1 t \tag{24-2}$$

and

$$V_2 = V \cos \omega_2 t \tag{24-3}$$

When these two signals are added together, they combine as follows:

$$V_1 + V_2 = V\left(\cos \omega_1 t + \cos \omega_2 t\right) \tag{24-4}$$

or

$$V_1 + V_2 = 2V \cos\left(\frac{\omega_2 - \omega_1}{2}\right)t \cos\left(\frac{\omega_2 + \omega_1}{2}\right)t \tag{24-5}$$

This is a signal having an average carrier frequency of $(\omega_2 + \omega_1)/2$ and amplitude modulation at frequency $(\omega_2 - \omega_1)/2$ as shown in Figure 24-8.

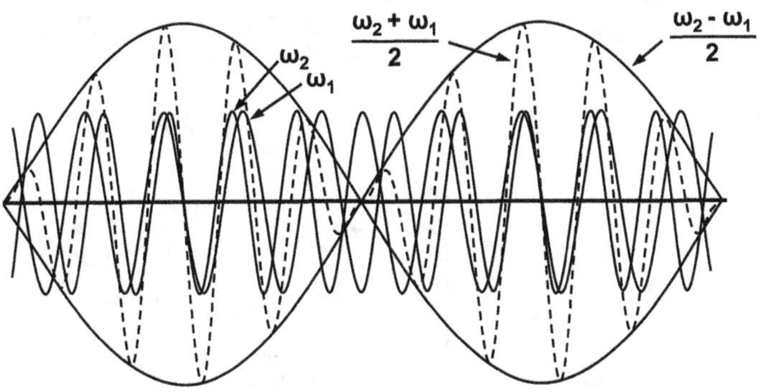

Figure 24-8 Two-carrier operation.

When the signal in Figure 24-8 is amplified by a TWT (or any other amplifier) in which saturation or AM/PM conversion occurs, the signal becomes distorted as shown in Figure 24-9. Harmonics of the modulation frequency and of the carrier frequency are generated. The total third harmonic signal, including the high-frequency variation may be written as:

$$V_{3rd} = 2\cos 3\left(\frac{\omega_2 - \omega_1}{2}\right)t \cos\left(\frac{\omega_2 + \omega_1}{2}\right)t \tag{24-6}$$

or as

$$V_{3rd} = \cos\left(2\,\omega_2 - \omega_1\right)t + \cos\left(2\,\omega_1 - \omega_2\right)t \tag{24-7}$$

These two cosine terms represent the third-order intermods, which are at frequencies close to the carrier frequencies as shown in Figure 24-10.

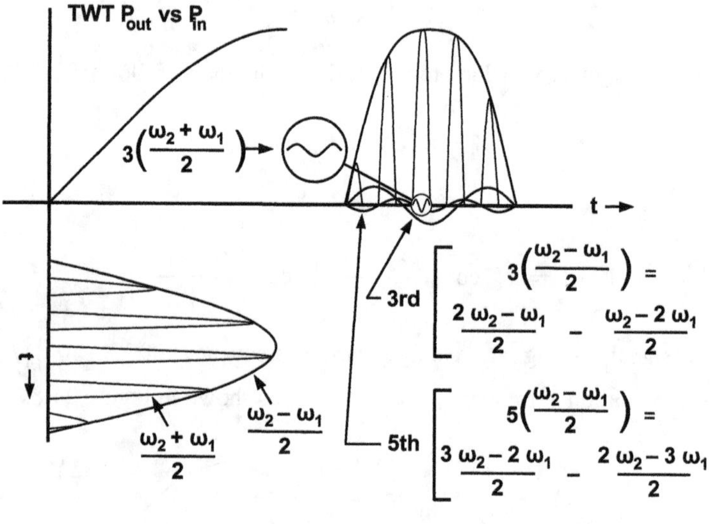

Figure 24-9 Harmonic signals generated during two-carrier operation.

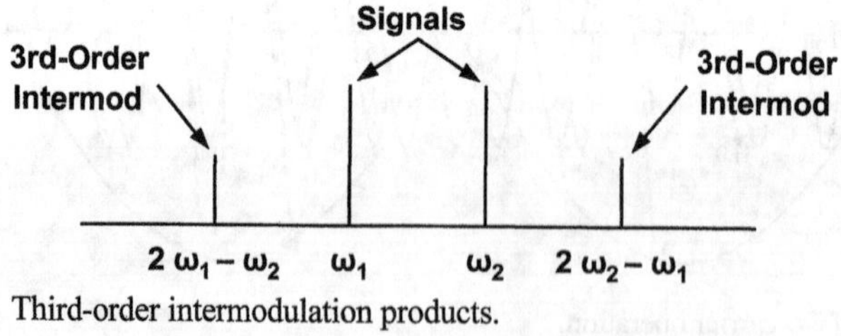

Figure 24-10 Third-order intermodulation products.

The total fifth harmonic signal including the high-frequency variation is:

$$V_{5th} = 2\cos 5\left(\frac{\omega_2 - \omega_1}{2}\right)t\,\cos\left(\frac{\omega_2 + \omega_1}{2}\right)t \tag{24-8}$$

or

$$V_{5th} = \cos\left(3\,\omega_2 - 2\,\omega_1\right)t + \cos\left(3\,\omega_1 - 2\,\omega_2\right)t \tag{24-9}$$

These two cosine terms the fifth-order intermods as shown in Figure 24-11.

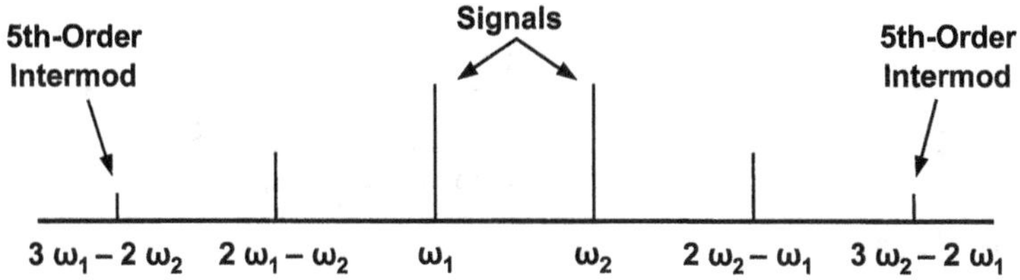

Figure 24-11 Fifth-order intermodulation products.

When a TWT approaches saturation and gain decreases, *gain compression* occurs. It is also possible, at input levels below saturation, for *gain expansion* to occur. Gain expansion (exaggerated for clarity) is compared with gain compression in Figure 24-12.

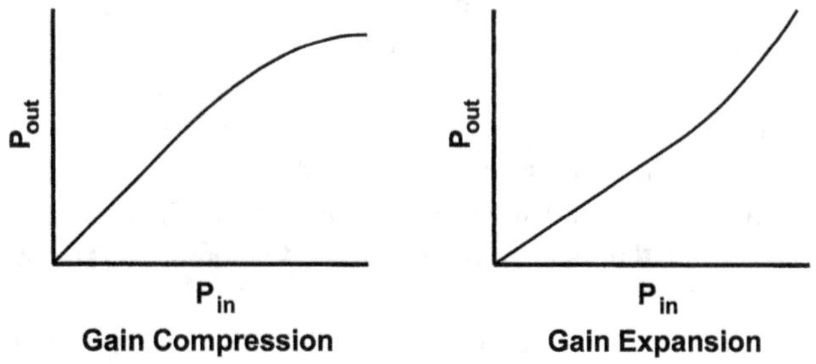

Figure 24-12 Comparison of gain compression and gain expansion.

Two causes for gain expansion (indicated in Figure 24-13) are the use of velocity steps or tapers and the use of a beam voltage greater than that which produces maximum gain (overvoltaging). If the beam is overvoltaged (the velocity parameter is too high), then the gain at small signal levels is less than optimum. As signal level is increased and more energy is extracted from the beam, then, as the beam slows down (and the velocity parameter decreases), gain increases.

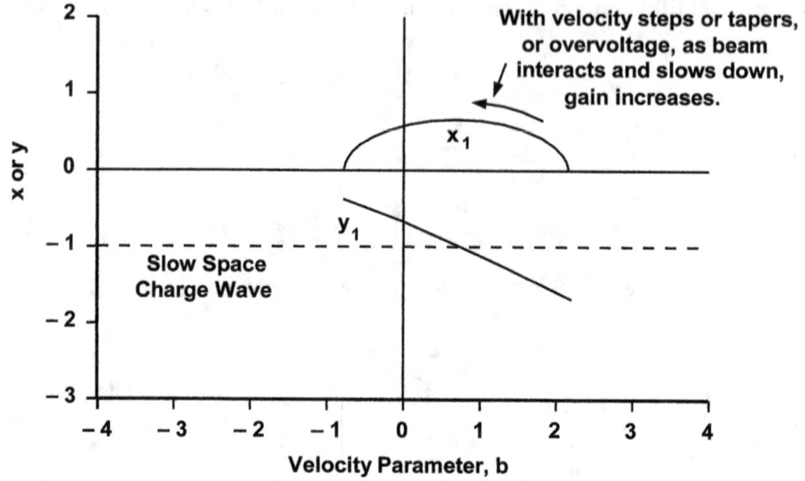

Figure 24-13 Reasons for gain expansion.

Similarly, when a velocity step or taper is used then, at small signal levels where little energy is extracted from the beam, the beam will be going too fast for optimum interaction in the step or taper region. As signal level is increased, the beam will slow down in the step or taper region and gain will increase.

The way in which gain expansion generates third-order intermods is indicated in Figure 24-14. For a sinusoidal input, the TWT output is distorted into a peaked shape. The third harmonic that is generated is 180° out of phase with that generated during gain compression and is the source of the third-order intermod. The affect of this phase difference is evident in Figure 24-15, which shows the third-order intermod for a TWT with a velocity taper and overvoltage [4]. As the drive power is increased so that operation moves from the gain expansion region to gain compression, the rate at which the third-order intermod increases varies significantly. This is in contrast to the linear increase in intermod level with increasing drive power that is usually experienced when only gain compression is present.

An example of the variation of intermod levels with drive when only gain compression is present is shown in Figure 24-16. In this case, when the tube is driven to saturation, third-order intermodulation products are each 8 to 10 dB below the carrier signals (two carriers of equal amplitude are assumed). By reducing the amplitudes of the drive signals so that the TWT operates well below saturated power output, the amplitudes of the intermodulation products drops linearly and rapidly with drive power level. In the example in Figure 24-16, with the TWT operated about 10 dB below saturation, the third-order intermodulation products drop to about 40 dB below the signal output level.

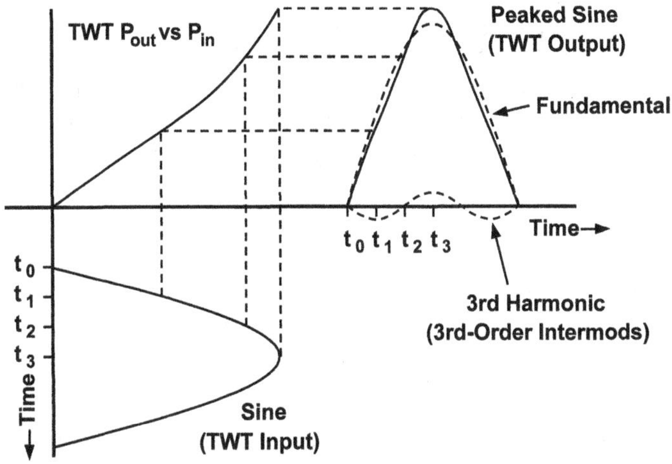

Figure 24-14 Effect of gain expansion on third-order intermodulation product.

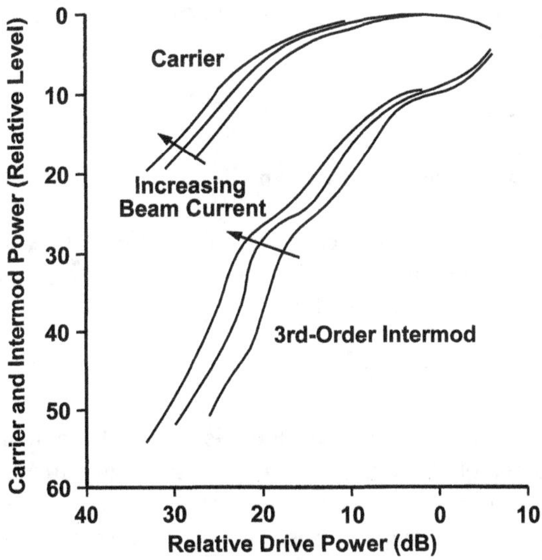

Figure 24-15 Third-order intermods for an overvoltaged TWT with a velocity taper.

The specification of intermodulation products is not straightforward. Usually, the amplitudes of the products are specified as the maximum allowable relative to the carrier signal levels, when these signals have equal amplitudes. The amplitude of the carrier signals is also usually specified relative to the saturated output power when the TWT is amplifying two signals of equal amplitude.

The method for specifying intermods illustrated in Figure 24-16 is called the *intercept point* method. This is based on an understanding of the rate at which the levels of the intermodulation products vary as drive level is varied. If one point is known, such as the intercept of the projection from the linear regions of the power curve for the carrier signals with the projection of the power curve for the third-order intermods, then the amplitudes of the intermods can be calculated for various drive levels.

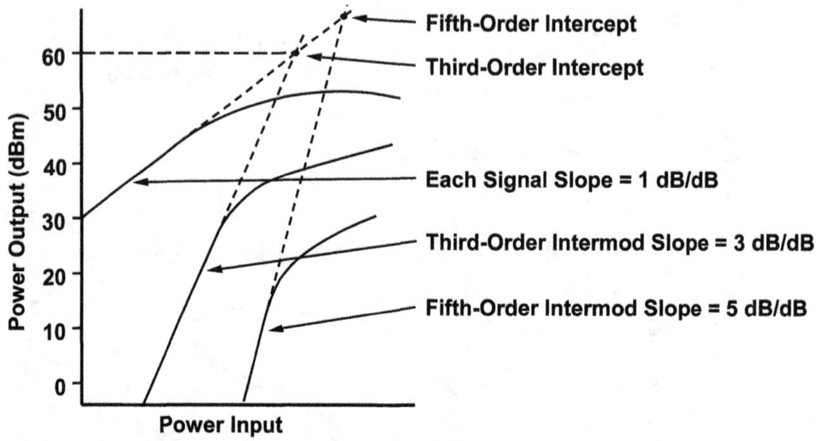

Figure 24-16 "Intercept point" method for specifying intermodulation distortion.

Figure 24-17 shows the intercept point for a TWT. This tube must operate about 10 dB below saturation (backed off 10 dB) for the third-order intermod to be 30 dB below the carrier (C/3IM = 30 dB).

24.2 Digital Communications

A major application for traveling wave tubes is in digital communications systems. They have two major advantages over solid-state amplifiers. They are more efficient and they are more reliable [5, 6].

To understand the demands placed on TWTs, it is necessary to understand some of the basic aspects of digital communications. First of all, information is transmitted as a series of ones and zeros. One or more of the properties of the transmitted signal may be varied (modulated) as shown in Figure 24-18 to produce the ones and zeros.

With frequency-shift keying (FSK) the frequency of a signal is switched between two values. One frequency represents a zero and the other represents a one. With amplitude-shift keying (ASK) the amplitude of the signal is switched between two values, and with phase-shift keying (PSK), phase is switched.

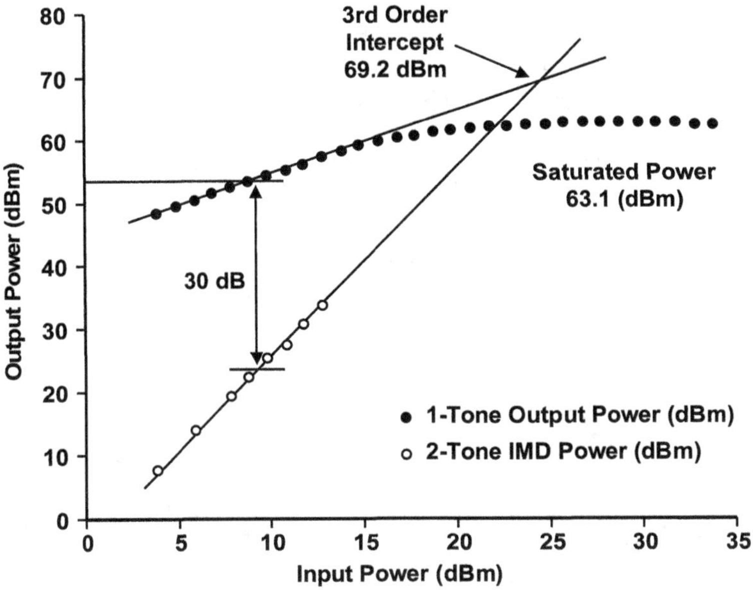

Figure 24-17 Third-order intercept for a 2-kW TWT. This tube must operate about 10 db backed off for C/3IM = 30 dB.

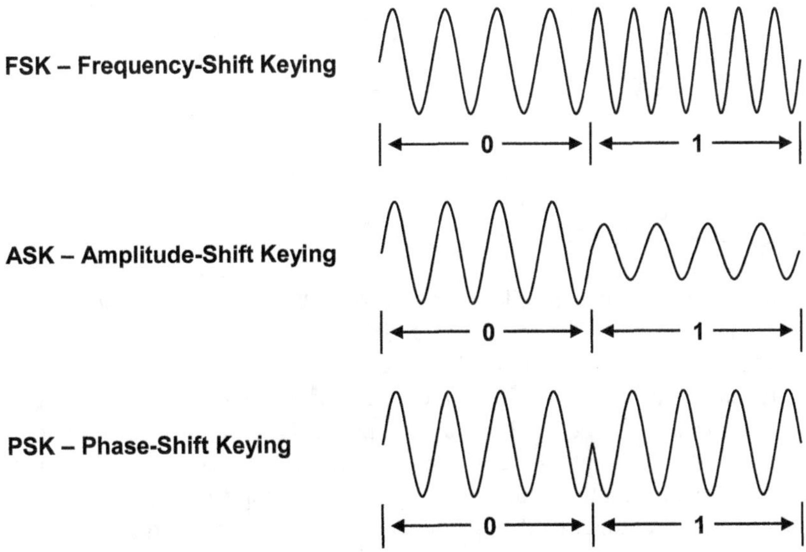

Figure 24-18 Signal properties that can be varied to transmit ones and zeros.

In the following sections, two modulation schemes, QPSK, which uses PSK and 16QAM, which uses PSK and ASK, will be discussed. The way in which PSK and ASK are implemented is as follows. As shown in Figure 24-19, a carrier signal generated by a local oscillator is split into two paths. One path contains a phase shifter that delays the signal 90° relative to the other path.

The signal containing the information to be transmitted is split into two independent components, *I* (in phase) and *Q* (quadrature). The *I* signal is used to modulate one of the carriers from the local oscillator and the *Q* signal is used to modulate the other. The two modulated carriers are then added together. The result is a digitally modulated signal, the amplitude and phase of which are determined by the amplitudes of the *I* and *Q* modulating signals.

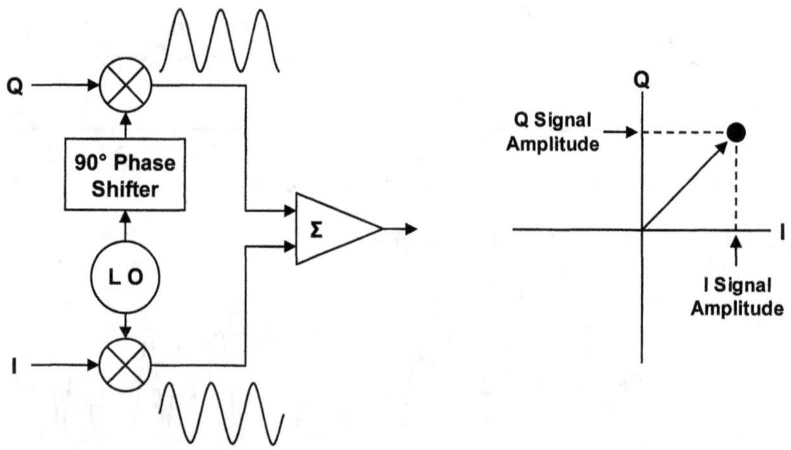

Figure 24-19 Circuit for producing an ASK and a PSK signal

24.2.1 QPSK and 16QAM

Data plots, referred to as constellation plots, for QPSK and 16QAM modulation techniques are shown in Figure 24-20. QPSK stands for quadrature phase-shift keying. With QPSK, there are four ideal positions (states) for the data points on the constellation plot. As shown in Figure 24-21, when QPSK is used, one signal serves as the phase reference and the other is phase shifted in increments that are multiples of 90°. Each phase state corresponds to information that is to be transmitted. For all practical purposes, from a TWT engineer's point of view, phase shifts occur at random. Also, phase shifts do not occur instantaneously, but instead take place over several cycles as shown in Figure 24-21.

When 16QAM is used, there are 16 states on the constellation diagram. In this case, there are two amplitude states at a phase of 45° (and at 135°, 225°, and 315°). There are also phase states at 15° and 75° into each quadrant of the constellation plot. The relative phases and amplitudes of the states in the first quadrant are shown in Figure 24-22.

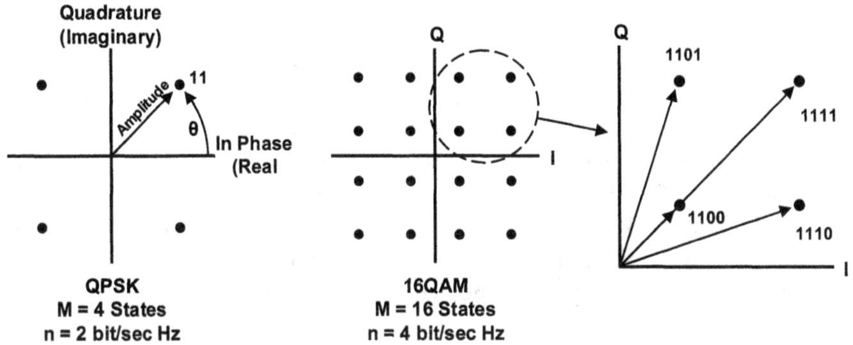

Figure 24-20 QPSK and 16QAM digital signal constellation plots.

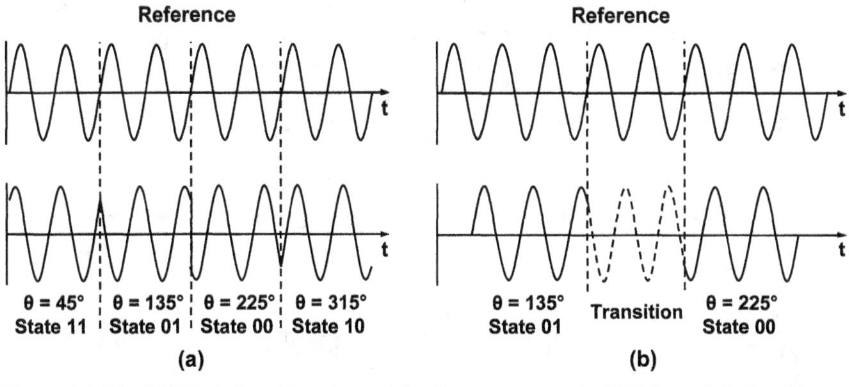

Figure 24-21 QPSK timing (a) and transition between states in QPSK modulation (b).

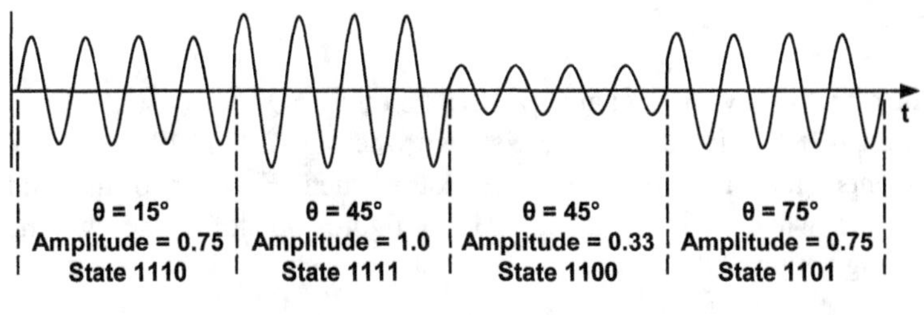

Figure 24-22 First quadrant of the constellation for 16QAM modulation.

24.2.2 Data Characteristics

Digital data are transmitted in "bits" that have a long enough duration to make it possible to identify the frequency, phase and amplitude. The rates at which bits are generated and transmitted are critical characteristics of a digital communications system.

A figure of merit for a digital transmission is E_b/N_o where E_b is the energy of the bit and N_o is the noise power density. E_b/N_o is the digital systems equivalent of the signal to noise ratio (SNR). The digital data rate, R, for a system is the rate at which bits are transmitted. The average signal power is E_bR.

Because of noise and amplifier characteristics, impairment of bits occurs. This is illustrated for QPSK in Figures 24-23 and 24-24. A bit with no impairment is shown in Figure 24-23(a). The amplitude and phase of the signal are exactly those required to place the bit on the target spot (State 11) on the constellation diagram. In Figure 24-23(b), the amplitude is impaired and in Figure 24-23(c) the phase is impaired but the bits are still identified as belonging to State 11. In Figure 24-23(d) phase has been badly distorted and so the signal is identified erroneously as belonging to State 01. In general, as shown in Figure 24-24, Gaussian noise and intermodulation distortions cause the signal to fall in a circular area about the target point. Amplitude distortion produces radial patterns and phase distortion produces angular patterns.

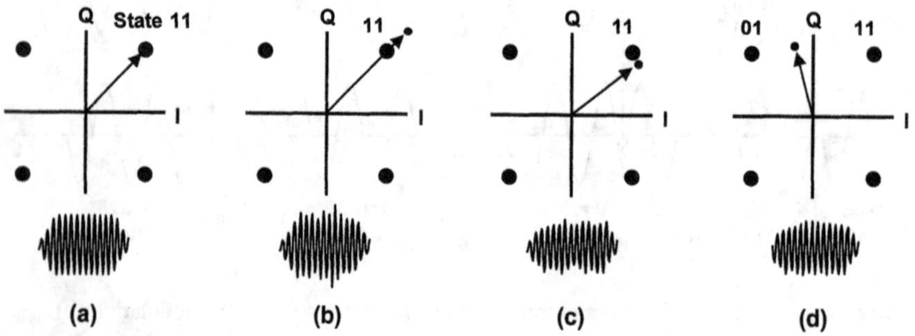

Figure 24-23 (a–d) Locations of data points on the constellation diagram for various impairments.

The effect of a weak signal (low SNR) on the 16QAM constellation is shown in Figure 24-25. No impairments are shown when E_b/N_o = 11 dB. There are a few impairments (indicated by the large dots) when E_b/N_o = 6 dB and many impairments when E_b/N_o = 2 dB. These figures illustrate the importance of maintaining a high SNR.

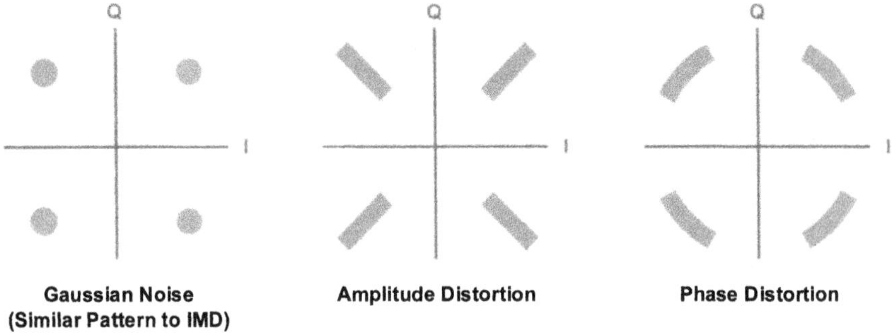

Figure 24-24 Impairment patterns produced by various distortions. (Adapted from: D. M. Goebel and W. L. Menninger, IVEC 2002. © 2002 IEEE.)

24.2.3 Amplifier Design to Reduce Distortion

From the foregoing discussion it is seen that the phase characteristics of the bits are extremely important. In QPSK, a phase error of > 45° will cause a signal to be interpreted in the wrong state. In 16QAM, phase is even more critical. Since phase states are only 30° apart, a phase error of > 15° will place the signal in the wrong state. Thus, it is clear that phase linearity is a very important characteristic of an amplifier.

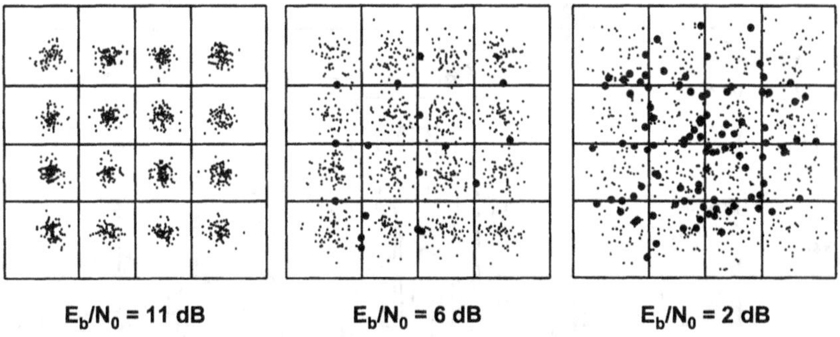

Figure 24-25 Constellation plots for 16QAM with various degrees of impairment. (Adapted from: D. M. Goebel and W. L. Menninger, IVEC 2002. © 2002 IEEE.)

Power and phase characteristics are shown for a conventional TWT in Figure 24-26(a) and for a TWT amplifier designed for phase linearity in Figure 24-26(b). The conventional design produced a phase shift of ~ 40° at saturation. For the TWT designed to minimize AM/PM, phase shift was < 10° near saturation. The design resulted in slight gain expansion and a nonuniform AM/AM curve [7].

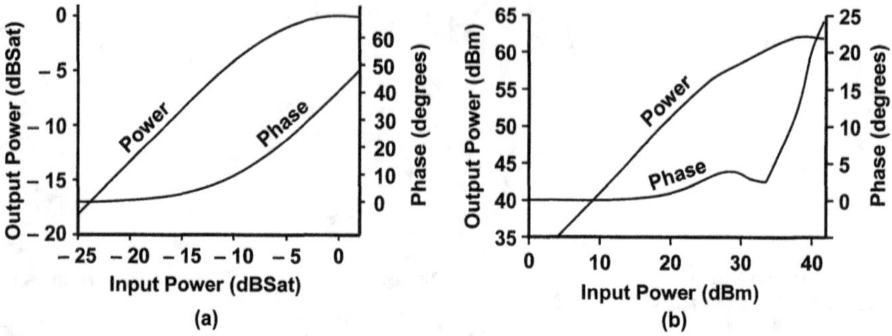

Figure 24-26 Comparison of the phase characteristics of a conventional TWT (a) and a TWT designed for phase linearity (b). (Adapted from: D. M. Goebel and W. L. Menninger, IVEC 2002. © 2002 IEEE.).

Even though TWTs are more efficient than solid-state amplifiers, their efficiency is far less than the maximum that they are capable of at saturation. This is because of the requirements for low intermodulation distortion. These requirements are in addition to the phase linearity requirements discussed earlier. Figure 24-27 shows the third- and fifth- order intermods produced by the eight-tone operation of a Boeing EDD 422H TWT at 150W of average power and continuous random phasing of the signals [8].

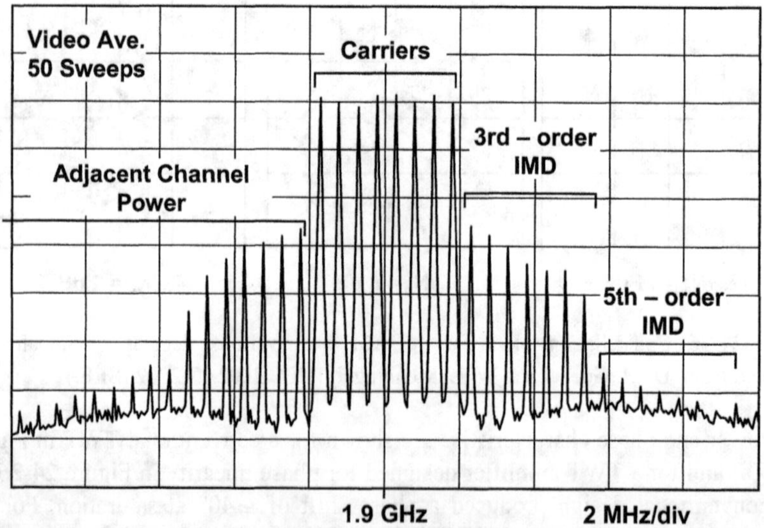

Figure 24-27 Sidebands generated by intermods of carrier frequencies. (Adapted from: D. M. Goebel and W. L. Menninger, IVEC 2002. © 2002 IEEE.)

The reductions in intermods that result from operating backed off are shown in Figures 24-28 and 24-29. When backed off operation is 9 dB below saturation, C/3IM is about 37 dBc. To achieve this level of performance, the helix design was optimized for intermod reduction. As shown in Figure 24-29, a tube with a standard helix had an intermod level about 7 dB higher than the tube with the optimized helix [8].

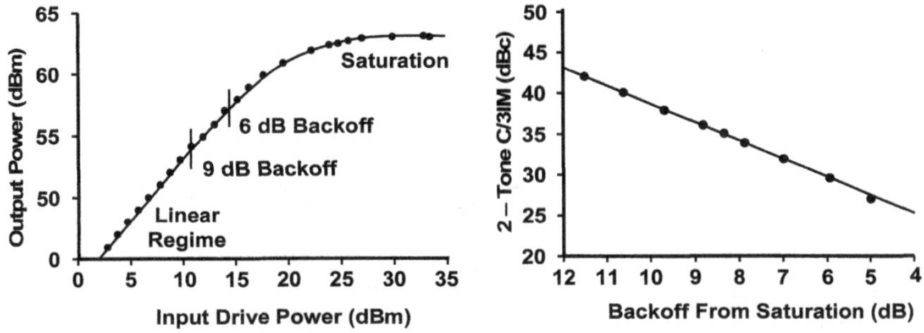

Figure 24-28 Intermod reduction that results from operating backed off. (Adapted from: D. M. Goebel and W. L. Menninger, IVEC 2002. © 2002 IEEE.)

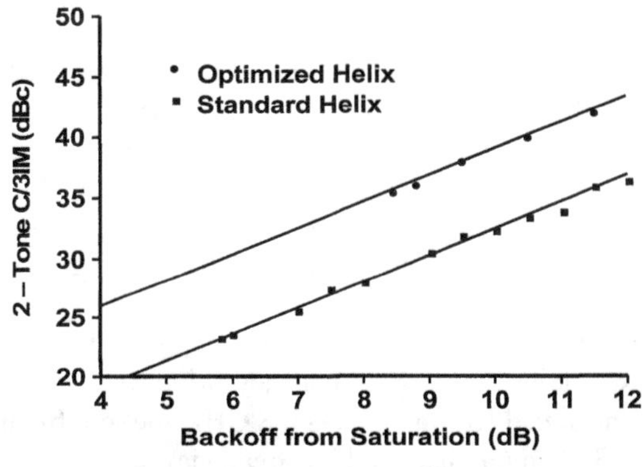

Figure 24-29 Reduction in intermod level by optimizing the helix design. (From: D. M. Goebel, et al., *IEEE Trans. Electron Devices*, 2001. © 2001 IEEE.)

To improve amplifier linearity and reduce intermodulation products, the technique [8] of predistorting the input signal to the amplifier is used (see Figure 24-30). Predistortion offsets the gain compression and phase shifts that cause intermod production. In addition, feedforward circuits [4] are often used to cancel the intermods from the TWT. The input signal and a small part of the amplifier output signal are fed into a correction amplifier, which amplifies the nonlinear signals from the TWT and mixes them out of phase with the TWT output to cancel the intermodulation products.

Predistortion can reduce the intermod level by 10 to 20 dB, and the feedforward circuit can reduce it by about 30 dB. Depending on the regulatory levels of intermods allowed, the main power amplifier must produce third-order intermod levels C/3IM of 20 to 40 dBc. This requires very linear performance by the TWT, which results from operating the tube well below saturation.

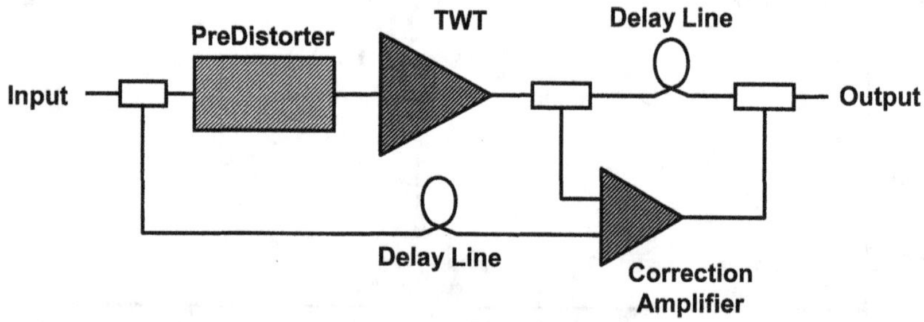

Figure 24-30 Linearizer circuit for minimizing intermodulation products. (Adapted from: D. M. Goebel and W. L. Menninger, IVEC 2002. © 2002 IEEE.)

24.3 SIGNAL CAPTURING

In the previous sections, it was assumed that the two signals applied to a TWT were amplified by the same amount. This is a correct assumption when the frequencies of the two signals are close together. However, when the two frequencies are widely separated but still within the bandwidth of the tube, the lower-frequency signal will be amplified more than the high-frequency signal. This phenomenon is referred to as the *capture effect* or as *signal capturing*. Figure 24-31 shows the effect of frequency difference on the power outputs of two signals of equal amplitude being amplified at the same time by a TWT. These results were measured at a power level for single frequency drive that was several dB below saturation. When the signal separation was 0.8 GHz, the low frequency signal was amplified by about 3 dB more than the high-frequency signal.

For the results in Figure 24-32 the frequency separation was constant but the drive power was varied (with equal signal amplitudes). At low drive levels, the gains for the two signals were equal. As saturation was approached, the gain for the low-frequency signal became larger than for the high-frequency signal and was about 10 dB higher at saturation.

Signal capturing is of little or no importance in applications of TWTs where the separation in frequencies of the signals is small. In electronic countermeasures applications, however, very broadband TWTs are sometimes used and the frequency separation of the signals can be large. In these cases, signal capturing does occur.

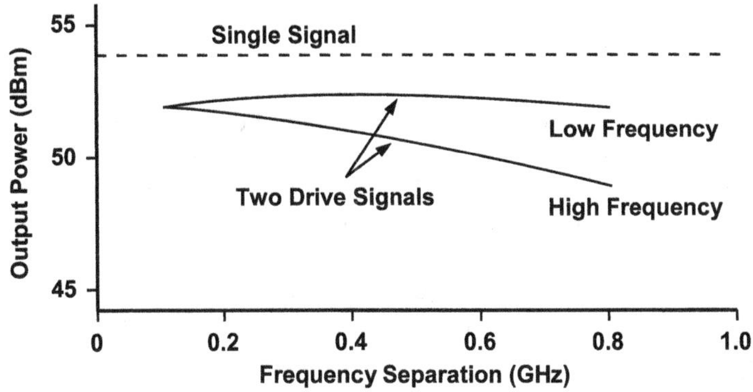

Figure 24-31 Effect of separation of signal frequencies on capturing. (From: R. True, AFTER Program Special Topics Course, November 1984.)

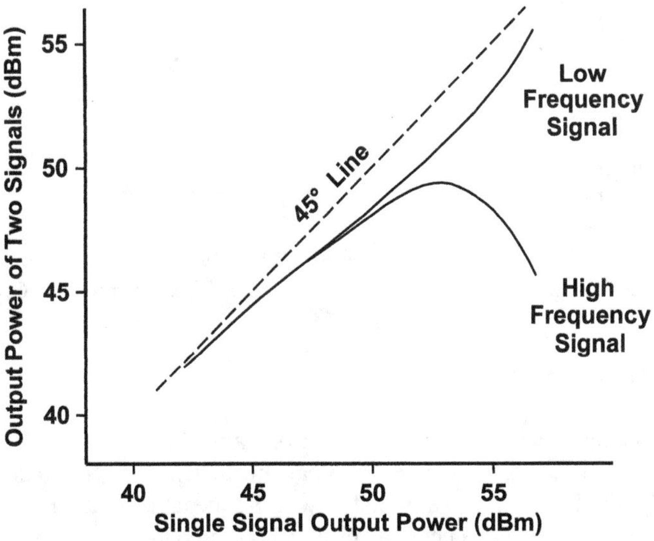

Figure 24-32 Saturation effect on signal capturing. (From: R. True, AFTER Program Special Topics Course, November 1984.)

24.4 VARIATIONS WITH FREQUENCY

When the gain versus frequency characteristics of a TWT are examined, two distinct types of gain variation are usually observed. One is a broadband variation for which the gain varies slowly as a function of frequency. The other is a relatively rapid variation of gain in narrow bandwidths.

24.4.1 Broadband Gain Variations

An example of the broadband variation in gain is shown in Figure 24-33. This gain characteristic was computed for small-signal operation of a coupled-cavity TWT [1]. The variation in gain with frequency results from the dispersion and impedance characteristics of the circuit along with the fact that gain varies with the electrical length of the circuit, which, in turn, varies with frequency. The imposition of a tight specification on gain flatness results in compromises in circuit design and beam conditions that reduce output power and efficiency. In an example [1], output power was a factor of 2.5 lower for the TWT with the required gain flatness than for a TWT designed for maximum efficiency.

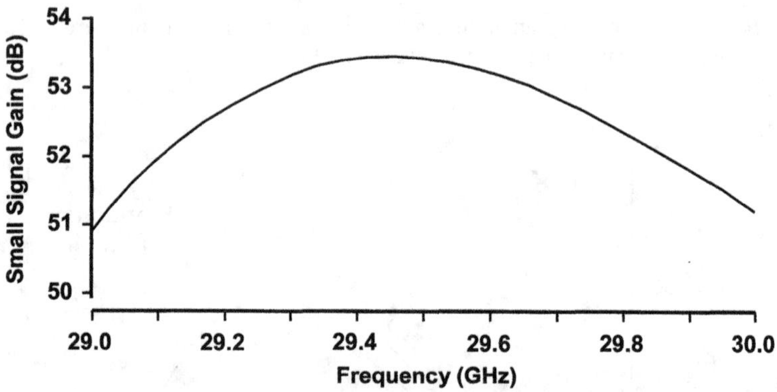

Figure 24-33 Small signal gain with no internal reflections. (From: M. L. Cabrales et al., "NASA Contractor Report 185146," January 1990.)

24.4.2 Narrowband Gain Variations

In narrow bandwidths, the dominant effect on gain is usually some feedback path within the TWT. For example, internal mismatches, which can never be completely eliminated, cause a periodic (with frequency) variation in gain. Figure 24-34 shows how this can occur. A mismatch at the output end of the slow wave structure causes the reflection of a small portion of the signal. This portion is then reflected by a mismatch at the input end of the structure. It then combines, in various phase relationships, with the desired growing wave. The result is that a "gain ripple" is produced as is shown in Figure 24-35. This particular curve was calculated for the millimeter-wave coupled-cavity TWT mentioned in a previous paragraph [4]. The frequency difference between gain peaks depends on the frequency of operation and the length of the RF circuit (which produce a closed-loop phase change of 360°) and is usually in the 100–300-MHz range.

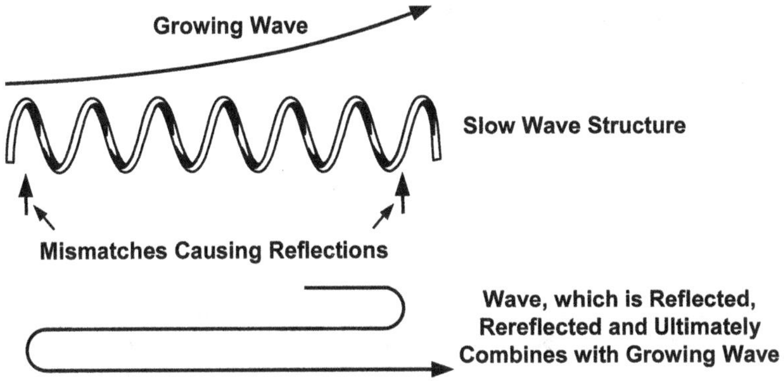

Figure 24-34 TWT section with two internal mismatches.

Internal reflections within a TWT are the primary cause for gain ripple. These reflections can occur at the following locations:

1. At the attenuator at the end of the input section of the TWT. These reflections are then rereflected at the input port to the TWT.
2. Within the center section of a TWT with two severs.
3. At the output port. These reflections are then rereflected at the sever or attenuator at the input end of the output section. This feedback path is usually the dominant path because gain is high and reflections from the output port vary with load characteristics. An isolator at the output helps control load reflections.
4. At imperfections in the slow wave structure. Periodically varying imperfections, for example at helix support rods, can be particularly detrimental in this respect.

It is also possible for returning electrons from the collector to provide a feedback path that produces gain ripple.

24.4.3 Phase Nonlinearities or Time Delay Distortion

Shown in Figure 24-36 are phase and time delay as functions of frequency. A linear increase of phase as a function of frequency is unavoidable because the phase length of a TWT depends on wavelength. The ripple portion of the phase curve depends on the same internal reflection phenomena that produce gain ripple. Time delay is the total time that it takes for a signal to pass through a TWT and is the derivative of phase delay. Thus, the same mechanisms that cause phase nonlinearities are responsible for time delay distortion.

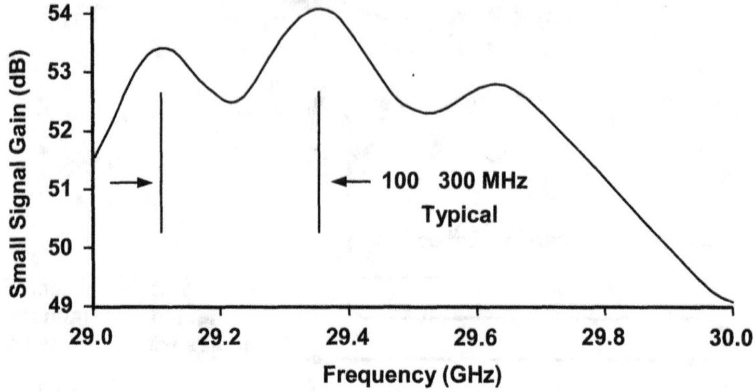

Figure 24-35 Gain ripple with 7% voltage reflections. (From: M. L. Cabrales et al., "NASA Contractor Report 185146," January 1990.)

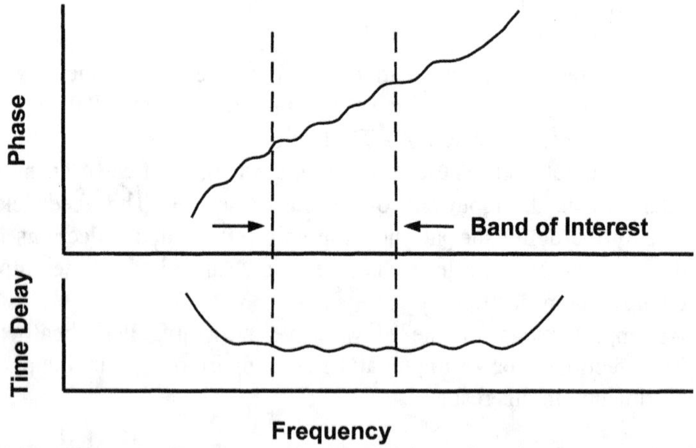

Figure 24-36 Phase nonlinearity and time delay distortion. (Adapted from: *L3 Communications TWT and TWTA Handbook.* 2007.)

The frequency band of interest is indicated in Figure 24-36 to illustrate the fact that, for most TWTs, the narrowband effects dominate the broadband effects. This is true because most TWTs operate over only a small portion of the passband of the circuit and so band edge effects are far from the frequency band of interest.

24.5 PUSHING AND PULLING

When an operating parameter, such as the beam voltage, varies, the RF performance of a tube is affected. The pushing figure is a measure of the change in a performance characteristic that occurs as a result of a change in an operating parameter. For example, the cathode- phase pushing figure is the change in phase of the output that results from a change in cathode voltage.

There is also a pulling figure and this has to do with the effect of the load on the tube. The load impedance has an effect on the power output and on the stability of the tube.

24.5.1 Amplitude Pushing

Amplitude pushing figures are relatively easy to determine. To measure amplitude pushing figures, the change in power output produced by a change in one of the tube operating parameters is measured. Typical specifications for the maximum values for the amplitude pushing figures for a tube with a gridded gun are as follows:

- *Cathode pushing*: 0.5 dB per 1% change in cathode voltage, V_a;
- *Grid pushing*: 0.5 dB per 1% change in grid voltage, V_g;
- *RF drive pushing*: 0.5 dB per 1 dB change in RF input power.

These pushing figures are all specified for operation at or very close to the nameplate operating conditions.

24.5.1.1 Amplitude Pushing with Gridless Guns

For constant efficiency, the RF power output, P_o, is proportional to the product of cathode current and voltage, that is

$$P_o \propto I\, V_a \tag{24-10}$$

but

$$I = p\, V_a^{3/2} \tag{24-11}$$

and, since the perveance is constant,

$$P_o \propto V_a^{5/2} \tag{24-12}$$

To find how amplitude pushing depends on cathode voltage, examine the change in P_o that occurs when V_a is varied, that is,

$$\Delta P_o \propto \frac{5}{2} V_a^{3/2}\, \Delta V_a \tag{24-13}$$

Now, dividing this equation by (24-12),

$$\frac{\Delta P_o}{P_o} = \frac{5}{2}\frac{\Delta V_a}{V_a} \tag{24-14}$$

To express this relation in terms of dB change in power output per 1% change in cathode voltage, remember that

$$\text{dB} = 10\log_{10}\frac{P_o + \Delta P_o}{P_o} = 10\log_{10}\left(1 + \frac{\Delta P_o}{P_o}\right) \tag{24-15}$$

but

$$\log_{10} = 0.434\ \ln \tag{24-16}$$

and

$$\ln\left(1 + \frac{\Delta P_o}{P_o}\right) \approx \frac{\Delta P_o}{P_o} \tag{24-17}$$

so

$$\text{dB} = 4.34\frac{\Delta P_o}{P_o} = 10.85\frac{\Delta V_a}{V_a} \tag{24-18}$$

As a result, for $\Delta V_a/V_a = 1\%$, the theoretical cathode pushing figure for a gridless gun is

$$\text{cathode pushing} = 0.11\ \text{dB} \ / \ 1\%\ \text{change in}\ V_a \tag{24-19}$$

The amplitude pushing that results from a change in RF drive level depends on the power transfer characteristic of the tube. If the tube is operating saturated, then, of course, RF drive pushing is zero. However, if the tube is operating below or above saturation, then RF drive pushing must be determined experimentally.

24.5.1.2 Amplitude Pushing with Gridded Guns

The calculation of the dependence of the amplitude pushing figure on the voltages in a gridded gun is not as straightforward as for the gridless gun. Of course, for constant efficiency, the RF power output is still proportional to the product of cathode current and voltage, but the current depends on grid voltage as well as cathode voltage. The way in which current varies with these voltages depends on the construction of the grid.

For example, if the wires of the grid are very close together, then the grid-cathode region behaves as a space charge limited diode and is shielded from the effects of the anode. As a result, current depends on the 3/2 power of grid voltage and is independent of anode voltage.

If, on the other hand, the wires of the grid are far apart, then the field established by the anode penetrates to the surface of the cathode and so the cathode voltage controls the current. The grid voltage has little effect on the current.

With the grid wire spacing that is used in practice, it is found that the current is almost linearly dependent on grid voltage and that it varies as the square root of cathode voltage, that is

$$I \propto V_g V_a^{1/2} \tag{24-20}$$

This means that

$$P_o \propto V_g \; V_a^{3/2} \tag{24-21}$$

To find how amplitude pushing depends on cathode voltage, examine the change in P_o that occurs when V_a is varied (with V_g held constant), that is,

$$\Delta P_o \left(V_g \text{ fixed} \right) \propto V_g \frac{3}{2} V_a^{1/2} \, \Delta V_a \tag{24-22}$$

Now, dividing this equation by (24-21),

$$\frac{\Delta P_o}{P_o} = \frac{3}{2} \frac{\Delta V_a}{V_a} \tag{24-23}$$

In this case, in terms of dB change in power output per 1% change in V_a, this becomes

$$\text{dB} = 4.34 \frac{\Delta P_o}{P_o} = 6.51 \frac{\Delta V_a}{V_a} \tag{24-24}$$

For $\Delta V_a / V_a = 1\%$

$$\text{cathode pushing} = 0.065 \text{ dB} \ / \ 1\% \text{ change in } V_a \tag{24-25}$$

and so the expected cathode pushing figure is well below the specification of 0.5 dB/1% given at the beginning of this section for a gridded gun.

To find how amplitude pushing depends on grid voltage, examine the change in P_o that occurs when V_g is varied (with V_a held constant), that is,

$$\Delta P_o \left(V_a \text{ fixed} \right) \propto \Delta V_g \, V_a^{3/2} \tag{24-26}$$

Now, dividing this equation by (24-21),

$$\frac{\Delta P_o}{P_o} = \frac{\Delta V_g}{V_g} \tag{24-27}$$

and so the pushing figure relating the change in power output in dB to a 1% change in grid voltage is

$$\text{grid pushing} = 0.0434 \text{ dB} \;/\; 1\% \text{ change in } V_g \tag{24-28}$$

and this is over an order of magnitude below the specified value of 0.5 dB/1%.

24.5.2 Phase Pushing

To measure phase pushing figures, it is necessary to make accurate measurements of relatively small changes in phase of the RF output when an operating parameter is changed. Typical specifications for the maximum values for the phase pushing figures are:

- *Cathode pushing*: 55° per 1% change in V_a;
- *Grid pushing*: 12° per 1% change in V_g;
- *Collector pushing*: 0.002° per 1 V change in collector voltage;
- *Filament pushing*: 10° per 1% change in filament voltage;
- *RF drive pushing*: 10° per 1 dB change in RF input power.

These pushing figures are all specified for operation at or very close to the nameplate operating conditions.

The calculation of phase pushing figures is not as straightforward as the calculation of amplitude pushing figures. For example, consider cathode pushing. The total phase length of the tube, φ, is proportional to the time that it takes for the RF signal to travel through the tube. That time is the length of the RF section divided by the phase velocity, v_p, of the signal on the circuit. As a result,

$$\phi \propto \frac{1}{v_p} \tag{24-29}$$

and the relation between a change in phase velocity and the resulting change in phase is

$$\frac{\Delta\phi}{\phi} = -\frac{\Delta v_p}{v_p} \tag{24-30}$$

For small signal levels, Pierce's theory gives the relation between a change in beam velocity and the resulting change in velocity of propagation of the signal on the circuit. The result is shown in Figure 24-37. Remember, b is Pierce's velocity parameter, which relates the beam velocity to the cold circuit phase velocity. Y_1 is the incremental propagation constant of the signal on the circuit. Notice that the slope of the y_1 versus b lines is nearly 1/2. The reason for this is that the velocity of the slow space charge wave (the bunch velocity) is controlled by the cold circuit phase velocity, which is constant, as well as by the beam velocity.

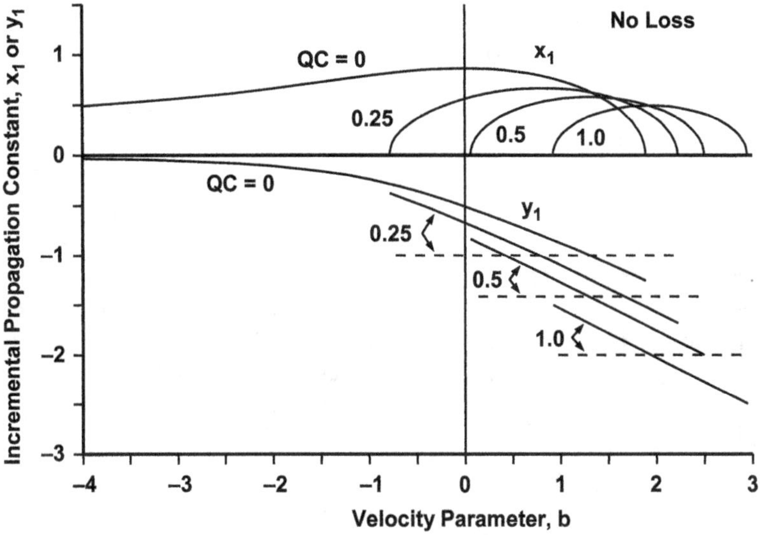

Figure 24-37 Effect of space charge on incremental propagation constants for growing wave.

Since the slope is 1/2, the change in signal velocity is 50% of the change in beam velocity. As a result,

$$\frac{\Delta\phi}{\phi} = -\frac{1}{2}\frac{\Delta u_o}{u_o} \tag{24-31}$$

Finally, $u_o \propto V_a^{1/2}$ and so

$$\frac{\Delta\phi}{\phi} = -\frac{1}{4}\frac{\Delta V_a}{V_a} \tag{24-32}$$

(Changes in velocity resulting from changes in beam density and beam size are being ignored. These are only a few percent of the cathode-induced changes.) As a result, the relation for the cathode phase-pushing figure is

$$\Delta\phi = -\frac{1}{4}\frac{\Delta V_a}{V_a}\phi = -\ 0.0025\,\phi \ \text{ degrees / 1\% change in } V_a \tag{24-33}$$

To use this relation, the phase length of the tube, φ, must be known. For traveling wave tubes, this phase length is usually ~ 5,000–10,000° and so the cathode phase pushing figure is usually

$$\Delta\phi = -0.0025\times 5{,}000 \text{ to } 10{,}000° = 12.5 \text{ to } 25° / 1\% \text{ change in } V_a \tag{24-34}$$

which is well below the maximum specified value of 55°/1%. It must be remembered that this result is based on Pierce's small signal theory. In practice, measured cathode phase- pushing figures are a factor of about two above this result.

Grid phase pushing occurs because, as grid voltage changes, beam current changes and this, in turn, changes the electron density in the beam. (Here, beam diameter will be assumed not to change.) As electron density changes, plasma frequency changes and so the velocity of the slow space-charge wave, v_{sscw}, changes. As was the case for cathode pushing (and from Figure 24-28), the resulting change in the velocity of the signal on the circuit is one half of the change in v_{sscw}. The velocity of the slow space-charge wave is

$$v_{sscw} = \frac{u_o}{1+\dfrac{\omega_q}{\omega}} \approx u_o\left(1-\frac{\omega_q}{\omega}\right) \tag{24-35}$$

This approximation is valid because ω_q/ω is usually about 0.1.

When ω_q changes, the resulting change in v_{sscw} is

$$\Delta v_{sscw} = -\ u_o\,\Delta\left(\frac{\omega_q}{\omega}\right) \tag{24-26}$$

and so

$$\frac{\Delta v_{sscw}}{v_{sscw}} = -\frac{\Delta\left(\omega_q/\omega\right)}{1-\omega_q/\omega} \approx -\Delta\left(\frac{\omega_q}{\omega}\right) \tag{24-37}$$

This means that

$$\frac{\Delta\phi}{\phi} = -\frac{\Delta v_p}{v_p} \approx \frac{1}{2}\Delta\left(\frac{\omega_q}{\omega}\right) \tag{24-38}$$

The plasma frequency is

$$\omega_q = F\,\omega_p = F\left(\eta\,\frac{\rho_o}{\varepsilon_o}\right)^{1/2} \propto \left(\frac{I}{u_o}\right)^{1/2} \propto \sqrt{I} \qquad (24\text{-}39)$$

Usually, for grid controlled guns, $I \propto V_g$ and so

$$\omega_q \propto \sqrt{V_g} \qquad (24\text{-}40)$$

The ratio of ω_q to ω is usually about 0.1. When V_g changes by 1%, ω_q changes by 1/2% and so ω_q/ω changes from 0.1 to 0.1005 or 0.05%. The change in phase is 1/2 of this or 0.025%. Since the total phase is 5,000 – 10,000°, the change in phase is 1.25 to 2.5° and so the grid pushing figure is

$$\Delta\phi = 1.25 \text{ to } 2.5° \,/\, 1\% \text{ change in } V_g \qquad (24\text{-}41)$$

which is much less than the maximum specified value of 12°/1%. Again, we must remember that this result is based on Pierce's small signal theory. In practice, measured grid phase pushing figures are a factor of two to three above this result.

Next, consider phase pushing by the filament voltage. Fluctuations or changes in filament voltage (and current) can have two effects on tube operation:

1. For slow variations in voltage, the cathode temperature changes. If the position of the cathode changes because of thermal expansion or contraction of the cathode support structure, then the cathode current changes and this, in turn, causes phase pushing. Also, if the cathode is not completely space-charge limited, emission will change with temperature and this will affect phase.
2. Fluctuations in filament current cause variations in the magnetic field generated by the filament. If the filament magnetic field links through the cathode surface, then the focusing of electrons in the gun and beam size are affected. As beam size fluctuates, the potential depression in the beam and the beam velocity vary, affecting the phase length of the tube.

It would appear that the collector should not have an effect on phase because the beam has completed its interaction with the circuit before entering the collector. However, fluctuations in the collector voltage can cause variations in the current reflected by the collector. Even a small current traveling backwards through a tube can carry a signal from output to input. As this signal combines with the input signal in various phase relations, the amplitude and phase of the resulting signal that is launched on the circuit vary. As a result, there is an effect on the phase and amplitude of the RF output.

The primary reason why the RF drive level affects phase is that, as drive is increased, more and more of the energy in the electron beam is converted to RF power output. As a result, the beam loses energy and its velocity decreases. As the velocity decreases, phase length increases.

24.5.3 Pulling

RF power generated within a microwave tube is transmitted to a load as shown in Figure 24-38. There may be (and, to some extent, always are) impedance mismatches at several places in the output transmission line where RF power is reflected back toward or away from the tube. Reflected power that enters the tube affects tube operation. This effect of the load on tube operation is referred to as "pulling."

The pulling effects that may be produced by reflected power are variations in the following:

1. Amplitude of the RF output power;
2. Phase of the RF output power;
3. For an oscillator, frequency of the RF output power.

Load reflections may also cause an amplifier to produce unwanted signals.

There is no straightforward way to predict the pulling effects of a load. Instead these effects must be measured by using loads that produce known voltage standing-wave ratios with known phases.

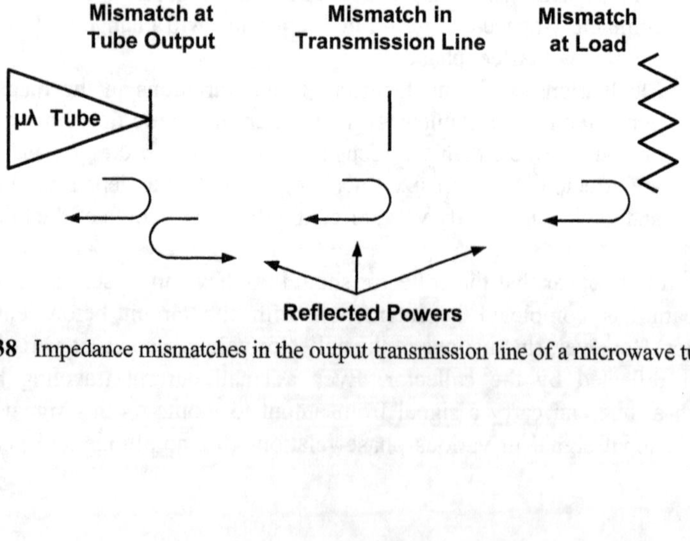

Figure 24-38 Impedance mismatches in the output transmission line of a microwave tube.

REFERENCES

[1] J. W. Hansen et al., "Systems aspects of communications TWTAs or how to deal with the tube manufacturer to your best advantage," Hughes Aircraft Company Electron Dynamics Division Applications Note, August 1982.

[2] H. G. Kosmahl and J. C. Peterson, "A TWT amplifier with a linear power transfer characteristic and improved efficiency, NASA Technical Memorandum 83590," *Tenth Communications Satellite Systems Conference*, Orlando, FL, 1984.

[3] J. W. Hansen, "Linearized TWTAs for low-distortion communications applications," *Microwave Power Tube Conference*, Monterey, CA, 1992.

[4] M. L. Cabrales et al., "Study of methods for the reduction of distortion in high-power traveling-wave tubes," NASA Contractor Report 185146, January 1990.

[5] W. L. Menninger et al., "70% efficient Ku-band and C-band TWTs for satellite downlinks," *IEEE Trans. Electron Devices,* Vol. 52, No. 5, 2005, pp. 673–678.

[6] R. Strauss, "Reliability of SSPAs and TWTAs," *IEEE Trans. Electron Devices,* Vol. 41, No. 4, 1994, pp. 625–626.

[7] D. M. Goebel and W. L. Menninger, "Design and performance of vacuum electron devices for digital communications," *Third IEEE IVEC*, 2002.

[8] D. M. Goebel et al., "Development of linear traveling wave tubes for telecommunications applications," *IEEE Trans. Electron Devices,* Vol. 48, No. 1, 2001, pp. 74–81.

Chapter 25

Breakdown and Protection

One of the most common and, perhaps, most poorly understood classes of failure mechanisms in microwave tubes centers on electrical breakdown. In the courses from which this book was originally developed, electrical breakdown was, at first, only briefly discussed. Because of the strong interest in breakdown phenomena expressed by those taking the course, that section was developed to cover major aspects of breakdown phenomena of concern to those who use microwave tubes.

Breakdown in microwave tubes may occur inside or outside the vacuum envelope at any of several locations as is indicated here and in Figure 25-1.

- Electron gun (between electrodes, or leads, or from electrodes or leads to ground);
- Collector (between electrodes, or leads, or from electrodes or leads to ground);
- High-power portions of the RF structure.

Often, it is not the tube that is at fault. For example, there may be a failure at any one of several locations in the RF system. If excessive RF power is reflected, the tube may be damaged for any one of several reasons. All too often the tube is thought to be at fault and is replaced before the original cause of the difficulty is corrected.

Whether or not breakdown will occur depends on two factors:

1. The applied field level and local field enhancement effects;
2. The breakdown field of the medium (gas, vacuum, liquid, or solid).

Very approximately, the breakdown fields for various media are as follows:

- Gas $\sim$ tens of V/cm to 10^5 V/cm depending on pressure and type;
- Vacuum $\sim 0.5\text{–}3 \times 10^5$ V/cm;
- Liquid $\sim 0.5\text{–}1.0 \times 10^6$ V/cm;
- Solid $\sim 0.5\text{–}1.0 \times 10^6$ V/cm.

At interfaces between solids, liquids and vacuum, or gas, the breakdown field may be significantly degraded. In addition, in a void in a solid or a liquid, the field is normally enhanced. The amount of enhancement depends on the shape and orientation of the void and the dielectric constants of the media, as indicated in Figure 25-2.

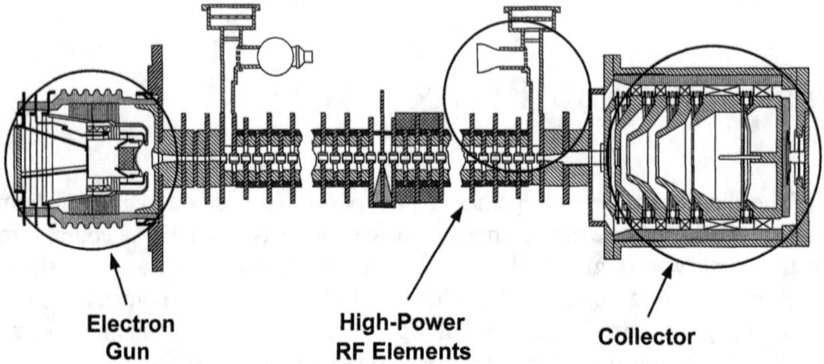

Figure 25-1 Regions in a microwave tube where electrical breakdown may occur.

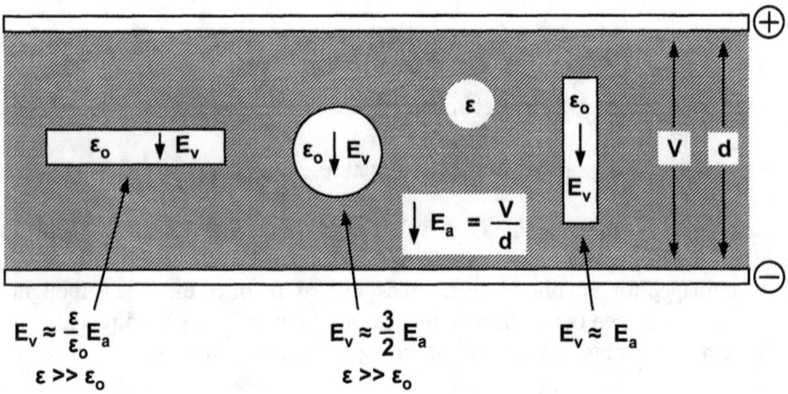

Figure 25-2 Electric field enhancement in voids in dielectric media.

The low breakdown fields of gas or vacuum relative to those of solids or liquids coupled with various field enhancement effects result in breakdown occurring, in most cases, in the vacuum or gas rather than in the solid or liquid. As a result, most of the breakdown discussions in this chapter will concentrate on vacuum and gas. First, a section is devoted to various field enhancement effects that reduce the value of the applied field at which breakdown occurs. Then, dc and RF breakdown in vacuum and in gas are covered. Breakdown in or across an insulator is sometimes a cause of failure and is also discussed.

25.1 FIELD ENHANCEMENT

The electric field between parallel plane electrodes is the voltage applied from one electrode to the other divided by the distance by which the electrodes are separated. In practice, it is common to calculate field levels in all sorts of configurations simply by dividing the applied voltage by the electrode separation distance. Fields calculated in this manner are often in error by a factor of from two to ten or more. Worse yet, because the power varies as the square of the field, the calculated power-handling capability is often in error by a factor ranging up to two orders of magnitude or more.

The parallel plane and coaxial conductors shown in Figure 25-3 provide a simple case for the comparison of field levels. A potential of 1,000V between electrodes is assumed, and equipotential profiles at 250V, 500V, and 750V are shown. For the parallel-plane electrode case, the potential profiles are equally spaced and the electric field is 1,000V/cm everywhere between the electrodes.

For the coaxial electrode case, the equipotential profiles near the center conductor are more closely spaced than in the parallel plane electrode case. The electric field at the surface of the center conductor is easy to calculate for this configuration and is 1560 V/cm, or over 50% above that for parallel plane electrodes.

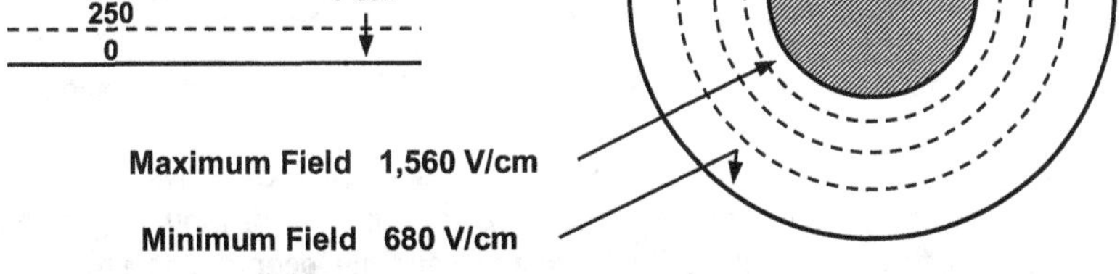

Figure 25-3 Potential distributions in parallel plane and 50-Ω coaxial electrode configurations.

If the maximum field was limited to 1,000 V/cm because of breakdown considerations, then the maximum voltage that could be applied to the coaxial electrodes would be 1,000/1.56, or 641V. In the TE11 mode (for which the potential distributions in Figure 25-3 are correct), the maximum RF power capability of the coaxial configuration would be (1/1.56)2, or 40% of that for the parallel-plane electrodes.

The factor by which the electric field exceeds the field for the parallel-plane electrode case is normally called the *field enhancement factor*, and is denoted by β. At the center conductor of the 50-Ω coaxial line, the field enhancement factor is 1.56.

Field enhancement factors have been calculated for many electrode configurations. Some of these are given in Figure 25-4. In most practical cases of interest, the electrode geometries are close enough to those shown in Figure 25-4 to permit a reasonable estimate of the value of β to be made. In some cases, values of β for two or more configurations can be combined to obtain an estimate for a configuration in question. For example, the value of β at the surface of a drop of condensation on the center conductor of a 50-Ω coaxial line is approximately 3 (β for a hemisphere on a plane) times 1.56 (β at the center conductor of a 50-Ω coaxial line), or about 4.7.

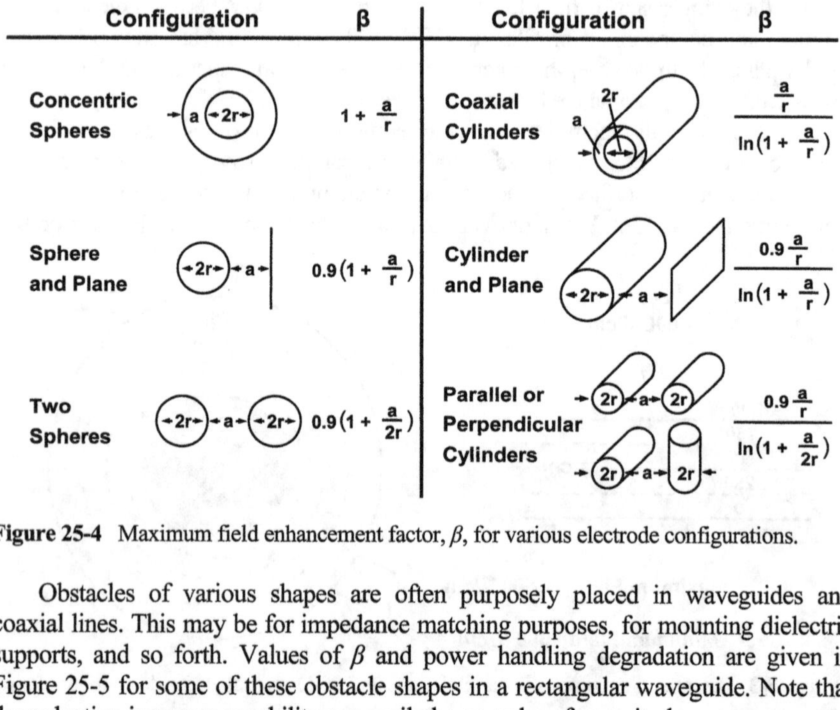

Figure 25-4 Maximum field enhancement factor, β, for various electrode configurations.

Obstacles of various shapes are often purposely placed in waveguides and coaxial lines. This may be for impedance matching purposes, for mounting dielectric supports, and so forth. Values of β and power handling degradation are given in Figure 25-5 for some of these obstacle shapes in a rectangular waveguide. Note that the reduction in power capability can easily be an order of magnitude or more.

The importance of field enhancement on electrical breakdown cannot be overemphasized. The most troublesome aspect of field enhancement is that, in many cases, its magnitude is either not known or is badly underestimated. The value of β of 1.56 calculated for the center conductor of a 50-Ω coaxial line is virtually meaningless if there are irregularities, such as scratches or machining marks, and so forth on the center conductor. These can easily have a β of 10 or more.

25.2 DC BREAKDOWN IN VACUUM [1–3]

Under ideal conditions, the breakdown field for vacuum exceeds that of all media including liquids and solids, and may be 10 MV/cm or higher. The upper limit results from field emission from the negative electrode. In practice, the field at which breakdown occurs is two to three orders of magnitude below the upper limit. For example, for average gap lengths (~1 cm), breakdown occurs at a few tens to a few hundreds of kV/cm.

For many vacuum tube applications, "Kilpatrick's criterion" [4] is used as the guideline to the maximum electric field that can be used in vacuum if breakdown is to be avoided. According to Kilpatrick, breakdown in vacuum varies as electrode spacing raised to the 0.75 power. Kilpatrick refers to field levels below this criterion as the no-spark region and to higher field levels as the possible-spark region.

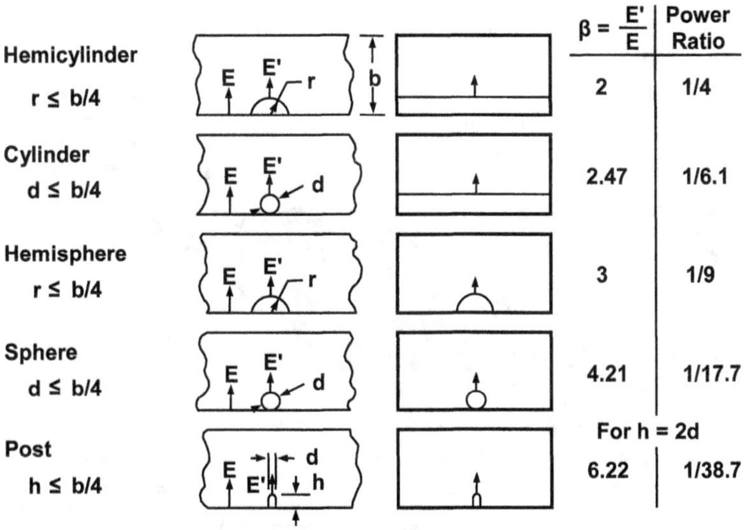

Figure 25-5 Values of β and power handling degradation for various obstacles in rectangular waveguides. (From: R. D. Wegenroth, Wheeler Laboratories Report 953P, June 1960.)

Figure 25-6 contains Kilpatrick's criterion and some results from the tube industry [5, 6]. The shaded areas indicate roughly the values of plate hold-off voltages for gridded tubes with cylindrical configurations and with thoriated tungsten (ThW) or oxide-coated cathodes. The data points represent the experimental breakdown characteristics of Pierce guns. The points marked by circles are for dc operation. The triangles and crosses represent long pulse and short pulse operation, respectively. Other results presented by Staprans [7] are in basic agreement with those given in Figure 25-6. Note that, for pulse operation, field levels somewhat above the Kilpatrick curve are sometimes used.

The Kilpatrick criterion is based on results for parallel plane electrodes while the tube results in Figure 25-6 are for various electrode configurations. Thus, varying degrees of field enhancement contribute to the scatter in the data. Apparently, when oxide coated cathodes are used, decomposition products form surface layers of reduced work function, greater roughness, different hardness, and so forth, which result in reduced breakdown voltage.

25.2.1 Electrode Phenomena Leading to Breakdown

The mechanisms responsible for breakdown in vacuum at field levels that are orders of magnitude below the field emission limit for a planar surface have been under investigation for decades. The electrode phenomena thought most likely to lead to vacuum breakdown in microwave tubes are shown in Figure 25-7. These phenomena and the way in which they cause breakdown are discussed in this section.

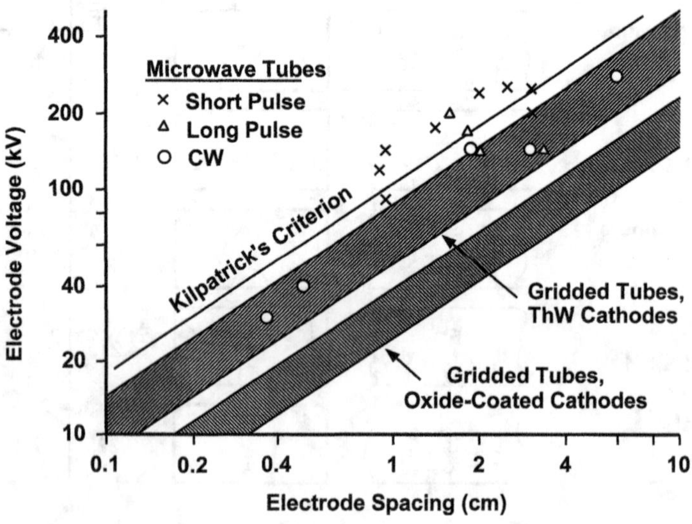

Figure 25-6 Kilpatrick's criterion and experimental breakdown characteristics of tubes.

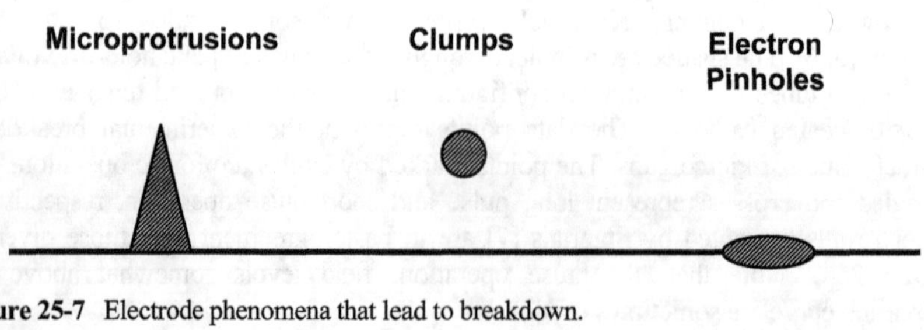

Figure 25-7 Electrode phenomena that lead to breakdown.

25.2.1.1 Breakdown Caused by Microprotrusions

To understand how microprotrusions can cause breakdown, it is necessary to investigate field emission phenomena. As was discussed in the chapter on cathodes, as the electric field applied to a cathodic electrode is increased to 10^7–10^8 V/cm (0.1–1.0 V/Å), electron emission increases very rapidly. The reason for this can be explained with the aid of Figure 25-8. At very high field levels, the potential barrier that electrons must overcome to escape from a metal becomes very narrow. It is so narrow, in fact, that when the wave function of an electron is examined, it is found that there is a finite probability that the electron may exist on the vacuum side of the barrier. In effect, electrons can pass through the barrier even though they don't have sufficient energy to overcome the barrier. This is known as the tunneling effect.

When a microprotrusion like that shown in Figure 25-9 is present, equipotential surfaces near the tip are crowded together so the electric field is higher than on nearby planar surfaces. If the field enhancement factor, β, at the tip is high enough (~ 100 or more) then field emission can occur at average field levels of ~ 10^5 V/cm.

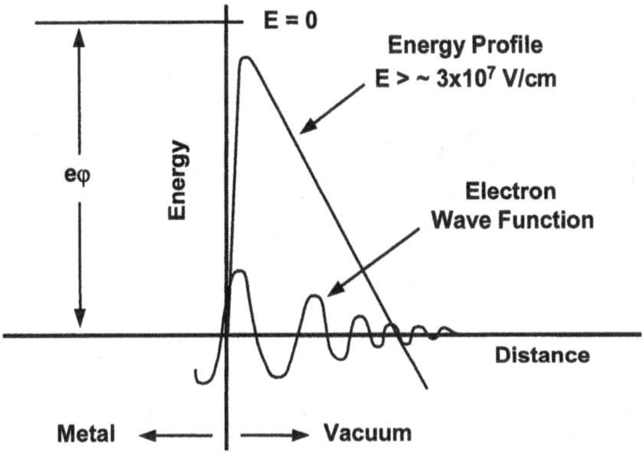

Figure 25-8 Electron potential energy profile with a very strong electric field.

The relation for current as a function of electric field when field emission is occurring is called the Fowler-Nordheim equation and is

$$J_o = C_1 E^2 e^{-C_2/E} \tag{25-1}$$

where C_1 and C_2 are constants that depend on work function.

Letting current density, $J_o = I/A$ where A is emitter area and $E = \beta V/d$ where β is the field enhancement and d is spacing, then

$$I = A\, C_1 \left(\frac{\beta V}{d}\right)^2 e^{-\frac{C_2 d}{\beta V}} \tag{25-2}$$

or

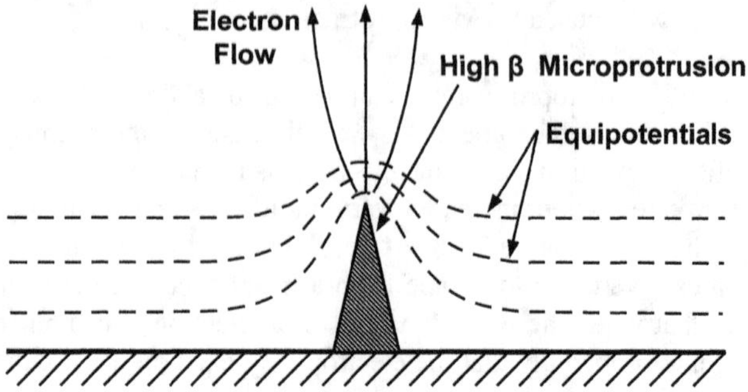

Figure 25-9 Equipotential concentration near the tip of a microprotrusion.

$$\ln\left(\frac{I}{V^2}\right) = \ln A C_1 \left(\frac{\beta}{d}\right)^2 - \frac{d C_2}{\beta}\left(\frac{1}{V}\right) \tag{25-3}$$

So a linear plot of ln I/V^2 versus $1/V$ as shown in Figure 25-10 indicates field emission.

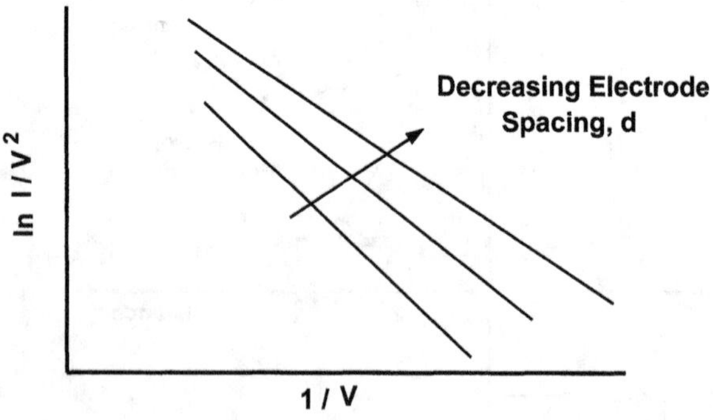

Figure 25-10 Fowler-Nordheim plots for field emission.

When electrical leakage is occurring, it is possible to determine if field emission is the cause or if conduction across an insulator surface is the cause. If conduction is the cause, then current is related to voltage by

$$I = C V \tag{25-4}$$

or

$$\frac{I}{V^2} = \frac{C}{V} \tag{25-5}$$

so

$$\ln\frac{I}{V^2} = \ln C + \ln\frac{1}{V} \tag{25-6}$$

Figure 25-11 shows an example of a Fowler-Nordheim plot for an electron gun in which it was apparent that electron conduction was causing leakage at low voltages and field emission was occurring at high voltages.

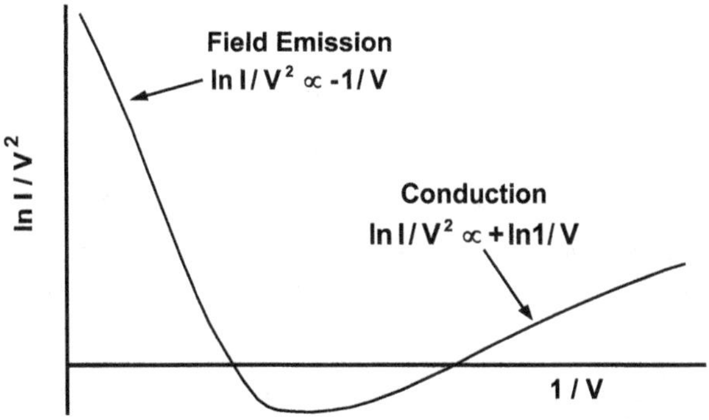

Figure 25-11 Fowler-Nordheim plot for an electron gun.

Microprotrusions can originate in several ways including the following:

1. Machining operations on some materials can produce sharp fibers.
2. A freshly polished surface can have filamentary structures lying in the plane of the surface. These can be erected normal to the surface by the strong electromechanical forces associated with a large electric field.
3. It is possible for microprotrusions to grow in the presence of large electric fields. This is a field activated surface diffusion process that is enhanced by heating the negative surface. Published data suggest that this effect is pronounced with electrode materials such as copper and aluminum that give unstable gaps.

For breakdown to be initiated by a microprotrusion, a microplasma must be generated. This is only possible if localized melting and vaporization of electrode material takes place. Outgassing can contribute to this process. As shown in Figure 25-12, there are two ways in which this can happen:

1. Cathode-initiated breakdown results from vaporization by current flow and resistive heating of the tip of a microprotrusion.
2. Anode-initiated breakdown occurs if the electron stream from the microprotrusion heats a spot on the anode to the point where melting and vaporization occur.

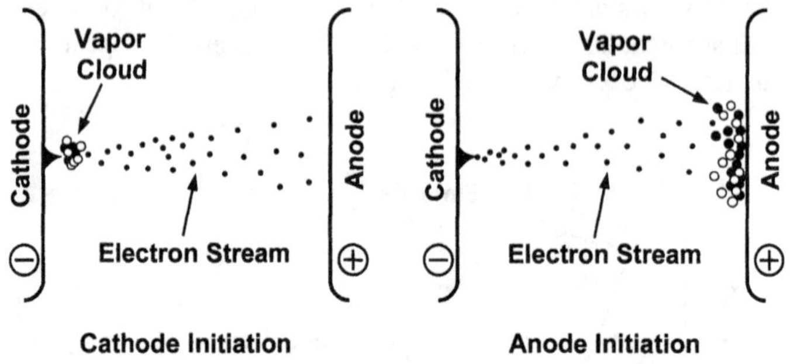

Figure 25-12 Breakdown initiation resulting from field emission from a microprotrusion. (Adapted from: *High Voltage Technology*, Chap. 4 by R. Hawley, Oxford University Press, 1968.)

As indicated in Figure 25-13, there can also be in-flight vaporization of a microparticle released from the anode hot spot.

Whether breakdown is cathode initiated or anode initiated depends on the field enhancement factor of the microprotrusion. Utsumi [8] found the high β emitters ($\beta > \sim 500$) (see Figure 25-14) usually lead to cathode-initiated breakdown while low β emitters ($\beta < \sim 30$) lead to anode-initiated breakdown. As shown in Figure 25-15, Utsumi also found that electrode separation played a role in determining whether breakdown was anode initiated or cathode initiated.

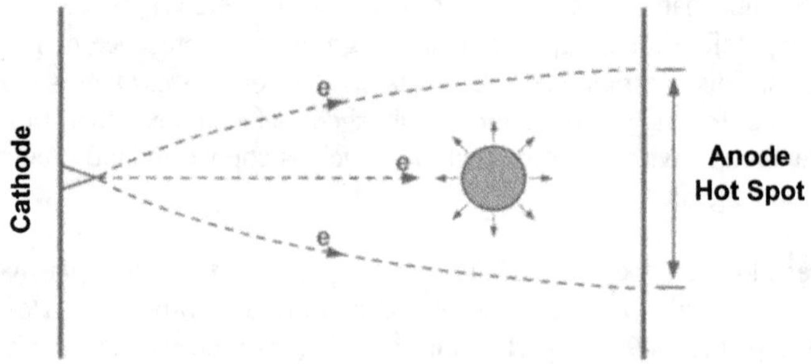

Figure 25-13 In-flight vaporization by electron bombardment of an anode microparticle.

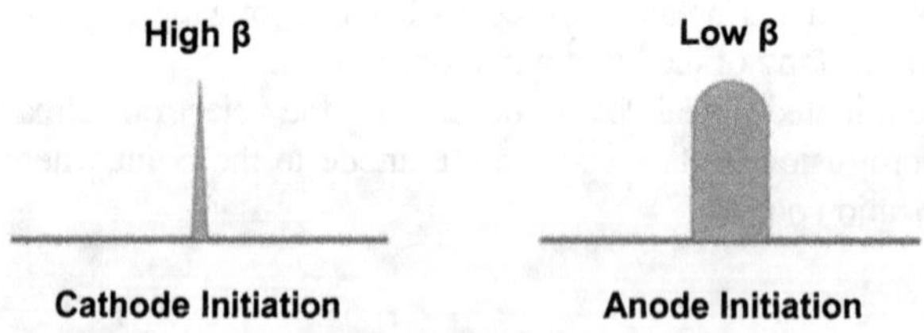

Figure 25-14 Emitters with high and low field enhancement factors.

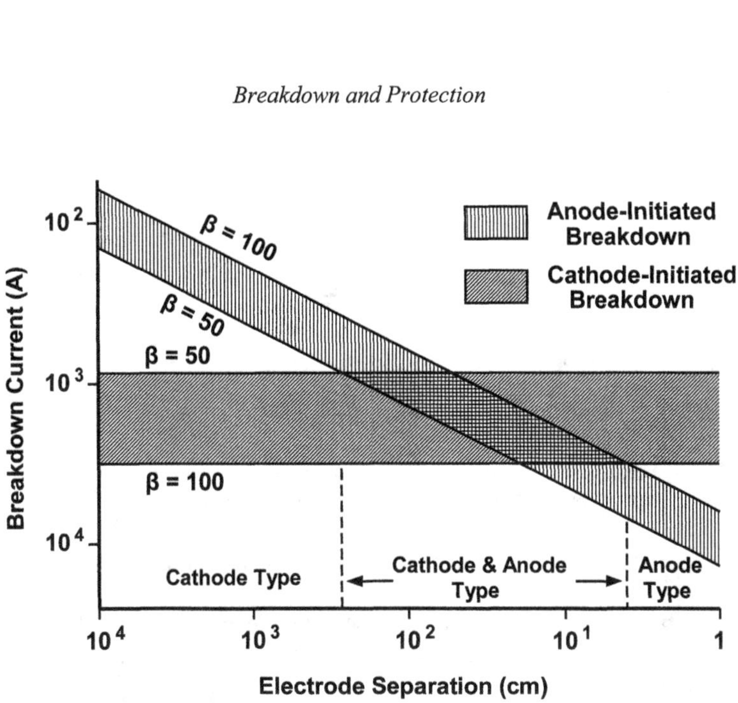

Figure 25-15 Electrode dominance in vacuum arc initiation. (From: T. Utsumi, *Technical Report* RADC-TR-66-712, January 1967.)

25.2.1.2 Breakdown Caused by Clumps

Cranberg [9] recognized that, in the presence of a high electric field, there are strong electro-mechanical forces. These forces tend to detach microparticles or clumps of microparticles from electrodes as shown in Figure 25-16. These clumps are charged to the same polarity as the parent electrode so they are also repelled. The resulting charged particles are accelerated to high velocities during the transit to the opposite electrode. On impact, particle energy is dissipated as heat and/or mechanical shock energy. If the impact energy is large enough, there can be sufficient vaporization of the particle and target to form a microplasma and, possibly, cause breakdown.

For large particles, breakdown can occur between a particle in transit and the electrode surface that it approaches. This can lead to breakdown of the gap.

When a particle strikes a surface, a crater can be produced. The splash rim of material surrounding the crater may have sharp protrusions that field emit and result in breakdown.

A very practical indication of the role of particles in breakdown may be had from a technique sometimes used by the tube industry, apparently to eliminate breakdown caused by particles. An electron gun to be highly stressed is placed in a vacuum chamber, high voltage is applied, and it is vibrated until arcing stops. The process may be repeated with varying orientations of the gun. Particles are observed to fall from the gun during the process, and it is presumed that they are the cause of arcing.

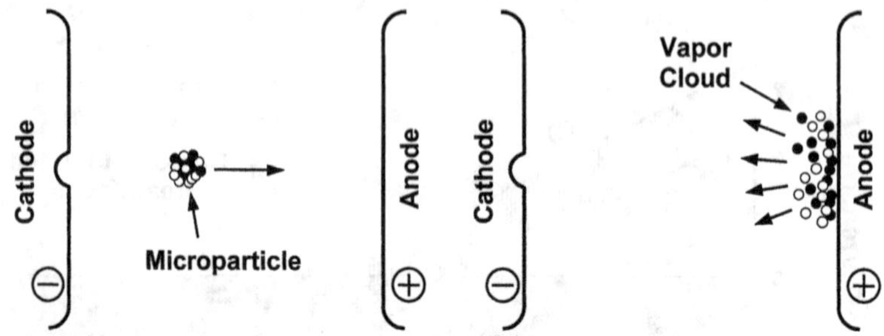

Figure 25-16 Breakdown initiation by a microparticle. (Adapted from: *High Voltage Technology,* Chapter 4 by R. Hawley, Oxford University Press, 1968.)

Examples of particles and sources of particles that can lead to breakdown are:

- Debris in the gun from the manufacturing process;
- Careless assembly procedures that cause chipping of insulators or other damage;
- Dust from inadequate cleanliness;
- Cathode evaporants that form loose films on various gun surfaces;
- Use of improper getters;
- Particles from heater potting material;
- Debris that enter the gun through the anode aperture;
- Embedded particles that are exposed by chemical etching.

In addition to these examples, as indicted in Figure 25-17, a shower of microparticles Figure 25-17(b) can be produced by the impact of a primary microparticle from the opposing electrode Figure 25-17(a). A shower of microparticles can also be produced by an arc in vacuum (to be discussed later in this chapter) and this can, subsequently, produce breakdown.

To find a relation for the conditions under which breakdown may be initiated by a clump, it is necessary to consider the energy, U, acquired by a particle that is accelerated across a gap. That energy is the product of the particle charge, q, and the accelerating voltage, V, that is

$$U = qV \tag{25-7}$$

If A is the cross-sectional area of the particle, the energy density on impact is

$$\frac{U}{A} = \frac{q}{A}V \tag{25-8}$$

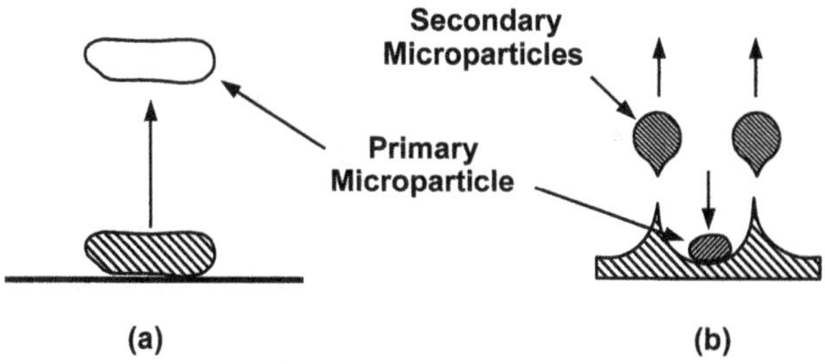

Figure 25-17 (a, b) A primary microparticle can produce secondary microparticles. (Adapted from: *High Voltage Vacuum Insulation*, by R. Latham, Academic Press, London, 1995.)

The charge density, q/A, acquired by the particle is proportional to the electric field, E, so

$$\frac{U}{A} \propto E\,V \tag{25-9}$$

Cranberg showed from experimental results that, in the limit when breakdown occurs,

$$E_{\text{lim}}\,V \approx 10^4 \tag{25-10}$$

where E is in kV/mm and V is in kV.

Figure 25-18 shows that the Cranberg limit applies to cathode-modulated Pierce guns. For anode-modulated tubes and dc-operated tubes the EV limit is closer to 10^3 $(\text{kV})^2$/mm. From Pierce gun design considerations, Durand and Shroff concluded that the relation between $E_{\text{lim}}V$ and V should be

$$E_{\text{lim}}\,V \sim V^{1.25} \tag{25-11}$$

This predicts that E_{lim} should increase with voltage. The dashed line in Figure 25-18 indicates the average of the $E_{\text{lim}}V$ products for Pierce guns and shows that there is, in fact, a trend for the maximum field to increase with increasing voltage.

It is interesting to observe that Kilpatrick's criterion (Figure 25-6) shows that, as spacing increases, the average electric field at which breakdown occurs decreases by about a factor of five (from ~ 200 kV/cm to ~ 40 kV/cm) as breakdown voltage increases from 20 kV to 400 kV. Over this same voltage range, (25-11) predicts that E_{lim} should increase by about a factor of two before breakdown occurs. Of course, we can understand from the Cranberg theory why E_{lim} can *increase* as the average field decreases. It is E_{lim} that is responsible for the charging of a clump and it is the gap voltage (and the average gap field) that accelerates the clump. The clump energy necessary to initiate breakdown is proportional to the product of the two fields and Thus, as one decreases, the other can increase.

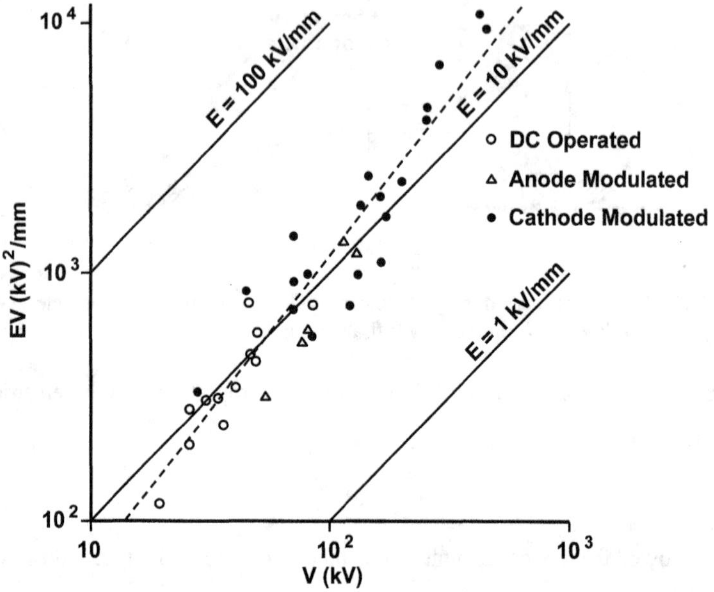

Figure 25-18 *EV* versus *V* for Pierce guns. (Adapted from: *High Voltage Vacuum Insulation*, Chapter 11 by A. J. Durand and A. M. Shroff. Academic Press London, 1995.)

Pulse length affects the maximum *EV* product as is shown in Figure 25-19. As pulse length increases, *EV* decreases. Durand and Shroff note that, when points corresponding to high values of *EV* are connected, almost all guns are below a limit given by

$$EV = \frac{100}{\text{pulse length}^{0.34}} \tag{25-12}$$

This limit must be taken into consideration during the design of electron guns. It should be noted that the squares in Figure 25-19 are the results of breakdown experiments on high- voltage guns from X-band klystrons at SLAC. These results show that it is possible to exceed the limit indicated by (25-12).

In Figures 25-18 and 25-19, most of the data indicate *EV* limits well below the Cranberg value of 10^4 (kV)^2/mm. There is the possibility that an explanation given by Latham may account for some of the low *EV* values. As shown in Figure 25-20 a particle may bounce back and forth between two electrodes and gradually acquire enough energy to cause breakdown. As a result, the gap voltage required to produce breakdown may be significantly below that needed for breakdown caused by a single-transit particle.

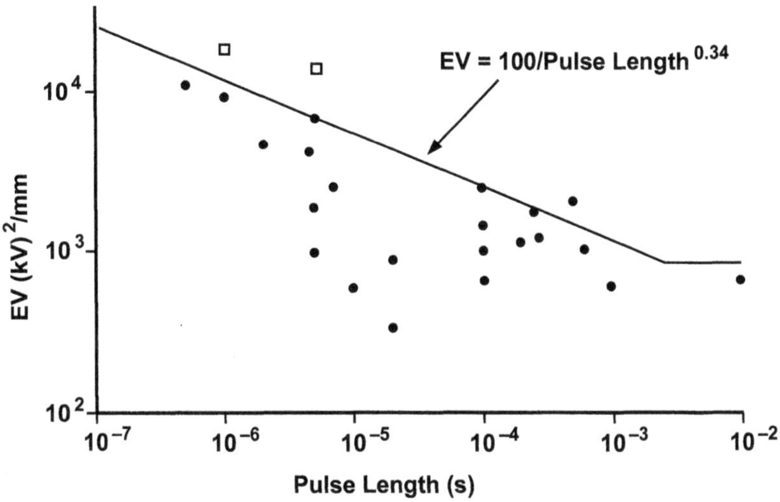

Figure 25-19 The effect of pulse length on *EV*. (Adapted from: *High Voltage Vacuum Insulation*, Chapter 11 by A. J. Durand and A. M. Shroff, Academic Press, London, 1995.)

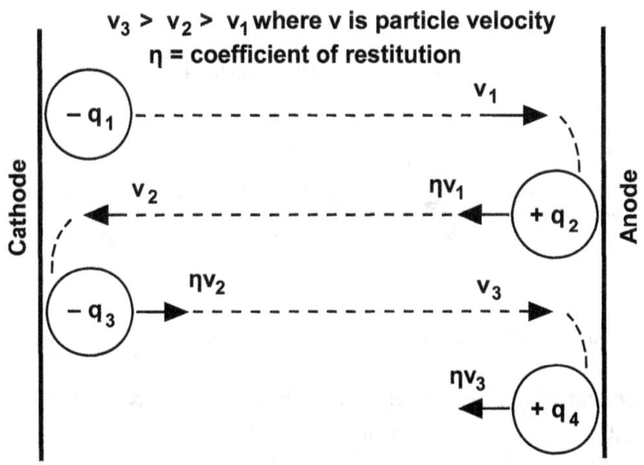

Figure 25-20 Particle bouncing and gradual energy accumulation proposed by Latham. (Adapted from: *High Voltage Vacuum Insulation*, by R. Latham, Academic Press, London, 1995.)

Durand and Shroff attributed the low *EV* values in Pierce guns to the presence of a cathode. Not only are the cathode and nearby electrodes very hot, there is the deposition of cathode evaporants (especially barium) to contend with. With attention to the following:

- Lowering the temperature of electrodes;
- Having well cleaned parts;
- Having well outgassed parts.

the EV limit may be increased.

25.2.1.3 Breakdown Caused by Electron Pinholes

The clump hypothesis for the initiation of breakdown appears to be valid. However, for breakdown thought to be initiated by microprotrusions:

1. Fowler-Nordheim plots predict unrealistic values of β (> 250 in some cases).
2. Often, no microprotrusions of the appropriate dimensions can be seen on the virgin electrodes.

As a result, in the mid-1970s the search for alternate emission mechanisms started. The result was the discovery of electron pinholes. In addition to the effect of loose particles, it has been demonstrated that a micron-sized particle like the insulating particle shown in Figure 25-21 that is attached to an electrode surface or partly imbedded in the surface can be the site for the initiation of breakdown.

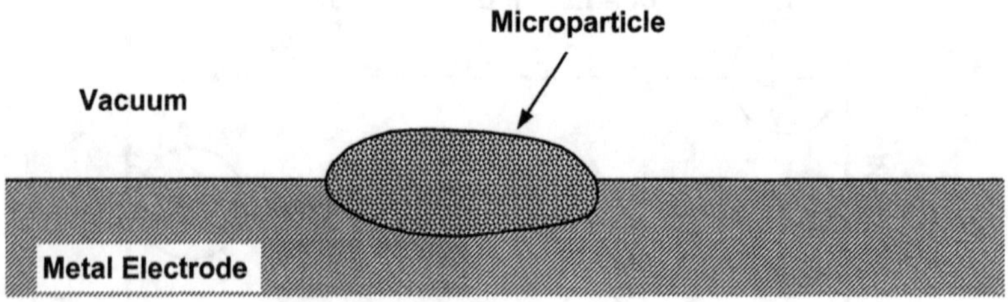

Figure 25-21 Partially embedded dielectric microparticle.

An excellent diagram for demonstrating the existence of electron pinholes is shown in Figure 25-22. The results of a scanning tunneling microscope (STEM) scan of an emitting site are shown. In the STEM process, a positively biased tungsten tip tracks an electrode surface at a constant distance from the surface. Spacing is determined by the tunneling current collected by the probe. As the probe approaches a pinhole with enhanced emitting properties ($\beta > 1$), the probe has to move away from the surface to maintain a constant current (with fixed bias). From the displacement of the probe from the surface Figure 25-22(a), the emitting properties of the surface Figure 25-22(b) can be determined. It has been shown that both the edge and the center of a particle can emit.

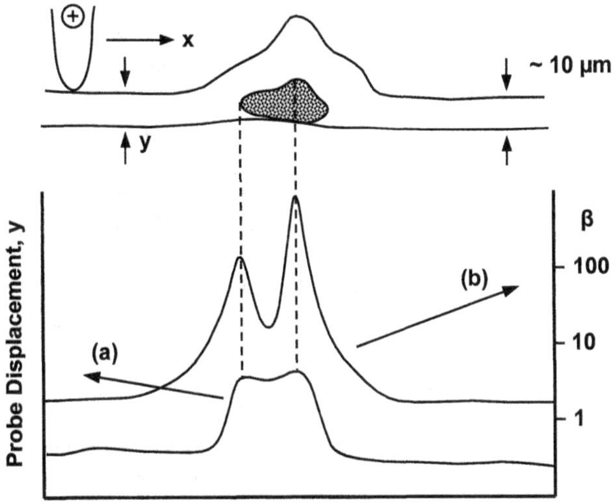

Figure 25-22 (a, b) Scanning tunneling microscope (STEM) analysis of an emission site. (Adapted from: *High Voltage Vacuum Insulation*, by R. Latham, Academic Press, London, 1995.)

The research of R.V. Latham [10] has provided a great deal of insight into the electron pinhole phenomena that cause this breakdown. The key results of Latham's studies (and the studies of others referenced by Latham) are summarized in Figure 25-23, which contrasts the properties of emission from a particulate site with that from a metallic whisker. In summary, the results show the following:

1. The basic emission process is associated with micron-sized particulate structures on the cathode surface. Frequently, such structures appear to be electrically insulated from the metal surface.
2. The material composition of the particles can be very varied, and apparently independent of the substrate cathode metal, with carbon being the single most common emission material, typically accounting for 20% of all sites.
3. In contrast to the linear Fowler-Nordheim characteristic of the metallic whisker, particulate emission sites have nonlinear characteristics, particularly at high fields.
4. Emission images obtained from individual sites indicate that the electrons originate from a number of apparently independent subsites that switch on and off randomly with mean frequencies that are dependent on pressure.
5. The electrons emitted from individual subsites have single-peak spectra that are symmetrical in shape, have large half-widths (typically > 300 meV), and exhibit a strong field-dependent shift from the Fermi level, which is characteristic of a nonmetallic emission mechanism. The key physical

implication of these data is that the emission process must involve the creation of hot electrons and that, consequently, the emission regime itself must include a dielectric or semiconducting medium.

6. The emission of electrons is associated with an electroluminescent effect that indicates the involvement of a semiconducting or insulating medium.

On the basis of the results listed above, two possible models shown in Figure 25-24 for the emission sites have been proposed. The M-I-V model involves a metal-insulator-vacuum regime. In the presence of a high electric field, electrons tunnel from the Fermi level of the metal into the conduction band of the insulator. They then move through the insulator under the influence of the penetrating field. Upon approaching the insulator-vacuum interface, the electrons experience very much higher field conditions, which cause them to be heated (accelerated) to temperatures in excess of 1,000°C. As a result, they are able to be quasi-thermionically emitted into the vacuum over the surface potential barrier at an energy that is significantly below the Fermi level. It is also assumed with this model that traps are present in the bulk of the insulator and these serve to control the flow of current at high field levels. This is the cause of the nonlinear Fowler-Nordheim characteristic shown in Figure 25-23.

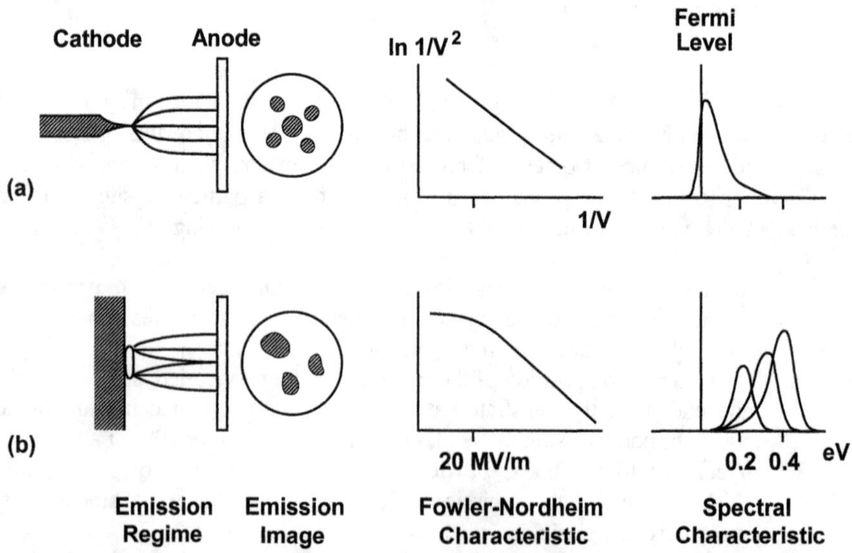

Figure 25-23 Comparison of emission properties of (a) a metallic whisker (manmade) and (b) a particulate site. (From: R. V. Latham and N. S. Xu, *Vacuum*, 1991.)

The metallic particle in contact with the insulating surface in the M-I-M-V model in Figure 25-24(b) serves to enhance the field at its contact point with the insulator. A conducting channel is formed in the region of the contact point and electrons are emitted into the metal flake as well as into the vacuum.

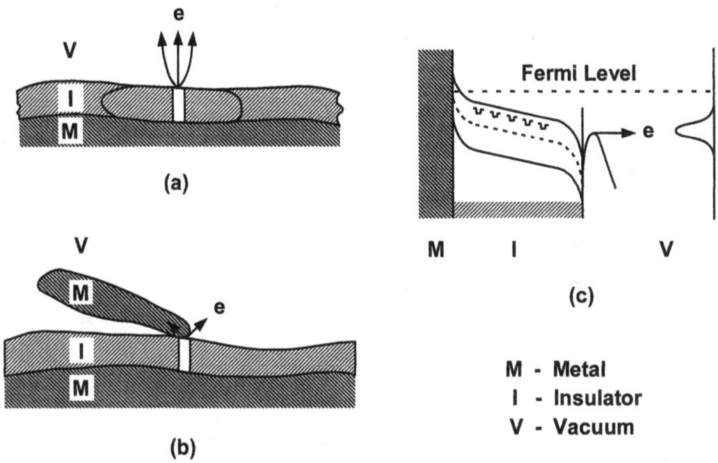

Figure 25-24 (a) M-I-V model, (b) M-I-M-V model, and (c) energy diagram of the conducting channel phenomenon. (From: R. V. Latham and N. S. Xu, *Vacuum*, 1991.)

25.2.2 Avoiding Breakdown

Four primary factors must be considered in avoiding breakdown:

1. Electrode geometry;
2. Electrode material;
3. Electrode surface preparation;
4. Electrode conditioning.

The area and spacing of electrodes along with field enhancement characteristics, all have strong effects on breakdown voltage. Obviously, radii should be as large as possible to minimize β. High β regions like slotted screw heads should be avoided or shielded. An increase in area will generally reduce breakdown voltage because more emitting sites and microparticles will be exposed.

For small spacings (< ~ 0.5 mm), breakdown tends to be field dependent. This indicates that breakdown is initiated by microprotrusion or electron pinhole phenomena. For large spacings (> ~ 2 mm), breakdown becomes voltage dependent. This indicates that breakdown becomes dominated by microparticle processes so cleanliness and the absence of particulate contamination are absolutely essential in microwave tube applications where spacings are almost always greater than 2 mm. In any event, electrode spacing should be as large as possible without overly compromising the operation of the device.

It has long been recognized that the choice of electrode material has a pronounced effect on breakdown voltage. It is generally agreed that, at one extreme there are materials like copper and nickel that are particularly susceptible to breakdown. At the other extreme, there are physically hard materials such as stainless steel, titanium, and molybdenum for which breakdown problems are minimized. However, hardness is not the only requirement because tungsten does not perform well.

Other than hardness, a second significant property of "good" electrode materials is their resistance to corrosion. The surfaces of these materials have a strong insulating oxide film, whereas, the surfaces of materials such as copper often have a semiconducting oxide film. While there is still controversy concerning breakdown mechanisms, these characteristics seem to be consistent with clump-initiated breakdown. For example, hard materials are less susceptible than soft materials to damage and the generation of secondary microparticles by impinging clumps. Semiconducting films are more likely to produce clump charging than insulating films. Increased clump charges result in increased clump velocities in the presence of high fields and increased surface damage on impact.

Concerning surface preparation, historically, the ideal electrode surface has been thought to be one that is free from microprotrusions and microparticles. This indicates the need for a surface treatment that includes:

1. A microscopic polishing procedure;
2. A final cleaning treatment.

Accordingly, the following procedures have historically been used to maximize breakdown voltage.

- Step 1: Take freshly machined electrodes and grind high-voltage surfaces with progressively finer grits.
- Step 2: After a thorough washing, polish to a 0.25-μm finish.
- Step 3: Ultrasonic clean with a degreasing agent, acetone, methanol, and distilled water.

Finally, great care must be taken to avoid recontamination during final assembly, so a clean room environment and procedures must be used. Still, surfaces prepared in this way will have a very complex surface microstructure so electrodes are often treated by using either a chemical or an electro-chemical polishing process. In addition, it has been shown that heating to ~ 1,100–1,200°C suppresses electron pinholes. (Of course, this is usually not possible with most microwave tube operations.)

Electrode conditioning is used to safely quench as many as possible of the sources of prebreakdown current. The simplest and most widely used technique is current conditioning, which is particularly suitable for initial treatment of electron guns. A high voltage supply is used with a current-limiting resistor to increase voltage in small steps so that the prebreakdown current is stabilized in each step. The time profiles of current and voltage for a microdischarge that does not result in breakdown are shown in Figure 25-25.

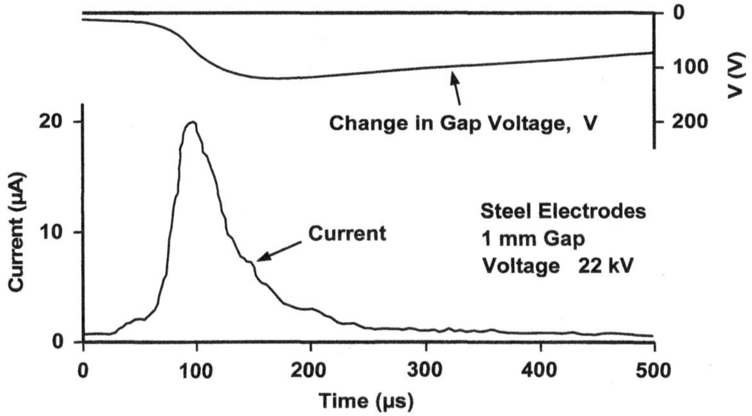

Figure 25-25 Microdischarge for 22 kV applied to a 1-mm gap between steel electrodes. (From: High Voltage Technology Seminar, Ion Physics Corporation, 1969.)

As high voltage is gradually applied to the electrodes (Figure 25-26), microdischarges are produced that "clean up" the electrodes. That is, microprotrusions are assumed to be thermally blunted or removed by the strong electromechanical forces. With a slowly increasing voltage, microparticles acquire a minimum charge and are not accelerated enough to cause breakdown. Instead, they are removed by continuous bouncing between electrodes or become welded to an electrode surface.

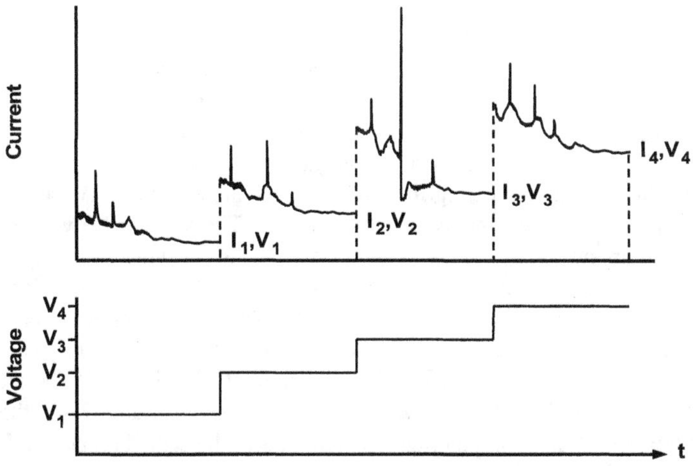

Figure 25-26 Voltage and current during conditioning. (Adapted from: *High Voltage Vacuum Insulation: The Physical Basis*, by R. V. Latham, Academic Press, 1981.)

In a high-voltage device, this conditioning process must be carried out when voltage is first applied and then may have to be repeated in an accelerated manner each time the device is turned on. It should also be noted that, even for a newly conditioned gap, the voltage hold-off capability degrades with time, as is shown by Figure 25-27.

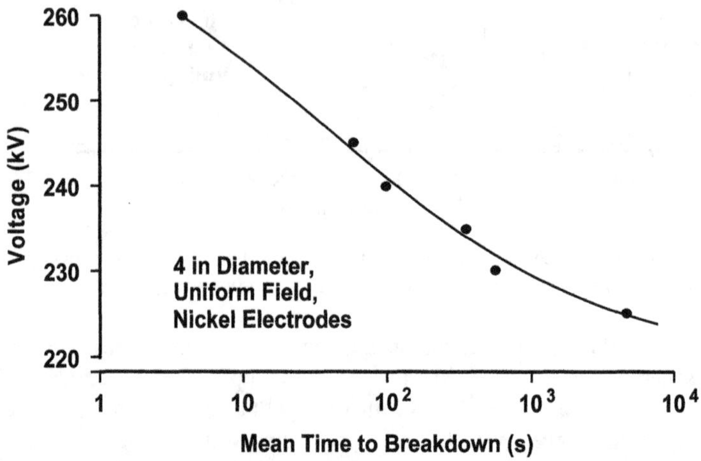

Figure 25-27 Breakdown characteristics for a conditioned vacuum gap. (From: High Voltage Technology Seminar, Ion Physics Corporation, 1969.)

25.2.3 Vacuum Arcs

Regardless of the initiation mechanism, dc breakdown in a vacuum leads to an arc on the negative electrode (at the current levels encountered under fault conditions in a microwave tube). This arc is normally known as a *vacuum arc.*

In a vacuum arc, the material of the negative electrode is vaporized and ionized by arc spots to form a conducting medium between the electrodes. Vacuum arcs can be established on any reasonably good conductor. The voltage between electrodes during a vacuum arc varies little with electrode spacing for typical (cm) spacings and is essentially the cathode-fall potential. This is because there is no appreciable anode-fall voltage and because the voltage drop in the plasma column is extremely small (~0.01 V/cm). The voltage drop is dependent on the material of the cathode and varies from 8 volts for mercury to 33 volts for cold steel. Figure 25-28 shows the relationship between the voltage at which a vacuum arc is sustained and the product of the boiling temperature and thermal conductivity of the cathode material. Barium is not included in this figure but is comparable to cadmium in arcing characteristics. As shown in Figure 25-29, the arcing voltage is a slowly varying function of current.

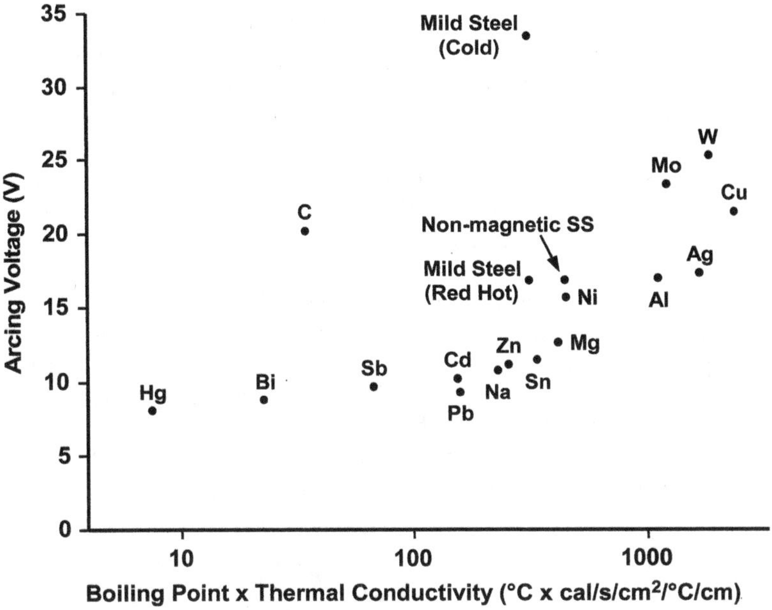

Figure 25-28 Voltage at which a vacuum arc is sustained for several commonly used metals. (Adapted from: M. P. Reece, *Proc. IEE*, 1963.)

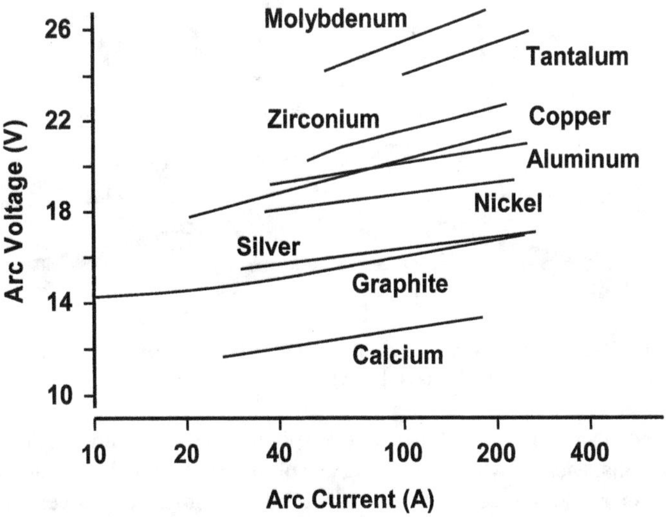

Figure 25-29 The dc volt-ampere characteristics of vacuum arcs. (From: W. D. Davis and H. C. Miller, *Jour. App. Phys.*, 1969.)

On a surface consisting of differing metals, a vacuum arc will operate on the low-arcing voltage metals. For example, if a calcium film partially covers a copper surface, a vacuum arc will remove the calcium film before attacking the copper. This results from the relatively low arcing voltage of calcium (about 12V) relative to that of copper (about 20V). Similarly, the evaporation products from thermionic cathodes will be preferentially removed from nearby grid or control electrode surfaces by a vacuum arc. This is because barium, strontium, calcium, and so forth all have similar arcing voltages, which are about one-half (or less) of those of the normal materials of grids and electrodes. As a side note, arcing may produce deposition of material on critical surfaces such as insulators

Unlike an arc in gas, which consists of one large spot anchored in position, many very small arc spots exist in a vacuum arc, as indicated in Figure 25-30. These spots travel at high speed (meters per second to tens of meters per second) with apparently random motion across the electrode surface. On a macroscopic scale, a surface subjected to prolonged vacuum arcing has the texture of a piece of frosted glass. On a microscopic scale, individual arc tracks somewhat resemble the tracks of insects crawling in random directions across a freshly painted surface.

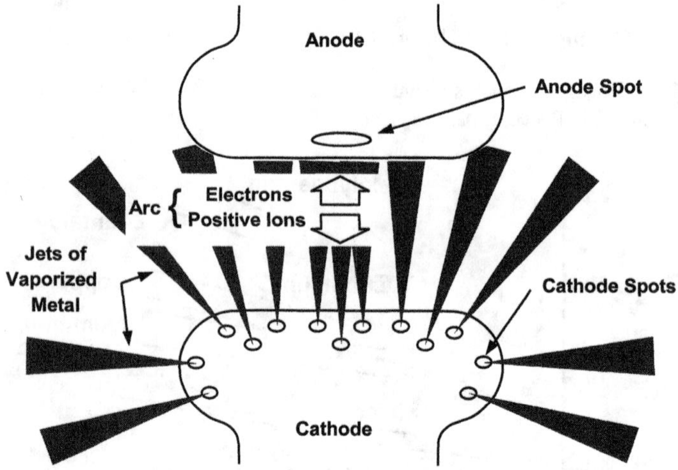

Figure 25-30 Details of a vacuum arc. (Adapted from: *Vacuum Arcs, Theory and Application*, J. M. Lafferty (ed.), John Wiley & Sons, 1980.)

It has been shown [11] that the cathode spot in the vacuum arc consists of a finite number of cells, each of which passes a current that is restricted to a narrow range of values. For example, Kesaev has shown that the current range of a cell in a mercury arc is approximately 0.08 to 0.1A. If the current through the arc is about 0.1A, then only one cell exists. If the current is raised above 0.1A, the cell divides into two cells. Further increases in current increase the number of cells until the total is equal to the current through the arc divided by the current per cell. If the current through a cell is reduced below about 0.08A, the cell extinguishes in a time that is thought to be less than 0.1 μs.

It is assumed that vacuum arcs on nearly all metals are cellular in nature. Possible exceptions are metals such as tungsten, which are capable of thermionic emission rates in excess of 10^6 A/cm^2 at temperatures slightly above their boiling points (as calculated using the Richardson-Dushman equation and room-temperature values of work functions). At these high temperatures, the computed evaporation rates lead to mass-to-charge ratios that are near the value of 0.1 mg/Coulomb that is observed in vacuum-arc experiments. As a result, if it is assumed that the cellular nature of the arc is necessary for the intense emission rates to occur on low-temperature materials, then cells may not necessarily exist on some high boiling point materials.

As the current through a vacuum arc is reduced, the number of cells decreases and the probability that all cells will extinguish simultaneously increases. As a result, the average arc lifetime becomes shorter and shorter. Figure 25-31 shows the interruption characteristics for several metals and shows, for example, that when the current through a tungsten (W) arc approaches 3A, the average lifetime approaches 10 μs.

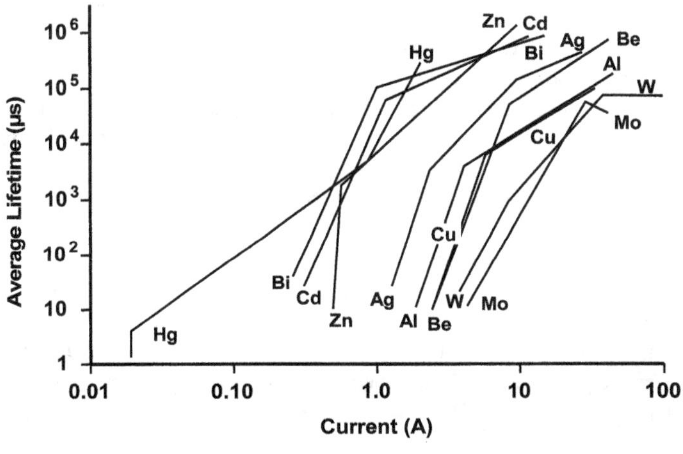

Figure 25-31 Average arc lifetime for pure electrode materials. (From: J. D. Cobine and G. A. Farrall, *Jour. App. Phys.*, December 1960.)

When a vacuum arc extinguishes, it does so very rapidly. In fact, the current is said to "chop," and the chopping time is usually less than 0.1 μs. The chopping characteristics of vacuum arcs are important to the designer of protection circuitry that is used with microwave tubes to prevent damage to the tube in the event that an arc occurs. For example, as the current is driven toward zero by a protection circuit, the arc will extinguish well before the current reaches zero. If there is significant inductance in the protection circuitry, the voltage generated by an arc that chops can be sufficient to cause insulation failure.

The rate at which material is removed from a surface by a vacuum arc varies somewhat from metal to metal but is generally about 0.1 mg per Coulomb of charge passed through the arc. Thus, if the time history of the current in a tube during an arc is known, the amount of erosion can be estimated with reasonable accuracy.

Similarly, if energy input to an arc in an electron gun is known, the erosion can be estimated. This is because the arcing voltage is restricted to a reasonably narrow range of values. For example, if a 20J arc occurs, the charge passed through the arc is estimated to be 1C using an average arcing voltage of 20V. For 1C of charge, 0.1 mg of material is removed. For a density of 10 gm/cm^3 (typical for most electron-gun metals), the volume of electrode material removed is 10^{-5} cm^3. For a 1 cm^2 area of electrode material, the thickness removed is only 10^{-5} cm or 0.1 μm. Thus, the amount of material removed by a 1C or a 20J arc is extremely small.

Unfortunately, the small amount of material removed by the 20J arc is not the only factor that must be considered. If the surfaces on which the vaporized arc material deposits are critical, such as insulators, leakage currents or insulator breakdown may result. Furthermore, gas molecules on the arced surfaces or nearby surfaces may be driven off by the arc, producing a gas burst. As an example, the release of a monolayer of gas from 1 cm^2 of area into a volume of 1 liter raises the pressure from 10^{-6} to 10^{-4} torr. This gas must be removed (by adsorption, an appendage pump, or a getter) before tube operation can be resumed.

The maximum permissible energy input during an arc in the gun is sometimes specified by the manufacturer. This energy is based largely on the manufacturer's experience. Maximum permissible energy levels from joules to tens of joules are not uncommon. Knowing the maximum permissible energy level, it is then possible for the tube user to design a protection circuit to interrupt the flow of energy before it becomes excessive.

Once a vacuum arc has extinguished, the voltage recovery rate can be extremely rapid. The recovery rate depends on the duration and intensity of the arc; however, a value of 10 kV/μs is not uncommon. Thus, it is often necessary to cause only a very brief interruption of current through a vacuum arc before voltage is reapplied to the electrodes. In fact, this principle has been used in the development of vacuum arc interrupters [12].

25.3 DC BREAKDOWN ON INSULATOR SURFACES

One of the weakest points in a vacuum device from an electrical breakdown point of view is the interface between a metal, an insulator, and a vacuum. This interface is often called the triple junction, and is illustrated in Figure 25-32. Because of electron launching mechanisms at the triple junction and surface effects on the insulator, the breakdown strength of the insulator (actually, breakdown of the insulator surface) is well below that of a vacuum gap or a solid insulator of equivalent length. Shannon,

Philip, and Trump suggested that through proper design, insulators in vacuum without conditioning should achieve 50 kV/cm, and conditioned insulators should be operable at 80 kV/cm [13].

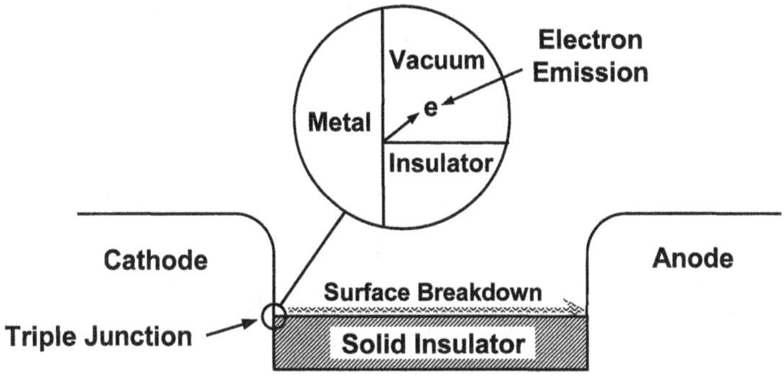

Figure 25-32 Triple junction, electron emission and breakdown across insulator surface.

At the triple junction there are at least two situations that might initiate breakdown. In one, the insulator is brazed to the metal surface. In the process of metallizing the surface of the insulator and brazing it to metal, it is virtually impossible to avoid the formation of a fillet of braze material on the surface of the insulator. This fillet has a razor-sharp edge with a high field enhancement factor. When voltage is applied, the high field can cause microprotuberances to grow, which further increases in the enhancement factor so field emission and, possibly, the generation of microparticles occur at relatively low voltages.

The second situation is when the insulator is mechanically held in place but not brazed. Then there are voids between the insulator and the metal along the triple junction. The effect of a void on equipotential surfaces is shown in Figure 25-33. The field enhancement factor in these voids may be as high as the relative dielectric constant of the insulator. For aluminum oxide, which is commonly used in these situations, the relative dielectric constant is about 9. This field enhancement factor, coupled with those of microprotuberances, is thought to lead to field emission and, possibly, microparticle generation.

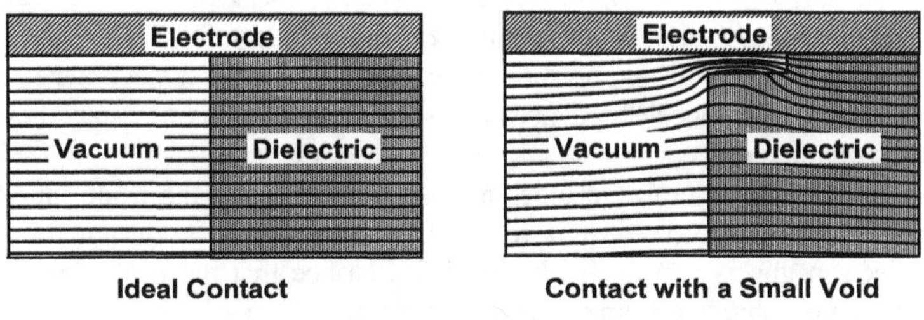

Figure 25-33 Effect of a small void on equipotential surfaces.

The second step in insulator flashover is thought to be positive charging of the insulator in the vicinity of the triple junction. This occurs, as indicated in Figure 25-34, because electrons that are field emitted from the triple junction strike the insulator surface causing secondary emission to occur. The secondary emission coefficient for most insulating materials is relatively high (values from 5 to 10 are common, especially if the primary electrons strike the surface at a grazing angle).

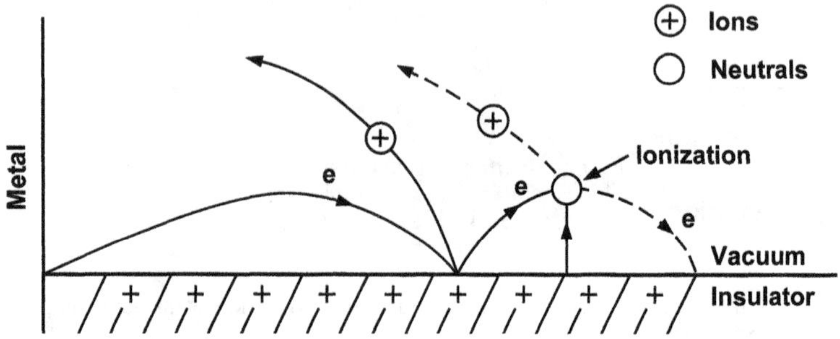

Figure 25-34 Events leading to insulator flashover. (Adapted from: *High Voltage Vacuum Insulation: The Physical Basis,* by R.V. Latham, Academic Press, 1981.)

The secondary emission process is regenerative in that the secondary electrons are accelerated by the field at the surface to produce additional secondaries. The result is that the insulator surface becomes positively charged in the triple junction region, and this increases the field and the rate of electron emission at the triple junction.

The next phase in the flashover process is thought to be desorption of gas molecules from the insulator surface and subsequent ionization as indicated in Figure 25-34. The positive ions are accelerated toward the cathode where they further enhance the field and, as a result, increase electron emission. Other processes possibly involving photon emission are thought to occur, and the result is rapid breakdown across the insulator surface.

The techniques that have been investigated for suppressing insulator flashover center on the prevention of the initial charging of the insulator in the region of the triple junction. As shown in Figure 25-35, if the angle of contact of the insulator with the electrode surface is about 45 degrees, the insulator flashover potential is about 5 times the flashover potential for normal incidence. The reason for this is thought to be that, since the electric fields are normal to the emitting electrode, electrons emitted at the triple junction are accelerated normal to the electrode. As a result, they don't strike the insulator surface.

It is common in microwave tubes to reduce the field at the triple junction by shaping the metal electrodes in this region. An example of the design for a high-voltage electron gun is shown in Figure 25-36. Notice that the triple junctions are almost completely shielded from equipotential surfaces.

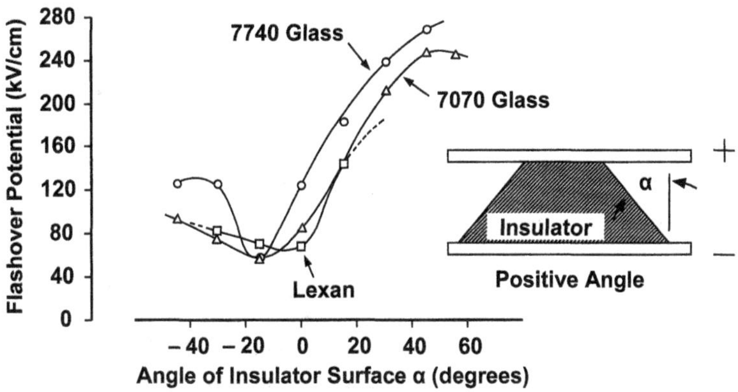

Figure 25-35 Effect of insulator geometry on insulator flashover. From A. Watson and J. Shannon, *Proc. Second Int. Symp. on Insulation of High Voltage in Vacuum,* September 1966.

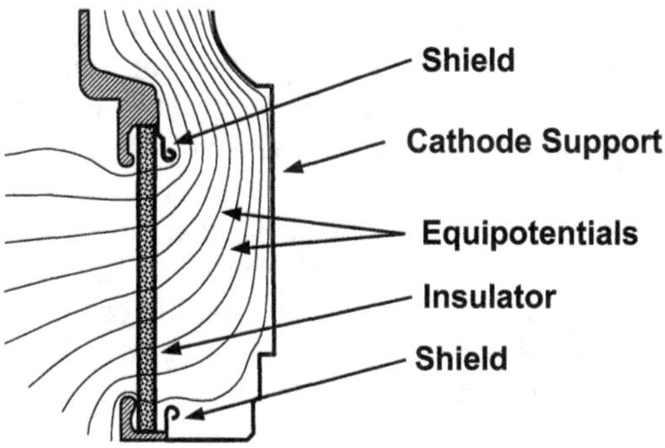

Figure 25-36 Electrode configuration for shielding the triple junctions in an electron gun.

25.4 RF BREAKDOWN IN VACUUM

Kilpatrick's criterion, discussed in Section 25.2, is often used as a guideline to the maximum RF voltage that can be applied between electrodes. Very often, however, another kind of discharge develops at voltage levels well below the Kilpatrick level. This discharge is called a *multipactor discharge*. In its simplest form, this discharge occurs when electrons move back and forth between two electrodes in synchronism with an RF field. As indicated in Figure 25-37, if the secondary emission coefficient of the electrodes is greater than unity (to be more specific, if the product of the secondary emission coefficients of the two electrodes is greater than unity), then the number of electrons involved in the process builds up with time. Many multipacting modes other than that shown in Figure 25-37 are possible. These will be described in this section.

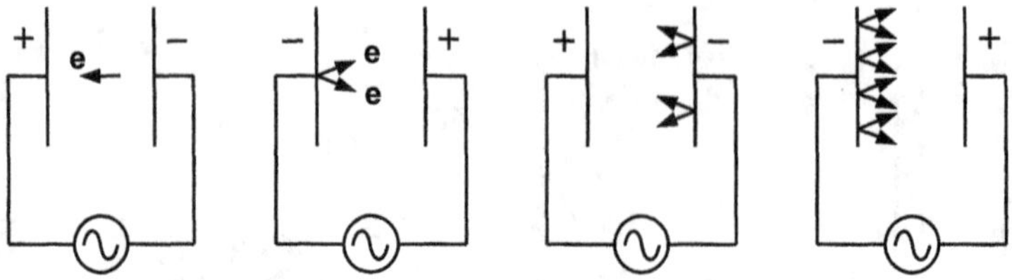

Figure 25-37 Illustration of simple two-surface multipactor.

During multipacting, large amounts of energy can be discharged in a small volume with the result that electrodes or RF windows can be damaged or destroyed. Even if excessive damage does not occur, multipacting can load an RF circuit so that the output power is reduced and modulated with a noiselike signal.

The phenomenon of secondary electron resonance, or multipactor, was first recognized and described by Farnsworth 75 years ago [14]. His primary interest in multipactor was its use in amplifying devices. With the exception of Vaughan's report in 1988 [15], not much has been published about multipactor in microwave tubes. As noted by Preist [16], "There is likewise very little to be heard on the subject of fleas, perhaps for the same reason, that undue familiarity with the subject may be taken to mean that one has problems." Still, as pointed out by Vaughan, multipactor is present in one form or another in nearly all high-power microwave tubes; often it is observable but harmless, occasionally beneficial, and sometimes disastrous. Thus, it is very important for microwave tube engineers to understand multipactor and the approaches and techniques that can be taken to control or eliminate it. In this section, multipacting modes are described along with some specific corrective actions that have been taken to eliminate multipactor.

25.4.1 Two-Surface Multipactor with No Magnetic Field

The theory of two-surface electric-field multipactor has been presented in various reports. The results given here were reported by Vaughan [15]. The configuration that he considered consists of two parallel electrodes driven by an RF voltage as shown in Figure 25-38.

Whether or not multipacting occurs depends primarily on the peak RF voltage, the frequency of the RF and the gap width. There are several combinations (zones) of voltage, frequency and gap width that can produce a multipactor discharge. In the $N = 1$ zone, electrons move back and forth between the surfaces in synchronism with each half cycle of the alternating RF. In the $N = 3$ zone, the electron transit time across the gap corresponds to 3/2 cycle of the applied RF voltage. This means that electrons start across the gap, stop and reverse direction of travel as the RF voltage reverses polarity, and then stop and reverse direction of travel again when the voltage reverses polarity before finally landing on the opposing electrode. In the $N = 5$ zone, the electron transit time across the gap corresponds to 5/2 cycle of the applied RF voltage.

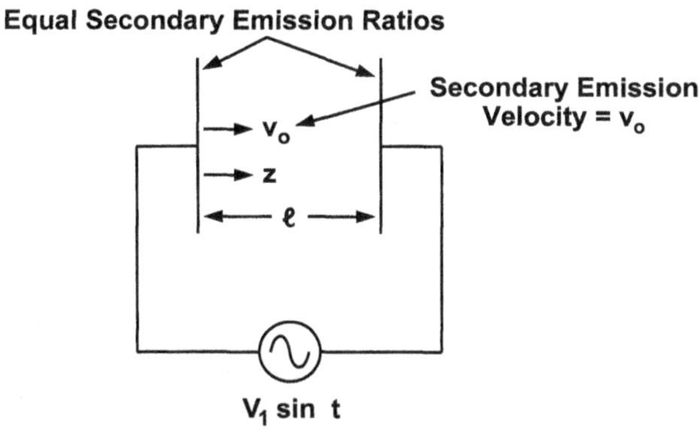

Figure 25-38 Gap configuration for analysis of two-surface electric-field multipactor.

The conditions that produce three of the multipactor zones are shown in Figure 25-39. For each zone, approximate contours of constant impact voltage are indicated by dashed lines. The lowest impact voltage is 100V. This is because, as Figure 25-40 shows, the secondary emission coefficient for typical materials drops below unity at impact energy levels below about 100 eV and so multipactor does not occur. At high impact voltages, the energy level at which the secondary emission ratio drops below unity is highly dependent on the angle of incidence of the impacting electrons. As shown in Figure 25-40, for normal incidence, the value is about 2,500 eV. The curve for impact at about 55° shows that the secondary emission coefficient drops below unity well above 4,000 eV.

For some forms of multipactor to be discussed later, the impact angle can approach 90° from normal. The energy of these electrons can be almost unlimited and the secondary emission coefficient will still be above unity.

The power level of the simple multipactor discussed in the previous paragraphs is usually too low to cause physical damage to the electrodes. However, the power level is sufficient to cause RF loading and, as a result, degradation of tube performance.

Vaughan points out that there are few general methods for suppressing multipactor. In general, a technique that employs a change in geometry or a change in operating conditions is preferred to one that is based on a surface treatment (such as a low secondary emission coating). This is because the change in geometry or operating conditions usually provides a permanent solution, whereas the surface treatment may deteriorate with time resulting in the return of conditions favorable to a multipactor.

An example of a geometrical solution is the one used in the output cavities of some klystrons. There, multipactor can occur between the drift tube tips as indicated in Figure 25-41. By cutting axial slots (sometimes referred to as fluting or serrating) the drift tube tips, the multipactor discharge can be suppressed. The reason is that the effective secondary emission coefficient of the drift tube tips is reduced to a value below unity. This occurs because many of the secondaries produced by the impacting electrons that fall into the slots are trapped in the slots.

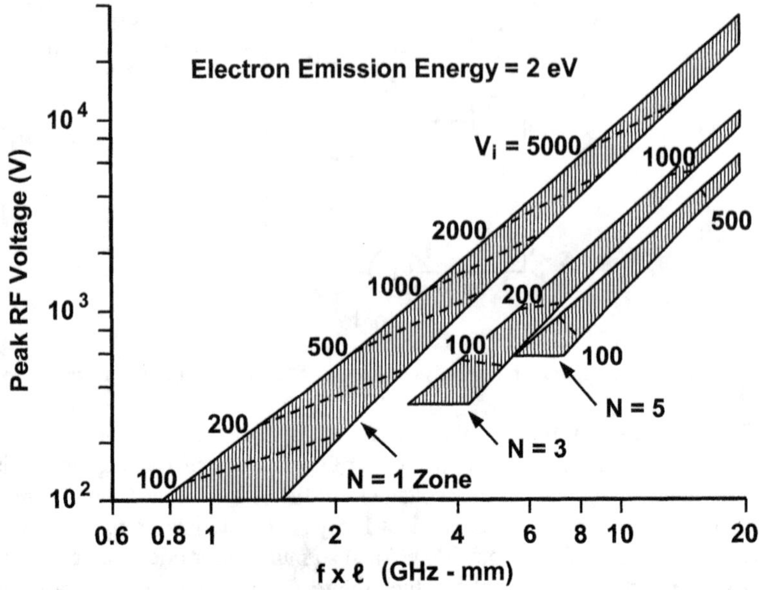

Figure 25-39 RF voltages at which two-surface electric-field multipactor discharges may occur. (From: J. R.M. Vaughan, *IEEE Trans. Electron Devices*, July 1988. © 1988 IEEE.)

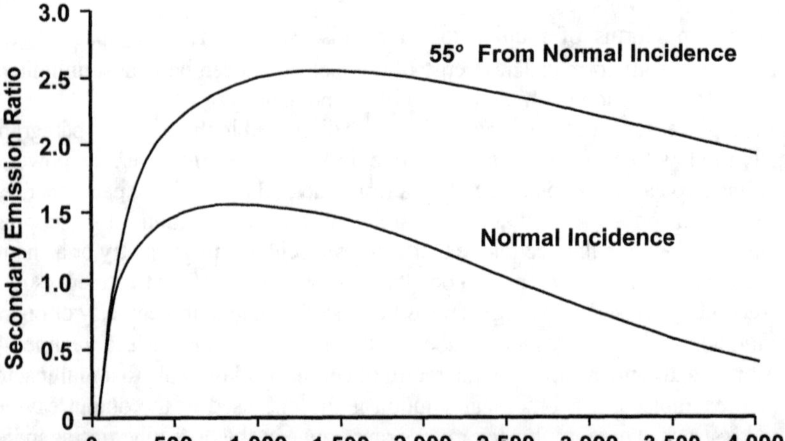

Figure 25-40 Typical variations of secondary emission ratio with impact energy for metal surfaces. (From: J. R. M. Vaughan, *IEEE Trans. Electron Devices*, July 1988. © 1988 IEEE.)

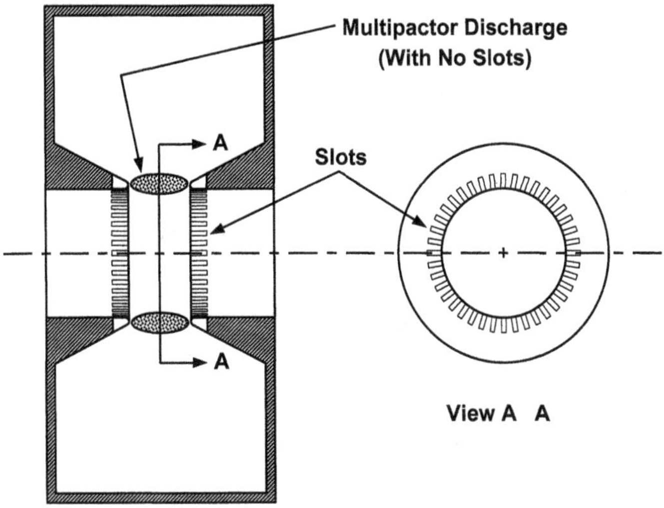

Figure 25-41 Use of slots in drift tube tips to suppress multipactor.

25.4.2 Two-Surface Multipactor in Combined Fields

When a transverse magnetic field is imposed on the parallel-plate electrode configuration considered Section 25.4.1, the electrons are deflected from straight-line paths to the curved trajectories shown in Figure 25-42. The most significant aspect of the change in trajectory shape is that the electrons can impact on the electrodes at an angle that can approach 90° from normal incidence. Thus, referring back to Figure 25-40, it is possible to see that the energies of the electrons can be tens of thousands of electron volts and the secondary emission coefficient of the surface will still remain above unity. Thus, the crossed-field multipactor can be a very high voltage and high-power discharge.

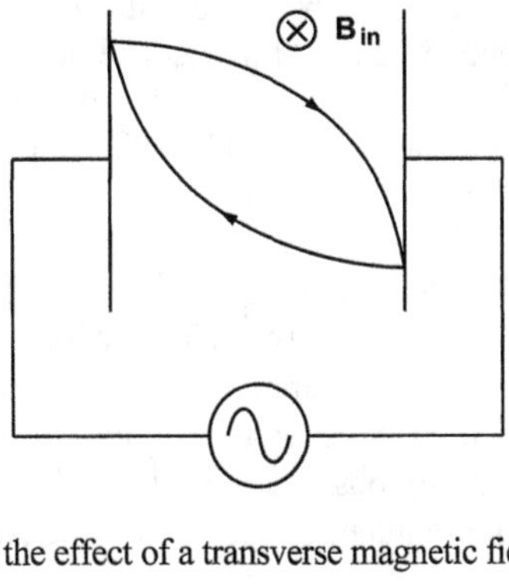

Figure 25-42 Illustration of the effect of a transverse magnetic field on electron trajectories.

Maximum secondary emission will occur when electron impact is tangential to the electrode surfaces and this is when we would expect the crossed-field discharge to be most intense. This will occur when an electron reaches the top of a cycloidal orbit in one-half cycle of the applied frequency and this, in turn, occurs when the applied frequency is equal to the cyclotron frequency, that is, when

$$\omega = \omega_c = \eta B \tag{25-13}$$

Vaughan [17] reported the case of crossed-field multipactor in the impedance transformer of a magnetron, the configuration of which is shown in Figure 25-43. It was thought that electrons generated by the multipactor at point A and missing the transformer surface at point B continued on to strike the window at point C causing window failure.

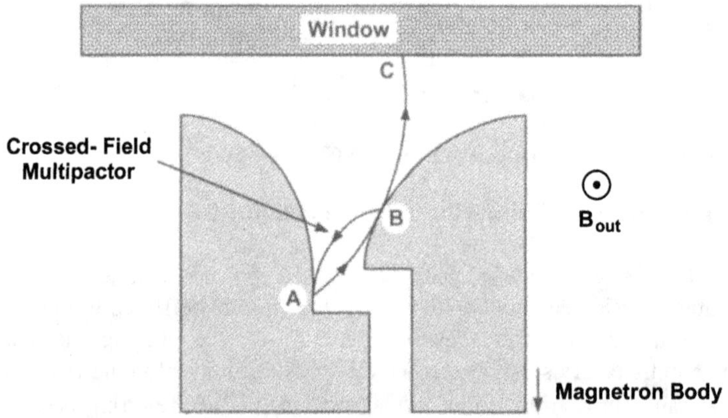

Figure 25-43 Possible crossed-field multipactor in magnetron impedance transformer. (From: J. R. M. Vaughan, *IEEE Trans. Electron Devices*, July 1988. © 1988 IEEE.)

The measured magnetic field in the region of the multipactor was 1,000 gauss and so the cyclotron frequency was 2,800 MHz. The operating frequency of the tube was 2,770 MHz and so it was nearly equal to the cyclotron frequency. X-ray photographs and good agreement between calculated impact velocities and observed X-ray hardness strongly indicated that the multipactor configuration was as shown in Figure 25-43.

The crossed-field multipactor was virtually eliminated by reducing the magnetic field in the region of the transformer by 90%. Vaughan also noted a substantial reduction in the intensity of the discharge when the gas pressure in the tubes was reduced. This was attributed to a reduction in ion formation in the multipactor discharge. It is thought that these ions neutralized the electron space charge forces in the discharge and so it was more intense than when the ions were absent.

25.4.3 Single-Surface Multipactor with No Magnetic Field

In the absence of a magnetic field, a multipactor discharge can occur on a single dielectric surface as is indicated in Figure 25-44 [16]. A dielectric is required because there must be a component of the electric field parallel to the surface. Electrons emitted from the surface (indicated by the vectors labeled with v_o in Figure 25-44) have a component of velocity perpendicular to the surface. The electric field accelerates the electrons in the direction parallel to the surface and, in the absence of any force toward the surface, the electrons drift away, as indicated by trajectories 1 and 2.

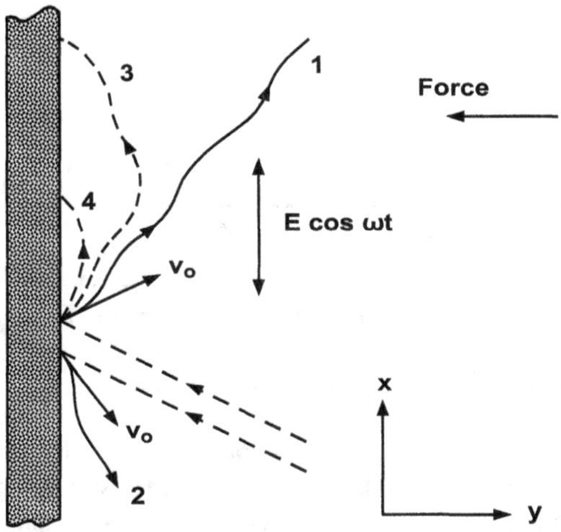

Figure 25-44 Electron motions near a dielectric surface with no magnetic field. (From: D. H. Preist, *Proc. 4th International Congress on Microwave Tubes*, 1963.)

If there is a force toward the dielectric surface, then electrons follow trajectories like 3 and 4 in Figure 25-44 and strike the surface. For electrons with sufficient energy and a surface with a secondary emission coefficient greater than unity, one or more secondaries will be emitted for each primary. Each of these can follow a trajectory like that shown in Figure 25-45 and produce additional secondaries. As a result, a multipactor discharge can build up on the dielectric surface and cause heating and failure of the dielectric.

The force shown in Figure 25-44 can arise when electrons (from a source such as the two-surface multipactor shown in Figure 25-43) strike the dielectric surface and cause electrons to be emitted. If the secondary emission coefficient is greater than unity, more electrons will leave the surface than arrive at it and so the surface will become positively charged. Thus, a force is generated that may enable the multipactor discharge to build up.

It should be noted that, unlike the two-surface multipactor discharges, there is no requirement with the single-surface discharge for a definite transit angle. Thus, the single-surface multipactor can exist over a large range of electric field strengths as long as the field is sufficient to accelerate electrons to impact velocities high enough so that the secondary emission coefficient is greater than unity.

The secondary emission coefficients of all of the dielectric materials used in microwave tubes is greater than unity. To suppress multipactor, it has been necessary to coat the dielectrics with low secondary yield films. Titanium-based films have been successfully used for this purpose [16, 18].

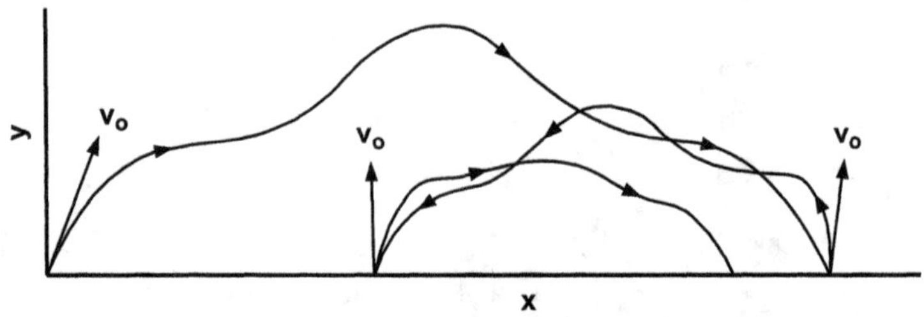

Figure 25-45 Electron motions during single-surface multipactor with no magnetic field. (From: D. H. Preist, *Proc. 4th International Congress on Microwave Tubes*, 1963.)

25.4.4 Single-Surface Multipactor in Combined Fields

An example of single-surface multipactor in combined fields on a metallic surface is shown in Figure 25-46 and an example on a dielectric surface is given in Figure 25-47. The electron trajectory shown in Figure 25-46 was computer generated [15]. This single-surface multipactor was observed using an X-ray pinhole camera in the output gap of a high-power klystron. The computation showed that electrons were leaving the cavity surface and returning one RF period later under the influence of the magnetic focusing field and the RF electric field. The electrons impinged tangentially on the cavity surface with an energy of about 35 keV to points displaced about 35° around the axis from their starting points. Thus, the whole discharge was rotating about the axis of the cavity. In the report of this multipactor [15] no technique was described for suppressing the discharge.

The multipactor shown in Figure 25-47 occurred on the inner surface of a dielectric cylinder used in the cavities of some klystrons [16]. This discharge is similar to the single- surface multipactor that occurs on a dielectric in the absence of a magnetic field. When the magnetic field is present, the force driving the electrons toward the surface of the dielectric is provided by the magnetic field in conjunction with the component of the initial electron velocity normal to the surface. The most intense discharge occurs when the cyclotron frequency is equal to the radio frequency. Magnetic fields stronger than the cyclotron field tend to prevent this form of multipactor.

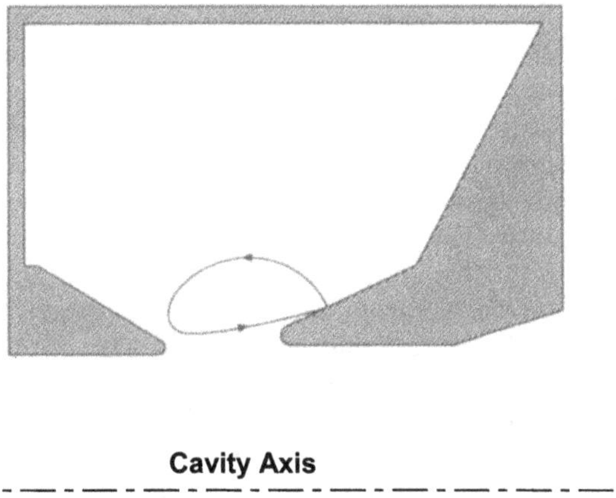

Figure 25-46 Electron trajectory in a multipactor discharge on a metallic surface in combined fields. (From: J. R. M. Vaughan, *IEEE Trans. Electron Devices*, July 1988. © 1988 IEEE.)

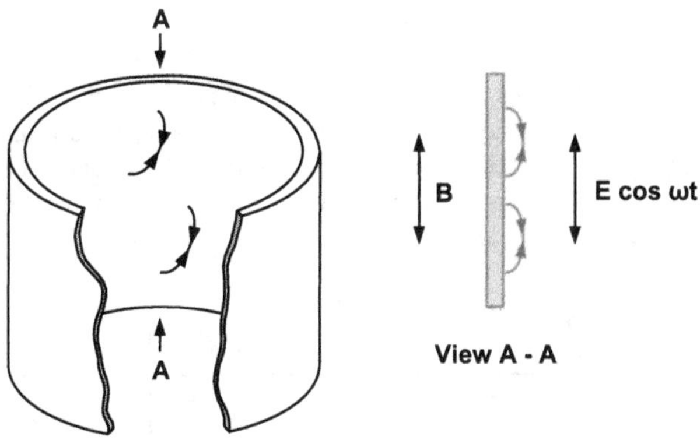

Figure 25-47 Electron motions in a multipactor discharge on a dielectric surface in aligned fields. (From: D. H. Preist, *Proc. 4th International Congress on Microwave Tubes*, 1963.)

25.5 RF BREAKDOWN OF INSULATORS

The multipactor discharges on dielectrics described in the previous section are discharges that occur on a surface. They may produce heating of the surface and that, in turn, may result in cracking and insulation failure. Vaughan [19] described another cause of window cracking that results from arcing to a conductive deposit on the vacuum side of the window. This problem was eliminated by preventing the formation of the conductive deposit.

Vaughan [19] has also described window punctures that resulted from the combined actions of an RF electric field and dc window charging on electrons deposited on the surface of a window. The window configuration was that shown in Figure 25-43 and the source of electrons was the crossed-field multipactor that occurred in the magnetron impedance transformer. The windows were generally observed to be charged positively between 5 and 12 kV on the *outside*. As a result, there was an electrostatic field of about 50 kV/cm through the window. Vaughan postulated that electrons striking the window formed a small cavity in the window surface and that the field at the tip of this cavity enhanced the electrostatic field in the window. The resulting dc force attracting electrons into the window, coupled with RF forces (produced by the RF electric field component parallel to the window surface) causing the electrons to oscillate sideways, produced a disintegrating action that resulted in electrons burrowing through the window. The result was the formation of a pinhole with a diameter on the order of 10 μm. This failure mode was eliminated by suppressing the crossed-field multipactor, which was the source of the electrons that initiated the failure process.

25.6 DC BREAKDOWN IN GAS [20–22]

Electrical discharges in gases are extremely complex. Several different types of discharges can exist, depending on the current and voltage applied to the gas. These various discharge types are indicated in Figure 25-48. Current and relative voltages shown in Figure 25-48 are representative values only and depend on factors such as gas type and the area of the electrodes used during testing.

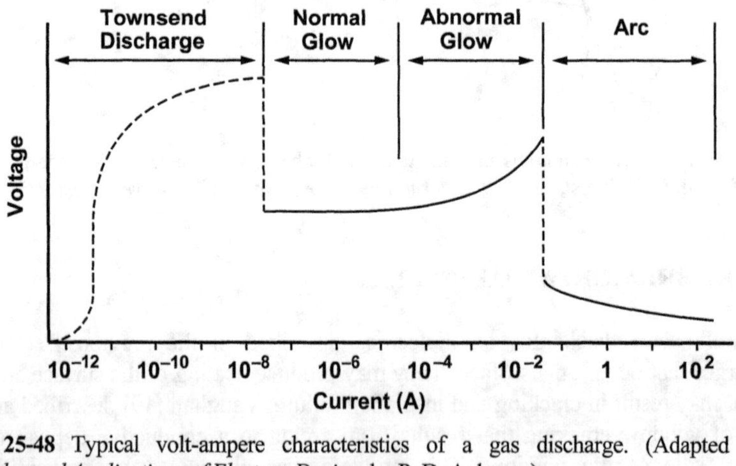

Figure 25-48 Typical volt-ampere characteristics of a gas discharge. (Adapted from: *Principles and Applications of Electron Devices* by P. D. Ankrum.)

When breakdown occurs, the discharge rapidly increases in intensity. Depending on circuit conditions (voltage and impedance), either a glow discharge or an arc may develop.

The breakdown voltage depends on the ease with which the gas is ionized and on the effectiveness of the ions in producing electrons when they strike the negative electrode. As a result, the breakdown voltage is a function of gas type and pressure, and electrode spacing and material.

The breakdown voltage is normally plotted in the manner shown in Figure 25-49. Breakdown voltage plots of this type are referred to as Paschen characteristics because of Paschen's law, which states that, for parallel plane electrodes in a particular gas at a particular temperature, the breakdown voltage is a function of the product of the gas pressure and the electrode spacing.

The shapes of the Paschen characteristics may be explained as follows: to the left of the minimum, the breakdown voltage increases as pressure and density decrease (with fixed electrode spacing). This is because the gas molecules are farther apart, and more electrons travel through the gas without producing ionization. In the limit, under high-vacuum conditions, there is essentially no ionization. Breakdown occurs because of electrode phenomena. DC breakdown in a vacuum is discussed in Section 25.2.

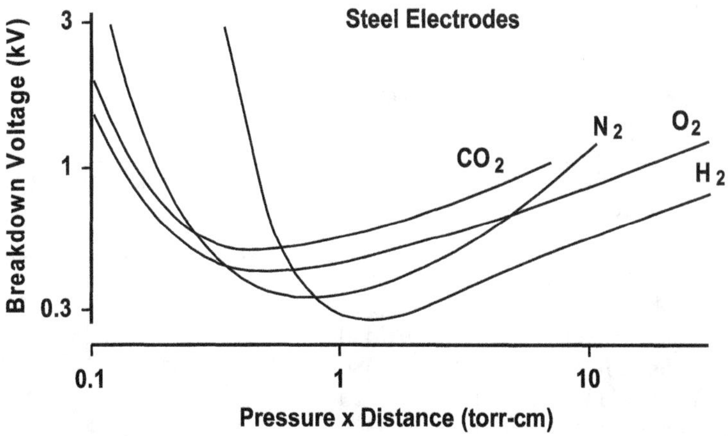

Figure 25-49 Paschen characteristics for some molecular gases. (From: M. J. Schonhuber, *IEEE Trans. P.A.S.* February 1969. © 1969 IEEE.)

To the right of the Paschen minimum, the breakdown voltage increases as pressure and gas density increase. In this case, gas molecules are closer together, and so a large electric field is required to give the electrons enough energy in the short distances between collisions to produce ionization. Thus, the breakdown voltage is high. At atmospheric pressure and above, the breakdown voltage for a 1-cm gap can be several tens and even hundreds of thousands of volts.

At intermediate pressures and electrode spacings, optimum conditions for ionization occur, and so there is a minimum in the Paschen curve. The minimum breakdown voltage for most gases is a few hundred volts and occurs at pressure × distance (torr-cm) near unity.

Breakdown voltages for several gases are listed in Table 25.1. The values given were obtained with a 1-cm gap at the Paschen minimum (the critical pressure-spacing product) and at atmospheric pressure. It should be noted that the freons and sulphur hexafluoride have high-breakdown voltages. These gases are referred to as electronegative gases. Their molecules require one or two electrons to fill the outer shell. As a result, they can easily capture free electrons and this results in the high-breakdown voltages for these gases.

Electronegative gases are expensive and tend to be somewhat hazardous, and so they are often mixed with more common gases such as air or dry nitrogen. As shown in Figure 25-50, even a small amount of sulphur hexafluoride (SF_6) in nitrogen significantly increases breakdown voltage.

The power supply or modulator for a microwave tube is normally capable of delivering a short-circuit current of at least several amperes and sometimes several kiloamperes. As a result, when breakdown occurs, a glow discharge rapidly transitions to an arc. Depending on the arc voltage and arc duration, damage can be significant.

Table 25.1

Breakdown Voltages for Several Commonly Used Gases

Gas	Volts (dc) (Paschen Min)	Kilovolts/cm (dc) (Atm. Pressure)
Air	315	33
Argon	280	4.8
Carbon Dioxide	430	28
Freon 14	480	32
Freon 115	430	90
Freon 116	500	—
Helium	189	1.63
Hydrogen	292	17
Nitrogen	265	32
Oxygen	440	—
Sulfur Hexafluoride	520	89

Source: W. G. Dunbar, "High Voltage Design Guide," Vol. IV, AFWAL-TR-82-2057.

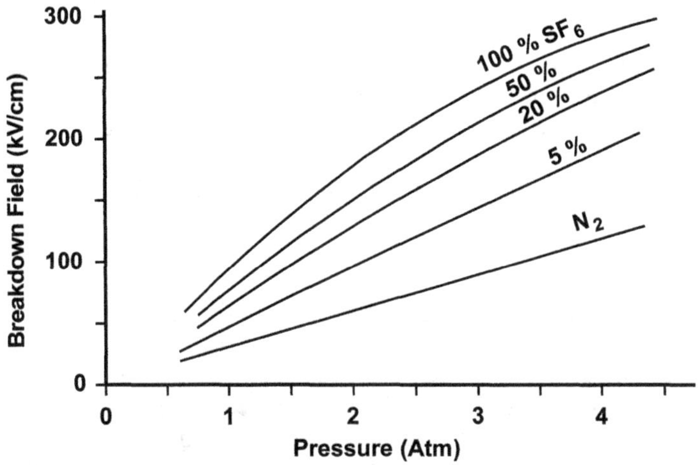

Figure 25-50 Effect on breakdown voltage of mixing SF6 with dry nitrogen. (From: N. H. Malik and A. H. Qureshi, *IEEE Trans. Elec. Ins.*, October 1980. © 1980 IEEE.)

The arc voltage depends on the voltage gradient of the arc column, which is largely dependent on gas type, arc current, and pressure. The arc-column voltage gradients for several common gases at atmospheric pressure are given in Figure 25-51. The combination of the column gradient of a few tens of volts/cm with the electrode drops results in arc voltages in roughly the 50- to 100-V range for arcs of average length (a few centimeters).

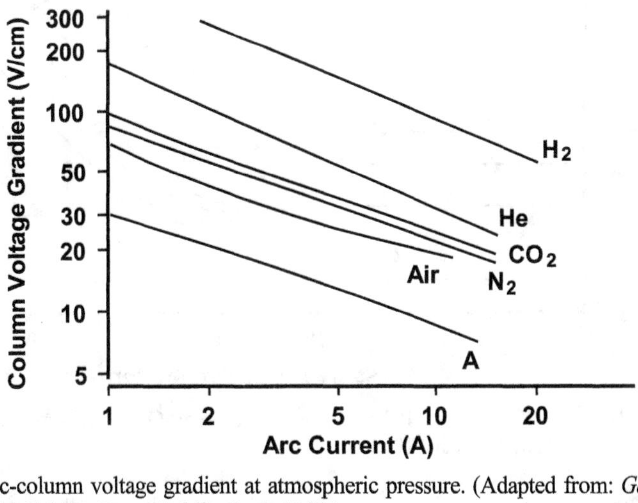

Figure 25-51 Arc-column voltage gradient at atmospheric pressure. (Adapted from: *Gaseous Conductors* by J. D. Cobine, Dover Publications, 1958.)

At the end of Section 25.3, estimates are made of the energy and charge levels at which excessive damage may occur during a dc arc inside the vacuum envelope of a tube. Unfortunately, it is not so easy to make comparable estimates for dc arcs outside the vacuum envelope. If the arc occurs between bare electrodes, it is possible that a substantial amount of energy (hundreds of joules or more) can be dissipated without causing excessive damage. If, however, the arc occurs in the vicinity of an insulating material, it is possible that the material may be damaged with only a few joules of energy dissipation.

25.7 RF BREAKDOWN IN GAS [21]

In the previous section, it was shown that the dc breakdown field for air at atmospheric pressure is about 30 kV/cm. At low frequencies (up to tens of Hz), the breakdown field remains at 30 kV/cm. As frequency is increased to the audio range and then into the RF range, the breakdown field drops and becomes highly dependent on frequency and electrode spacing, as well as pressure.

The unified plot given in Figure 25-52 was developed by Woo [23] for 50-Ω coaxial transmission lines and gives breakdown power as functions of frequency, pressure, and spacing under laboratory conditions. This plot appears complex at first glance, but it is relatively straightforward once a person becomes familiar with it.

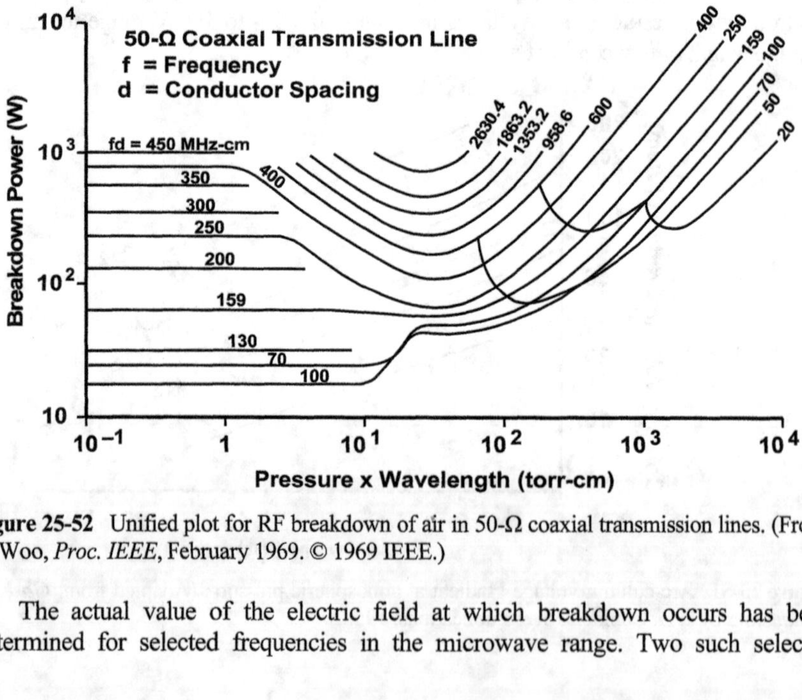

Figure 25-52 Unified plot for RF breakdown of air in 50-Ω coaxial transmission lines. (From: R. Woo, *Proc. IEEE*, February 1969. © 1969 IEEE.)

The actual value of the electric field at which breakdown occurs has been determined for selected frequencies in the microwave range. Two such selected frequencies are shown in Figures 25-53 and 25-54. In both of these cases (992 MHz and 9.4 GHz), extrapolations to atmospheric pressure indicate that the breakdown field is about 20 kV/cm, or comparable to the dc breakdown field level of about 30 kV/cm. The characteristic diffusion length in Figures 25-53 and 25-54 is a measure of the size and shape of the cavity in which the measurements were made.

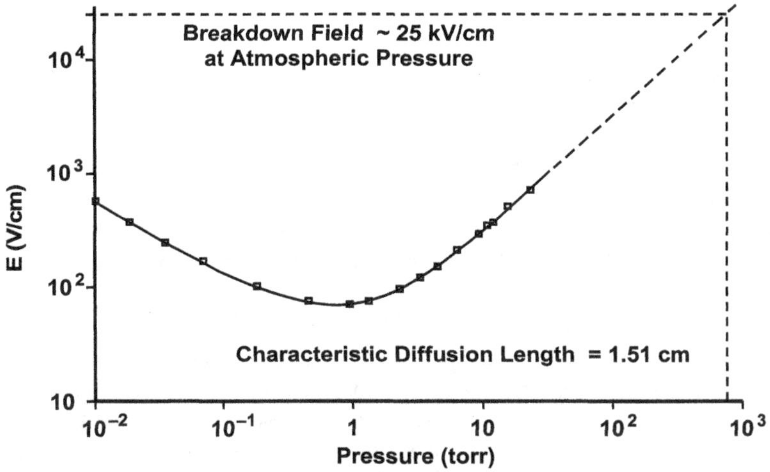

Figure 25-53 Breakdown in air at 992 MHz. (Adapted from: *Gaseous Electronics*, Volume I, "Electric Discharge," M. H. Hirsh and H. J. Oskam (eds.), Academic Press, 1978.)

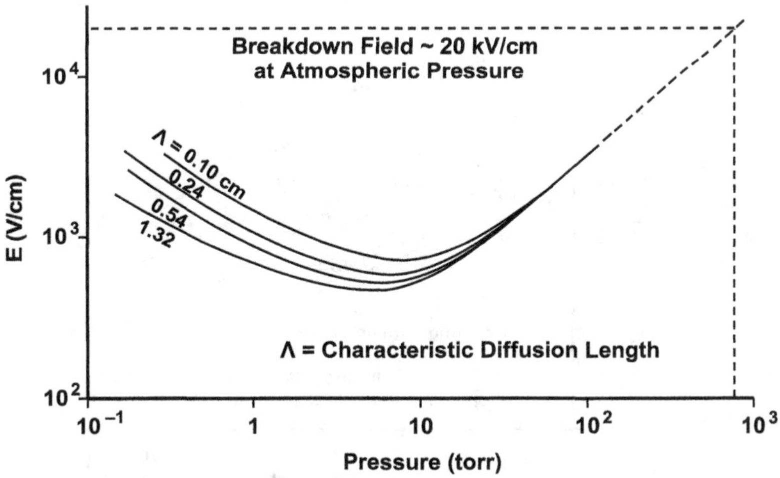

Figure 25-54 Breakdown in air at 9.4 GHz. (Adapted from: *Gaseous Electronics*, Volume I, "Electric Discharge," M. H. Hirsh and H. J. Oskam (eds.), Academic Press, 1978.)

The information in Figures 25-52, 25-53, and 25-54 would lead one to believe that extremely high-power levels can be transmitted through waveguides and coaxial lines of relatively modest size. For example, an extrapolation of Woo's unified plots to atmospheric pressure in a 7/8-inch coaxial line and a frequency of 1 GHz indicates that the breakdown power is about 200 kW. Still, as shown in Figure 25-55, the power ratings for a standard 7/8-inch coaxial line are 43 kW peak and only about 1,250W average. Some of the degradation results from field enhancement associated with supporting elements for the center conductor. More important, for the average power case, heating, especially of the center conductor and gas, is a major factor limiting the power rating.

A situation similar to that in coaxial lines exists in waveguides as shown in Figure 25-56. The ratings in the lefthand plot are for a conductor temperature rise of 42.4°C, and those for the righthand plot are for a temperature rise of 110°C. Based on a breakdown field of 20 kV/cm, it would be expected that WR 90, which has a small dimension of about 1 cm, should be able to transmit megawatts of power. Instead, the rating is in the kilowatt range.

The power ratings for the waveguides in Figure 25-56 decrease by a factor of about three when the conductor temperature rise increases from 42.4°C to 110°C. Part of this decrease in rating may be attributed to the decrease in gas density in the waveguide as temperature is increased while pressure is held constant. This is equivalent to moving to the left in the high pressure region on the Paschen curve toward lower breakdown voltages. The reduction in pressure at constant temperature in moving to the left on the Paschen curve is equivalent to a reduction in gas density. This reduction in power capability with temperature is illustrated more clearly in Figure 25-57.

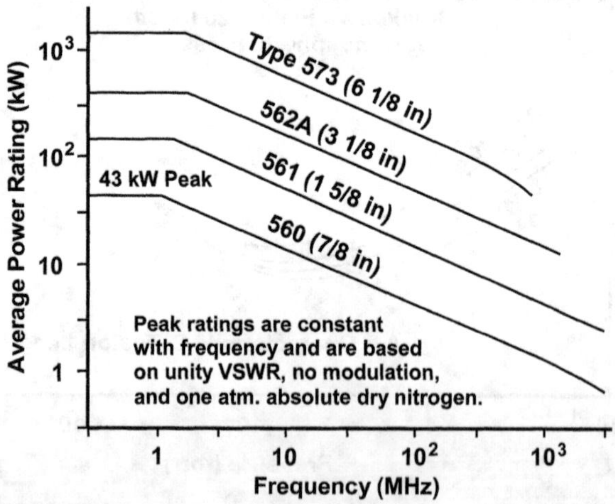

Figure 25-55 Average power ratings for standard rigid lines. (Adapted from: *Microwave Engineers Handbook*, T. Saad (ed.), Artech House, 1971.)

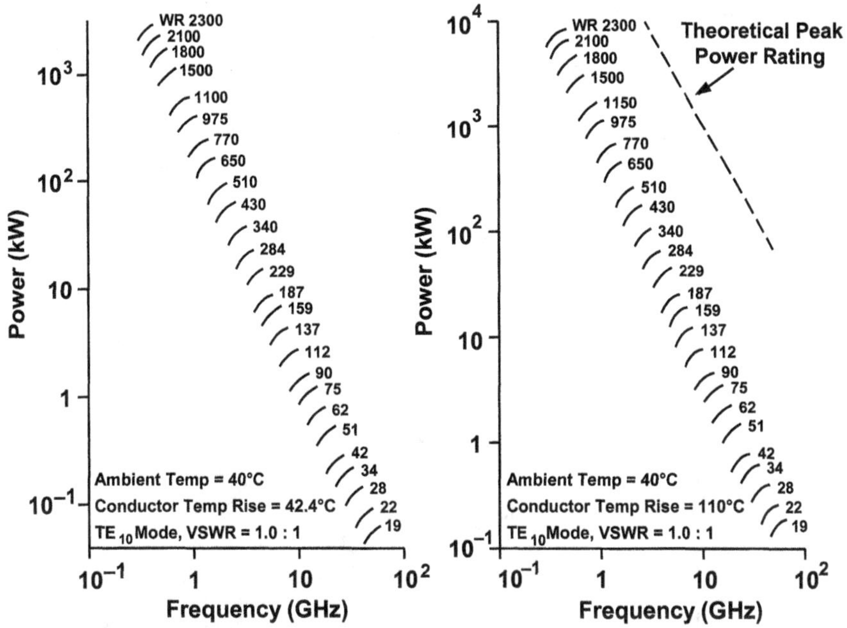

Figure 25-56 Average power rating for copper rectangular waveguide. (Adapted from: *Microwave Engineers Handbook*, T. Saad (ed.), Artech House, 1971.)

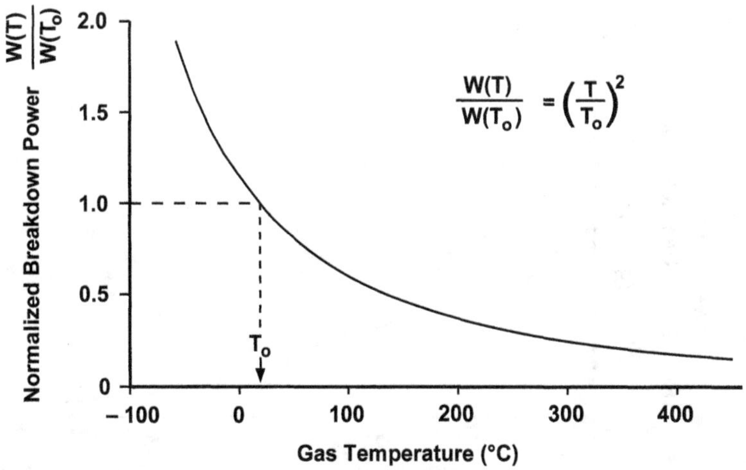

Figure 25-57 Waveguide breakdown power as a function of temperature at fixed pressure. (From: R. H. Stone and R. M. White, "Final Report TIS R61 ELM 205 General Electric Co." January 1961.)

As would be expected, a change in pressure in a waveguide or a coaxial line has an extremely large effect on power handling capability. As an example, the change in pressure from sea level to about 50,000 ft is given in Figure 25-58, and the corresponding reduction in waveguide power capacity is shown in Figure 25-59. Note that at 50,000 ft, the power capacity drops to about 3% of the sea level capacity. If an RF system is pressurized, a significant increase in power handling capability can be realized as shown in Figure 25-60.

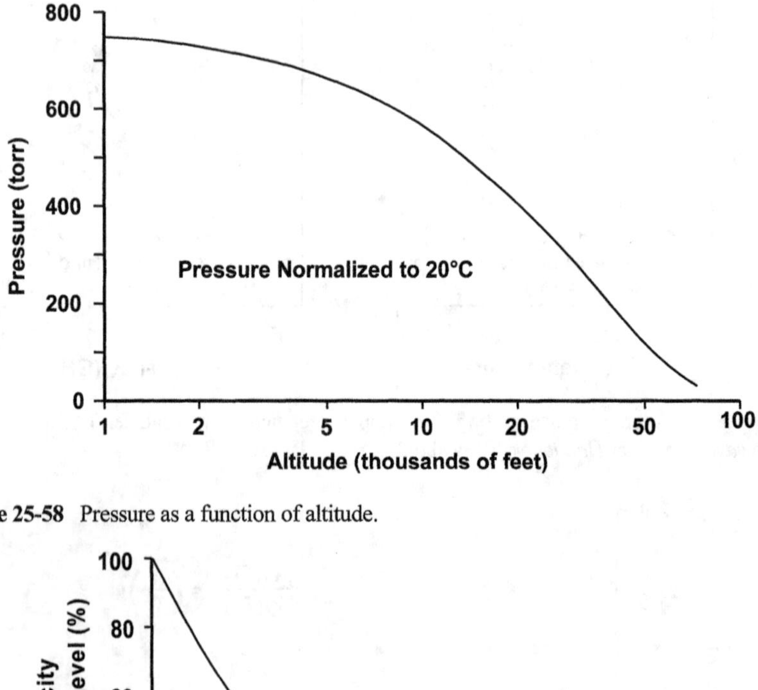

Figure 25-58 Pressure as a function of altitude.

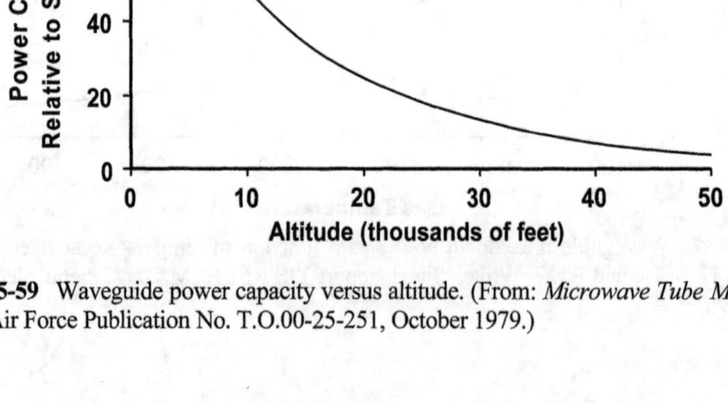

Figure 25-59 Waveguide power capacity versus altitude. (From: *Microwave Tube Manual*, by Varian, Air Force Publication No. T.O.00-25-251, October 1979.)

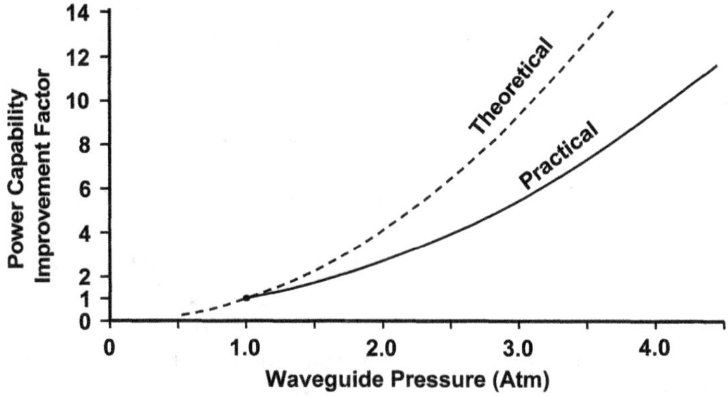

Figure 25-60 Improvement in power capability with pressurization. (From: *Microwave Tube Manual*, by Varian, Air Force Publication No. T.O.00-25-251, October 1979.)

In Section 25.7, the use of electronegative gases for increasing breakdown voltage was discussed. As shown in Figure 25-61, similar improvements are possible in RF components. The improvement in power handling capability with pure sulphur hexafluoride (SF_6) at a pressure of 20 psi is a factor of about 16. Even a small percentage of SF_6 mixed with air produces a significant increase in power capability. Other electronegative gases such as freon 12 and freon 114 produce similar improvements in power handling capability.

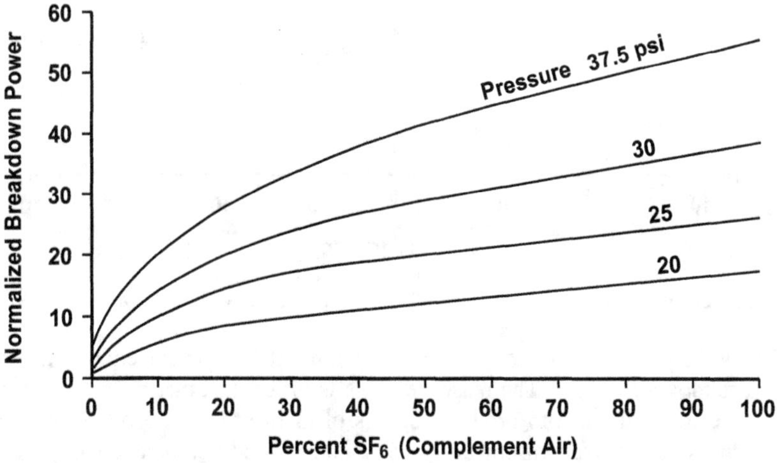

Figure 25-61 Breakdown power for mixtures of SF6 and air normalized to air at 20 psi. (From: V. C. Vannicola and L. L. Stevens, Air Force Report No. RADC-TDR-64-67, March 1964.)

In pulsed RF systems, the peak power capability increases for at least two reasons. The first has already been pointed out in connection with Figure 25-55. That is, the heating effect is decreased. The other reason is based on the fact that it takes time for a gas to break down. First, there is the time required for the breakdown process to start. This is called the formative time. Second, there is the time required for the discharge to develop to the point where appreciable RF power is absorbed or reflected.

The formative time for air can be determined from Figure 25-62. Here, P is gas pressure, E is electric field, and τ is the formative time. For a 1-μs pulse, and atmospheric pressure, the $P\tau$ product is slightly less than 10^{-3} torr-sec, which is near the right side of Figure 25-62. Thus, at pulse lengths substantially shorter than 1 μs, the formative time is long enough that the breakdown field increases. An example of the dependence of breakdown power on pulsewidth is given in Figure 25-63. Note that the reduction in power capability when the pulse length is increased from 1 to 3 μs is about 30%.

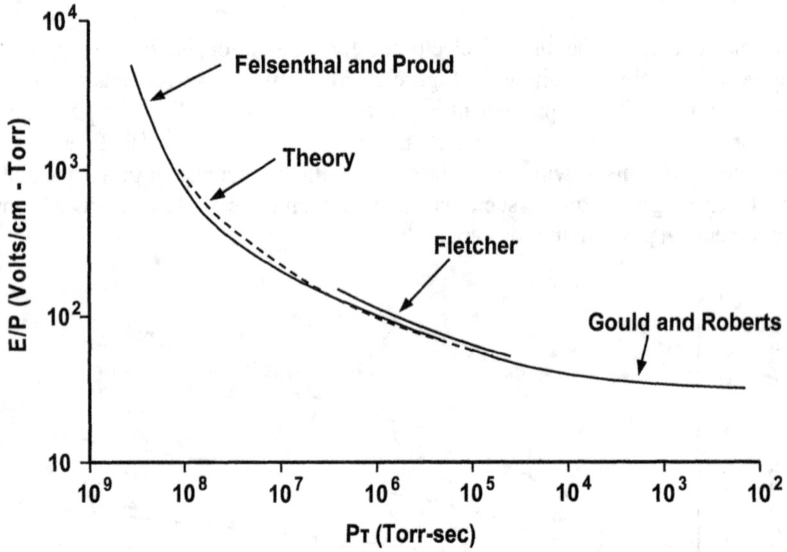

Figure 25-62 Formative time in air. (From: P. Felsenthal and J. M. Proud, *Phys. Rev.*, 1965.)

The voltage standing wave ratio (VSWR) in the RF output system of a tube may cause breakdown to occur. The reason for this is, of course, that the reflected wave adds to the forward wave to produce regions of increased field level. A VSWR of 1.5:1 reduces the breakdown power by 35%, and 2:1 reduces breakdown power by 50%. If a system must be designed for operation with a significant VSWR, then the normal waveguide or coaxial line rating must be reduced to prevent breakdown as indicated by Figure 25-64.

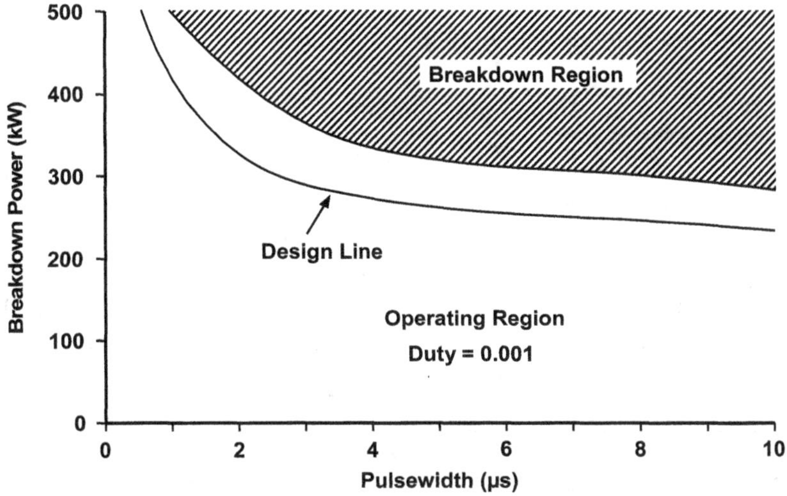

Figure 25-63 Effect of pulsewidth on breakdown power in waveguide. (From: *Microwave Tube Manual*, by Varian, Air Force Publication No. T.O.00-25-251, October 1979.)

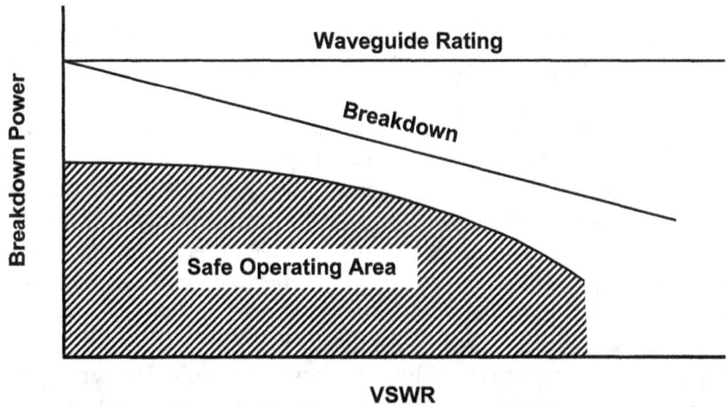

Figure 25-64 Effect of VSWR on RF power system rating. (From: *Microwave Tube Manual*, by Varian, Air Force Publication No. T.O.00-25-251, October 1979.)

If breakdown does occur in the RF output system connected to a microwave tube, it is probable that the resulting arc will reflect most of the incident power back toward the tube. As a result, the VSWR will be very high and may produce breakdown at other points along the RF system, as is indicated in Figure 25-65. In this case, an arc at plane I produces a standing wave and causes the electric field to exceed the breakdown level of a critical element such as the output window of a tube located at plane IV.

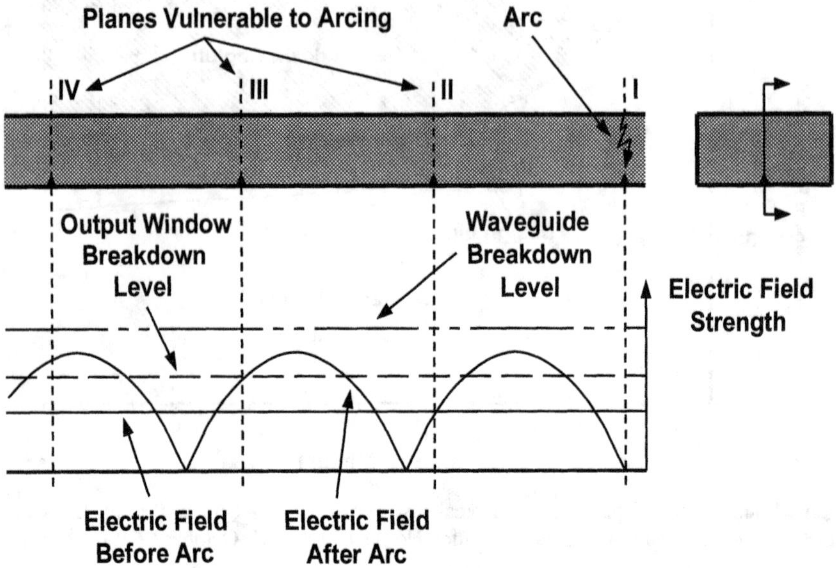

Figure 25-65 Arc propagation in a waveguide. (From: R. H. Stone and R. M. White, Final Report TIS R61 ELM 205 General Electric Co., January 1961.)

An isolator is often used to protect a microwave tube from excessive reflected RF power. The isolator may also impede or even stop a traveling arc. The effect of a ferrite isolator on VSWR is shown in Figure 25-66. While the isolator is effective in protecting the tube, it will not prevent arcing in the remainder of the RF system.

In some microwave systems, an attempt is made to reduce arcing problems and arc damage by using a detector of some type. Reflected power can be sensed with a directional coupler and measured with a threshold detector that generates a signal to turn off the microwave tube if excessive reflections occur.

A waveguide arc can be sensed by detecting visible arcs. The sensing portion of the detector is usually mounted in a 90° waveguide bend located between the tube output flange and the waveguide transmission line. The output signal from the sensing device is normally used to turn off the high-power tube.

In another technique sometimes used to protect an output window, a metal probe is inserted through, but insulated from, the waveguide near the output window. When an arc occurs and moves toward the window, the electrical circuit between the probe and the waveguide is closed, so that a relay can be activated to turn off the tube.

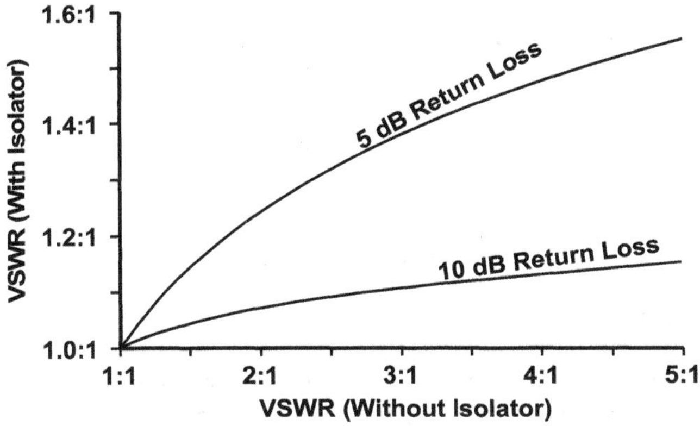

Figure 25-66 Effect of ferrite isolator on VSWR. (From: *Microwave Tube Manual*, by Varian, Air Force Publication No. T.O.00-25-251, October 1979.)

25.8 FAULT DETECTION AND TUBE PROTECTION

There are several types of faults that occur in microwave tubes. Two that often occur are:

1. Excess body (interception) current;
2. Excess reflected RF power.

Excess body current can result from an arc in the gun or from any one of many factors that upset the electron beam dynamics and cause too much of the beam to strike the RF circuit or the anode of the electron gun. Excess reflected RF power is normally caused by a problem in the RF circuit attached to the output of the tube. Even though the tube is often not at fault, excess reflected power may cause it to fail and so the tube must be protected.

25.8.1 Excess Body Current

The most common location for an arc in a microwave tube is in the electron gun and the arc that is formed is a vacuum arc. There is also the possibility that an arc may occur outside the vacuum envelope in the gun region. When an arc occurs, excess current flows to the gun anode (or to the anode lead outside the vacuum envelope) and, if the anode is connected to the body of the tube (as it often is), then the current returns to the power supply through the tube body and so the current is referred to as body current.

Body current caused by an arc is usually distinguishable from body current caused by beam interception because it is usually much larger in magnitude and occurs much more rapidly. In designing circuits that will protect a tube from excess body current, it is necessary to design for the fast rising, large amplitude current of the arc as well as for the (usually) more gradual increase of beam interception current to an unacceptable level. The more difficult aspect of the circuit design is usually the arc protection part and that is what we will concentrate on here. In designing arc detection and tube protection circuits, the two arc characteristics that must be known are:

1. Arc voltage;
2. Maximum permissible arc energy.

Early in this section it was shown that the arc voltage for an arc in air is about 50V. In Section 25.3, it was noted that the voltage of a vacuum arc can range from 8V for mercury to 33V for cold steel. The average voltage for most commonly used metals is about 20V. To be safe, the protection circuitry should be designed for the higher voltage (50V) arc.

The energy, W_a, delivered to an arc is, of course, the product of the power to the arc, P_a, and the duration of the arc, τ_a so

$$W_a = P_a \tau_a \tag{25-14}$$

But the power is the product of the arc voltage (50V) and the arc current, I_a, so

$$W_a = 50 I_a \tau_a \tag{25-15}$$

Knowing the circuit design for the tube power supply, I_a can be calculated and so the duration of the arc, τ_a, must be limited by the protection circuitry to limit arc energy to the specified level. That is, the protection circuitry must respond in time

$$\tau_a < \frac{W_a}{50 I_a} \tag{25-16}$$

For example, if the maximum arc energy specification is 2 joules and the short-circuit current from the power supply is 1,000A (this high current is entirely possible when a supply using large filter capacitors is used), then

$$\tau_a < 40\ \mu\text{s} \tag{25-17}$$

Thus, the time response requirement on the protection circuitry is usually very modest. In fact, arc energies can easily be limited to a small fraction of a joule by limiting short circuit current and using a switching circuit that reacts within a few microseconds.

25.8.2 Excess Reflected RF Power

Figure 25-67 shows some of the elements of an RF system that can cause excessive reflected RF power. In the cases shown, an impedance mismatch occurs because of a mechanical problem in the RF system. Arcing in the output RF system can also produce a very large impedance mismatch and almost total power reflection.

If the power reflection occurs on the output side of the isolator and the isolator is not defective then the reflected power will not reach the tube and thus, of course, the tube will be protected. Usually, a directional coupler (which is not shown in Figure 25-67) is used to measure reflected power and turn off the system in the event of excessive reflected power.

If the power reflection occurs on the input side of the isolator, or the isolator is defective (or there is no isolator and no directional coupler for measuring reflected power), then excessive power may reach the microwave tube. This may cause failure of the output window or failure of internal tube elements such as the attenuator in a traveling wave tube. In some systems an optical detector is placed in the RF output line near the tube window to shut the system off if an arc occurs.

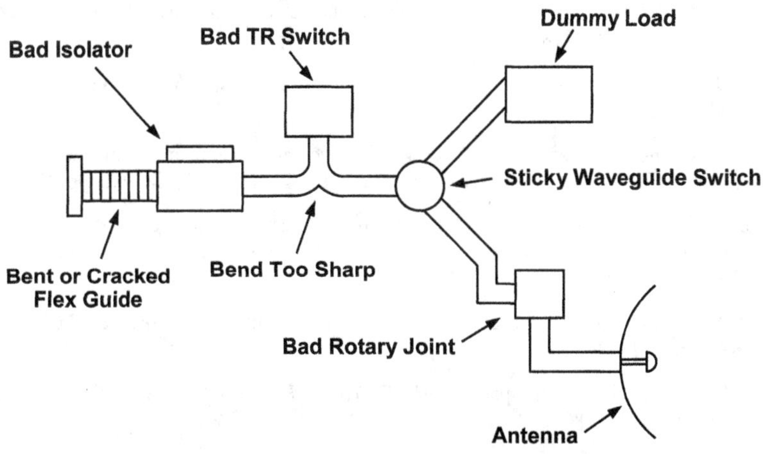

Figure 25-67 Common problem areas in RF systems. (From: *Microwave Tube Manual*, by Varian, Air Force Publication No. T.O.00-25-251, October 1979.)

REFERENCES

[1] J. M. Lafferty, *Vacuum Arcs, Theory and Applications*, New York: John Wiley and Sons, 1980.

[2] R. V. Latham, *High Voltage Vacuum Insulation: The Physical Basis*, New York: Academic Press, 1981.

[3] R. V. Latham, *High Voltage Vacuum Insulation: Basic Concepts and Technical Practice*, New York: Academic Press, 1995.

[4] W. D. Kilpatrick, "Criterion for vacuum sparking designed to include both RF and dc," *Rev. Sci. Inst.*, Vol. 28, No. 10, 1957, pp. 824–826.

[5] Machlett Laboratories, "Machlett pulse tubes for high voltage, high power, video and RF pulsing," June 1964.

[6] A. Staprans, "Voltage breakdown limitations of electron guns for high-power microwave tubes," *Proc. Second International Symposium on Insulation of Voltages in Vacuum*, Cambridge, MA, 1966.

[7] A. Staprans, "Electron gun breakdown," *High Voltage Workshop*, Monterey, CA, February 1985.

[8] T. Utsumi, "Physical processes in electrical breakdown in vacuum," *Technical Report* RADC-TR-66-712, January 1967.

[9] L. Cranberg, "Initiation of electrical breakdown in vacuum," *Jour. of App. Phys.*, Vol. 23, No. 5, May 1952, pp. 518–522.

[10] R. V. Latham, and N. S. Xu, "Electron pin holes: the limiting defect for insulating high voltages by vacuum, a basis for new cold cathode electron sources," *Vacuum*, Vol. 42, No. 18, 1991, pp. 1173–1181.

[11] I. G. Kesaev, *Cathode Processes in the Mercury Arc*, New York: Consultants Bureau, 1964.

[12] A. S. Gilmour, Jr., and D. L. Lockwood, "The interruption of vacuum arcs at high dc voltages," *IEEE Trans. Electron Devices*, Vol. ED-22, No. 4, April 1975, pp. 173–180.

[13] J. P. Shannon, S. F. Philip, and J. G. Trump, "Insulation of high voltages across solid insulators in vacuum," *Jour. of Vacuum Science and Technology*, Vol. 2, No. 5, September-October 1965, pp. 234–238.

[14] P. T. Farnsworth, "Television by electron image scanning," *J. Franklin Inst.*, Vol. 218, 1934, pp. 411–444.

[15] J. R. M. Vaughan, "Multipactor," *IEEE Trans. Electron Devices*, Vol. 35, No. 7, July 1988, pp. 1172–1180.

[16] D. H. Preist, "Multipactor motions in microwave tubes," *Proc. 4th International Congress on Microwave Tubes*, Eindhoven, The Netherlands: Centrex, 1963, pp. 34–40.

[17] J. R. M. Vaughan, "Magnetron anode erosion in the presence of cathodes containing oxides," *IEEE Trans. Electron Devices*, Vol. ED-17, No. 4, April 1970, pp. 377.

[18] R. W. Bierce, J. Jasberg, and J. V. Lebacqz, "The Stanford Two Mile Accelerator," Ch. 10, R. B. Neal, (ed.), New York: Benjamin, 1968.

[19] J. R. M. Vaughan, "Some High-Power Window Failures," *IRE Trans. Electron Devices*, Vol. ED-8, No. 4, July 1961, pp. 302–308.

[20] J. D. Cobine, *Gaseous Conductors, Theory and Engineering Applications*, New York: Dover Publications, 1958.

[21] M. N. Hirsh and H. J. Oskam, *Gaseous Electronics, Volume 1, Electrical Discharges*, New York: Academic Press, 1978.

[22] P. D. Ankrum, *Principles and Applications of Electron Devices*, Scranton, PA: International Textbook Company, 1959.

[23] R. Woo, "RF voltage breakdown in coaxial transmission lines," *Proc. IEEE*, Vol. 57, No. 2, February 1969, pp. 254–256.

Appendix A

Useful Constants and Conversions

Constants

Boltzmann constant	k	1.38×10^{-23} J/K
electronic charge	e	1.6×10^{-19} C
electron rest mass	m	9.1×10^{-31} kg
electron charge-to-mass ratio	η	1.76×10^{11} C/kg
permittivity of vacuum	ε_o	8.85×10^{-12} F/m
permeability of vacuum	μ_o	$4\pi \times 10^{-7}$ H/m
Planck constant	h	6.625×10^{-34} J-s
speed of light in vacuum	c	3×10^{8} m/s
hydrogen atom mass	m_h	$1{,}836 \times m$(electron)

Conversions

1 eV	=	1.6×10^{-19} J
1 Tesla	=	10^4 Gauss
room temp. (20°C) kT	=	1/40 eV

Appendix B

Vacuum Technology

If a microwave tube is to operate properly, a high or ultrahigh vacuum must be maintained throughout its life. As a result, special techniques, processes, and materials must be used. The emphasis in this appendix is on vacuum technology. Subjects discussed are:

- Units of measurement;
- Ranges of operation;
- Sources of gas;
- Vacuum systems;
- Vacuum pumps;
- Vacuum gauges;
- Bakeout;
- Microwave tube materials;
- Fabrication techniques.

Several excellent books on vacuum technology are listed with the references at the end of this appendix [1–11].

B.1 UNITS OF MEASUREMENT

Many years ago, the most commonly used unit of pressure was the millimeter of mercury (mmHg). Atmospheric pressure is 760 mmHg. Now, the accepted units of pressure are the torr and the newton/m^2 (pascal). The relation between these units is

$$1 \text{ torr} \approx 1 \text{ mmHg} \approx 1.33 \times 10^2 \text{ pascal}$$

Another unit of pressure sometimes used at relatively low vacuum (high pressure) levels is the micron:

$$1 \text{ micron} = 10^{-6} \text{ meters of Hg} = 10^{-3} \text{ mmHg} \approx 10^{-3} \text{ torr}$$

Still another unit of pressure used in European literature is the mbar (millibar):

$$1.33 \text{ mbar} = 1 \text{ torr}$$

In microwave tube work, it is rarely and perhaps never necessary to know pressure to an accuracy better than a factor of two or so. As a result, it is safe to remember that

$$1 \text{ torr} \approx 1 \text{ mmHg} \approx 1 \text{ mbar}$$

B.2 RANGES OF OPERATION

In evacuating a microwave tube, the pressure changes from atmospheric to a value as low as 10^{-8} to 10^{-9} torr or lower. Thus, the pressure is reduced by over eleven orders of magnitude. It is common to divide this very wide pressure variation into four ranges:

- 760 torr to ~1 torr – Rough vacuum
- 1 torr to ~10^{-3} torr – Medium vacuum
- 10^{-3} torr to ~10^{-7} torr – High vacuum
- < 10^{-7} torr – Ultrahigh vacuum

In most electron tubes, the pressure is at the low end of the high vacuum range or the high end of the ultrahigh vacuum range (i.e., from 10^{-7} to 10^{-9} torr). In some high-power klystrons that have been in operation for > 50,000 hrs, the pressure is even lower.

At atmospheric pressure, the gas density is about 2.7×10^{19} molecules/cm^3. As shown in Figure B-1, density drops linearly with decreasing pressure and, at 10^{-8} torr, is about 3×10^8 molecules/cm^3. While this density appears to be high, 3×10^8 molecules will form a monolayer on an area of only 10^{-6} cm^2. The time required to form a monolayer is also shown in Figure B-1. At a pressure of 10^{-8} torr, a monolayer is formed in a about 3 min.

Mean free path length is the mean distance traveled by a molecule between collisions. Figure B-1 shows that, at 10^{-8} torr, the mean free path length is ~10^4 m. So the average molecule could pass through a 1-m vacuum chamber 10,000 times without colliding with another molecule. At 1 torr however, the mean free path length is ~10^{-4}-m so molecules collide with each other more frequently than with the walls of a vacuum system.

At relatively high pressures, where molecules collide with each other far more frequently than with vacuum chamber walls, gas flow is referred to as viscous flow. At low pressures, where molecules collide with vacuum system walls more frequently than with each other, gas flow is referred to as molecular flow. In the viscous flow and molecular flow regions, the analysis of gas flow is relatively straightforward. In the pressure range from ~10^{-2} torr to 10^{-4} torr, a transition from viscous to molecular flow occurs, and gas flow behavior is difficult to analyze.

The effects of viscous and molecular flow in vacuum system design are illustrated in Figure B-2. In the viscous flow region, molecules gradually migrate in the direction of the pump because of the change in pressure, ΔP. With molecular flow, molecules collide only with the walls of the chamber and only occasionally are directed toward the pump. As a result, a large pump orifice is required if a high pump speed is to be achieved. For example, if 1-inch (2.5-cm) tubulation was used to connect the vacuum system to the pump, the speed would be limited to a few liters per second. If a 6-inch (15-cm) diameter orifice was used, the speed could be well over 1,000 liters per second.

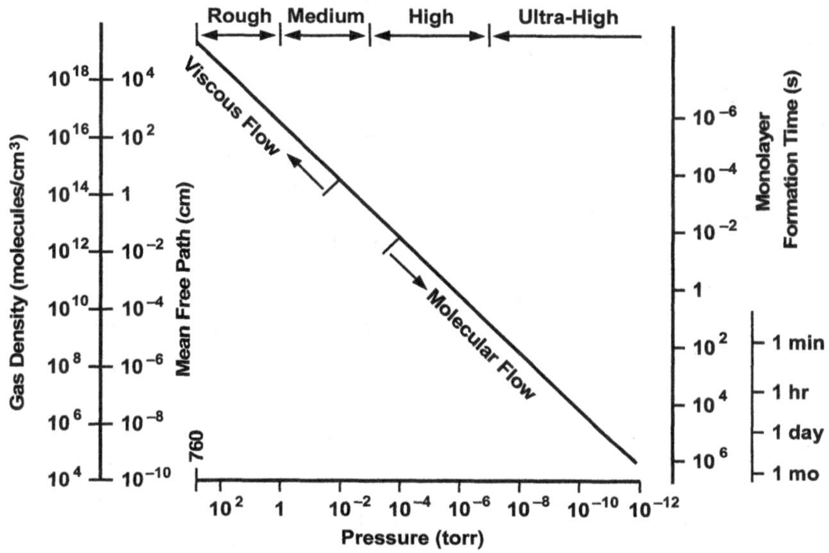

Figure B-1 Relationship of several quantities defining the degree of vacuum.

B.3 SOURCES OF GAS

If the only gas source in a container that is to be evacuated is the gas located in the volume of the container, then the container could be pumped out very quickly. In addition, if the pump being used is a "perfect pump" that is, regardless of pressure, the pumping speed is unchanged and the pump contributed no gas to the system, then the ultimate pressure would be unlimited. Unfortunately, there are many gas sources in any container that is to be evacuated and there is no perfect pump. The gas sources are identified in Figure B-3 and described next.

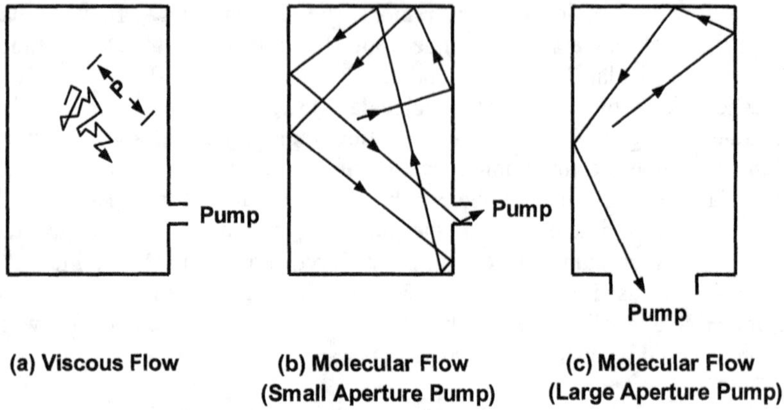

Figure B-2 (a–c) Comparison of viscous and molecular flow.

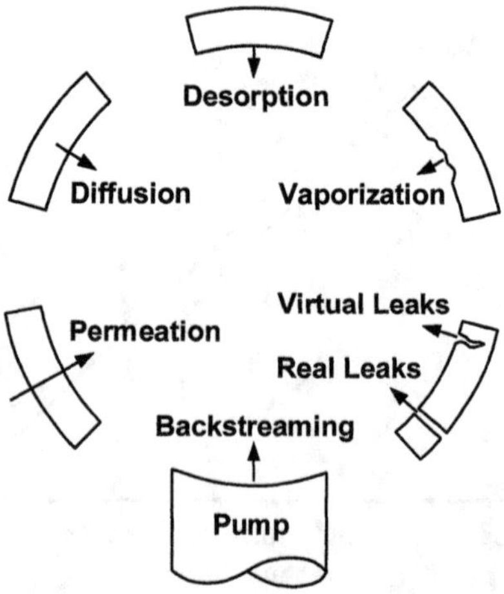

Figure B-3 Possible sources of gas in a vacuum system. (Adapted from: *A User's Guide to Vacuum Technology*, J. F. O'Hanlon, John Wiley & Sons, Inc., 2003.)

B.3.1 Backstreaming

Backstreaming is the transport of gas from a vacuum pump back into the chamber being pumped. Perhaps the best examples of pumps for which backstreaming is sometimes a problem are the oil-sealed mechanical pump and the diffusion pump. The oil in the mechanical pump vaporizes and the oil molecules can diffuse throughout the vacuum system being pumped. In the diffusion pump, oil is the working fluid that produces the pumping action, but the oil vapors can diffuse toward the system being pumped. When either or both of these pumps are used, backstreaming occurs. The backstreaming gas consists of molecules of the pumping fluid and its fractions. A baffle can be used between the pump and the chamber being evacuated to reduce backstreaming, but still some molecules of pumping fluid migrate to the chamber being evacuated.

B.3.2 Permeation

Permeation results from an electronic interaction between a gas and a solid. This interaction results in the movement of gas molecules from atom to atom through the solid. The rate of gas flow due to permeation depends on the gas and the solid and varies by orders of magnitude with different gas-solid combinations. Permeations for several gas-metal systems are shown in Figure B-4.

The only appreciable permeation rate of interest for vacuum-tube design is hydrogen through iron and iron alloys. Iron is part of the vacuum envelope in many PPM-focused TWTs. The concentration of H_2 in atmospheric air is too low to be of concern. However, hydrogen released from the corrosion of iron by water can be of concern. As a result, surface treatments of iron parts, such as electroplating, which prevent corrosion, will minimize the generation and permeation of hydrogen.

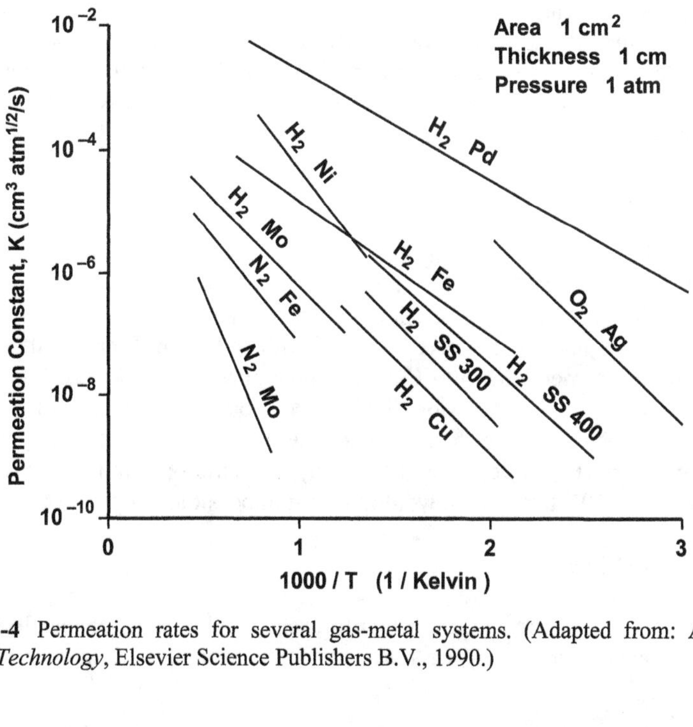

Figure B-4 Permeation rates for several gas-metal systems. (Adapted from: A. Roth, *Vacuum Technology*, Elsevier Science Publishers B.V., 1990.)

B.3.3 Diffusion

In general, diffusion is the movement of one material through another. In high and ultrahigh vacuum chambers, we are concerned mostly about the diffusion of gases through the metallic walls of the chamber to the interior surface of the walls of the chamber. There, the gas is desorbed and contributes to the gas that must be removed from the system. The rate at which gas diffuses through metals is an exponential function of temperature so the use of heat (baking) to accelerate the diffusion process is often used.

B.3.4 Desorption

Gas molecules adhere to surfaces within a vacuum chamber and are said to be adsorbed on the surfaces. The source of this gas can be the atmosphere within the chamber or diffusion or permeation from the walls of the chamber. Gas-free surfaces in vacuum can become covered with gas molecules very quickly. For example, Figure B-1 shows that, at a pressure of 10^{-6} torr, the time required to form a monolayer of gas on a gas free surface is only one second. When adsorbed gas molecules are released from a surface they are said to be desorbed. The rate of desorption is an exponential function of temperature so baking is very effective in removing desorbed gas.

Many adsorbed gasses are too tightly bonded to surfaces to be set free by baking. These can be desorbed by electron impact. As a result, many microwave tubes are operated with a pump attached for several hundred hours with the electron beam being used to assist in the outgassing process.

B.3.5 Vaporization

Great care is taken in selecting low–vapor pressure materials for use in the construction of microwave tubes because the operation of the tubes is very sensitive to the presence of gas. As examples of metals to be avoided, consider brass, which contains zinc, or screws plated with cadmium. First of all, the vapor pressure of these metals at 500°C (the temperature at which tubes are usually baked) is ~1 torr so the interior of the tube would be coated with a zinc or cadmium film. Secondly, in those parts of the tube operating at a temperature of ~200°C or more (the collector and the gun), the zinc or cadmium gas pressure would be ~10^{-4}–10^{-5} torr, which would make it impossible to operate the tube.

B.3.6 Virtual Leaks

Virtual leaks occur when gases, which are chemically or mechanically trapped inside the vacuum envelope, are released. Some examples of sources of virtual leaks are:

1. A weld joint that is made on the outside (atmospheric side) of the vacuum envelope. Dirt or other contaminants, which are trapped between the welded parts on the inside (vacuum side) of the joint, release gas that slowly leaks into the vacuum chamber.
2. A screw that is used to fasten a part in an electron gun. The threads of the screw have not been properly vented and so gas that is trapped and not completely driven out during bakeout is slowly released.
3. An inadequately outgassed part inside the vacuum envelope. Gas from the part slowly diffuses into the vacuum chamber.
4. During thermal cycling, a microscopic crack that develops on the inner (vacuum side) of a ceramic insulator. This crack exposes a tiny gas-filled void and the gas slowly leaks into the vacuum chamber.

Virtual leaks are avoided by following good design practice, eliminating gas traps, and properly processing parts. There is no external test that can detect a virtual leak. The use of a getter or an appendage pump is an inexpensive and straightforward way to prevent gas problems in the event that a virtual leak occurs.

B.3.7 Real Leaks

A real leak results from a pore in the wall of the vacuum envelope. Examples of pores are voids caused by impurities or inclusions in the wall material or cracks caused by thermal or mechanical stress.

The leak rate for air as a function of pore size (radius and length) is shown in Figure B-5. The slope of 1/3 indicates molecular flow. A hole size on the order of at least 3 molecular diameters (~10A or 10^{-7} cm) is required for pore leakage. Smaller holes are likely to be plugged by large molecules. Thus, pore leakage is unlikely to occur below a leak rate of about 10^{-15} torr-liters/sec for a 1-mm-long pore. (1 torr-liter/sec ≈ 1 cm^3/sec at standard temperature and pressure). At this leak rate, the pressure in a dormant TWT with an internal volume of 0.025 liter and without a getter or pump would increase to about 10^{-6} torr in one year. At this pressure, there is high likelihood that proper tube operation would not be possible. Thus, even the smallest pore leak is unacceptable for most applications.

B.4 VACUUM SYSTEMS

There is no single vacuum pump that is capable of evacuating a chamber from atmospheric pressure down to the ultrahigh vacuum range. Instead, as is shown in Figure B-6, at least two pumps are required. One, the roughing pump or fore pump, evacuates the system from atmospheric pressure down to the 10^{-2}–10^{-4} torr range. The other, the high- vacuum pump, reduces the pressure to the high to ultrahigh vacuum range.

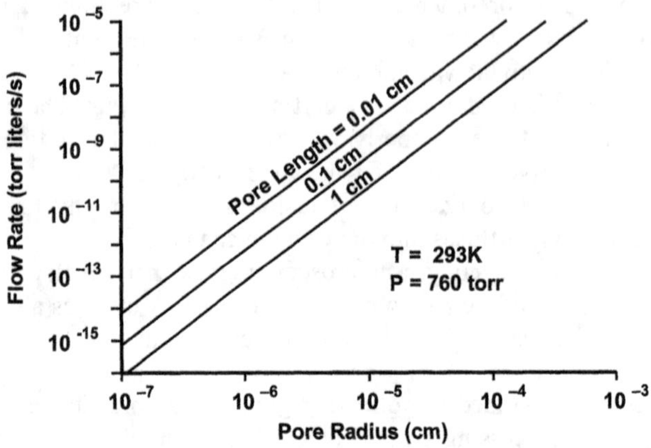

Figure B-5 Flow rate of air as a function of pore size.

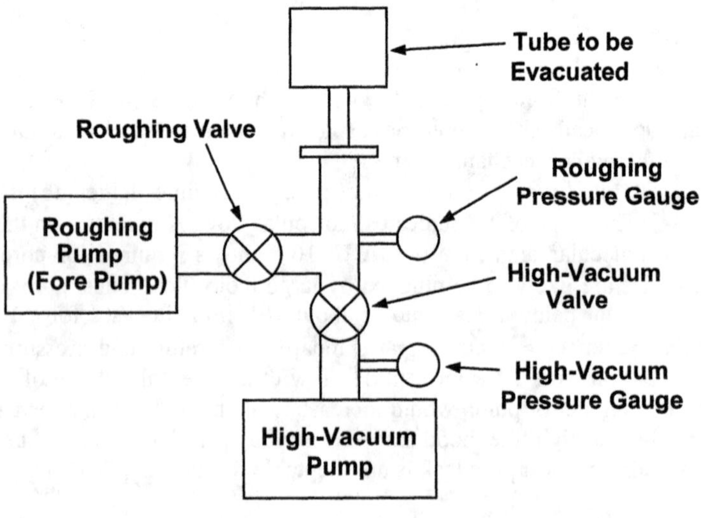

Figure B-6 Basic high-vacuum system.

The operation of the system is straightforward. If the entire system is at atmospheric pressure, then, with both valves open, the roughing pump is started. When the pressure in the system has decreased to the pressure at which the high-vacuum pump can start operating (usually ~10^{-2}–10^{-4} torr) then the high-vacuum pump is started and the roughing pressure valve is closed. If all goes well (there are no leaks for example) then the high-vacuum pump will reduce the pressure to the high to ultrahigh vacuum range.

Very often, the entire system is not started at atmospheric pressure. Instead, initially, the high-vacuum valve is closed and the high-vacuum pump is operating (it was evacuated and started during a previous pump-down cycle). Then, the roughing pump is started, the tube to be evacuated is pumped to the 10^{-2}–10^{-4} torr range, the roughing valve is closed, and the high-vacuum valve is opened.

From the discussion in this section, it is clear that vacuum pumps may be divided into two major categories: roughing pumps and high-vacuum pumps. The pumps most commonly used in the fabrication and operation of microwave tubes are described in Sections B.5 and B.6.

B.5 ROUGHING PUMPS

There are four roughing pumps that are frequently used in microwave tube manufacturing:

- Oil filled mechanical pump;
- Scroll pump;
- Sorption pump;
- Venturi pump.

B.5.1 Oil-Filled Mechanical Pumps

An example of the oil-filled mechanical pump is the sliding vane rotary pump, the essential elements of which are illustrated in Figure B-7. A rotating cylinder containing spring-loaded vanes transfers gas from the inlet port to the output port. A high-grade vacuum oil is used in the assembly containing the rotating cylinder and vanes. The oil serves as a lubricant and also provides a vacuum seal between moving parts.

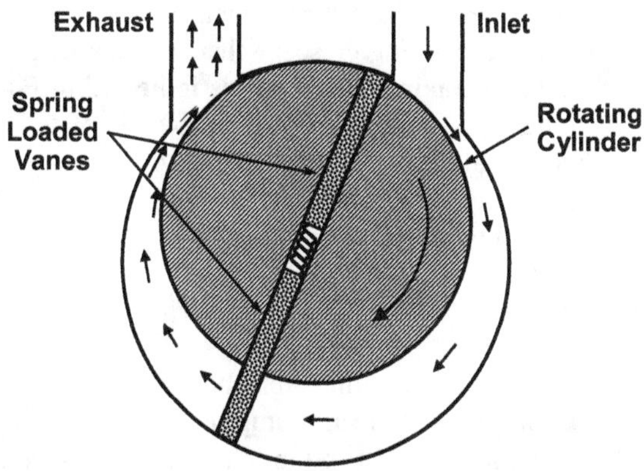

Figure B-7 Essential elements of a sliding vane rotary pump.

A problem with mechanical pumps is the pumping of condensable vapors such as water. If these vapors condense during the compression cycle, they may emulsify with the pump oil and degrade the lubricating properties. Also, the ultimate pressure capability of the pump is reduced because the vapors are released from the pump oil in the inlet side of the pump. This problem may be avoided with the use of a gas ballast device that operates during the compression cycle. Pumping speed curves for sliding vane rotary pumps are shown in Figure B-8. The ultimate pressure is ~10^{-2}–10^{-4} torr.

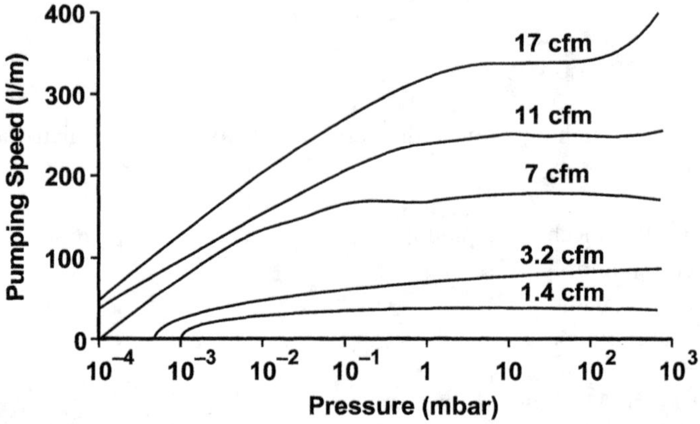

Figure B-8 Pumping speeds of sliding vane rotary pumps. (Adapted from: Varian vacuum products brochure.)

B.5.2 Scroll Pumps

Scroll pumps are oil-free and require no lubricant or sealant. The scroll pump uses two plates, each of which has a spiral-shaped pattern impressed in it. One of these plates is stationary and fits inside the other, which moves with an orbital motion. As the movable scroll orbits, moving crescent-shaped gas pockets are formed that compress the gas that they contain. This action is indicated in Figure B-9. First a large pocket of gas is trapped (in a), then as the movable plate orbits, (b through h), the pocket becomes smaller and smaller. Finally, (in i and j), the gases are expelled at the center of the fixed plate. A pumping speed curve for a Varian triscroll pump is shown in Figure B-10.

B.5.3 Sorption Pumps

A sorption pump is capable of evacuating a chamber from atmospheric pressure to the medium vacuum range. A typical pump is shown in Figure B-11. The operating principle is the physical adsorption of gases on the surfaces of molecular sieves. Zeolite, an alkali alumino-silicate, is frequently used.

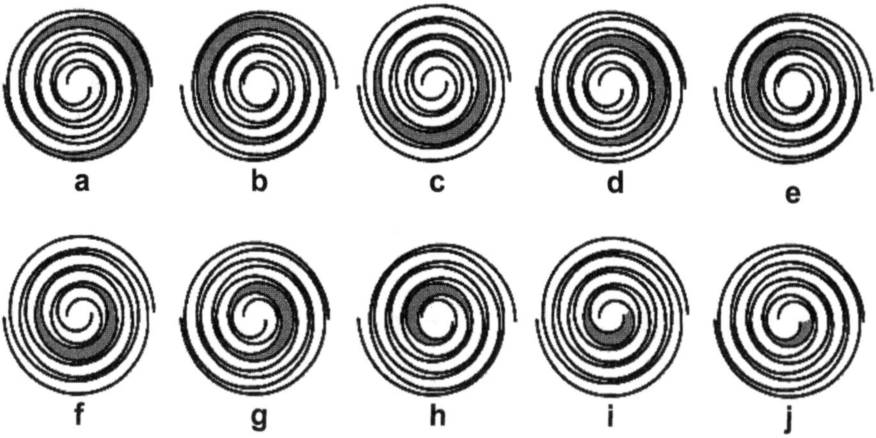

Figure B-9 Gas compression in a scroll pump.

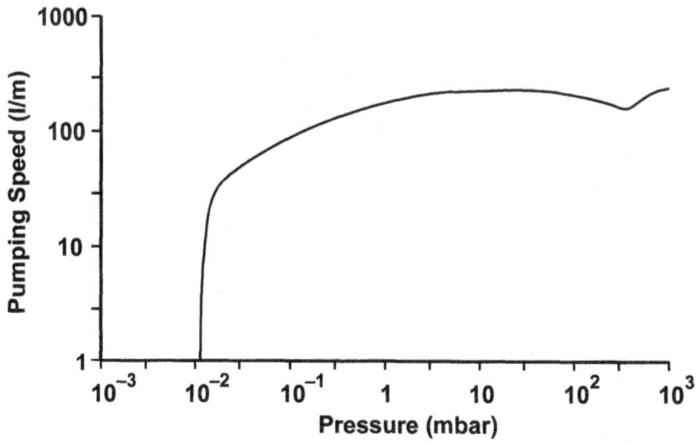

Figure B-10 Pumping speed for a scroll pump. (Adapted from: Varian vacuum products brochure.)

The surface area of the pores is about 10^3 m^2 per gram. The pore diameter of zeolite 13X (about 13Å) is approximately the size of water vapor, oil vapor, and large gas molecules. Large molecules are, therefore, readily adsorbed. Small molecules such as hydrogen, helium, and neon are very poorly adsorbed.

The adsorption of gases is dependent on pressure and temperature as shown in Figure B-12. Note that the difference between the quantity of nitrogen adsorbed and the quantity of neon or helium adsorbed is at least four orders of magnitude.

In practice, hydrogen and helium are normally present in the gases to be evacuated from a microwave tube. This is because a hydrogen atmosphere is used during brazing processes and an inert gas atmosphere containing helium is frequently used for inert gas welding. Since hydrogen and helium are normally present, it is necessary to be cautious in using a sorption pump. Because the pump operates mostly in the viscous flow range, hydrogen and helium are dragged along with the gases readily pumped by the zeolite. As long as the pumping process takes place fairly quickly, hydrogen and helium can be pulled into the sorption pump, and then a valve can be closed before the hydrogen and helium diffuse back into the chamber being pumped.

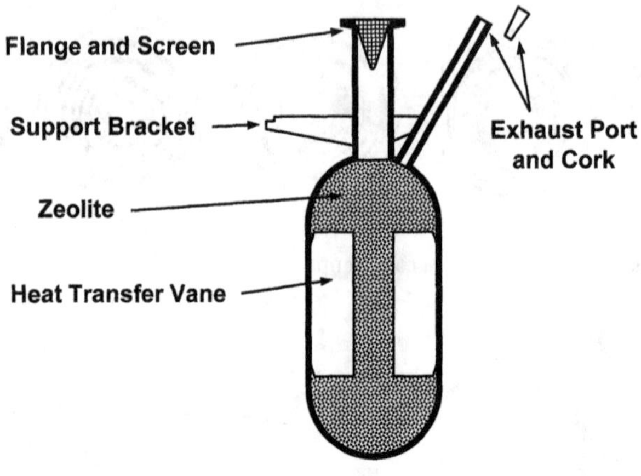

Figure B-11 Typical sorption pump.

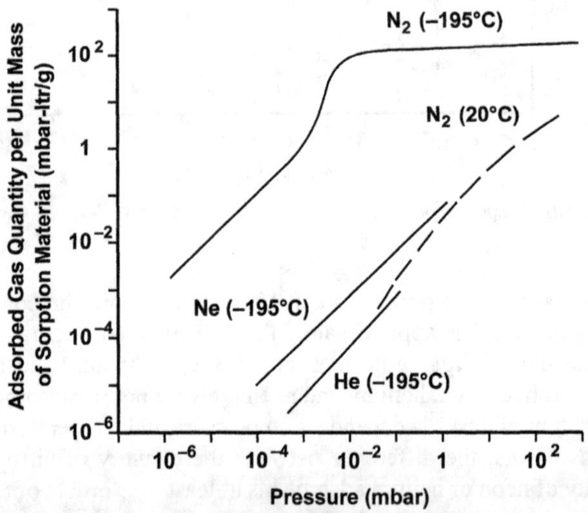

Figure B-12 Adsorption isotherms of zeolite 13X for nitrogen, neon, and helium. (Adapted from: *Product and Vacuum Technology Reference Book* by Leybold Heraeus, Inc.)

Sorption pumps are immersed in liquid nitrogen to achieve large pumping capacities. To rejuvenate zeolite after pumping, the pump is warmed to room temperature. Gas is released from the zeolite and escapes from the pump through an exhaust port. If a great deal of water vapor or other condensables has been pumped, it may be necessary to bake the zeolite for a few hours at 200°C or more.

B.5.4 Venturi Pumps

The operation of the sorption pump can be improved by using a venturi pump (Figure B-13) to remove 50–80% of the gas in the system before the sorption pump is used.

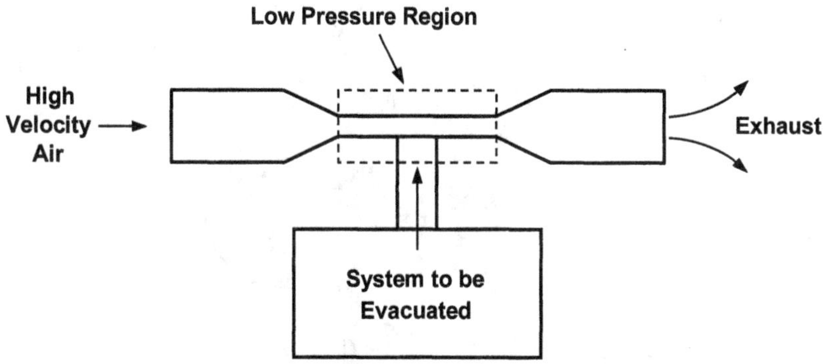

Figure B-13 Venturi pump.

B.6 HIGH-VACUUM PUMPS

There are five commonly used high-vacuum pumps in the microwave tube industry:

- Diffusion pump;
- Ion pump;
- Turbomolecular pump;
- Cryogenic pump;
- Nonevaporable getter.

B.6.1 Diffusion Pumps

A diffusion pump (Figure B-14) is normally used in conjunction with a rotary pump to achieve pressures in the high or ultrahigh vacuum range. A very high quality silicon oil is vaporized by a heating element in the bottom of the pump.

The oil vapors are formed into jets. Collisions of oil molecules with gas molecules cause the gas to diffuse toward the exhaust port. A pressure difference (between inlet and exhaust) of approximately six orders of magnitude can be produced with a diffusion pump. The pump is normally water cooled and so the oil is condensed and returns to the oil bath in the bottom of the pump.

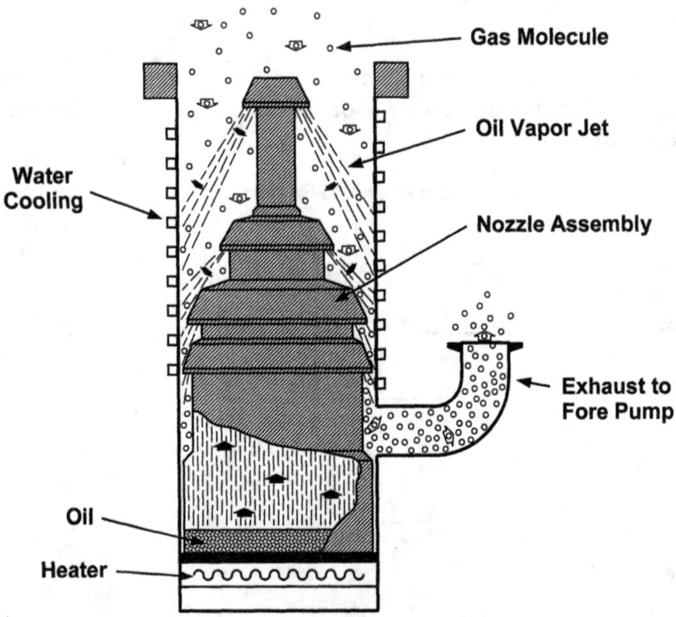

Figure B-14 Oil-filled diffusion pump. (Adapted from: *Product and Vacuum Technology Reference Book* by Leybold Heraeus, Inc.)

Typical pumping speed and throughput for a diffusion pump are shown in Figure B-15. Note that speed is constant over a wide range of pressures.

At very low pressures, the ultimate vacuum level is limited primarily by diffusion of oil molecules back into the chamber being pumped. This is referred to as backstreaming and was discussed previously in this appendix. A baffle is sometimes used between the pump and the chamber to limit backstreaming. Metallic elements are arranged as indicated in Figure B-16 so that there is no line-of-sight path through the baffle. Baffles are normally cooled, sometimes by liquid nitrogen, to enhance the condensation of oil vapors. With a baffle it is sometimes possible to reduce the ultimate pressure by over two orders of magnitude.

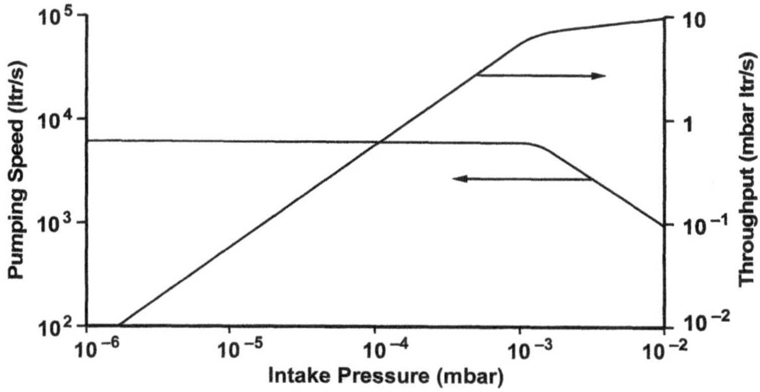

Figure B-15 Typical pumping speed and throughput for a diffusion pump.

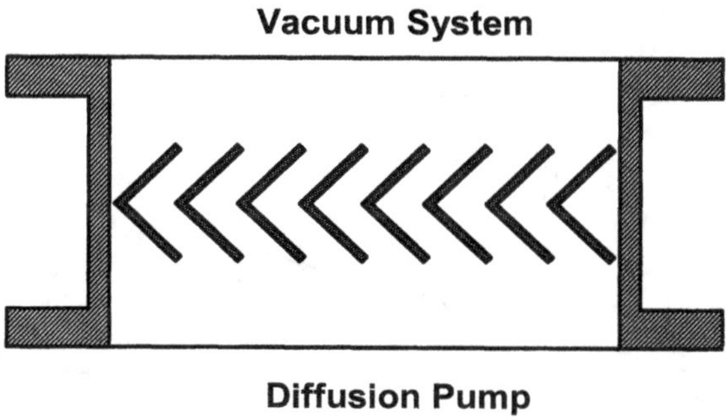

Figure B-16 Chevron-type baffle.

B.6.2 Ion Pumps

The configuration of an ion pump is shown in Figure B-17. Ion pumps are capable of evacuating a chamber to the ultrahigh vacuum range. They require a pressure in the low end of the medium vacuum range (~10^{-2} torr) before they can be started. Often, an sorption pump is used to generate this medium vacuum level. Once an ion pump has been started, it can operate without assistance from other pumps.

The operation of an ion pump is indicated in Figure B-18. A cellular anode structure is located between two planar cathodes. A magnetic field is applied perpendicular to the cathode planes and parallel to the cell axes. Electrons spiral around the magnetic field lines and ionize gas molecules. Electrons released during ionization participate in further ionization of other gas molecules. During collisions, electrons give up energy and gradually migrate to the anode. The positive ions are accelerated to the negative electrode, which is fabricated from titanium. The ions interact with the titanium, and at least three pumping actions occur as indicated in Figure B-19.

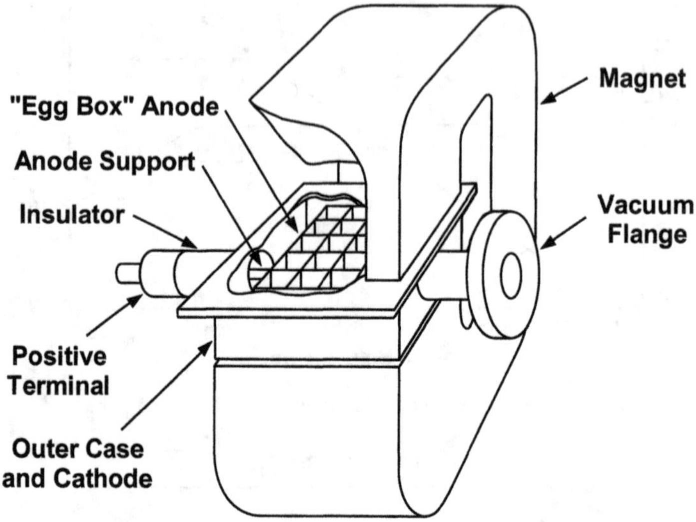

Figure B-17 Typical ion pump.

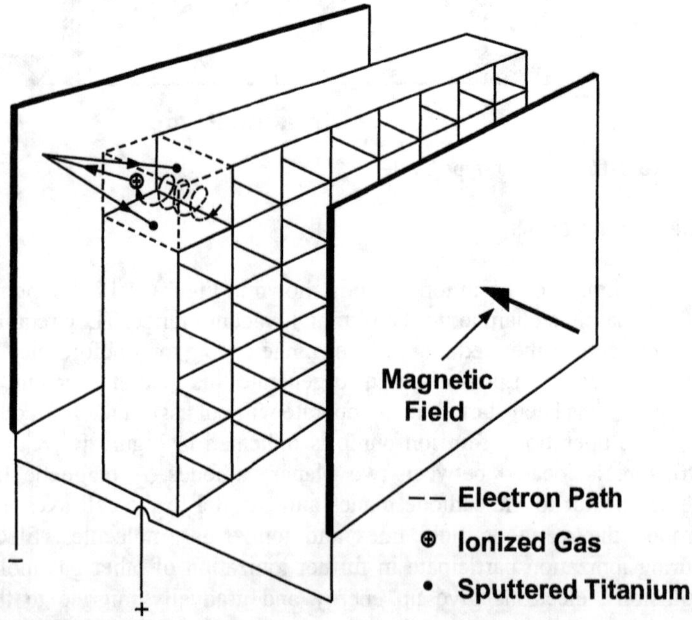

Figure B-18 Operation of diode-type ion pump.

1. Ions sputter titanium cathode material, which deposits and acts as a getter film, adsorbing reactive gases (such as nitrogen, oxygen, and hydrogen).
2. The energy of some ions is sufficient for them to be buried in the electrode surfaces, where they are adsorbed by ion implantation. This process is effective for all types of ions including those of rare or inert gases.
3. Enhanced pumping of noble gases is possible with a triode ion pump (Figure B-20) because sputtered titanium from the cathodes buries gas atoms on the target electrode.

The pumping speed of an ion pump depends on the type of gas being pumped. Compared with air, the speed for other gases is

- Hydrogen — 150 to 200%;
- Oxygen — 80%;
- Methane — 100%;
- Other light hydrocarbons — 80 to 120%;
- Argon — 30%;
- Helium — 28%.

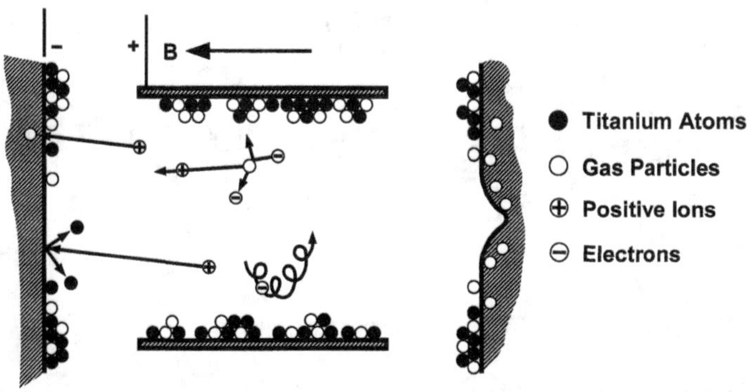

Figure B-19 Pumping processes in a diode-type ion pump. (Adapted from: *Product and Vacuum Technology Reference Book* by Leybold Heraeus, Inc.)

The voltage of operation is typically about 3.5 kV for small pumps and about 7 kV for large pumps. Ion pumps as small as 0.2 l/s and as large as several thousand liters per second are available. A small pump is often used as an appendage pump, which is an integral part of a microwave tube. An example of a coupled-cavity TWT with an appendage ion pump is shown in Figure B-21. As shown in Figure B-22, pump current is directly proportional to pressure and, after calibration, may be used as a measure of pressure.

The magnetic flux density varies somewhat with the size of the pump and is 0.125 tesla (1,250 gauss) for an 8 l/s ion pump. The stray magnetic field of the pump is of concern in electron beam applications and so care must be taken in locating the pump.

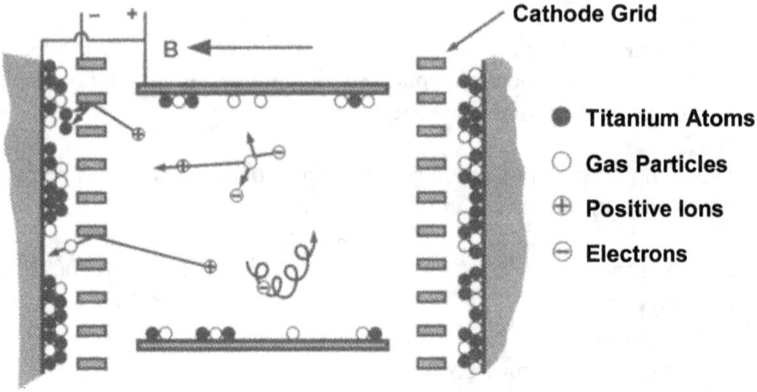

Figure B-20 Pumping processes in a triode-type ion pump. (Adapted from: *Product and Vacuum Technology Reference Book* by Leybold Heraeus, Inc.)

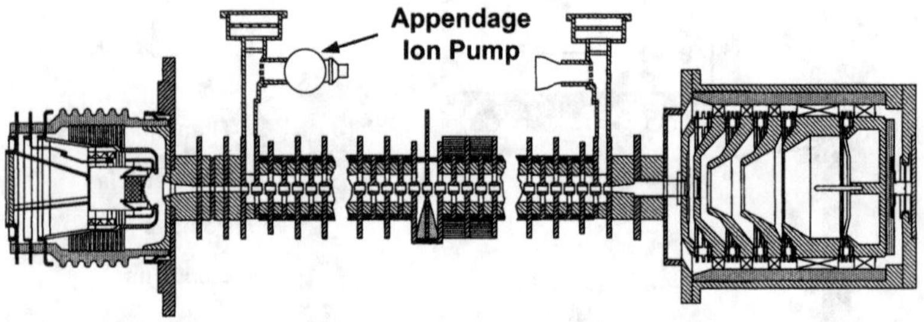

Figure B-21 Coupled-cavity TWT with an appendage pump.

B.6.3 Turbomolecular Pumps

A turbomolecular pump is like a jet engine. A rotor, spinning at a high speed, compresses gas by imparting momentum to impinging molecules. As indicated in Figure B-23, many rotor discs are used. The gas channels for the discs are in series, so that, even though the pressure gradient generated by each disc is small, the pump is capable of maintaining a high pressure ratio. The pumps operate at rotor speeds up to ~80,000 rpm. They have large compression ratios for hydrogen and ultimate pressures in the 10^{-10} torr range. The fore pressure is ~10^{-2} torr so oil-free backing pumps such as the scroll pump can be used.

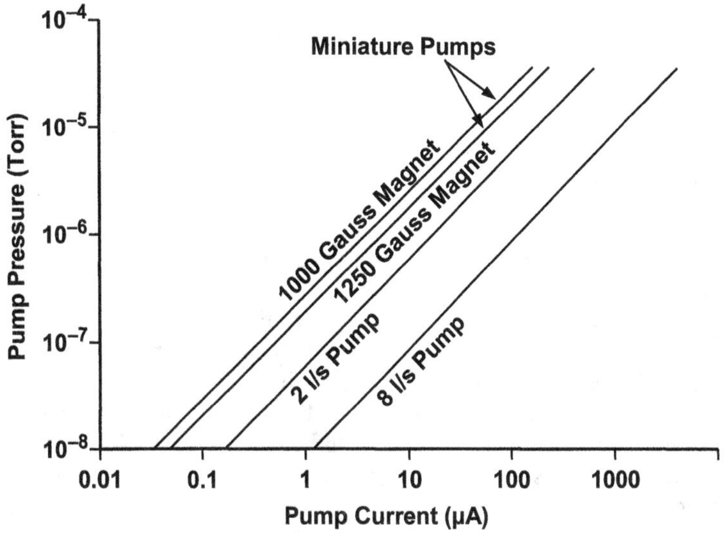

Figure B-22 Ion pump pressure versus current.

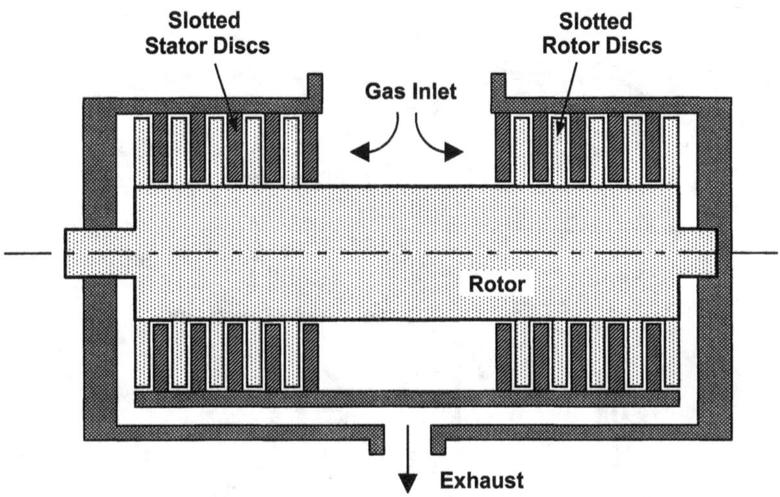

Figure B-23 Turbomolecular pump.

B.6.4 Cryogenic Pumps

Cryogenic pumps operate on the principle that gases can be condensed at low temperatures and have extremely low vapor pressures. An example of a cryogenic pump is shown in Figure B-24. This pump operates in the 10^{-3}–10^{-10} torr range.

Pumping speeds for some common gases are:

- Water — 4,000 liters/second;
- Air — 1,500 liters/second;
- Hydrogen — 2,500 liters/second;
- Argon — 1,200 liters/second.

In operation, the 80K array traps water and hydro-carbon vapors. The 15K array condenses nitrogen, oxygen, and argon. The specially processed charcoal in this array traps helium, hydrogen, and neon.

Cryogenic pumps require periodic regeneration to return them to normal operation capabilities. Regeneration removes trapped gases in a process similar to defrosting a refrigerator. The frequency with which a pump must be regenerated depends, of course, on the pump capacity and the gas load (operating pressure). In some cases it is suggested that a pump should be regenerated on a regular schedule that coincides with the system maintenance, weekend system shutdown, and so forth. Typically, the frequency of regeneration ranges from a few hundred hours for high gas load operations to many thousand hours at low pressures (below 10^{-5} torr).

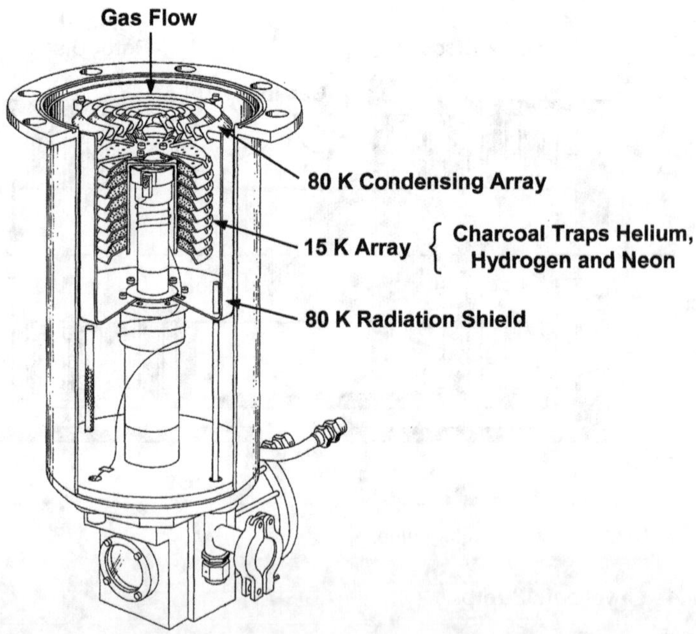

Figure B-24 The Cryo-Torr 8 high-vacuum pump. (Adapted from: CTI-Cryogenics *Cryo-Torr Manual.*)

B.6.5 Nonevaporable Getters

The phenomenon of gettering has long been used in vacuum tubes. In conventional getters, a chemically active metal is evaporated to form a film on the inner surface of the vacuum envelope. Gases that react with the metal film are said to be gettered. Conventional getters have not found application in high-power microwave tubes primarily because the evaporated film may cause electrical leakage or breakdown in high- voltage structures.

Nonevaporable getters are now commonly used in microwave tubes. The ST175 getters available from SAES GETTERS/USA Inc., are particularly attractive for use. These getters utilize a mixture of titanium and molybdenum powders in a weight ratio of 9:1. The material is lightly pressed and sintered to form a structure that is porous and has good mechanical stability. It is possible to use these getters in holders made of nichrome as is indicated in Figure B-25.

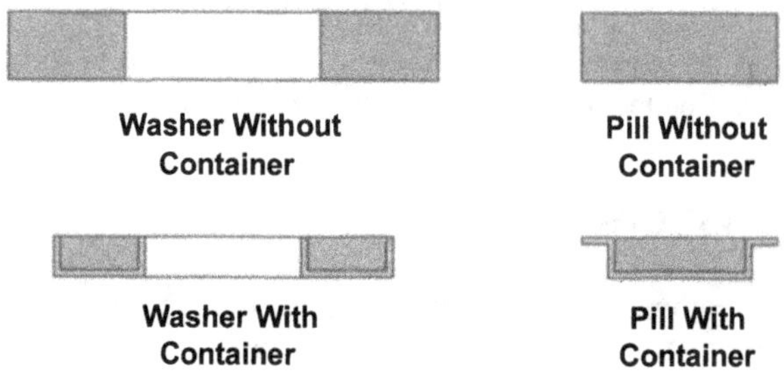

Figure B-25 SAES ST175 getters without and with nichrome containers. (Courtesy of SAES GETTERS/USA, Inc.)

The ST175 can be satisfactorily activated by heating to 500°C for 10 minutes in vacuum ($< 10^{-3}$ torr) so these getters can be conveniently used during the bakeout of tubes. As an example, the results of sorption tests for CO and H_2 activated in this manner are shown in Figure B-26.

Some of the microwave tube experience with nonevaporable getters installed in an electron gun is as follows:

1. Appendage ion pump current, which might sometimes be high, is virtually zero.
2. Arcing in the electron gun is greatly reduced.
3. After extended periods of storage, voltage can be applied immediately to a tube without gassing or arcing.
4. The getters are recoverable. After activation and then exposure to air, getters can be reactivated.

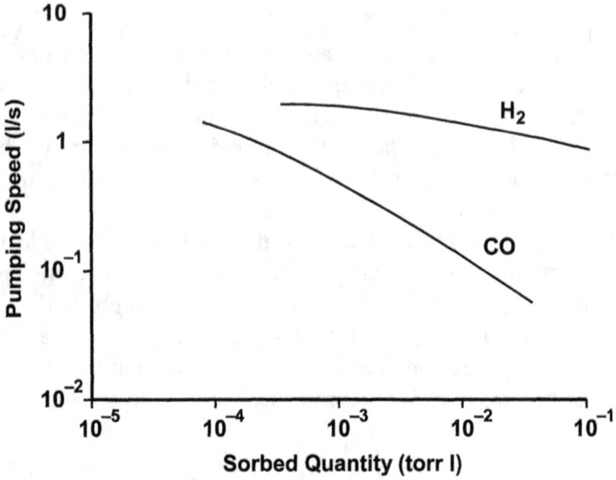

Figure B-26 H_2 and CO sorption characteristics for 1.5 cm OD × 0.6 cm ID × 0.2 cm thick SAES ST175 getter. (Courtesy of SAES GETTERS/USA, Inc.)

B.7 VACUUM GAUGES

The gauges most commonly used in the microwave tube industry for measuring pressures in the rough to ultrahigh vacuum ranges are:

- Thermocouple gauge — Rough and medium vacuum
- Ionization gauge — High and ultrahigh vacuum
- Ion pump current — High and ultrahigh vacuum

The use of ion-pump current as a measure of pressure was discussed in Section B.6.2. Thermocouple and ionization gauges will be described in this section.

B.7.1 Thermocouple Gauge

The thermocouple gauge is one of several gauges that use the thermal conductivity of gas to provide the pressure indication. As indicated in Figure B-27, a filament of wire heated by an electric current is mounted in an envelope attached to the vacuum system. The temperature of the wire depends on heat loss through the gas, which depends on gas pressure. The wire temperature is monitored with a thermocouple, which is calibrated to indicate pressure. Thermocouple gauges operate over the pressure range of ~ 10 to 10^{-3} torr.

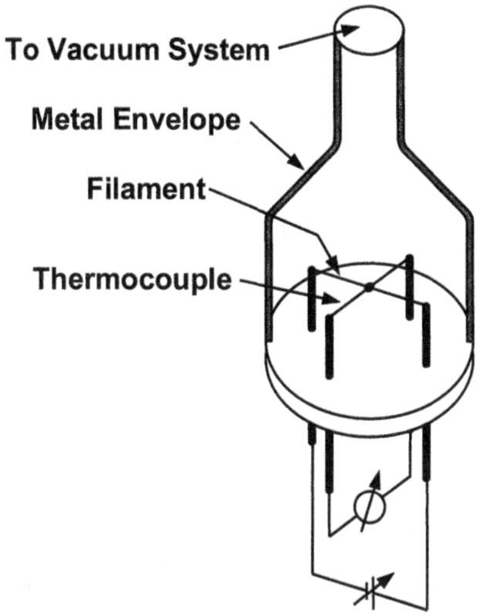

Figure B-27 Thermocouple gauge. (Adapted from: *Handbook of Thin Film Technology* by L. I. Maissel and R. Glang, McGraw-Hill, Inc., 1970.)

B.7.2 Ionization Gauge

Ionization gauges of the type shown in Figure B-28 are commonly used to measure pressures in the high and ultrahigh vacuum ranges. Electrons, accelerated by a positive grid, produce ions that are collected by an ion collector. The resulting ion current is a measure of gas pressure. These gauges are sometimes referred to as inverted triode ion gauges, or Bayard-Alpert gauges. In these gauges, a very small diameter wire is used as the ion collector. This is to minimize the X-rays produced when electrons strike the positive grid. Some of these X-rays impinge on the collector wire, causing it to release electrons. The resulting flow of electron current away from the collector causes an error in pressure indication.

B.8 BAKEOUT

Bakeout is the process of heating a tube to facilitate evacuation. As pointed out previously in this appendix, the rates of diffusion and the desorption of gases vary exponentially with temperature, so heat is used to aid in the evacuation process. The bakeout temperature commonly used in the microwave tube industry is ~500–525°C. Typically, bakeout takes place in a system like that shown in Figure B-29. The tube to be evacuated is connected to a vacuum pump and heated until the desired vacuum level is obtained.

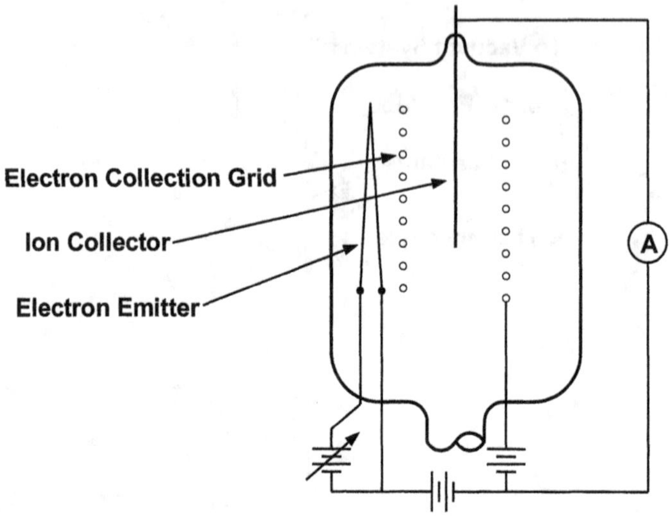

Figure B-28 Ionization gauge.

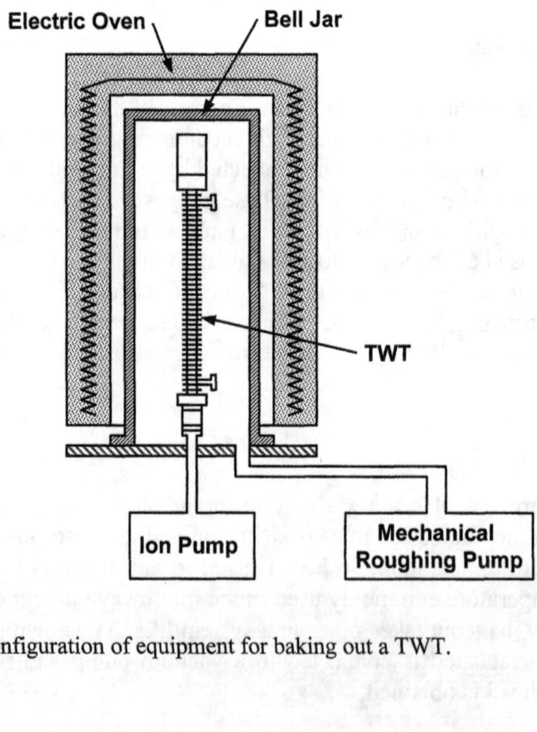

Figure B-29 Configuration of equipment for baking out a TWT.

Since the tube is very hot (at 500°C the entire tube glows a dull red color) it is necessary to protect the outside from oxidation. Not only does oxidation look bad, but it could damage vacuum joints, change critical mechanical dimensions, and cause flaking of material. Oxidation is prevented by providing a partial vacuum around the outside of the tube during the bakeout process with a mechanical roughing pump.

B.9 MICROWAVE TUBE MATERIALS

For proper performance, microwave tubes must operate in the high to ultrahigh vacuum range. At best, an ion pump or getter may be used as an integral part of a tube to help maintain the vacuum level. At worst, no pump is used, and the tube may be required to operate properly after several years in storage. The most important factor in achieving and maintaining an acceptable vacuum level is the use of the correct materials for fabricating a tube. Many of the materials commonly used are listed in Figure B-30.

	Aluminum (O_2)	Barium	Beryllium (O_2)	Boron (Nitride)	Calcium	Carbon	Chromium	Copper	Gold	Iridium	Iron	Manganese	Molybdenum	Nickel	Osmium	Palladium	Platinum	Rhenium	Rhodium	Scandium	Silver	Strontium	Thorium	Titanium	Tungsten	Vanadium	Zirconium
Heater	■																	■							■		
Cathode	■	■	■		■					■				■	■		■			■		■	■		■		
Grids						■		■	■				■												■		
Anode								■																			
Cavities								■																			
Helix								■	■				■					■							■		
Support	■		■	■		■																					
Attenuator						■																					
Collector						■		■																			
Envelope							■	■			■			■													
Brazes								■	■			■	■			■	■		■		■						■
Polepieces											■															■	
Window	■		■			■																					
Getter		■																						■		■	■

Figure B-30 Some materials used inside the vacuum envelope of microwave tubes.

The primary factor considered in selecting a metal for use inside the vacuum envelope is vapor pressure. The charts in Figure B-31, B-32, and B-33 give vapor pressure as a function of temperature for the metallic elements.

As an example of the use of these charts, consider zinc (see Figure B-31). At low temperatures, the vapor pressure is reasonable (10^{-9} torr at 100°C). At a somewhat elevated temperature (400°C), the vapor pressure is only 10^{-1} torr. This shows why brass (which contains zinc) must not be used in a vacuum system that may be heated. If brass was used and baked at 400°C or higher, zinc vapor would permeate the entire system.

The curve for barium in Figure B-33 is significant in considering the operation of cathodes. At a cathode operating temperature of 1,000°C, the vapor pressure of barium is over 1 torr. Thus, excess barium (over ~ monolayer) on the cathode surface evaporates very rapidly. This evaporated barium may eventually deposit on insulating surfaces and cause electrical leakage or breakdown. To remove excess barium, a new cathode is often placed in vacuum and operated at a high temperature before it is placed in a tube.

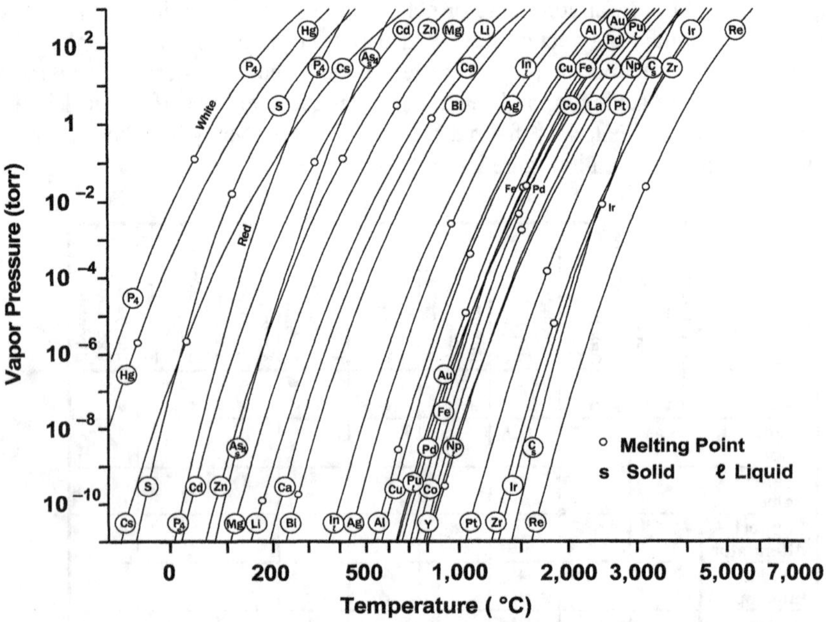

Figure B-31 Vapor pressures of the elements. (Adapted from: R. E. Honig and D. A. Kramer, *RCA Review*, June 1969, copyright 1969 by Radio Corporation of America.)

Nonmetallic materials, such as those shown in Figure B-34, are sometimes used in demountable vacuum systems. These materials severely limit the pressure and temperature at which the system can operate.

When a nonmetallic material is required in a system that must be baked, one of a very small number of materials is selected. If the material is part of the vacuum envelope, then either Al_2O_3 (aluminum oxide), BeO (beryllium oxide), glass or, in special cases, CVD diamond is selected. These materials have extremely low vapor pressures. Inside the vacuum envelope, in addition to Al_2O_3 or BeO, boron nitride or diamond is used.

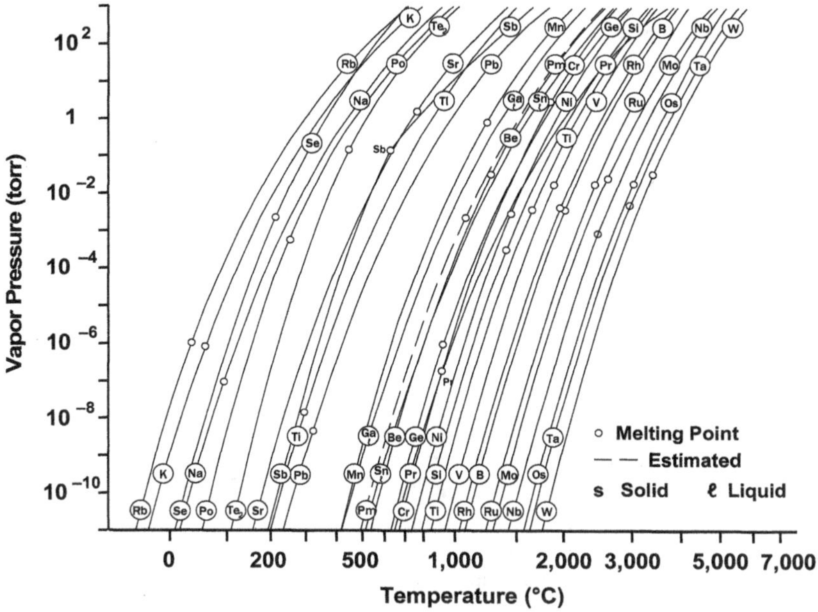

Figure B-32 More vapor pressures of the elements.

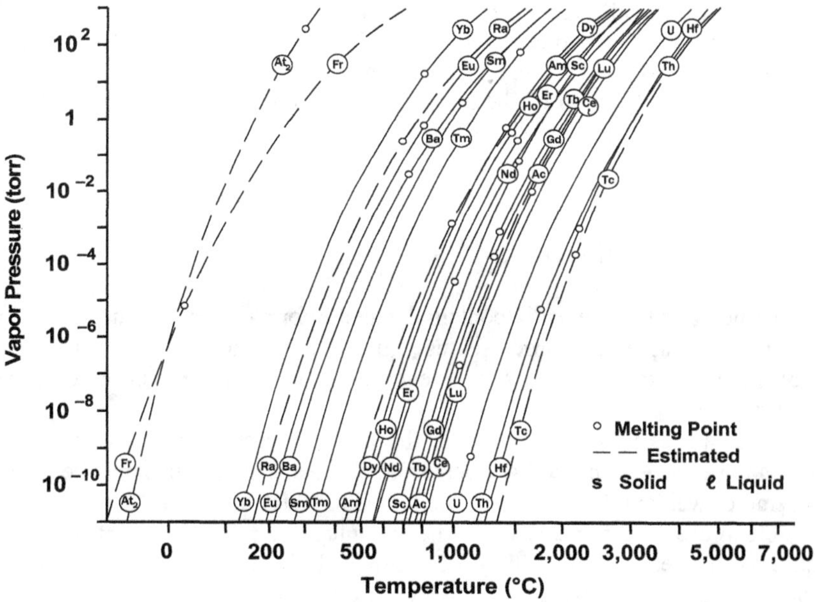

Figure B-33 More vapor pressures of the elements.

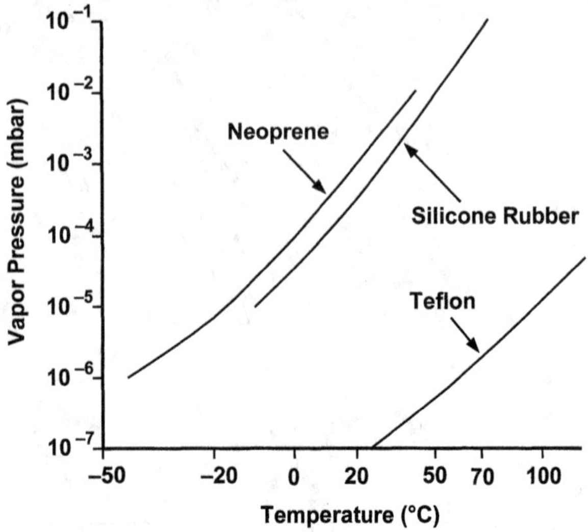

Figure B-34 Vapor pressure of nonmetallic sealing materials. (Adapted from: *Product and Vacuum Technology Reference Book* by Leybold Heraeus, Inc.)

Some liquids may have extremely low vapor pressures. Examples are diffusion pump oils. Still, there is no liquid suitable for use in the high or ultrahigh vacuum environment of a microwave tube. It is also appropriate to point out that no lubricants are suitable for use in a microwave tube. Thus, in the spin tuned magnetron, where a portion of the anode rotates at 4,000 rpm, no lubricants are used. Instead, metals are carefully selected, treated and precisely machined so that galling and excessive wear do not occur.

B.10 JOINING TECHNIQUES

The references for this appendix contain detailed information concerning many of the joining techniques and processes used in the microwave tube industry. It is possible here to describe briefly only a few of the important aspects of microwave tube fabrication.

First of all, as was pointed out in the previous section, it is essential that materials to be used are properly selected and cleaned. Then, metals may be joined by brazing or welding. Ceramics may be joined to metals by brazing. In no case is it possible to use solid fluxes because they may become trapped in joints and then produce virtual leaks, that is, they very slowly leak out of joints and produce contamination.

B.10.1 Brazing

From O'Hanlon [3], "The joining process of two metal parts with a third one having a lower melting point is known generally as *soldering*. When the solder has a melting point lower than 400°C, the process is known as soft soldering; if the solder melts above 500°C, the process is known as hard soldering.

Brazing is defined as the metal joining process in which melted filler metal is drawn by *capillary attraction* into the space between the closely adjacent surfaces of the parts to be joined (British Standard 499/1952 [12]). The temperatures required for brazing are above 500°C (American Welding Society [13]).

It must be remembered that, during the evacuation of a microwave tube, bakeout temperatures in excess on 500°C are normally used. Thus, to avoid the possibility of melting, softening, or excessive vaporization, filler materials must be used that melt at temperatures well above 500°C. It is common, for example, to use filler materials that melt at about 900°C. The *Brazing Handbook* [14] contains descriptions of brazing filler materials and their uses.

The temperatures required for brazing are very high and it is important to properly control the variation of temperature with time. The reasons for this can be seen by referring to Figure B-35. During stage 1, the parts are preheated as fast as possible consistent with avoiding warping and thermal shock. Thermal equilibrium is established during stage 2. It is important for all parts to reach the solidus temperature at the same time. Stage 3 is the brazing step. The final temperature is adjusted to obtain the desired viscosity of the filler. The time spent in the liquidus region is minimized. Finally, the parts are cooled as quickly as possible. The most critical portion of cooldown is dropping the temperature below the solidus temperature.

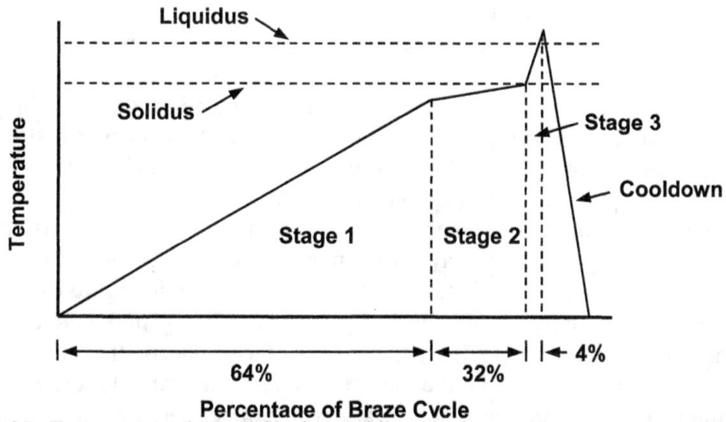

Figure B-35 Temperature–time profile for avoiding brazing problems. (From: R. English, Microwave Tube Manufacture lectures, Crane Division, Naval Surface Warfare Center.)

Some of the problems that occur if temperature is not properly controlled during brazing are shown in Figure B-36. The ideal braze joint is the one with the full fillet. If either temperature or time are inadequate, then the filler material may not be completely melted and so the underbrazed condition results. If the parts are not properly cleaned or the joint tolerance is incorrect, then voids may occur which can result in leaks or virtual leaks. If the temperature is too high or the time that the parts are held at a high temperature is too long, then the filler may alloy with the parts being brazed and cause erosion.

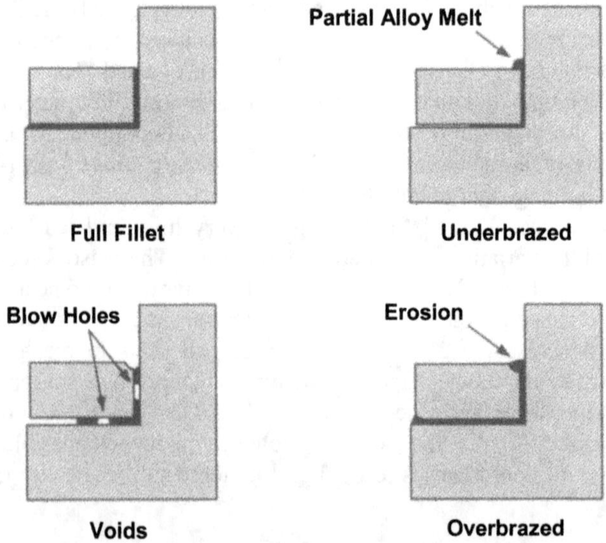

Figure B-36 Examples of braze problems. (From: R. English, Microwave Tube Manufacture lectures, Crane Division, Naval Surface Warfare Center.)

Because of the critical nature of the brazing operation, there are only a few processes that can be used. The most common are oven brazing and induction brazing. Fluxes cannot be used so brazes must be performed in vacuum or in a reducing atmosphere of hydrogen. Hydrogen is often used rather than a vacuum because brazing can be performed more rapidly. In some cases, such as brazing of stainless steel, extremely dry hydrogen must be used to prevent oxidation. An induction heating system for brazing in hydrogen is shown in Figure B-37.

Ceramic-to-metal seals are made by first metallizing the ceramic and then brazing the metallic part to the metallized surface. One of the most important factors to be considered is the thermal expansion of the parts. Figure B-38 shows how thermal expansion coefficients vary with temperature. When ceramics and metals having thermal expansion coefficients that differ significantly are used, great care must be taken in the design of the seal. In some cases the metal element is designed to be flexible in the region of the braze. In a butt ceramic-to-metal seal, as shown in Figure B-39, the metal is sandwiched between ceramic pieces that are massive enough to force the metal to expand and contract with the ceramic.

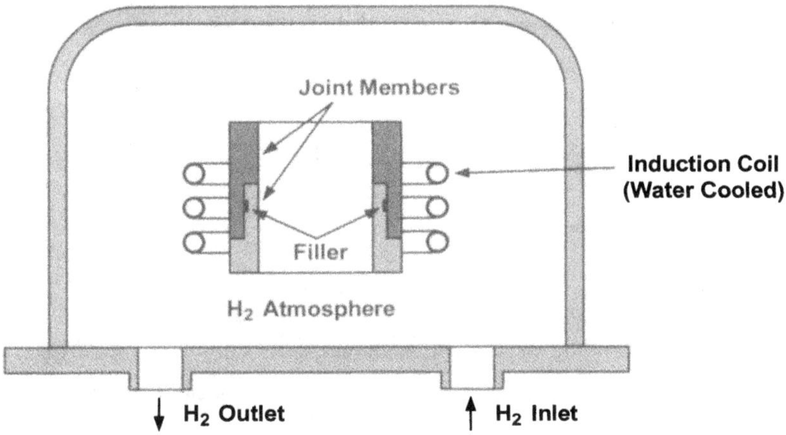

Figure B-37 Brazing by induction heating in hydrogen.

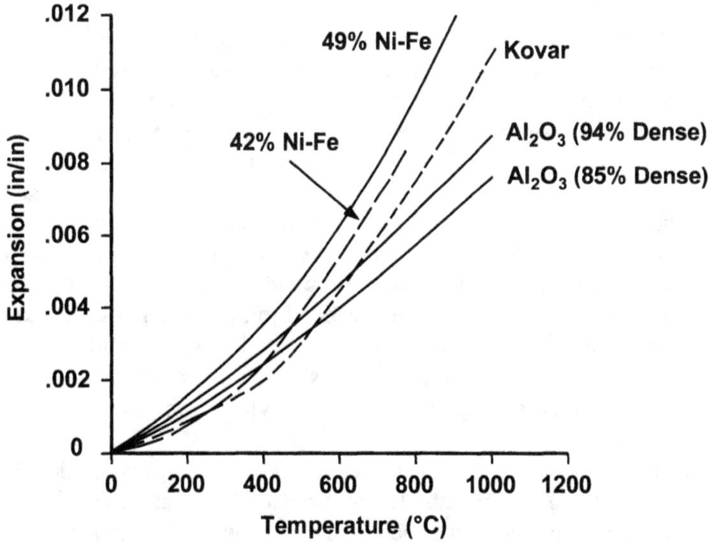

Figure B-38 Thermal expansion characteristics. (Adapted from: *Handbook of Electron Tube and Vacuum Techniques* by Fred Rosebury, copyright 1965 by Addison-Wesley Publishing Company, Inc.)

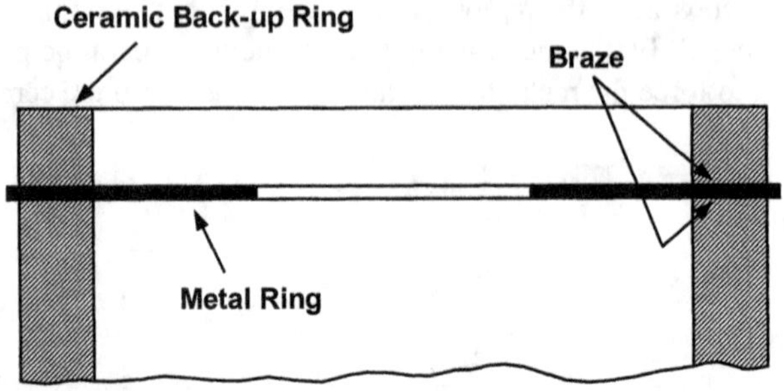

Figure B-39 Butt ceramic-to-metal seal.

B.10.2 Welding

Several types of welding are used in the manufacture of microwave tubes:

- Tungsten inert gas (TIG);
- Resistance;
- Electron beam;
- Laser.

TIG welding can be used manually or in automated systems A drawing of a manual inert gas welding torch is given in Figure B-40. With TIG welding, the heat produced by an arc melts the metal of the joint between workpieces and joins them together. Important parameters in the welding process are the current, which controls the depth and speed of the weld and current types (dc positive or negative or ac). Either argon or helium or a mixture of the two is the gas that is used. The gas must be very pure with impurities of 50 ppm at a maximum.

Care must be taken in designing weld joints if adequate strength is to be obtained and gas and dirt traps are to be avoided. Examples of the precautions that must be taken in the formation of a welded joint between a tube and a flange are shown in Figure B-41. Whenever possible, the weld should be made on the vacuum side of the joint. Then, weld slag or other contaminants that may be trapped between the mating surfaces of the cylinder and flange will be outside of the vacuum seal.

Resistance welding (sometimes called spot welding) is, perhaps, the simplest method for joining two metals. Using a device like that depicted in Figure B-42, high pressure is applied to the metal parts to be welded and then a current pulse is passed through the parts causing them to melt and fuse together.

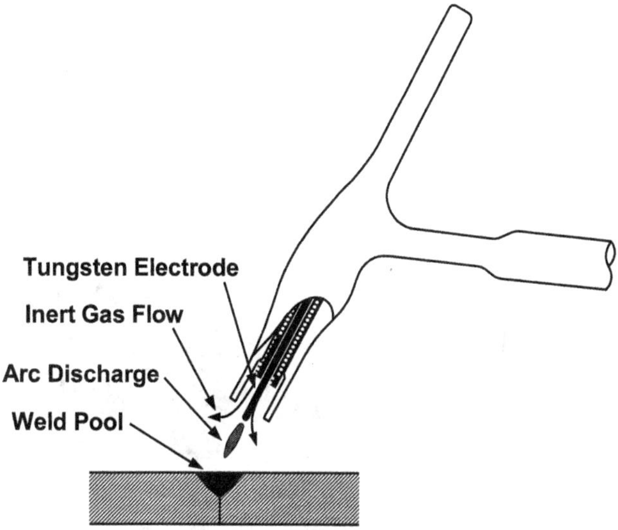

Figure B-40 Schematic of tungsten inert gas (TIG) welding torch. (Adapted from: *Handbook of Thin Film Technology* by L. I. Maissel and R. Glang. © 1970 McGraw-Hill, Inc.)

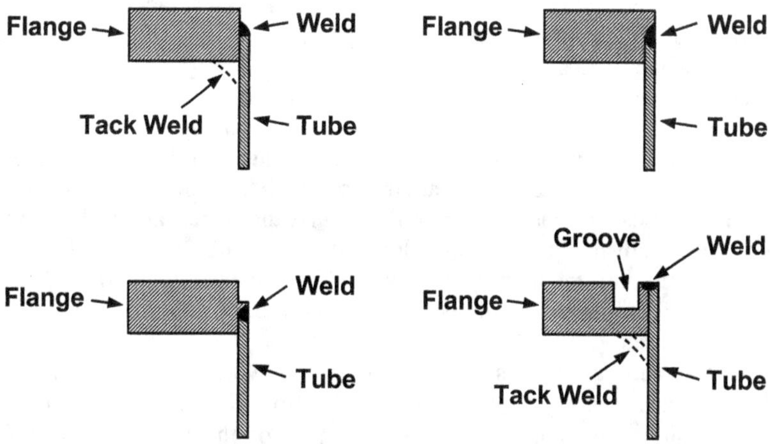

Figure B-41 Common tube-to-flange weld designs. (Adapted from: *Handbook of Thin Film Technology* by L. I. Maissel and R. Glang. © 1970 McGraw-Hill, Inc.)

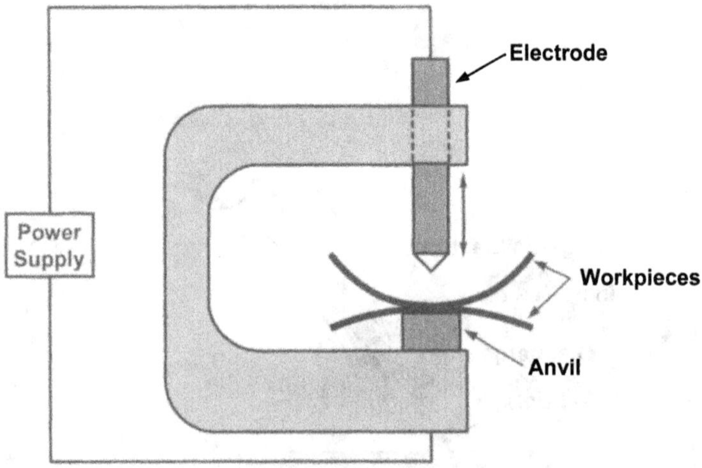

Figure B-42 Apparatus for spot welding metal parts. (Adapted from: R. English, Microwave Tube Manufacture lectures, Crane Division, Naval Surface Warfare Center.)

Electron beam welding is generally performed in vacuum but can be done in an atmosphere. A beam accelerated to 0.3 – 0.7 times the speed of light is used to form deep narrow welds. The region affected by the weld is small so welding close to parts affected by heat is possible. Disadvantages are that a vacuum requirement is usually required, equipment costs are high, and close-fitting joints are required because of the narrow beam width that is used.

Laser welding is accomplished with either a pulsed or a CW light beam. It is often used for welds of small parts in hard-to-reach locations such as the connection in a TWT of the last turn of a helix to the center conductor of a coaxial connector. It is also used when complex weld patterns are required. Compared with electron beam welding, smaller spot sizes are possible and the region affected by heat is smaller. One disadvantage of laser welding is that a large fraction of the light beam is reflected until the metals being welded have melted. Other disadvantages are that it is difficult to define the beamwidth and precise alignments and control of gaps are required.

After assembly and processing of a metal-to-ceramic vacuum tube has been completed, it is necessary to separate the tube from the pumping station. (An appendage pump may remain as an integral part of the tube.) The technique universally used to separate the tube from the pumping station is illustrated in Figure B-43. Soft copper tubulation is used to make the vacuum connection to the pumping station. This tubulation is molecularly clean on the internal surfaces because it has been baked along with the microwave tube (at 450°C or higher).

By squeezing the tubulation with a special tool, the internal mating surfaces of the copper can be forced to flow together to form a cold weld. In its crudest form, the tool is similar to a manually operated bolt cutter with the cutting edges replaced by cylindrical steel bars.

Because the manual squeeze-off (pinch-off) operation with the tool that is much like a bolt cutter is highly operator dependent, an automated system is normally employed.

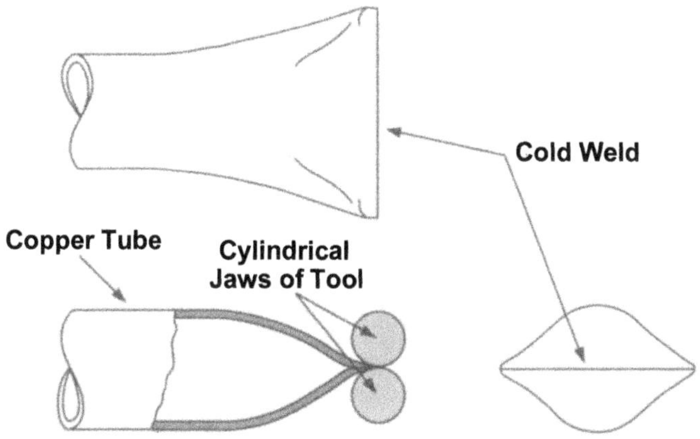

Figure B-43 Method used to squeeze-off (pinch-off) copper tubulation. (Adapted from: *Power Travelling Wave Tubes* by J. F. Gittins, American Elsevier Publishing Company, Inc., 1965.)

REFERENCES

[1] *Vacuum Physics and Technology, Methods of Experimental Physics*, Vol. 14, New York: Academic Press, 1979.

[2] S. Dushman, and J. M. Lafferty, *Scientific Foundations of Vacuum Techniques*, New York: John Wiley and Sons, 1962.

[3] J. F. O'Hanlon, *A User's Guide to Vacuum Technology*, New York: John Wiley and Sons, 2003.

[4] M. Pirani, and J. Yarwood, *Principles of Vacuum Engineering,* New York: Reinhold, 1963.

[5] R. W. Roberts, and T. A. Vanderslice, *Ultra High Vacuum and Its Applications*, Englewood Cliffs, NJ: Prentice Hall, 1963.

[6] A. Roth, *Vacuum Technology*, Amsterdam, the Netherlands: Elsevier Science Publishers B.V., 1990.

[7] A. Roth, *Vacuum Sealing Techniques*, Woodbury, NY: AIP Press, 1994.

[8] M. Knoll, *Materials and Processes of Electron Devices*, Berlin: Springer-Verlag, 1959.

[9] W. H. Kohl, *Materials and Techniques for Electron Tubes*, New York: Reinhold, 1960.

[10] F. Rosebury, *Handbook of Electron Tube and Vacuum Techniques,* Reading, MA: Addison-Wesley, 1965.

[11] R. English, "Lectures on Microwave Tube Manufacture," Crane Division of the Naval Surface Warfare Center.

[12] British Standard 499/1952.

[13] American Welding Society.

[14] *Brazing Handbook*, Miami, FL: American Welding Society, 1991.

Appendix C

Magnetics

Magnetic fields are used for focusing electron beams in nearly all microwave tubes. These fields may be generated by permanent magnets or electromagnets. In a permanent magnet, electron motion within the material produces the magnetic field. In an electromagnet, current passing through a wire produces a magnetic field surrounding the wire. Ferromagnetic materials, permanent magnets, and electromagnets are discussed in this appendix [1–5].

C.1 MAGNETIC QUANTITIES

In considerations of magnetics and magnetic materials, three quantities are of primary interest:

- Magnetic field intensity, H, is one of the elementary fields in nature.
- Magnetic flux density, B, is the density of magnetic flux lines produced by the magnetic field.
- Magnetic permeability, μ, is the property of a medium that determines how much flux is produced by a certain field level. Flux density is related to field intensity by

$$B = \mu H \tag{C-1}$$

The units commonly used for H, B, and μ are given in the following table along with conversion factors. The value for μ in air (free space), μ_o, is

$$\mu_o = 4\,\pi \times 10^{-7} \ \ \text{H/m} \tag{C-2}$$

It is common to use the term *relative permeability*, μ_r, which is the ratio of the permeability of a material to that of free space. That is,

$$\mu_r = \frac{\mu}{\mu_o} \tag{C-3}$$

Quantity	*MKS Unit (a)*	*CGS Unit (b)*	*Conversion*
H	Amperes/meter	Oersted	$(a) = 79.577 \times (b)$
B	Tesla	Gauss	$(a) = 10^{-4} \times (b)$
μ	Henrys/meter	Gauss/Oersted	$(a) = 1.257 \times 10^{-6} \times (b)$

C.2 MAGNETIC CIRCUITS

It is convenient, when thinking about magnetic circuits, to relate magnetic quantities to analogous quantities in electrical circuits. Figure C-1 indicates how magnetic and electric circuits are related. *MMF* is the magnetomotive force and is analogous to voltage. The source of the magnetomotive force can be either a permanent magnet or an electromagnet. Φ is the magnetic flux and is analogous to the current and reluctance is analogous to resistance.

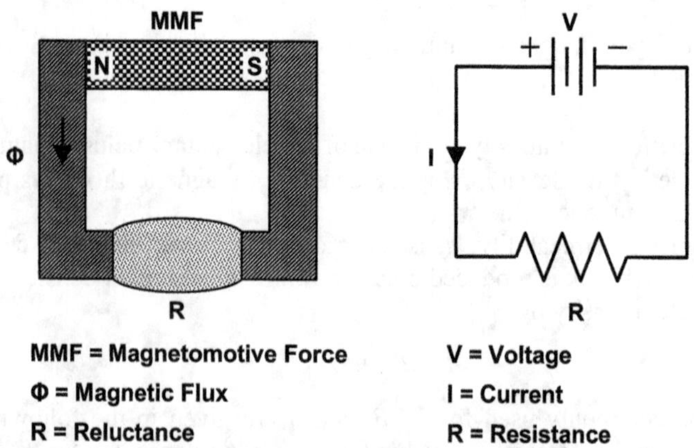

Figure C-1 Comparison of magnetic and electric circuits.

Permeance, denoted by *P*, is the inverse of reluctance, is analogous to electrical conductance and is often used rather than reluctance. *MMF*, Φ, *P*, and *R* are related to *H* and *B* by

$$MMF = \int H \cdot dl \tag{C-4}$$

$$\Phi = \int B \cdot dA \tag{C-5}$$

$$R = \frac{1}{P} = \frac{MMF}{\Phi} \tag{C-6}$$

Just as resistance is related to the physical characteristics of a medium by $R = \ell / \sigma A$, where σ is conductivity, reluctance is related by $R = \ell / \mu A$ where μ is permeability.

Much can be learned by examining the magnetic circuit in Figure C-1. Consider Figure C-2 and assume that there is no flux leakage outside the magnetic circuit and also assume that the flux density in the gap, G, is uniform. The problem is to find expressions for the flux density, B_g, in the gap and for the operating conditions of the magnet.

Let H_m be the magnetic field strength of the magnet, let H_{Fe} be the magnetic field in iron pole pieces that connect the magnet to the gap, and let H_g be the magnetic field in the gap. In terms of field quantities, the magnetic circuit equation is as follows:

$$\int H\,dl = \int_A^B H_{Fe}\,dl + \int_B^C H_G\,dl + \int_C^D H_{Fe}\,dl + \int_D^A H_m\,dl = 0 \tag{C-7}$$

which, assuming a constant H_{Fe} in the pole pieces with lengths ℓ_1 and ℓ_2, becomes

$$H_m T = H_{Fe}\ell_1 + H_{Fe}\ell_2 + H_g G \tag{C-8}$$

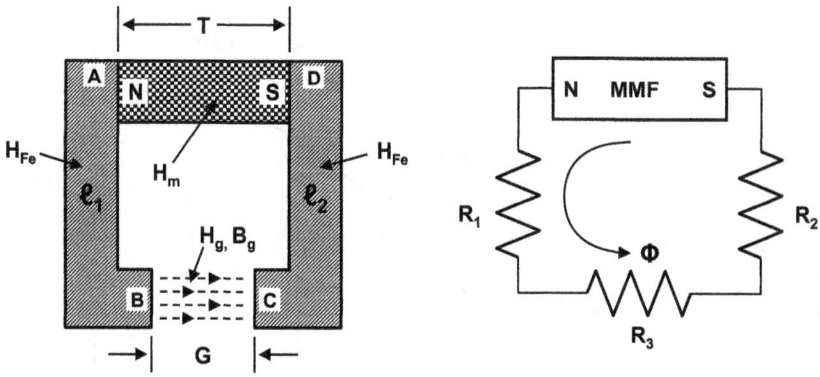

Figure C-2 Simple magnetic circuit.

In terms of circuit quantities, the magnetic circuit equation is

$$MMF = \Phi(R_1 + R_2 + R_3) \tag{C-9}$$

The reluctances, R_1, R_2 and R_3 are given by:

$$R_1 = \frac{\ell_1}{\mu_{Fe} A_1}, \quad R_2 = \frac{\ell_2}{\mu_{Fe} A_2}, \text{ and } R_3 = \frac{G}{\mu_o A_g} \tag{C-10}$$

where μ_{Fe} is the permeability of iron and A is the cross-sectional area. Since $\mu_{Fe} >> \mu_o$ and $MMF = H_m T$,

$$H_m T = \Phi \frac{G}{\mu_o A_g} = B_g G / \mu_o \tag{C-11}$$

or

$$B_g = \mu_o H_m \frac{T}{G} \tag{C-12}$$

Thus, if H_m is known, the flux density in the gap can be found.

Since magnetic flux is constant, $B_g = B_m A_m / A_g$, and so

$$\frac{B_m}{H_m} = \mu_o \frac{T}{A_m} \frac{A_g}{G} = \frac{T}{A_m} P_g \tag{C-13}$$

where P_g is the permeance of the gap. This relation gives the slope of a line, called the load line, on the B-H characteristic for the magnet material. This is illustrated in Figure C-3. The demagnetizing curve is discussed in the following section.

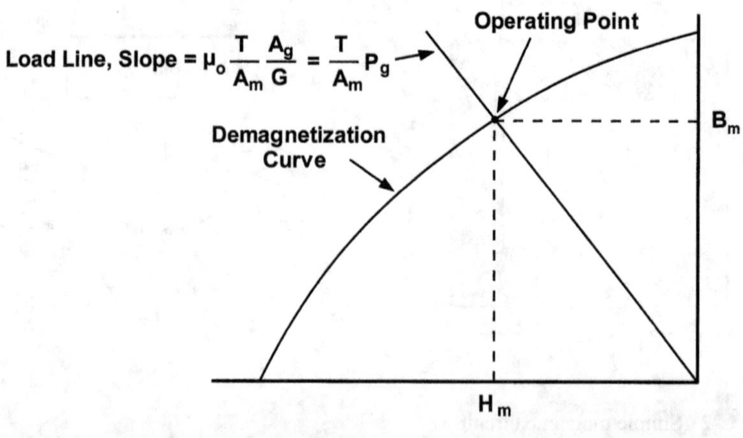

Figure C-3 Illustration of magnetic load line.

C.3 MAGNETIC MATERIALS

Magnetic materials can be divided into three categories. These are:

- $\mu_r < 1$ diamagnetic
- $\mu_r > 1$ paramagnetic
- $\mu_r >> 1$ ferromagnetic

Diamagnetism is present in all materials. The establishment of magnetic flux is opposed by the material so that the relative permeability is less than unity. This effect is explained by the laws of Faraday and Lenz. When a magnetic field is applied, an electric field is produced. This electric field acts on the electron orbits in a material to produce microscopic reaction currents. The reaction currents produce a magnetic field that opposes the applied field. Diamagnetism is an extremely small effect. In most materials, it is overridden by paramagnetism or ferromagnetism.

Paramagnetic materials have a relative permeability slightly greater than unity. In these materials, the application of an external magnetic field aligns permanent magnetic dipoles of atoms, ions, or molecules with the applied field, thus enhancing the field.

In ferromagnetic materials, a specialized condition exists that greatly aids the applied magnetic field. This condition involves the electronic structures of atoms and the distances between atoms. The most stable configuration of the electrons is when their spin magnetic moments are parallel. The result is that complete alignment of the spins occurs within regions called *magnetic domains*. The application of an external field both aligns domains and shifts domain boundaries to produce an enormous field.

C.3.1 Ferromagnetic Materials

Ferromagnetic materials are highly nonlinear. The relation between B and H must be determined experimentally. This is done with a circuit like that in Figure C-4.

From Ampere's law, the magnetic field, H, in a magnetic sample is proportional to the current, I. From Faraday's law, the voltage, V, induced in a winding is proportional to the time rate of change of magnetic flux linking the winding. The integral of voltage is, therefore, proportional to the flux and the magnetic flux density, B.

A signal proportional to B is applied to the vertical input of an oscilloscope. A signal proportional to H is applied to the horizontal input. The result displayed on the oscilloscope is known as a hysteresis curve and is illustrated in Figure C-5.

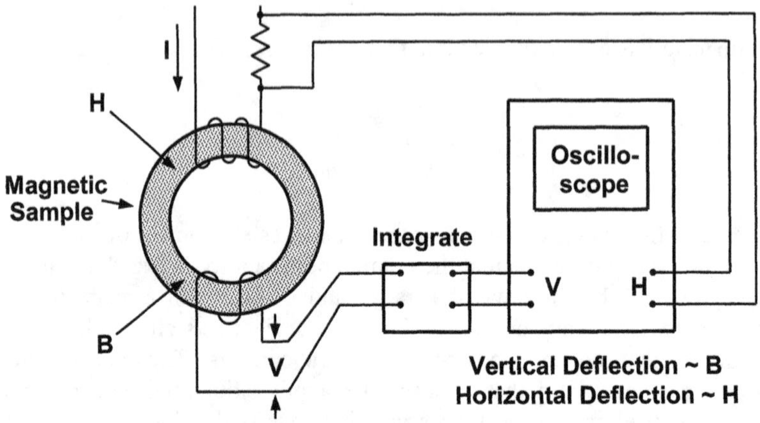

Figure C-4 Circuit for determining the magnetic characteristic of materials.

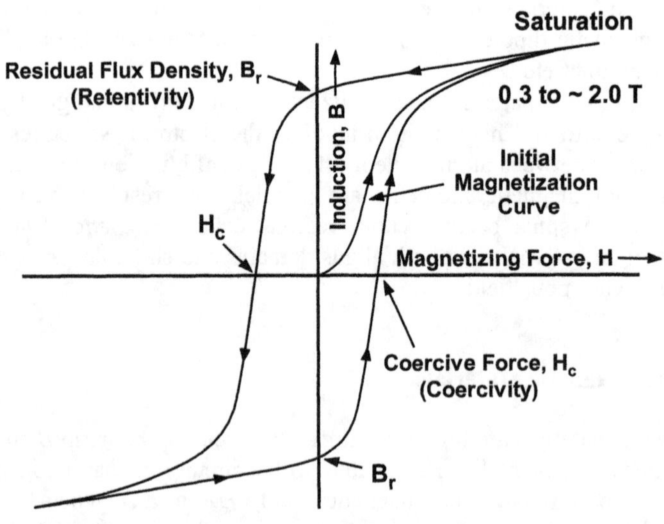

Figure C-5 Hysteresis curve resulting from use of circuit shown in Figure C-7.

When current is increased, the right-hand portion of the hysteresis curve is followed. When current is decreased, the left-hand portion is followed. At very large current excursions, the shape of the hysteresis curve is relatively independent of the current, and a loop known as the saturation loop is traversed.

Several important characteristics of magnetic materials can be illustrated with the hysteresis loop. Some of these are shown in Figure C-5. When a magnetic field is first applied to a demagnetized sample, the initial magnetization curve is followed. When H is zero, there is a residual flux density, B_r, in the material called the *retentivity*. The flux density, B, is reduced to zero when a field, H_c, called the *coercive force* or *coercivity*, is applied.

If the alternating magnetizing current is gradually increased from zero, small hysteresis loops are followed which gradually increase in size as shown in Figure C-6(a). The tips of these loops form the normal magnetization curve. This is very similar in shape to the initial magnetization curve. This curve is a reproducible characteristic for a magnetic material. Figure C-6(b) shows that a material can be demagnetized by gradually reducing the alternating magnetic field to zero.

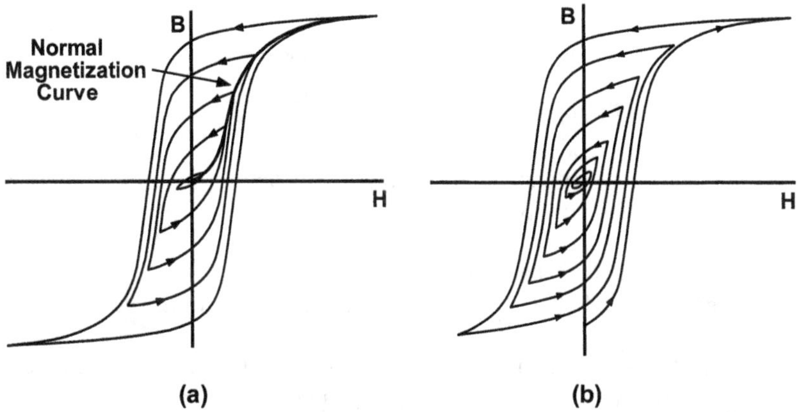

Figure C-6 (a) Magnetization and (b) demagnetization with a gradually changing alternating current.

The relative permeability quoted for a material is based on the slope of the normal magnetization curve as shown in Figure C-7. The maximum value is normally used. It is significant to note that, at very low values of the magnetic field, the permeability is low. In fact, for most magnetic materials, the permeability at low values of H is so low that they will not shield against stray magnetic fields. This is sometimes of importance in the electron gun of a tube when it is necessary to shield the cathode from stray magnetic flux. At high values of H, the permeability is also low. In fact, when a material is driven well into saturation, the permeability approaches the permeability of free space. This must be kept in mind in the design of magnetic pole pieces to be used in regions of high magnetic field. The concentration of flux lines at corners of pole pieces, for example, can cause saturation and distortion of the field pattern.

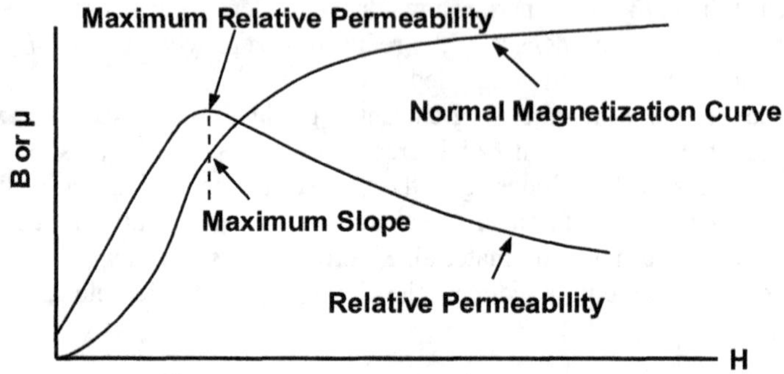

Figure C-7 Relation of relative permeability to normal magnetization curve.

C.3.2 Normal and Intrinsic Hysteresis Curves

The current used for applying a magnetizing force to the sample in the test described in Figure C-4 does two things:

1. Orients the domains and magnetizes the sample;
2. Generates a field that would be there in the absence of the sample.

As a result, to find the true (intrinsic) hysteresis characteristic of the sample, the additional field that must be supplied must be subtracted. The result is that two hysteresis loops shown in Figure C-8 must be considered. These are:

1. The normal induction loop that contains the additional field;
2. The intrinsic induction loop which is a characteristic of only the metal.

The flux density for the normal loop is

$$B = B_i + \mu_o H \tag{C-14}$$

where B_i is the magnetic flux density for intrinsic induction (which results from the magnetization of the material). Figure C-9 is useful in explaining the difference between normal and intrinsic induction.

In Figure C-9(a), the situation is the same as in Figure C-4. That is, the current produces a magnetizing force that magnetizes the sample (produces intrinsic flux density B_i) and produces the free-space flux density, $\mu_o H$. In Figure C-9(b), the magnetizing force (current carrying coil) has been removed so H is 0 and the sample has been left in a magnetized state with a flux density, B_r, which is the residual flux density (retentivity).

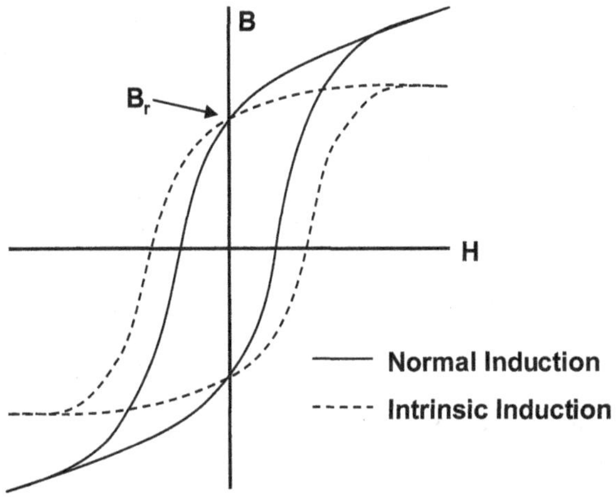

Figure C-8 Normal and intrinsic hysteresis loops.

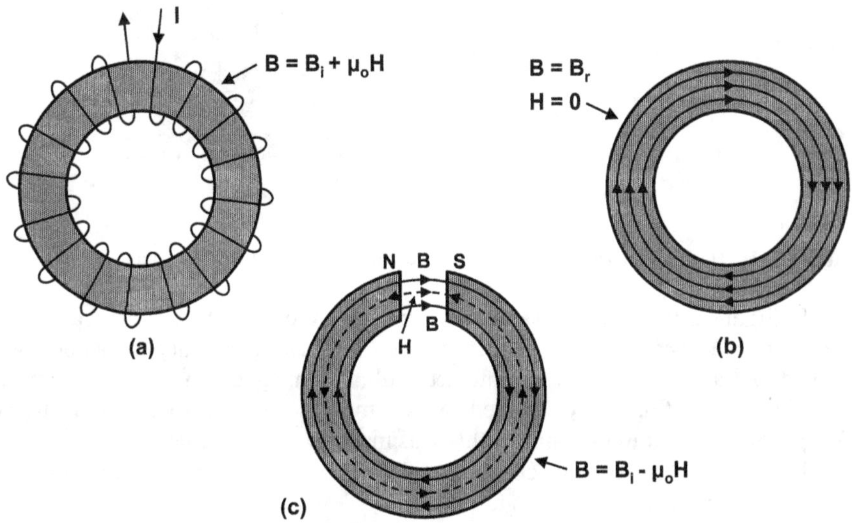

Figure C-9 (a–c) Normal and intrinsic inductance effects in magnetic material with a gap.

In Figure C-9(c), there is a gap in the magnetic sample. When the gap is cut, north and south magnetic poles are formed in the faces of the magnet. Within the gap, the magnetic field, H, is from north to south and is in the same direction as the flux density, B. The flux density produced by H in the gap is $\mu_o H$. In the magnet, the field, H, is also from north to south but is in the direction *opposite* to B.

The flux density produced by H in the magnet is $-\mu_o H$ Thus, H tends to demagnetize the magnet. This effect is the same as reversing the current through the coil in Figure C-9(a). This reverses the field generated by the coil and moves the operating point on the B–H curve to the left of the B axis as shown in Figure C-10. This is the second quadrant of the hysteresis loop. The curve in this region is called the demagnetization curve. Here, the field is opposite to B so demagnetization occurs. From Figure C-9, it is seen that the flux density in the gap is $B_i + \mu_o H$ so it is the normal B–H curve that is used for magnet design and for determining demagnetization.

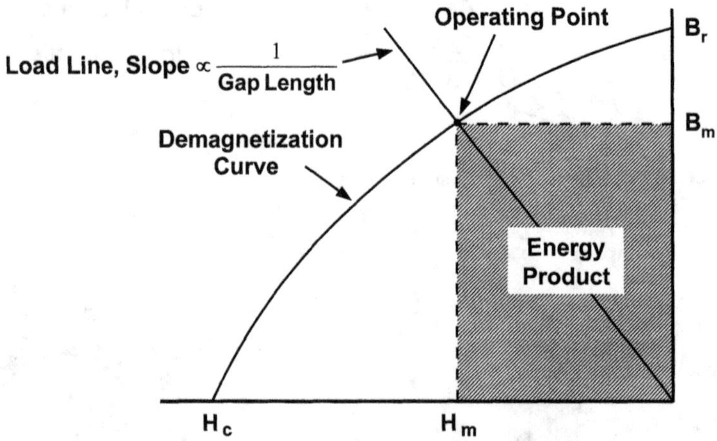

Figure C-10 Demagnetizing curve in second quadrant of the hysteresis loop.

In Figure C-9 when the magnetizing coil was removed, the sample was left in a magnetized state with a flux density of B_r. Thus, a permanent magnet was formed, but all of the flux was contained within the magnetic material, so there is no way that the flux could be used, for example, to exert a force on electrons. When a gap was cut in the sample, flux was made available for use.

Now, two considerations concerning the design of permanent magnet circuits are immediately evident from Figure C-10:

1. For optimum utilization of the magnet, the gap must be long enough to provide a reasonable gap volume in which interactions with the flux density can take place. On the other hand, the gap must not be so long that the flux density becomes too low.
2. A material with a coercivity, $-H_c$, as large as possible as well as with a large retentivity, B_r, should be used to maximize the gap length (and gap volume) for a given flux density.

C.3.3 Energy Product

Magnet materials for TWTs are selected primarily on the basis of the magnetic energy that they can produce in an air (or vacuum) gap. A second important consideration, which will be discussed later, is temperature stability. The energy per unit volume, w, in a magnetic field is

$$w = \int \mathbf{H} \bullet d\,\mathbf{B} \tag{C-15}$$

Thus, ignoring fringe effects, the total energy, W_g, in the magnetic field in the gap of a magnet is w multiplied by the volume of the gap, or

$$W_g = \ell_g\, A H_g\, B \tag{C-16}$$

where A is the cross-sectional area of the gap. Because $H_m \ell_m = H_g \ell_m$,

$$W_g = \ell_m\, A H_m B \tag{C-17}$$

but this is the energy in the field within the magnet. Thus, the energy delivered by the magnet is the energy established in the gap. This is analogous to saying that the power delivered by an electrical circuit is the power received by the load. The energy delivered by the magnet is proportional to the cross-hatched area in Figure C-10. This area is referred to as the energy product of the magnet. To maximize the energy product (and maximize the energy established in the gap), the gap length, l_g, must be such that the shaded area is maximized. The variation of energy product with the position of the operating point is shown in Figure C-11.

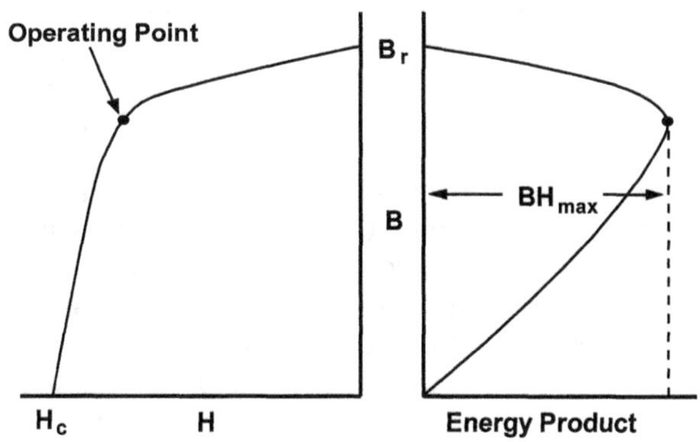

Figure C-11 Demagnetization curve and energy product.

C.3.4 Rare Earth Magnet Materials

Rare earth magnets are the magnets of choice for most modern microwave tube applications. The most common material used is samarium cobalt. The advantage of samarium cobalt over some other materials that have been used in the past is shown by the demagnetization curves in Figure C-12. The energy product for samarium cobalt is approximately four times that of alnico V, which is a magnet material that was in common use in microwave tubes 30 years ago. The coercive force is over an order of magnitude greater than that of alnico V, with the result that the samarium cobalt magnets are extremely stable.

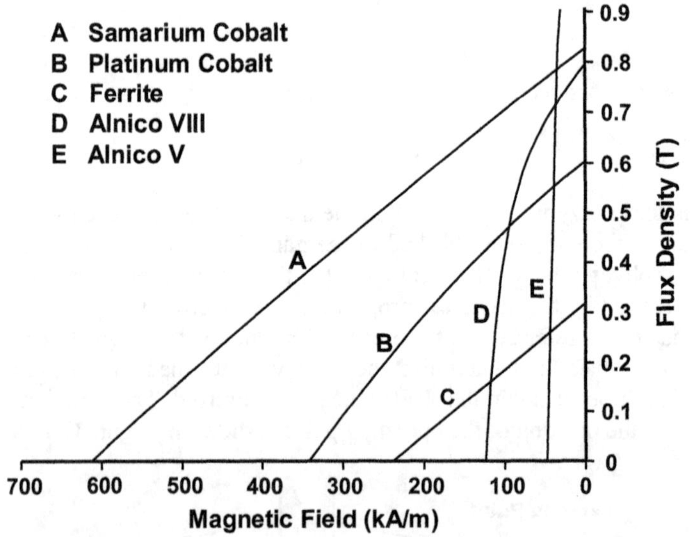

Figure C-12 Comparison of rare earth cobalt magnets with conventional magnets. (Adapted from: "Rare Earth Cobalt Magnets", Varian Brochure, 1972.)

Permanent magnets on microwave tubes must be able to operate over a wide temperature range with no significant change in characteristics. Conventional samarium cobalt ($SmCo_5$) has a field strength temperature coefficient of about 0.04%/°C. Electron Energy Corporation developed a series of Sm_2Co_{17} magnet grades that are temperature compensated so that they have a near-zero temperature coefficient. This was done by substituting gadolinium for a portion of the samarium the alloy [6].

Figure C-13 shows the demagnetization curves at various temperatures for one of a series of Sm_2Co_{17} high-temperature magnets. These magnets have an energy product of 20 MGOe. Notice that the normal demagnetization plots are extremely linear. This simplifies the analysis of magnetic circuits in which these magnets are used.

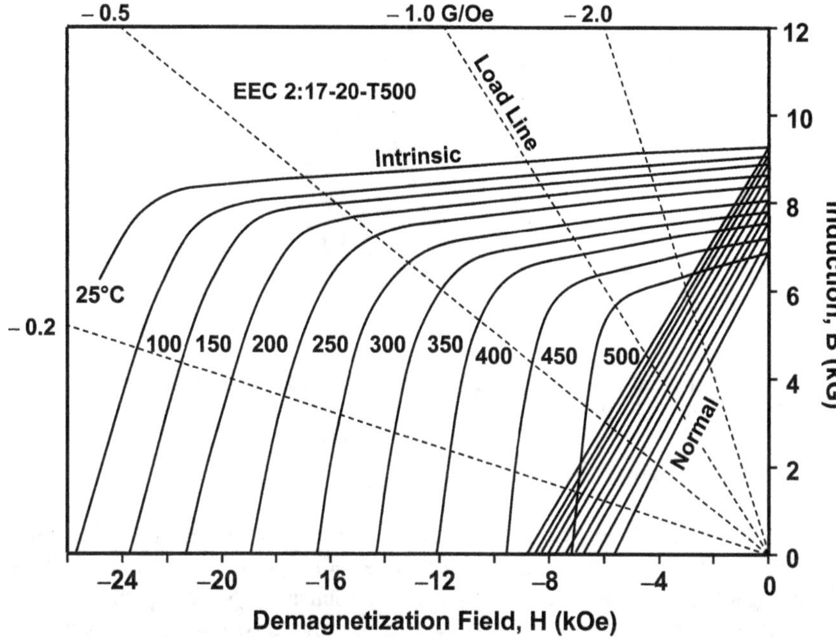

Figure C-13 Demagnetization curves for samarium cobalt Sm_2Co_{17} high-temperature magnets. (Adapted from: J. F. Liu and M. H. Walmer, *IEEE Trans. Electron Devices*, Vol. 52, No. 5, 2005. © 2005 IEEE.)

C.4 PERMANENT MAGNETS [5]

C.4.1 Straight Field Magnets

Historically, permanent magnets were sometimes used for electron beam focusing in TWTs and klystrons. These magnets were large and heavy and could produce only limited magnetic fields. Figure C-14 shows the magnet used on the SLAC 2422 klystron [7]. The magnet was barrel shaped, produced a magnetic flux density of about 0.06T, and was designed for a gap length of 16.5 inches. The inside diameter was 8.5 inches which provided space for the insertion of lead shielding and for the easy removal of the tube from the magnet. The maximum outside diameter was approximately 18 inches and the total weight was nearly 800 pounds. Permanent magnets were used because they made it possible to eliminate 240 focusing coils, focusing supplies, and water interlocks.

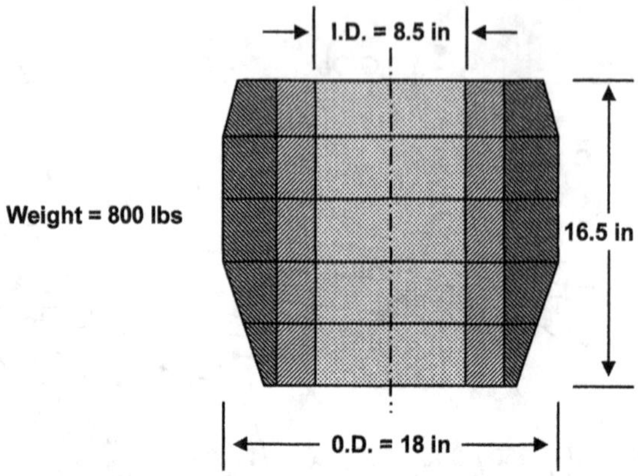

Figure C-14 Permanent magnet used on the SLAC 2422 klystron. (Adapted from: J. V. Lebacqz, SLAC-PUB-87, 1965.)

C.4.2 Periodic Permanent Magnets

In the majority of applications of microwave tubes, it is necessary to minimize weight and size. For these applications, periodic permanent magnet (PPM) focusing is used. The weight reduction of the focusing system, when compared with a solenoid or a unidirectional magnet, can be one to two orders of magnitude. The reason for this weight reduction can be explained by first noting that a solenoid or unidirectional magnet establishes a magnetic field in a volume far larger than that required for focusing the electron beam. Since the weight of the magnet is roughly proportional to the magnetic energy stored in the volume surrounding the magnet, much of the weight of the magnet results from the production of unused magnetic field. Figure C-15 [8] is useful in explaining why the PPM weight is so low. Assume first that the magnet in Figure C-15(a) has the most efficient design (magnet weight is minimized) to produce a magnetic field to focus an electron beam as it travels through the magnet. The most efficient design is the one that nearly optimizes the energy product for the magnet.

Next, assume that it is desired to focus the same beam over a distance that is three times the length of the magnet in Figure C-15(a). This may be accomplished with the magnet in Figure C-15(b) but, to produce the same magnetic field that was produced by the magnet in Figure C-15(a), it is necessary to increase all dimensions of the magnet by a factor of three. (Remember that the dimensions of a permanent magnet are established to nearly maximize the energy product. A magnet configuration other than that shown in Figure C-15(b) would result in either a reduction in magnetic field or an increase in magnet weight.) As a result, the magnet in Figure C-15(b) weighs 3^3 or 27 times as much as the magnet in (a), and the energy stored in the field surrounding the magnet in Figure C-15(b) is 27 times that for Figure C-15(a). By comparison, if three magnets like that in (a) were stacked with like poles together as shown in Figure C-15(c), then a field sufficient to focus the beam could be produced over the same length as for magnet Figure C-15(b). The total weight and energy stored for the magnets in Figure C-15(c) would be one-ninth of the weight and energy stored for the magnet in Figure C-15(b).

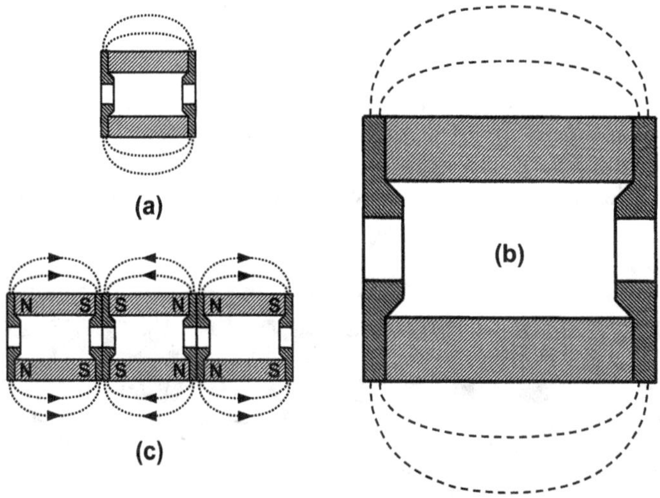

Figure C-15 (a–c) Size (and weight) reduction resulting from the use of periodic permanent magnets. (Adapted from: J. T. Mendel et al., *Proc. IRE*, May 1954. © 1954 IRE (now IEEE).)

If the magnet in Figure C-15(b) was replaced by n smaller magnets aligned like those in Figure C-15(c), the weight of the resulting structure would be $1/n^2$ times that of the larger magnet. In practice, the weight saving is not this great. This is because the field on the axis of the periodic structure does not appear as a square wave with abrupt reversals occurring at magnet interfaces. Instead, the field may be more nearly sinusoidal and the peak value must be increased to produce the same effective field as would be produced by the square wave. As a result, the magnet sizes must be increased to produce the required effective field.

Since magnet weight is roughly proportional to $1/n^2$, and since it is usually desirable to minimize magnet and tube weight, it is reasonable to ask what limits the number of magnets that is used. Qualitatively, this question can be answered by referring to Figure C-16 in which an electron beam focused by a periodic permanent magnet structure is illustrated. First, the interior dimension of the magnets has to be large enough to surround the electron beam and (not shown in Figure C-16) the RF circuit. Second, each magnet and the gap between pole pieces must be long enough so that magnetic field lines of sufficient strength for focusing penetrate into the electron beam. Third, as shown in Chapter 7, the wrong choice of magnet length and period of the magnet structure can lead to excessive beam ripple or beam divergence. Finally, other considerations to be discussed later in this appendix lead to the use of *double-period focusing* or, possibly, to *long-period focusing*.

In PPM structures like that in Figure C-16, ring-shaped magnets are used to form magnetic lenses that periodically focus the electron beam traveling along the center of the structure. Periodic magnetic focusing of an electron beam is discussed in Chapter 7.

A problem with the ring-shaped magnets illustrated in Figure C-16 is that each must be split in order to be placed on the tube. This is one of the factors leading to nonuniformities in the magnetic fields produced by these structures. To obtain acceptable tube performance, it is often necessary to compensate for the nonuniformities by placing magnetic shims on the outside of the magnetic assembly.

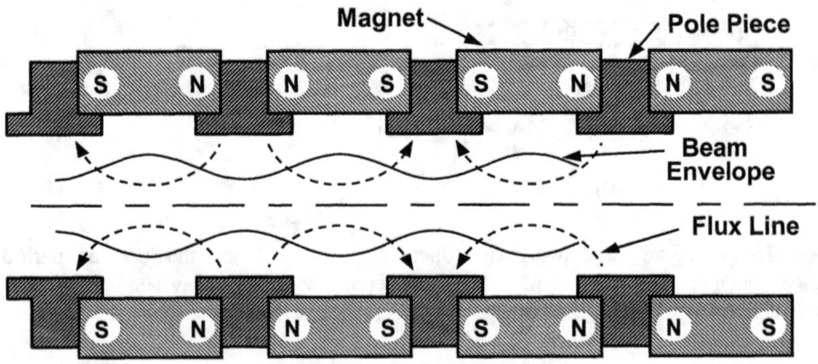

Figure C-16 Periodic permanent magnet structure.

A technique for avoiding some of the problems of ring-shaped magnets is shown in Figure C-17. Several (often four) pill-shaped magnets are placed between magnetic pole pieces that surround the axis of the tube. The magnets can be magnetized to the desired level and then placed on the tube without being split. Field uniformity is increased (as compared with the ring-shaped magnets). In addition, magnet cost is reduced because magnets of one size can be used on tubes having a wide range of sizes.

C.4.3 Double-Period and Long-Period Focusing

Double-period focusing is used in coupled-cavity TWTs. In those tubes, as shown in Figure C-18, some cavity walls also serve as pole pieces that concentrate the magnetic flux in the electron beam. Other cavity walls serve as flux guides to shape the magnetic field in the beam.

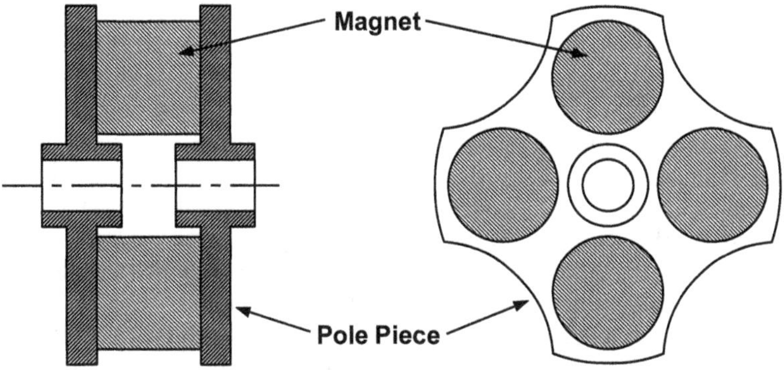

Figure C-17 A PPM configuration using pill-shaped magnets.

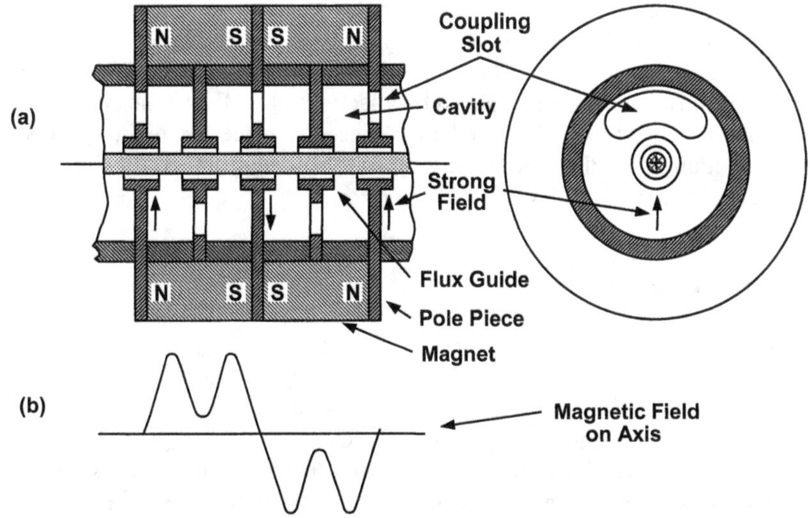

Figure C-18 (a, b) Configuration of magnets, pole pieces, and cavities for double-period focusing.

Slots in the cavity walls are used to couple the RF field from cavity to cavity. Often, the slots in adjacent walls are on opposite sides of the axis. The slots produce asymmetries in the focusing field, which produce transverse forces that deflect the electron beam. By arranging the magnets as shown in Figure C-18, the transverse forces produced by adjacent pole pieces are in opposite directions and so the deflections of the beam are in opposite directions. This prevents the beam from gradually being forced off axis as it moves from pole piece to pole piece.

Next, consider the rationale for long-period focusing (LPF) [9]. A single permanent magnet can be used to produce a uniform field to focus a beam. If a reversal of the otherwise uniform field takes place in a very short distance, the electron beam won't "know" that the field reverses. The rotation of the beam will reverse but, assuming no cathode flux, the focusing force will remain constant. With many reversals and very short distances between reversals, many smaller magnets could be used, each one of which had its polarity reversed from the previous magnet. The advantage in doing this would be the reduction in the weight of the magnets. PPM focusing is the limit, in terms of weight reduction but not necessarily in terms of practicality and cost. Long-period focusing is the compromise between minimum weight and minimum cost.

A primary application proposed for long-period focusing is high-frequency TWTs which have very small-diameter electron beams and, with PPM focusing, require many very small magnets. For example, 50 or more small ring magnets are required for a millimeter-wave TWT. The cost of parts and of labor to assemble the parts is very high. Phillips et al. proposed the use of long-period focusing to reduce the parts count, the cost of parts, and the cost of labor to assemble the parts [9]. A portion of an LPF structure and the associated magnetic field produced by it is shown in Figure C-19. With this configuration, compared with PPM focusing, the number of ring magnets is reduced by a factor of 4. In the single period shown, a pair of saturable flux guides and four spacers replace the eight spacers that would be required by PPM focusing. Phillips et al. demonstrated the feasibility of LPF in tests on a 200-watt TWT.

C.5 POLE PIECES

In many of the preceding figures, reference to pole pieces has been made. The pole pieces must be capable of carrying the flux from the magnets to the electron beam. In low-frequency tubes, where dimensions are large and flux densities are modest, the flux densities in the pole pieces are modest. However, in high-frequency devices, where dimensions are small, the flux density in the pole pieces can be high enough so that saturation occurs. When saturation occurs, the flux passes through the pole piece material as if it was made of air and the field available for focusing the beam is degraded. To avoid this problem, the saturation flux density of the pole pieces must be as high as possible and the pole piece must be designed so that the flux density is below the saturation value.

It is tedious but reasonably straightforward to estimate the flux density in the pole pieces and to determine if there is a danger of saturation [10, 11]. The flux density can be estimated if the operating point of the magnet is known and the dimensions of the magnetic circuit are known. To find the operating point of the magnet, the load line must be determined. To plot the load line, it is necessary to know the permeance of the magnet configuration.

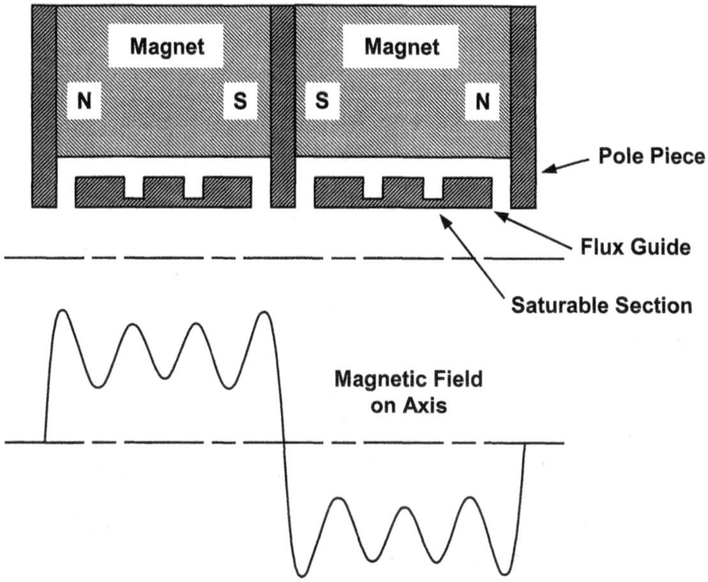

Figure C-19 Example of long-period focusing. (From: R. M. Phillips, et al. *Final Technical Report* RADC-TR-90-110, June 1990.)

As an example, the total permeance in Figure C-20(a) is

$$P_T = P_1 + P_2 + P_3 \tag{C-18}$$

where P_1 is the path between ferrules, P_2 is the leakage path inside the ferrules and P_3 is the leakage path outside the pole pieces. The load line is given by

$$\frac{B_m}{H_m} = \frac{T}{A} P_T \tag{C-19}$$

where T is the magnet length and A is the cross-sectional area.

Permeance depends very much on the position of the magnets relative to the pole pieces as indicated in Figure C-20. The configuration in Figure C-20(a) is representative of those used on helix TWTs. In coupled-cavity TWTs, the configuration is more like that shown in Figure C-20(b) or C-20(c). The pole pieces are the walls of the cavities and the magnets are at a radial distance that is large enough to make room for the cavities. As a result, P_4 is a significant permeance. Permeances P_5 and P_6 depend largely on flux requirements and thermal considerations.

The calculation of some of the permeances is straightforward. For example, referring to the dimensions in Figure C-21,

$$P_1 = \mu_o \frac{\left(r_{f2}^2 - r_{f1}^2\right)}{g} \quad \text{(C-20)}$$

On the other hand, permeances P_2 and P_3 are not so easy to determine. To find these, boundary value problems must be solved using Fourier series involving Bessel functions as well as sine and cosine functions.

After the permeances have been determined and the magnet load line has been established, the magnetic field that the pole pieces must support can be found. With the magnetic field and the pole piece dimensions, magnetic flux densities can be calculated. This is when the magnetic characteristics of the pole pieces must be known. As shown in Figure C-22, the saturation flux density for iron, which is commonly used, is about 21 kG. The highest saturation flux density material is vanadium permendur which, in Figure C-22, is shown to be about 22 kG but is sometimes reported to be as high as 23 kG [12].

Before leaving the subjects of pole pieces and permanent magnets, we will briefly discuss the effect of magnetic objects on the focusing of an electron beam. In connection with Figure C-16, it was pointed out that ring-shaped magnets have to be broken in order to be placed on a tube. As a result, nonuniformities in the focusing field for the beam occur and magnetic shims are placed on the outside of the tube in an effort to compensate for the nonuniformities.

It is important to realize that magnetic materials of any kind that are brought into close vicinity of a permanent magnet focused tube can affect the focusing of the beam. Magnetic tools are particularly dangerous, but even a small magnetic scrap like that shown in Figure C-23 can affect the beam trajectory and possibly destroy the tube.

C.6 ELECTROMAGNETS

A wire carrying a current produces a magnetic field as shown in Figure C-24. The direction of the field lines in relation to the direction of current flow can be remembered with the aid of the familiar right-hand rule. (The thumb points in the direction of conventional current and the fingers point in the direction of the field lines.) For microwave tube applications, the current carrying wire is formed into a solenoid, as shown in Figure C-25.

Very close to each turn of the solenoid, the flux lines are nearly circular. However, at greater distances, the flux in the region between adjacent turns cancels, so the flux lines around each pair of turns are nearly elliptical. At still greater distances, there is further cancellation of opposing flux lines and the pattern in the center of Figure C-25 occurs. Near the middle of the solenoid, the flux lines are parallel and the flux density is constant. Near the ends of the solenoid, the flux lines diverge and the flux density decreases as indicated by the dashed line in Figure C-26. This large decrease in field near the ends of the solenoid is not acceptable for most microwave tube applications.

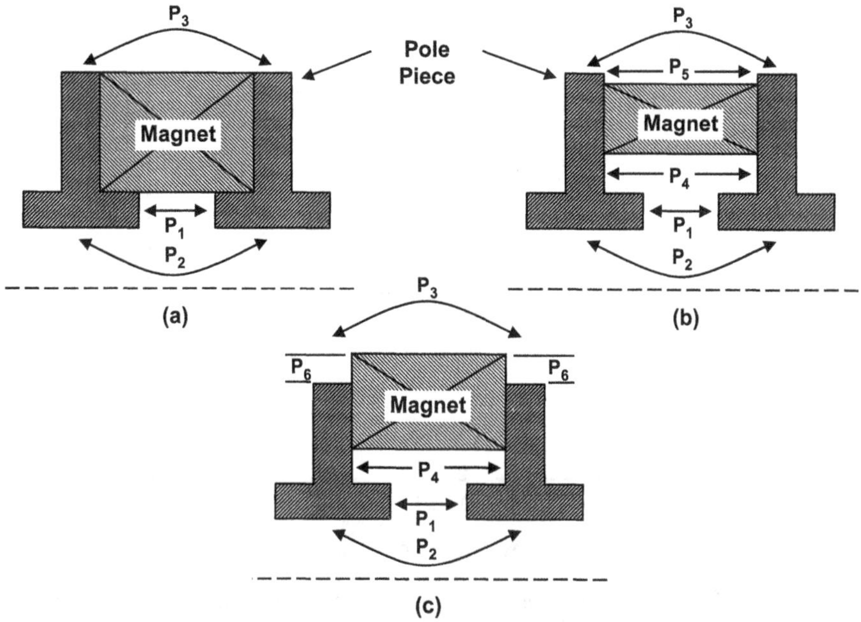

Figure C-20 (a–c) Some common magnet and pole piece configurations. (Adapted from: M. Santra et al., *IEEE Trans. Electron Devices*, Vol. 56, No. 5, 2009. © 2009 IEEE.)

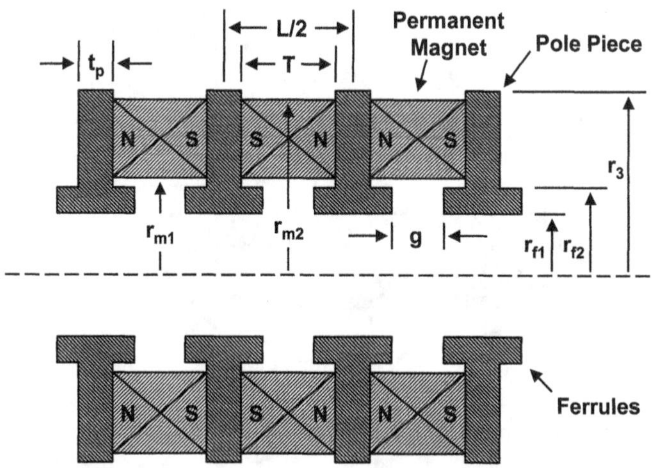

Figure C-21 Dimensions required for the determination of the flux densities in pole pieces. (Adapted from: M. Santra et al., *IEEE Trans. Electron Devices*, Vol. 56, No. 5, 2009. © 2009 IEEE.)

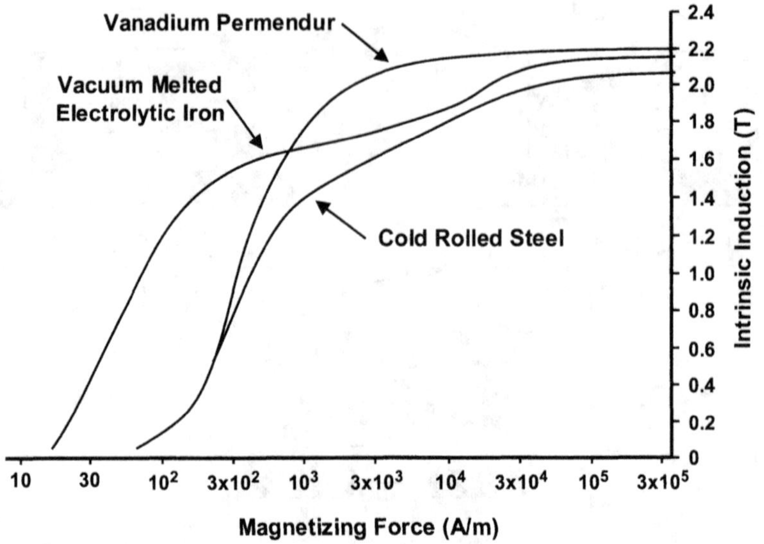

Figure C-22 DC magnetization curves for some pole piece materials. (Adapted from: DC Magnetization Curves, General Electric Co., 1956.)

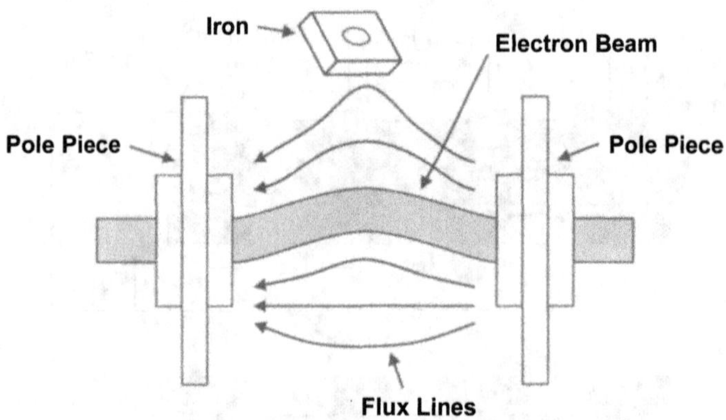

Figure C-23 Distortion of magnetic focusing field by a magnetic scrap.

To enhance uniformity of the field on axis where the electron beam is to flow, magnetic pole pieces, as shown in Figure C-27, are used. The pole pieces have a high relative permeability ($\mu_r \approx 5{,}000$) and their effect is to strengthen the field near the ends of the solenoid as shown by the solid curve in Figure C-26. The magnetic behavior of the pole pieces is described in the following section.

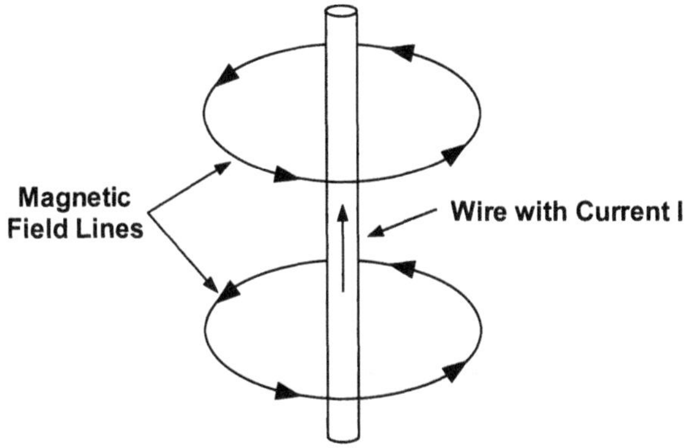

Figure C-24 Magnetic field around a wire carrying a current.

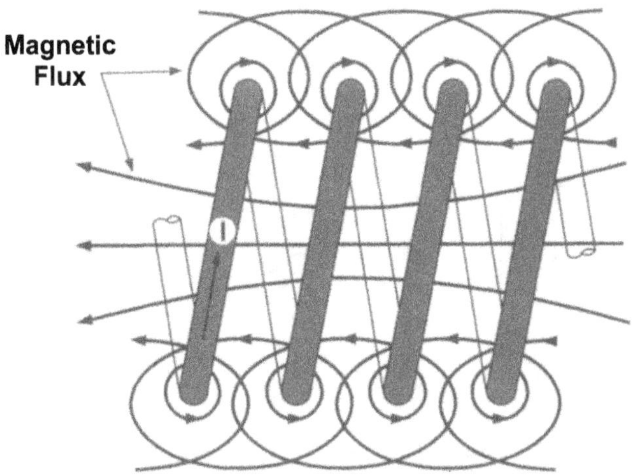

Figure C-25 Portion of a solenoid showing magnetic flux linkages.

Figure C-28 shows a typical solenoid for a TWT or klystron. Individual coils of the solenoid are separated to provide space for RF feedthroughs, cooling lines, and so forth. Normally, a solenoid is surrounded by a magnetic shield, or magnetic bars, as shown in Figure C-28, to provide a return path for flux lines. These magnetic shields do not enhance the magnetic flux density on the axis of the solenoid. Instead, they confine the magnetic flux lines surrounding a solenoid to a localized region near the solenoid. This helps minimize the possibility that a magnetic object, such as a tool, will accidentally be pulled toward the tube being focused by the solenoid, thus causing damage to, or destruction of, the tube. The magnetic shields or bars also help to minimize the interactions of the solenoid with nearby electrical circuits.

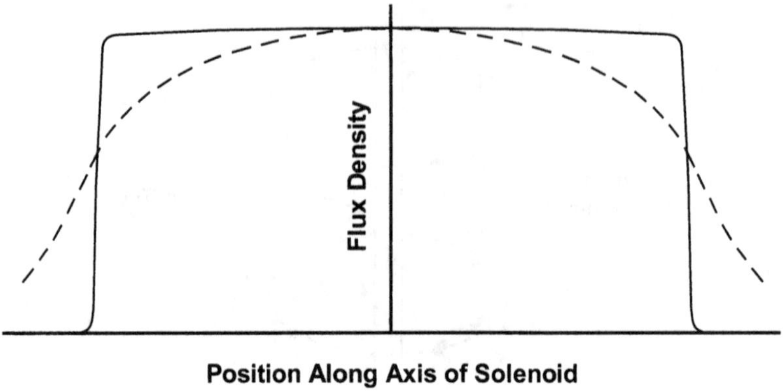

Figure C-26 Solenoid flux density without (dashed line) and with (solid line) magnetic pole pieces.

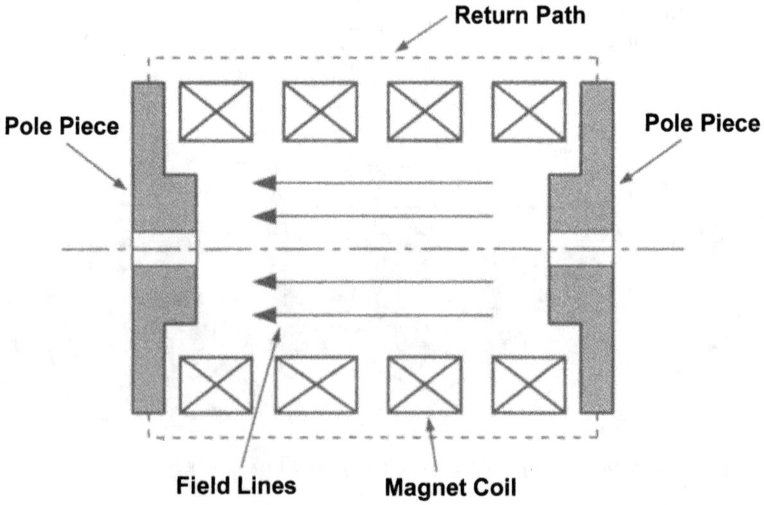

Figure C-27 Focusing solenoid consisting of coils and magnetic pole pieces to improve field uniformity.

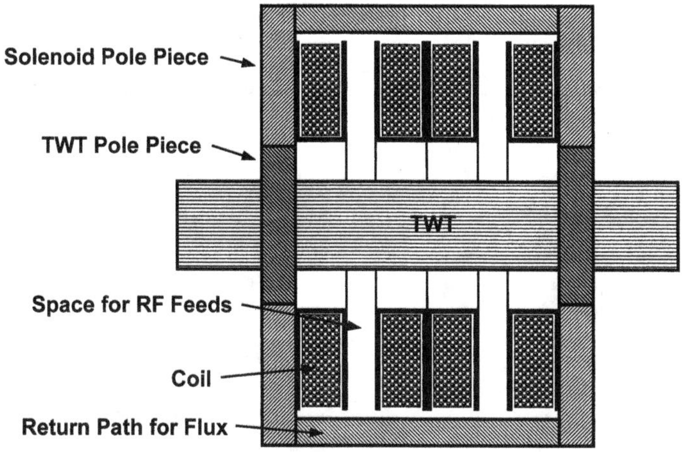

Figure C-28 Typical configuration of a solenoid and a microwave tube.

REFERENCES

[1] W. H. Hayt, Jr., *Engineering Electromagnetics*, New York: McGraw-Hill, 1981.

[2] J. D. Krauss and K. R. Carver, *Electromagnetics*, New York: McGraw-Hill, 1973.

[3] L. M. Magid *Electromagnetic Fields, Energy, and Waves*, New York: McGraw-Hill, 1972.

[4] C. G. Skitek, and S. V. Marshall, *Electromagnetic Concepts and Applications*, Englewood Cliffs, NJ: Prentice Hall, 1982.

[5] R. J. Parker and R. J. Strudders, *Permanent Magnets and Their Applications*, New York: John Wiley and Sons, 1962.

[6] J. F. Liu and M. H. Walmer, "Thermal stability and performance data for SmCo 2:17 high-temperature magnets on PPM focusing structures," *IEEE Trans. Electron Devices,* Vol. 52, No. 5, 2005, pp. 899–902.

[7] J. V. Lebacqz, *High Power Klystrons*, Vol. SLAC-PUB-87, Stanford Linear Accelerator Center, Stanford, CA, 1965.

[8] J. T. Mendel, C. F. Quate, and W. H. Yocom, "Electron beam focusing with periodic permanent magnet fields," *Proc. IRE*, Vol. 42, No. 5, May 1954, pp. 800–810.

[9] R. M. Phillips, S. Snyder, and M. Baird, "Advanced electron beam focusing (long period focusing)," *Final Technical Report RADC-TR-90-119*, June 1990.

[10] F. Sterzer and W. W. Siekanowicz, "Design of periodic permanent magnets for focusing of electron beams," *RCA Review,* Vol. 18, No. 1, 1957, pp. 39–59.

[11] M. Santra, L. Kumar, and J. Balakrishnan, "An improved analysis of PPM focusing structures including the effect of magnetic saturation in the iron pole pieces," *IEEE Trans. Electron Devices,* Vol. 56, No. 5, 2009, pp. 974–980.

[12] S. Sprague, *Lamination Steels, Third Edition*, Electric Motor Education and Research Foundation, 2007.

Glossary

AM near-carrier noise: AM noise and spurious outputs at frequencies close to the carrier frequency, usually 100 Hz to 500 kHz relative to the carrier. Components not related to the presence of the carrier (signal) are usually caused by ripple on power supply voltages. Interactions in the TWT are usually responsible for components that exist only when a carrier (signal) is present. (*See also* near-carrier noise.)

Amplitude pulling factor: The amount of gain change experienced when the VSWR of the input source or the load is changed.

Amplitude pushing factor: The amount of gain change experienced as the voltage on a particular tube element is varied, usually expressed in dB/volt.

AM/AM conversion: *amplitude modulation to amplitude modulation conversion* describes a type of distortion in microwave amplifiers. It is the change in the output RF voltage produced by variations in input signal level, usually expressed in dB/dB. An ideal amplifier has an AM/AM conversion of 1 dB/dB. As the amplifier approaches saturation, the amplification process becomes nonlinear and gain compression occurs.

AM/PM conversion: *amplitude modulation to phase modulation conversion* describes a type of distortion in microwave amplifiers. It is the change in phase angle of the output RF voltage produced by variations in input signal level, usually expressed in degrees/dB. An ideal amplifier has an AM/PM conversion of 0 degrees/dB.

AM/PM transfer: *amplitude modulation to phase modulation transfer* describes a type of distortion in microwave amplifiers. It is the change in phase of the output RF voltage on one signal produced by variations in input signal level on a signal at another frequency, usually expressed in degrees/dB.

Amplitude shift keying (ASK): A form of digital modulation in which the amplitude of the carrier signal is switched between two states.

***b*:** Pierce's velocity parameter, b is a measure of actual beam velocity relative to synchronous beam velocity.

Backstreaming: (1) A condition in which electrons reflected from the collector or secondary electrons from the collector travel "backward" toward the electron gun. This is an undesirable effect, distorting the primary electron beam and any modulation that may be present. (2) A condition in which gas particles from a vacuum pump travel "backward" into the chamber being pumped.

Backward wave oscillator (BWO): A wideband voltage tunable oscillator related to a traveling wave tube. It uses a broadband circuit similar to a TWT. Regeneration exists because the backward wave mode on the circuit propagates RF energy in a direction opposite to that of the electron stream. Oscillation occurs when the loop gain is equal to or greater than unity. High-power TWTs can experience undesired backward wave oscillations when the backward wave mode has sufficient regeneration.

Barium: The primary emitting constituent of a dispenser cathode.

Beam efficiency: (also called RF efficiency) The RF output power divided by the beam power.

Beam power: The product of cathode voltage and cathode current.

Body current: The part of the electron current from the gun that does not arrive at the collector. This current is intercepted by the RF circuit and other electrodes.

Body-helix protection: A combination of circuit elements in the TWT power supply, ranging from a simple resistive network to a complex "crowbar" device. The circuitry is designed to prevent damage to the TWT as the result of arcing or unusually high body current.

Brillouin field: *See* Brillouin focusing.

Brillouin flow: Electron flow resulting from the use of Brillouin focusing.

Brillouin focusing: A magnetic focusing scheme in which the magnetic field that provides focusing is parallel to the direction of electron flow and the electron gun is shielded from the magnetic field. The magnetic field required to focus an ideal beam is called the *Brillouin field.*

Bunching: The result of the interaction of an electron beam with an RF field. Electrons which are accelerated by the field move toward those that have been decelerated, forming a high-density region usually referred to as a *bunch*.

C: Pierce's gain parameter, C is a measure of the effectiveness of the circuit in a TWT in interacting with the electron beam.

Cathode loading: The current density at the emitting surface of the cathode, usually expressed in amperes per square centimeter.

Cavity: A microwave-resonant circuit, usually consisting of a hollow metallic enclosure in which RF fields can be established at one or more frequencies.

Centipede structure: A fundamental forward wave coupled-cavity structure. Adjacent cavities are coupled by a multiplicity of dual coupling loops.

Charge density: The amount of electric charge contained in a unit of volume, usually expressed in coulombs per cubic meter.

Circuit efficiency: The efficiency with which the slow wave circuit of a microwave tube delivers the RF power generated in the tube to the output connector.

Cloverleaf structure: A fundamental forward wave coupled-cavity structure. Each cavity has the shape of a four-leaf clover. Adjacent cavities are magnetically coupled by radial slots and are rotated 45°.

Coaxial magnetron injection gun (CMIG): An electron gun used in a coaxial gyrotron.

Cold match: The input or output RF impedance match of a TWT obtained when the TWT is not operating (the TWT is electrically "cold").

Collector: The electrode that collects the spent electron beam after it has been used to provide power for microwave amplification or oscillation. The kinetic energy in the spent electron beam is converted to thermal energy (heat) in the collector or is recovered and returned to the power supply by the collector.

Collector efficiency: The efficiency with which the collector recovers power from the spent beam.

Confined flow: A type of beam focusing similar to Brillouin flow except that a small amount of magnetic flux is caused to thread the cathode. More field is required than for Brillouin flow, but the beam is less affected by RF bunching (less scalloping is introduced when RF is applied to the tube and body current is affected less by the application of RF drive). Confined flow is sometimes referred to as *immersed flow*.

Confocal mirrors: For identical mirrors, the radius of curvature the mirrors is the distance between the mirrors.

Contrawound helix: A helix slow-wave structure where two helices, wound in opposite directions, are superimposed into a single structure. This circuit offers substantially higher power than a conventional helix with some sacrifice of bandwidth.

Control grid: An electrode, usually a gridlike structure, mounted between the cathode and the anode to control the flow of electrons. An appropriate negative voltage (with respect to the cathode) reduces the electron flow (beam current) to zero or cutoff and an appropriate positive (or, in some cases, less negative) voltage allows current to flow. This electrode is usually some sort of mesh structure.

Coupled-cavity tube: A TWT with a slow-wave structure made up of a number of "cavities" electrically coupled by means of coupling holes, slots, or loops. This circuit is capable of very high-power operation.

Crossed-field device: A microwave tube in which the electric field between the cathode and anode is perpendicular to the magnetic field. The resulting average motion of the electrons emitted from the cathode is nearly perpendicular to both the electric and magnetic fields and parallel to the direction of propagation of energy on a nearby slow wave circuit.

Current density: The current per unit area. For electrons in an electron beam, the current density is the beam current divided by the cross-sectional area of the beam. The current density at the cathode surface (cathode loading) is equal to the current divided by the surface area of the cathode and is usually expressed in amperes per square centimeter.

Chemical vapor deposited diamond (CVD): Used as the output window for RF power on high-power high frequency microwave tubes and gyrotrons.

***d*:** Pierce's loss parameter, d is a measure of the reduction in gain that occurs in a TWT as the result of circuit loss.

dBc: The level of a spurious output relative to the main signal (carrier) expressed in decibels below the carrier power level.

dBm: Power level in decibels above a milliwatt; 60 dBm is a power level of 60 dB above a milliwatt and is a kilowatt.

dBW: Power level in decibels above a watt; 60 dBW is a power level of 60 dB above a watt and is a megawatt.

Denison mode converter: A modification of the Vlasov converter, which is compact and can be placed between the gyrotron cavity and the output window.

Density modulation: The process of controlling the density of electrons emitted from a cathode, usually, by a grid or other control electrode located near the cathode.

Depressed collector: Collector depression is the process of applying a negative potential (with respect to the tube body or "ground") to the collector to reduce the velocity of electrons entering the collector. This reduces the kinetic energy of the electrons and so less energy is converted to heat and X-rays when the electrons impinge on the electrodes of the collector. The energy recovered from the electrons is returned to the power supply and so the conversion efficiency is greatly improved. A collector designed to operate in this mode is called a *depressed collector*. Depressed collectors can have one or more depressed stages. Multiple-stage depressed collectors use velocity sorting techniques to direct the high-velocity electrons to stages having the greatest depression and slow electrons to the stages having the least depression.

Dispenser cathode: A cathode that has a body made of porous tungsten. The pores contain the active materials that enhance electron emission. When the cathode is heated to its operating temperature, the material in the pores migrates toward the emitting surface, continuously replenishing the supply of active material at the emitting surface.

Dispersion: A term used to describe the variation in phase velocity of an RF wave with frequency. A nondispersive circuit propagates RF energy at a phase velocity that is independent of frequency.

Drift tube: A section of metal tubing in which an electron beam "drifts" without being accelerated or decelerated by applied electric fields.

Drive: A term to indicate the RF input or RF signal to an electronic device.

Dual mode: Usually refers to devices capable of operating efficiently in both a low power CW mode and a high-power pulse mode.

Duty cycle: The fraction of the time during which a device is operating. The term *duty cycle* is usually used with pulsed operation and is the product of pulsewidth and pulse repetition frequency.

Dynamic range for linear operation: The range of operation of an amplifier in which the power output is nearly proportional to the power input. The lower limit is the input drive level at which the signal input is equal to the noise input plus the noise figure. The upper limit is the input drive level that produces 1 dB or less of gain compression.

Dynamic velocity taper (DVT): A variation in the pitch of the helix of a TWT to improve efficiency and linearity. (*See also* tapered velocity.)

Efficiency: The RF output power divided by the sum of all power provided by the power supplies. (*See also* overall efficiency).

Electronic efficiency: The efficiency with which dc beam power is converted to RF power at the desired frequency within a microwave tube.

Electronic power conditioner (EPC): A power supply/modulator usually associated with space-qualified TWTAs. In addition to providing power, EPCs often include protection circuits, circuits for receiving commands, and circuits for providing telemetry data.

Energy product: A measure of the magnetic energy produced in the gap in a permanent magnet. The energy product is the product of B and H at the magnet operating point.

Equalizer: A passive device providing selective loss and/or phase shift over an operating band such that the net gain and/or phase shift of the equalized amplifier matches a required profile.

External Q: A measure of the affect on the Q of a resonant circuit by an external circuit attached to that resonant circuit. External Q is the reciprocal of the difference between the reciprocals of the loaded and the unloaded Q.

Ferrule: A drift tube tip in the cavity of a klystron or coupled-cavity TWT.

Fine grain: A term used to refer to small variations in gain or phase of an amplifier.

Focus electrode: An element in the electron gun that is used to focus the electrons into a well-defined beam. It is sometimes called the *beam-forming electrode* (BFE).

Folded helix circuit: A helix slow-wave structure where two helices, wound in the same direction, are superimposed into a single structure. This circuit is sometimes called a *bifilar helix* and offers substantially higher power than a conventional helix with some sacrifice of bandwidth.

Folded waveguide circuit: Another name for a coupled-cavity circuit.

Frequency shift keying (FSK): A form of digital modulation in which the frequency of the carrier signal is switched between two states.

Fundamental backward wave circuit: A slow-wave circuit used in coupled-cavity TWTs. The fundamental mode of the circuit is a backward wave mode. The circuit is operated in a forward space-harmonic mode.

Fundamental forward wave circuit: A slow-wave circuit used in high-power coupled-cavity TWTs. The fundamental mode of the circuit is a forward wave mode.

Fundamental Gaussian mode content (FGMC): The fraction of the mode content in the output beam from a gyrotron that is Gaussian.

Gain: The ratio of output voltage, current, or power to the input voltage, current, or power, respectively, in an amplifier stage, receiver, or system. Power gain is usually expressed in dB.

Gain compression: The decrease in gain of an amplifier between the small signal gain and the gain at saturation. Typically, gain compression is 6 to 8 dB in TWTs.

Gated emission: Emission from a cathode that is controlled (gated), usually with a grid adjacent to the cathode.

Getter: A device that, after activation, absorbs or adsorbs gases within an electron tube.

Grid: An electrode mounted between the cathode and the anode to control the flow of electrons. When used to control the amount of electron current in a TWT or klystron, the grid is called a *control grid.* The grid electrode may be a cylindrical-shaped ring or a section of wire screen or mesh.

Group delay: The distortion that results when the time delay of a signal being processed is not constant as a function of frequency. Areas of interest are linear delay, parabolic delay, and the ripple component. The linear component is the difference in delay between two frequencies within the specified bandwidth. Parabolic delay is determined by fitting the equation of a parabola to a plot of delay versus frequency. Ripple is the maximum peak-to-peak variation relative to a plot of the smoothed curve.

Group velocity: The velocity at which energy advances along an RF circuit. For a coaxial transmission line with no dielectric, the group velocity is the speed of light. For waveguide, the group velocity is less than the speed of light. The group velocity cannot be greater than the speed of light.

Gyrotron: An oscillator or amplifier that is capable of generating extremely high powers at frequencies of tens of gigahertz to terahertz.

Gyro-monotron or gyro-oscillator: A gyrotron oscillator.

Gyro-klystron: A gyrotron amplifier.

Gyro-twystron: A gyrotron amplifier.

Gyro-TWT: A gyrotron amplifier.

Harmonic drive: The inclusion of phase-conditioned harmonic power in the input RF signal to reduce harmonic power and harmonic capture, and to improve efficiency at the low end of the operating band of a TWT.

Harmonic interaction: The effect of the harmonic content of the RF input signal on the beam modulation. This is generally undesirable and usually reduces the fundamental power output.

Heater: An electric heating element for raising an indirectly heated cathode to the operating temperature.

Hot match: The RF impedance match at the input or output of a TWT, obtained when the TWT is operating (electrically "hot").

Hub: *See* Ferrule.

Hull cutoff condition: The combination of perpendicular electric and magnetic fields that causes an electron in a smooth bore magnetron to just graze anode.

Hysteresis loop: *See* intrinsic hysteresis loop and normal hysteresis loop.

Inverted magnetron injection gun (ICMIG): An electron gun used on coaxial gyrotrons.

Immersed flow: *See* Confined flow.

Inline slot structure: A fundamental backward wave coupled-cavity structure in which adjacent cavities are electromagnetically coupled by two slots spaced 180° apart in the wall between the cavities. The slots in the walls of adjacent cavity are aligned.

Insertion loss: The reduction in signal strength that results when a passive device is installed (inserted) into an RF transmission line. Insertion loss is caused by power being absorbed in the device and by reflections from discontinuities at the interfaces between the device and the transmission line. Usually expressed in decibels.

Intercepting grid: A control grid that is not shielded from electrons emitted by the cathode. The grid intercepts approximately 10 to 15% of the emission current when a positive potential is applied.

Intercept point: On a plot of intermodulation distortion data as a function of RF input drive, the carrier data at small signal drive levels has a slope of 1:1, third-order intermodulation products have a slope of 3:1 and fifth-order intermodulation products have a slope of 5:1. Extrapolating the slopes for the carriers and the third-order products produces an intercept point. The output power at this point is called the *third-order intercept point*. Similarly, the fifth-order intercept point is the point at which the carrier slope intercepts the fifth-order product slope. (*See also* Intermodulation distortion.)

Interdigital line: A slow-wave circuit composed of a comblike structure with alternate segments being connected together at one end and remaining segments connected together at the opposite end.

Intermodulation distortion: Distortion resulting from the production of frequencies that are the sum of, and the difference between, frequencies contained in the applied waveform.

Intrinsic hysteresis loop: The magnetization characteristic of a ferromagnetic metal.

Intrinsic efficiency: Efficiency calculated by dividing the RF output power from a microwave tube by the sum of all power provided with the exception of heater power and that provided to the solenoid for focusing the beam.

Ion oscillations: Fluctuations in microwave tube operation resulting from ion effects in the electron beam. These fluctuations are commonly observed as variations in RF power output or as RF signals removed from the carrier frequency by the frequency of ion oscillations in the electron beam.

Iridium: A transition metal sometimes used with dispenser cathodes for enhancing emission.

Klystron: A velocity-modulated microwave tube that uses the interaction between an electron beam and the RF energy in cavities to provide signal amplification. In a klystron, the RF interaction takes place at discrete locations along the electron beam. Common types of klystrons are the reflex klystron (an oscillator having only one cavity), two-cavity klystron amplifiers and oscillators, and multicavity klystron amplifiers.

Leakage current: Undesirable flow of current between electrodes often resulting from conductive paths through or over the surface of an insulator or from unwanted electron emission from electrode surfaces.

Linear-beam tube: A microwave tube (usually a klystron or TWT) in which a long cylindrical electron beam is used.

Linearizer: A device that improves the linearity of the AM/AM conversion curve of an amplifier (causing the slope of power output versus power input to be nearly 1 dB/dB over a wide range of drive levels). Linearizers are usually designed to also reduce the AM/PM conversion (causing the phase shift to remain nearly constant over a wide range of drive levels). The use of a linearizer reduces the intermodulation distortion.

Loaded Q: The Q of a resonant circuit when there is external coupling to the circuit.

Loss buttons: In coupled-cavity TWTs, lossy elements used to provide frequency selective loss to inhibit the excitation of higher order modes.

Magnetron: A crossed-field microwave oscillator containing a concentric cathode and anode containing resonant cavities. An axial magnetic field causes electrons to orbit between the cathode and the anode. The RF voltages across gaps in the resonators modulate the velocities of the electrons. This causes the orbiting electrons to form into "spokes" that rotate around the tube axis. As the spokes rotate past the resonator gaps, they induce currents that excite the cavities. The RF voltages build up to large levels. High-power output is obtained at high efficiency.

Metal-ceramic: A term applied to tube assemblies that employ ceramics as a part of their vacuum walls. Electrical isolation between elements is achieved by using ceramic insulators. The assembly is integrated by brazing the ceramic and metal parts.

Micron: A measure of pressure. One micron is 10^{-3} torr.

Mode interference: *See* interfering mode.

M-type tube: A crossed-field tube.

Multimode: Having the capability of operating efficiently at more than one power level. (*See also* dual mode.)

Multioctave: Capable of operating satisfactorily over a frequency range of two or more octaves.

Multipactor: An electron-RF field interaction in which electrons oscillate back and forth in synchronism with an RF voltage applied between electrodes. Secondary electrons produced when primary electrons strike the electrodes cause the number of oscillating electrons to increase with time. Multipactor is usually an undesirable effect and can occur across cavity or drift tube gaps, in output waveguides or on the surface of the ceramic output window.

Multistage collector: A collector with several segments, each successive segment being "depressed" more than the preceding segment. This enhances the efficiency of the collector in recovering power from the spent electron beam.

Near-carrier noise: AM noise, PM noise, and spurious outputs at frequencies close to the carrier frequency, usually 100 Hz to 500 kHz relative to the carrier. Components not related to the presence of the carrier (signal) are usually caused by ripple on power supply voltages. Interactions in the TWT are usually responsible for components that exist only when a carrier (signal) is present.

Noise figure: The ratio of the signal-to-noise on the output of a device to the signal-to-noise on the input. It is important because it indicates the amount of noise the amplifier contributes to the signal, and it is an absolute indicator of the sensitivity of the device. Noise figure is usually expressed in decibels, and is abbreviated NF or F.

Noise power: The noise generated by a device or amplifier when measured at the output port when the input port is terminated and no RF drive is applied.

Noise power ratio (NPR): A measure of intermodulation when an infinite number of carriers is being amplified. The input to the amplifier is a well-defined bandwidth of random noise with a deep notch in the center. Intermodulation in the amplifier produces intermodulation products that partially fill the notch. The ratio of noise power in the passband to the noise power in the notch, measured at the output of the amplifier, is the NPR.

Nonintercepting grid: A control grid that is mechanically shielded from the cathode by a "shadow grid" in such a way that few electrons are intercepted by the control grid, even when a positive voltage is applied to the control grid (with respect to the cathode).

Normal hysteresis loop: The magnetization characteristic of a ferromagnetic metal plus the field that would be there in the absence of the metal.

Octave: The interval between two frequencies having a ratio of 2:1.

Osmium: A transition metal sometimes used with dispenser cathodes for enhancing emission.

Outgassing: A term used to describe the release of various gasses from the internal surfaces of a vacuum device during processing and testing.

O-type tube: A linear-beam tube.

Overall efficiency: The ratio of RF output power from a microwave tube to the total of all power inputs to the tube.

Overdrive: An input signal level greater than that required for saturation, resulting in decreasing output power.

Oxide-coated cathode: A cathode with an emissive surface consisting of oxides of alkaline-earth metals (usually barium, calcium, and strontium).

Penultimate cavity: The next to the last cavity of a multicavity klystron.

Perveance: A numerical constant relating current and voltage in a space-charge–limited diode. The relation for perveance, usually denoted by P, is

$$P = \frac{I}{V^{3/2}}$$

Perveance depends only on the geometry of the diode. When the voltage is the cathode-to-anode voltage in the electron gun, the perveance is called the *gun perveance.* When the voltage is the cathode-to-circuit voltage, the perveance is called the *beam perveance.*

Phase linearity: A term referring to the degree of deviation from a straight line of the phase versus frequency characteristic of a device.

Phase-pulling factor: The amount by which a change in input or output VSWR changes the phase shift in the TWT.

Phase-pushing factor: The amount by which a change in the voltage on an electrode changes the phase shift in the TWT, usually expressed in degrees/volt.

Phase tracking: The closeness or similarity of the phase characteristics of a number of devices. This is an important consideration when power combining the outputs from two or more devices.

Phase velocity: The velocity at which phase fronts for an electromagnetic wave propagate. For a coaxial transmission line with no dielectric, the magnitude of the phase velocity is the speed of light. For a waveguide, the magnitude of the phase velocity is greater than the speed of light. Slow-wave circuits are designed to provide a phase velocity having a magnitude much less than the speed of light.

PM near-carrier noise: PM (phase modulation) noise and spurious output at frequencies close to the carrier frequency, usually 100 Hz to 500 kHz relative to the carrier. Components not related to the presence of the carrier (signal) are usually caused by ripple on power supply voltages. Interactions in the TWT are usually responsible for components that exist only when a carrier (signal) is present. (*See also* near-carrier noise.)

Pole piece: A section of ferromagnetic material used to shape the magnetic flux distribution from a solenoid (electromagnet) or permanent magnet.

Power combining: A scheme whereby the outputs from two or more amplifiers are combined to provide a greater output power than available from a single amplifier.

Power curve: A plot of output power versus input drive or input frequency. (For power curve versus input drive, *see also* AM/AM conversion.)

Phase shift keying (PSK): A form of digital modulation in which the phase of the carrier signal is switched between two states.

Periodic permanent magnet focusing (PPM): A method of focusing a TWT where permanent magnets of opposite polarity are placed side by side along the length of the tube.

Pulse-up ratio: The ratio, usually expressed in dB, between the CW power level and the pulse-power level in a dual mode device.

***QC*:** Pierce's space charge parameter, QC is a measure of the strength of space charge forces causing debunching relative to the strength of the circuit field that is bunching the beam.

Quadrature phase shift keying (QPSK): A form of digital modulation in which the phase of the carrier signal is switched between four states.

Quasi-optical: Basic elements are optical but effects of finite wavelength are accounted for.

Reike diagram: A Smith chart based load diagram for microwave oscillators, particularly klystrons and magnetrons.

Residual gas: The gas remaining in an evacuated vacuum tube.

RF efficiency: (also called beam efficiency) The RF output power divided by the beam power.

Rhenium: A transition metal sometimes used with dispenser cathodes for enhancing emission.

Ring-bar tube: A TWT with a slow-wave structure composed of ringlike segments connected by straight straps or "bars." This device is capable of higher power levels than a conventional helix tube at a significant reduction in bandwidth.

Ring-loop tube: Similar to a ring-bar tube, except that the bars connecting the ringlike segments are replaced by curved straps.

Samarium cobalt: A high-energy product magnet material commonly used on microwave tubes.

Saturated gain: The gain of an amplifier operating at saturation.

Saturated power output: The power output of an amplifier operating at saturation.

Saturation: A term used to describe that point on the RF power output versus RF power input characteristic where an increase in input power does not produce an increase in output power.

Screen grid: A grid structure placed between the control grid and anode to reduce the capacitive coupling between those two electrodes. In a multimode electron gun, it is the element used to control beam current from the edge of the cathode.

Secondary emission: Emission of electrons resulting from the bombardment of a surface by high energy electrons.

Serrodyne: An operating mode for a TWT in which the input signal is translated to a new frequency at the output. A sawtooth voltage is superimposed on the dc voltage between the cathode and the slow wave circuit, causing the phase shift in the TWT to change in a linear fashion as a function of time. The amplitude of the sawtooth voltage can be adjusted to provide 360° of phase shift during each cycle, causing the output of the TWT to be offset by the frequency of the sawtooth waveform (the "flyback" time of the sawtooth must be very short so the TWT "thinks" that the phase is being offset continuously).

Shadow grid: A grid structure placed between the cathode and control grid and electrically connected to the cathode. This element shields the control grid from interception.

Shot noise: Noise power generation in a vacuum tube resulting from the random nature of the emission of electrons.

Single-stage collector: A TWT or klystron collector that has only one element for collecting electrons. The stage may be operated at ground potential or may be depressed. (*See also* depressed collector, multistage collector.)

Slow-wave circuit: Any structure that "slows" the phase velocity of an RF wave below the velocity of light to establish synchronism or near synchronism between the velocities of the wave and an electron beam.

Small-signal gain: The gain of an amplifier operating at small-signal levels.

Space charge: The electrical charge of electrons in a vacuum environment (as opposed to electrons within the crystal structure of a metal).

Space-charge–limited operation: Operation of a cathode at an emission density high enough so that the cathode current is limited by the space-charge field of the electron cloud near the cathode surface.

Spent electron beam: The electron beam exiting from the interaction region of a linear-beam tube.

Spherical diode: A diode constructed to duplicate the spherical geometry of a typical electron gun. Used for the evaluation and analysis of electron emission, analysis of electron gun designs, and for realistic cathode life testing.

Stagger tuning: A method of tuning the cavities of a klystron, also called *broadband tuning*. One or more of the cavities is tuned to a frequency(s) above and one or more is tuned to a frequency(s) below the center of the frequency band to provide approximately uniform amplification over a band of frequencies.

Staggered-slot structure: A fundamental backward wave coupled-cavity structure in which adjacent cavities are electromagnetically coupled by one opening or "slot" in the wall between the cavities. The slots are rotated 180° from cavity wall to cavity wall.

Synchronous tuning: The tuning of all cavities of a klystron to the same frequency.

Synchronous voltage: The beam voltage in a TWT at which the electron beam has the same velocity as the phase velocity of the RF wave on the circuit.

Tapered termination: A gradual increase in the amount of loss applied to a slow-wave structure to control reflections within a TWT.

Tapered velocity: A change in the pitch of a helix, the height of cavities, or some other axial dimensions of a slow-wave circuit to change the phase velocity of the RF wave. This is done because the extraction of kinetic energy from the electron beam causes the beam to slow down near the output of the TWT. The taper helps to maintain synchronism between the RF wave and the electron beam, which improves the efficiency of the TWT. (*See also* DVT.)

Temperature-limited operation: Operation of a cathode under conditions such that the cathode current depends on cathode temperature.

Thermionic emission: The emission of electrons from a surface as the result of heating of the surface.

Tetrode: A vacuum tube having four elements, usually a cathode, control grid, screen grid, and anode. In multimode TWTs, *tetrode* is a term describing the electron gun (cathode, shadow grid, control grid, and screen grid).

Torr: A unit of pressure. Atmospheric pressure is 760 torr. Most linear-beam microwave tubes operate at pressures below 10^{-7} torr.

Transfer curve: A curve or family of curves for which output power is plotted as a function of input drive power at one or more beam voltages. Sometimes referred to as a *gain curve*. (*See also* AM/AM transfer.)

Triode: A three-element vacuum tube having a cathode, a control grid, and an anode.

Traveling wave tube (TWT): A microwave tube using a broadband circuit in which an electron beam interacts continuously with a guided electromagnetic field to amplify microwave frequencies.

Traveling wave tube amplifier (TWTA): A combination of a power supply, a modulator (for pulsed systems), and a traveling wave tube, often packaged in a common enclosure.

Twystron: A high-power microwave tube having an input circuit consisting of klystron cavities and a cloverleaf coupled-cavity structure for the output circuit. A twystron has a nearly flat power output characteristic over a bandwidth as high as 10% or more.

Vacuum envelope: Any structure containing or capable of containing a high vacuum environment. The "body" structure of a microwave tube is usually referred to as the *vacuum envelope*.

Velocity modulation: The modulation of the velocities of electrons in a microwave tube as a result of their interaction with RF electric fields.

Velocity resynchronization: Any method for changing the axial velocity of an RF wave or of an electron beam to improve the synchronism between that wave and an electron beam. (*See also* tapered velocity, velocity jump.)

Velocity step: *See* tapered velocity.

Velocity taper: *See* tapered velocity.

Vlasov mode converter: A smooth surface circular waveguide with a launching cut and reflectors that form a Gaussian-like beam in a gyrotron.

Work function: The minimum energy, usually expressed in electron volts (eV), required for an electron to be thermionically emitted from a cathode.

Whispering gallery mode: An electromagnetic wave mode in a gyrotron, TE_{mn} mode, where $m >> n$.

About the Author

A. S. Gilmour, Jr., received a B.E.E., an M.S.E.E., and a Ph.D. from Cornell University. He was an assistant professor of electrical engineering at Cornell when he received a two-year appointment as professor in residence at the Cornell Aeronautical Laboratory in Buffalo, New York. He was then named head of the Wave Electronics Section at the laboratory. This was followed by a two-year assignment as manager of the Electronics Department of Sanders Associates in Williamsville, New York. He was then appointed professor of electrical and computer engineering at the State University of New York at Buffalo. He served as chairman for several years and is now professor emeritus.

Dr. Gilmour has been actively involved in the research, design, development and application of microwave tubes for over five decades. His klystron and TWT work resulted in a nearly perfect Brillouin electron beam. He made the first detailed measurements of the RF current and velocity distributions in Brillouin beams. He has participated in the design or improvement of transmitters for a large number of military microwave systems. He was instrumental in the establishment of a NASA/USAF program to develop high-efficiency TWTs for electronic countermeasures systems. During the past three decades, he has worked extensively with Government personnel to correct deficient tube and system designs in order to reduce failure rates and enhance performance.

Dr. Gilmour has over 200 publications, presentations, and patents to his credit. Among the most significant are the compilation and publication of the *Conference Records of the Power Modulator Symposia* (over 5,000 pages), and the publication of the books *Microwave Tubes* and *Principles of Traveling Wave Tubes*. His first book, *Microwave Tubes,* was based on the five-day course on microwave tubes that he developed and presented at various military, industrial, and educational organizations. *Principles of Traveling Wave Tubes* was based on the eight-day course on traveling wave tubes that he developed and presented two to three times at each of the major microwave tube companies in the United States as well as at the Crane Division of the Naval Surface Warfare Center.

Dr. Gilmour was a member of the Program Committees for the fourteenth through twentieth Power Modulator Symposia and was chairman in 1988. He was responsible for hosting the Twelfth Power Modulator Symposium in Buffalo, New York. For nearly a decade, he served on program committees for High Voltage

Workshops and was chairman in 1990. He was a member of the Committee on Pulsed Power Sources of the Air Force Studies Board, Assembly of Engineering, National Research Council, and National Academy of Sciences. He served on the program committees for International Pulsed Power Conferences for over a decade.

Dr. Gilmour is a fellow member of the Institute of Electrical and Electronic Engineers. His background includes certification as an electromagnetic compatibility engineer and as a professional engineer in the State of New York.

Index

A

B

C

D

E

F

G

H

I

J

K

L

M

N

P

V

W

Recent Titles in the Artech House Microwave Library

Active Filters for Integrated-Circuit Applications, Fred H. Irons

Advanced Techniques in RF Power Amplifier Design, Steve C. Cripps

Automated Smith Chart, Version 4.0: Software and User's Manual, Leonard M. Schwab

Behavioral Modeling of Nonlinear RF and Microwave Devices, Thomas R. Turlington

Broadband Microwave Amplifiers, Bal S. Virdee, Avtar S. Virdee, and Ben Y. Banyamin

Computer-Aided Analysis of Nonlinear Microwave Circuits, Paulo J. C. Rodrigues

Designing Bipolar Transistor Radio Frequency Integrated Circuits, Allen A. Sweet

Design of FET Frequency Multipliers and Harmonic Oscillators, Edmar Camargo

Design of Linear RF Outphasing Power Amplifiers, Xuejun Zhang, Lawrence E. Larson, and Peter M. Asbeck

Design Methodology for RF CMOS Phase Locked Loops, Carlos Quemada, Guillermo Bistué, and Iñigo Adin

Design of RF and Microwave Amplifiers and Oscillators, Second Edition, Pieter L. D. Abrie

Digital Filter Design Solutions, Jolyon M. De Freitas

Discrete Oscillator Design Linear, Nonlinear, Transient, and Noise Domains, Randall W. Rhea

Distortion in RF Power Amplifiers, Joel Vuolevi and Timo Rahkonen

EMPLAN: Electromagnetic Analysis of Printed Structures in Planarly Layered Media, Software and User's Manual, Noyan Kinayman and M. I. Aksun

An Engineer's Guide to Automated Testing of High-Speed Interfaces, José Moreira and Hubert Werkmann

Essentials of RF and Microwave Grounding, Eric Holzman

FAST: Fast Amplifier Synthesis Tool—Software and User's Guide, Dale D. Henkes

Feedforward Linear Power Amplifiers, Nick Pothecary

Foundations of Oscillator Circuit Design, Guillermo Gonzalez

Frequency Synthesizers: Concept to Product, Alexander Chenakin

Fundamentals of Nonlinear Behavioral Modeling for RF and Microwave Design, John Wood and David E. Root, editors

Generalized Filter Design by Computer Optimization, Djuradj Budimir

High-Linearity RF Amplifier Design, Peter B. Kenington

High-Speed Circuit Board Signal Integrity, Stephen C. Thierauf

Intermodulation Distortion in Microwave and Wireless Circuits, José Carlos Pedro and Nuno Borges Carvalho

Introduction to Modeling HBTs, Matthias Rudolph

Klystrons, Traveling Wave Tubes, Magnetrons, Crossed-Field Amplifiers, and Gyrotrons, A. S. Gilmour, Jr.

Lumped Elements for RF and Microwave Circuits, Inder Bahl

Lumped Element Quadrature Hybrids, David Andrews

Microwave Circuit Modeling Using Electromagnetic Field Simulation, Daniel G. Swanson, Jr. and Wolfgang J. R. Hoefer

Microwave Component Mechanics, Harri Eskelinen and Pekka Eskelinen

Microwave Differential Circuit Design Using Mixed-Mode S-Parameters, William R. Eisenstadt, Robert Stengel, and Bruce M. Thompson

Microwave Engineers' Handbook, Two Volumes, Theodore Saad, editor

Microwave Filters, Impedance-Matching Networks, and Coupling Structures, George L. Matthaei, Leo Young, and E.M.T. Jones

Microwave Materials and Fabrication Techniques, Second Edition, Thomas S. Laverghetta

Microwave Mixers, Second Edition, Stephen A. Maas

Microwave Network Design Using the Scattering Matrix, Janusz A. Dobrowolski

Microwave Radio Transmission Design Guide, Second Edition, Trevor Manning

Microwaves and Wireless Simplified, Third Edition, Thomas S. Laverghetta

Modern Microwave Circuits, Noyan Kinayman and M. I. Aksun

Modern Microwave Measurements and Techniques, Second Edition, Thomas S. Laverghetta

Neural Networks for RF and Microwave Design, Q. J. Zhang and K. C. Gupta

Noise in Linear and Nonlinear Circuits, Stephen A. Maas

Nonlinear Microwave and RF Circuits, Second Edition, Stephen A. Maas

QMATCH: Lumped-Element Impedance Matching, Software and User's Guide, Pieter L. D. Abrie

Practical Analog and Digital Filter Design, Les Thede

Practical Microstrip Design and Applications, Günter Kompa

Practical RF Circuit Design for Modern Wireless Systems, Volume I: Passive Circuits and Systems, Les Besser and Rowan Gilmore

Practical RF Circuit Design for Modern Wireless Systems, Volume II: Active Circuits and Systems, Rowan Gilmore and Les Besser

Production Testing of RF and System-on-a-Chip Devices for Wireless Communications, Keith B. Schaub and Joe Kelly

Radio Frequency Integrated Circuit Design, Second Edition, John W. M. Rogers and Calvin Plett

RF Bulk Acoustic Wave Filters for Communications, Ken-ya Hashimoto

RF Design Guide: Systems, Circuits, and Equations, Peter Vizmuller

RF Measurements of Die and Packages, Scott A. Wartenberg

The RF and Microwave Circuit Design Handbook, Stephen A. Maas

RF and Microwave Coupled-Line Circuits, Rajesh Mongia, Inder Bahl, and Prakash Bhartia

RF and Microwave Oscillator Design, Michal Odyniec, editor

RF Power Amplifiers for Wireless Communications, Second Edition, Steve C. Cripps

RF Systems, Components, and Circuits Handbook, Ferril A. Losee

The Six-Port Technique with Microwave and Wireless Applications, Fadhel M. Ghannouchi and Abbas Mohammadi

Solid-State Microwave High-Power Amplifiers, Franco Sechi and Marina Bujatti

Stability Analysis of Nonlinear Microwave Circuits, Almudena Suárez and Raymond Quéré

Substrate Noise Coupling in Analog/RF Circuits, Stephane Bronckers, Geert Van der Plas, Gerd Vandersteen, and Yves Rolain

System-in-Package RF Design and Applications, Michael P. Gaynor

TRAVIS 2.0: Transmission Line Visualization Software and User's Guide, Version 2.0, Robert G. Kaires and Barton T. Hickman

Understanding Microwave Heating Cavities, Tse V. Chow Ting Chan and Howard C. Reader

For further information on these and other Artech House titles, including previously considered out-of-print books now available through our In-Print- Forever® (IPF®) program, contact:

Artech House Publishers
685 Canton Street
Norwood, MA 02062
Phone: 781-769-9750
Fax: 781-769-6334
e-mail: artech@artechhouse.com

Artech House Books
16 Sussex Street
London SW1V 4RW UK
Phone: +44 (0)20 7596 8750
Fax: +44 (0)20 7630 0166
e-mail: artech-uk@artechhouse.com

Find us on the World Wide Web at: www.artechhouse.com

www.ingramcontent.com/pod-product-compliance
Lightning Source LLC
LaVergne TN
LVHW011632200926
848484LV00010B/2535